FISHES AND FORESTRY

FISHES AND FORESTRY

Worldwide Watershed Interactions
and Management

EDITED BY

T.G. NORTHCOTE

and

G.F. HARTMAN

Blackwell
Science

© 2004 Blackwell Science Ltd
a Blackwell Publishing company

Editorial offices:
Blackwell Science Ltd, 9600 Garsington Road, Oxford OX4 2DQ, UK
   Tel: +44 (0)1865 776868
Iowa State Press, a Blackwell Publishing Company, 2121 State Avenue, Ames, Iowa 50014-8300, USA
   Tel: +1 515 292 0140
Blackwell Publishing Asia Pty Ltd, 550 Swanston Street, Carlton, Victoria 3053, Australia
   Tel: +61 (0)3 8359 1011

First published 2004

Library of Congress Cataloging-in-Publication Data is available

ISBN 0-632-05809-9

A catalogue record for this title is available from the British Library

Set in 10/12½ pt Sabon
by Sparks Computer Solutions Ltd, Oxford
http://www.sparks.co.uk
Printed and bound in Great Britain using acid-free paper by
MPG Books Ltd, Bodmin, Cornwall

For further information on Blackwell Publishing, visit our website:
www.blackwellpublishing.com

Contents

PART VIII: EFFECTING BETTER FISH–FORESTRY INTERACTIONS

The colour plate section follows page 362.

Preface

Having started a fish–forestry interaction course in the Faculty of Forestry, University of British Columbia early in the 1970s, one of us (TGN) soon realized that the students needed a book that assembled the complex breadth and depth of the subject. Somewhat later, attempts to develop such a book failed for a number of reasons. The focus was to be largely on salmonid fishes in the Pacific northwestern area of North America, integrating results from relevant major research and management studies started in the 1950s to 1970s in Oregon, Washington, British Columbia and Alaska. Various conferences, workshops and other publications in this geographical area now have covered separately much of this focus and these have been extensively referred to in appropriate Parts and chapters of this book.

Over the last several decades, both editors of this book have travelled extensively and have worked on inland fish, fisheries and forested habitats, especially in Latin America, Scandinavia and New Zealand (TGN), as well as in Africa, Scandinavia and New Zealand (GFH). In 1995 after the Pacific Science Congress in Beijing, we toured aquatic and some forested regions of central and far western China under the auspices of the Chinese Academy of Sciences. We realized that major problems in fish–forestry interaction were not only to be found in the Pacific Northwest, nor only with respect to salmonids, but occurred widely over forested regions of the world and involved many other important groups of fishes in both inland and estuarine–coastal waters.

In late 1999 via Dr D.T. Crisp who was then completing a book on the ecology, conservation and rehabilitation of trout and salmon, we learned that Blackwell Science might be interested in publishing a book on fish–forestry interaction. They approached us and together we worked up an outline that gained their approval.

We each brought into the project some similar and some different interests, background and expertise. That of TGN was largely in university teaching and research, first in limnology and fish ecology/migration, and then as well in fish–forestry interaction. That of GFH spanned university teaching and research in fish ecology, but also fisheries and wildlife management in various parts of Canada, and direction of a long-term major research project dealing with impacts of forestry practices on a coastal stream ecosystem – Carnation Creek in British Columbia.

From the onset we wanted this book to approach the subject from a worldwide viewpoint. Several of our colleagues argued against this, saying that such coverage would just not be possible. We appreciated their concern for two reasons. In respect to depth of coverage for topics such as forest ecology, stream ecology and lacustrine ecology, there are tens of thousands of relevant publications for each of these disciplines.

In respect to breadth of coverage it was, as expected, very difficult to find contributors from all areas who were able to complete chapters, or in some cases even available and prepared to attempt to do so. Regarding the latter we ran into several disappointments. We had hoped to have a separate chapter on hydrology–geomorphology but the authors for this belatedly defaulted so this coverage in synoptic form is given in Chapter 13. We knew that we must have coverage of Russia/Siberia and of Africa. GFH had contacts in both these areas and had worked intensively in several regions of Africa. After many attempts he could not find anyone in either area that would take on the task. We also knew that eastern and southern USA should be covered but our lead candidate there had to drop out because of serious family illness. For China, another key area, we had committed chapter authors, but the lead one eventually defaulted late in the sequence.

To fill these and other geographical gaps as well as possible, we have added brief coverage in a section of the final chapter where we review general insights from a worldwide summary of the subject. Even then there are other regions that probably should have been included. However, we had concern about how much we could put between two covers, as surely did our publishers. Also as retirees, there was a limit to how far we could push ourselves and maintain our good personal relationship, and perhaps more importantly the humour of our respective spouses. In the course of requesting chapter authors and in the editing of them, we made many fine friends, most only by e-mail. Even so it has been a great pleasure to get to know even superficially so many exceptional people from all around the world.

The Contents section provides an outline of the major Parts and chapters of the book. At the end of the introductory first chapter we give a rationale for this organization.

Acknowledgements

We are most grateful for the help provided throughout preparation of this book from the David Suzuki Foundation and Otto Langer, as well as from Casimer Lindsey, Gordon Miller, George Pattern, Bill Pollard and Ann Thompson. Over this whole period Heather Northcote and Helen Hartman have had to put up with missed holidays as well as two cranky old men. We thank them for their endurance, patience, and especially for their good advice – much of which we followed. Josie Severn of Blackwell Publishing provided counsel and understanding during several critical periods.

List of contributors

C.A.R.M. Araujo-Lima INPA, Manaus, CP 478, AM 69011-970, Brazil

J.E. Barker Department of Forest Sciences, University of British Columbia, 2424 Main Mall, Vancouver, BC, V6T 1Z4, Canada

W. Barrella PUCSP, Sorocaba, CP 33-45770, SP 18030-230, Brazil

D.G. Bengen Faculty of Fisheries and Marine Science, Bogor Agricultural University (IPB), Bogor, Indonesia

R.E. Bilby Weyerhaeuser Company, WTC 1A5, PO Box 9777, 32901 Weyerhaeuser Way S, Federal Way, Washington 98001, USA

G. Bull Faculty of Forestry, University of British Columbia, Vancouver, BC, V6T 1Z4, Canada

C.J. Cederholm Washington Department of Natural Resources, Land Management Division, 1111 Washington Street SE, PO Box 47016, Olympia, Washington 98504-7016, USA

K.D. Clarke Fisheries and Oceans Canada, Science, Oceans, and Environment Branch, PO Box 5667, 1 White Hills Road, St John's, NF, A1C 5X1, Canada

D.T. Crisp 21A Main Street, Mochrum, Newton Stewart, Wigtownshire DG8 91Y, Scotland

R.A. Cunjak Canadian Rivers Institute, New Brunswick Cooperative Fish & Wildlife Research Unit, Department of Biology, Faculty of Forestry and Environmental Management, Bag Service 45111, Fredericton, NB, E3B 6E1, Canada

R.A. Curry Canadian Rivers Institute, New Brunswick Cooperative Fish & Wildlife Research Unit, Department of Biology, Faculty of Forestry and Environmental Management, Bag Service 45111, Fredericton, NB, E3B 6E1, Canada

J.H. Dick Independent Environmental Impact Assessment Consultant, 2117 Sandowne Road, Victoria, BC, V8R 3J2, Canada

I.M. Dutton The Nature Conservancy (TNC), Indonesia Programme, Jakarta, Indonesia

M.J. Duncan National Institute of Water and Atmospheric Research Ltd, PO Box 8602, Christchurch, New Zealand

T. Eriksson Department of Aquaculture, Swedish University of Agricultural Sciences, S-901 83 Umeå, Sweden

W.D. Erskine Senior Specialist, Soil, Water and Fish, Environmental Management and Forest Practices Directorate, State Forests of New South Wales, Locked Bag 23, Pennant Hills, NSW 2120 and Honorary Research Associate, Division of Geography, School of Geosciences, University of Sydney, NSW 2006, Australia

K.D. Fausch Department of Fishery and Wildlife Biology, Colorado State University, Fort Collins, CO 80523, USA

R.M. García-Núñez Research Professor, Universidad Autónoma Chapingo, 56230 Chapingo, Mexico

G.J. Glova National Institute of Water and Atmospheric Research Ltd, PO Box 8602, Christchurch, New Zealand

E.R. Hall Department of Civil Engineering, University of British Columbia, 2324 Main Mall, Vancouver, BC, V6T 1Z4, Canada

J.D. Hall Department of Fisheries and Wildlife, Oregon State University, 104 Nash Hall, Corvallis, OR 97331-3803, USA

K.J. Hall Department of Civil Engineering, University of British Columbia, 2324 Main Mall, Vancouver, BC, V6T 1Z4, Canada

J.H. Harris Cooperative Research Centre for Freshwater Ecology, RiffleRun, Bootawa Road, Tinonee, NSW 2430 and Honorary Research Associate, Division of Geography, School of Geosciences, University of Sydney, NSW 2006, Australia

G.F. Hartman Fisheries Research & Education Services, 1217 Rose Ann Drive, Nanaimo, BC, V9T 3Z4, Canada

M.C. Healey Institute of Resources and Environment, 436E-2206 East Mall, University of British Columbia, Vancouver, BC, V6T 1Z3, Canada

B.J. Hicks   Centre for Biodiversity and Ecology Research, Department of Biological Sciences, University of Waikato, Private Bag 3105, Hamilton, New Zealand

N. Higuchi   INPA, Manaus, CP 478, AM, 69011-970, Brazil

S.G. Hinch   Department of Forest Sciences, 3022 Forest Sciences Centre, University of British Columbia, Vancouver, BC, V6T 1Z4, Canada

D.L. Hogan   Research Branch, BC Ministry of Forests, PO Box 9519, Stn Prov Govt, Victoria, BC, V8W 9C2, Canada

M. Inoue   Department of Biology and Earth Sciences, Faculty of Science, Ehime University, Bunkyo-cho 2-5, Matsuyama 790-8577, Japan

M. Karagosian   Department of Wildlife, Fish, and Conservation Biology, University of California, Davis, CA 95616, USA

J.P. Kimmins   Department of Forest Sciences, University of British Columbia, Vancouver, BC, V6T 1Z4, Canada

K.V. Koski   NOAA Fisheries Laboratory, 11305 Glacier Highway, Juneau, AK 99801-8626, USA

F. Krogstad   Research Associate, College of Forest Resources, University of Washington, Seattle, Washington 98195, USA

J. Leggett   Ministry of Water, Land & Air Protection, 400-640 Borland Street, Williams Lake, BC, V2G 4T1, Canada

C.D. Levings   Fisheries and Oceans Canada, 4160 Marine Drive, West Vancouver, BC, V7V 1N6, Canada

N. McCubbin   N. McCubbin Consultants Inc., 140 Fisher's Point, Foster, Quebec, J0E 1R0, Canada

T.E. McMahon   Ecology Department, Lewis Hall, Montana State University, Bozeman, MT 59717, USA

K. Martin-Smith   Research Fellow, School of Zoology, University of Tasmania, Australia

K. Moore   Moore Resource Management, Box 1029, Queen Charlotte City, BC, V0T 1S0, Canada

K.R. Munkittrick   Canada Research Chair in Ecosystem Health Assessment, Canadian Rivers Institute and Department of Biology, University of New Brunswick, St John, NB, E2L 4L5, Canada

M.L. Murphy   17355 Glacier Highway, Juneau, AK 99801-8331, USA

F. Nakamura   Graduate School of Agriculture, Department of Forest Science, Hokkaido University, Sapporo 060-8589, Japan

T.G. Northcote   Departments of Zoology and Forest Sciences, University of British Columbia, c/o RR2 S77B C2, Summerland, BC, V0H 1Z0, Canada

C.A. Paszkowski   Department of Biological Science, University of Alberta, Edmonton, Alberta, T6G 2E9, Canada

A. Peter   EAMAG/ETH, Seestrasse 6047, Kastanienbaum, Switzerland

M. Rask   FGFRI, Evo Fisheries Research Station, FIN-16970 Evo, Finland

N.H. Ringler   Department of Environmental and Forest Biology, State University of New York, College of Environmental Science and Forestry, Syracuse, NY 13210, USA

A. Sánchez-Vélez   Research Professor, Universidad Autónoma Chapingo, 56230 Chapingo, Mexico

P. Schiess   McMc Resources Professor of Forest Engineering, College of Forest Resources, University of Washington, Seattle, Washington 98195, USA

G.J. Scrimgeour   Alberta Conservation Association, PO Box 40027, Baker Centre, Edmonton, Alberta, T5J 4M9, Canada

D.A. Scruton   Fisheries and Oceans Canada, Science, Oceans, and Environment Branch, PO Box 5667, 1 White Hills Road, St John's, NF, A1C 5X1, Canada

R.J. Steedman   Ontario Ministry of Natural Resources, Center for Northern Forest Ecosystem Research, Thunder Bay, Ontario, P7B 5E1, Canada

W.M. Tonn   Department of Biological Science, University of Alberta, Edmonton, Alberta, T6G 2E9, Canada

P.J. Tschaplinski   Research Branch, BC Ministry of Forests, PO Box 9519, Stn Prov Govt, Victoria, BC, V8W 9C2, Canada

M.K. Young   USDA Forest Service, Rocky Mountain Research Station, 800 East Beckwith Avenue, Missoula, MT 59801, USA

Part I
Introduction

Chapter 1
An introductory overview of fish–forestry interactions

T.G. NORTHCOTE AND G.F. HARTMAN

What are fish–forestry interactions?

Many species of fish occupying inland waters of the world during all or part of their lifespan reside in watersheds that were or still are surrounded by forests and are dependent in major ways upon such cover. The complexity of interactions between fishes and forests has only been recognized in the last few decades. These interactions are multifaceted, dynamic processes involving most inland surface waters (streams, rivers, marshes, lakes, reservoirs, estuaries), forests, subsurface waters, geology and soils, climate and its changes, and the biotic components of the relevant ecosystems. The latter range from bacteria to birds and mammals that interact in many ways with fish. The interactions also include the aspects of forestry tied to human development, economics, population growth, and even philosophies.

Forestry is an important activity in many regions of the world. However, it is only one of the many factors affecting fish abundance and long-term survival in watersheds. Overfishing, dams, hatcheries, aquaculture, genetic stock alterations and others must be considered and evaluated. Climate change, anthropogenic or not, now also seems to have had major impacts on Pacific salmon abundance during the last three centuries (Brown 2000; Finney et al. 2000), and on other freshwater fishes as well. And forest ecosystems are also greatly affected – see the special section in BioScience 51: 709–79 (2001) and technical comment abstracts in Science 299: 1015 (2003) (www.sciencemag.org./cgi/content/full/299/5609/10152a,&b).

The rate of forest removal is high though variable in many areas of the world. A survey of the status of the world's forests by the Food and Agriculture Organization of the United Nations in March 2001 suggests that global rates of forest loss decreased in the 1990s to an annual average net loss of 9 million hectares or 0.2% of the global total. There was almost immediate disputation of the survey validity (Stokstad 2001), with strong suggestions that it understated the rate of loss. The Brazilian Amazon has the world's highest absolute rate of deforestation, conservatively averaging nearly 2 million hectares per year from 1995 to 1999 (Laurance et al. 2001a; see also Silveira 2001; Laurance et al. 2001b; Williams 2003). About 70% of Thailand was forested by 1936, dropping to 25% by 1982 (Ramitanondh 1989), and is significantly less now (Campbell & Parnrong 2001), with 30% of a national park alleged to have been illegally logged. A similar problem occurs in Indonesia (Jepson et al. 2001). Only some 10% of remaining forests in parts of the Mekong basin now are commercially valuable (UNEP, Global Environment Outlook-2000). The annual area harvested for timber in

Canada rose steadily from about 4000 km² in 1920 to over 10,000 km² in 1995 (May 1998), still a relatively low rate of about 0.43% per annum, but one pushing into less productive higher elevations and more northern areas of the country. The question of sustainability must be considered, locally and globally. Deforestation reversal is possible, as in Europe where the forest area has increased by more than 10% since the early 1960s (UNEP, Global Environment Outlook-2000). However, afforestation rates in many areas such as Africa are far less than those of deforestation.

Watershed ecosystems are highly complex. Many years of research are required to gain some understanding of processes in a small drainage and much more to understand large systems. In the small Carnation Creek watershed (about 10 km²) in south coastal British Columbia, interactive processes between forests, logging and fishes were complex, time-lagged and variable depending on drainage features, tree species, logging patterns and fish species (four salmonids and two cottids) with their different life history stages (Hartman & Scrivener 1990).

Consider then the complexities that must underlie fish and forestry interaction in the Amazon River basin, some 6.5 million km² in area, with the richest fish fauna in the world. Over 1300 species (Lowe-McConnell 1987), many poorly known and probably many more still undescribed, live there within some 40% of the world's remaining tropical rainforest. Furthermore, interactive complexities via invertebrate fluxes between forests and streams, representing up to 44% of the annual energy budget of their fish and bird predators, can shift seasonally in timing and extent of flow. Thus human-induced change in stream riparian habitat may cause serious effects on reciprocal energy flow between land and water environments (Willson *et al.* 1998; Nakano & Murakami 2001; Power 2001; Fausch *et al.* 2002). Salmon and other anadromous fish returning from the sea bring vital nutrients back to stream systems (Ben-David *et al.* 1998). Clearly in both small and large watersheds there are many complex and important interactions between fishes and forests. If the further complex of effects that arise from the many-faceted activities of forestry are imposed on watersheds, an extremely diverse array of responses may be expected. The nature and extent of these will depend on the physical severity of the impacts, the hydrological and geomorphological features of the drainage, and the characteristics of the biota.

Where can fish–forestry interactions occur?

As a first approach to outlining the global distribution of fish–forestry interactions, we summarize the world land cover occupied by forests (Fig. 1.1A). In places these interactions may extend into areas below those that are forested as a result of downstream progression of impacts, or even above by climatic impacts of lowland deforestation on upper montane cloud forests (Lawton *et al.* 2001). Furthermore shrublands and woody savannas may be harvested for fuel wood or small construction timber although not technically classified as being forested.

Globally (Anonymous 1999; see also Olson *et al.* 2001), forest/woodland contributes the highest percentage of land cover (27.5%), followed by a cropland/natural vegetation mosaic (19.2%), shrubland (14.2%) and grass/savanna (14.0%), with the

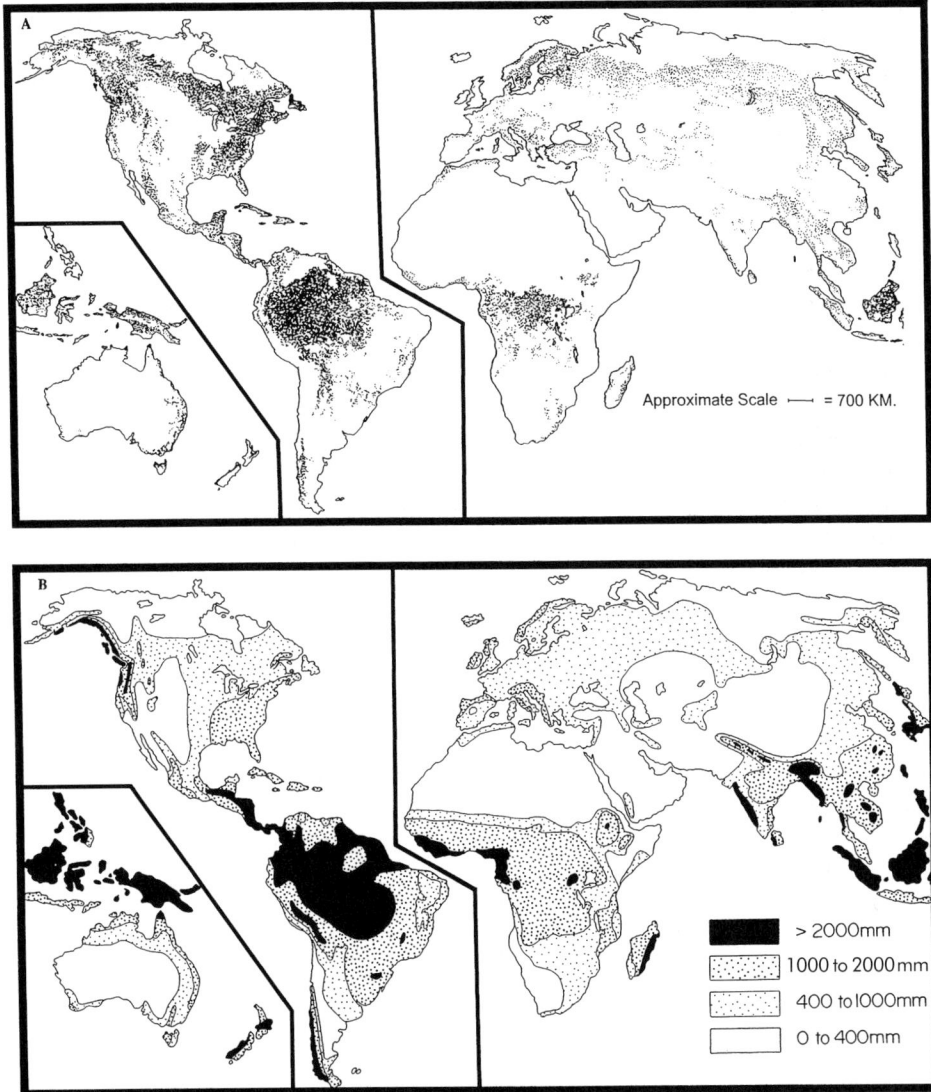

Fig. 1.1 (A) World distribution of major forested land cover areas (stippled), combining five forest categories. Adapted and generalized from *The Times Atlas of the World*, Tenth Comprehensive Edition (Anonymous 1999), which should be consulted for any detailed consideration of specific regions. Based on 1992/1993 satellite imagery at a ground resolution of 1 km and shown in the Goode Interrupted Homolosine Projection. See also Stokstad (2001) for a world forest distribution map showing deforestation and afforestation areas >0.5% per year. (B) World distribution of four ranges of mean annual precipitation.

remainder wetland, urban, and barren snow or ice. On a continental comparison of forest/woodland cover, North and South America have the highest and about equal amounts – some 43%. Next in sequence are Australia (about 27%), Eurasia (about 25%) and Africa (about 24%). A slightly different pattern emerges if only forest cover is considered (Table 1.1). Europe has a large forest area, exceeded slightly by Central and South America, with the latter areas having the highest percentage of forest cover, 47.1%. North America has some 457 million hectares of forest, or about a quarter of its land base. Africa and Asia each have over 500 million hectares in forest, about 18% and 16% of their total land areas respectively. Overall, some 26.6% of the world's total land area is forested. Half of the world's closed canopy forest lies in Russia, Canada and Brazil (Holden 2001a).

A broad band of forest cover extends from western Alaska across North America and from eastern Norway across much of Sweden, Finland and Russia/Siberia to the latter's eastern margins (Fig. 1.1A). Vegetation cover above 40° N is spreading as a result of global warming (Holden 2001b). Apart from those areas, however, much of the remainder of the world in a similar northern latitude band is rather sparsely forested. The forested areas from western Mexico down through Central America are not followed at similar latitudes across the world's lands until one reaches Myanmar (Burma) (note that here and elsewhere, place, country and region names follow Anonymous (1999)), Thailand, Cambodia, Laos and Vietnam. In the southern hemisphere (and extending partly into the southernmost region of the northern hemisphere in South America), is the largest concentration of the world's forested land (Fig. 1.1A). At approximately similar latitudes in western and central Africa there is another sizable forested area, picked up again in Sumatra, the areas of Malaysia, Indonesia and New Guinea. Throughout the remainder of South America and Africa (including Madagascar) there are scattered bands of roughly north–south oriented forest lands. Similar bands are also found along the eastern margins of Australia (including northern Tasmania), and along mainly the western side of the South Island of New Zealand, as well as in some north and central parts of the North Island.

Table 1.1   Forested areas[a] in major regions of the world

Region	Forest area (million ha)	Forest area as % of total land area
Central, South America	950.0	47.1
Europe	933.4	41.3
North America	457.1	24.9
Africa	520.2	17.7
Asia	503.0	16.4
Oceania	90.7	10.7
World	3454.4	26.6

[a]Reorganized regionally from Forestry Commission Facts and Figures 1998–99, Edinburgh, Scotland; originally published by FAO/UN in *State of the World's Forests* 1999. (Courtesy Dr D.T. Crisp.)

What fishes may be involved?

There are 34 orders of fishes with freshwater representatives (Nelson 1994) relevant here (Table 1.2) that include over 10,000 freshwater species. Two of these orders have only one species (the Australian lungfish, the North American bowfin) and another, the cods, has only one fully freshwater species, although other species of cod sometimes enter river estuaries or ascend further upstream (Scott & Crossman 1973).

In contrast, there are four orders of freshwater fishes each with well over 1000 species (Table 1.2) – the cyprinids, the catfishes, the percids and the characins – followed by the cyprinodonts with over 800 species. Together these five major groups account for nearly 90% of the relevant freshwater fish species.

All 34 orders of freshwater fishes have species living at least during part of their life in rivers or streams (Table 1.2), and at least 25 of these occur in lakes or swamps, along with 26 orders with estuarine representatives. Clearly the three major types of inland waters have strong representation from a very diverse assemblage of fishes that may be exposed to whatever effects forestry practices have on these different inland water habitats. At least 20 of the orders of freshwater fishes include species having considerable importance for local food use, for major commercial fisheries, and for aquaculture or aquarium culture purposes (Table 1.2).

Central and South America and Africa each support over 3000 freshwater species (Table 1.2), with representatives from 23 and 22 different orders respectively. Eurasia has over 2000 freshwater species (20 different orders), with over 800 in North America (25 orders) and nearly 800 in insular Oceania (19 orders). Australia and New Zealand together have 187 freshwater species (23 orders).

In general, there would seem to be a strong potential for forestry interactions to have effects on a very large number of freshwater fish species. Many are of considerable importance to man. They occur over a broad range of the world's forested lands, and are found in various types of inland waters, from very small streams to large rivers and lakes to estuaries.

There is considerable overlap between forested areas of the world and precipitation greater than 100 cm per year (cf. Figs 1.1A and B). The boreal zone is an exception; most cold zone forests have annual precipitation between 25 and 100 cm. In Central and South America 41 of 46 families of fish using inland habitats have distributions that are overlapped 70% or more by dense forest and rainfall >100 cm per year (Table 1.3). In the insular Oceania region, 17 of 34 families have distributions that are overlapped 70% or more by dense forest and precipitation >200 cm annually. In North America the distributions of anadromous salmonids (family Salmonidae) are overlapped by dense forest and precipitation >100 cm per year, and 32 of 34 families there have distributions in dense to sparse forest zones. In Eurasia 5 of 54 families have distributions that are overlapped 70% or more by dense forest, and all but two families occur where precipitation exceeds 25 cm per year. In Africa, 3 of 33 families have distributions that are overlapped 70% or more by dense forest zones. Most of the fish families considered have distributions within dense forest zones, but many also extend their distribution into more sparse forest cover where precipitation may be as low as 25 cm per year.

Table 1.2 Freshwater habitats used and approximate number of fish species in 34 orders as recorded mainly by Nelson (1994) for major regions of the world. S,R = streams, rivers; L,S = lakes, swamps; E = estuaries.

Orders	S,R	L,S	E	North America[a]	Eurasia[b]	Central and South America	Africa[c]	Australia and New Zealand	Insular Oceania[d]	Totals[e]
1. Lampreys	+	+	+	27	17	1		3		48
2. Ground sharks	+	+	+			1+	1+	2+	1+	5+
3. Rays, sawfishes	+		+			18	1+	4	1+	24+
4. Lungfish[f]	+							1		1
5. Lungfish[g]	+					1	4			5
6. Polypterids	+	+					10			10
7. Acipenserids[h]*	+	+	+	9	10+					19+
8. Gars	+	+	+	7		1+				8+
9. Bowfin	+	+		1						1
10. Osteoglossids*	+	+	+	2	5	3	204	2	7	223
11. Tarpons	+	+	+	1	1	1	1	1	1	6
12. Eels*	+	+	+	1	1		4+	5+	2+	13+
13. Herrings*	+	+	+	2	10+	2	27	8	10+	59+
14. Gonorhynchids*		+	+				28	1	1	30
15. Cyprinids[i]*	+	+	+	337	1732	1	477	6	429	2982
16. Characins*	+	+	+	1		885+	208			1094+
17. Catfishes*	+	+	+	45	203+	1403	429	14	33	2127+
18. Gymnotids	+	+				62				62
19. Esocids[j]*	+	+	+	7	6					13

20. Osmerids[k]*	+	+	+	+	+	+	4	13	6	1	29	3	56
21. Salmonids[l]*	+	+	+	+	+		38	27	6	2	7		80
22. Percopsids	+						9						9
23. Cods*	+	+					2	1					3
24. Toadfishes			+						6				6
25. Mullets*	+				+	+	2+				1+	1+	4+
26. Atherinids[m]*	+	+	+	+	+	+	20+	14+	14+	14+	23	135	220+
27. Belonids[n]	+	+	+	+	+	+	1	23	1+	1+	4+	23+	53+
28. Cyprinodontids[o]*	+	+	+	+	+	+	73	16	389	285	3	60+	826+
29. Gasterosteids[p]*	+	+	+	+	+	+	14+	12+	8+	8+	2	8+	52+
30. Synbranchids[q]	+	+	+	+	+	+	3	43+	5	51	2	5+	109+
31. Scorpaenids[u]*	+	+		+	+		11	72		2	2		85
32. Percids*	+	+	+	+	+	+	210	121	324	1538	62	60	2315
33. Flatfishes*	+	+	+	+	+	+	2	1	1	2	4	3	13
34. Puffers*			+	+	+	+			5+	10+	1	10+	26+
Totals							829+	2328+	3144+	3306+	187+	793+	10587+

*Indicates that species within the order have considerable importance for food, commercial or recreational fisheries, or for aquaculture, aquarium culture.

aIncluding Mexico; bincluding China; cincluding Madagascar; dSumatra, Java, + chain out to Timor, Borneo, Philippines, New Guinea (Plate 2, Anonymous 1999); enot necessarily the same as in 'species using freshwater' in Nelson (1994); fAustralian; gSouth American, African; hsturgeon + paddlefish; iminnows, gyrinocheilids, suckers, loaches; jpikes + mudminnows; ksmelts, noodlefishes, antipodean smelts, galaxids; lwhitefishes, ciscoes, graylings, salmon, trout, char, etc.; mmedakas and related subfamilies, needlefishes, halfbeaks; orivulines, killifishes, topminnows, valenciids, anablepids, poeciliids, goodeids, pupfishes; psticklebacks, pipefishes, indostomids; qswamp-eels, chaudurids, spiny eels; usculpins, oilfishes, abyssocottids.

Table 1.3 Number of cases for major world regions where freshwater fish families[a] occur primarily in inland or coastal areas, and where their distributions are overlapped 70% or more by forested lands (Fig. 1.1A)

Region	Number of cases[b]		Number of cases
	Inland	Coastal	≥ 70% overlap by forests
North America	33	16	1
Central, South America	46	14	41
Africa	33	18	3
Eurasia	54	24	5
Australia, New Zealand	18	22	2
Oceania	34	22	17

[a]As given for 157 freshwater fish family distributions by Berra (1981).
[b]Columns will not total 157 because some families are represented in several regions and others only in one.

Why a book on worldwide fish–forestry interactions?

In addition to at least three bibliographies dealing in part or whole with fish–forestry interaction (Gibbons & Salo 1973; Blackie *et al.* 1980; Macdonald *et al.* 1988), there are a number of publications which cover some aspects of the subject of interaction between fishes and forestry practices. A paper by Salo (1967) contained some important early insights into logging impacts and the ability of salmon to recover from habitat disturbance. One of the most comprehensive works, Meehan (1991), grew out of an earlier series of reports examining relationships between forest and range management and salmonid fishes in western North America, and Murphy (1995) provides more recent coverage. Others are from symposia or conferences on special aspects of the subject (Krygier & Hall 1971; Hartman 1982; Meehan *et al.* 1984; Salo & Cundy 1987; Chamberlin 1988; Laursen 1996; Brewin & Monita 1998). Some deal largely with special aspects of forestry effects and interactions mainly with salmonids in the Pacific Northwest of North America (Brown 1980; Toews & Brownlee 1981; Maser & Sedell 1994), while others consider in great depth and detail a single watershed (Hartman & Scrivener 1990; Candell & Guvå 1992; Hogan *et al.* 1998; Näslund 1999). Many of the earlier studies on watershed deforestation gave little or no emphasis to fish, such as that on the Hubbard Brook Experimental Forest in eastern USA, updated by Martin *et al.* (2000). Exceptions are 5 of the 18 papers summarized by Carignan & Steedman (2000) dealing with a temperate North American assembly of aquatic ecosystem responses to episodic deforestation and acid rain.

For various reasons, the subject to date has concentrated largely on salmonid fishes. These may be severely affected by forestry activities; see for example the relevant section in Crisp (2000). We wanted to extend consideration as much as possible to non-salmonid fishes and to other temperate, subtropical and tropical inland waters of the world, believing that in doing so interesting parallels might be shown and informative differences might emerge. To our knowledge there is no book attempting to give a broad world coverage of fish–forestry interactions based on an integrated ecological understanding of forests, fishes and the inland waters involved.

We feel strongly that relevant professional workers (forestry, fisheries and others) need a comprehensive synthesis of the subject. So too we believe do senior undergraduates, and graduate students (as well as some of their professors), and also the natural resource and environmentally concerned public in general.

What is the historical background of fish–forestry interactions?

Forest removal for timber began millennia ago in Greece and Italy (Abramovitz & Mattoon 1999), and elsewhere in Europe in the Middle Ages (Elliott *et al.* 1998; Williams 2003). The effects on fish in watersheds exposed to such activities were largely unrecorded. So too were those effects associated with later major forest cutting for firewood, agriculture, housing or boat construction in many other parts of the world such as northern Europe, North America and South America. Historically, and even up to the present day in some regions, forest removal has long preceded the development of understanding or concern about its impacts on aquatic systems and especially fish. In the Great Lakes region of North America large white pine trees up to 2 metres in diameter were so abundant in the mid-1800s that loggers considered that those smaller than 1 metre across were 'undersized' (Abramovitz & Mattoon 1999; see also Abrams 2001). Large-scale logging of white pine has been put forward as one of the three main causes for the extirpation of Arctic grayling in watersheds draining into the Great Lakes (Vincent 1962). But the shape and structure of North American forests probably was altered considerably over millennia before the arrival of recent European settlers, as a result of climate changes and early Eurasian colonists (Kloor 2000).

Large-scale old-growth forest logging with major effects on the flora and fauna of watersheds, and possibly also climate, is still carried out in boreal and temperate forests of Canada (Schindler 1998), western Europe and Eurasia. Deforestation by logging and other human activities has occurred in parts of South America, as well as in many of the world's tropical forests (Kummer & Turner 1994; Skole *et al.* 1994; Cochrane *et al.* 1999; Curran *et al.* 1999; Nepstad *et al.* 1999; Gascon *et al.* 2000; Wuethrich 2000; Holden 2002). Indeed almost half of the world's former forest cover is now gone (Abramovitz & Mattoon 1999), with minimal chances of major replacement.

There are also effects related to log transport, especially in streams and rivers. In Sweden man-made river 'floods' by sudden release of water stored behind small dams started in the 1500s and 1600s, and between 1881 and 1965 nearly 200 million logs were floated down one river alone (Birger 1966). This practice must have had serious effects on stream biota including fish but studies on these were not started until the late 1800s (Müller 1962). Similar releases for downstream log transport were used in many parts of North America (Sedell *et al.* 1991) and New Zealand well into the 1900s. Railway and road transport, especially in the construction and maintenance phases, can seriously affect watershed conditions and fish (Cederholm *et al.* 1982; Reid & Dunne 1984; Paulsen & Fisher 2001). The many effects of log processing (milling for lumber, pulping for paper) have a history spanning well over a century, as have many of the silvicultural practices during forest renewal and protection from disease and pests. Application of pesticides to control forest insects dates back close to half a century (Fettes 1962; Warner & Fenderson 1962). Furthermore, forestry-related fire suppression has had unfavourable long-term ecological effects (Holling & Meffe 1996; Kloor 2000), despite the claim attributed to G.W. Bush that restrictions on logging should be relaxed because fewer forests would mean fewer forest fires (Anonymous 2002; see also Malakoff 2002).

Section/chapter organization and rationale

It is the above broad array of topics that we will cover in the next seven Parts (see Contents). Chapters 2–5 provide an overview of the ecology of forests, watersheds (streams and rivers), lakes and estuaries. Then we summarize key information on the biology of fish within these interconnected systems in Chapters 6–9, as a first essential for fish protection and management in forested landscapes. In the next three chapters (10–12) we describe the activities in which the forest industry engages to obtain wood fibre, transport it, process it and regenerate new stands of timber. Then follows a review of the effects that these forestry activities – including pulp and paper mills – have on watersheds, lakes and estuaries, along with their food webs leading to fish (Chapters 13–16). Next there are a series of chapters that we have solicited to describe in more depth the fish–forestry interactions in a number of major regions around the world. Finally we consider the regulation–management–education interfaces reflected in the regional chapters and speculate on means to effect better fish–forestry interactions (Chapters 31–34).

The book has aimed at a broad coverage. Because of this, many of the chapters – such as those on the ecology of forests, streams, lakes and estuaries, which might in themselves each fill a single book – are not treated in great depth. The references selected should lead to more in-depth consideration.

Acknowledgements

We thank Drs D.T. Crisp, J.D. Hall, C.C. Lindsey and P.H.W. Mylechreest for their careful and helpful reviews of the manuscript. Ms G. Edwards kindly made available use of Anonymous (1999) before we could obtain our own copy.

References

Abrams, M.D. (2001) Eastern white pine versatility in the presettlement forest. *BioScience*, **51**, 967–79.

Abramovitz, J.N. & Mattoon, A.T. (1999) Reorienting the forest products economy. In: *State of the World*, (ed. L. Starke), pp. 60–77, 206–213. W.W. Norton & Co., New York.

Anonymous (1999) *The Times Atlas of the World*, 10th Comprehensive Edition. Times Books, division of Random House, New York.

Anonymous (2002) Scorecard, G.W. Bush. *Macleans*, 2 September 2002, 13.

Ben-David, M., Hanley, T.A. & Schell, D.M. (1998) Fertilization of terrestrial vegetation by spawning Pacific salmon: the role of flooding and predator activity. *Oikos*, **83**, 47–55.

Berra, T.M. (1981) *An Atlas of the Distribution of the Freshwater Fishes of the World.* University of Nebraska Press, Lincoln.

Birger, N. (1966) *Samverken med en Flod. Ångermanälvens Flottningförening från Förstadierna til Sandslån Sommaren 1996*. Victor Pettersons Bokindustri AB, Stockholm.

Blackie, J.R., Ford, E.D., Horne, J.E.M., Kinsman, D.J.J., Last, F.T. & Moorhouse, P. (1980) Environmental effects of deforestation. An annotated bibliography. *Freshwater Biological Association Occasional Publication No. 10*, Ambleside, UK.

Brewin, M.K. & Monita, D.M.A. (eds) (1998) *Forest-fish Conference: Land Management Practices Affecting Aquatic Ecosystems*. Natural Resources Canada, Canadian Forest Service, Information Report NOR-X-356.

Brown, G.W. (1980) *Forestry and Water Quality*. Oregon State University Book Stores, Corvallis, OR, USA.

Brown, K. (2000) Pacific salmon run hot and cold. *Science*, **290**, 685–6.

Campbell, I.C. & Parnrong, S. (2001) Limnology in Thailand: present status and future needs. *Verhandlungen Internationale Vereinigung für Theoretische und Angewandte Limnologie*, **27**, 2135–41.

Candell, L.-G. & Guvå, L. (1992) *Från spelflotte till Ådalen III Angermanälven*. CEWE-Förlaget, Bjästa, Sweden.

Carignan, R. & Steedman, R.J. (2000) Impacts of major watershed perturbations on aquatic ecosystems. *Canadian Journal of Fisheries and Aquatic Sciences*, **57** (Suppl.2), 1–4.

Cederholm, C.J., Reid, L.M., Edie, B.G. & Salo, E.O. (1982) Effects of forest road erosion on salmonid spawning and gravel composition and populations of the Clearwater River, Washington. In: *Habitat Disturbance and Recovery: Proceedings of a Symposium*, pp. 1–17. California Trout, San Francisco, CA, USA.

Chamberlin, T.W. (ed) (1988) *Proceedings of the Workshop: Applying 15 Years of Carnation Creek Results*. Pacific Biological Station, Nanaimo, British Columbia.

Crisp, D.T. (2000) *Trout and Salmon: Ecology, Conservation and Rehabilitation*. Blackwell Science, Oxford.

Cochrane, M.A., Alencar, A., Sculze, M.D., *et al.* (1999) Positive feedbacks in fire dynamic of closed canopy tropical forests. *Science*, **284**, 1832–5.

Curran, L.M., Caniago, I., Paoli, G.D., *et al.* (1999) Impact of El Niño and logging on tree canopy recruitment in Borneo. *Science*, **286**, 2184–8.

Elliott, S.R., Coe, T.A., Helfield, J.M. & Naiman, R.J. (1998) Spatial variation in environmental characteristics of Atlantic salmon (*Salmo salar*) rivers. *Canadian Journal of Fisheries and Aquatic Sciences*, **55** (Suppl. 1), 267–80.

Fausch, K., Power, M.E. & Murakami, M. (2002) Linkages between stream and forest food webs: Shigeru Nakano's legacy for ecology in Japan. *Trends in Ecology and Evolution*, **17**, 429–34.

Fettes, J.J. (1962) Forest aerial spraying – dosage concepts and avoidance of hazard to fish and wildlife. *Proceedings of the Fifth World Forestry Congress (1960)*, **2**, 924–9.

Finney, B.P., Gregory-Eaves, I., Sweetman, J. Douglas, M.S.V. & Smol, J.P. (2000) Impacts of climatic change and fishing on Pacific salmon abundance over the past 300 years. *Science*, **290**, 795–9.

Gascon, C., Williamson, G.B. & da Fonseca, G.A.B. (2000) Receding forest edges and vanishing reserves. *Science*, **288**, 1356–8.

Gibbons, D.R. & Salo, E.O. (1973) An annotated bibliography of the effects of logging on fish of the western United States and Canada. *USDA Forest Service General Technical Report PNW-10*, Portland, OR, USA.

Hartman, G.F. (ed) (1982) *Proceedings of the Carnation Creek Workshop, a 10 Year Review.* Pacific Biological Station, Nanaimo, British Columbia.

Hartman, G.F. & Scrivener, J.C. (1990) Impacts of forestry practices on a coastal stream eco-system, Carnation Creek, British Columbia. *Canadian Bulletin of Fisheries and Aquatic Sciences*, **223**, 1–148.

Hogan, D.L., Tschaplinski, P.J. & Chatwin, S. (eds) (1998) *Carnation Creek and Queen Charlotte Islands Fish/Forestry Workshop: Applying 20 Years of Coast Research to Management Solutions.* British Columbia Ministry of Forests, Victoria, Canada.

Holden, C. (2001a) Savable forest. *Science*, **293**, 1587.

Holden, C. (2001b) Greenhouse is here. *Science*, **293**, 1987.

Holden, C. (2002) World's richest forest in peril. *Science*, **295**, 963.

Holling, C.S. & Meffe, G.K. (1996) Command and control and the pathology of natural resource management. *Conservation Biology*, **10**, 328–37.

Jepson, P., Jarvie, J.K., MacKinnon, K. & Monk, K.A. (2001) The end for Indonesia's lowland forests? *Science*, **292**, 859–61.

Kloor, K. (2000) Returning America's forests to their 'natural' roots. *Science*, **287**, 573–5.

Krygier, J.T. & Hall, J.D. (eds) (1971) *Forest Land Uses and Stream Environment.* Oregon State University, Corvallis, OR.

Kummer, D.M. & Turner, B.L. (1994) The human causes of deforestation in southeast Asia. *BioScience*, **44**, 323–8.

Laurance, W.F., Cochrane, M.A., Bergen, S., *et al.* (2001a) The future of the Brazilian Amazon. *Science*, **291**, 438–9.

Laurance, W.F., Fearnside, P.M., Cochrane, M.A., D'Angelo, S., Bergen, S. & Delamonica, P. (2001b) Response. *Science*, **292**, 859–61.

Laursen, S.B. (ed) (1996) *At the Water's Edge: The Science of Riparian Forestry.* University of Minnesota, Extension Service, St Paul, MN.

Lawton, R.A., Nair, U.S., Pielke, R.A. & Welch, R.M. (2001) Climatic impact of tropical lowland deforestation on nearby montane cloud forests. *Science*, **294**, 584–87.

Lowe-McConnell, R.H. (1987) *Ecological Studies in Tropical Fish Communities.* Cambridge University Press, Cambridge, UK.

Macdonald, J.S., Miller, G. & Stewart, R.A. (1988) The effects of logging, other forest industries and forest management practices on fish: an initial bibliography. *Canadian Technical Report of Fisheries and Aquatic Sciences*, No. 1622.

Malakoff, D. (2002) Arizona ecologist puts stamp on forest restoration debate. *Science*, **292**, 2194–96.

Martin, C.W., Hornbeck, J.W., Likens, G.E. & Buso, D.C. (2000) Impacts of intensive harvesting on hydrology and nutrient dynamics of northern hardwood forests. *Canadian Journal of Fisheries and Aquatic Sciences*, **57** (Suppl. 2), 19–29.

Maser, C. & Sedell, J.R. (1994) *From the Forest to the Sea.* St Lucie Press, Delray Beach, FL.

May, E. (1998) *At the Cutting Edge.* Key Porter Books, Toronto.

Meehan, W.R. (ed) (1991) *Influences of Forest and Rangeland Management on Salmonid Fishes and their Habitats.* American Fisheries Society Special Publication 19, Bethesda, MD.

Meehan, W.R., Merrell, T.R. & Hanley, T.A. (eds) (1984) *Fish and Wildlife Relationships in Old-growth Forests.* Proceedings of a Symposium, American Institute of Fishery Research Biologists, Juneau, AK, USA.

Müller, K. (1962) *Flottningens inerkan på fisket.* Druckerei H. Guntrum II. K.G., Schlitz/Hessen, Germany.

Murphy, M.L. (1995) Forestry impacts on freshwater habitat of anadromous salmonids in the Pacific Northwest and Alaska – requirements for protection and restoration. US Department of Commerce, National Oceanic and Atmosphere Administration, NOAA Coastal Ocean Program, *Decision Analysis Series 7*, NOAA Coastal Ocean Office, Silver Spring, MD.

Nakano, S. & Murakami, M. (2001) Reciprocal subsidies: dynamic interdependence between terrestrial and aquatic food webs. *Proceedings of the National Academy of Science USA*, **98**, 166–70.

Näslund, I. (ed.) (1999) *Fiske, skogsbruk och vattendrag-nyttjande i ett uthålligt perspektiv. Erfarenheter från fornskning I Ammerans dalgång*. Fiskeriverkets Försöksstation, Kälarne, Sweden.

Nelson, J.S. (1994) *Fishes of the World*, 3rd edn. John Wiley & Sons, New York.

Nepstad, D.C., Veríssimo, A., Alencar, A., *et al.* (1999) Large-scale impoverishment of Amazonian forests by logging and fire. *Nature*, **398**, 505–8.

Olson, D.M., Dinerstein, E. & Wikramanayaki, E.D., *et al.* (2001) Terrestrial ecoregions of the world: a new map of life on earth. *BioScience*, **51**, 933–38.

Paulsen, C.M. & Fisher, T.R. (2001) Statistical relationship between parr-to-smolt survival of Snake River spring-summer chinook salmon and indices of land use. *Transactions of the American Fisheries Society*, **130**, 347–58.

Power, M.E. (2001) Prey exchange between a stream and its forested watershed elevates predator densities in both habitats. Commentary. *Proceedings of the National Academy of Science USA*, **98**, 14–15.

Ramitanondh, S. (1989) Forests and deforestation in Thailand: a pandisciplinary approach. In: *Culture and Environment in Thailand*, (ed. M. Shari), pp. 23–50. Editions Duang Kamol for The Siam Society, Bangkok.

Reid, L.M. & Dunne, T. (1984) Sediment production from forest road surfaces. *Water Resources Research*, **20**, 1753–61.

Salo, E.O. (1967) Study of the effects of logging on pink salmon in Alaska. *Proceedings, Society of American Foresters*, Washington, DC.

Salo, E.O. & Cundy, T.W. (eds) (1987) *Streamside Management: Forestry and Fishery Interactions*. Contribution No. 57. College of Forest Resources and Institute of Forest Resources, University of Washington, Seattle, WA.

Schindler, D.W. (1998) A dim future for boreal waters and landscapes. *BioScience*, **48**, 157–64.

Scott, W.B. & Crossman, E.J. (1973) *Freshwater Fishes of Canada*. Bulletin 184. Fisheries Research Board of Canada, Ottawa.

Sedell, J.R., Leone, F.N. & Duval, W.S. (1991) Water transportation and storage of logs. In: *Influences of Forest and Rangeland Management on Salmonid Fishes and their Habitats*, (ed. W.R. Meehan), pp. 325–368. American Fisheries Society Special Publication 19, Bethesda, MD, USA.

Silveira, J.P. (2001) Development of the Brazilian Amazon. *Science*, **292**, 1651–52.

Skole, D.L., Chomentowski, W.H., Salas, W.A. & Nobre, A.D. (1994) Physical and human dimensions of deforestation in Amazonia. *BioScience*, **44**, 314–22.

Stokstad, E. (2001) U.N. report suggests slowed forest losses. *Science*, **291**, 2294.

Toews, D.A.A. & Brownlee, M.J. (1981) *A Handbook for Fish Habitat Protection on Forest Lands of British Columbia*. Canada Department of Fisheries and Oceans, Vancouver, BC.

Vincent, R.E. (1962) *Biogeographical and ecological factors contributing to the decline of Arctic grayling, Thymallus arcticus Pallas, in Michigan and Montana*. PhD thesis, University of Michigan, Ann Arbor.

Warner, K. & Fenderson, O.C. (1962) Effects of DDT spraying for forest insects in Maine trout streams. *Journal of Wildlife Management*, **26**, 86–93.

Williams, M. (2003) *Deforesting the Earth*. University of Chicago Press, Chicago.

Willson, M.F., Gende, S.M. & Marston, B.H. (1998) Fishes and the forest. *BioScience,* **48**, 455–62.

Wuethrich, B. (2000) Combined insults spell trouble for rainforests. *Science*, **289**, 35–37.

Part II
Ecology of the Systems

Chapter 2
Forest ecology
The study of the ecological (spatial), biological and temporal diversity of forest ecosystems

J.P. KIMMINS

Introduction

The topic of forest ecology is too broad to be summarized adequately in this chapter, so this is not attempted. Rather the chapter focuses on the key issue of forest stewardship and sustainability: managing the dynamics of forest ecosystems in terms of a balance between ecosystem disturbance and post-disturbance ecosystem development. For a more detailed treatment, see Kimmins (1997, 2003a).

Forest ecology is the study of forest ecosystems – ecosystems in which ecological processes are dominated by trees. Trees are frequently, but not always, the dominant plant life form in humid, precipitation-rich climates that have adequate summer warmth, and forests are therefore intimately related to the rivers, streams and other aquatic habitats that occur or arise in such climatic areas. Because of the intimacy of the forest–water interconnections, most of the ecosystems that feed river systems either are forested, or were forested before land clearance for agriculture and other land uses. Their physical, chemical and biological characteristics are, or were previously, determined largely by the presence of live and dead trees. The fish and other aquatic organisms of forest streams – the first, second and third order streams that arise in and drain forested landscapes – are, to a great extent, forest organisms. The characteristics of their aquatic environment and the majority of their food supply and nutrients, for part or all of their life cycle, originate as forest plant biomass and litterfall. Understanding, predicting and managing the hydrology and ecology of these water bodies and the aquatic plants, animals and microbes they support must, therefore, be based on an understanding of the ecology of forest ecosystems.

The world's population doubled between 1960 and 2000 – an increase of three billion. Fortunately, the rate of human population growth is slowing, but it is still expected that this century will see another three to five billion people adding to the pressure that our species is putting on the planet's forests, water and fish resources (Lutz *et al.* 2001; UN 2001a,b; IIASA 2001). Even after the population has peaked and possibly declined somewhat, pressure on the world's forests and forest resources is expected to continue to increase for a considerable period of time as per capita standard of living rises (Wackernagel & Rees 1996). It is therefore imperative that humans establish a more sustainable relationship to their environment and its resources now, before more damage is done to the forests and the water they yield in areas where the population continues to increase.

The biophysical issues of ecosystem sustainability, productivity, biodiversity and carbon budgets, and the overarching issue of climate change, have dominated the discussion about the environment for the past decade. However, resolution of the present conflict between the human species and the global environment is fundamentally a political and social rather than a natural science issue. While it is unlikely that we will find a sustainable solution without a sound ecological foundation for the evolution of policy and practice, as much or more work is needed on the social, cultural, economic and political aspects of our environmental problems without which all the natural science in the world will not achieve our objectives (see, for example, Lugo 1995). The review of some of the key concepts and issues of forest ecology in this chapter is made in the context of contributing to this ecological foundation, but it is not considered that such information will, on its own, provide the solution to fish, water and watershed issues.

This chapter examines the three key dimensions of forest ecosystems that are important from both a conservation and a management perspective: ecological diversity (the ecological stage, set by the climate, topography, geology, soil and physical disturbance factors), temporal diversity (the ecological play, representing ecosystem disturbance and the ecosystem process of post-disturbance recovery and development), and biological diversity (the ecological actors, determined by the ecological stage and the ecological play). In the face of 6 billion people headed for 9 to 11 billion, and with the threat of human-induced climatic change, ecosystem response to natural and human-induced disturbance has become the key issue and is, therefore, the focus of this chapter.

Forest ecosystem attributes

Forest ecology is the study of forest ecosystems. A *forest* is a terrestrial ecosystem in which the ecological processes, the physical and chemical conditions, and the availability of plant, microbial and animal habitats are determined by the dominance of the system by trees. A forest may refer to a stand-level ecosystem (e.g. 1–100 ha), or to a local (e.g. 100–1000 ha) or regional landscape (e.g. 1000–100,000 ha or more). A forest stand is defined as: an aggregation of trees occupying a specific area and sufficiently uniform in composition, age, arrangement and condition so that it is distinguishable from the forest in adjoining areas. Stands are the basic management unit in silviculture (Dunster & Dunster 1996). Forested landscapes can, and generally do, contain aquatic and terrestrial areas that are not occupied by trees, but the overall ecological, hydrological and biological character of the landscape still reflects the dominant role of trees.

A *terrestrial ecosystem* is any system of biological and physical components and processes of energy and material exchange that exhibits the following five attributes: structure, function, interconnectedness, complexity and change over time.

Aquatic ecosystems have similar attributes, but have structural, functional and temporal differences in comparison with terrestrial ecosystems, depending on the type of aquatic ecosystem involved. Large lake ecosystems differ considerably from forest ecosystems, whereas first order forest streams share more in common with their terrestrial counterparts. Many aquatic ecosystems exhibit less change or slower change over time than many terrestrial ecosystems.

For a detailed treatment of ecosystem function, physical factors, and population and community structures and processes, see Kimmins (2003a). The major focus in the following discussion is on the attribute of change over time.

Structure

Forest ecosystems exhibit a variety of structural characteristics of which the mass and spatial arrangement of live and dead plant biomass are the most important. They determine habitat for forest animals and microbes, and affect soil and hydrological processes. The vertical and horizontal variability in plant structure is a major component and determinant of biological diversity. Multiple plant canopy levels (trees, shrubs, herbs, bryophytes, epiphytes and climbers), the diversity of sizes of standing dead trees (snags) and decomposing logs on the ground (coarse woody debris or CWD), and the horizontal variation (the spatial variability) in vertical live plant structure, snags and CWD are the major components of forest community structure.

Other components of structure are the soil and the animal and microbial communities. While not strictly speaking 'structure', microclimate has traditionally been considered to be part of the structural ecosystem attribute. However, the creation of microclimate can be considered a functional aspect of forest ecosystems.

Function

The creation of biomass by plants is the single most important functional process in terrestrial ecosystems. It creates the energy (food) supply that drives most animal and microbial populations, and is the dominant foundation for biotic community food webs. Plant biomass also creates habitat for animals and microbes, and produces acids and other chemicals that are important in geological weathering and soil development, and thus in nutrient cycling.

The creation of plant biomass requires nutrients. As a consequence, a major function of ecosystems is regulation of the inputs of nutrient chemicals into ecosystems, the mobilization of unavailable nutrient and other elements from inorganic and organic sources, and the circulation, concentration and retention of nutrients within the ecosystem.

A third important function of terrestrial ecosystems is the regulation of the hydrological cycle. Precipitation inputs to forests are largely determined by air mass movements and seasonal variations in temperature interacting with topographic features, and by systems of variation in atmospheric pressure. Plant biomass generally plays a modest role, although effects such as fog drip and snow accumulation can be significant in some forests. In contrast, trees play a major role in regulating the loss of water back to the atmosphere through canopy interception loss, evaporation and transpiration. They also regulate snowmelt, surface runoff, infiltration into soil, water storage in and movement through the soil, and the quantity, quality and regimen of water moving out of the terrestrial ecosystem into aquatic ecosystems.

In forest ecosystems the creation of an organic forest floor, the maintenance of mineral soil organic matter and mineral soil structure, and the creation of root channels that allow for rapid conduction of water from the soil surface to groundwater, are important contributions to forest hydrology and the character of associated water bodies. The production of coarse woody debris (CWD: decomposing logs) that limits surface runoff on slopes and creates and sustains stream habitat diversity is another important role of the forest.

The creation of microclimates as a plant-mediated modification of local climate is also a major functional role of ecosystems.

Interactions and interconnectedness between ecosystem components

A key attribute of any system is the interactions between its structural components and functional processes. An assemblage of soil, plants, animals and microbes that are not interacting, if this were possible, is not an ecosystem. Only when these structural components are arranged in such a way that they interact to function in biomass creation, energy flow, material exchanges, hydrological control and microclimatic regulation do they constitute an ecosystem. There is a wide range of interactions, from the association of soil fungi with the tree roots that is critical in tree nutrition and access to soil water, to interactions between minor vegetation and trees, and the action of pathogens that result in the death of individual trees and create snags and CWD. It is the interconnectedness of the structural components and functional processes of forest ecosystems that defines their characteristics and response to disturbance, and constitutes the ecosystem. This interconnectedness requires that forests be managed as a system rather than managing the component parts or values separately.

Complexity

It is an inevitable feature of a system that has many structural components, many functional processes and a wide range of interactions, that it is difficult to predict with acceptable accuracy the characteristics and change over time of such a system, or any one component or process thereof, unless one has a sound knowledge of the overall system structure, function and interactions (Chapter 3 in Kimmins 2003a). There is a close (but not always linear) and negative correlation between unaccounted for complexity and the ability to predict future states of a system. This gives an apparent stochasticity and unpredictability to complex forest ecosystems when knowledge of the system is incomplete. This is frequently the reason why management for individual ecosystem components, such as timber, sometimes has unexpected outcomes. The incomplete knowledge and resultant low predictability contribute to the belief by some that forest ecosystems behave chaotically, cannot be predicted and therefore cannot be managed, a belief that reflects the lack of knowledge of ecosystems and their structures and processes on the part of such believers more than any fundamental characteristic of a forest ecosystem. However, complexity in the face of incomplete knowledge and understanding certainly renders the management of forest ecosystems to achieve particular objectives difficult.

Ecosystem complexity includes characteristic levels of functional redundancy that enables ecosystems to function normally while undergoing change that periodically removes or adds individual species and structures.

Change over time

Note that the following discussion of change is very general. As with all generalizations, there will be many examples from around the world where these generalized concepts will be incorrect. In fact there are only two words in ecology and forest management that appear to be almost inevitably true: *It Depends*. For example, the relationship between forest ecosystem disturbance and measures of biodiversity can be very different between some tropical and some temperate or boreal forests.

Ecosystems are living systems, not static entities. While they include important physical, inanimate structural components and processes, they are dominated by living components and biotic processes. Implicit in all living systems is change over time, and this biotic change results in change in the physical components and processes of the ecosystem, complementing abiotically induced physical changes.

Ecosystems are not analogous to human-created engineering systems such as mechanical watches or moon rockets. These engineering systems may fail if individual pieces are removed and their structure and component parts change, unless functional redundancy has been designed into the system. Ecosystem structure, species composition, function, degree of interconnectedness and complexity are continually changing as a result of 'natural' (non-human) and human-caused disturbance and the processes of post-disturbance ecosystem development or recovery. (Humans are part of nature, but I have followed the convention here of 'natural change' being a reference to non-human processes – see Peterken (1996) for a useful discussion of 'natural'.)

Change in ecosystems, which is broadly referred to as *succession*, is the result of many processes:

(1) Physical factors, which produce *allogenic* succession, such as fire, wind, snowstorm, flood, landslide, or changes induced by climate change. These processes are controlled by forces external to the ecosystem in question and they cause ecosystem disturbance (see definition later in chapter) – they alter the rates and patterns of ecosystem change expected from autogenic succession (see below).

(2) Biotic processes, which disturb ecosystems to initiate new sequences of biotic communities and physical conditions. Such *biogenic* successional processes include insect and disease epidemics of native species, invasions of non-native species, and human disturbance of the forest. Biogenic succession is differentiated from autogenic succession (below) in that biogenic factors cause ecosystem disturbance whereas autogenic processes do not. The distinction between biogenic succession and autogenic succession is based on the definition of ecological disturbance given later in the chapter. In biogenic succession, the sequence, rate and direction of successional change are determined by biotic factors, processes or events that alter the rates and pathways of succession that are expected as a result of autogenic succession.

(3)   Internal ecosystem processes, referred to as *autogenic* succession. These are the population and community processes of invasion and colonization of a particular biotic community by additional native species, the environmental alteration (e.g. in light levels and availability of soil nutrients) caused by the plant community, and the subsequent exclusion (by competition or other processes) of the original species in the community. These processes collectively result in a sequence of biotic communities that successively occupy and are replaced over time in a particular ecosystem. This sequence of communities is accompanied by alterations of the physical environmental conditions that are generated by the community.

Change in ecosystems is frequently a necessary component of long-term ecosystem stability and 'healthy' functioning (Pickett *et al.* 1992; Attiwill 1994a, b; Rogers 1996; Kimmins 1996, 2000a, 2003a; Perry and Amaranthus 1997; Perera *et al.* in press). Change occurs as plants establish, mature, die and are replaced by new individuals of the same or different species. As this predictable and inevitable change in plant populations and communities occurs, there is change in soils, microclimates and hydrology; in animal and microbial habitats and communities; and in the rates of ecosystem processes. The change in soils that occurs both as a result of ecosystem disturbance and during the course of autogenic succession is frequently necessary for, or beneficial to, the subsequent plant community, and therefore to the associated animals and microbes.

In the absence of biotic change, parasites, predators and diseases of particular plant or animal species may build up to epidemic levels. Successions of plant, animal and microbial communities (temporal diversity) prevent this and help to maintain these relationships at levels that do not disrupt ecosystem function excessively. Thus, if there is such a thing as a 'balance of nature', it is the changes in ecosystem conditions and species composition over time that result in non-declining patterns of ecosystem change within the limits of the historical range of variation. Local extirpations of species with geographically extensive ranges are often part of the long-term stability and health of regional meta-populations. However, such local extirpations can lead to species extinctions of 'endemic' species that have very limited geographical ranges, such as is the case in some tropical forest species.

By periodically replacing whole biotic communities, succession increases biological diversity (both landscape scale diversity and stand-level temporal diversity; it may temporarily increase or decrease other measures of stand-level diversity). Similarly, disturbance is positively related to many measures of biodiversity if it acts to initiate new successional sequences. This topic is explored further below.

Disturbance, whether allogenic or biogenic, can be a key process in maintaining the productivity and historical character of many temperate and boreal forest ecosystems, and in some types of forest is essential for their long-term persistence, as noted in the examples given below (see also Attiwill 1994a, b; Kimmins 1996, 2003a; Perera *et al.* in press). The accumulation of woody material (CWD) in the forest floor and soil of a forest certainly has some habitat and hydrological benefits, but over time it can, in some forest ecosystem types, reduce the availability of nitrogen and the rate of nutrient cycling (e.g. Pastor & Post 1986; Keenan *et al.* 1993). This is because the higher levels of lignins, tannins and other complex organic chemicals that are present in woody material, and which soil microbes are unable to decompose rapidly, reduce the rate of forest floor mineralization and nutrient circulation. This impairs tree nutrition, which reduces tree leaf area and consequently reduces tree photosynthesis and biomass creation. The lower tree leaf area allows more light to penetrate the forest canopy, increasing the growth of the understory. While this may benefit wildlife species that are dependent on herbs and shrubs, it increases competition for soil water and nutrients between trees and these other plant life forms, further reducing tree growth. Sometimes this increased competition renders the trees more susceptible to diseases, parasites and insects. Sometimes disturbances can lead to stand replacement, reducing competition between trees and other plant types and the dominance of the soil by decaying wood, both leading to an increase in ecosystem productivity; sometimes they can have the opposite effect.

It is commonly thought that trees are the most competitive and evolutionarily successful plant life form in forests. After all, trees dominate all other types of plant in closed canopy forests. However, in many cool, humid climates, shrub species belonging to the Heather family (the *Ericaceae*) or comparable southern taxa (e.g. *Ipacridaceae*) are competitively superior to trees in the long-term absence of disturbance. Because these shrubs can reproduce vegetatively by rhizomes, tolerate low levels of nutrients and develop very dense fine root systems, they may be able to reduce nitrogen availability to trees. As such, shrubs may be well adapted to colonize gaps in the forest that are created when individual trees die and restrict tree regeneration in these gaps. In the long-term absence of disturbance, this can gradually convert some types of closed forest into shrub-dominated woodland or even shrub heathland (Dammann 1971; Malcolm 1975; Messier & Kimmins 1990; Fraser 1993; Prescott & Weetman 1994; Titus *et al.* 1995). Sphagnum moss may similarly take over northern forests by reducing soil temperatures and access to essential soil nutrients by the trees and/or by altering soil moisture conditions (Heilman 1966, 1968; Banner *et al.* 1983). It is severe, stand-replacing disturbance, such as fire, landslides, insect epidemics or human-induced ecosystem disturbance events that periodically tip the competition balance back in favour of trees in such ecosystems. Where such processes are operating, the long-term maintenance of closed forest conditions may require periodic disturbance at some particular spatial scale, severity and frequency. The critical combination of these disturbance characteristics that will sustain particular forest conditions and attributes, including biological diversity and ecosystem function, can be identified through the concept of *ecological rotation* (discussed below).

Important diversities in forests: part of the complexity attribute of ecosystems

'Diversity' has become a scientifically important as well as a politically correct concept in forestry and conservation, as well as in other aspects of society. But what is it? There are three major dimensions of diversity that must be addressed in designing sustainable forestry.

Ecological diversity – the 'ecological stage'

The ecological stage determines what species, what type of forest ecosystem, and what natural disturbance–ecosystem recovery sequences can be expected. The ecological stage is defined by the regional and local climate, by the geology, soil and topography, and by the characteristic regimes of physical disturbance such as wind, fire, snow, flood and landslide or avalanche (allogenic factors). These physical factors determine the frequency, scale and severity of physical ecosystem disturbance and change, and help to determine the rate of processes of ecosystem recovery by autogenic succession. They determine which species (ecological actors) can 'perform' on the ecological stage and thus which 'ecological play' (disturbance–recovery successional sequences of species, community structure and function) will occur.

Ecological diversity determines to a great extent the spatial diversity of forests – how they vary from place to place. It is described by ecological site classification schemes such as the biogeoclimatic classification of British Columbia (Meidinger & Pojar 1991).

Biological diversity – the 'ecological play'

Biological diversity is the genetic, taxonomic, structural, functional and adaptational diversity of organisms, populations and communities. Within the ecological framework set by ecological diversity, the various measures of biological diversity are determined by the type, severity, frequency and scale of physical and/or biological disturbance to the ecosystem interacting with the adaptations of the species involved and their population and community processes. Disturbance is defined here as: *any allogenic or biogenic factor or event that alters the rate, pattern and pathway of ecosystem change that is expected to result from the autogenic successional processes operating in the ecosystem in question* (c.f. Kimmins in press).

Disturbance removes or alters population levels of individual species depending on their genetically controlled tolerances and requirements. It alters the frequency of different genotypes within populations. It affects the population and community processes that collectively determine which species arrive and colonize a disturbed area, which species subsequently invade and displace these early biotic communities, and which species will be able to resist further invasion and create at least temporarily self-replacing communities.

There are many measures of biological diversity, and each of these can be evaluated at several (traditionally, three) different spatial scales: local stand level (*alpha*), local landscape level (*beta*) and regional landscape level (*gamma*). Beta diversity is the variation in alpha diversity measures between stands in the local landscape due to differences in soil and disturbance history. Gamma diversity is the variation in alpha and beta diversity measures along climatic gradients across a region.

Temporal diversity – changes in the species list, the community structure, the ecosystem processes and the physical/chemical conditions in the ecosystem over time

The levels of all the measures of biodiversity change over time. This temporal diversity is most obvious at the stand (alpha) level, but can also be seen as a shifting mosaic of changing ecosystem conditions and characteristics at the local (beta) and regional (gamma) landscape scales. Temporal diversity is linked to the concept of stability. Stability at the stand level is non-declining patterns of change, which involves a balance between the frequency and severity of ecosystem change and the resilience of the ecosystem – its rate of recovery (discussed further below). At the local and regional landscape levels, stability is a shifting mosaic of changing stand-level conditions. At the regional level the overall character of the mosaic may be relatively constant, whereas at the local landscape level it may change over time if the forest is subject to periodic large-scale disturbance. Accompanying the change in the structure and function of the biotic community over time there is change in the soil and microclimatic conditions.

Accompanying the changes in biodiversity measures over time, there are changes in soil, microclimate and hydrological attributes and processes. The complexity of biodiversity is explored in Bunnell & Huggard (1999).

Forests and sustainability: the concept of ecological rotation

The concepts of sustainability and ecosystem change initially appear to be incompatible. How can a system be sustainable if it is constantly changing? The answer lies in the widely accepted concept that forest ecosystems are dynamic, and that 'stability' at the stand level means a non-declining pattern of change. But how can one assess whether a particular forest ecosystem disturbance is part of a non-declining pattern and is therefore sustainable, or is part of a directional and non-sustainable change? An answer can be found in the concept of ecological rotations (Kimmins 1974).

An ecological rotation is the time taken for a particular ecosystem condition, structure, composition or other attribute to return to pre-disturbance levels following disturbance, or for recovery to some new, socially desired condition. Ecological rotations are defined by the severity (degree of ecosystem change) and extent (spatial scale, which influences the processes of recovery) of the disturbance, interacting with the rate of recovery of ecosystem conditions – the rate of autogenic succession. Where the recovery rate is slow, ecological rotations will be longer than where the rate is fast, for any particular severity and spatial scale of disturbance. Where the degree of disturbance-induced ecosystem change is high, the ecological rotation will be longer than where the change is less, for any particular rate of ecosystem recovery. If the severity of disturbance is high, stability must involve less frequent disturbance; where the severity is low, the ecosystem can be sustained in the face of frequent disturbance (Kimmins 1974, 2003a).

Ecosystem sustainability at the stand level cannot be judged by any one of: disturbance severity, disturbance scale, disturbance frequency or ecosystem resilience (rate of autogenic succession). It is only when all of these factors are considered as an interacting system of determinants that sustainability of local forest ecosystems, and thus concepts such as stewardship and ethical forestry, can be assessed. Simple visual observation of short-term, stand-level, ecosystem change following a disturbance provides a very unreliable basis for evaluating sustainability, biodiversity and stewardship questions. Yet most of the debate about forest stewardship is currently based on such 'snapshot' evaluations of local ecosystem change, to the exclusion of a dynamic evaluation of ecosystem recovery and the frequency of future disturbance events over ecologically meaningful time (multiple decades or centuries) and spatial scales (landscape as well as stand).

Definition of what severity and which scale of disturbance are consistent with ecosystem function, biodiversity, social and other considerations will depend on what ecosystem conditions, functions and values are to be sustained, what type of ecosystem one is dealing with, and what the historical relationship has been between disturbance and the desired functions and values (i.e. the historical 'natural' disturbance regime) in the ecosystem in question.

In some temperate and northern coniferous-dominated forest ecosystems, shade-intolerant, disturbance-adapted, early successional deciduous hardwood tree species play a very important, although relatively temporary, ecological role. Their deciduous nature can result in warmer soils in cold forest climates. Their litterfall is often more decomposable than coniferous litterfall, resulting in increased soil animal and bacterial activity, and greater rates of decomposition and nutrient cycling. This can improve soil structure and soil fertility, resulting in higher organic production in subsequent, conifer-dominated seral stages, and may result in better infiltration of water into the soil during heavy rain or rapid snowmelt events. Some such trees (e.g. alder and birch) establish relationships with micro-organisms that convert inert atmospheric nitrogen gas into organic nitrogen, and thus increase nitrogen availability. This can acidify the soil, causing accelerated weathering of primary soil minerals, which releases other essential plant nutrients (e.g. potassium, magnesium, phosphorus, sulphur, micronutrients) and other chemicals (e.g. calcium) that play important roles in soil development and nutrition.

Many of these early seral deciduous hardwood species are fast growing, relatively short lived, and subject to stem decay, which creates an early supply of snags and stem cavities for cavity nesting birds and mammals. These hardwoods may also be resistant to soil fungi that decay the roots of, and kill, conifer trees, or render the conifers susceptible to wind- or snow-induced mortality. Periods of ecosystem occupancy by such hardwoods induced by stand-replacing disturbance can maintain lower levels of these conifer root and stem diseases. Except during epidemics of coniferous insects, the hardwoods are often associated with higher sustained populations of insects than adjacent conifer stands, providing food for birds and for aquatic organisms as the insect larvae fall from streamside trees. Mixed species stands of disturbance-dependent, early seral deciduous hardwoods and later successional evergreen conifers can result in higher levels of ecosystem productivity than either group of trees alone because of ecological niche diversification. Clearly, these disturbance-related hardwoods are an important component of the landscapes where they occur, suggesting the need for appropriate severity and frequency of management-induced ecosystem disturbance if natural disturbance events are not sustaining desired abundance of such hardwoods in the landscape. The benefits of mixtures can be achieved either in time (alternating hardwood and conifer seral stages; temporal mixed woods) or in space (mixed species stands; spatial mixed woods).

High severity, large spatial scales and/or high frequencies of ecosystem disturbance will maintain landscapes in early seral biotic communities to the exclusion of ecosystem structures, functions and values that are associated with late seral forests. Conversely, low severity, small-scale and/or infrequent disturbance regimes will sustain late seral conditions and values to the exclusion of the early seral stages (Kimmins in press). Depending on the desired values and the relationship between ecosystem sustainability and disturbance, either of these successional extremes may be consistent with management objectives. Generally this will not be the case for reasons given above. More commonly, a shifting landscape pattern of different seral stages and ecological conditions will be needed to maintain desired diversities of values and conditions across a forested landscape, and will permit the temporal sequences of seral stages within particular ecosystems that constitute the true 'balance of nature'. For many forest ecosystems, attempts to hold the ecosystem in any one successional stage through repeated management cycles will fail to sustain desired measures of ecosystem productivity, 'health', biological diversity and ecosystem function, and will fail to emulate past natural disturbance regimes and the resultant natural range of variation in ecosystem conditions. However, some unmanaged forest ecosystems do exist naturally in a narrow range of early, mid or late seral conditions as a result of the disturbance regimes to which nature subjects them.

Which of these two approaches will emulate natural disturbance most closely will depend on the forest type and its natural disturbance history. As a consequence, all the different silvicultural systems of forest stand management will normally have relevance somewhere in the diversity of forest types that exist in most forest regions (Kimmins in press). None of the classic silvicultural systems mimic natural disturbance precisely, largely because they *all* take tree stems away, whereas many 'natural' disturbances leave much of the stem mass on site. However, 20–30% of the biomass of a mature tree is typically left on site following clear-cutting in the form of stumps, roots, bark, branches, treetops and broken or decayed pieces of stem. The importance of decaying root systems for the growth of the next tree crop has been demonstrated (Van Lear *et al.* 2001), as has the important role of stumps in post-disturbance collembola abundance and diversity (Marshall *et al.* 1998). Within the range of available silvicultural (tree management) systems it is possible to identify one that will emulate a particular disturbance most closely. Although we have inadequate empirical evidence because of our generally short period of forest management, it would appear that employing different management-induced disturbance severities at different frequencies over time would sustain the greatest range of seral stages and the associated biological diversity (Kimmins in press). Using any one level of management-induced disturbance across ecologically variable landscapes will result in lower levels of diversity.

Achievement of the right balance between disturbance and ecosystem recovery requires an understanding of ecosystem components and processes, and how these vary spatially and over time (i.e. ecosystem-level forest ecology). Such knowledge can be found in most standard forest ecology textbooks (e.g. Kimmins 2003a).

Application of forest ecology knowledge in sustainable forest management

Successful management of forests to sustain desired values and conditions requires knowledge of ecological, biological and temporal diversity.

Ecological site classification and biotic inventories: identifying ecological and biological diversity

The first step in designing forest management strategies (policies) and selecting appropriate tactics (practices) is to identify the types of forest ecosystem (the ecological diversity) in the management area in question through the process of ecosystem and site classification (e.g. the biogeoclimatic or BEC classification of British Columbia; Pojar *et al.* 1987; Meidinger & Pojar 1991; Kimmins 2003a). Climates, soils, geology, topography and physical disturbance regimes should be identified, followed by an inventory of the biotic communities that are occupying this diversity of ecological conditions. The biotic inventory should then be interpreted in terms of its relationship to the physical environment, its temporal patterns of change (the successional relationships in the area in question), and the ecosystem disturbance regimes that have historically affected the area.

Having established this ecological inventory and understanding, the next step is to identify the range in values and ecosystem conditions that exist or have existed across the landscape and are ecologically possible today and in the future, and which portion of this range is desired and socially acceptable. This leads to a statement of the management objectives in terms of spatial patterns and temporal sequences of change in the desired conditions and values. Having defined the objectives, combinations of management-induced and anticipated 'natural' disturbance regimes are defined that are consistent with the ecology and sociology of the desired spatial and temporal forest patterns (e.g. Lieffers *et al.* 1996; Bergeron *et al.* 1999).

Determination of appropriate management-induced disturbance regimes: stand dynamics vs succession

Where historical disturbance has occurred with a frequency that is similar to the time-scale of timber production, and at a spatial scale that is socially acceptable, emulation of natural disturbance may be an appropriate forest management paradigm (c.f. Denslow 1995; Lugo 1995). Where past disturbance was infrequent but large-scale and severe, natural events may not be a socially acceptable template for management (Kimmins in press).

Discussions of emulation of natural disturbance have sometimes assumed that the different natural disturbance types are unique, quantifiable events that produce a characteristic and limited range of ecosystem impacts. In reality, most types of natural disturbance can result in a wide range of severity, spatial scale and ecosystem change. Fire, wind and insects can kill virtually every tree on hundreds of thousands of hectares, from ridge tops to valley bottoms, including riparian areas. Alternatively, they may produce scattered, small-scale mortality that merely reduces stand density and within-stand competition, and may leave riparian stands intact. Any intermediate level of tree mortality and spatial patterns of disturbance is possible. Landslides tend to produce a narrow range of severity (generally high) and scale of disturbance (generally quite small to medium). Similarly, diseases generally produce only scattered patch disturbance rather than replacing entire forest stands over extensive areas, and the severity of ecosystem disturbance is often quite low, unless it predisposes the area to an increased risk of fire and wind disturbance. This suggests that in emulating natural disturbance the full range of traditional silvicultural systems, modified where appropriate to retain individual trees or groups of trees (variable retention silviculture – Franklin *et al.* 1997, 2001; Mitchell & Beese 2002), should be employed as appropriate for the values and ecosystem types concerned, from clear-cutting to individual tree selection harvesting.

Stand-replacing natural or management-induced disturbance results in essentially even-aged forests. These may be single tree species (monoculture) stands because of one or more of availability of seeds, suitability of seedbeds, tolerance of the disturbance-induced microclimate, seed predation, herbivory, seedling diseases or other factors. Alternatively, they can be multi-species stands. Low severity, small-scale disturbance can result in uneven-aged, multi-canopy layer monoculture stands of shade-tolerant tree species, or mixed communities of intermediate to high shade tolerance if seed sources of this diversity of tree species is available. Post-disturbance planting can result in mixed or single species stands depending on what is planted and the natural regeneration that augments the planted trees. Planting generally results in even-aged stands, but can result in uneven-aged stand structures if there is continuous recruitment of shade-tolerant natural regeneration.

Even-aged stands, whether 'natural' or produced through forest management by clear-cutting, patch-cutting, seed tree or shelterwood systems, are generally dense at early ages (a large number of trees per hectare), and this density declines as the stand gets older due to competition-related stand self-thinning. This process takes the stand through four recognizable phases:

(1) *Stand initiation*, in which populations of trees of one or more species establish following the period of competition reduction (and, frequently, soil disturbance) produced by stand-replacing disturbance or the later phases of development of the previous seral stage.

(2) *Stem exclusion*, the phase following crown closure (when the canopies of individual small trees, touch, overlap and intercept most of the incoming light) in which light competition kills understory vegetation and less competitive trees, stand density declines and the remaining trees increase in stem and canopy size.

(3) *Understory re-initiation*, the phase when dying trees are large enough to leave rather persistent gaps in the canopy and the wind-induced swaying of taller trees causes branch breakage, both of which increase light penetration to the ground and permit the development of understory vegetation. This phase also facilitates

the establishment of seedlings of the more shade-tolerant tree species of the next seral stage if there is a source of their seed. Thus, the understory re-initiation phase of one seral stage is also the stand initiation phase of the subsequent seral stage.

(4)   *Old growth*. This is the phase during which the individuals of the initial population/community of trees are approaching maximum size and longevity, experiencing physiological weakening and increased physical damage (e.g. wind, snow), disease and/or insect damage, and exhibiting other symptoms of old age. The abundance of large snags and large CWD produced by the survivors of the initial cohort reaches its maximum for that seral stage. At this development phase of the first cohort, the population of the next seral stage is developing through its stem exclusion and possibly also its understory re-initiation phase, depending on the longevity of the earlier seral species. Thus, there is a continuing overlap of the various stand development phases of the different seral stages as autogenic succession proceeds.

These four phases of stand development occur in each seral stage of ecosystem development (i.e. succession). Stand development and succession are related but different phenomena. If there is no invasion of more shade-tolerant species during the understory re-initiation and old growth phases of a particular seral stage, then there is no succession – only stand development. If the loss of canopy cover in the old growth phase is sufficient, and no new species invade, seedlings of the present overstory cohort may establish to produce a self-replacing seral stage, which is referred to as a climax. Alternatively, canopy gaps may be colonized by shrubs and/or herbs, leading to a progressive loss of closed forest cover (Kimmins in press). Closed forest is not necessarily the climax condition in the long-term absence of disturbance. A useful review of both stand development and succession in Vancouver Island forests of British Columbia can be found in Trofymow & MacKinnon (1998). A broader review of the old growth condition is presented in Kimmins (2003b).

One of the most interesting aspects of stand development from a wildlife, stream and fisheries perspective is the production of snags and CWD. Because the stem exclusion phase produces relatively small dead stems, it is not as useful for wildlife habitat and the development of stream diversity as the understory re-initiation and old growth phases that produce larger diameter snags and CWD. Similarly, shrubs and tree regeneration that are important for stream bank stability and fish habitat of low order streams are poorly developed in the dark, stem exclusion phase, which is therefore also of reduced value for foraging species of terrestrial wildlife. Forests that do not develop beyond the stem exclusion phase because of high frequency of natural disturbance, or relatively short rotation timber management, thus tend to have lower wildlife and fish habitat values. This is why it is important to manage streamside (riparian) stands either on much longer rotations or with reduced tree density through thinning, or to exclude tree harvesting entirely from these areas, to ensure that features of understory re-initiation and old growth phases of stand development are sustained either continuously or periodically over an appropriate proportion of the stream length. Minimal disturbance of riparian areas may lead to the loss of deciduous hardwoods as noted above. The desire to have both late stand development phases and/or late seral stages in the riparian zone, but to also have some stream sections with early seral hardwoods, suggests a diversity of disturbance regimes along the watercourse.

Timber management, whether it involves clear-cutting or low disturbance partial harvesting, has typically truncated both the sequence of seral stages and the series of stand developmental phases within a seral stage. Where this occurs, some measures of biological diversity of the managed forest can be expected to decline (see papers in Gillam & Roberts 1995). Variation in silvicultural systems, types and severities of ecosystem disturbance over time within a particular stand can prevent this problem (Kimmins in press).

Partial harvesting and uneven-age stands are visually more acceptable to many people than clear-cutting and even-age management, and are currently 'politically correct' and publicly popular. However, as noted above, such low disturbance severity harvesting can eliminate or greatly reduce the abundance of early seral stages and deciduous hardwoods and promote late seral conifer forests in areas where conifers form the climax ecosystem community. It can eliminate or reduce early seral conifers or shade-intolerant hardwoods from climax, shade-tolerant hardwood forests. Partial harvesting typically eliminates or reduces competition-related tree mortality because the frequently repeated harvesting reduces stand density and competition. Thus, while partially harvested stands may look nice, they can be starved of snags and CWD, as are short rotation even-aged stands. In contrast, even-age stands created by clear-cutting, patch-cutting, shelterwood or natural disturbance at lower frequencies (longer cutting cycles) that are not thinned (intermediate harvests) and are allowed to go through stem exclusion and understory re-initiation phases of stand development can be a rich source of small and medium-sized snags and CWD. This, together with legacies from previous stands, may be one of the reasons for the relative lack of statistical difference in biodiversity measures in the different forest ages described in papers in Trofymow & MacKinnon (1998). Studies in chronosequences of very wet and drier forest types on west and east Vancouver Island, British Columbia, reported differences in function and some aspects of biodiversity between regenerating (3–9 years post clear-cut), immature (32–43 years post clear-cut), mature (66–99 years post stand replacing disturbance) and 'old growth' (176 to >450 years post disturbance). However, for several variables the only statistically significant differences were between the regenerating and 'old growth' forests because of the high degree of variation in what were defined as 'old growth' stands.

It is obviously necessary to consider both stand dynamics and successional relationships in deciding on the silvicultural system that is needed to sustain desired values and functions. Sullivan & Sullivan (2001) and Sullivan *et al.* (2001) review the effects of variable retention harvests on diversity of stand structure and small mammals.

Promoting diversity in forest ecosystems

Because diversity in everything (spatial and temporal) seems to be a good idea, a mosaic of different severities, scales, types and frequencies of disturbance across a landscape, and a diversity of disturbance regimes over time in any one stand, may be the optimum strategy for sustainable forest management. Such a strategy will result in increased diversity in stream habitats, from stream exposure, to early seral conditions, to various later seral stages, and different phases of stand development within each stage. The universal application of one system (e.g. the past use of even-age, monoculture stands with clear-cutting nearly everywhere) is unlikely to result in the desired diversity. But the banning of clear-cutting, suggested by some, and its replacement by any one other system everywhere (e.g. low severity, small-scale, frequent disturbance produced by partial harvesting that produces fewer snags and CWD) constitutes the same error and will also fail to sustain habitat and ecosystem diversity. Neither approach respects and reproduces the diversity we observe in most unmanaged forests. However, the previous statement notwithstanding, some unmanaged forests do approximate these two ends of the disturbance spectrum. Consequently, both large-scale clear-cutting and large-scale low disturbance partial harvesting will be appropriate from an ecological and sustainability perspective somewhere in a region characterized by high ecological diversity. Either may be socially unacceptable under specific circumstances.

As noted above, a recent development in natural disturbance emulation is variable retention or stand structure management silviculture (Franklin *et al.* 1997, 2001; Arnott & Beese 1997; Mitchell & Beese 2002). Reflecting the fact that larger scale and more severe 'natural' disturbance events frequently leave undisturbed patches and living individual trees scattered within the disturbance area, harvesting systems are now being designed to create such retention patches. These have aesthetic value, and may have long-term wildlife habitat values depending on how they are designed; aggregated retention may be a biologically superior strategy and less subject to wind damage than dispersed retention, although the latter may be aesthetically more pleasing. They also have the merit of space-for-time substitution. By retaining ecosystem elements that have longer ecological rotations within a matrix of forest managed for values that have shorter ecological rotations, one can sustain within a given management unit ecological conditions and functions that require both short- or long-term successional or stand development sequences for their renewal.

Variable retention systems have many merits and are used to address both the social values of aesthetics and economic timber production, and the environmental values of certain measures of biodiversity and wildlife habitat. However, there can be conflicts between these two sets of values. Patterns of retention that are optimal for wildlife may not be optimal for aesthetics or timber production, and vice versa. Careful analysis at both stand and spatial landscape scales over one or more timber production rotation timescales is needed to establish the desired balance of values and identify the trade-offs between the different values. The best option may be to adopt a wide spectrum of variable retention intensities, designs and patterns across the landscape to ensure that all values are represented. Any single way of doing variable retention everywhere is generally to be avoided. However, by using stream and fish habitat values and considerations as the framework on which to build a variable retention system, many of the desired objectives can be achieved.

In comparing different silvicultural systems and harvest systems, one often finds that the public interpretations of management-induced disturbance are inaccurate. Clear-cutting has been widely condemned as destructive of ecosystems and not comparable to the effects of stand-replacing wildfire or insect epidemics. It is certainly true that there are fundamental differences between clear-cutting and these 'natural' events (Keenan & Kimmins 1993; McRae *et al.* 2001), just as there are between partial harvesting and natural disturbance. A major difference is that clear-cutting and partial harvesting remove tree stems, but there are other important differences. Because harvesting frequently disturbs only a small percentage of the soil and minor vegetation and regeneration, it is often a much less severe ecosystem disturbance than wildfire. Harvesting that only removes commercial-sized stems leaves 20–30% of the tree biomass and causes much less loss of nutrients than a severe fire. Harvesting only the tree stems leaves the most nutrient-rich biomass components, only removing the biomass that has the lowest concentrations of the limiting nutrients – usually nitrogen and phosphorus. Fire generally has the opposite nutrient loss effect, removing the foliage and branches and the upper forest floor that can, in a severe fire, account for the majority of the readily available and actively circulating nitrogen. Harvesting that includes removal of branches and foliage as well as stems (whole tree harvesting) causes nutrient losses more comparable to, but still lower than, those caused by fire because it does not remove the forest floor, whereas fire generally does to some degree (Wei *et al.* 1997).

If tree harvesting leaves significant shrub and herb cover and seedlings/saplings of more shade-tolerant tree species, and if it results in very little soil disturbance, it may fail to produce the level of ecosystem disturbance needed to ensure new successional sequences and facilitate the natural regeneration of early seral, disturbance-dependent species. Even the visually dramatic clear-cutting may be a much less severe disturbance than some natural disturbance events. In fact, it can, and sometimes does, accelerate successional development, rather than being 'destructive' and retarding it. Alternatively, in some ecosystems it can lead to poorly stocked, shrub-dominated woodland unless the area is promptly planted and the seedlings are released from shrub and/or herb competition.

This great diversity in the successional effects of clear-cutting and other silvicultural/harvesting systems reflects the different phases of stand development and the different seral stages at which the harvesting takes place, as well as different harvesting techniques, time of year, differences in climate and soil, and various other factors. Traditionally, the inadequate severity of disturbance to soil or minor vegetation by some clear-cutting has been rectified by burning, mechanically scarifying the site, or the use of sheep or herbicides to manage plant competition, and by the planting of desired tree species. These disturbances resynchronize the microclimate, soil conditions and competition components of the ecosystem so that they are suitable for desired tree species; planting overcomes the sometimes very exacting seedbed and microclimatic conditions needed for natural regeneration, and problems of lack of seed source, seed predation and seed pathology.

Clearly, there is need to undertake careful analysis of the successional consequences of different silvicultural and harvesting systems and ensure that the level of ecosystem disturbance matches the ecology of the desired values. This includes applying disturbance regimes that ensure the variety of seral conditions required to produce the desired ecosystem conditions. For example, frequent low-severity harvesting or natural disturbance in riparian environments can result in either closed, late seral forest or open shrub/herb woodland, to the exclusion of earlier seral tree species. In coastal

British Columbia, this may reduce the abundance of deciduous hardwood species, and the long-term production of snags and CWD – all of which are important to stream function and diversity. Where natural disturbance provides sufficiently frequent and severe disturbance in riparian stands, these possibly undesirable outcomes will not occur.

Forest diversity and forest management: implications for streams and fish

What are the implications of the apparent universality of diversity for the management of forests with respect to streams and fish habitat?

The single most important conclusion is that regulation-based management of diverse forests is unlikely to be successful unless the regulations are specific to the ecological (climate, soil, geology, topography, physical disturbance regimes) and biological measures of diversity. They must be sensitive to the particular aquatic and terrestrial species involved, and must be applied flexibly enough to account for this diversity. There are differences in the hydrological and fish habitat consequences of forest disturbance between factors such as:

- snowpack-dominated and rain-dominated hydrological regimes
- different climatic types, soils and geologies
- different topographies, stream orders and gradients
- differences in CWD, bed load and sediment characteristics.

These require site- and situation-specific regulations with respect to riparian management, percentages of watersheds in stands of different tree age, stand development stages and seral conditions, and the types of disturbance that are consistent with maintaining water quality. For example, aquatic invertebrates were found to respond differently to sedimentation events between streams draining fine textured and coarser textured surficial materials (Heise 2002).

To be sustainable of desired values, forest management must be based on a recognition and classification of the multiple dimensions of forest diversity, both spatial and temporal, and a linkage between this classification and forest management policies and practices. Regulation-based forestry applied without the ability to recognize and interpret different ecosystem situations is very unlikely to be successful. This emphasizes the importance of adequate experience and education of forest policy-makers and managers, and of experience and training of forest workers. Ecologically based management guide books can be a great help in rendering regulations site- and situation-specific, but will never replace on-site judgement by knowledgeable and experienced field personnel. Thus, education and training must be strongly biased towards field instruction and experience. Theoretical knowledge is important to prepare an individual to 'read' the forest, just as learning the vocabulary and grammatical rules is important preparation for communicating and reading the literature of a foreign language. However, there is no substitute for the actual speaking and reading. As David Henry Thoreau, Aldo Leopold (1966), Daniel Botkin (1990, 2001) and most other ecologists have asserted, there is no substitute in learning to understand and respect forests to actually being in and observing the forest itself. Forestry education, and the education of hydrologists, aquatic ecologists, biologists, forest managers, environmentalists and conservation scientists, must have a heavy emphasis on fieldwork. Lecture room theory is only a preparation for the more important activity of learning in the field.

The spatial diversity of forests is of critical importance for watershed management that will sustain streams and fish habitat. Moist and fertile riparian areas have a dominant role in determining stream chemistry – they generally have relatively nutrient-rich canopy drip and litterfall, abundant leaf litterfall, rapid nutrient cycling, and a high potential for nutrient leakage into streams following stand-replacing disturbance. However, the rapid re-growth of trees and minor vegetation on such sites following disturbance re-establishes nutrient-retaining mechanisms (Vitousek & Reiners 1975) and these riparian ecosystems are generally an efficient filter for nutrients being leached from upslope ecosystems. If the soil is very moist, there may be active denitrification, reducing the probability that any nitrate nitrogen not taken up by riparian vegetation reaches the stream. So efficient is this biological filter that streams draining young forests that are actively accumulating biomass may be starved of nitrogen, phosphorus and other nutrients. The flush of nutrients into streams following riparian stand disturbance (the assart flush: Kimmins 2003a) may appreciably increase instream net primary production and subsequent food chain energy flow (e.g. Perrin 1981; Perrin *et al.* 1987; Perrin & Richardson 1997), but this may simply balance the short-term reduction in leaf litter and throughfall nutrient inputs.

Disturbance to upper slope ecosystems, especially to steep 'headwall' sites, will generally have little effect on stream water quality because of the filtering effect of lower slope and riparian ecosystems, as long as water remains within the soil profile. However, such sites may have slope instability problems some years after disturbance as stumps and roots decay (O'Loughlin 1974; Swanston 1974).

Slope failures can lead to mass wasting that deposits soil materials into streams irrespective of riparian buffers. Accumulations of logging debris in summer dry ravines and channels can lead to debris avalanches during extreme wet season precipitation or rain-on-snow events. Both these processes are important over long timescales as mechanisms of delivering CWD and gravel to streams, but they can result in short- to medium-term fish habitat damage, and harvest-related acceleration of these processes can have very negative impacts on streams and fish.

Disturbance of mid-slope stands will often pose a lower risk of undesirable changes to streams and fish habitat than changes to headwall or riparian stands. Generally, on gentler slopes, and with more closed nutrient cycles, harvest or natural disturbance to these mesic or 'zonal' sites will have modest implications for streams unless there are issues of groundwater temperature and chemistry that are not ameliorated in the riparian ecosystems, or where debris accumulations in summer dry channels contribute to debris torrents initiated higher on the slope.

Where possible, management-related disturbance should be designed to reflect the characteristics of the different types of ecosystem within a watershed.

Tools with which to apply knowledge of forest ecology in forest ecosystem and watershed management

If the material in this chapter, which is only a small fraction of the field of forest ecology, has left you frustrated with the level of complexity and a feeling that it cannot easily be applied, the chapter has achieved part of its objective. Another part is to suggest how to address the problems of complexity and the issue of application.

The first tool is ecological site classification, which provides a way of cataloguing the diversity of our knowledge on a site-specific basis and identifying in the field the types of forest ecosystem that will respond in a particular way to a particular disturbance event or regime. The second tool is ecologically based models that can use the memory and computational speed of computers to forecast possible forest futures under alternative management scenarios. While models will always be less satisfactory than the results of well-designed, funded and monitored field experiments, with 6.2 billion people in the world (2003), headed towards 9–11 billion within the time it would take to complete rotation-length field experiments, we cannot afford to wait for empirical evidence. While such research must always be continued, it must be complemented by models that combine empirical experience with research-based knowledge about ecological processes. Such models can be used to make scientifically credible forecasts of the outcomes of different forest management choices while we wait patiently for the results of the long-term field research.

By combining models of forest ecosystem function and management (e.g. FORE-CAST: Kimmins *et al.* 1999; Seely *et al.* 1999) with hydrology models, watershed models, fish habitat models (e.g. Heinzelmann 2002), digital elevation and slope stability models, regional timber supply models and wildlife habitat suitability models, it is now possible to produce meta-models with which to explore possible ranges of outcomes of different ways in which forest stands, small watersheds and regional landscape watersheds could be managed. Such scenario analysis makes possible the exploration of trade-offs between different environmental and social values at different spatial scales and timescales (Messier *et al.* 2003).

Society has ethical responsibilities to the present human generation to provide it with the diversity of forest-related values that it needs and wants. It also has an equal responsibility to pass on to future generations the forests and associated values that we believe our descendants will want and need. The question is, how to balance these two ethical imperatives? Only by projecting our current knowledge over ecologically significant temporal and spatial scales by using meta-models can we evaluate what constitutes sustainable management and stewardship of forests in terms of many different values, including fish and water. The trend in modelling is to use stand-level ecosystem management models such as FORECAST to drive landscape-level models and produce visual images of stand and landscape conditions (Kimmins 2001). FORECAST produces movies from the annual snapshots, enabling users to rapidly evaluate the possible outcomes, for many different social and environmental values, of different stand management practices. FORECAST is used to drive a watershed model (POSSIBLE FOREST FUTURES) and a large landscape timber supply (ATLAS) and habitat supply model (SIMFOR) in a meta-model that also produces movies. Linkage of FORECAST to hydrology models and a fish habitat model (Heinzelmann 2002) is in process.

Concluding statement

This chapter has not attempted to provide a rigorous review of forest ecology. This is not possible even in a lengthy textbook, because forest ecology spans botany, zoology, microbiology, geology, pedology, hydrology and climatology, and then adds ecological layers as the interactions of these components of the ecosystem are considered – ecosystem structure, function, interactions, complexity (diversity) and change over time (disturbance and autogenic succession). What I have attempted to do is to illustrate the critical importance of approaching forest management and conservation from an ecosystem concept at a range of spatial and temporal scales, and encompassing both ecological (physical) and biological diversity. Management and conservation must respect nature as it is and not as we might wish it to be for one or more of convenience, simplicity or emotion-based reasons. The use of ecologically based meta-models based on the best available understanding of forest ecosystems and linked to ecological site classification systems that address the spatial diversity of forests will help us to define which is the best way to manage forests to meet our biophysical objectives which must then be linked to our social objectives.

References

Arnott, J.T. & Beese, W.J. (1997) Alternative to clearcutting in BC coastal montane forests. *The Forestry Chronicle*, **73**, 670–8.

Attiwill, P.M. (1994a) The disturbance of forest ecosystems: the ecological basis for conservation management. *Forest Ecology and Management*, **63**, 247–300.

Attiwill, P.M. (1994b) Ecological disturbance and the conservative management of eucalyptus forests in Australia. *Forest Ecology and Management*, **63**, 301–46.

Banner, A., Pojar, J. & Rouse, G.E. (1983) Postglacial paleoecology and successional relationships of a bog woodland near Prince Rupert, British Columbia. *Canadian Journal of Forest Research*, **13**, 938–47.

Bergeron, Y., Harvey, B., Leduc, A. & Gauthier, S. (1999) Forest management guidelines based on natural disturbance dynamics: stand and forest-level considerations. *The Forestry Chronicle*, **75**, 49–53.

Botkin, D.B. (1990) *Discordant Harmonies: A New Ecology for the Twenty First Century*. Oxford University Press, New York.

Botkin, D.B. (2001) *No Man's Garden. Thoreau and a New Vision for Civilization and Nature*. Island Press, Washington, DC.

Dammann, A.W.H. (1971) Effect of vegetation changes on the fertility of a Newfoundland forest site. *Ecological Monographs*, **41**, 253–70.

Denslow, J.S. (1995) Disturbance and diversity in tropical rain forests: the density effect. *Ecological Applications*, **54**, 962–8.

Dunster, J. & Dunster, K. (1996) *Dictionary of Natural Resource Management*. UBC Press, Vancouver, BC.

Franklin, J.F., Berg, D.R., Thornburgh, D.A. & Tappeiner, J.C. (1997) Alternative silvicultural approaches to timber harvesting: variable retention harvest systems. In: *Creating a Forestry for the 21st Century. The Science of Ecosystem Management*, (eds K.A. Kohm & J.F. Franklin), pp. 111–39. Island Press, Washington, DC.

Franklin, J.F., Spies, T.A., Van Pelt, R., *et al.* (2001) Disturbances and structural development of natural forest ecosystems with silvicultural implications, using Douglas-fir forests as an example. *Forest Ecology & Management*, **155**, 399–423.

Fraser, L. (1993) *The influence of salal on planted hemlock and cedar saplings on northern Vancouver Island*. MSc thesis, University of British Columbia, Vancouver.

Gilliam, F.S. & Roberts, M.R. (1995) Plant diversity in managed forests (Collection of papers). *Ecological Applications*, **5**(4).

Goodman, D. (1975) The theory of diversity-stability relationships in ecology. *Quarterly Review of Biology*, **50**, 237–66.

Heilman, P.E. (1966) Change in distribution and availability of nitrogen with forest succession on north slopes in interior Alaska. *Ecology*, **47**, 825–31.

Heilman, P.E. (1968) Relationship of availability of phosphorus and cations to forest succession and bog formation in interior Alaska. *Ecology*, **49**, 331–6.

Heinzelmann, F.S.P. (2002) *A modelling framework to predict relative effects of forest management strategies on coastal stream channel morphology and fish habitat*. PhD thesis, University of British Columbia, Vancouver.

Heise, B. (2002) Assessing the effectiveness of logging road deactivation using stream silt and aquatic invertebrates. In: *Proceedings of the 2002 Interior Watershed Conference*, Kamloops, BC. BC Ministry of Forests/BC Forest Continuing Studies Network, Victoria, BC.

Houghton, J.T., Jenkins, G.J. & Ephraums, J.J. (1990) *Climate Change*. IPCC Scientific Assessment, WMO-UNEP. Cambridge University Press, Cambridge, UK.

Houghton, R.A., Hobbie, J.E., Melillo, J.M., *et al.* (1983) Changes in the carbon content of terrestrial biota and soils 1860 and 1980: a net release of CO_2 to the atmosphere. *Ecological Monographs*, **53**, 236–62.

Huston, M.A. (1994) *Biological Diversity*. Cambridge University Press, Cambridge, UK.

IIASA (2001) *IIASA Population Projection Results* http://www/iiasn.act.at/Research/POP/docs/Population_Projections-Results.html

Keenan, R.J. & Kimmins, J.P. (1993) Ecological effects of clearcutting. *Environmental Reviews*, **1**, 121–44.

Keenan, R.J., Prescott, C.E. & Kimmins, J.P. (1993) Mass and nutrient content of woody debris and forest floor in western red-cedar and western hemlock forests on northern Vancouver Island. *Canadian Journal of Forest Research*, **23**, 1052–9.

Kimmins, J.P. (1974) Sustained yield, timber running, and the concept of ecological rotation: a British Columbia view. *The Forestry Chronicle*, **50**, 27–31.

Kimmins, J.P. (1996) Importance of soil and role of ecosystem disturbance for sustained productivity of cool temperate and boreal forests. *Soil Science Society of America Journal*, **60**, 1643–54.

Kimmins, J.P. (1997b) *Balancing Act. Environmental Issues in Forestry*, 2nd edn. University of BC Press, Vancouver.

Kimmins, J.P. (1999) Biodiversity, Beauty and the Beast – are beautiful forests sustainable, are sustainable forests beautiful, and is 'small' always ecologically desirable? *The Forestry Chronicle*, **75**, 955–60.

Kimmins, J.P. (2000) Respect for Nature: An essential foundation for sustainable forest management. In: *Ecosystem Management of Forested Landscapes; Directions and Implementation*, (eds R.G. D'Eon, J. Jackson & E.A. Ferguson), pp. 3–24. UBC Press, Vancouver.

Kimmins, J.P. (2001) Visible and non-visible indicators of forest sustainability: beauty, beholders and belief systems. In: *Forests and Landscapes. Linking Ecology, Sustainability and Aesthetics*, (eds S.R.J. Sheppard & H.W. Harshaw), pp. 43–56. CABI Publishing, UK.

Kimmins, J.P. (2003a) *Forest Ecology. A Foundation for Sustainable Forest Management and Environmental Ethics in Forestry*, 3rd edn. Prentice Hall, Upper Saddle River, NJ.

Kimmins, J.P. (2003b) Old growth forests. An ancient and stable sylvan equilibrium, or a relatively transitory ecosystem condition that offers people a visual and emotional feast? Answer – It depends. *The Forestry Chronicle*, **79**, 429–40.

Kimmins, J.P. (in press) Emulation of natural forest disturbance. What does this mean, and how closely does management-induced disturbance duplicate the ecological effects of non-human forest ecosystem disturbance processes? In: *Emulating Natural Forest Landscape Disturbances: Concepts and Applications,* (eds A.H. Perera, L.J. Buse & M. Weber), Columbia University Press, New York.

Kimmins, J.P., Mailly, D. & Seely, B. (1999) Modelling forest ecosystem net primary production: the hybrid simulation approach used in FORECAST. *Ecological Modelling*, **122**, 195–224.

Leopold, A. (1966) *A Sand County Almanac, With Essays on Conservation from Round River.* Ballantine Books, New York.

Lieffers, V.J., MacMillan, R.B., MacPherson, D., Branter, K. & Stewart, J.D. (1996) Semi-natural and intensive silviculture systems for the boreal mixedwood forest. *The Forestry Chronicle*, **72**, 286–92.

Lugo, A.E. (1995) Management of tropical biodiversity. *Ecological Applications*, **5**, 956–61.

Lutz, W., Sanderson, W. & Scherbov, S. (2001) The end of world population growth. *Nature*, **412**, 543–5.

McRae, D.J, Duchesne, L.C., Freedman, B., Lynham, T.J. & Woodley, S. (2001) Comparisons between wildfire and forest harvesting, and their implications in forest management. *Environmental Reviews*, **9**, 223–60.

Malcolm, D.C. (1975) The influence of heather on silviculture practice – an appraisal. *Scottish Forestry*, **2**, 14–24.

Marshall, V.G., Setälä, H. & Trofymow, J.A. (1998) Collembolan succession and stump decomposition in Douglas-fir. *Northwest Science*, **72**, 84–5.

May, R.M. (1973) *Stability and Complexity in Model Ecosystems.* Princeton University Press, Princeton, NY.

Meidinger, D.V. & Pojar, J. (1991) *Ecosystems of British Columbia*, Special Series No. 6. BC Ministry of Forests, Victoria, BC.

Messier, C. & Kimmins, J.P. (1990) Nutritional stress in *Picea sitchensis* plantations in coastal British Columbia: the effects of *Gaultheria shallon* and declining soil fertility. *Water Air Soil Pollution*, **54**, 257–67.

Messier, C., Fortin, J., Schmiegelow, F., *et al.* (2003) *Modelling Tools to Assess the Sustainability of Forest Management Scenarios.* Networks of Centres of Excellence-Sustainable Forest Management Networks, Edmonton, Alberta.

Mitchell, S.J. & Beese, W.J. (2002) The retention system: reconciling variable retention with the principles of silvicultural systems. *The Forestry Chronicle*, **78**, 397–403.

Naeem, S. (1998) Species redundancy and ecosystem reliability. *Conservation Biology*, **12**, 39–45.

Oliver, C.D. & Larson, B.C. (1990) *Forest Stand Dynamics.* McGraw-Hill, New York.

O'Loughlin, C.L. (1974) A study of tree root strength deterioration following clearcutting. *Canadian Journal of Forest Research*, **4**, 104–13.

Orians, G.H. (1975) Diversity, stability and maturity in natural ecosystems. In: *Unifying Concepts in Ecology,* (eds W.H. van Dobben & R.H. Lowe-McConnell), pp. 139–50. W. Junk b.v. Publishers, The Hague.

Pastor, J. & Post, W.M. (1986) Influence of climate, soil moisture and succession on forest carbon and nitrogen cycles. *Biogeochemistry*, **2**, 3–27.

Perera, A.H., Buse, L.J. & Weber, M. (in press) *Emulating Natural Forest Landscape Disturbances: Concepts and Applications.* Columbia University Press, New York.

Perrin, C.J. (1981) *On the summer regulation of nitrogen and phosphorus transport in a small stream in southwestern British Columbia.* MSc thesis, University of British Columbia, Vancouver.

Perrin, C.J. & Richardson, J.S. (1997) N and P limitation of benthos abundance in the Nechako River, British Columbia. *Canadian Journal Fish Aquatic Science*, **54**, 2574–83.

Perrin, C.J., Bothwell, M.I. & Slavery, P.A. (1987) Experimental enrichment of a coastal stream in British Columbia: effects of organic and inorganic additives on autotrophic pheriphyton production. *Canadian Journal Fish Aquatic Science*, **44**, 1247–56.

Perry, D.A. & Amaranthus, M.P. (1997) Disturbance, recovery and stability. In: *Creating a Forestry for the 21st Century. The Science of Ecosystem Management*, (eds K.A. Kohm & J.F. Franklin), pp. 31–56. Island Press, Washington, DC.

Pickett, S.T.A. & White, P.S. (1985) *The Ecology of Natural Disturbances and Patch Dynamics.* Academic Press, Orlando, FL.

Pickett, S.T.A, Parker, V.T. & Fielder, P.L. (1992) The new paradigm in ecology: implications for conservation biology above the species level. In: *Conservation Biology*, (eds P.L. Fielder & S.K. Jain). Chapman & Hall, New York.

Pimm, S.L. (1984) The complexity and stability of ecosystems. *Nature*, **307**, 321–6.

Pojar, J., Klinka, K. & Meidinger, D.V. (1987) Biogeoclimatic ecosystem classification in British Columbia. *Forestry Ecology Management*, **22**, 119–54.

Price, K., Pojar, J., Roburn, A., Brewer, L. & Poirier, N. (1998) Windrown or clearcut – what's the difference? Structure, processes and diversity in successional forests of coastal British Columbia. *Northwest Science*, **72** (Special issue no. 2), 30–3.

Rogers, P. (1996) Disturbance ecology and forest management: a review of the literature. *USDA Forest Service Intermountain Research Station, General Technical Report*, INT-GTR-336.

Salim, E. & Ullsten, O. (1999) Our forests, our future. *Report of the World Commission on Forests and Sustainable Development*. Cambridge University Press, Cambridge, UK.

Schulz, E.D. & Mooney, H.A. (1993) *Biodiversity and Ecosystem Function*. Springer Verlag, New York.

Seely, B., Kimmins, J.P., Welham, C. & Scoullar, K. (1999) Defining stand-level sustainability, exploring stand-level stewardship. *Journal of Forestry*, **97**, 4–10.

Spies, T.A. (1997) Forest stand structure, composition and function. In: *Creating a Forestry for the 21st Century. The Science of Ecosystem Management*, (eds K.A. Kohm & J.F. Franklin), pp. 11–30. Island Press, Washington, DC.

Spies, T.A. & Franklin, J.F. (1991) The structure of natural young, mature and old-growth forests in Oregon and Washington. In: *Wildlife and Vegetation of Unmanaged Douglas-fir Forests, General Technical Report*, PNW-GTR-285, (tech. Coordinators L.F. Ruggiero, K.B Aubry, A.B. Carey & M.H. Huff), pp. 91–109. USDA Forest Service, Portland, OR.

Sullivan, T.D. & Sullivan, D.S. (2001) Influence of variable retention harvests on forest ecosystems. II. Diversity and population dynamics of small mammals. *Journal of Applied Ecology*, **38**, 1234–52.

Sullivan, T.P., Sullivan, D.S. & Lindgren, P.M.F. (2001) Influence of retention harvests on forest ecosystems. I. Diversity of stand structure. *Journal of Applied Ecology*, 38, 1221–33.

Swanston, D.N. (1974) Slope stability problems associated with timber harvesting in mountainous regions of the western United States. *USDA Forest Service General Technical Report*, PNW-21.

Tilman, D. & Downing, J.A. (1994) Biodiversity and stability of grasslands. *Nature*, **367**, 363–5.

Titus, B.D., Sidhu, S.S. & Mallik, A.U. (1995) A summary of some past studies on *Kalmia angustifolium* L. A problem species in Newfoundland forestry. *Canadian Forest Service, Newfoundland and Labrador Region, St John's, Info. Rept.* N-X-296.

United Nations (2001a) *World Population 2000*. United Nations Population Division, Department of Economic and Social Affairs. Publ. ST/ESA/SLR.A/197.

United Nations (2001b) *World Population Prospects. The 2000 Revision. Highlights Population Division.* Department of Economic and Social Affairs, Publ. ESA/P/WP/65.

Van Lear, D.H., Kapeluck, P.R. & Carroll, W.D. (2001) Productivity of loblolly pine as affected by decomposing root systems. *Forest Ecology and Management*, **138**, 435–43.

Vitousek, P.M. & Reiners, W.A. (1975) Ecosystem succession and nutrient retention: a hypothesis. *BioScience*, **25**, 376–81.

Wackernagel, M. & Rees, W.E. (1996) *Our Ecological Footprint: Reducing Human Impact on Earth.* New Society Publishers, Gabriola Island, BC.

Wei, X., Kimmins, J.P., Peel, K. & Steen, O. (1997) Mass and nutrients in woody debris in harvested and wildfire-killed lodgepole pine forests in the central interior of British Columbia. *Canadian Journal of Forest Research*, **27**, 148–55.

Woodwell, G.M. & Smith, H.H. (1969) *Diversity and Stability in Ecological Systems. Brookhaven Symposia in Biology* #22. Brookhaven National Laboratory, Biology Department, Upta, NY. 11973 BNL 50175 (C-56) (Biology and Medicine-TID-4500).

Chapter 3
Elements of stream ecosystem process

G.F. HARTMAN AND R.E. BILBY

Introduction

Stream ecosystems around the world reflect an enormous diversity in size, ecological features and processes. Viewed in some ways they are all different, considered in other ways they are all the same (Cummins *et al.* 1984). This chapter identifies major physical and biological processes common to stream systems so that this knowledge may be brought to bear on the management of forests.

Nature of stream systems

Role of geology, landform, climate and time

The past geological history of stream systems reflects primary influences of geological conditions, landform and climate, and secondary effects of the passage of time and concurrent forest evolution. These factors influence sediment and wood input to streams, which interact with hydrological conditions to determine the structure of the channel (Fig. 3.1).

In the Stillaguamish River basin, Benda *et al.* (1992) showed how geomorphic processes have controlled the evolution of salmon habitat during the past 14,000 years. The geomorphic processes included high levels of sediment transport and deposition during glacial retreat followed by downcutting of channels through these deposits. Forest establishment then provided organic materials to the channels that enabled the creation of productive salmon habitat. The best salmonid habitat has formed on glacial terraces during the last 5000–6000 years.

Stream conditions and fish habitat

The processes in Fig. 3.1 indicate primary influences on the structure of stream channels. Structure is also determined by location in the drainage system, i.e. headwater, transport or depositional zone. As the dimensions of the channel increase downstream, its features are changed by the nature of these functions within it. Headwater sections, composed of small tributaries, are characterized by water and sediment release to the larger main, high order, river channel (see Hewlett (1982) for an explanation of stream

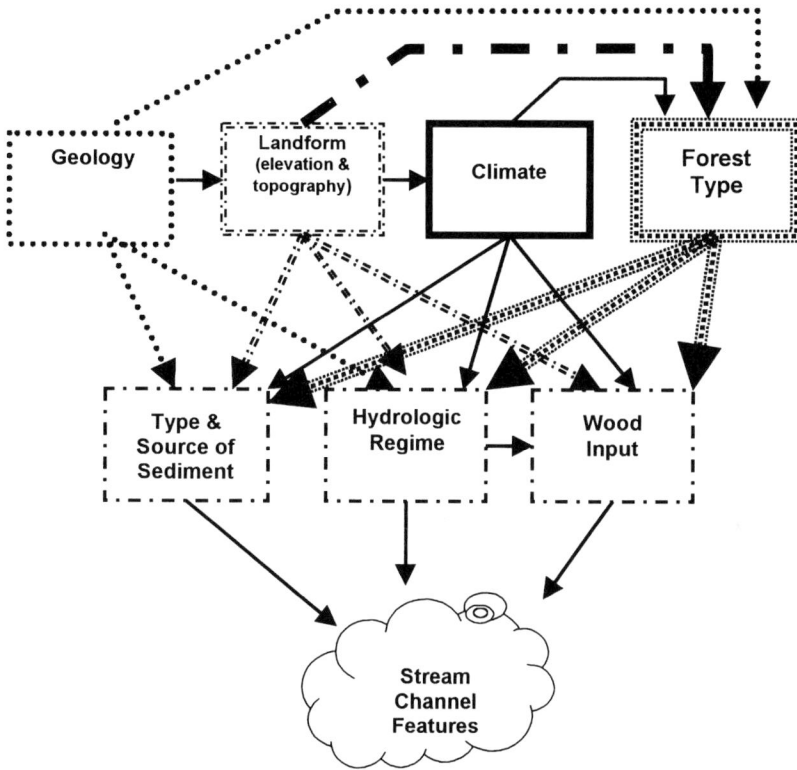

Fig. 3.1 Major factors affecting the formation and character of stream channels.

order). The main channel is characterized by material transport through it. Sections in or near the estuary are zones of deposition.

Within the diverse channel types, many features interact to determine the characteristics of fish habitat (Fig. 3.2):

- Combinations of river or stream conditions of width, depth, channel configuration, streambed features and hydraulic complexity determine local structure.
- Riparian and up-slope forests determine, in part, cover conditions for fish, and they link the terrestrial and aquatic habitats.
- Riparian forests contribute litter and terrestrial insects to the stream system. Such material supports 'heterotrophic' production processes within the system.
- Sunlight, and stream and soils nutrients, provide the basis for 'autotrophic' production, i.e. production based on photosynthetic processes within the stream.

Any fish habitat is defined by some combination of these physical features, that include or influence water temperature, oxygen availability and food production processes (Fig. 3.2).

Despite the diversity that the combinations of conditions might cause, stream ecosystems all share common processes. These processes include linkages with the

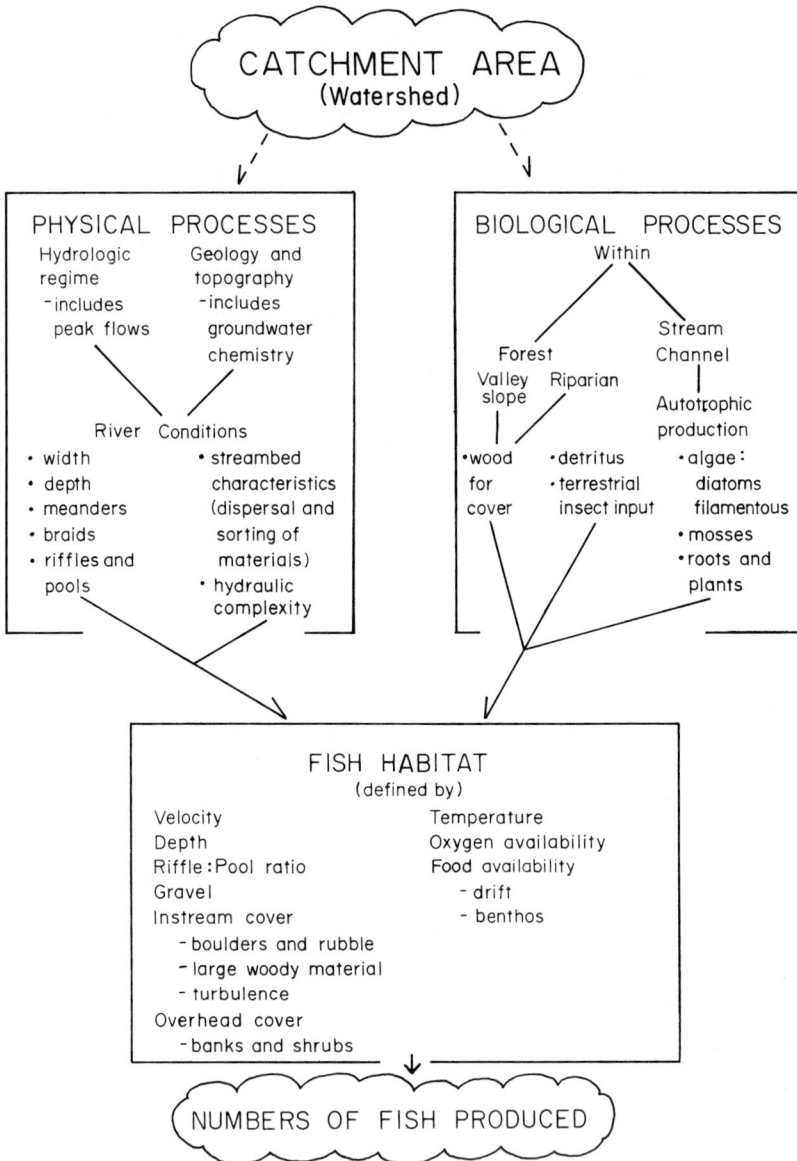

Fig. 3.2 Schematic relationships of physical and biological features and processes that combine to produce fish habitat and influence numbers of fish produced (J.H. Mundie, pers. comm.).

terrestrial ecosystem, input and transport of inorganic and organic materials, and fundamental characteristics of trophic system organization. Hydrological processes provide an integrating function in all watersheds as they link geological, climatic and forest components and influence water quality, channel conditions and trophic processes. Swanston (1991) described these critical connections, and Jones & Mulholland (2000) described important groundwater and hyporheic water behaviour and their relationships to the stream and biotic processes. Hydrological conditions change with seasons and storm events, and may disturb and re-configure channel features.

Natural disturbances exert important control over physical habitat change in watersheds. They can alter channel features, availability of large woody material and food production processes. Hydrological disturbances can range from low level events, that can occur at weekly intervals, to major floods that occur at 100–1000-year intervals. Soil movements can range from small slumps of a few cubic metres to large landslides (Swanston 1991). Disturbance may have short-term negative effects on biotic processes but long-term positive effects in 're-setting' the system. Major disturbances may replace woody debris, re-sort and clean spawning gravel, and re-construct side channels.

In the Pacific Northwest of North America, salmonids have evolved to survive within a range of natural disturbances of particular frequencies and severity (Reeves *et al.* 1995). However, human alterations of watershed processes, that lead to severe repeated disturbances, may force the system into a new ecological trajectory. As such, there may be no opportunity for recovery and conditions may be incompatible for survival of the fish species of interest.

The roles of wood and sediment

Large woody material (LW) within stream channels is an obvious indication of the linkage between stream and forested landscape. It modifies transport and storage of sediment, and affects channel form. It influences movement and transformation of litter and breakdown particles, and provides habitat for aquatic biota (Bilby & Bisson 1998). Wood in streams may also affect the characteristics of the riparian forest (Fetherston *et al.* 1995).

The role of wood in streams is strongly influenced by channel size and type and by characteristics of the surrounding forest. Abundance of LW typically decreases with increasing stream size (Bilby & Ward 1989). It tends to be distributed randomly in small streams but is often accumulated in jams in large ones (Swanson *et al.* 1982). Historically, wood accumulations in major rivers of northwestern North America were very large, and some reached several kilometres in length (Sedell & Luchessa 1982). Clumping of wood in large channels is a product of their greater transport capacity. This results in larger average diameter, length and volume of pieces with increasing stream size because smaller pieces are transported out of the system (Bilby & Ward 1989).

LW abundance varies with the characteristics of riparian forests in different parts of North America. Wood abundance and average piece size in streams along the temperate, wet, northwestern Pacific coast of North America are among the highest in the world, reflecting the high tree density and large size (Harmon *et al.* 1986). Amounts of LW in streams in forests in interior regions of western North America and eastern North America are lower due to lower density and smaller trees (Bilby & Bisson 1998).

LW is delivered to streams both by continuous processes and by relatively rare, episodic events (Keller & Swanson 1979; Bisson *et al.* 1987). Tree mortality due to competition, disease, bank erosion and windthrow generates small amounts of wood at frequent intervals (Rot *et al.* 2000). Rate of input of wood to the channel via these processes often varies as a function of successional stage and species in the riparian stand (Likens & Bilby 1982; Grette 1985). Landslides, avalanches, catastrophic windthrow, fire or severe floods occur infrequently but add massive amounts of wood to the channel in a single event (Orr 1963; Keller & Swanson 1979; McGarry 1994; Benda *et al.* 1998). These processes account for most of the wood in streams in watersheds prone to such events (Benda *et al.* 1998). Wood input from fires depends on fire frequency, which may range from decades to centuries (Agee 1988). In floodplain areas, with large erodible channels, severe floods widen channels, transport fallen trees and add large amounts of LW to the stream (Keller & Swanson 1979).

Leaching, fragmentation, microbial decay and fluvial transport all eliminate LW from streams (Keller & Swanson 1979). Leaching plays a minor role in wood decomposition (Harmon *et al.* 1986), but fragmentation assisted by microbial decay (Aumen 1985) is of major importance. These processes occur, however, at very slow rates in streams and wood can persist for very long periods (Franklin *et al.* 1981). In streams on the western coast of Canada, debris jams may form with depositional areas above them and eroded channel reaches below them. This build-up of jams and upstream depositional areas, and its subsequent breakdown occurs on about a 50-year cycle (Hogan & Bird 1998). Wood from conifer trees may last for over a century in small Oregon streams (Swanson & Lienkaemper 1978). In some cases wood can persist for many centuries. Pieces of wood in a Pacific Northwest river in North America have been dated at over 1000 years old (Hyatt & Naiman 2001) and wood up to 2000 years old has been found in a Tasmanian river (Nanson *et al.* 1995).

In small streams, large wood often has a major impact on channel form (Montgomery & Buffington 1993). It forms waterfalls (Heede 1972), creates pools (Robison & Beschta 1990) and increases variability in channel width (Zimmerman *et al.* 1967). In small streams LW frequently spans the channel forming 'log steps' that create depositional sites (Heede 1972; Montgomery & Buffington 1993). The percentage of small stream pools containing or associated with LW were 80% in southwest Washington and northern Idaho (Bilby 1984; Sedell *et al.* 1985), 86% in a northern California stream (Lisle & Kelsey 1982) and 27–45% in small British Columbia (BC) streams (Fausch & Northcote 1992).

As channel size increases and gradient decreases, LW becomes less frequent (Fig. 3.3), and has less influence on channel form and pool creation (Bilby & Ward 1989; Montgomery & Buffington 1993). In large channels occasional LW accumulations may increase channel width, promote sediment deposition and encourage the development of meander cutoffs and secondary channels (Keller & Swanson 1979). The depositional sites are less frequent but larger than in small streams (Bilby & Ward 1989). They may be very large in some cases (Abbe & Montgomery 1996), increasing as additional wood is collected. Establishment of trees and shrubs on sediment deposits provides additional stability and encourages further deposition. Eventually it may coalesce with the riparian forest (Fetherston *et al.* 1995).

Wood removal from streams dramatically increases sediment transport. Removal of wood released 5250 m³ of sediment from a 250-m reach of a stream in the Oregon

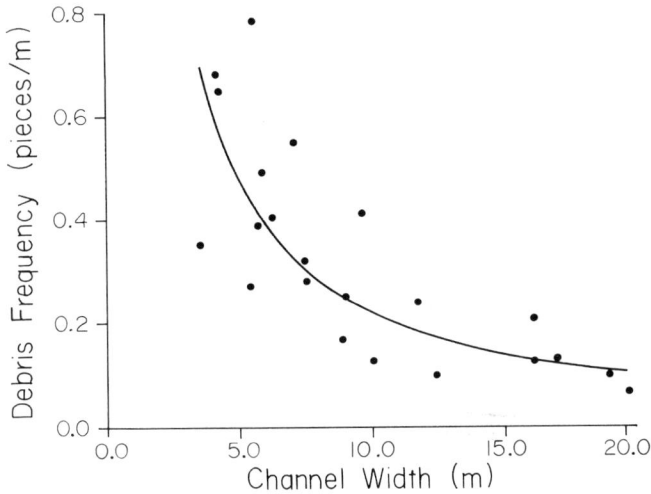

Fig. 3.3 Frequency of large wood pieces in relation to mean channel width for streams in western Washington. $\log_{10}$ debris frequency = $-1.12 \log_{10}$ channel width + 0.46: $r^2 = 0.69$, N = 22. Redrawn with permission from Bilby & Ward (1989).

Coast Range (Beschta 1979), and 60% of the stored sediment from a 100-m reach in a northern California stream (MacDonald & Keller 1983). Removal of LW from a 175-m reach of a small stream in New Hampshire doubled the rate of sediment export the following year (Bilby 1981).

Wood in streams traps large amounts of particulate organic matter (Naiman & Sedell 1979). This material constitutes a significant proportion of a stream's nutrient capital (Bilby 1981) and it is an important food source for many stream-dwelling invertebrates (Vannote *et al.* 1980). Wood removal or addition alters nutrient availability (Bilby & Likens 1980) and can have a direct impact on the biological productivity of the system (Wallace *et al.* 1995). The carcasses and reproductive products deposited by the spawning salmon make an important nutrient contribution to the productivity of Pacific Northwest streams (Bilby *et al.* 1998; Wipfli *et al.* 1998). Woody debris retained about 60% of the coho salmon carcasses in a number of western Washington streams (Cederholm *et al.* 1989).

Wood in stream channels is directly utilized by macroinvertebrates as substrate for egg laying and rearing, and as a food source (Harmon *et al.* 1986). Abundance is often high in the litter accumulations associated with LW (Merritt & Cummins 1978). Members of at least five orders of stream macroinvertebrates use wood for some aspect of their life history (Dudley & Anderson 1982). Addition of LW to a stream in North Carolina resulted in a change in community composition, a 2.1-fold increase in biomass, and a 24-fold increase in invertebrate abundance (Wallace *et al.* 1995). In streams with fine-grain beds, wood provides a stable substrate for species that are incapable of dealing with a frequently shifting bottom (Cudney & Wallace 1980). Biomass of insects on wood in a southeastern US stream was 5–10 times that on the sandy streambed (Benke *et al.* 1984).

Much of the research on the relationship between wood and fish has focused on salmon in watersheds along the Pacific coast of North America (Sedell *et al.* 1984; Murphy *et al.* 1985; Fausch & Northcote 1992). Wood creates cover and retains gravel, providing nest-building habitat. Pools created by LW are favoured habitats of juvenile coho salmon and older age classes of cutthroat trout and steelhead (Bisson *et al.* 1987). Simulated LW complexes served as cover for steelhead (*Oncorhynchus mykiss*) parr and coho (*O. kisutch*) fry (Shirvell 1990), and as refuge habitat for juvenile coho (McMahon & Hartman 1989).

Wood in streams increases their capacity to support fish. 'Wood-rich' stream reaches in southern British Columbia supported standing stocks of juvenile coho salmon and cutthroat trout five-fold higher than sites with little wood (Fausch & Northcote 1992). In southeastern Alaska, the average coho salmon density in 'wood-poor' streams was only 25% of that in streams with high wood volumes (Murphy *et al.* 1985). Fish abundance usually decreases following wood removal from channels (Lestelle 1978; Bryant 1983; Dolloff 1986) and sometimes increases following its addition (Ward & Slaney 1979; House & Boehne 1986).

The perception that LW accumulations blocked fish migration led to extensive efforts to remove wood from channels (Merrell 1951). Recent evaluations have indicated that blockage by wood is rarely a problem except, perhaps, during periods of low stream-flow (Bilby & Bisson 1998). Historically, wood accumulations are estimated to have blocked fish from 5% to 20% of potential habitat (Sedell *et al.* 1985).

Physical characteristics and vegetation of riparian areas tend to be more heterogeneous than they are up-slope. Inundation, soil saturation, and physical disturbance of the streamside vegetation from flooding and ice jams contribute to this heterogeneity (Agee 1988). LW in riparian areas provides an important location for tree seedling establishment (Harmon *et al.* 1986) by elevating them, reducing competition and keeping moisture-intolerant species above wet soil. Along large rivers in the Pacific Northwest, over 80% of the riparian conifer regeneration occurs on woody debris (McKee *et al.* 1984; Thomas *et al.* 1993).

Trophic processes: influences

The production of plant life, micro-organisms and invertebrates is a critical determinant of fish population size. While physical and biological processes come together to determine the nature of fish habitat (Fig. 3.2), four important conditions within the stream affect the conditions for trophic processes. Current, stream substrate (affected by LW), temperature and oxygen influence both fish habitat and food production. Their effects and roles are described in more detail in Minshall (1984) and Allan (1995). Some of these features influence others, e.g. temperature influences oxygen availability and demand; current influences the size distribution of sediment particles along a stream.

'Current' is referred to in Allan (1995) as the 'defining feature that unites all rivers and streams'. Current transports nutrients and food to plants and animals and removes wastes produced by them. Intra-gravel water movement is required to remove wastes from around fish eggs and invertebrates within the streambed.

Continual directional flow of water imposes forces on all life forms on or above the streambed. Stream-dwelling organisms have, however, evolved strategies to deal with such forces. Some species of small organisms live in the low velocity boundary layer between the streambed and the moving water. The hydraulic determinants of aquatic insect habitats and the problems of small organisms living in lotic environments were examined by Newbury (1984) and Statzner *et al.* (1988). Stream insects prevent their displacement by current in several ways. Blackfly larvae spin mats of silk on stones and attach themselves to them with hooks that ring their posterior end. Some mayfly nymphs have depressed body forms and overlapping gills that form a 'vacuum' disc that permits them to stick to smooth stones in the current. Others burrow into the streambed or build attached cases to avoid strong currents (Minshall 1984). Net structures of caddis larvae occupying rapid flow habitat are rigidly supported and streamlined, while those of species living in slower moving water are weaker and unsupported (Edington 1968). Species of fish are adapted to flows in streams by their body shape and swim-bladder development (Allan 1995). A species of dace that occupied relatively fast-flowing water had a smaller swim-bladder volume than one that occurred in slower water (Gee & Northcote 1963).

Although stream current imposes constraints on organisms living in it, it also facilitates the transport of dissolved organic material, particulate matter and drifting food organisms throughout a stream or river.

Substrate of the streambed is where organisms find food and shelter and compete for space. The main types of substrate are stones, sand, wood or plants (Allan 1995). The presence of detritus (Egglishaw 1964), the sizes of streambed particles, their stability and heterogeneity are important determinants of bottom type preference by organisms. Numbers of invertebrates within the upper 25–30 cm of the streambed may be very large but different in various habitats; e.g. 24,079 to 44,765/m^2 in two riffles, 26,966/m^2 in a pool and 57,271/m^2 in spawning channel gravel (Mundie 1971).

Particle size preferences may change during the life cycle of an insect (Egglishaw 1969). In tests of ten species of insect, Cummins & Lauff (1969) concluded that while current (water velocity), temperature and chemicals determine the general range of insects, particle size and food supply exert primary influence on their micro-distribution. There are several comprehensive reviews of the relationships between aquatic insects and their substrates (Cummins 1966; Cummins *et al.* 1966; Minshall 1984; Allan 1995).

Substrate complexity and porosity affect preferences of some insects (Minshall 1984). Sand is regarded as a poor substrate because it has low porosity and it may be unstable. In spite of this, there is a variety of sand-dwelling macroinvertebrates that are small, abundant and live interstitially at considerable depth in the streambed (Allan 1995). Embedded silt, which reduces porosity, has a negative effect on the inhabitability of the substrate in rocky stream bottoms. However, the loose silt in the bottom of a pool may contain high densities of invertebrates.

The relationship between invertebrate numbers and substrate type is variable because many factors other than substrate affect abundance. More stable substrates, such as wood, often support greater densities of macroinvertebrates than unstable substrates such as sand and silt (Benke *et al.* 1984). Plant surfaces often support high densities of invertebrates. However, these higher densities on plants may be an artifact of the two-dimensional way in which density is expressed. Plants provide a surface area far in excess of the streambed area on which they are located (Minshall 1984), and they may be covered with diatoms and bacteria that attract feeding invertebrates.

Stream water temperature regimes potentially affect all life functions and processes of lotic organisms. Stream temperature regimes vary with latitude, elevation, season, time of day and location in the watershed. In streams in eastern USA, total annual degree-day accumulation was about 7000 at 32° latitude and about 3400 at 45° latitude (Vannote & Sweeney 1980). In White Clay Creek, Pennsylvania, groundwater and spring seepage kept stream temperatures nearly constant all year round. The range of seasonal variability, within a stream system, tends to be low in first order springbrooks, higher in second order streams, and higher yet downstream through the third order zones (Vannote & Sweeney 1980; Allan 1995).

Shading from streamside forest canopy reduces diel temperature fluctuation during summer, but may increase it during winter (Beschta *et al.* 1987). Such effects have an important bearing on stream trophic processes. Water temperature affects growth and body size of stream animals (Allan 1995). Among salmon, it controls food conversion efficiency (Brett 1976; Brett *et al.* 1982), energetics and metabolic processes (Brett 1995), migration and, in extreme situations, survival (Brett 1995). For young coho salmon, a temperature change can re-set the timing of a sequence of life stages (Holtby & Scrivener 1989). Water temperature may affect growth, metabolism, reproduction, emergence and distribution of stream insects. Although insects have evolved to survive conditions of temperature fluctuation in streams, they show little ability to compensate or acclimate to environmental temperatures beyond that to which they are adapted (Vannote & Sweeney 1980).

Oxygen, which is a requirement for most organisms, is usually near saturation in unpolluted flowing water. The oxygen levels that support aquatic organisms are very low compared with that experienced by terrestrial forms. Water may contain from zero to 14.6 ppm. However, the atmosphere around us contains about 200,000 ppm oxygen. Furthermore, an increase in water temperature causes a reduction in maximum oxygen saturation level. The capacity of freshwater to retain dissolved oxygen decreases from 14.6 ppm at 0°C to 7.63 ppm at 30°C. Moreover, as the temperature of water rises, the metabolic demands for oxygen increase for most freshwater organisms within it. These circumstances cause aquatic biologists to be concerned about conditions, created by forestry or any other industry, that affect oxygen availability in freshwater.

Many kinds of stream organisms depend upon current to supply oxygen at their gill or body surfaces and some have behaviour patterns that compensate for reduction or absence of current (Allan 1995).

Autotrophic and heterotrophic production

Autotrophic processes, from photosynthesis within the stream, and heterotrophic processes from litter input, make energy available to an array of consumer organisms at different trophic levels. In the following section we examine features of these processes. Chapter 13 reviews the effects of forestry activities upon them. The riparian forest influences both of the above energy processes. It controls radiant energy flux at the stream surface, and consequently affects the level of photosynthesis. It is the source of litter that enters the stream, and terrestrial insects that fall into it. Riparian trees may intercept groundwater nutrients (Hill 2000), and during rainfall, materials that have accumulated on the canopy may be washed into the stream.

The influence of the riparian forest on a stream decreases with increasing channel size. As the channel widens, with increasing stream order, the fraction of the stream receiving sunlight increases, stimulating plant growth within the channel. Concurrently, the relative importance of input of terrestrial organic matter declines. As a result, the relative importance of the energy processes shift, in a continuum, from heterotrophy to autotrophy (Vanotte *et al*. 1980). Interlocking autotrophic and heterotrophic energy processes are set within and influenced by the complex physical environment. They are linked at their foundation by solar energy, and are functionally connected inasmuch as the same consumers depend upon them.

Autotrophic organisms convert carbon dioxide and nutrients into organic matter using energy from sunlight in photosynthetic processes. Vascular aquatic plants, algae, diatoms, mosses, some bacteria and protists are autotrophs in lotic environments (Allan 1995). Each of these types requires different habitats and each sends energy via different routes through the stream ecosystem (Murphy & Meehan 1991).

The role of autotrophic production in streams may have been underestimated (Minshall 1978). Production of diatoms is rapid and frequently several times greater than the standing crop (Allan 1995). It can support a large consumer biomass (Murphy & Meehan 1991). Even in forested streams, where autotrophic production might be expected to be low, 50 of 75 invertebrate taxa had algae, especially diatoms, in their diet (Allan 1995).

Light, nutrient availability, water temperature, stream scouring, and grazing, especially by insects, control the standing crop of stream autotrophs (Hynes 1970). Grazers can alter standing crop, photosynthetic activity and species composition of plants. Their effects may occur concurrently with other forms of control (Allan 1995). The production of periphyton (attached algae) in Walker Branch Creek, Tennessee, was light-limited, but biomass standing crop was controlled by insect grazing (Steinman 1992). The amount of algal grazing may vary from one stream system to the next. It can change seasonally (Allan 1995) and can result in <4% to >64% removal of the standing crop (Lamberti & Moore 1984).

In a stream in coastal BC, periphyton production was limited primarily by phosphorus and secondarily by light conditions (Stockner & Shortreed 1976). Other studies, ranging from in situ experiments (Pringle & Bowers 1984), to channel studies (Mundie *et al*. 1991), and river enrichment work (Peterson *et al*. 1985), provided results consistent with those of Stockner & Shortreed (1976, 1978).

Sunlight is required for diatom growth, but its ultraviolet component may suppress it. This may appear contradictory, because in normal stream bottom habitat exposed to ultraviolet light, algal standing crop may be more abundant than it is in unexposed areas. This occurs because the ultraviolet light suppresses the grazers even more than it suppresses the algae (Bothwell *et al*. 1994).

Periphyton may be lost due to over-growth, smothering, die-off and sloughing, or direct scouring by the current. The final spring freshet and first autumn freshet delineated the growing season during which significant algal biomass accumulated in Carnation Creek. When flows were high, scouring loss was high (Shortreed & Stockner 1983). High discharges may also carry sediment that removes algal growth (Allan 1995). Extreme discharges also tumble the stones and grind off the algae.

The organic matter from outside of the stream may arrive in many forms. Within the stream, it may be categorized by particle size and origin (Allan 1995). The ratio of carbon to nitrogen in a particular fragment of leaf may change through time as it is ingested, excreted and re-colonized with bacteria and fungus. Coarse particulate organic matter (CPOM) includes leaves, twigs, woody debris, dying macrophytes, fruit, seeds, animal faeces and fragments of aquatic plants. CPOM may also come from salmon carcasses (Cederholm *et al.* 2000) that provide a direct food source for invertebrates and fish. Marine-derived nutrients from the salmon carcasses may influence stream primary production processes. Stream ecosystems with salmon, or other anadromous fishes that die after spawning, may be unique in regard to this nutrient source. In streams containing large populations of spawning salmon, the contribution of nutrients and organic matter to the stream ecosystem is so important that the salmon must be regarded as habitat elements in themselves.

Fine particulate organic matter (FPOM) is generally derived from CPOM. Colonization of CPOM by fungi and bacteria initiate its decomposition to finer particles. FPOM also originates from faeces of small consumers, from dissolved organic material (DOM) accumulated by microbial uptake or flocculation, from sloughed-off algae, or from the forest floor and stream bank. DOM comes from groundwater, surface inflow, leachate washed from the foliage as rain falls through it, and extracellular release from aquatic plants and algae.

Riparian-based litter entering the coastal streams of western Canada is the dominant component in heterotrophic production. Seasonal input of leaves and other litter varies according to the species and age composition of vegetation involved (Murphy & Meehan 1991). Autumn leaf input is characteristic of most streams in temperate deciduous, desert and grassland areas (Anderson & Sedell 1979). During autumn, in a third order BC coastal stream, > 10 metric tonnes of litter per kilometre entered the stream. This was supplemented by conifer needle input from May to December (Neaves 1978).

The contribution of litter input varies according to stream size and location. It is relatively important in first to third order streams, which are strongly influenced by the riparian vegetation (Vannote *et al.* 1980). Such litter input is important in the overall function of a stream system because these smaller channels constitute close to 70% of the accumulated stream length in the Pacific Northwest (Naiman *et al.* 1992). Leaf litter input is lower in first and second order alpine and grassland streams. In small tropical streams litter input is the main source of energy. However, where the canopy structure is complex and multi-storied, litter may be intercepted by certain types of plants. In such circumstances litter distribution may be patchy along the length of the stream (Covich 1988).

Leaves that fall into the water undergo a series of chemical and physical changes (Kaushik & Hynes 1971; Barlocher 1985; Allan 1995). During the first 24–48 hours, up to 25% of leaf weight is lost through leaching of soluble carbohydrates and polyphenols (Kaushik & Hynes 1971; Allan 1995). Following leaching, bacteria and fungi colonize leaf litter. They reduce its weight as they consume the organic matter. However, they add protein as they colonize the remaining leaf tissue (Kaushik & Hynes 1971). The fungi and bacteria, on the surface of the ingested CPOM, provide much of the nutrient value derived by stream insects (Vanotte *et al.* 1980). These insects, in turn, constitute the primary food for vertebrate predators in the stream.

The rate at which leaves break down is dependent upon the stream water chemistry, temperature, and physical and chemical characteristics (C/N ratio) of the leaves (Park 1976). Leaves with high nutrient content broke down rapidly (Kaushik & Hynes 1971; Allan 1995). Those with tough cuticles (conifer needles) resisted fungal invasion and broke down slowly. The breakdown times for woody plant leaves ranged in half-life from 60 to 500 days (Webster & Benfield 1986).

Invertebrates utilize various behavioural and morphological adaptations to obtain and consume the various types of organic matter present in streams (Anderson & Sedell 1979; Lamberti & Moore 1984; Merritt *et al.* 1984; Allan 1995). Anderson & Sedell (1979) recognized a number of main feeding groups: shredders, collectors, grazers, gougers and predators (secondary consumers). Many of the taxa exhibited multiple feeding behaviours. Shredders break the leaf and litter material down by fragmenting and consuming parts of it along with the fungi and bacteria upon it. Scrapers remove algae, detritus and biofilm material from rocks and other solid objects (Allan 1995). Filter-feeders obtain fine particulate organic matter with spun nets, or body parts modified to form filters (Allan 1995). Collector-gatherers feed upon the organic detritus on the surface of sediments (Lamberti & Moore 1984). Some elmid beetle and caddisfly larvae feed upon wood (Allan 1995), although relatively few species of lotic invertebrates utilize this low-nutrient material as their primary food source.

The physical conditions of the habitat and the kinds and amounts of available food determine the relative abundance of the feeding groups of insects (Fig. 3.4). Consequently the different groups change their abundance in relation to stream order (Figs 3.4 and 3.5).

Invertebrate drift

Downstream movement of living organisms in suspension is called 'drift'. Allan (1995) provides excellent description and discussion of the topic. There is no specific taxonomic composition to stream drift (Hynes 1970). Some taxa are found in the drift only during periods of elevated discharge, suggesting that they were dislodged from the streambed by current. Others are found in the drift at all flow levels and display distinct diurnal patterns, indicating that this is volitional behaviour. Ephemeroptera, Diptera, Plecoptera and Trichoptera are common insect groups in drift, and amphipods and isopods are also frequently reported (Allan 1995).

Most species of insects that volitionally enter the drift do so at night (Hynes 1970; Ciborowski 1983; Allan 1984; Forrester 1994). Some exceptions do occur.

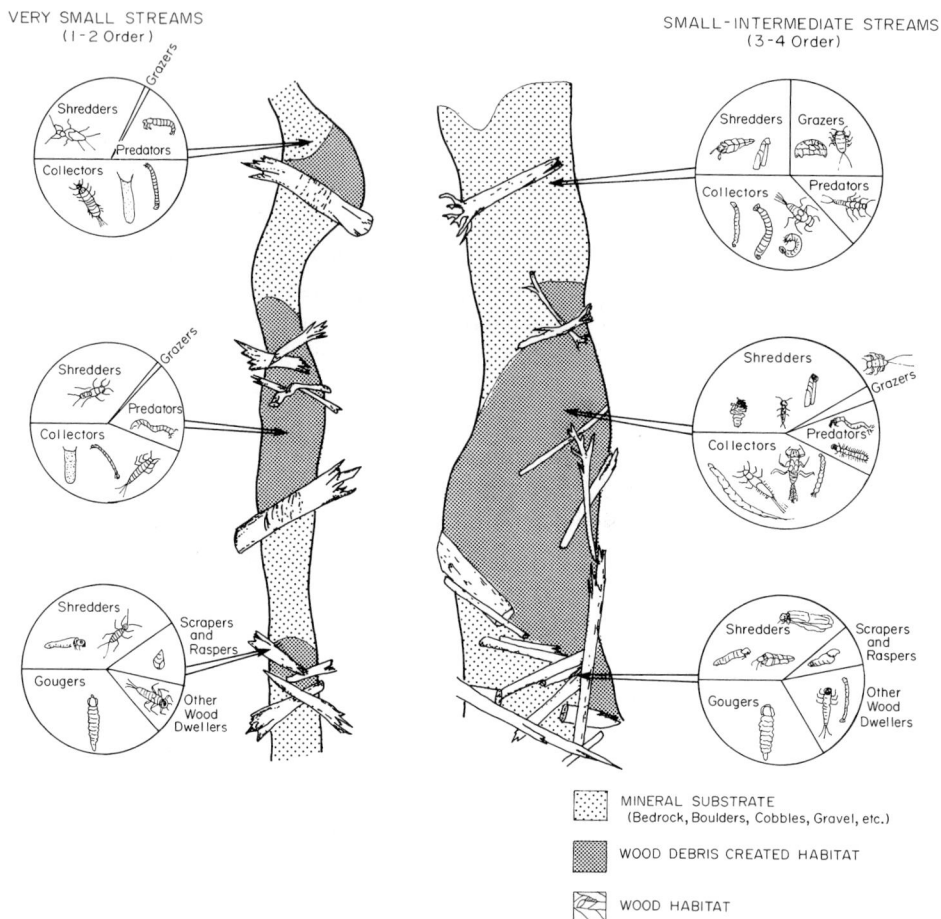

Fig. 3.4 Relative abundance of different feeding groups of invertebrates, on various substrates, in a second (left) and a third-fourth order (right) stream. Reproduced with permission from the *Annual Review of Entomology*, Vol 24 © 1979 by Annual Reviews www.AnnualReviews.org

Chironomids are described as aperiodic, and water mites and some Trichopterans are day-active (Allan 1995). Behaviour patterns are not universal, and Kerby *et al.* (1995) found no distinct diel pattern of drift among simuliids, chironomids and caenid mayflies in four small streams in Queensland, Australia.

Nocturnal drift may have evolved as a predator-avoidance mechanism (Flecker 1992; Forrester 1994). Within a series of streams, drift became more nocturnal with increasing risk of predation. Drift was primarily nocturnal in Andean piedmont streams that were occupied by predacious fish, but it was not nocturnal in streams where predators were absent (Flecker 1992). Insect size, food availability and injury have also been cited as directly influencing drift (Allan 1978, 1984; Kerby *et al.* 1995).

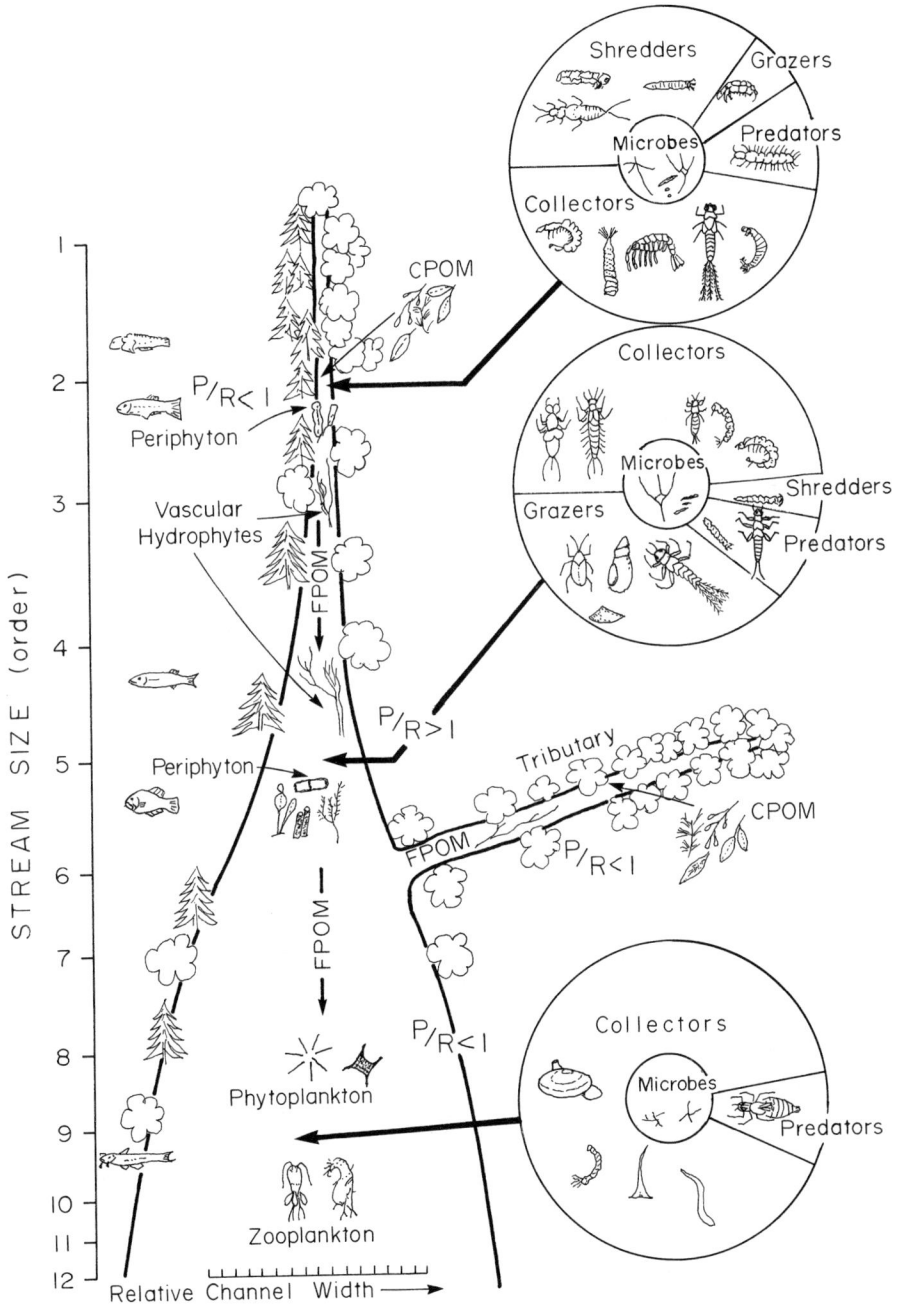

Fig. 3.5 Change in structural and functional features with changing stream size in a hypothetical system. Re-drawn with permission from Vannote *et al.* (1980).

Continuous to pulsed systems: ranges of conditions

The processes that we have described occur within a wide range of watershed types in which the overall flow is primarily linear. Many streams and rivers, however, overflow their banks, flood the valley floor and undergo an annual transition to lake-like condition and then back to that of a single channel. The relationships of the river to the forest, and the key processes underlying fish production in these two types of system are different.

River continuum concept

For many running water systems, there is a linear physical and ecological continuity from the headwaters downstream as small streams expand into large rivers (Vannote *et al.* 1980). The river continuum concept (RCC) deals with the connected processes of formation, delivery, transport and use of CPOM, FPOM, DOM; with the nature of invertebrate assemblages and energetics; and with the changes in these attributes with increasing stream size (Fig. 3.5). From a detailed study of first to seventh order drainages, in four different regions of North America, conclusions about river processes were consistent with the RCC (Minshall *et al.* 1983). The gradual change in ecosystem function (Vannote *et al.* 1980; Minshall *et al.* 1983) is driven by longitudinal changes in the climatic and geographic controls that affect sediment and nutrient supply, runoff pattern and geomorphic responses. Progressive downstream changes in riparian conditions determine light, litter input and physical storage via debris accumulations. The biology of the system responds to these predictable changes in physical habitat and type and amount of organic matter available.

The RCC provides a useful framework within which to compare stream systems and consider land use impacts. However, it has been criticized because it is not applicable in some situations. Research workers in New Zealand expressed concern about the universality of RCC (Winterbourn *et al.* 1981, 1984; Barmuta & Lake 1982). Shredders were absent, food specialization among insects was minimal, and organic biofilm layers (periphyton and microbes in an organic matrix that encrusts the streambed substrate) were more important than terrestrially derived CPOM and FPOM in some New Zealand streams. Litter input was more important, in the lower reaches of large rivers, than predicted by the RCC (Meyer & Edwards 1990; Allan 1995). The idea of the overall continuum is useful, but the framework must be viewed as a sliding scale that can shift upstream or down depending upon the overall environmental situation (Minshall *et al.* 1983). The RCC does not apply fully when the whole adjacent valley floor is inundated seasonally and lake-like conditions are produced.

Flood pulse systems

In large rivers with extensive floodplains a 'flood pulse' occurs within which the ratio of lentic to lotic area increases seasonally as floodplains are inundated. Where regular seasonal inundation occurs, the flood pulse determines the floodplain biota. Storage of organic matter and nutrients and nutrient recycling within the system occur primarily on the floodplain, and biological components of the system are adapted to this pattern (Junk *et al.* 1989).

Numerous species of fish within flood pulse systems such as the Amazon River have evolved patterns of feeding and movement that are appropriate for flood pulse conditions (Goulding 1980; Bayley 1995). Agricultural or logging activities that remove flooded forest, where fish may spend half of their time foraging, will dramatically affect feeding and growth of these fish. However, forestry impacts may not be transmitted downstream as directly as in non-pulsed rivers.

Small drainages may have floodplains that, seasonally, experience partial or full-cover flooding. Food production may be high at these sites, and fish may develop life history strategies that include the seasonal occupation of these locations. Brown (1985) described such conditions for a small stream on the western coast of BC.

Many large rivers in North America were originally connected to their floodplains (Sedell & Froggatt 1984) and at one time exhibited flood pulse conditions. Other large rivers, e.g. the Moise River, with a confined channel in a drainage area of 20,000 km^2, have never exhibited floodplain inundation pulses and have features that are consistent with the RCC (Sedell *et al.* 1989).

We suggest that across the spectrum of watersheds, both large and small, there is a complete range of conditions from full flood pulse to none. Regardless of the nature of the system, flood pulse or river continuum, forest management must be predicated on understanding such broad river dynamics as well as the knowledge of associated biotic processes.

Concluding comments

Watershed ecosystems are influenced by an array of interconnected physical and biological processes, and the balance among them varies with climate and geographic setting. Within any setting there is a complex of conditions that are most suitable for biotic processes there. A holistic understanding of background physical conditions, ecological processes, and their relative importance within watersheds, will provide the best framework within which knowledge and experience may be applied for the protection or restoration of watersheds. If the nature and location of forestry activities are known, experienced managers and scientists will be able to anticipate the type of physical impacts that may arise from them. They may also be able to predict how forestry impacts may propagate through the system. The precise responses of various components of the biotic community may be more difficult to predict, but understanding how the system functions is a critical part of the exercise of predicting biological response to land use activities.

There are a number of key watershed processes that must be maintained if the systems are to be productive. The relative importance of these processes will vary regionally due to differences in climate and geology, and they may vary locally according to watershed aspect, elevation and mean gradient. Regardless, disruption of these processes will compromise the productivity and resilience of aquatic communities inhabiting them.

A long history of research in regions such as western North America and Europe permits an understanding of watershed processes that provides foundation for specific management requirements. In parts of the world where forestry activities have recently begun to expand, basic watershed research programmes are a necessity if aquatic habitats and fishery resources are to be maintained.

Acknowledgements

It is our pleasure to acknowledge G. Miller and G. Pattern of the Pacific Biological Station, DFO, Library. They helped to find literature for this and three other chapters in the book. G. Miller not only knows where every volume is in the library, but he also knows what is in most of them. We appreciate the fine review work by J.H. Mundie and T.G. Northcote. A. Thompson helped in preparation of the manuscript, and J.C. Scrivener, H.I. Hartman and W.G. Hartman assisted us with proofreading it. We had help in figure preparation from B. Knight, whose time was kindly made available by Weyerhaeuser. Our thanks to all.

References

Abbe, T.B. & Montgomery, D.R. (1996) Large woody debris jams, channel hydraulics, and habitat formation in large rivers. *Regulated Rivers*, **12**, 201–21.

Agee, J.K. (1988) Successional dynamics in forest riparian zones. In: *Streamside Management: Riparian Wildlife and Forestry Interactions*, (ed K.J. Raedeke), pp. 31–43. Contribution No. 59. Institute of Forest Resources, University of Washington, Seattle, WA.

Allan, J.D. (1978) Trout predation and the size composition of stream drift. *Limnology and Oceanography*, **23**, 1231–7.

Allan, J.D. (1984) The size composition of invertebrate drift in a Rocky Mountain stream. *Oikos*, **43**, 68–76.

Allan, J.D. (1995) *Stream Ecology: Structure and Function of Running Waters*. Kluwer Academic Publishers, Dordrecht.

Anderson, N.H. & Sedell, J.R. (1979) Detritus processing by macroinvertebrates in stream ecosystems. *Annual Review of Entomology*, **24**, 351–77.

Aumen, N. G. (1985) *Characterization of lignocellulose decomposition in stream wood samples using ^{14}C and ^{15}N techniques*. PhD dissertation, Oregon State University.

Barlocher, F. (1985) The role of fungi in the nutrition of stream invertebrates. *Botanical Journal of the Linnean Society*, **91**, 83–94.

Barmuta, L.A. & Lake, P.S. (1982) On the value of the river continuum concept. *New Zealand Journal of Marine and Freshwater Research*, **16**, 227–9.

Bayley, P.B. (1995) Understanding large river-floodplain ecosystems. *BioScience*, **45**, 153–8.

Benda, L., Beechie, T.J., Wissmar, R.C. & Johnson, A. (1992) Morphology and evolution of salmonid habitats in a recently deglaciated river basin, Washington State, USA. *Canadian Journal of Fisheries and Aquatic Sciences*, **49**, 1246–56.

Benda, L.E., Miller, D.J., Dunne, T., Reeves, G.H. & Agee, J.K. (1998) Dynamic landscape systems. In: *River Ecology and Management: Lessons from the Pacific Coastal Ecoregion*, (eds R.J. Naiman & R.E. Bilby), pp. 261–88. Springer-Verlag, New York.

Benke, A.C., Van Arsdall, T.C., Gillespie, D.M. & Parrish, F.K. (1984) Invertebrate productivity in a subtropical blackwater river: the importance of habitat and life history. *Ecological Monographs*, **54**, 25–63.

Beschta, R.L. (1979) Debris removal and its effect on sedimentation in an Oregon Coast Range stream. *Northwest Science*, **53**, 71–7.

Beschta, R.L., Bilby, R.E., Brown, G.W., Holtby. L.B. & Hofstra, T.D. (1987) Stream temperature and aquatic habitat: fisheries and forestry interactions. In: *Streamside Management: Forestry and Fishery Interactions*, (eds E.O. Salo & T.W. Cundy), pp. 191–232. Contribution No. 57. College of Forest Resources, University of Washington, University of Washington Institute of Forest Resources, Seattle, WA.

Bilby, R.E. (1981) Role of organic debris dams in regulating the export of dissolved and particulate matter from a forested watershed. *Ecology*, **62**, 1234–43.

Bilby, R.E. (1984) Post-logging removal of woody debris may affect stream channel stability. *Journal of Forestry*, **82**, 609–13.

Bilby, R.E. & Bisson, P.A. (1998) Function and distribution of large woody debris in Pacific coastal streams and rivers. In: *River Ecology and Management: Lessons from the Pacific Coastal Ecoregion*, (eds R.J. Naiman & R.E. Bilby), pp. 324–46. Springer-Verlag, New York.

Bilby, R.E. & Likens, G.E. (1980) Importance of organic debris dams in the structure and function of stream ecosystems. *Ecology*, **61**, 1107–13.

Bilby, R.E. & Ward, J.W. (1989) Changes in characteristics and function of woody debris with increasing size of streams in western Washington. *Transactions of the American Fisheries Society*, **118**, 368–78.

Bilby, R.E., Fransen, B.R., Bisson, P.A. & Walter, J.K. (1998) Response of juvenile coho salmon and steelhead to the addition of salmon carcasses to two streams in southwest Washington, USA. *Canadian Journal of Fisheries and Aquatic Sciences*, **55**, 1909–18.

Bisson, P.A., Bilby, R.E., Bryant, M.D., *et al.* (1987) Large woody debris in forested streams in the Pacific Northwest: past, present, and future. In: *Streamside Management: Forestry and Fishery Interactions*, (eds E.O. Salo & T.W. Cundy), pp. 143–90. Contribution Number 57. Institute of Forest Resources, University of Washington, Seattle, WA.

Bothwell, M.L., Sherbot, D.M. & Pollock, C.M. (1994) Ecosystem response to solar ultraviolet-B radiation: influence of trophic-level interactions. *Science*, **265**, 97–100.

Brett, J.R. (1976) Feeding metabolic rates of young sockeye salmon, *Oncorhynchus nerka*, in relation to ration levels and temperature. Department of Environment, Canada. *Fisheries and Marine Service, Technical Report* 675.

Brett, J.R. (1995) Energetics. In: *Physiological Ecology of Pacific Salmon*, (eds C. Groot, L. Margolis & W.C. Clarke), pp. 3–68. UBC Press, Vancouver.

Brett, J.R., Clarke, W.C. & Shelbourn, J.E. (1982) Experiments on thermal requirements for growth and food conversion efficiency of juvenile chinook salmon, *Oncorhynchus tshawytscha*. *Canadian Fisheries and Aquatic Sciences Technical Report* 1127.

Brown, T.G. (1985) *The role of abandoned stream channels as over-wintering habitat for juvenile salmonids*. Thesis submitted in partial fulfillment of the requirements for the degree of Master of Science, University of British Columbia, Vancouver.

Bryant, M.D. (1983) The role and management of woody debris in west coast salmonid nursery streams. *North American Journal of Fisheries Management*, **3**, 322–30.

Cederholm, C.J., Houston, D.B., Cole, D.L. & Scarlett, W.J. (1989) Fate of coho salmon (*Oncorhynchus kisutch*) carcasses in spawning streams. *Canadian Journal of Fisheries and Aquatic Science*, **46**, 1347–55.

Cederholm, C.J., Johnson, D.H., Bilby, R.E., *et al.* (2000) *Pacific Salmon and Wildlife: Ecological Contexts, Relationships, and Implications for Management*. Washington Department of Natural Resources, Business Systems Support Division, Olympia, WA.

Ciborowski, J.J. (1983) Influence of current velocity, density, and detritus on drift of two mayfly species (*Ephemeroptera*). *Canadian Journal of Zoology*, **61**, 119–25.

Covich, A. (1988) Geographical and historical comparisons of neotropical streams: biotic diversity and detrital processing in highly variable habitats. *Journal of the North American Benthological Society,* 7, 361–86.

Cudney, M. D. & Wallace, J.B. (1980) Life cycles, microdistribution and production dynamics of six species of net-spinning caddisflies in a large southeastern (U.S.A.) river. *Holarctic Ecology,* 3, 169–82.

Cummins, K.W. (1966) A review of stream ecology with special emphasis on organism-substrate relationships. In: *Organism-substrate Relationships in Streams,* (eds K.W. Cummins, C.A. Tryon & R.T. Hartman), pp. 2–51. Special Publication No. 4. Pymatuning Laboratory of Ecology, University of Pittsburgh.

Cummins, K.W. & Lauff, G.H. (1969) The influence of substrate particle size on the microdistribution of stream macrobenthos. *Hydrobiologia,* 34, 145–81.

Cummins, K.W., Tryon, C.A. & Hartman, R.T. (1966) Organism-substrate relationships in streams. In: *Symposium held in the Pymatuning Laboratory of Ecology, July 16–17, 1964.* Special publication No. 4. Pymatuning Laboratory of Ecology, University of Pittsburgh.

Cummins, K.W., Minshall, G.W., Sedell, J.R., Cushing, C.E. & Petersen, R.C. (1984) Stream ecosystem theory. *Internationale Vereinigung für Theoretische und Angewandte Limnologie,* 22, 1818–27.

Dolloff, C.A. (1986) Effects of stream cleaning on juvenile coho salmon and Dolly Varden in southeast Alaska. *Transactions of the American Fisheries Society,* 115, 743–55.

Dudley, T.L. & Anderson, N.H. (1982) A survey of invertebrates associated with wood debris in aquatic habitats. *Melanderia,* 39, 1–21.

Edington, J.M. (1968) Habitat preferences in net-spinning caddis larvae with special reference to the influence of water velocity. *Journal of Animal Ecology,* 37, 675–92.

Egglishaw, H.J. (1964) The distributional relationship between the bottom fauna and plant detritus in streams. *Journal of Animal Ecology,* 33, 463–76.

Egglishaw, H.J. (1969) The distribution of benthic invertebrates on substrata in fast-flowing streams. *Journal of Animal Ecology,* 38, 19–33.

Fausch, K.D. & Northcote, T.G. (1992) Large woody debris and salmonid habitat in a small coastal British Columbia stream. *Canadian Journal of Fisheries and Aquatic Science,* 49, 682–93.

Fetherston, K.L., Naiman, R.J., & Bilby, R.E. (1995) Large woody debris, physical process, and riparian forest development in montane river networks of the Pacific Northwest. *Geomorphology,* 13, 133–44.

Flecker, A.S. (1992) Fish predation and the evolution of invertebrate drift periodicity: evidence from neotropical streams. *Ecology,* 73, 438–48.

Forrester, G.E. (1994) Diel patterns of drift by five species of mayfly at different levels of fish predation. *Canadian Journal of Fisheries and Aquatic Sciences,* 51, 2549–57.

Franklin, J. F., Cromack, K., Denison, W., *et al.* (1981) *Ecological Characteristics of Old-growth Douglas-fir Forests.* USDA Forest Service, Pacific Northwest Forest and Range Experiment Station. Portland, OR. *General Technical Report* PNW-118.

Gee, J.H. & Northcote, T.G. (1963) Comparative ecology of two sympatric species of dace (*Rhinichthys*) in the Fraser River system, British Columbia. *Journal of the Fisheries Research Board of Canada,* 20, 105–18.

Goulding, M. (1980) *The Fishes and the Forest: Explorations in Amazonian Natural History.* University of California Press, Berkeley, CA.

Grette, G.B. (1985) *The abundance and role of large organic debris in juvenile salmonid habitat in streams in second growth and unlogged forests.* Masters dissertation, University of Washington, Seattle, WA.

Harmon, M.E., Franklin, J.E, Swanson, F.J., *et al.* (1986) Ecology of coarse woody debris in temperate ecosystems. *Advances in Ecological Research,* 15, 133–302.

Heede, B.H. (1972) Influences of a forest on the hydraulic geometry of two mountain streams. *Water Resources Bulletin,* 8, 523–30.

Hewlett, J.D. (1982) *Principles of Forest Hydrology*. University of Georgia Press, Athens, GA.

Hill, A.R. (2000) Stream chemistry and riparian zones. In: *Streams and Ground Waters*, (eds J.B. Jones & P.J. Mulholland), pp. 83–110. Academic Press, San Diego.

Hogan, D.L. & Bird, S.A. (1998) Forest management and channel morphology in small coastal watersheds: results from Carnation Creek and the Queen Charlotte Islands. In: *Forest-fish Conference: Land Management Practices Affecting Aquatic Ecosystems*, (eds M.K. Brewin & D.M. Monita), pp. 209–26. Canadian Forest Service, Northern Forestry Centre, Information Report NOR-X-356, Edmonton.

Holtby, L.B. & Scrivener, J.C. (1989) Observed and simulated effects of climatic variability, clear-cut logging, and fishing on numbers of chum salmon (*Oncorhynchus keta*) and coho salmon (*O. kisutch*) returning to Carnation Creek, British Columbia. In: *Proceedings of the National Workshop on Effects of Habitat Alterations on Salmonid Stocks*, (eds C.D. Levings, L.B. Holtby & M.A. Henderson), pp. 62–81. Canadian Special Publication of Fisheries and Aquatic Sciences 105.

House, R.A. & Boehne, P.L. (1986) Effects of instream structures on salmonid habitat and populations in Tobe Creek, Oregon. *North American Journal of Fisheries Management*, **6**, 38–46.

Hyatt, T.L. & Naiman, R.J. (2001) The residence time of large woody debris in the Queets River, Washington, USA. *Ecological Applications*, **11**, 191–202.

Hynes, H.B.N. (1970) *The Ecology of Running Waters*. University of Toronto Press, Toronto.

Jones, J.B., & Mulholland, P.J. (2000) *Streams and Ground Waters*. Academic Press, New York.

Junk, W.J., Bayley, P.B. & Sparks, R.E. (1989) The flood pulse concept in river-floodplain systems. In: *Proceedings of the Large River Symposium (LARS)*, (ed. D.P. Dodge), pp. 110–27. *Canadian Special Publication of Fisheries and Aquatic Sciences* 106.

Kaushik, N.K & Hynes, H.N.B. (1971) The fate of dead leaves that fall into streams. *Archiv für Hydrobiologie*, **68**, 465–515.

Keller, E.A. & Swanson, F.J. (1979) Effects of large organic material on channel form and fluvial processes. *Earth Surface Processes*, **4**, 361–80.

Kerby, B.M., Bunn, S.E. & Hughes, J.M. (1995) Factors influencing invertebrate drift in small forest streams, South-eastern Queensland. *Marine and Freshwater Research*, **46**, 1101–8.

Lamberti, G.A. & Moore, J.W. (1984) Aquatic insects as primary consumers. In: *The Ecology of Aquatic Insects*, (eds V.H. Resh & D.M. Rosenberg), pp. 164–95. Praeger Scientific, New York.

Lestelle, L.C. (1978) *The effects of forest debris removal on a population of resident cutthroat trout in a small, headwater stream*. Masters dissertation, University of Washington, Seattle, WA.

Likens, G.E. & Bilby, R.E. (1982) Development, maintenance and role of organic debris dams in streams. In: *Sediment Budgets and Routing in Forested Drainage Basins*, Forest Service General Technical Report, PNW-141, (eds F.J. Swanson, R.J. Janda, T. Dunne, & R.N. Swanston), pp. 122–8. US Department of Agriculture.

Lisle, T.E. & Kelsey, H.M. (1982) Effects of large roughness elements on the thalweg course and pool spacing. In: *American Geomorphological Field Group Field Trip Guidebook*, (ed. L.B. Leopold), pp. 134–5. American Geophysical Union, Berkeley, CA.

MacDonald, A. & Keller, E.A. (1983) Large organic debris and anadromous fish habitat in the coastal redwood environment: the hydrologic system. *Technical Completion Report OWRT Project B-213-CAL*. Water Resources Center, University of California, Davis, CA.

McGarry, E.V. (1994) *A quantitative analysis and description of the delivery and distribution of large woody debris in Cummins Creek, Oregon*. Masters thesis, Oregon State University.

McKee, A., LeRoi, G. & Franklin, J.F. (1984) Structure, composition and reproductive behavior of terrace forests, South Fork Hoh River, Olympic National Park. In: *Ecological Research in National Parks of the Pacific Northwest*, (eds E.E. Starkey, J.F. Franklin & J.W. Mathews), pp. 22–9. National Park Service Cooperative Study Unit, Corvallis, OR.

McMahon, T.E. & Hartman, G.F. (1989) Influence of cover complexity and current velocity on winter habitat use by juvenile coho salmon (*Oncorhynchus kisutch*). *Canadian Journal of Fisheries and Aquatic Science*, **46**, 1551–7.

Merrell, T.R. (1951) Stream improvement as conducted in Oregon on the Clatskanie River and tributaries. *Fish Commission, Oregon Research Briefs*, **3**, 41–7.

Merritt, R.W. & Cummins, K.W. (1978) *An Introduction to the Aquatic Insects of North America*. Kendall/Hunt Publishing Company, Dubuque, IA.

Merritt, R.W., Cummins, K.W. & Burton, T.M. (1984) The role of aquatic insects in the processing and cycling of nutrients. In: *Aquatic Insect-substratum Relationships*, (eds V.H. Resh & D.M. Rosenberg), pp.134–63. Praeger Scientific, New York.

Meyer, J.L. & Edwards, R.T. (1990) Ecosystem metabolism and turnover of organic carbon along a blackwater river continuum. *Ecology*, **71**, 668–77.

Minshall, G.W. (1978) Autotrophy in stream ecosystems. *BioScience*, **28**, 767–71.

Minshall, G.W. (1984) Aquatic insect-substratum relationships. In: *The Ecology of Aquatic Insects*, (eds V.H. Resh & D.M. Rosenberg), pp. 358–400. Praeger Scientific, New York.

Minshall, G.W., Petersen, R.C., Cummins, K.W., *et al*. (1983) Interbiome comparisons of stream ecosystem dynamics. *Ecological Monographs*, **53**, 1–25.

Montgomery, D.R. & Buffington, J.M. (1993) Channel classification, prediction of channel response, and assessment of channel condition. *Draft Report to the Sediment, Hydrology, and Mass Wasting Committee of the Washington State Timber/Fish/Wildlife Agreement*. Department of Geological Sciences and Quaternary Research Center, University of Washington, Seattle, WA.

Mundie, J.H. (1971) Sampling benthos and substrate materials, down to 50 microns in size, in shallow streams. *Journal of the Fisheries Research Board of Canada*, **28**, 849–60.

Mundie, J.H., Simpson, K.S. & Perrin, C.J. (1991) Responses of stream periphyton and benthic insects to increased dissolved inorganic phosphorus in a mesocosm. *Canadian Journal of Fisheries and Aquatic Sciences*, **48**, 2061–72.

Murphy, M.L. & Meehan, W.R. (1991) Stream ecosystems. In: *Influences of Forest and Rangeland Management on Salmonid Fishes and their Habitats*, (ed. W.R. Meehan), pp. 17–46. Special Publication 19. American Fisheries Society, Bethesda, MD.

Murphy, M.L., Koski, KV., Heifetz, J., Johnson, S.W., Kirchofer, D. & Thedinga, J.F. (1985) Role of large organic debris as winter habitat for juvenile salmonids in Alaska streams. In: *Proceedings; Western Association of Fish and Wildlife Agencies, 1984*, pp. 251–62.

Naiman, R.J. & Sedell, J.R. (1979) Benthic organic matter as a function of stream order in Oregon. *Archiv für Hydrobiologie*, **87**, 404–22.

Naiman, R.J., Beechie, T.J., Benda, L.E., *et al*. (1992) Fundamental elements of ecologically healthy watersheds in the Pacific Northwest coastal ecoregion. In: *Watershed Management: Balancing Sustainability and Environmental Change*, (ed. R.J. Naiman), pp. 127–88. Springer-Verlag, New York.

Nanson, G.C., Barbetti, M. & Taylor, G. (1995) River stabilization due to changing climate and vegetation during the late Quaternary in western Tasmania, Australia. *Geomorphology*, **13**, 145–58.

Neaves, P.I. (1978) Litter, export, decomposition, and retention in Carnation Creek, Vancouver Island. *Canadian Fisheries and Marine Service, Technical Report*, 809.

Newbury, R. (1984) Hydraulic determinants of aquatic insect habitats. In: *The Ecology of Aquatic Insects*, (eds V.H. Resh & D.M. Rosenberg), pp. 323–57. Praeger Scientific, New York.

Orr, P.W. (1963) *Windthrown Timber Survey in the Pacific Northwest, 1962*. USDA Forest Service, Pacific Northwest Region, Portland, OR.

Park, D. (1976) Carbon and nitrogen levels as factors influencing fungal decomposers. In: *The Role of Terrestrial and Aquatic Organisms in Decomposition Processes*, (eds J.M. Anderson & A. Macfadyen), pp. 41–59. Blackwell Scientific, Oxford.

Peterson, B.J., Hobbie, J.E. & Hershey, A.E. (1985) Transformation of a tundra river from heterotrophy to autotrophy by addition of phosphorus. *Science*, **229**, 1383–6.

Pringle, C.M. & Bowers, J.A. (1984) An in situ substratum fertilization technique: diatom colonization on nutrient-enriched, sand substrata. *Canadian Journal of Fisheries and Aquatic Sciences*, **41**, 1247–51.

Reeves, G.H., Benda, L.E., Burnett, K.M., Bisson, P.A. & Sedell, J.R. (1995) A disturbance-based ecosystem approach to maintaining and restoring freshwater habitats of evolutionarily significant units of anadromous salmonids in the Pacific Northwest. In: *Evolution and the Aquatic Ecosystem: Defining Unique Units in Population Conservation*, (ed. J.L. Nielsen), pp. 334–49. *American Fisheries Society Symposium*, **17**.

Robison, E.G. & Beschta, R.L. (1990) Characteristics of coarse woody debris for several coastal streams of southeast Alaska, USA. *Canadian Journal of Fisheries and Aquatic Science*, **47**, 1684–93.

Rot, B.W., Naiman, R.J. & Bilby, R.E. (2000) Stream channel configuration, landform, and riparian forest structure in the Cascade Mountains, Washington. *Canadian Journal of Fisheries and Aquatic Sciences*, **57**, 699–707.

Sedell, J.R. & Froggatt, J.L. (1984) The importance of streamside forests to large rivers: the isolation of the Willamette River, Oregon, U.S.A., from its floodplain by snagging and streamside forest removal. *Internationale Vereinigung für Theoretische und Angewandte Limnologie*, **22**, 1828–34.

Sedell, J.R. & Luchessa, K.J. (1982) Using the historical record as an aid to salmonid habitat enhancement. In: *Acquisition and Utilization of Aquatic Habitat Inventory Information*, (ed. N.B. Armantrout), pp. 210–23. Western Division, American Fisheries Society.

Sedell, J.R., Yuska, J.E., & Speaker, R.W. (1984) Habitats and salmonid distribution in pristine, sediment-rich river valley systems: S. Fork Hoh and Queets River, Olympic National Park. In: *Fish and Wildlife Relationships in Old-growth Forests*, (eds W.R. Meehan, T.R. Merrell, Jr. & T.A. Hanley), pp. 33–46. American Institute of Fishery Research Biologists. Juneau, AK.

Sedell, J.R., Swanson, F.J. & Gregory, S.V. (1985) Evaluating fish response to woody debris. In: *Proceedings of the Pacific Northwest Stream Habitat Workshop, October 10–12, 1984*, (ed. T.J. Hassler), pp. 224–45. Humboldt State University, Arcata, CA.

Sedell, J.R., Richey, J.E. & Swanson, F.J. (1989) The river continuum concept: a basis for the expected ecosystem behavior of very large rivers. In: *Proceedings of the International Large River Symposium (LARS)*, (ed. D.P. Dodge), pp. 49–55. Special Publication, 106. Canadian Fisheries and Aquatic Sciences.

Shirvell, C.S. (1990) Role of instream rootwads as juvenile coho salmon (*Oncorhynchus kisutch*) and steelhead trout (*O. mykiss*) cover habitat under varying streamflows. *Canadian Journal of Fisheries and Aquatic Sciences*, **47**, 852–61.

Shortreed, K.S. & Stockner, J.G. (1983) Periphyton biomass and species composition in a coastal rainforest stream in British Columbia: effects of environmental change caused by logging. *Canadian Journal of Fisheries Aquatic Sciences*, **40**, 1887–95.

Statzner, B., Gore, J.A. & Resh, V.H. (1988) Hydraulic stream ecology: observed patterns and potential applications. *Journal of the North American Benthological Society*, **7**, 307–60.

Steinman, A.D. (1992) Does an increase in irradiance influence periphyton in a heavily-grazed woodland stream? *Oecologia*, **91**, 163–70.

Stockner, J.G. & Shortreed, K.R. (1976) Autotrophic production in Carnation Creek, a coastal rainforest stream on Vancouver Island, British Columbia. *Journal of the Fisheries Research Board of Canada*, **33**, 1553–63.

Stockner, J.G. & Shortreed, K.R. (1978) Enhancement of autotrophic production by nutrient addition in a coastal rainforest stream on Vancouver Island. *Journal of the Fisheries Research Board of Canada*, **35**, 28–34.

Swanson, F.J. & Lienkaemper, G.W. (1978) *Physical Consequences of Large Organic Debris in Pacific Northwest Streams.* General Technical Report PNW-69. USDA Forest Service, Pacific Northwest Forest and Range Experiment Station. Portland, OR.

Swanson, F.J., Gregory, S.V., Sedell, J.R. & Campbell, A.G. (1982) Land-water interactions: the riparian zone. In: *Analysis of Coniferous Forest Ecosystems in the Western United States,* (ed. R.L. Edmonds), pp. 267–91. Hutchinson Ross, Stroudsburg, PA.

Swanston, D.N. (1991) Natural processes. In: *Influences of Forest and Rangeland Management on Salmonid Fishes and their Habitats,* (ed. W.R. Meehan), pp. 139–79. American Fisheries Society Special Publication 19.

Thomas, J.W., Raphael, M.G., Anthony, R.G., *et al.* (1993) *Viability Assessments and Management Considerations for Species Associated with Late-Successional and Old-Growth Forests of the Pacific Northwest: The Report of the Scientific Analysis Team.* USDA Forest Service, Portland, OR.

Vannote, R.L. & Sweeney, B.W. (1980) Geographic analysis of thermal equilibria: a conceptual model for evaluating the effect of natural and modified thermal regimes on aquatic insect communities. *American Naturalist,* **115,** 667–95.

Vannote, R.L., Minshall, G.W., Cummins, K.W., Sedell, J.R. & Cushing, C.E. (1980) The river continuum concept. *Canadian Journal of Fisheries and Aquatic Sciences,* **37,** 130–7.

Wallace, J.B., Webster, J.R. & Meyer, J.L. (1995) Influence of log additions on physical and biotic characteristics of a mountain stream. *Canadian Journal of Fisheries and Aquatic Science,* **52,** 2120–37.

Ward, B.R. & Slaney, P.A. (1979) *Evaluation of In-Stream Enhancement Structures for the Production of Juvenile Steelhead Trout and Coho Salmon in the Keogh River: Progress 1977 and 1978.* Fisheries Technical Circular 45. British Columbia Fish and Wildlife Branch, Victoria, BC.

Webster, J.R. & Benfield, E.F. (1986) Vascular plant breakdown in freshwater ecosystems. *Annual Reviews in Ecological Systematics,* **17,** 567–94.

Winterbourn, M.J., Rounick, J.R. & Cowie, B. (1981) Are New Zealand stream ecosystems really different? *New Zealand Journal of Marine and Freshwater Research,* **15,** 321–8.

Winterbourn, M.J., Cowie, B. & Rounick, J.S. (1984) Food resources and ingestion patterns of insects along a west coast, South Island, river system. *New Zealand Journal Marine and Freshwater Research,* **18,** 43–51.

Wipfli, M.S., Hudson, J. & Caouette, J. (1998) Influence of salmon carcasses on stream productivity: response of biofilm and benthic macroinvertebrates in southeastern Alaska, USA. *Canadian Journal of Fisheries and Aquatic Sciences,* **55,** 1503–11.

Zimmerman, R.C., Goodlet, J.C. & Comer, G.H. (1967) The influence of vegetation on channel forms of small streams. In: *Symposium on River Morphology.* pp. 255–75. Publication No. 75. International Association of Scientific Hydrology.

Chapter 4
Fundamentals of lake ecology relevant to fish–forestry interactions

T.G. NORTHCOTE

Introduction

The science of limnology – the study of inland waters – has its roots steeped in lakes. From the Greek word *limnos* (lake, pool, swamp) and first used in the late 1800s by F.A. Forel to title his monographic works on Lake Geneva, limnology has gradually broadened to include all inland waters, be they fresh or salty, lentic or lotic, surface or subterranean. Nowadays most limnological texts include some sections on streams and rivers (e.g. Wetzel 2001).

Nevertheless, much of the consideration given to the inland waters of forested landscapes and their interactions with fish has focused largely on flowing waters. This might be expected in the Pacific northwestern parts of North America where stream rearing and sea-running salmonids are so important. Lacustrine subjects, however, have been given little coverage in a number of significant reviews, bibliographies and books related to fish and forestry interactions (Gibbons & Salo 1973; Blackie *et al.* 1980; Toews & Brownlee 1981; Salo & Cundy 1987; Meehan 1991; Murphy 1995; Brewin & Monita 1998; Näslund 1999).

Oceans make up more than 97% of water in the biosphere, with polar ice and glaciers 2% and groundwater less than 0.3%. Of the remaining inland surface waters (231.2 thousands of cubic kilometres), lakes contribute nearly 99.5% (Wetzel 2001). For example, Lake Baikal contains about a fifth of the world's freshwater volume. However, most lakes are small and are thus vulnerable to human activities in their tributaries and around their shorelines. Even Lake Baikal has been affected by logging activities and pulpmill pollutants (Northcote 1972a), and these practices continue (Goldman 1994).

Lakes within the watershed of a stream–river system can alter the latter in many ways. They can change both summer and winter temperatures as well as discharge downstream. They intercept sediment and large woody debris but may produce particulate foods important to the biota in downstream river reaches. For some stream and river migratory fishes they provide spawning habitats for adults and rearing and feeding habitats for juveniles.

Lakes seldom are 'standing bodies' of water, fresh or otherwise. They are often moving internally as a result of complex though important processes. They are highly active and responsive inland aquatic ecosystems affected in major ways by human activities, including forestry, not only within their watersheds but often from their airsheds and even by global climatic changes arising in part from anthropogenic causes.

Lakes as products of their watersheds and airsheds

The notion that lakes were largely closed systems where most of the important processes affecting their functioning and productivity occurred internally grew out of the popular view, held almost to the mid-1900s, that lakes were 'microcosms' to be regarded as 'closed communities'. In recent decades we have found all too painfully that lakes are anything but closed. Limnologists and the public have been forced to face the many problems surrounding lake eutrophication, acidification, increased ultraviolet radiation and global climate changes. Lakes whether small or large are now regarded as products of their watersheds and airsheds. Indeed recently developing views further strengthen lake connections to landscape ecology (Turner 1998), with broad spatial scales, controlling factors and comparisons extending out from lakes themselves into catchments, lake districts, physiographic regions, and even continents (Tonn *et al.* 1990).

Positioning of lakes within watersheds has important implications on how markedly they may be affected by watershed inputs. This is nicely illustrated by the series of Okanagan Valley basin lakes (Stockner & Northcote 1974), where one of the uppermost in the chain – Wood Lake – has long been subject to cultural eutrophication whereas one of the lowermost in the chain – Vaseux Lake – by its positioning down in the series is 'protected' from the more serious effects of eutrophication. Both are relatively small and shallow lakes (Table 4.1) but with greatly different drainage areas and residence times, those of Wood Lake being small and long respectively, while those of Vaseux are large and short. Furthermore, the five lakes upstream from Vaseux (Fig. 4.1) serve as partial nutrient traps, three of them much larger and deeper, whereas Wood Lake has only one small and shallow basin lake upstream from it. So apart from the many other aspects of watershed characteristics that affect lake functions, positioning within watersheds must be considered.

Watershed landscape topography also may be important. Two small eutrophic lakes less than a kilometre apart in south-central British Columbia responded very differently to factors producing overwinter oxygen depletion and resultant fish mortality (Halsey 1968; Northcote 1980). Corbett Lake, the upper and deeper one (mean depth 7.0 m), in a small side valley protected from effective autumnal wind mixing, experienced frequent fish winter-kills. In contrast, Courtney Lake, the lower and shallower one (mean depth 4.9 m), was much more open to prevailing wind direction, mixed

Table 4.1   Comparative morphometric and hydrologic features[a] of two Okanagan basin lakes of differing watershed positioning (see Fig. 4.1)

Lake	Position in watershed	Drainage area (km²)	Surface area (km²)	Maximum depth (m)	Residence time (years)	Eutrophic condition
Wood	Upper	184	930	34	19.8	High
Vaseux	Lower	6425	275	27	0.03	Lower

[a]From Northcote (1973), Stockner & Northcote (1974) and additional polar planimetry.

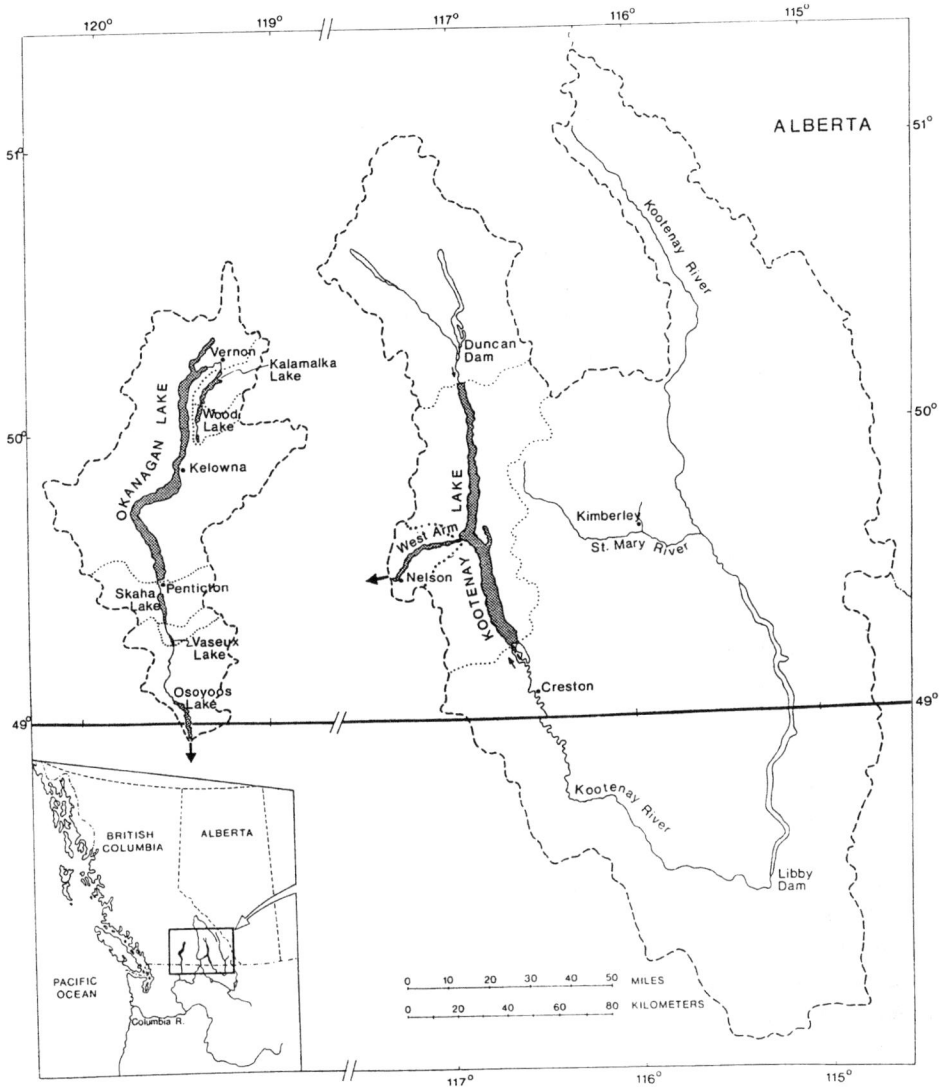

Fig. 4.1 Kootenay and Okanagan basin lakes showing their watersheds (– -) and major sub-basins (...). Inset shows lake location in western North America. Adapted from Northcote (1980).

thoroughly, reoxygenated in autumn and never had fish kills. Artificial autumn circulation of Corbett Lake subsequently prevented fish kills. Landscape topography and the position of lakes accounted for major divergences in responses to drought (Webster *et al.* 1996). Watershed features such as area, slope and percentage of wetlands affected water chemistry, as shown by D'Arcy & Carignan (1997) for lakes in Quebec, Canada.

Lake watershed morphometry (size and shape), as well as positioning in relation to the lake outlet, have effects on their internal production processes. Large Kootenay

Lake (417 km^2) in southeastern British Columbia has a watershed over 100 times the size of its surface area with about 80% of the catchment draining into the southern end (Fig. 4.1). Phosphate enrichment from a fertilizer plant over 350 km upstream in a small headwaters stream resulted in severe eutrophication of this large lake from the mid-1950s to the early 1970s (Northcote 1972b, 1973). The combined effect of reduced fertilizer input and construction of a large impoundment (Fig. 4.1) in the middle reaches of the inflowing Kootenay River brought about, by the late 1970s, a marked reversal in nutrient loading to Kootenay Lake (Northcote 1980; Ashley *et al.* 1999).

The importance of major drainage sub-basin location in relation to the outlet of lakes is shown by contrasting differences between two small lakes, Deer in British Columbia and Tutira in New Zealand. The largest sub-basin in Deer Lake (nearly half of the total watershed and supplying about 36% of the lake's annual phosphorus load) is located so as to pass its discharge through much of the lake system (Northcote & Luksun 1992), whereas by far the largest sub-basin of Lake Tutira is located close to its outlet, reducing its potential to effect eutrophication of the lake (Northcote 1980).

Lakes in watersheds collect, alter, and sometimes concentrate, effects of various human activities in their watersheds – deforestation by logging or reforestation being but two of many (Burgis & Morris 1987). Nutrients and other materials may be trapped temporarily in their bottom waters.

Theoretical water retention times (or the converse – flushing rates) have great significance. This is especially evident for the large Kootenay Lake whose major watershed sub-basins drain through the relatively shallow and narrow West Arm (Fig. 4.1). The main lake has a long residence time – some 566 days – and after a few years of high phosphate enrichment became eutrophic. Its West Arm with a residence time of 5.5 days and as brief as 1.6 days during the high early summer runoff period (Northcote 1973) never showed severe effects of eutrophication in the 1960s. In Kamloops Lake, British Columbia (56 km^2 in surface area and 73 m mean depth), seasonal changes in discharge of its major inlet river varied residence time from a late winter high of 212 days to a spring runoff period of only 11 days (Ward 1964, 1966).

Airsheds as well as watersheds of lakes can have large effects on water quality, biota and overall functional processes. Many of the accumulative effects of acid precipitation, stratospheric ozone depletion and climatic warming are reviewed by Schindler (1998) for the world's several million boreal lakes. These 'big three' of global human stressors to natural ecosystems along with other human activities on lakes and their watersheds are summarized in Fig. 4.2.

Lake mean depth has been used as a morphometric parameter to classify lake trophic status – high mean depth indicative of low productivity (oligotrophic) and low mean depth of high productivity (eutrophic) ones. Earlier approaches in this area are found in Rawson (1955), with modifications by Northcote & Larkin (1956). Lake surface area in relation to mean depth and hence volume obviously has major effects on retention time and thereby production, as do shoreline slope, littoral zone area and volume, in relation to that of the whole lake. Lake water level is normally high in the runoff season, and depending on lake edge morphometry, can inundate large areas of shallow shorelines, increasing inputs of nutrients and terrestrial invertebrates as well as shoreline-attached algae (periphyton) and rooted aquatic plant (macrophyte)

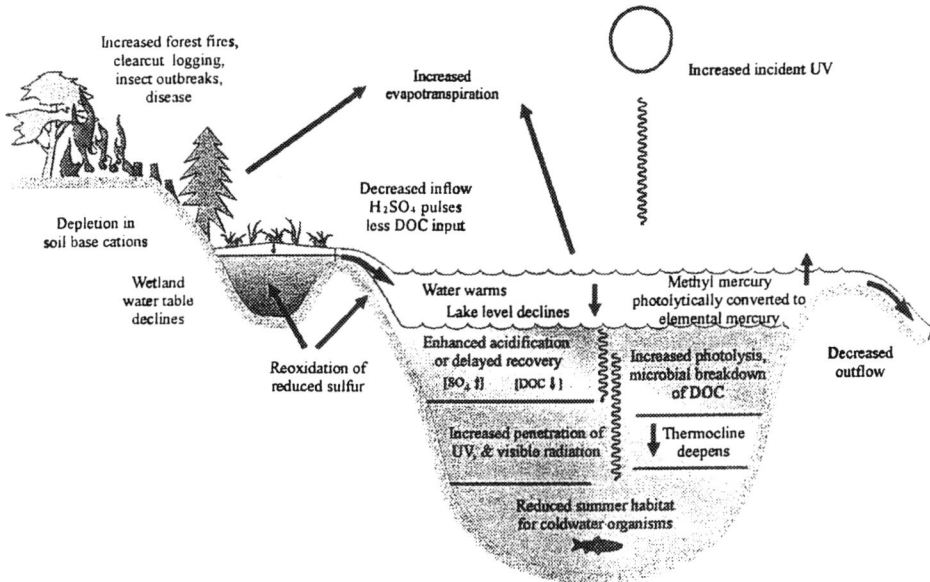

Fig. 4.2 Combined effects of climatic warming, acid precipitation, stratospheric ozone depletion and major human activities on boreal lakes and their watersheds. Sulphate entering the wetlands and lakes with acid precipitation is stored and reduced as sulphur when water tables are high. When climate warming and drought cause water tables to lower, the sulphur is reoxidized, causing acid pulses to enter lakes and streams. Both drought and acidification cause dissolved organic carbon (DOC) to decrease. Adapted from a colour figure in Schindler (1998).

production. Such interactions are important in both temperate and tropical lakes, especially those situated in large river floodplains (Engle & Melack 1993).

For a temperate seepage lake in northern Wisconsin, USA, Hagerthey & Kerfoot (1998) showed that high groundwater discharge sites around the shoreline were significantly greater in phosphorus concentrations and epibenthic algal biomass than were those at low flow sites. Groundwater as well as surface water inputs may vary spatially around lake shorelines, as well as seasonally and annually, further complicating production dynamics, especially in floodplain lakes (Lesack 1995).

The importance of interfaces for productivity and biodiversity

Many of the processes controlling productivity and biodiversity within lakes occur at or near interfaces. There are two key interfaces for any lake – at its surface between air and water, and at its bottom between water and whatever material encloses the water – rock, gravel, sand, or silt. Its upper air–water interface is the site of several highly important processes. Here radiation from the sun drives primary photosynthetic production throughout the upper water column (euphotic zone) via planktonic algae (phytoplankton), algae attached to surfaces (periphyton), and rooted aquatic plants (macrophytes). Their abundance and diversity varies spatially (usually being highest in the upper onshore areas), seasonally, and among the three groups. Long wave solar radiation heats the upper layers of lakes and warm near-surface waters are circulated to varying depths into the lake mainly by wind action at the air–water interface, although other processes are involved usually to a lesser degree. Surface wave action is responsible for much of the gaseous exchanges of oxygen, carbon dioxide and nitrogen, so necessary to biological production within lakes. Much of the planktonic invertebrate community occurs in the upper warmed layers.

The second key interface in any lake is that between the water and the lake bottom. Here, especially in the warm seasons of temperate lakes and throughout much of the year in tropical lakes, the important processes of bacterial decomposition and nutrient regeneration occur, although these also take place in the water column itself. At the water–bottom interface are found the benthic invertebrate communities, some living at or close to this interface and others up to 30 cm or more into the softer sediments, again often in highest abundance and diversity in nearshore areas. Benthic pathways are a neglected but integral part of lake food webs (Vadeboncoeur *et al.* 2002).

These two interfaces overlap along the shoreline, usually creating a region of highest biological productivity and diversity to be found anywhere in a lake. But shorelines are also the very location where many human activities have been concentrated for centuries if not millennia. For some of the large European lakes such as Maggiore in northern Italy and Constance in Germany/Switzerland/Austria it is difficult to find even a few metres of natural lake shoreline remaining. And in the large Okanagan Lake in Canada where sizable human populations have only existed for less than a century, about 80% of the lake shoreline in several long stretches has been modified in varying ways by man (Northcote & Northcote 1996), with major reduction in their productivity and diversity. Lake shorelines at or close to the entry of streams and rivers may be greatly affected by sediment inputs as well as other particulate and dissolved materials, some from natural but others from human-generated sources. This is evident in Lake Malawi and elsewhere in Africa (pers. comm. G.F. Hartman), and certainly in Lake Titicaca, Peru (Northcote 1992). The composition of surficial fine-grained sediment, important for macrophyte growth around the littoral zone of a large Quebec/Vermont lake, could be predicted by a model including fetch, depth and shoreline slope (Petticrew & Kalff 1991). Wave disturbance and substrate slope affect sediment characteristics of small lakes (Cyr 1998).

Several morphometric parameters have been used in quantifying the relative amount of shoreline bordering lakes of differing shape, size and bottom slope. Most commonly used are shoreline length, shoreline development (the degree to which lake shape departs from being circular), littoral area and littoral slope; see Håkanson (1981).

There now are means to differentiate between littoral and pelagic food webs in lakes using stable carbon isotopes. In Canadian Shield lakes ^{13}C enrichment is clearly evident in littoral versus pelagic consumers and this may well be a characteristic feature of lakes worldwide (France 1995). Littoral habitat has long been viewed as an important variable controlling the abundance and production of fish, with macrophyte presence being a key factor (Hinch *et al.* 1991). In North American Great Lakes a fish production index is significantly greater in littoral habitats with submerged macrophytes than without them (Randall *et al.* 1996). Submerged trees and coarse woody debris in lake littoral zones are important spawning sites for some fish (Colgan & Ealey 1973), and as cover to significantly lower predation risk for other fish (Tabor & Wurtsbaugh 1991). Coarse woody debris occurs in the littoral zone of many of the 260,000 Ontario lakes (Guyette & Cole 1999), as well as in lakes and reservoirs in several of the large limnological regions of British Columbia (Northcote & Larkin 1956). Relatively little seems to be known about the ecological function of woody debris in lakes compared with that in flowing waters. Its role in reservoirs is reviewed by Northcote & Atagi (1997). Many of the woody debris pieces of eastern white pine (*Pinus strobus*) in Ontario lake littoral zones have been there for several hundred years, but no mature ones have fallen into a study lake there for the last 100 years (Guyette & Cole 1999), as a consequence of all the large trees around the forest–lake ecotone being harvested in the late 1800s and early 1900s (see also Abrams 2001).

Stratification, destratification and mixing

In temperate areas most lakes cycle through a fairly predictable sequence of full spring mixing followed by a gradual build-up of summer thermal stratification into an upper warm mixing layer above a sharp temperature/density gradient, and a lower cool partially non-mixing layer (Weyhenmeyer & Meili 2000). The latter may become severely depleted in dissolved oxygen and thereby not habitable by many species of fish. In autumn upper layers may cool to the point where strong wind action can mix the lake so it once again can circulate freely. If ice cover forms in late autumn or winter the lake of course can no longer mix readily and slight negative thermal stratification occurs beneath the ice. Strong oxygen depletion in the lower layers can develop (Mathias & Barica 1980), to the degree that overwinter or early spring fish mortalities may result, sometimes just in the first few days after ice break-up. There are numerous exceptions and modifications to this pattern even as outlined for temperate lakes, and certainly for subtropical and tropical lakes. These variations are explained in most limnological texts. Tropical lakes may not show large thermal variation vertically, but this can occur in density, and may restrict deep mixing so that lower layers become deoxygenated (Hecky *et al.* 1994). Elsewhere mixing can be restricted by morphometry, wind sheltering by trees and other factors (Whitmore *et al.* 1991). Temperature and atmospheric pressure effects on dissolved oxygen content of inland waters are well known (Mortimer 1981). Lake shape, nutrient concentration and loading are important in affecting the degree of anoxia in lakes (Nürnberg 1995a) and fish species richness (Nürnberg 1995b).

As noted previously, lakes are not static water bodies. Wind-driven waves on their surface cause parallel foam streaks to develop at wind speeds of about 2–7 metres per second (Langmuir 1938). Between and beneath these streaks are cells of converging and diverging water circulation which aggregate planktonic algae and small invertebrates, forming high prey densities for predators such as fish to exploit. Furthermore, when thermally stratified, lakes (especially long narrow ones) under periodic wind stress can develop large internal stratification slopes and thereby currents (internal seiches). These complex oscillations have important implications both chemically and biologically on the distribution and production of lake biota (Ostrovsky *et al.* 1996). Sometimes they result in macroinvertebrate and fish mortalities (Emery 1970). In early spring or late autumn, onshore temperature and hence density differences with off-shore water can lead to formation of a thermal bar which can trap onshore pollutants from rivers, cities or industries (Spain *et al.* 1976).

Light penetration and its ecological significance

The quantity and quality of radiation that reaches various regions in a lake directly or indirectly affects nearly every process occurring there. The main radiation source is sunlight, and its quantity and quality have been altered by passage through the earth's atmosphere before it reaches lake surfaces, although by far the greatest changes occur within lake waters themselves. No matter how transparent the lake water, less than 20% of long wave radiation remains after 1 metre depth, and only about half or less of the ultraviolet radiation. Discolouring stains such as in the brown water of bog lakes have little effect on longer wavelengths but virtually cut off the ultraviolet wavelengths. Stains in lake waters result from dissolved inorganic compounds (irons producing yellows or reds, calcium carbonate producing greens) and from dissolved or colloidal organic matter including lignin–protein complexes, 'humic' carboxylic acids, as well as polysaccharides, carbohydrates and saturated fatty acids. Suspended particles in lake waters greatly reduce the quantity of subsurface transmission and, depending on their size, differentially affect light quality – small particles reducing short wavelengths most and large particles being nearly non-selective.

Turbidity and transparency of lakes results from inorganic suspended material such as glacial flour and silt, and from various types of organic matter including plant and animal plankton. The settling rate of suspended matter depends on whether or not it is live or dead, on the specific gravity and size of the particles, on their shape (surface area to volume ratio), and on water viscosity, which varies seasonally with temperature.

Measurement of photosynthetically active radiation in lakes is usually done with a radiometer covering the 400–700-nm range. Water transparency of lakes has routinely been measured by the maximum depth to which a 20-cm diameter white or black and white quadrant disc (Secchi disc) can be observed from the surface. Values of about 40 m have been reported for Lake Baikal in Siberia and in a few Swedish lakes, with those in the 10–20-m range common for clear, relatively unproductive lakes, and often much less than 1 m in highly productive lakes during dense algal blooms or in highly unproductive lakes with heavy glacial flour or silt loads. Because these vary with run-off, there can be large seasonal and spatial differences in lakes (Northcote *et al.* 1999). Approximate factors have been used to convert Secchi depth readings to 1% surface illumination, the maximum depth of the euphotic zone depth (Koenings & Edmundson 1991). In lakes where there is little interference from inorganic particulates, Secchi disc transparency tracks changes in planktonic algal biomass, as was the case in Lake Washington, USA (Edmondson 1972). Phytoplankton production can be regulated by inorganic turbidity, greatly limiting light penetration in some temperate and tropical lakes (Lind *et al.* 1992; Northcote *et al.* 1999).

Increases up to 10–20% per decade in incident ultraviolet B (UV-B) radiation (280–320 nm) at temperate latitudes have resulted from stratospheric ozone depletion over the last decade and a half (Williamson 1995). The depth of UV-B penetration is only a few metres in all but the clearest lakes. In some regions brown-staining dissolved organics are characteristic of lake waters and these may further reduce UV-B depth penetration, giving protection to aquatic biota, although not always (Arts *et al.* 2000; Kaczmarska *et al.* 2000). In higher elevation clear waters one might expect greater impacts from increased UV radiation, as was suggested for littoral attached diatoms (Vinebrooke & Leavitt 1996). The effects of UV-B radiation on lentic freshwater ecosystems are complex with multiple indirect as well as direct ones, and with synergistic responses from other large-scale perturbations (Williamson 1995; Schindler 1998). In New Zealand lakes, aquatic plants such as *Isoetes* and *Chara* were stressed by UV radiation, but not *Potamogeton* (Rae *et al.* 2001).

Nutrient sources, needs and uptakes

Production of biotic communities in lakes, from bacteria to aquatic vertebrates, is in large part controlled by nutrient supply and availability – dissolved forms of nitrate, phosphate, silicate and bicarbonate along with a number of microelements. In recent decades the importance of several major supply sites for nutrients has become evident. These include surface and groundwater inputs of lake drainage basins (even from their headwater reaches, Peterson *et al.* 2001), airborne inputs, and internal inputs from lake waters, biota (including fish) and bottom sediments.

Concern over excessive algal blooms in the 1960s, mainly in the areas where dense human populations were present, resulted in the identification of phosphorus and nitrogen inputs as key causes. Loading models were developed to predict when lakes might become seriously affected or by how much inputs should be reduced to bring about major improvements (Vollenweider 1968).

In some lake waters microelement deficiencies involving molybdenum, zinc, iron and others have been identified (Evans & Prepas 1997; Emmenegger *et al.* 2001). In the early 1970s, marked within-lake differences (partly a result of residence time), in expression of high phosphate loading on algal populations were recognized, as in the West Arm and main basin of Kootenay Lake (Northcote 1973). Vollenweider (1976) then included retention time in his phosphate loading formula. Further improvements and refinements have followed (Kenny 1990; McCullough 1998; Hatch *et al.* 1999; Devito *et al.* 2000; Guildford & Hecky 2000).

Effects of lakeshore forest cover on planktonic algal production were shown for Castle Lake in California, USA (Goldman 1961). Most of its western shoreline was composed of an *Abies–Pinus* open forest, whereas over half the eastern shoreline was covered by nitrogen-fixing *Alnus*. Planktonic algal production was significantly higher off the eastern than the western shoreline, presumably a result of higher nitrate inputs. Especially in regions of dense riparian cover by deciduous trees, terrestrial airborne litterfall from at least 10 m away from the shoreline may contribute significantly to lake carbon, nitrogen and phosphorus nutrient budgets (Sebetich & Horner-Neufeld 2000).

Suspended inorganic sediments entering lakes can have significant negative effects via light limitation on phytoplankton, periphyton and macrozooplankton response to nutrient loading, especially phosphorus (Ennis 1975; Cuker & Hudson 1992; Northcote *et al.* 1999).

It became apparent soon after application of Vollenweider type and other nutrient loading models for prediction or restoration of lake water and biotic quality that internal sources and recycling of nutrients must be included in these models. In temperate lakes which mix more or less completely in spring and autumn but which deoxygenate in their lower layers over summer and winter, there can be major phosphorus inputs from lake bottom sediments (Nürnberg 1987; Murphy *et al.* 2001). In addition, bacterial breakdown of organic material from many biotic sources within the water column must be considered, including fish (Persson 1997). In some lakes night-time convective circulation may be an important means for movement of littoral phosphorus into deeper offshore pelagic waters (Stefan *et al.* 1989; James & Barko 1991). As a further complication, the balance between nitrogen and phosphorus inputs (the N:P ratio) has been an important issue in lake nutrient response (Levine & Schindler 1992; Guildford & Hecky 2000; Downing *et al.* 2001).

Though most of the work on lake nutrient dynamics has concerned temperate systems, there have been valuable studies on both deep and shallow tropical lakes (Edmond *et al.* 1993; Carignan & Planas 1994; Lehman & Branstrator 1994; Carignan & Vaithiyanathan 1999).

Major players in lake communities and their functional interactions

There are three major groups of primary producers using solar radiation as an energy source in conjunction with their photosynthetic pigments and available dissolved nutrients, mainly but not exclusively phosphorus, nitrogen and carbon (chiefly in the form of bicarbonate), to form plant sugars and eventually other complex compounds. The first group is planktonic algae (phytoplankton) along with the blue-green cyanobacteria and in some lakes also photosynthetic purple sulphur bacteria. Secondly there are the algae attached to different surfaces (periphyton), often given other names depending on the type of surfaces to which they are attached – epiphytic algae being those attached to rooted aquatic plants (macrophytes), epipelic algae those growing on fine organic sediments, epilithic algae being those growing on rock surfaces, and so on. Thirdly there are the rooted aquatic plants, some of which may be completely submerged and living on the bottom down to several metres if there is sufficient light, others with both submerged and floating leaves, and still others with emergent stems and leaves. The relative contribution of these three major groups of primary producers to overall production varies greatly within lakes (seasonally and spatially) and between lakes. Too frequently, published measurements of primary production include only the phytoplanktonic component because techniques for obtaining it via ^{14}C uptake are relatively easy, well worked out and standardized.

Secondary production in lakes comes from a wide range in sizes and types of planktonic invertebrate animals including protozoans up to some insects, as well as from invertebrates living in or on the lake bottom (the benthos), and the fish and other vertebrate groups living both in the water column and/or at the lake bottom. The sizes and feeding (trophic) relationships of these vary greatly over the life cycles of the species involved and among the aquatic ecosystems being considered. In general, lake secondary production measurements come from estimation of numbers, biomass and growth rates and can be fraught with difficulties.

The usual view of food web structure has been that phytoplankton, mainly diatoms and flagellates, were grazed upon by intermediate and larger sized zooplankton and these in turn by larval, juvenile and some species of adult fish. Other pathways are now becoming evident and include consumption of very small algae (0.2–2 µm, picoplankton) by protozoans, the presence of a microbial food web using detritus, and a recycling of this organic matter into higher trophic levels, referred to as the 'microbial loop' (Carrick *et al.* 1991; Hart *et al.* 2000).

Lake type and succession: natural and anthropogenic

Lakes, in the long run, are self-destroying aquatic depressions in the landscape. They accumulate sediments from their drainage basin inputs, from their edges by erosive wave action, and from their internal organic production, often at rates of only a very few millimetres annually, but over the centuries they gradually fill in. They become shallower and richer in nutrients, their productivity increases, as does their rate of sediment accumulation. Thereby even rather large unproductive oligotrophic lakes may become productive eutrophic ones in a few millennia. This natural process of eutrophication and succession does not always proceed at a constant rate. It is subject to sometimes rapid changes and reversals, as is evidenced by analyses of sediment profiles, in terms of physical, chemical and microfossil evidence contained therein.

A wide range of human activities in and around lake airsheds, watersheds and shorelines can greatly accelerate the normally slow process of lake eutrophication. This in general is called cultural eutrophication and forestry may be one of the activities involved (see Chapter 14). Major deforestation in the watershed of the small oligotrophic Blelham Tarn in the English Lake District about a thousand years ago resulted in its first sharp increase in sedimentation rate since deglaciation, followed by another in the 1500s associated with the start of agricultural fertilization, and with the largest increase occurring after sewage enrichment in the 1900s (Pennington *et al.* 1977). In the large North American Lake Erie, phosphorus loading increased slowly after forest clearance and early watershed settlement in the late 1800s, but exponentially from the late 1940s to the mid-1970s (Schelske & Hodell 1995). It has declined considerably with phosphate abatement programmes, a nice demonstration that cultural eutrophication can be reversed. Over the last hundred years most lakes in Connecticut, USA, with watersheds remaining about 80% forested have not changed in their trophic status, whereas those becoming about 25% residential have had the greatest change (Silver *et al.* 1999).

Prediction of trophic status, conditions and fish production

Effective management of lake ecosystems in the face of multiple impacts from natural disturbance and from human activities and interventions demands the ability to make general predictions, often without detailed information on many of the factors involved or on how they interact together or sequentially. Predictive models designed to provide help in this regard have a long history in lake limnology. More recently, the role of prediction has become explicit (Peters 1986; Håkanson & Peters 1995; Håkanson 1996; Bondini 2000).

Thienemann (1927) proposed that for north German lakes those with a mean depth >18 m were oligotrophic and those <18 m were eutrophic. Rawson (1939) extended that suggestion in his frequently reproduced chart showing the inter-related factors affecting lake productivity, many of which had strong morphometric interactions. During the Second World War the Canadian government asked him to develop a method to predict sustainable commercial fish production in the many unexploited northern lakes, which he did in large part by using their mean depths as a predictive indicator. Later he used that same parameter to develop his sharply hyperbolic curves (high productivity in kilograms per hectare at low mean depth and low productivity at high mean depth) for zooplankton and zoobenthos as well as fish in a number of large North American lakes (Rawson 1955). Similar shaped hyperbolic curves related biomasses of the same three biotic components in their abundance relationships to mean depth in some 100 British Columbia lakes ranging widely in size (Northcote & Larkin 1956). However, they showed that a better productivity predictor for British Columbian lakes was nutrient level as indicated by their total dissolved solid (TDS) contents. In the mid-1960s the above two predictors – mean depth and TDS – were combined into a morphoedaphic index by dividing TDS mg/L by mean depth (m), both expressed in $\log_{10}$ units (Ryder 1965). This index was expanded to cover north temperate North American lakes, Finnish lakes, temperate and south temperate USA reservoirs, as well as tropical African lakes (Ryder *et al.* 1974) and later modified (Ryder 1982; Jackson *et al.* 1990; Downing *et al.* 1990; Rempel & Colby 1991).

There are now means to predict the cyclic annual surface water temperatures of lakes (Shuter *et al.* 1983; Matuszek & Shuter 1996) and the timing of spring stratification (Demers & Kalff 1993). Decades ago Patalas (1961) showed for 50 Polish lakes that summer thermocline depths could be predicted within about 20% by a 4.4 multiplication of the square root of their effective length, a function 'reinvented' by Ragotzkie (1978) and applied with modifications elsewhere in the world (Patalas 1984; Quiros & Drago 1985; Gorham & Boyce 1989). Freeze-up temperatures of lake surfaces also can be estimated from their fetch with adjustments for small and large lakes (Ragotzskie 1978). Even by the early 1950s, the period of simple uninodal internal seiches could be predicted with reasonable accuracy (Mortimer 1952), and this ability was later refined to include far more complex internal movements of stratified lakes, along with recognition of their ecological significance.

Hypolimnial oxygen deficits, with their implications for phosphorus release, have been predictable for at least two decades (Cornett & Rigler 1979; Nürnberg 1988;

Molot *et al.* 1992). Also there is a model to predict lake water colour (Håkanson 1993).

Because phosphorus loading was recognized over three decades ago as being key to many problems of cultural eutrophication (Vollenweider 1968), models were developed shortly thereafter to obtain approximate but useful estimates of its loading into lakes (Patalas 1972; Patalas & Salki 1973) as well as its in-lake concentrations (Dillon & Rigler 1974; Dillon & Kirchner 1975; Prepas & Rigler 1981; Vighi & Chiaudani 1985). From estimates of phosphorus loading or its winter/spring concentration at first vernal circulation, one could then predict forthcoming summer phytoplanktonic algal biomass (Oglesby & Schaffner 1978; Smith 1982; Riley & Prepas 1985). Cyanobacteria blooms and dominance in lakes often have been associated with low N:P ratios but their prediction now seems more strongly correlated with total concentration of these nutrients (Downing *et al.* 2001). Furthermore, fish-mediated nutrient recycling and its effects on zooplankton grazing of phytoplankton may favour cyanobacteria dominance (Attayde & Hansson 2001). Fee (1979) obtained a significant regression (r = 0.94) between annual phytoplankton production and epilimnial area divided by its volume, and considered that mean epilimnial depth was more important in this regard than whole lake mean depth. Crustacean zooplankton abundance of lakes seemed positively related to planktonic algal biomass and the annual load of total phosphorus (Patalas 1972).

Fish production in lakes is a function of their mean biomasses and their growth rates. Growth rate and biomass are affected by various factors such as temperature, pH, nutrient concentration and primary productivity (Downing & Plante 1993). Lake littoral habitats with a high percentage of bottom covered by submerged macrophytes show significantly greater fish production indices than those with low macrophyte cover (Randall *et al.* 1996), but the index is also significantly related to phosphorus, slope of the littoral area, and fish species richness.

Effects of some major man-made perturbations

Some of these such as watershed changes, UV-B increases, and nutrient loading additions have already been noted. Two large topics – lake acidification and global climate change – will be considered here briefly.

Acid rain and its impacts

Serious effects on inland waters from acid rain require the appropriate mix of geological, climatic and geographical factors (Fig. 4.3). This includes:

(a) hard, low-weathering geology of granites and gneisses with thin, low-nutrient soils, and consequent low buffering capacity;
(b) moderate to high rainfall regions with rapid runoff, low solution of nutrients;
(c) location in an air mass drift path from heavy industry or highly urbanized centres.

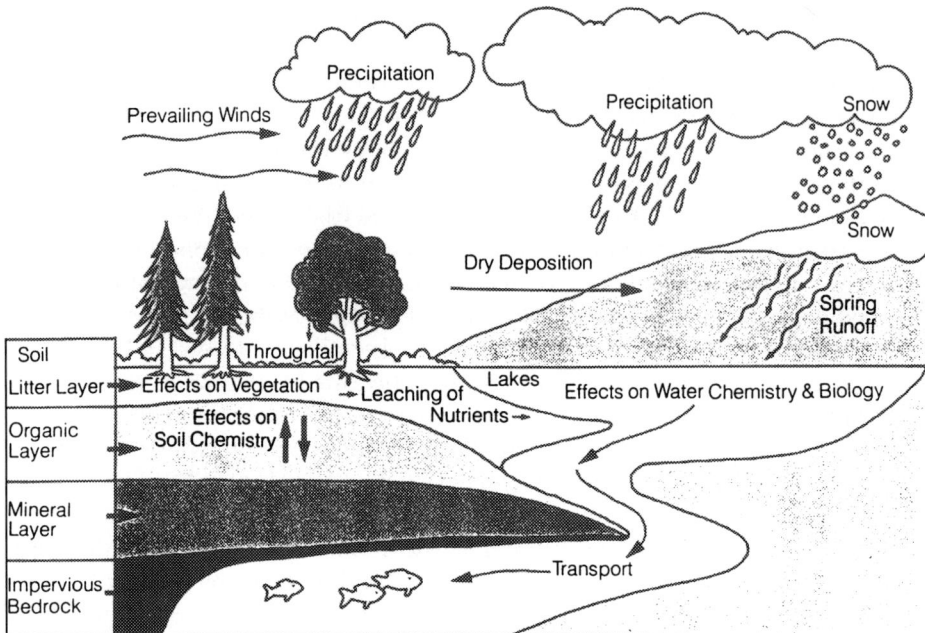

Fig. 4.3 The airshed[a] and watershed[b] involvment in lake acidification, highly simplified and generalized. Adapted from Anonymous (1980). [a]Prevailing winds from heavily industrialized and/or urbanized centres; [b]in hard rock, low weathering geological areas with thin, low-nutrient soils that lead to poorly buffered waters.

Because of (a) and (b) above, lake waters have low total dissolved solid content (usually well below 50 mg/L), low alkalinity and buffering capacity. The levels of pH may initially be only slightly acidic (6–6.5) but are very sensitive to external change because of their low HCO_3 buffering. As a result of low solubility of surrounding basin rocks in such regions and because of high rainfall, most dissolved materials enter surface waters from the atmosphere in rain, snow or dry fall, although snowmelt and spring runoff can produce short-term severe 'pulses' of acidic water to lakes. However, biological processes within lakes may be important natural sources of acid neutralizing capacity in low alkalinity systems (Schindler *et al.* 1986). Areas prone to acidification problems occur in parts of Britain, Norway and Sweden as well as the Precambrian Shield area of Canada, northeastern and western USA up into coastal British Columbia, although the latter areas are to some degree 'protected' by relatively clean airflow off the Pacific Ocean.

Acid rain brings about major physical and chemical changes in lake waters. Sulphate content rises, as in the Swedish areas of acidification where it has about doubled since the 1950s. Even more seriously, the key nutrients, nitrates and phosphates, characteristically low in areas subject to acidification, are further reduced by three effects of low pH – bacterial decomposition inhibition and thereby a lowering of nutrient cycling rates, nitrogen fixation inhibition and phosphate precipitation. Heavy metal trace elements may increase because of their high content in precipitation associated with industrial activities and because high acidity promotes increased leaching of metals, especially aluminium, from rocks and soils in the watershed (Beamish & van Loon 1977). There is also a sharp increase in water transparency, often because increasing acidity coagulates and precipitates out brown-staining organic compounds.

Acidification has many impacts on biotic components of lake ecosystems. Phytoplankton primary producers show reduced biomass (often <1 mg/L chlorophyll a) and species number with a shift to acidophilic species. *Sphagnum* and other mosses are highly tolerant of acidification so they can flourish, further increasing acidity by their cell wall exchange processes and secretions. The zooplankton community is reduced in abundance, especially the larger forms such as *Daphnia,* along with a decline in species diversity. Some species such as small *Bosmina* seem tolerant of high acidity. Benthic invertebrates undergo a reduction in both abundance and diversity with molluscs, mayflies and crayfish being very sensitive but apparently not stoneflies and dragonflies, and with some air-breathers (gyrinid beetles and water boatmen) becoming abundant. Fish sensitivity to acidification varies greatly with species but many salmonids and some cyprinids such as ide (*Leuciscus rutilus*) are very vulnerable. Fish population changes begin to occur at about pH 5.5 with reproduction becoming only sporadically successful, and recruitment declines. Older fish initially show an increase in growth rate, but as lakes become more acid only a few old fish are left, and these then show decreased growth.

Many Atlantic salmon and brown trout populations in southernmost Norway were eliminated by low pH levels there, although some survived in drainages with limestone rocks, for example, near Oslo. In Ontario, Canada, some populations of bass, lake trout, brook trout and whitefish gradually started to disappear in the late 1950s and 1960s with suckers (catostomids) being the last to die off. Low pH appears to affect osmoregulation, permeability and other membrane processes, especially of eggs. It also may alter metallic ion concentrations and permeability with aluminium being implicated.

Attempts to correct lake acidification have shown the usual range found in environmental problems from 'treating the patient' – liming of lakes and their watersheds (see Eriksson *et al.* 1983), to 'attacking the source' – banning use of high sulphur content coal and oil as well as scrubbing flue gases to remove sulphur. Due mainly to reduced acidic fallout from industry many Swedish lakes are now recovering (pers. comm. S. Holmgren, G. Milbrink). Some recovery of a forest ecosystem in New Hampshire, USA, subject to long-term effects of acid rain, has been reported (Likens *et al.* 1996) but their recovery will be delayed because of large calcium and magnesium losses from the soil complex. An assessment of lake pH changes in southeastern Canada (Jefferies *et al.* 2000) suggests that some 76,000 lakes and 970,000 ha of lake area there will remain damaged unless there are additional reductions in SO_2 emissions to those required by the Canada/USA Air Quality Agreement – an unlikely scenario in the present political arena in both countries.

Global climate change effects on lakes

Global warmings and related changes have occurred several times in the past geological history of the Earth but the latest, starting in the 1860s, has stimulated much discussion and more speculation as to causes (Hansen *et al.* 2002). It now seems that both natural (solar, volcanic) and anthropogenic forcings are involved (Stott *et al.* 2000; Zwiers & Weaver 2000). Latitudinally in North America greatest summer effects on lakes probably will occur from about 48° to 52° N (Schindler *et al.* 1990), but there is little doubt that there will be winter effects on them as well, notably shortening the duration of ice cover (Anderson *et al.* 1996; Magnuson *et al.* 2000).

A detailed and comprehensive coverage of climate warming effects on lake ecosystems in the central boreal forest region and elsewhere in Canada has been assembled (Schindler *et al.* 1990, 1996; Schindler 2001) with highlights summarized in Table 4.2. They found significant 20-year trends in over a dozen important physical and chemical variables, with several other trends including biological ones, although declining to assign these changes definitely to the effects of global climate change. Overall the climatic and hydrological effects of increasing temperature, wind speed and evaporation trigger an interlocking set of lake changes which drive many of the key in-lake limnological processes.

Some 25,000 lakes in the Mackenzie River delta north of the Canadian boreal forest may be vulnerable to water level changes associated with climate warming (Marsh & Lesack 1996). A 3°C increase in mean July epilimnetic temperature over the recent 16 years in an Alaskan arctic lake with no increase in food availability for young-of-year lake trout may reduce their early survival if the warming trend continues (McDonald *et al.* 1996). Global climate change will also have effects on subtropical and tropical inland waters and their fishes (McDowall 1992; Hambright *et al.* 1994; Lehman 1998).

Closing comment

Contrary to earlier notions, modern limnological studies have shown that lakes are not closed ecosystems. Lakes via their surfaces and shorelines, as well as from the watersheds, airsheds and groundwaters supplying them, can be affected in major ways by a wide range of human activities which surely include forestry. Some of the largest lakes in the world – either in volume (Lake Baikal), or in surface area (North American Great Lakes) – have been affected by forestry practices in several ways. Not surprisingly, the multitude of small lakes in forested areas of the world are very vulnerable to these effects, and ecological processes of such lakes in many deforested and even reforested regions already have been severely altered with response times for correction requiring decades to centuries in some cases. These will be examined in detail in Chapter 14.

Acknowledgements

Drs D.T. Crisp, S. Holmgren and G. Milbrink gave many helpful suggestions, especially in shortening of the manuscript, as did Dr T. Murphy. I am indebted to Dr F.J. Ward for his detailed review of the manuscript and suggestions for improvement in wording.

Table 4.2 A 20-year period (late 1960s to1990) of climate warming effects on lake physical (1–13), chemical (1–6) and biological (1–4) characteristics (mean annual) in the central boreal forest Experimental Lakes Area of northwestern Ontario, Canada

Variables	Change	Trend significant
Physical		
1. Air temperature (°C)	increase 0.08 per year	yes
2. Whole lake temperature (°C)	increase up to 0.09 per year	yes
3. Wind speed ice-free period (km/h)	increase c. 5 km/h	apparently
4. Thermocline depth (m)	increase up to 0.08 per year	yes
5. Summer thermal capacity (degree-days)	increase up to 26.0 per year	yes
6. Precipitation (mm)	decrease −10.5 per year	yes
7. Evaporation (mm)	increase 8.8 per year	yes
8. Runoff (mm)	decrease −10.8 per year	yes
9. 1st order tributary no flow period (days)	increase up to 0.68 per year	yes
10. Lake residence time (years)	increase	yes
11. Ice-free season (days)	increase 0.8 per year	yes
12. Lake transparency (Secchi) (m)	increase	apparently
13. Lake 1% surface light depth (m)	increase up to 0.20 per year	yes
Chemical		
1. Stream major ion export to lake (Na, K, Mg, Ca) (kg/ha)	decrease	yes or nearly so
2. Precipitation major ion export to lake (%)	increase	not tested
3. Stream total dissolved; total P export to lake (kg/ha)	decrease −0.0011; −0.0017 per year	yes
4. Stream total dissolved; total N export to lake (kg/ha)	slight decrease	no
5. Lake major ion concentration (Na, K, Mg, SO$_4$, Cl)	increase	not tested
6. Lake nutrient concentration: total dissolved P, Si	decrease	yes
Biological		
1. Phytoplankton biomass, epilimnion (µg/L)	increase	apparently
2. Phytoplankton production total, sub-epilimnial	variable	no
3. Phytoplankton diversity: Simpson's index	increase	apparently
4. Cold summer fish refugia	slight decrease	probably no

Summarized from data in Schindler *et al.* (1990, 1996).

References

Abrams, M.D. (2001) Eastern white pine versatility in the presettlement forest. *BioScience*, **51**, 967–79.

Anderson, W.L., Robertson, D.M. & Magnuson, J.J. (1996) Evidence of recent warming and El Niño-related variation in ice breakup of Wisconsin lakes. *Limnology and Oceanography*, **41**, 815–21.

Anonymous (1980) *The Case against Acid Rain. A Report on Acidic Precipitation and Ontario Programs for Remedial Action*. Ontario Ministry of Environment, Toronto, Canada.

Arts, M.T., Robarts, R.D., Kasai, F., *et al.* (2000) The attenuation of ultraviolet radiation in high dissolved organic carbon waters of wetlands and lakes on the northern Great Plains. *Limnology and Oceanography*, **45**, 292–9.

Ashley, K., Thompson, L.C., Sebastian, D., Lasenby, D.C., Smokorowski, K.E. & Andrusak, H. (1999) Restoration of kokanee salmon in Kootenay Lake, a large intermontane lake, by controlled application of limiting nutrients. In: *Aquatic Restoration in Canada*, (eds T. Murphy & M. Munawar), pp. 127–69. Ecovision World Monograph Series, Backhuys Publishers, Leiden, The Netherlands.

Attayde, J.L. & Hansson, L.A. (2001) Fish-mediated nutrient recycling and the trophic cascade in lakes. *Canadian Journal of Fisheries and Aquatic Sciences*, **58**, 1924–31.

Beamish, R.J. & van Loon, J.C. (1977) Precipitation loading of acid and heavy metals to a small lake near Sudbury, Ontario. *Journal of the Fisheries Research Board of Canada*, **34**, 649–58.

Blackie, J.R., Ford, E.D., Horne, J.E.M., Kinsman, D.J.J., Last, F.T. & Moorhouse, P. (1980) *Environmental Effects of Deforestation, An Annotated Bibliography*. Freshwater Biological Association Occasional Publication No. 10, Titus Wilson & Son, Kendal, England.

Bondini, A. (2000) Reconstructing trophic interactions as a tool for understanding and managing ecosystems: application to a shallow eutrophic lake. *Canadian Journal of Fisheries and Aquatic Sciences*, **57**, 1999–2009.

Brewin, M.K. & Monita, D.M.A. (1998) *Forest–Fish Conference: Land Management Practices Affecting Aquatic Ecosystems*. Natural Resources Canada, Canadian Forest Service, Information Report NOR-X-356.

Burgis, M.J. & Morris, P. (1987) *The Natural History of Lakes*. Cambridge University Press, Cambridge, UK.

Carignan, R.A. & Planas, D. (1994) Recognition of nutrient and light limitation in turbid mixed layers: three approaches compared in the Parana floodplain (Argentina). *Limnology and Oceanography*, **39**, 580–96.

Carignan, R. & Vaithiyanathan, P. (1999) Phosphorus availability in the Parana floodplain lakes (Argentina): influence of pH and phosphate buffering by fluvial sediments. *Limnology and Oceanography*, **44**, 1540–8.

Carrick, H.J., Fahnenstiel, G.L., Stoermer, E.F. & Wetzel, R.G. (1991) The importance of zooplankton–protozoan trophic couplings in Lake Michigan. *Limnology and Oceanography*, **36**, 1335–45.

Colgan, P. & Ealey, D. (1973) Role of woody debris in nest site selection by pumpkinseed sunfish, *Lepomis gibbosus*. *Journal of the Fisheries Research Board of Canada*, **30**, 853–6.

Cornett, R.J. & Rigler, F.H. (1979) Hypolimnetic oxygen deficits: their prediction and interpretation. *Science,* **205**, 580–1.

Cuker, B.E. & Hudson, L. (1992) Type of suspended clay influences zooplankton response to phosphorus loading. *Limnology and Oceanography,* **37**, 566–76.

Cyr, H. (1998) Effects of wave disturbance and substrate slope on sediment characteristics in the littoral zone of small lakes. *Canadian Journal of Fisheries and Aquatic Sciences*, **55**, 967–76.

D'Arcy, P. & Carignan, R. (1997) Influence of catchment topography on water chemistry in southwestern Quebec Shield lakes. *Canadian Journal of Fisheries and Aquatic Sciences,* **54**, 2215–27.

Demers, E. & Kalff, J. (1993) A simple model for predicting the date of spring stratification in temperate and subtropical lakes. *Limnology and Oceanography,* **38**, 1077–80.

Devito, K.J., Creed, I.F., Rothwell, R.L., & Prepas, E.E. (2000) Landscape controls on phosphorus loading to boreal lakes: implications for the potential impacts of forest harvesting. *Canadian Journal of Fisheries and Aquatic Sciences,* **57**, 1977–84.

Dillon, P.J. & Kirchner, W.B. (1975) The effects of geology and land use on the export of phosphorus from watersheds. *Water Research,* **9**, 135–48.

Dillon, P.J. & Rigler, F.H. (1974) A test of a simple nutrient budget model predicting the phosphorus concentration in lake waters. *Journal of the Fisheries Research Board of Canada,* **31**, 1771–8.

Downing, J.A. & Plante, C. (1993) Production of fish populations in lakes. *Canadian Journal of Fisheries and Aquatic Sciences,* **50**, 110–20.

Downing, J.A., Plante, C. & Lalonde, S. (1990) Fish production correlated with primary productivity not the morphoedaphic index. *Canadian Journal of Fisheries and Aquatic Sciences,* **47**, 1929–36.

Downing, J.A., Watson, S.B. & McCauley, E. (2001) Predicting Cyanobacteria dominance in lakes. *Canadian Journal of Fisheries and Aquatic Sciences,* **58**, 1905–8.

Edmond, J.M., Stallard, R.F., Craig, H., Craig, V., Weiss, R.F. & Coulter, G.W. (1993) Nutrient chemistry of the water column of Lake Tanganyika. *Limnology and Oceanography,* **38**, 725–38.

Edmondson, W.T. (1972) Nutrients and phytoplankton in Lake Washington. In: *Nutrients and Eutrophication: The Limiting Nutrient Controversy,* (ed G.E. Likens), Limnology and Oceanography Special Symposium 1.

Emery, A.R. (1970) Fish and crayfish mortalities due to an internal seiche in Georgian Bay, Lake Huron. *Journal of the Fisheries Research Board of Canada,* **27**, 1165–8.

Emmenegger, L., Schoenenberger, R., Sigg, L. & Sulzberger, B. (2001) Light-induced redox cycling of iron in circumneutral lakes. *Limnology and Oceanography,* **46**, 49–61.

Engle, D.L. & Melack, J.M. (1993) Consequences of riverine flooding for seston and the periphyton of floating meadows in an Amazon floodplain lake. *Limnology and Oceanography,* **38**, 1500–20.

Ennis, G.L. (1975) Distribution and abundance of benthic algae along phosphate enriched Kootenay Lake, British Columbia. *Verhandlungen Internationale Vereinigung für theoretische und angewandte Limnologie,* **19**, 562–70.

Eriksson, F., Hornstrom, E., Mossberg, P. & Nyberg, P. (1983) Ecological effects of lime treatment of acidified lakes and rivers in Sweden. *Hydrobiologia,* **101**, 145–64.

Evans, J.E. & Prepas, E.E. (1997) Relative importance of iron and molybdenum in restricting phytoplankton biomass in high phosphorus saline lakes. *Limnology and Oceanography,* **42**, 461–72.

Fee, E. (1979) A relation between lake morphometry and primary productivity and its use in interpreting whole-lake eutrophication experiments. *Limnology and Oceanography,* **24**, 401–16.

France, R.L. (1995) Differentiation between littoral and pelagic food webs in lakes using stable carbon isotopes. *Limnology and Oceanography,* **40**, 1310–13.

Gibbons, D.R. & Salo, E.O. (1973) *An Annotated Bibliography of the Effects of Logging on Fish of the Western United States and Canada.* USDA Forest Service General Tech. Rept. PNW-10, Portland, OR.

Goldman, C.R. (1961) The contribution of alder trees (*Alnus tenuifolia*) to the primary productivity of Castle Lake, California. *Ecology,* **42**, 282–8.

Goldman, C.R. (1994) Baikal: the Greatest Great Lake. In: *1994 Yearbook of Science and the Future.* pp. 208–25. Encyclopedia Britannica, Chicago, IL.

Gorham, E. & Boyce, F.M. (1989) Influence of lake surface area and depth upon thermal stratification and the depth of the summer thermocline. *Journal of Great Lakes Research*, **15**, 233–45.

Guildford, S.J. & Hecky, R.E. (2000) Total nitrogen, total phosphorus, and nutrient limitation in lakes and oceans: is there a common relationship? *Limnology and Oceanography*, **45**, 1213–23.

Guyette, R.P. & Cole, W.C. (1999) Age characteristics of coarse woody debris (*Pinus strobus*) in a lake littoral zone. *Canadian Journal of Fisheries and Aquatic Sciences*, **56**, 496–505.

Hagerthey, S.E. & Kerfoot, W.C. (1998) Groundwater flow influences on the biomass and nutrient ratios of epibenthic algae in a north temperate seepage lake. *Limnology and Oceanography*, **43**, 1227–42.

Håkanson, L. (1981) *A Manual of Lake Morphometry*. Springer-Verlag, Berlin.

Håkanson, L. (1993) A model to predict lake water colour. *Internationale Revue Gesamten Hydrobiologie*, **78**, 107–37.

Håkanson, L. (1996) Predicting important lake habitat variables from maps using modern modelling tools. *Canadian Journal of Fisheries and Aquatic Sciences*, **53** (Suppl. 1), 364–82.

Håkanson, L. & Peters, R.H. (1995) *Predictive Limnology Methods for Predictive Modelling*. SPB Academic Publishing, Amsterdam, The Netherlands.

Halsey, T.G. (1968) Autumnal and over-winter limnology of three small eutrophic lakes with particular reference to experimental circulation and trout mortality. *Journal of the Fisheries Research Board of Canada*, **25**, 81–99.

Hambright, K.D., Gophen, M. & Serruya, S. (1994) Influence of long-term climatic changes on the stratification of a subtropical, warm monomictic lake. *Limnology and Oceanography*, **39**, 1233–42.

Hansen, J., Ruedy, R., Sato, M. & Lo, K. (2002) Global warming continues. *Science*, **295**, 275.

Hart, D.R., Stone, L. & Berman, T. (2000) Seasonal dynamics of the Lake Kinneret food web: the importance of the microbial loop. *Limnology and Oceanography*, **45**, 350–61.

Hatch, L.K., Reuter, J.E. & Goldman, C.R. (1999) Relative importance of stream-borne particulate and dissolved phosphorus fractions to Lake Tahoe phytoplankton. *Canadian Journal of Fisheries and Aquatic Sciences*, **56**, 2331–9.

Hecky, R.E., Bugenyi, F.W.B., Ochumba, P., *et al.* (1994) Deoxygenation of the deep water of Lake Victoria, East Africa. *Limnology and Oceanography*, **39**, 1476–80.

Hinch, S.G., Collins, N.C. & Harvey, H.H. (1991). Relative abundance of littoral zone fishes: biotic interactions, abiotic factors, and post-glacial colonization. *Ecology*, **72**, 1314–24.

Jackson, D.A., Harvey, H.H. & Somers, K.M. (1990) Ratios in aquatic sciences: statistical shortcomings with mean depth and the morphoedaphic index. *Canadian Journal of Fisheries and Aquatic Sciences*, **47**, 1788–95.

James, W.F. & Barko, J.W. (1991) Littoral-pelagic phosphorus dynamics during night-time convective circulation. *Limnology and Oceanography*, **36**, 949–60.

Jeffries, D.S., Lam, D.C.L., Wong, I. & Moran, M.D. (2000) Assessment of changes in lake pH in southeastern Canada arising from present levels and expected reductions in acidic deposition. *Canadian Journal of Fisheries and Aquatic Sciences*, **57** (Suppl. 2), 40–9.

Kaczmarska, I., Clair, T.A., Ehrman, J.M., MacDonald, S.L., Lean, D. & Day, K.E. (2000) The effect of ultraviolet B on phytoplankton populations in clear and brown temperate Canadian lakes. *Limnology and Oceanography*, **45**, 651–63.

Kenny, B.C. (1990) *On the Dynamics of Phosphorus in Lake Systems*. Environment Canada NHRI Paper No. 45, IWD Scientific Series No. 182, Saskatoon, Saskatchewan.

Koenings, J.P. & Edmundson, J.A. (1991) Secchi disk and photometer estimates of light regimes in Alaskan lakes: effects of yellow color and turbidity. *Limnology and Oceanography*, **36**, 91–105.

Langmuir, I. (1938) Surface motion of water induced by wind. *Science*, **87**, 119–23.

Lehman, J.T. (1998) *Environmental Change and Response in East African Lakes*. Kluwer Academic, Dordrecht, The Netherlands.

Lehman, J.T. & Branstrator, D.K. (1994) Nutrient dynamics and turnover rates of phosphate and sulphate in Lake Victoria, East Africa. *Limnology and Oceanography*, 39, 227–33.

Lesack, L.F.W. (1995) Seepage exchange in an Amazon floodplain lake. *Limnology and Oceanography*, 40, 598–609.

Levine, S.N. & Schindler, D.W. (1992) Modification of the N:P ratio in lakes by in situ processes. *Limnology and Oceanography*, 37, 917–35.

Likens, G.E., Driscoll, C.T. & Buso, D.C. (1996) Long-term effects of acid rain: response and recovery of a forest ecosystem. *Science*, 272, 244–5.

Lind, O.T., Doyle, R., Vodopich, D.S. & Trotter, B.G. (1992) Clay turbidity: regulation of phytoplankton production in a large, nutrient-rich tropical lake. *Limnology and Oceanography*, 37, 549–65.

McCullough, G.K. (1998) The contribution of forest litterfall to phosphorus inputs to Lake 239, Experimental Lakes Area, northwestern Ontario. In: *Forest–Fish Conference: Land Management Practices Affecting Aquatic Ecosystems*, (eds M.K. Brewin & D.M.A. Monita), pp. 159–68. Proceedings of the Forest-Fish Conference, Calgary, Alberta. Natural Resources Canada, Canadian Forest Service Information Report NOR-X-356.

McDonald, M.E., Hershey, A.E. & Miller, M.C. (1996) Global warming impacts on lake trout in arctic lakes. *Limnology and Oceanography*, 41, 1102–8.

McDowall, R.M. (1992) Global climate change and fish and fisheries: what might happen in a temperate oceanic archipelago like New Zealand? *GeoJournal*, 28, 29–37.

Magnuson, J.J., Robertson, D.M., Benson, B.J., *et al.* (2000) Historical trends in lake and river ice cover in the Northern Hemisphere. *Science*, 289, 1743–6.

Marsh, P. & Lesack, L.F.W. (1996) The hydrologic regime of perched lakes in the Mackenzie Delta: potential responses to climate change. *Limnology and Oceanography*, 41, 849–56.

Mathias, J.A. & Barica, J. (1980) Factors controlling oxygen depletion in ice-covered lakes. *Canadian Journal of Fisheries and Aquatic Sciences*, 37, 185–94.

Matuszek, J.E. & Shuter, B.J. (1996) An empirical method for the prediction of daily water temperatures in the littoral zone of temperate lakes. *Transactions of the American Fisheries Society*, 125, 622–7.

Meehan, W.R. (1991) *Influences of Forest and Rangeland Management on Salmonid Fishes and their Habitats*. AFS Special Publication 19, Bethesda, MD.

Molot, L.A., Dillon, P.J., Clark, B.J. & Neary, B.P. (1992) Predicting end-of-summer oxygen profiles in stratified lakes. *Canadian Journal of Fisheries and Aquatic Sciences*, 49, 2363–72.

Mortimer, C.H. (1952) Water movements in lakes during summer stratification: evidence from the distribution of temperature in Windermere. *Philosophical Transactions of the Royal Society of London, Series B*, 236, 355–404.

Mortimer, C.H. (1981) The oxygen content of air-saturated fresh waters over ranges of temperature and atmospheric pressure of limnological interest. *Mitteilungen Internationale Vereinigen Limnologie*, 22.

Murphy, M.L. (1995) *Forestry Impacts on Freshwater Habitat of Anadromous Salmonids in the Pacific Northwest and Alaska – Requirements for Protection and Restoration*. US Dept Comm. Nat. Ocean. Atmosp. Admin. Coastal Ocean Program, Decision Analysis Series No. 7. Silver Spring, MD.

Murphy, T., Lawson, A., Kumagai, M. & Nalewajko, C. (2001) Release of phosphorus from sediments in Lake Biwa. *Limnology*, 2, 119–28.

Näslund, I. (1999) *Fiske, skogsbruk och vattendrag – nyttjande i ett uthålligt perspektiv*. Fiskeriverkets försökstation, Kalarne, Sweden.

Northcote, T.G. (1972a) Pollution in Russia: a look at Lake Baikal. *Canadian Conservationist*, Summer 1972, 22–4.

Northcote, T.G. (1972b) Kootenay Lake: man's effects on the salmonid community. *Journal of the Fisheries Research Board of Canada*, 29, 861–5.

Northcote, T.G. (1973) *Some Impacts of Man on Kootenay Lake and its Salmonoids.* Great Lakes Fishery Commission Tech. Rept. No. 25.

Northcote, T.G. (1980) Morphometrically conditioned eutrophy and its amelioration in some British Columbia lakes. In: *Developments in Hydrobiology*, Vol. 2, (eds J. Barica & L.R. Mur), pp. 75–85. Dr W. Junk b.v. Publishers, The Hague, The Netherlands.

Northcote, T.G. (1992) Eutrophication and pollution problems. In: *Lake Titicaca. A Synthesis of Limnological Knowledge*, (eds C. Dejoux & A. Iltis), pp. 551–61. Kluwer Academic Publishers, Dordrecht, The Netherlands.

Northcote, T.G. & Atagi, D.Y. (1997) *Ecological Interactions in the Flooded Littoral Zone of Reservoirs: The Importance and Role of Submerged Terrestrial Vegetation with Special Reference to the Nechako Reservoir of British Columbia, Canada.* BC Ministry of Environment, Lands and Parks, Skeena Fisheries Report SK-111.

Northcote, T.G. & Larkin, P.A. (1956) Indices of productivity in British Columbia lakes. *Journal of the Fisheries Research Board of Canada*, **13**, 515–40.

Northcote, T.G. & Luksun, B. (1992) Restoration and environmental sustainability of a small British Columbia urban lake. *Water Pollution Research Journal of Canada*, **27**, 341–64.

Northcote, T.G. & Northcote, H. (1996) Shoreline marshes of Okanagan Lake: are they habitats of high productivity, diversity, scarcity and vulnerability? *Lakes & Reservoirs: Research and Management*, **2**, 157–61.

Northcote, T.G., Fillion, D.B., Salter, S.P. & Ennis, G.L. (1999) *Interactions of Nutrients and Turbidity in the Control of Phytoplankton in Kootenay Lake, British Columbia, Canada, 1964 to 1966.* Report to Columbia Basin Fish & Wildlife Compensation Program, BC Hydro/BC Environment, Nelson, BC, Canada.

Nürnberg, G.K. (1987) A comparison of internal phosphorus loads in lakes with anoxic hypolimnia: laboratory incubation versus in situ hypolimnetic phosphorus accumulation. *Limnology and Oceanography*, **32**, 1160–4.

Nürnberg, G.K. (1988) Prediction of phosphorus release rates from total and reductant-soluble phosphorus in anoxic lake sediments. *Canadian Journal of Fisheries and Aquatic Sciences*, **45**, 453–62.

Nürnberg, G.K. (1995a) Quantifying anoxia in lakes. *Limnology and Oceanography*, **40**, 1100–11.

Nürnberg, G.K. (1995b) The anoxic factor, a quantitative measure of anoxia and fish species richness in central Ontario lakes. *Transactions of the American Fisheries Society*, **124**, 677–86.

Oglesby, R. & Schaffner, W. (1978) Phosphorus loadings to lakes and some of their responses. Part 1. A new calculation of phosphorus loading and its application to New York lakes. *Limnology and Oceanography*, **23**, 120–34.

Ostrovsky, I., Yacobi, Y.Z., Walline, P. & Kalikhman, I. (1996) Seiche-induced mixing: its impact on lake productivity. *Limnology and Oceanography*, **41**, 323–32.

Patalas, K. (1961) Wind- und morphologiebedingte Wasserbegungs-typen als bestimmender Factor für die Intensität des Stoffkrieslaufes in nordpolnischen. *Verhandlungen Internationale Vereinigung für theoretische und angewandte Limnologie*, **14**, 59–64.

Patalas, K. (1972) Crustacean plankton and eutrophication of St. Lawrence Great Lakes. *Journal of the Fisheries Research Board of Canada*, **29**, 1451–62.

Patalas, K. (1984) Mid-summer mixing depths of lakes at different latitudes. *Verhandlungen Internationale Vereinigung für theoretische und angewandte Limnologie*, **22**, 97–102.

Patalas, K. & Salki, A. (1973) Crustacean plankton and eutrophication of lakes in the Okanagan Valley, British Columbia. *Journal of the Fisheries Research Board of Canada*, **30**, 519–42.

Pennington, W., Cranwell, P.A., Haworth, E.Y., Bonny, A.P. & Lishman, J.P. (1977) Interpreting the environmental record in the sediments of Blelham Tarn. *FBA Annual Report*, **45**, 37–47.

Persson, A. (1997) Phosphorus release by fish in relation to external and internal load in a eutrophic lake. *Limnology and Oceanography*, **42**, 577–83.

Peters, R.H. (1986) The role of prediction in limnology. *Limnology and Oceanography*, **31**, 1143–59.

Peterson, B.J., Wollheim, W.M., Mulholland, P.J. *et al.* (2001) Control of nitrogen export from watersheds by headwater streams. *Science*, **292**, 86–90.

Petticrew, E.L. & Kalff, J. (1991) Predictions of surficial sediment composition in the littoral zone of lakes. *Limnology and Oceanography*, **36**, 384–92.

Prepas, E.E. & Rigler, F.H. (1981) A test of a simple model to predict short-term changes in the phosphorus concentration in lake water. *Verhandlungen Internationale Vereinigung für theoretische und angewandte Limnologie*, **21**, 187–96.

Quiros, R. & Drago, E. (1985) Relaciones entre variables fisicas, morfometricas y climaticas en lagos patagonicos. *Revista de la Asociacion de Ciencias Naturales de Litoral*, **16**, 181–99.

Rae, R., Hawes, I., Hanelt, D. & Howard-Williams, C. (2001) Ultraviolet light: is it harming New Zealand's lakes? *Water & Atmosphere*, **9**, 10–12.

Ragotzkie, R.A. (1978) Heat budgets of lakes. In: *Lakes: Chemistry, Geology, Physics*, (ed. A. Lehrman), pp. 1–19. Springer-Verlag, New York.

Randall, R.G., Minns, C.K., Cairns, V.W. & Moore, J.E. (1996) The relationship between an index of fish production and submerged macrophytes and other habitat features at three littoral areas in the Great Lakes. *Canadian Journal of Fisheries and Aquatic Sciences*, **53** (Suppl. 1), 35–44.

Rawson, D.S. (1939). Some physical and chemical factors in the metabolism of lakes. In: *Problems of Lake Biology*. Publication of the American Association for the Advancement of Science 10.

Rawson, D.S. (1955) Morphometry as a dominant factor in the productivity of large lakes. *Verhandlungen Internationale Vereinigung für theoretische und angewandte Limnologie*, **12**, 164–75.

Rempel, R.S. & Colby, P.J. (1991) A statistically valid model of the morphoedaphic index. *Canadian Journal of Fisheries and Aquatic Sciences*, **48**, 1937–43.

Riley, E.T. & Prepas, E.E. (1985) Comparison of the phosphorus-chlorophyll relationships in mixed and stratified lakes. *Canadian Journal of Fisheries and Aquatic Sciences*, **42**, 831–5.

Ryder, R.A. (1965) A method for estimating the potential fish production of north-temperate lakes. *Transactions of the American Fisheries Society*, **94**, 214–18.

Ryder, R.A. (1982) The morphoedaphic index – use, abuse and fundamental concepts. *Transactions of the American Fisheries Society*, **111**, 154–64.

Ryder, R.A., Kerr, S.R., Loftus, K.H. & Regier, H.A. (1974) The morphoedaphic index, a fish yield estimator – a review and evaluation. *Journal of the Fisheries Research Board of Canada*, **31**, 663–88.

Salo, E.O. & Cundy, T.W. (1987) *Streamside Management: Forestry and Fishery Interactions*. Institute of Forest Resources, University of Washington, Seattle.

Schelske, C.L. & Hodell, D.A. (1995) Using carbon isotopes of bulk sedimentary organic matter to reconstruct the history of nutrient loading and eutrophication in Lake Erie. *Limnology and Oceanography*, **40**, 918–29.

Schindler, D.W. (1998) A dim future for boreal waters and landscapes. *BioScience*, **48**, 157–64.

Schindler, D.W. (2001) The cumulative effects of climate warming and other human stresses on Canadian freshwaters in the new millennium. *Canadian Journal of Fisheries and Aquatic Sciences*, **58**, 18–29.

Schindler, D.W., Turner, M.A., Stainton, M.P. & Linsey, G.A. (1986) Natural sources of acid neutralizing capacity in low alkalinity lakes of the Precambrian Shield. *Science*, **232**, 844–7.

Schindler, D.W., Beaty, K.G., Fee, E.J., *et al.* (1990) Effects of climatic warming on lakes of the central boreal forest. *Science*, **250**, 967–70.

Schindler, D.W., Bayley, S.E., Parker, B.R., *et al.* (1996) The effects of climate warming on the properties of boreal lakes and streams of the Experimental Lakes Area, northwestern Ontario. *Limnology and Oceanography*, 41, 1004–17.

Sebetich, M.J. & Horner-Neufeld, G. (2000) Terrestrial litterfall as a source of organic matter in a mesotrophic lake. *Verhandlungen Internationale Vereinigung für theoretische und angewandte Limnologie*, 27, 2225–31.

Shuter, B.J., Schlesinger, D.A. & Zimmerman, A.P. (1983) Empirical predictors of annual surface water temperature cycles in North American lakes. *Canadian Journal of Fisheries and Aquatic Sciences*, 40, 1838–45.

Silver, P.A., Lott, A.M., Cash, E., Moss, J. & Marsicano, L. (1999) Century changes in Connecticut, U.S.A., lakes as inferred from siliceous algal remains and their relationships to land-use change. *Limnology and Oceanography*, 44, 1928–35.

Smith, V. (1982) The nitrogen and phosphorus dependence of algal biomass in lakes: an empirical and theoretical analysis. *Limnology and Oceanography*, 27, 1101–11.

Spain, J.D., Wernert, G.M. & Hubbard, D.W. (1976) The structure of the spring thermal bar in Lake Superior. *Journal of Great Lakes Research*, 2, 296–306.

Stefan, H.G., Horsch, G.M. & Barko, J.W. (1989) A model for the estimation of convective exchange in the littoral region of a shallow lake during cooling. *Hydrobiologia*, 174, 225–34.

Stockner, J.G. & Northcote, T.G. (1974) Recent limnological studies of Okanagan Basin lakes and their contribution to comprehensive water resource planning. *Journal of the Fisheries Research Board of Canada*, 31, 955–76.

Stott, P.A., Tett, S.F.B., Jones, G.S., Allen, M.R., Mitchell, J.F.B. & Jenkins, G.J. (2000) External control of 20th century temperature by natural and anthropogenic forcings. *Science*, 290, 2133–7.

Tabor, R.A. & Wurtsbaugh, W.A. (1991) Predation risk and the importance of cover for juvenile rainbow trout in lentic systems. *Transactions of the American Fisheries Society*, 120, 728–38.

Thienemann, A. (1927) Der Bau des Seebeckens in seiner Bedeutung für den Ablauf des Lebens im See. *Verhandlungen der Zoologische-Botanischen Gesellschaft in Oesterreich*, 77, 87–91.

Toews, D.A.A. & Brownlee, M.J. (1981) *A Handbook for Fish Habitat Protection on Forest Lands in British Columbia*. Fisheries and Oceans Canada, Vancouver.

Tonn, W.M., Magnuson, J.J., Rask, M. & Toivonen, J. (1990) Intercontinental comparison of small-lake fish assemblages: the balance between local and regional processes. *The American Naturalist*, 136, 345–75.

Turner, M.G. (1998) Landscape ecology, living in a mosaic. In: *Ecology*, (eds S.I. Dodson, T.F.H. Allen, S.R. Carpenter, *et al.*), pp. 78–122. Oxford University Press, New York.

Vadeboncoeur, Y., Vander Zanden, M.J. & Lodge, D.M. (2002) Putting the lake back together: reintegrating benthic pathways into lake food web models. *BioScience*, 52, 44–54.

Vighi, M. & Chiaudani, G. (1985) A simple method to estimate lake phosphorus concentrations resulting from natural, background, loadings. *Water Research*, 19, 987–91.

Vinebrooke, R.D. & Leavitt, P.R. (1996) Effects of ultraviolet radiation on periphyton in an alpine lake. *Limnology and Oceanography*, 41, 1035–40.

Vollenweider, R.A. (1968) *Scientific Fundamentals of the Eutrophication of Lakes and Flowing Water, with Particular Reference to Phosphorus and Nitrogen as Factors in Eutrophication*. OECD Technical Report DAS/CS1/68.27.

Vollenweider, R.A. (1976) Advances in defining critical loading levels for phosphorus in lake eutrophication. *Memorie dell'Istituto Italiano di Idrobiologia*, 33, 53–83.

Ward, F.J. (1964) *Limnology of Kamloops Lake*. Bulletin XVI, International Pacific Salmon Fisheries Commission, New Westminster, BC, Canada.

Ward, F.J. (1966) Initiation of vernal heating in Kamloops Lake, B.C. *Verhandlungen Internationale Vereinigung für theoretische und angewandte Limnologie*, 16, 111–16.

Webster, K.E., Kratz, T.K., Bowser, C.J. & Magnuson, J.J. (1996) The influence of landscape position on lake chemical response to drought in northern Wisconsin. *Limnology and Oceanography*, **41**, 977–84.

Wetzel, R.G. (2001) *Limnology*, 3rd edn. Academic Press, San Diego.

Weyhenmeyer, G.A. & Meili, M. (2000) Hypolimnetic lake sediments in frequent motion. *Verhandlungen Internationale Vereinigung für theoretische und angewandte Limnologie*, **27**, 2317–22.

Whitmore, T.J., Brenner, M., Rood, B.E. & Japy, K.E. (1991) Deoxygenation of a Florida lake during winter mixing. *Limnology and Oceanography*, **36**, 577–84.

Williamson, C.E. (1995) What role does UV-B radiation play in freshwater ecosystems? *Limnology and Oceanography*, **40**, 386–92.

Zwiers, F. & Weaver, A.T. (2000) The causes of 20th century warming. *Science*, **290**, 2081–3.

Chapter 5
Fundamental aspects of estuarine ecology relevant to fish–forestry interactions

T.G. NORTHCOTE AND M.C. HEALEY

Introduction

Estuaries form the lowermost link in the chain of aquatic environments stretching from the uppermost reaches of watersheds to the ocean. They are the transitional ecotone between freshwater and marine habitats and often provide critical areas for both marine and freshwater fishes (Healey 1982; Monaco *et al.* 1989; Nelson *et al.* 1991; Roman *et al.* 2000). The general definition of an estuary proposed by Pritchard (1967) as 'a semi-enclosed body of water which has free connection with the sea and within which seawater is measurably diluted with freshwater derived from land drainage' is still commonly recognized (e.g. Hobbie 2000). Estuaries of significance to fish ecology occur where a freshwater inflow to the sea creates a localized region of intermediate salinity and buoyancy-driven circulation sufficiently large to influence the distribution and abundance of fishes.

Estuaries are among the most biologically productive ecosystems in the world, with primary production in the order of 1000 $gCm^{-2}y^{-1}$, higher than comparable areas of forest or farmland (Bell *et al.* 2000). Estuaries have also attracted much human economic activity, and with their associated river systems, frequently serve as communication gateways to continental hinterlands. As a result, many estuaries have been highly altered and degraded. Degradation can be caused not only by human activity within the estuary itself but also by human activity upstream, the impacts of which propagate downstream (Alongi 1998; Jackson *et al.* 2001). Furthermore, there can be ocean-driven impacts such as exotic species introduction from offshore ballast water release by ships, or large-scale changes in nearby ocean regimes (Dr C. Levings, pers. comm.).

In this chapter we briefly review some fundamental aspects of estuarine structure and function. Our emphasis is on aspects of estuarine ecology that are most likely to be affected by forestry-related activities both within the estuary and upstream or offshore. Readers needing a more comprehensive review of estuarine ecology should consult Alongi (1998), Hobbie (2000) or Day *et al.* (1989). Our review will be biased toward temperate zone estuaries as these are the best studied (Ayvazian 1999). Far northern and southern estuaries have rarely been studied. Only a few of the thousand or so estuaries in Australia and New Zealand have received close attention (Bell *et al.* 2000; Saenger & Holmes 2000). Earlier studies of West African tropical estuaries are covered by Pillay (1967) and a general worldwide coverage can be found in Lowe-McConnell (1987). Nordstrom & Roman (1996) discuss European, West African, Australian,

New Zealand, USA and Central American estuaries, focusing on processes affecting shorelines. Case studies of estuaries from the Americas, Africa, India, Southeast Asia and Australia are provided in Blaber (1997). Allanson & Baird (1999) discuss many of the 300 South African estuaries.

Physical and chemical characteristics

The shape and size of estuaries depend on many processes controlling watershed formation and evolution, as well as those affecting sea level change and coastal erosion/deposition. For more pristine estuaries, present features can be a result of events occurring as far back as Pleistocene or even Cretaceous times (Cooper *et al.* 1999). Heavily glaciated coastal landscapes (e.g. fjords) are often characterized by a diversity of types of estuaries formed by glacial scouring of bedrock, reshaping and widening of pre-glacial valleys, and deposition of sediments that formed barrier beaches and islands when sea level rose (Emmett *et al.* 2000; Roman *et al.* 2000). Coastlines characterized by active tectonics are likely to have steep-sided estuaries with limited wetland development and little shallow water nursery habitat for fishes, although these can be modified by sedimentation rates. Where there is a gently sloping coastal plain, river deltas and wetlands can be extensive, as in northeast Asia (Korean mudflats, mouth of the Chang Jiang River), and provide large areas of nursery habitat (Deegan *et al.* 1986; Dame *et al.* 2000). Human activity can have profound effects on the physical character of estuaries with concomitant effects on biota (e.g. Healey 1994; Nordstrom & Roman 1996). Typical man-made physical alterations include narrowing, straightening and deepening of drainage channels, dyking and filling of intertidal flats and marshes, removing vegetation and adding impervious surfaces, and installing groynes and barriers to impede or direct sediment transport and deposition.

Estuaries experience large and sometimes rapid changes in physical and chemical characteristics driven by tides, winds, temperature and changes in river discharge. The latter can be the major factor controlling hydrology and morphology, along with bottom and mouth topography. For example, early summer peak flow of over $14,000\ m^3\ s^{-1}$ in the Fraser River in western Canada turns its estuary into a freshwater-dominated system, whereas during winter low flows of $650\ m^3\ s^{-1}$ moderately saline water extends 30 km upstream and tidal fluctuations some 100 km upstream (Northcote & Larkin 1989). Tidal influence extends over 160 km up the Guyana River (Lowe-McConnell 1987). The St John River estuary in eastern Canada is a freshwater-dominated system with a drainage area to surface ratio of 230:1, yet saline water occasionally reaches 60 km inland (Metcalfe *et al.* 1976). When discharge in the European Rhone and Ebre rivers is below average (1700 and $424\ m^3\ s^{-1}$ respectively), salt wedges are established and their deltas become depositional. If discharge is greater than average, however, the salt wedge is pushed out into the Mediterranean Sea and the deltas become erosional (Ibanez *et al.* 1997). In the Amazon, the interface between fresh and salt water moves up and down river over 200 km with changing discharge, tidal back-up is evident 800 km upstream and the river's muddy plume is visible 200 km out into the Atlantic (Lowe-McConnell 1987).

These seasonal and episodic changes in salinity distribution have a profound influence on the distribution, abundance and productivity of fishes. Estuaries typically demonstrate a species minimum at some point along the gradient from fresh to salt water (Day *et al.* 1989). Wagner (1999) found a species richness depression for fish assemblages in lower Chesapeake Bay tributaries at salinities of 8–10 ppt and a peak associated with the tidal freshwater interface (salinities of 0–2 ppt). In general, fishery yield from estuaries is positively related to river discharge (Deegan *et al.* 1986; Wilbur 1993). The effects can often be masked by changes in other factors such as nutrient loading and sediment load and, in some instances, the correlation is negative rather than positive (e.g. FAO 1995). The mechanism underlying these correlations is not well understood, although Friedland *et al.* (1996) have shown how juvenile nursery habitat for one species (Atlantic menhaden, *Brevoortia tyrannus*) is structured by seasonal patterns of discharge that affect the location of phytoplankton maxima.

Water residence time in estuaries is closely linked to freshwater discharge and can vary from a few days to more than a year (Miller & McPherson 1991; Allanson & Winter 1999). Residence time greatly affects biogeochemical and ecological processes driving production in lower levels of estuarine food webs as well as its temporal variability and spatial heterogeneity (Jay *et al.* 2000). Structural attributes of estuaries also contribute to water retention, especially retention of fine particulate materials that may be sources of nutrients or toxic materials. Many authors have emphasized the importance of structural aspects of estuaries, including large woody debris from upstream in the watershed. Structure affects patterns of erosion and deposition of particulate material and can also provide refuges from predators for estuarine fishes (Naiman & Sibert 1978; Gonor *et al.* 1988; Everett & Ruiz 1993; Maser & Sedell 1994). All of these processes and attributes can be dramatically altered by human activities in the estuary or watershed that affect the amount and seasonality of discharge, sediment loads, and nutrient and toxic inputs – one of them being forestry activity.

For many estuaries nutrients, particularly nitrogen (N), are the primary limitation on productivity. Nutrient sources and transport vary greatly among estuarine systems (Jordan *et al.* 1991). In some estuaries, such as the Nanaimo River estuary in British Columbia, Canada, riverine inputs of N and phosphorus (P) are low most of the year (Naiman & Sibert 1979). For the outer Fraser River estuary, the major system on the Canadian west coast, entrainment of oceanic water is considered the principal source of nutrients for phytoplankton growth (Harrison *et al.* 1991). Indeed, the two-layered circulation pattern of estuaries, with low salinity water moving seaward at the surface and high salinity water moving landward at depth and being entrained into the surface flow, is considered a major factor in the high productivity of estuaries around the world. In other estuaries, such as the St Lawrence in eastern Canada, river transport is a major source of nutrients, with monthly amounts ranging up to 6.72×10^3 tons for N, 0.48×10^3 tons for P and 6.73×10^3 tons for silicate (Si) (Sinclair *et al.* 1976; Greisman & Ingram 1977). For many US east coast estuaries, land-derived N sources, especially from sewage, may predominate (Valiela *et al.* 1992; McClelland & Valiela 1998; Bowen & Valiela 2001). Headwater streams in North America can control N export even down to estuaries (Peterson *et al.* 2001). Human inputs of nutrients are also implicated in the eutrophication of estuaries on the northern coast of the Mediterranean (FAO 1995). In Long Island Sound, 95% of the N input comes from anthropogenic sources, two-thirds from sewage and one-third from atmospheric deposition (Howarth 2000). Loading of N and P to the broad estuary and nearby coastal waters of the Mississippi River has doubled since the 1950s, although Si has dropped by half. Nutrient pollution, particularly from N, is considered the largest problem now facing estuaries in the USA (Bricker *et al.* 1999; Anonymous 2000).

Nutrient budgets in estuarine marshes show large seasonal changes, with periods of nutrient import alternating with periods of export. For example, in winter to summer the salt marsh of the York River tributary in lower Chesapeake Bay, USA, acts as a nutrient sink with very active ammonia uptake in spring to early summer. In late summer, however, if low oxygen concentrations develop during slack water, the marsh becomes a net exporter of ammonia (Wolaver *et al.* 1980). Great Sippewissett Marsh in east coast USA also shows a strong seasonal cycle of import and export of ammonia – high import during July and early August, high export from mid-August to December, and a balance for the rest of the year (Nordstrom & Roman 1996).

As indicated above, the nutrient dynamics of estuaries are complex with variable nutrient inputs from riverine, atmospheric and oceanic sources, combined with changing patterns of uptake, storage and recycling by estuarine organisms and sediments. The availability of nutrients in estuarine waters and their continual replenishment underlies the high primary productivity of estuaries. High production at the bottom of the food web is reflected in high fisheries yields, ranging from 20 to 500 $kg\,ha^{-1}\,y^{-1}$, and a positive correlation between primary production and fish yield (Alongi 1998). Impacts of land use within the watershed on estuarine nutrient dynamics can be profound, particularly where riverine inputs are a significant component of estuarine nutrient loading (FAO 1995).

Biological characteristics

The rich abundance of life in estuaries rises from a foundation of organic carbon. There are three main sources of organic carbon: (1) in situ production by green plants, (2) detrital carbon (dissolved [DOC] and fine particulate [FPOC] and coarse particulate [CPOC]) brought downstream to the estuary from the watershed, and (3) both living and detrital carbon brought into the estuary from oceanic sources (Fig. 5.1). Critical biological processes at the base of the food web that make carbon available to consumers are photosynthesis by green plants and remineralization of detrital carbon by bacteria in the microbial loop (Fig. 5.1).

Organic carbon from terrestrial sources, brought downstream as DOC, FPOC and CPOC, can be very large (Fig. 5.1). Globally, inputs of terrestrial river-borne carbon are estimated to be 198 million tons DOC and 172 million tons POC. In river-dominated estuaries, high amounts of carbon can be delivered to the estuary. For example, the Fraser River in western Canada discharges some 472,000 tons of carbon per year to its estuary, most of it in dissolved form (Healey & Richardson 1996). This represents 295 $g\,C\,m^{-2}\,y^{-1}$ delivered to 1600 km^2 of the Fraser plume in the Strait of Georgia off the Fraser River mouth. Riverine organic carbon inputs, therefore, are roughly equal

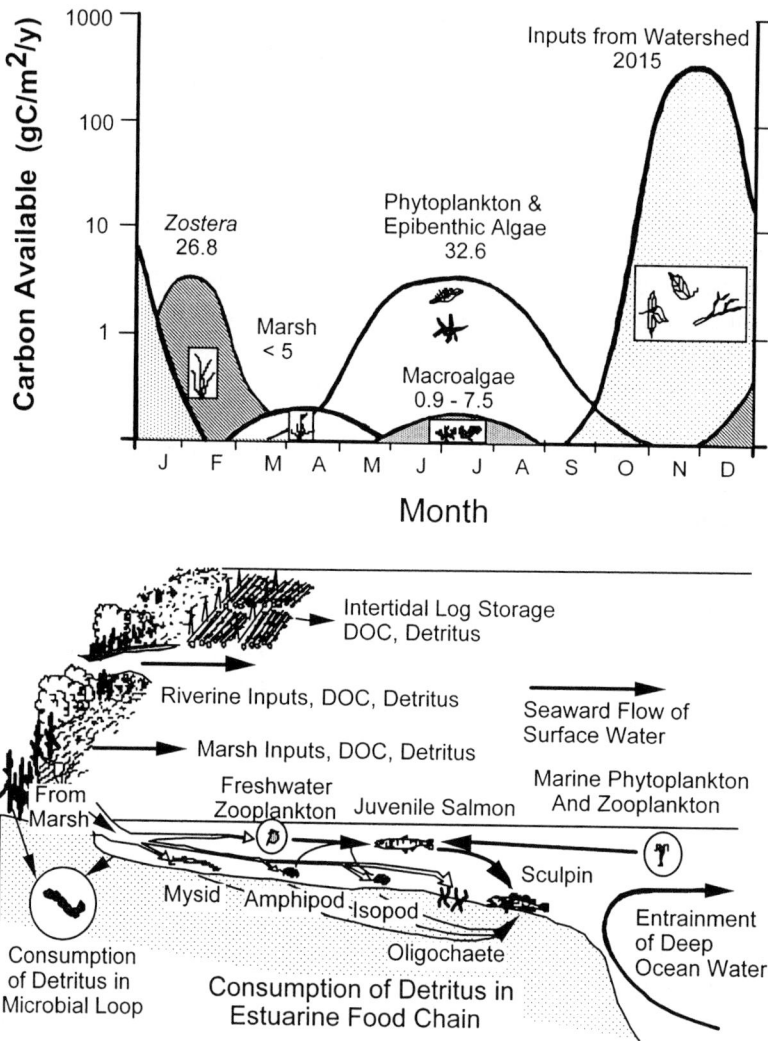

Fig. 5.1 Upper: Seasonal changes in carbon sources coming into the Nanaimo River estuary, southwestern British Columbia, Canada. Numbers with sources show total $gC\,m^{-2}\,y^{-1}$ from the source. For marsh, total production is $>500\ gC\,m^{-2}\,y^{-1}$ but little is transported to the estuary. Adapted from Naiman & Sibert (1978). Lower: The detritus-based food web typical of tidal marsh habitat in the Fraser River estuary, British Columbia, Canada. Adapted from Kistritz (1978) and Dorcey *et al.* (1978).

to the primary production estimated for the area by Harrison *et al.* (1983). For the Nanaimo River estuary, Naiman & Sibert (1979) estimated that riverine inputs of organic carbon were approximately 2000 $gC\,m^{-2}\,y^{-1}$, mostly as DOC (Fig. 5.1). This compares with 64 $gC\,m^{-2}\,y^{-1}$ autochthonous production. For the Great Whale River in the Canadian Arctic, Hudon (1994) estimated that 8.8 tons (ash-free dry weight, about 50% carbon) of particulate organic material (POM) >0.5 mm and 1021 million insects were delivered to the estuary during spring break-up. Particulate organic matter of this size fraction represents only about 2–5% of the amount of DOC, so that actual river contribution of carbon to its estuary is probably in the order of 100–200 tons during break-up. This does not include the fraction of DOC in insects, a potential direct food source for fishes. Impoundment of rivers may reduce such inputs of organic carbon to estuaries, as does deforestation of their watersheds.

Tidally and seasonally flooded marshes of estuaries are also a potential source of organic carbon (Fig. 5.1). Primary production in marsh and mangrove communities is often very high, ranging up to 3000 $gC\,m^{-2}\,y^{-1}$ (Alongi 1998). Little marsh and mangrove plant production is consumed as living material. Dead plant material in these systems is consumed there, buried in sediments, or transported to other parts of the estuary. Welsh *et al.* (1982) estimated that marshes contributed between about 2% and 60% of organic carbon loads to several estuaries in New England, USA, but most values were <5%. Alongi (1998) gives export rates from marshes of 9–379 $gC\,m^{-2}\,y^{-1}$, similar to those from mangrove forests. There are fewer measures of carbon flux from the ocean, but this could be large due to tidal exchange, particularly in macrotidal open estuaries.

Most of the organic carbon carried downstream in rivers or from tidal wetlands is considered to be highly refractory and not a good source of food for estuarine organisms. It is, however, a suitable substrate for bacterial growth and can provide nourishment, especially for estuarine benthic invertebrates. Bacterial communities of estuaries, both in the water column and sediments, are usually much more abundant than upstream in freshwater or in the adjacent ocean (Barnes 1984). Free bacteria numbered $4 \times 10^9\,L^{-1}$ in the upper St Lawrence estuary of eastern Canada and decreased exponentially along the salinity gradient to $0.5 \times 10^9\,L^{-1}$ at salinities >20 ppt(Painchaud *et al.* 1996). A similar relationship is reported for the Fraser River estuary in British Columbia (Harrison *et al.* 1983). Abundance appeared to be largely under hydrodynamic control in the St Lawrence but was also affected by microheterotrophic grazing. In the Hudson River estuary of eastern USA, bacteria ranged in abundance from 10^9 to $10^{10}\,L^{-1}$ over the year with a summer maximum (Findlay *et al.* 1991). Measurements made in the Mississippi River plume suggest that in summer 10–58% of phytoplankton production is sufficient to support the bacterial population but in winter the bacteria are dependent on other sources of carbon, presumably from riverine inputs (Alongi 1998). This echoes the suggestion by Naiman & Sibert (1979) that different carbon sources may be important at different seasons, with riverine inputs particularly important in winter (Fig. 5.1).

The role of bacteria in remineralization of organic matter has long been recognized but the importance of the microbial loop in food webs of aquatic communities has only recently come to the fore (Alongi 1998). As Jackson *et al.* (2001) suggest, however, the importance of this pathway may have been enhanced by overexploitation of top predators. Decomposition and remineralization of organic carbon materials in estuaries follows a three-step process. First, soluble organic substances are leached out and so provide food for bacteria and some other estuarine microfauna. Second, bacteria colonize the remaining organic particles, breaking them down into sugars and proteins but leaving larger polymers and other complexes to settle into sediments.

Finally, heterotrophic consumption by larger estuarine organisms goes on throughout this process with bacteria and some components of the organic detritus comprising the food of these consumers. In the York River estuary, Texas, USA, heterotrophic protists fed equally well on heterotrophic bacteria and cyanobacteria, suggesting that the protists oxidized much of the total bacterial biomass and thus were important nutrient regulators in this ecosystem (Eldridge & Sierachi 1993). Both aerobic and anaerobic processes are given in Kennish (1990).

Primary production in estuaries is distributed among pelagic phytoplankton, benthic microalgae, seagrasses, tidal marsh plants and mangrove forests. The combined production from these sources can amount to several $kgC.m^{-2}.y^{-1}$ but their relative importance varies substantially among estuaries. Phytoplankton are important energy fixers for zooplankton grazers and suspension-feeding benthic invertebrates. Benthic micro- and macro-algae and seagrasses provide food for a wide range of organisms from bottom-dwelling invertebrates to fish and birds. Marshes and mangrove forests have their own communities of consumers but at times can export much organic material to the open water. Many consumer organisms invade marshes and mangrove forests tidally and seasonally or during particular life stages and feed there.

Phytoplankton biomass varies seasonally in temperate estuaries with maximum values in late spring and early summer. Seasonal cycles of biomasses are much less pronounced in tropical estuaries. Annual phytoplankton production ranges from <10 to >500 $gC m^{-2} y^{-1}$ depending on turbidity and nutrient conditions (Kennish 1990). Phytoplankton biomass in upper San Francisco Bay, western USA is low (2–3 mg chlorophyll a per m³) during spring periods of high river discharge when water residence time in the estuary is short. It peaks in late summer and autumn (>30 mg chlorophyll a per m³) when river discharge is low and residence time long (Alpine & Cloern 1992). In the tidal freshwater portion of the Hudson River estuary, eastern USA, net phytoplankton production is strongly light-limited and algal cells in the mixing regime can spend up to 18 hours per day below the 1% light level (Cole *et al.* 1992). Salinity, diffuse light attenuation coefficients and ammonia concentration explained 71% of variability in phytoplankton biomass in the Neuse River estuary, eastern USA (Pinckney *et al.* 1997). In highly turbid estuaries, phytoplankton compensation depths may be <1 m and productivity is light-limited. In many estuaries, however, production is nutrient-limited, often by nitrogen (Parker *et al.* 1975; Boynton *et al.* 1982; Lewitus *et al.* 1998), but sometimes by phosphorus (D'Elia *et al.* 1986).

Benthic microalgal production in estuaries typically ranges from 50 to 150 gC $m^{-2} y^{-1}$. In some east coast US and European estuaries production from this source can exceed 2000 $gC m^{-2} y^{-1}$ and may be greater than that in the river upstream, especially where light (shading by tree canopy) or nutrient limitation occurs (Stockner & Shortreed 1976). Benthic macroalgae also can be highly productive but distributions tend to be very patchy. Production estimates for the various intertidal species range from 35 to 4400 $gC m^{-2} y^{-1}$ but their contribution to overall production may be small because they occupy limited areas (Alongi 1998), except on large mudflats.

Seagrass communities are among the most productive of estuarine autotrophs, with rates of above-ground production for different species ranging from 35 to 6700 gC $m^{-2} y^{-1}$ (most from 150 to 500 $gC m^{-2} y^{-1}$). Below-ground production is seldom measured but can be as high as 35% of total production for some species (Alongi 1998). Tropical seagrasses tend to be more productive, and production to biomass ratios decline with increasing latitude from 9–15:1 in the tropics to 1–7:1 in temperate latitudes. Grasses also provide a substrate for microalgal colonization and these and other epiphytes can contribute between 2% and 60% of total primary production in seagrass meadows. As with macroalgae, seagrass contribution to total estuarine production is highly variable because of their patchy distribution.

Tidal marshes and mangrove forests are, on a per unit area basis, often the most productive communities in estuaries. Above-ground primary production for salt marsh species in continental USA ranges from 500 to 7500 $gCm^{-2}y^{-1}$. Their below-ground primary production can be as large or larger than above ground, with values in the USA ranging from 279 to 7600 $gCm^{-2}y^{-1}$ (Alongi 1998). Above-ground production in mangrove forests averages higher than in salt marshes, with values ranging from 990 to 3900 $gCm^{-2}y^{-1}$ in southeast Asia and China. Macroalgae, microalgae and epiphytes can contribute as much as half to primary production in marshes and mangrove forests, but this is often overlooked.

The contribution of the different autotrophic communities to total primary production has been estimated for several estuaries in the USA, Australia and Europe (Table 5.1). For many east coast USA estuaries, marshes contributed from 74% to 86% of total primary production, and in Australia, mangroves contributed 85% of the total. In Europe, by contrast, marshes contributed only 0–20% of total production while phytoplankton made up 52–75% of the total. Benthic microalgae also were important in European estuaries, all of which had extensive intertidal mud and sand flats. Other autotrophic communities were relatively unimportant in these estuaries.

Table 5.1 Contribution (%) of various autotrophic communities to total primary production in a number of estuaries in the USA, Australia and Europe

Estuary	Phytoplankton	Microalgae	Macroalgae	Marsh/ mangrove	Epiphyte/ neuston	Seagrass
South Carolina	4	9	9	74	4	–[a]
New York	6	6	9	79	+[b]	–
Massachusetts	2	10	–	86	2	–
Australia	4	3	–	85	7	–
Ems-Dollard	60	30	0	10	–	0
Wadden Sea	65	30	1	2	–	2
Grevelingen	58	23	2	1	–	15
Veerse Meer	52	20	25	0	–	3
Oosterschelde	75	20	2	2	–	0
Westerschelde	45	35	0	20	–	0

Adapted from data in Alongi (1998).
[a]Here and elsewhere, not measured.
[b]Trace (<1%) contribution.

The combination of productivity sources makes estuaries a rich feeding environment for many filter-feeding and grazing organisms. Typically, zooplankton and benthic invertebrates form an intermediate filter- or deposit-feeding link between phytoplankton primary producers, POC and larger consumers, although some larger forms can also be predatory. Estuaries are often energetic environments, with strong currents induced by tides and river discharge. Getting into and remaining in an estuary can be a significant problem for small pelagic organisms like zooplankton or the larval forms of larger invertebrates and fishes. Over 60 years ago, Rogers (1940) pointed out in a study of the St John River estuary in the Canadian Maritimes that a zooplankton community could be maintained there in spite of strong river outflow by either remaining in the deeper saltwater layer or by using diel vertical migration to alternate between outflowing surface waters and inflowing deeper waters. Abundance of pelagic zooplankton is strongly seasonal in temperate estuaries, with highest abundance typically following the spring phytoplankton bloom. Zooplankton production in temperate estuaries usually ranges between 5 and 10 $gC m^{-2} y^{-1}$ (Kennish 1990).

The benthic invertebrate fauna of estuaries has been classified by size: microfauna (<0.1 mm), meiofauna (0.1–0.4 mm) and macrofauna (>0.4 mm); by their location: epifauna, infauna and interstitial fauna; and by their feeding: deposit feeders, suspension feeders, herbivores, carnivores, scavengers, parasites. Benthic invertebrates are a diverse group, with over 100 species of macrofauna identified in the sediments of Puget Sound, northwestern USA, and in the Gulf of Nicoya, Costa Rica (Vargas 1996; Emmett *et al.* 2000). Meiofauna reach average densities of about 10 million m^{-2} in estuaries and production can be >20 $gC m^{-2} y^{-1}$ (Kennish 1990). A striking characteristic of some estuaries is the high proportion of introduced species among the soft bottom benthos. In San Francisco Bay, California, USA, for example, introduced species represent the majority of organisms and constitute over 95% of the benthic biomass (Emmett *et al.* 2000).

Estuarine fish communities are composed of marine, freshwater and true estuarine species. Although the latter number is relatively small, their contribution to the total is variable. Roman *et al.* (2000) note that the proportion of estuarine fishes classified as resident or seasonally resident is high (80–100%) in north temperate estuaries, but low (15–50%) in subtropical ones. Total species found in estuaries also tends to be high in the tropics, with 121 species in Terminos Lagoon, Mexico (Yanez-Arancibia & Linares 1980), 214 in the Gulf of Nicoya, Costa Rica (Vargas 1996) and 128 in West African estuaries (Pillay 1967). By contrast, 37 species were recorded from the Fraser River estuary, western Canada (Northcote & Larkin 1989), 58 from Great Bay-Little Egg Harbour, New Jersey, USA (Able *et al.* 1996) and 50 from the Thames (Mann 1972). Able *et al.* (1996) also found that the number of species and the proportion of resident species varied strongly among habitat types. Fewest species and highest proportion of resident species were found in marsh habitats, with increasing number of species and a decreasing proportion of resident species as one moved from intertidal creek habitats to subtidal creek habitats and subtidal shorelines.

Estuaries in temperate regions have long been recognized as important spawning and nursery areas for commercially valuable marine fish species. At least two species of Pacific salmon (*Oncorhynchus* spp.) may spawn intertidally in estuaries (Groot

1989). Some species, such as Atlantic menhaden and Pacific salmon (*Oncorhynchus* spp.), depend strongly on estuaries as juvenile nurseries (e.g. Northcote 1976; Dorcey *et al.* 1978; Healey 1982; Warlen 1994), with young chinook salmon (*O. tshawytscha*) spending up to a month or more growing in such habitat (Levy & Northcote 1982). Even small estuaries have an important role in the production of juvenile coho salmon (*O. kisutch*) as shown by Tschaplinski (1987). Blaber (1997) argued that the temperate zone paradigm of fish-estuarine dependence could not be generalized to tropical estuaries, although some do provide important nursery habitats for marine species. Estuaries also provide habitat for adults of a number of economically important invertebrate fisheries such as those for crabs and shrimp. Food web relationships and biophysical processes affecting fish production from estuaries have received considerable attention but remain controversial. Fish and invertebrate production in estuaries has been linked to freshwater discharge, primary production and area of intertidal marsh, all factors heavily influenced by human activities both in estuaries and upstream (Day *et al.* 1989; FAO 1995; Alongi 1998; Kremmer *et al.* 2000). Most relationships are based on correlations and the mechanisms underlying them are not well worked out.

Of considerable significance to the models of fish production in estuaries is the supporting organic carbon base. As described earlier, many different plant communities contribute to the organic carbon load in estuaries. The importance of these sources in the food webs leading to economically valuable fish and invertebrates has been hotly debated. Recent studies using stable isotopes are now providing better information on carbon pathways to secondary and tertiary consumers. These data indicate that all carbon sources contribute to estuarine food webs, although the relative importance of different sources varies considerably among estuaries (Alongi 1998). No single carbon source (e.g. phytoplankton) can be said to dominate carbon pathways leading to top predators.

Concluding remarks

Because they are the ecotone between marine and riverine environments, estuaries are important in the production dynamics of species from both realms. Any alteration or degradation of estuarine ecosystems can potentially have effects that extend well beyond the borders of the estuary itself. Estuaries are also highly vulnerable and greatly threatened by human activities throughout the watersheds that discharge into the estuary. Impoundment or diversion of river flows, changes in sediment load, changes in riparian vegetation, land use changes, discharge of xenobiotics upstream, all have effects that propagate downstream. Within the estuary itself, changes in land use, channelization, dredging and filling of marshes and other shallow water habitats, discharge of pollutants all impact on ecological processes. Changes in ocean regimes and importation of exotic species from offshore releases also can influence these open systems. Many of these impacts can be generated by forest practices and, in some parts of the world, forestry can be a major cause of degradation of estuarine productivity. These and others will be reviewed in Chapter 15.

Acknowledgements

Drs P.J. Harrison and C. Levings provided many helpful suggestions in their reviews at the manuscript stage.

References

Able, K.W., Witting, D.A., McBride, R.S., Rountree, R.A. & Smith, K.J. (1996) Fishes of polyhaline estuarine shores in Great Bay–Little Egg harbor, New Jersey: a case study of seasonal and habitat influences. In: *Estuarine Shores: Evolution, Environments and Human Alterations*, (eds K.F. Nordstrom & C.T. Roman), pp.335–53. John Wiley & Sons, New York.

Allanson, B. & Baird, D. (1999) *Estuaries of South Africa*. Cambridge University Press, Cambridge, UK.

Allanson, B.R. & Winter, D. (1999) Chemistry. In: *Estuaries of South Africa*, (eds B.R. Allanson & D. Baird), pp. 53–89. Cambridge University Press, Cambridge, UK.

Alongi, D.M. (1998) *Coastal Ecosystem Processes*. CRC Press, Boca Raton, FL.

Alpine, A.E. & Cloern, J.E. (1992) Trophic interactions and direct physical effects control phytoplankton biomass and production in an estuary. *Limnology and Oceanography*, **37**, 946–55.

Anonymous (2000) *Clean Coastal Waters: Understanding and Reducing the Effects of Nutrient Pollution*. US National Research Council, National Academy Press.

Ayvazian, S.G. (1999) *Fish and Fisheries of Tropical Estuaries*, (S.J.M. Blaber) (1997) Chapman & Hall, New York. Reviewed in: *Transactions of the American Fisheries Society*, **128**, 183–4.

Barnes, R.S.K. (1984) *Estuarine Biology*. Edward Arnold, London.

Bell, R., Green, M., Hume, T. & Gorman, R. (2000) What regulates sedimentation in estuaries? *NIWA Water and Atmosphere*, **8**, 13–16.

Blaber, S.J.M. (1997) *Fish and Fisheries of Tropical Estuaries*. Chapman & Hall, New York.

Boynton, W.R., Kemp, W.M. & Keefe, C.W. (1982) A comparative analysis of nutrients and other factors influencing estuarine phytoplankton production. In: *Estuarine Comparisons*, (ed. V.S. Kennedy), pp. 69–90. Academic Press, New York.

Bowen, J.L. & Valiela, I. (2001) The ecological effects of urbanization of coastal watersheds: historical increases in nitrogen loads and eutrophication of Walquoit Bay estuaries. *Canadian Journal of Fisheries and Aquatic Sciences*, **58**, 1489–500.

Bricker, S.B., Clement, C.G., Pirhall, D.E., Orlando, S.P. & Farrow, D.G.C. (1999) *National Estuarine Eutrophication Assessment: Effects of Nutrient Enrichment in the Nation's Estuaries*. Special Projects Office and the National Centers for Coastal Ocean Science, National Ocean Service and Atmospheric Administration.

Cole, J.J., Caraco, N.F. & Peierls, B.L. (1992) Can phytoplankton maintain a positive carbon balance in a turbid freshwater, tidal estuary? *Limnology and Oceanography*, **37**, 1608–17.

Cooper, A., Wright, I. & Mason, T. (1999) Geomorphology and sedimentology. In: *Estuaries of South Africa*, (eds B. Allanson & D. Baird), pp. 5–25. Cambridge University Press, Cambridge, UK.

Dame, R., Alber, M., Allen, D., *et al.* (2000) Estuaries of the south Atlantic coast of North America: their geographic signatures. *Estuaries*, **23**, 793–819.

Day, J.W., Hall, C.A., Kemp, M.W. & Yanez-Arancibia, A. (1989) *Estuarine Ecology*. John Wiley & Sons, New York.

Deegan, L., Day, J.W., Gosselink, J.G., Yanez-Arancibia, A., Chavez, G.S. & Sanchez-Gil, P. (1986) Relationship among physical characteristics, vegetation distribution and fisheries yield in Gulf of Mexico estuaries. In: *Estuarine Variability*, (ed. V.S. Kennedy), pp. 83–100. Academic Press, New York.

D'Elia, C.F., Saunders, J.G. & Boynton, W.R. (1986) Nutrient enrichment studies in a coastal plain estuary: phytoplankton growth in large-scale, continuous cultures. *Canadian Journal of Fisheries and Aquatic Sciences*, **43**, 397–406.

Dorcey, A.H.J., Northcote, T.G. & Ward, D.V. (1978) Are the Fraser marshes essential to salmon? *Westwater Research Centre Lecture Series 1*. University of British Columbia, Vancouver.

Eldridge, P.M. & Sierachi, M.E. (1993) Biological and hydrodynamic regulation of the microbial food web in a periodically mixed estuary. *Limnology and Oceanography*, **38**, 1666–79.

Emmett, R., Llanso, R., Newton, J., *et al.* (2000) Geographic signatures of North American west coast estuaries. *Estuaries*, **23**, 765–92.

Everett, R.A. & Ruiz, G.M. (1993) Coarse woody debris as a refuge from predation in aquatic communities: an experimental test. *Oecologia-Heidelberg*, **93**, 475–86.

FAO (1995) Effects of riverine inputs on coastal ecosystems and fishery resources. *FAO Fisheries Technical Paper 349*. FAO Marine Resources Service, Fishery Resources Division.

Findlay, S., Pace, M.L., Lints, D., Cole, J.J., Caraco, N.F. & Peierls, B. (1991) Weak coupling of bacterial and algal production in a heterotrophic system: the Hudson River estuary. *Limnology and Oceanography*, **36**, 268–78.

Friedland, K.D., Ahrenholz, D.W. & Guthrie, J.F. (1996) Formation and seasonal evolution of Atlantic menhaden juvenile nurseries in coastal estuaries. *Estuaries*, **19**, 105–14.

Gonor, J.J., Sedell, J.R. & Benner, P.A. (1988) What we know about large trees in estuaries, in sea and coastal beaches. In: *From the Forest to the Sea: A Story of Fallen Trees*, (eds C. Masser, R.F. Tarrant, J.M. Trappe & J.F. Franklin), pp. 83–112. US Forest Service Technical Report, PNW-GTR-229.

Greisman, P. & Ingram, G. (1977) Nutrient distribution in the St. Lawrence estuary. *Journal of the Fisheries Research Board of Canada*, **34**, 2117–23.

Groot, E.P. (1989) *Intertidal spawning of chum salmon: saltwater tolerance of the early life stages to actual and simulated intertidal conditions*. MSc thesis, University of British Columbia.

Harrison, P.J., Fulton, J.D., Taylor, F.J.R. & Parsons, T.R. (1983) Review of the biological oceanography of the Strait of Georgia: pelagic environment. *Canadian Journal of Fisheries and Aquatic Sciences*, **40**, 1064–94.

Harrison, P.J., Clifford, P.J., Cochlan, W.P., *et al.* (1991) Nutrient and plankton dynamics in the Fraser River plume, Strait of Georgia, British Columbia. *Marine Ecology Progress Series*, **70**, 291–304.

Healey, M.C. (1982) Juvenile Pacific salmon in estuaries: the life support system. In: *Estuarine Comparisons*, (ed. V.S. Kennedy), pp. 315–42. Academic Press, New York.

Healey, M.C. (1994) Effects of dams and dikes on fish habitats in two Canadian river deltas. In: *Global Wetlands, Old World and New*, (ed. W. Mitsch), pp. 385–98. Elsevier, Amsterdam.

Healey, M.C. & Richardson, J.S. (1996) Changes in the productivity base and fish populations of the Lower Fraser River (Canada) associated with historical changes in human occupation. *Archiv für Hydrobiologie*, Supplement **113**, 279–90.

Hobbie, J.E. (2000) *Estuarine Science*. Island Press, Washington, DC.

Howarth, R.W. (2000) *Estuary Restoration and Maintenance: The National Estuary Program*, (M.J. Kennish) (2000) CRC Press, Boca Raton, FL. Reviewed in: *Limnology and Oceanography*, **45**, 1889.

Hudon, C. (1994) Biological events during ice breakup in the Great Whale River (Hudson Bay). *Canadian Journal of Fisheries and Aquatic Sciences*, **51**, 2467–81.

Ibanez, C., Pont, D. & Prat, N. (1997) Characterization of the Ebre and Rhone estuaries: a basis for defining and classifying salt-wedge estuaries. *Limnology and Oceanography*, **42**, 89–101.

Jackson, J.B., Kirby, M.X., Berger, W.H., *et al.* (2001) Historical overfishing and the recent collapse of coastal ecosystems. *Science*, **293**, 629–38.

Jay, D.A., Geyer, W.R. & Montgomery, D.A. (2000) An ecological perspective on estuarine classification. In: *Estuarine Science*, (ed. J.E. Hobbie), pp. 149–76. Island Press, Washington, DC.

Jordan, T.E., Correll, D.L., Miklas, J. & Weller, D.E. (1991) Nutrients and chlorophyll at the interface of a watershed and estuary. *Limnology and Oceanography*, **36**, 251–67.

Kennish, M.J. (1990) *Ecology of Estuaries, Vol. II, Biological Aspects*. CRC Press, Boca Raton, FL.

Kistritz, R.U. (1978) An ecological evaluation of Fraser Estuary tidal marshes: the role of detritus and the cycling of elements. *Westwater Technical Report 15*. University of British Columbia, Vancouver.

Kremmer, J.N., Kemp, W.M., Giblin, A.E., Valiela, I., Seitzinger, S.P. & Hofmann, E.E. (2000) Linking biogeochemical processes to higher trophic levels. In: *Estuarine Science: A Synthetic Approach to Research and Practice*, (ed. J.E. Hobbie), pp. 299–345. Island Press, Washington, DC.

Levy, D.A. & Northcote, T.G. (1982) Juvenile salmon residency in a marsh area of the Fraser River estuary. *Canadian Journal of Fisheries and Aquatic Sciences*, **39**, 270–6.

Lewitus, A.J., Koepfler, E.T. & Morris, J.T. (1998) Seasonal variation in the regulation of phytoplankton by nitrogen and grazing in a salt-marsh estuary. *Limnology and Oceanography*, **43**, 636–46.

Lowe-McConnell, R.H. (1987) *Ecological Studies in Tropical Fish Communities*. Cambridge University Press, Cambridge, UK.

McClelland, J.W. & Valiela, I. (1998) Linking nitrogen in estuary producers to land-derived sources. *Limnology and Oceanography*, **43**, 577–85.

Mann, K.H. (1972) Case history: the River Thames. In: *River Ecology and Man*, (eds R.A. Oglesby, C.A. Carlson & J.A. McCann), pp. 215–32. Academic Press, New York.

Maser, C. & Sedell, J.R. (1994) *From the Forest to the Sea: The Ecology of Wood in Streams, Rivers, Estuaries and Oceans*. St Lucie Press, Delray Beach, FL.

Metcalfe, C.D., Dadswell, M.J., Gillis, G.F. & Thomas, M.L.H. (1976) *Physical, Chemical and Biological Parameters of the St John Estuary, New Brunswick, Canada*. Technical Report of the Fisheries Research Board of Canada 686.

Miller, R.L. & McPherson, B.F. (1991) Estimating estuarine flushing and residence times in Charlotte Harbor, Florida, via salt balance and a box model. *Limnology and Oceanography*, **36**, 602–12.

Monaco, M.E., Nelson, D.M., Czapla, T.E. & Pattillo, M.E. (1989) *Distribution and Abundance of Fishes and Invertebrates in Texas Estuaries*. ELMR Rept. No. 3. NOAA/NOS Strategic Assessments Division, Rockville, USA.

Naiman, R.L. & Sibert, J.R. (1978) Transport of nutrients and carbon from the Nanaimo River to its estuary. *Limnology and Oceanography*, **23**, 1183–93.

Naiman, R.L. & Sibert, J.R. (1979) Detritus and juvenile salmon production in the Nanaimo estuary III: importance of detrital carbon to the estuary ecosystem. *Journal of the Fisheries Research Board of Canada*, **36**, 504–20.

Nelson, D.M., Monaco, M.E., Irlandi, E.A., Settle, L.R. & Coston-Clements, L. (1991) *Distribution and Abundance of Fishes and Invertebrates in Southeast Estuaries*. ELMR Rept. No. 9. NOAA/NOS Strategic Environmental Assessments Division, Rockville, Madison, USA.

Nordstrom, K.F. & Roman, C.T. (1996) *Estuarine Shores*. John Wiley & Sons, Chichester, England.

Northcote, T.G. (1976) Biology of the Lower Fraser and ecological effects of pollution. In: *The Uncertain Future of the Fraser River*, (ed. A.H.J. Dorcey), pp. 85–119. University of British Columbia Press, Vancouver.

Northcote, T.G. & Larkin, P.A. (1989) The Fraser River: a major salmonine production system. In: *Proceedings of the International Large River Symposium*, (ed. D.P. Dodge), pp. 172–204. Canadian Special Publication of Fisheries and Aquatic Sciences 106.

Painchaud, J., Lafaivre, D., Therriault, J-C. & Legendre, L. (1996) Bacterial dynamics in the upper St. Lawrence estuary. *Limnology and Oceanography*, **41**, 1610–18.

Parker, R.R., Sibert, J. & Brown, T.J. (1975) Inhibition of primary productivity through heterotrophic competition for nitrate in a stratified estuary. *Journal of the Fisheries Research Board of Canada*, **32**, 72–7.

Peterson, B.T., Wollheim, W.M., Mulholland, P.J., *et al.* (2001) Control of nitrogen export from watersheds by headwater streams. *Science*, **292**, 86–90.

Pillay, T.V.R. (1967) Estuarine fishes in West Africa. In: *Estuaries*, (ed. G. Lauff), pp.639–46. American Association for the Advancement of Science Publication No. 83, Washington, DC.

Pinckney, J.L., Millie, D.F., Vinyard, B.T. & Paerl, H.W. (1997) Environmental controls on phytoplankton bloom dynamics in the Neuse River estuary, North Carolina, U.S.A. *Canadian Journal of Fisheries and Aquatic Sciences*, **54**, 2491–501.

Pritchard, D.W. (1967) What is an estuary: physical viewpoint. In: *Estuaries*, (ed. G. Lauff), pp. 3–5. Publication No. 83, American Association for the Advancement of Science, Washington, DC.

Rogers, H.M. (1940) Occurrence and retention of plankton within the estuary. *Journal of the Fisheries Research Board of Canada*, **5**, 164–71.

Roman, C.T., Jaworski, N., Short, F.T., Findlay, S. & Warren, R.S. (2000) Estuaries of the Northeastern United States: habitat and land use signatures. *Estuaries*, **23**, 743–64.

Saenger, P. & Holmes, N. (2000) Australian estuaries: comparisons around a continent. *Rivers for the Future*, **Spring**, 20–22.

Sinclair, M., El-Sabh, M. & Brindle, J-R. (1976) Seaward nutrient transport in the lower St. Lawrence estuary. *Journal of the Fisheries Research Board of Canada*, **33**, 1271–7.

Stockner, J.G. & Shortreed, K. (1976) Autotrophic production in Carnation Creek, a coastal rainforest stream on Vancouver Island, British Columbia. *Journal of the Fisheries Research Board of Canada*, **33**, 1553–63.

Tschaplinski, P.J. (1987) *Comparative ecology of stream-dwelling and estuarine juvenile coho salmon (Oncorhynchus kisutch) in Carnation Creek, Vancouver Island, British Columbia*. PhD thesis, University of Victoria, British Columbia.

Valiela, I., Foreman, K., La Montaine, M., *et al.* (1992) Coupling of watersheds and coastal waters: sources and consequences of nutrient enrichment in Waquoit Bay, Massachusetts. *Estuaries*, **15**, 443–57.

Vargas, J.A. (1996) Ecological dynamics of a tropical mudflat community. In: *Estuarine Shores*, (eds K.F. Nordstrom & C.T. Roman), pp. 355–71. John Wiley & Sons, Chichester, UK.

Wagner, C.M. (1999) Expression of the estuarine species minimum in littoral fish assemblages of the lower Chesapeake Bay tributaries. *Estuaries*, **22**, 304–12.

Warlen, S.M. (1994) Spawning time and recruitment dynamics of larval Atlantic menhaden, *Brevoortia tyrannus*, into a North Carolina estuary. *Fishery Bulletin*, **92**, 420–33.

Welsh, B.L., Whitlatch, R.B. & Bohlen, W.F. (1982) Relationship between physical characteristics and organic carbon sources as a basis for comparing estuaries in southern New England. In: *Estuarine Comparisons*, (ed. V.S. Kennedy), pp. 53–67. Academic Press, New York.

Wilbur, D.H. (1993) The influence of the Apalachicola River flows on blue crab, *Callinectes sapidus*, in north Florida. *Fishery Bulletin*, **92**, 180–8.

Wolaver, T.G., Wetzel, R.L., Zieman, J.C. & Webb, K.L. (1980) Nutrient interactions between salt marsh mudflats and estuarine water. In: *Estuarine Perspectives*, (ed. V.S. Kennedy), pp. 123–33. Academic Press, New York.

Yanez-Arancibia, A. & Linares, F.A. (1980) Fish community structure and function in Terminos Lagoon, a tropical estuary in the southern Gulf of Mexico. In: *Estuarine Perspectives*, (ed. V.S. Kennedy), pp. 465–82. Academic Press, New York.

Part III
Fish Biology and Ecology

Chapter 6
Fish life history variation and stock diversity in forested watersheds

T.G. NORTHCOTE

Introduction

In salmonid fishes there is overwhelming evidence that within a watershed and within one lake there may be great diversity in a number of life history characteristics for species that should be known and carefully considered in both forestry and fishery management practices. Furthermore, it is becoming clear that some of this diversity is under genetic control, having developed through reproductive isolation even within a dozen generations in sockeye salmon (Barton 2000; Hendry *et al.* 2000). In other species both environmental and genetic controls may be involved (Northcote 1969, 1981; Kelso *et al.* 1981; Northcote & Kelso 1981; Jonsson & Jonsson 1999). Salmonids are not all that special in this regard, for such differentiation occurs in sticklebacks (Pennisi 2000; Rundle *et al.* 2000), and in other fishes such as smelt and ayu (Jonsson & Jonsson 1993), shad (Glebe & Leggett 1981) and sunfish (Ehlinger 1990; Robinson & Wilson 1996).

Consequently we need to review some of the more outstanding differences evident in the life history characteristics of salmonid fishes, give some examples for non-salmonids which are not so well studied in this regard, and examine the extent to which this diversity occurs in both temperate and tropical forested regions of the world.

Salmonid life history variation

General features

The family Salmonidae, collectively known as salmonids, includes three subfamilies (Nelson 1994): the Coregoninae with about 32 species (whitefishes, ciscoes and the inconnu), the Thymallinae with about five species (graylings) and the Salmoninae with about 30 species (lenok, huchen, chars, Atlantic salmon and brown trouts, Pacific trouts, Pacific salmon, and about four other isolated species).

Many terms, scientific and popular, have been applied to the sequence of salmonid life history stages, some 18 of which are defined briefly in Table 6.1. Salmonids in some of these stages may be found in a wide variety of inland waters ranging from the uppermost reaches of small streams on down through the receiving rivers and lakes to estuaries, out into coastal seawaters, and far offshore. Others are found only in freshwater habitats or occasionally in estuaries.

Table 6.1 Generalized salmonid life history stages, terms and habitat sequences (see Crisp (2000) for full term definitions, habitats used)

Stage	Term definition	Main habitats used
1. Adults	Non-sexually mature males, females	Streams, rivers, lakes, estuaries, coastal and offshore seas
2. Spawners	Sexually maturing or mature males, females	Estuaries, rivers, streams, lakes
3. Jacks	Precocious males in spawning run at least a year before normal	As in 2
4. Kelts	Spent spawners on physiological return to mature adult state	As in 1
5. Redd	Egg nest(s), often in gravel pit(s) dug by female	Streams, rivers, lakes, rarely estuaries
6. Milt	Free-running sperm fluid	
7. Roe	Eggs in female abdominal cavity	
8. 'Green' eggs	Fertilized, within 30 s of water contact	As in 5
9. 'Hardened' eggs	After 30 s, outer egg membrane forms protective, elastic capsule	As in 5
10. Eyed eggs	Developing eyes and vertebral column evident	As in 5
11. Hatching	Outer egg membrane breaks open releasing larval young	As in 5
12. Alevins	Free-moving larvae, 'sac-fry' as yolk sac still evident	As in 5
13. Emergence	Late alevins or early 'swim-up' fry, leaving hatching site	As in 5
14. Fry[a]	Free-swimming and feeding young	As in 5
15. Parr	Late stage fry adopt feeding stations, establish and defend territories (Crisp 2000)	Streams, rivers, lakes, estuaries
16. Fingerlings	North American term for 1+ or older juveniles, 'finger' size	As in 15
17. Smolts	Late stage parr undergoing external and internal physiological changes for seaward migration	As in 15
18. Grilse	One-sea-winter fish; may return to spawn or remain longer in sea (Crisp 2000)	Coastal seas, estuaries

[a]North American salmonid biologists use 'fry' to indicate the free-swimming stage for longer than given by Allan & Ritter (1977), often referring to 0+ fry as young in their first year of life.

A generalized salmonid life history cycle could start with adults spawning in streams, lakes or rivers, sometimes at their mouths where they empty over bars into saltwater. Female salmon, trout and char usually prepare a redd in streams, rivers, and sometimes in lakes, although lake spawners may not, nor do grayling, whitefishes or ciscoes.

Except for Pacific salmon, which die after first spawning (not for Japanese masu and amago salmon where some may spawn over several years), other salmonids may spawn several times before dying. Spawning time is usually between early spring and summer for Pacific trouts, huchen and graylings, but in autumn to winter for most other salmonids.

Egg development, hatching and emergence times are temperature-dependent and proceed more rapidly in spring than in autumn spawners. Development rate is also species and genetic stock dependent, being selected for specific local conditions.

On emergence the fry may stay near their site of substrate exit, becoming parr and eventually adults, or they may migrate varying distances downstream, upstream (or both sequentially), along shorelines, offshore, or seaward – again depending on species and stock. Further complex juvenile and subadult migratory stages may occur, as outlined in Chapter 7.

When maturity approaches, adults (especially males) change slightly or greatly in external appearance (colouration, head shape, dentition, tubercles), and in internal physiology, again species and stock dependent. They undertake migrations of short up to very long distances to appropriate spawning habitat, often that used by their parents. Female egg number (and size) is fish size, species and stock dependent, ranging from less than 50 in small headwater trouts (Northcote & Hartman 1988) and small sized chars (Jonsson *et al.* 1988), to nearly 24,000 for large huchen (Holčík *et al.* 1988).

Further information on general life history features of many salmonids can be found in Scott & Crossman (1973), Holčík *et al.* (1988), Groot & Margolis (1991), Northcote & Ennis (1994), Northcote (1995) and Groot (1996). But it is the locality and stock-specific patterns of life history information that should be used to guide fish–forestry interaction decisions. The range in these patterns for salmonids is summarized in Table 6.2, where examples of four categories of life history and stock diversity are evident in nearly all nine of the genera considered.

The whitefishes and related species

Two of the coregonine genera – *Prosopium* and *Stenodus* – occur mainly in northern North America and parts of northern Asia, and in at least some of the species have examples of differentiation for most major life history/stock diversities (migratory behaviour, reproductive stages, feeding ecology and seasonal habitat use). Some nine species of *Coregonus*, widely distributed in parts of northern North America, Europe and Asia, together show examples of all categories.

For riverine mountain whitefish (*Prosopium williamsoni*) in the Sheep River system of southwestern Alberta, Canada, there is a small resident fraction of the population and a larger migratory fraction (Davies & Thompson 1976) which exemplify all four

Table 6.2 A summary of life history variation and stock diversity presence (+) or absence (−) among genera in the family Salmonidae throughout their world distribution

Genus[a]	World location	Life history/stock diversity				References
		A	B	C	D	
Prosopium (3)	North America, NE Asia	+	+	+	+	20, 24, 26
Coregonus (9)	North America, Europe, N Asia	+	+	+	−	17, 18, 24, 31
Stenodus (1)	North America, NW Asia	+	+	+	+	8, 24
Thymallus (2+)	North America, Europe, Asia	+	+	+	+	23, 25
Brachymystax (1)	Asia – Korea	+	−	+	+	22, 24
Hucho (4)	Europe, Asia – Japan	+	+	+	+	7, 24
Salvelinus (11)	N. Hemisphere, South America, Africa, New Zealand, Australia	+	+	+	+	2, 4, 10, 11, 15, 24, 28, 29
Salmo (5)	North America, N. Europe, S. America, Africa, New Zealand, Australia	+	+	+	+	9, 11, 12, 13, 14, 15, 16, 19, 21, 24, 27, 28, 30
Oncorhynchus (11)	North America, Europe, Siberia, Japan, Africa, India, Pakistan, New Zealand, Australia	+	+	+	+	1, 3, 5, 6, 11, 24, 32, 33, 34

Key supporting references are given below; others are in the text of this chapter or in references of those cited.
A = migratory behaviour, B = reproductive ecology, C = feeding ecology, D = seasonal habitat use.
[a]Number of species in parentheses; life history/stock diversity information may not be available for all.
1, Burger *et al.* (2000); 2, Curry *et al.* (1997); 3, Dunham *et al.* (1999); 4, Gislason *et al.* (1999); 5, Hendry & Quinn (1997); 6, Hendry *et al.* (2002); 7, Holčik *et al.* (1988); 8, Howland *et al.* (2000); 9, Hutchings & Jones (1998); 10, Jensen (1994); 11, Jonsson & Jonsson (1993); 12, Jonsson & Jonsson (1999); 13, Jonsson *et al.* (1999); 14, Jonsson *et al.* (2001); 15, Jonsson & Jonsson (2001); 16, Koljonen *et al.* (1999); 17, Kristofferson & Clayton (1990); 18, Lindström (1970); 19, Marschall *et al.* (1998); 20, McCart (1970); 21, Näslund *et al.* (1998); 22, Nikolsky (1963); 23, Northcote (1995); 24, Northcote (1997); 25, Northcote (2000a); 26, Northcote & Ennis (1994); 27, Riddell & Leggett (1981); 28, Ryan (1993); 29, Skúlason *et al.* (1996); 30, Stewart *et al.* (2002); 31, Svärdson (1998); 32, Teel *et al.* (2000); 33, Woody *et al.* (2000); 34, Young (1999).

life history categories outlined in Table 6.2 – see Northcote & Ennis (1994) for the complex sequence of habitats used. For five Swedish lacustrine *Coregonus* forms, in different lakes there may be lake littoral spawners, stream spawners or lake pelagic spawners. In several of 10 lakes studied, two and sometimes three different subpopulations were using these different spawning habitats (Lindström 1970). Inconnu (*Stenodus leucichthys*) mainly in the Alaskan section of the Yukon River have both restricted and long distance migratory populations with different spawning, feeding and wintering habitats (Alt 1977). In the Mackenzie River system of northwestern Canada there are both freshwater and anadromous populations of inconnu that differ markedly in timing and distance of their reproductive migrations (Howland *et al.* 2000).

The graylings

The widespread evidence for different stocks of Arctic grayling was summarized from North American studies for populations in northeastern British Columbia (Northcote 1993). Later the review was broadened to compare life history characteristics between Arctic and European grayling (Northcote 1995), and subsequently updated (Northcote 2000a) – see also Ibbotson *et al.* (2001). Overall the studies reviewed suggest that there are clear stock differences in timing and location of spawning sites, feeding sites, growth rates and migratory behaviour, even though at times different stocks may intermingle or migrate from one part of a river system to another. Of the wealth of detailed information coming from research on European grayling in rivers and streams of Austria (Uiblein *et al.* 2000), one of their most important features is said to be 'the occurrence of adaptations to distinct environmental conditions in different stocks'. The stocks respond differentially to different negative effects caused by human activities in the various river reaches.

The salmon, trout and char

Of the about 30 species of Salmoninae, Groot (1996) has provided a comprehensive life history review for 15 of the more important species in three genera – *Oncorhynchus* (Pacific salmon and Pacific trouts), *Salmo* (Atlantic trouts) and *Salvelinus* (chars – some species of which confusingly are given the common name of trout, e.g. brook trout, lake trout, bull trout). The ecological complexity of freshwater life histories for five species of Pacific salmon and two species of trout in coastal British Columbia was reviewed by Scrivener *et al.* (1998); see especially their Fig. 1. Furthermore Willson (1997) has covered variation in life histories of these three salmonine genera reviewing differences in degree of anadromy, age of maturation, frequency of reproduction, body size and fecundity, and seasons of breeding. Here only selected features will be presented along with more recent information. Five salmonines within the genera *Brachymystax* and *Hucho* give evidence of most recognized life history/stock diversity categories (Table 6.2), where they occur from Europe and Asia to Japan and Korea, as do many species within the other three genera (*Salvelinus, Salmo, Oncorhynchus* – Table 6.2 and Northcote 1997).

Perhaps more than any other salmonid, Arctic char show very high levels of morphological, life history and ecological variability both among lakes and rivers and also within them (Hindar & Jonsson 1982, 1993; Jonsson *et al.* 1988; Jensen 1994; Hammar 1998; Jonsson & Jonsson 2001). The latter review the amazing diversity of life history characteristics to be found in different morph systems of Arctic char in Canada, Iceland, Scotland, England, Norway and Sweden. Control of char life history variability is in itself variable among systems, sometimes resulting largely from environmental conditions and in others under more dominant genetic regulation (Skúlason *et al.* 1996; Gislason *et al.* 1999). But many of the other species of *Salvelinus* also show marked life history variability with different seasonal alterations between spawning, feeding and wintering habitats (Northcote 1997), and even the so-called 'lake trout' *Salvelinus namaycush* may have populations living year-round in northern rivers (Olver 1991).

Over six decades ago local variability in life history and behavioural characteristics were noted in Atlantic salmon (*Salmo salar*) and the possibility was raised of their being under genetic control (White & Huntsman 1938). The extent, causes and significance of this variability are still being investigated both in Atlantic salmon (Riddell & Leggett 1981; Ryan 1993; Hutchings & Jones 1998; Marschall *et al.* 1998; Koljonen *et al.* 1999; Stewart *et al.* 2002) and in brown trout (Jonsson & Jonsson 1993; Näslund *et al.* 1998; Jonsson *et al.* 1999, 2001; Carlsson & Nilsson 2000).

The Pacific trouts also show great life history and stock variability with seasonal races (summer and winter) in the anadromous form of rainbow–steelhead (Withler 1966; Smith 1969; Nakamoto 1994; Hendry *et al.* 2002). Complex lakeward movements by stocks of young rainbow trout from inlet and outlet lake tributaries (Northcote 1969) are under both environmental and genetic controls (Kelso & Northcote 1981; Kelso *et al.* 1981). Cutthroat trout, both stream-dwelling (Dunham *et al.* 1999) and lake-dwelling forms (Gresswell *et al.* 1994), exhibit marked differences in life history patterns.

Life history variations in the Pacific salmon have been well covered by Groot & Margolis (1991) and by Groot (1996), so only a few earlier studies not cited there or in Willson (1997) of interest here will be noted, along with some appearing more recently. Although Pacific salmon are usually autumn spawners, there are stocks which enter spawning habitats unusually early in summer such as the sockeye in the Gitnadoix system, lower Skeena River, British Columbia. Stocks of other species may spawn as late as April; chum salmon, for example (Wickett 1964). Smoker & Fukashima (1997) discuss determinants of spawning timing for pink salmon. Male life history variation in age of spawner return occurs in several species of Pacific salmon and differs significantly among populations (Young 1999), as do timing and speed of estuarine movement during seaward migration as smolts (Moser *et al.* 1991; McMahon & Holtby 1992). Juvenile chinook salmon life history is highly variable. Some stocks spend from a few days to a few months in natal streams ('ocean type') before seaward migration to estuaries. Others stay a year or more in streams ('stream type'), not necessarily their natal ones (Murray & Rosenau 1989; Levings & Lauzier 1991; Scrivener *et al.* 1994). In British Columbia there appear to be two genetically different groups of chinook

– coastal and interior – which largely coincide with the above juvenile ocean and stream-type life history patterns (Teel *et al.* 2000). New Zealand introduced chinook salmon apparently came from a single California, USA population, but in about 20 generations have adapted to local conditions in the four major rivers used, showing a suite of life history differences (Quinn & Unwin 1993). The importance of life history adaptive divergence has been emphasized in sockeye populations from Lake Washington and Frazer Lake, USA (Hendry & Quinn 1997; Burger *et al.* 2000). Spawning by sockeye salmon in large rivers with no lakes nearby has not been considered a common trait, but Eiler *et al.* (1992) show 11 such stocks (two in southwestern Siberia, seven in Alaska, and two in British Columbia), and report in detail on the Alaskan/British Columbia Taku River spawning stock, which represents up to 60% of the total returns to the system and is genetically distinct. Lakeshore spawning, although not a common trait in sockeye, contributes on average over 30% of the sockeye spawning runs in glacially turbid Tustumena Lake, Alaska (Burger *et al.* 1995). This genetic diversification has occurred within the 2000 years since it has become fully deglaciated (Burger *et al.* 1997).

Non-salmonid life history variation

Life history variation is high and extensive within salmonid fish species as we have seen, but the phenomenon also occurs in many other major groups of fishes including at least five additional orders of those in Chapter 1 (Table 1.2 – cyprinids, osmerids, cyprinodontids, gasterosteids and percids).

Cyprinids, with relatively high fecundity and a large number of individual reproductive cycles in comparison with salmonids, might be expected to show more genetic diversity and life history variability (Mitton & Lewis 1989). Indeed a 45-year-old dam on the Rhone River in France resulted in significant genetic structure and differentiation developing in populations of chub (*Leuciscus cephalus*) and roach (*Rutilus rutilus*) localized upstream and downstream from the barrier (Laroche *et al.* 1999). They reported genetic isolation among chub populations 100 km apart in the Rhone River. Such close isolation did not occur there among the roach populations. Hanfling & Brandl (1998) also show considerable genetic and morphological variation in the chub, both within and across central European drainages. Such variation also occurs in two other species of *Leuciscus* (Coelho *et al.* 1995), in the stoneroller minnow (*Campostoma anomalum*) (Heithaus & Laushman 1997), and in other cyprinids (Tibbets & Dowling 1996).

Sympatric divergence of trophic ecotypes of osmerid smelt has been reported in northeastern North America (Taylor & Bentzen 1993). For cyprinodontids, genetic variability occurs in the guppy (*Poecilia reticulata*) with several factors involved (Shaw *et al.* 1994). It also occurs through space and time in the mosquitofish *Gambusia* (McClenaghan *et al.* 1985). Gasterosteid species, especially *Gasterosteus aculeatus*, show remarkably different trophic phenotypes cohabiting in coastal lakes (Larson & McIntyre 1993; Schluter & McPhail 1993), having developed repeatedly in such waters of British Columbia in the few thousand millennia since deglaciation. Ecological, morphological and life history variation is evident in several species of percids including yellow perch (Aalto & Newsome 1990), in centrarchids such as pumpkinseed (Robinson *et al.* 1993), and in darters (Heithaus & Laushman 1997). The latter authors discuss how much of the genetic variation responsible for the differences is apportioned within and among populations as well as rivers. Robinson & Wilson (1994) give examples for 14 freshwater families, over 90 species in all.

Tropical freshwater fishes, especially those found in large species flocks, may show amazing sophistication of trophic divergence within species. A common cichlid scale-eater in Lake Tanganyika occurs in pale and dark colour morphs. This species exhibits seven different predatory hunting behaviours, several being colour morph-specific (Nshombo 1994). One morph uses female mimicry of an endemic killifish to attract its males, which then become victims! Several other scale-eating cichlids in Lake Tanganyika have dextral or sinstral mouth openings as an adaptation for efficient removal of prey scales, with the ratio of right and left 'handedness' being stabilized near unity in every population examined (Hori 1993; Kohda *et al.* 1997). Profound differences in behaviour and social organization of a lek-breeding Tanganyikan cichlid, *Cyathopharynx furcifer*, are of importance in demonstrating the importance of predator-mediated evolutionary mechanisms (Rossiter & Yamagashi 1997). Some 90% of the fish species in the three largest East African lakes – Malawi, Tanganyika and Victoria – are cichlids and most are endemic, so the evidence and hypotheses to account for such high speciation rates demand the attention given to them (Seehausen 2000).

The cyprinodontid species flock in the littoral zone of Lake Titicaca (Peru/Bolivia) show remarkable diversity in size, trophogastric morphology, feeding and habitat distribution (Northcote 2000b). Much more relevant information on tropical freshwater fish interactions and life history variability can be found in Lowe-McConnell (1987) for an African equatorial forest river (Zaire), a seasonal floodplain river (Niger), several African large lakes (Turkana, Victoria, Malawi, Tanganyika), the Amazon and other floodplain rivers of South America, and for several Far Eastern forest rivers and reservoirs, including some in Australia.

Concluding comment

Both salmonid and non-salmonid fishes, widely distributed among many of the forested watersheds of the world, display a remarkable range in life history variability and stock specificity. The various traits outlined are evident not only among species but within species (stocks) of those carefully examined. They occur in a single lake and between different reaches of a river or even a small stream. Clearly then for effective and responsible management both by forestry and fishery professionals, the extent and nature of this diversity needs to be recognized, known locally, and then applied appropriately in their various management practices.

Acknowledgements

I am most grateful to Drs Bror and Nina Jonsson who reviewed this chapter and provided many helpful suggestions and references. Dr J.D. McPhail agreed to review the chapter, as did Drs C.C. Lindsey and G.F. Hartman, both of whom provided many useful suggestions.

References

Aalto. S.K. & Newsome, G.E. (1990) Additional evidence supporting demic behaviour of a yellow perch (*Perca flavescens*) population. *Canadian Journal of Fisheries and Aquatic Sciences*, **47**, 1959–62.

Allan, I.R.H. & Ritter, J.A. (1977) Salmon terminology. *Journal du Conseil International pour l'Exploration se la Mer*, **37**, 293–9.

Alt, K.T. (1977) Inconnu, *Stenodus leucichthys*, migration studies in Alaska 1961–74. *Journal of the Fisheries Research Board of Canada*, **34**, 129–33.

Barton, N. (2000) The rapid origin of reproductive isolation. *Science*, **290**, 462–3.

Burger, C.V., Finn, J.E. & Holland-Bartels, L. (1995) Pattern of shoreline spawning by sockeye salmon in a glacially turbid lake: evidence for subpopulation differentiation. *Transactions of the American Fisheries Society*, **124**, 1–15.

Burger, C.V., Spearman, W.J. & Cronin, M.A. (1997) Genetic differentiation of sockeye salmon subpopulations from a geologically young Alaska lake system. *Transactions of the American Fisheries Society*, **126**, 926–38.

Burger, C.V., Scribner, K.T., Spearman, W.J., Swanton, C.O. & Campton, D.E. (2000) Genetic contribution of three life history forms of sockeye salmon to colonization of Frazer Lake, Alaska. *Canadian Journal of Fisheries and Aquatic Sciences*, **57**, 2096–111.

Carlsson, J. & Nilsson, J. (2000) Population genetic structure of brown trout (*Salmo trutta* L.) within a northern boreal forest stream. *Heriditas*, **132**, 173–81.

Coelho, M.M., Brito, R.M., Pacheco, T.R., Figueiredo, D. & Pires, A.M. (1995) Genetic variation and divergence of *Leuciscus pyrenaicus* and *L. carolitertii* (Pisces, Cyprinidae). *Journal of Fish Biology*, **47** (Suppl.), 243–58.

Crisp, D.J. (2000) *Trout and Salmon*. Blackwell Science, Oxford.

Curry, R.A., Brady, C., Noakes, D.L.G. & Danzmann, R.G. (1997) Use of small streams by young brook trout spawned in a lake. *Transactions of the American Fisheries Society*, **126**, 77–83.

Davies, R.W. & Thompson, G.W. (1976) Movements of mountain whitefish (*Prosopium williamsoni*) in the Sheep River watershed, Alberta. *Journal of the Fisheries Research Board of Canada*, **33**, 2395–401.

Dunham, J.B., Peacock, M.M., Rieman, B.E., Schroeter, R.E. & Vinyard, G.L. (1999) Local and geographic variability in the distribution of stream-living Lahonton cutthroat trout. *Transactions of the American Fisheries Society*, **128**, 875–89.

Ehlinger, T.J. (1990) Habitat choice and phenotype-limited feeding efficiency in bluegill: individual differences and trophic polymorphisms. *Ecology*, **71**, 886–96.

Eiler, J.H., Nelson, B.D. & Bradshaw, R.F. (1992) Riverine spawning by sockeye salmon in the Taku River, Alaska and British Columbia. *Transactions of the American Fisheries Society*, **121**, 701–8.

Gislason, D., Ferguson, M.M., Skúlason, S. & Snorrason, S.S. (1999) Rapid and coupled phenotypic and genetic divergence in Icelandic Arctic char (*Salvelinus alpinus*). *Canadian Journal of Fisheries and Aquatic Sciences*, **56**, 2229–34.

Glebe, B.D. & Leggett, W.C. (1981) Temporal, intra-population differences in energy allocation and use by American shad (*Alosa sapidissima*) during the spawning migration. *Canadian Journal of Fisheries and Aquatic Sciences*, **38**, 795–805.

Gresswell, R.E., Liss, W.J. & Larson, G.L. (1994) Life-history organization of Yellowstone cutthroat trout (*Oncorhynchus clarki bouvieri*) in Yellowstone Lake. *Canadian Journal of Fisheries and Aquatic Sciences*, **51** (Suppl. 1), 298–309.

Groot, C. (1996) Salmonid life histories. In: *Principles of Salmonid Culture*, (eds W. Pennel & B.A. Barton), pp. 97–230. Elsevier Science, Amsterdam.

Groot, C. & Margolis, L. (1991) *Pacific Salmon Life Histories*. UBC Press, Vancouver.

Hammar, J. (1998) *Evolutionary Ecology of Arctic Char* (Salvelinus alpinus (*L.*)). Doctoral dissertation, Acta Universitatis Uppsaliensis, Uppsala.

Hanfling, B. & Brandl, R. (1998) Genetic and morphological variation in a common European cyprinid, *Leuciscus cephalus*, within and across central European drainages. *Journal of Fish Biology*, **52**, 706–15.

Heithaus, M.R. & Laushman, R.H. (1997) Genetic variation and conservation of stream fishes: influence of ecology, life history and water quality. *Canadian Journal of Fisheries and Aquatic Sciences*, **54**, 1822–36.

Hendry, A.P. & Quinn, T.P. (1997) Variation in adult life history and morphology among Lake Washington sockeye salmon (*Oncorhynchus nerka*) populations in relation to habitat features and ancestral affinities. *Canadian Journal of Fisheries and Aquatic Sciences*, **54**, 75–84.

Hendry, A.P., Wenburg, J.K., Bentzen, P., Volk, E.C. & Quinn, T.P. (2000) Rapid evolution of reproductive isolation in the wild: evidence from introduced salmon. *Science*, **290**, 516–18.

Hendry, M.A., Wenburg, J.K., Myers, K.W. & Hendry, A.P. (2002) Genetic and phenotypic variation through the migratory season provides evidence for multiple populations of wild steelhead in the Dean River, British Columbia. *Transactions of the American Fisheries Society*, **131**, 418–34.

Hindar, K. & Jonsson, B. (1982) Habitat and food segregation of dwarf and normal Arctic char (*Salvelinus alpinus*) from Vangsvatnet Lake, western Norway. *Canadian Journal of Fisheries and Aquatic Sciences*, **39**, 1030–45.

Hindar, K. & Jonsson, B. (1993) Ecological polymorphism in Arctic char. *Biological Journal of the Linnean Society*, **48**, 63–74.

Holčík, J., Hensel, K., Nieslanik, J. & Skacel, L. (1988) *The Eurasian Huchen, Hucho hucho*. Dr. W. Junk Publishers, Dordrecht, The Netherlands.

Hori, M. (1993) Frequency-dependent selection in handedness of scale-eating cichlid fish. *Science*, **260**, 216–19.

Howland, K.L., Tallman, R.F. & Tonn, W.M. (2000) Migration patterns of freshwater and anadromous inconnu in the Mackenzie River system. *Transactions of the American Fisheries Society*, **129**, 41–59.

Hutchings, J.A. & Jones, M.E.B. (1998) Life history variation and growth rate thresholds for maturity in Atlantic salmon, *Salmo salar*. *Canadian Journal of Fisheries and Aquatic Sciences*, **55** (Suppl.1), 22–47.

Ibbotson, A.T., Cove, R.J., Ingraham, A., *et al.* (2001) A review of grayling ecology, status and management practice. *Technical Report W245*. Environment Agency, Bristol, UK.

Jensen, A.J. (1994) Growth and age distribution of a river-dwelling and lake-dwelling population of anadromous Arctic char at the same latitude in Norway. *Transactions of the American Fisheries Society*, **123**, 370–6.

Jonsson, B. & Jonsson, N. (1993) Partial migration: niche shift versus sexual maturation in fishes. *Reviews in Fish Biology and Fisheries*, **3**, 348–65.

Jonsson, B. & Jonsson, N. (2001) Polymorphism and speciation in Arctic charr. *Journal of Fish Biology*, **58**, 605–38.

Jonsson, B., Skúlasson, S., Snorrason, S.S., *et al.* (1988) Life history variation of polymorphic Arctic char (*Salvelinus alpinus*) in Thingvallavatn, Iceland. *Canadian Journal of Fisheries and Aquatic Sciences*, **45**, 1537–47.

Jonsson, B., Jonsson, N., Brodlkorb, E. & Ingebrigtsen, P.-J. (2001) Life-history traits of brown trout vary with size of small streams. *Functional Ecology*, **15**, 310–17.

Jonsson, N. & Jonsson, B. (1999) Trade-off between egg mass and egg number in brown trout. *Journal of Fish Biology*, **55**, 767–83.

Jonsson, N., Naesje, T.F., Jonsson, B., Saksgard, R. & Sandlund, O.T. (1999) The influence of piscivory on life history traits of brown trout. *Journal of Fish Biology*, **55**, 1129–41.

Kelso, B.W. & Northcote, T.G. (1981) Current response of young rainbow trout from inlet and outlet spawning stocks of a British Columbia lake. *Verhandlungen Internationale Vereinigen für theoretische und angewandte Limnologie*, **21**, 1214–21.

Kelso, B.W., Northcote, T.G. & Wehrhahn, C.F. (1981) Genetic and environmental aspects of the response to water current by rainbow trout (*Salmo gairdneri*) originating from inlet and outlet spawning streams of two lakes. *Canadian Journal of Zoology*, **59**, 2177–85.

Kohda, M., Hori, M. & Nshombo, M. (1997) Inter-individual variation in foraging behaviour and dimorphism. In: *Fish Communities in Lake Tanganyika*, (eds H. Kawanabe & M. Nagoshi), pp. 123–36. Kyoto University Press, Japan.

Koljonen, M.-L., Jansson, H., Paaver, T., Vasin, O. & Koskiniemi, J. (1999) Phylogeographic lineages and differentiation pattern of Atlantic salmon (*Salmo salar*) in the Baltic Sea with management implications. *Canadian Journal of Fisheries and Aquatic Sciences*, **56**, 1766–80.

Kristofferson, A.H. & Clayton, J.W. (1990) Subpopulation status of lake whitefish (*Coregonus clupeaformis*) in Lake Winnipeg. *Canadian Journal of Fisheries and Aquatic Sciences*, **47**, 1484–94.

Laroche, J., Durand, J.D., Bouvet, Y., Guinand, B. & Brohon, B. (1999) Genetic structure and differentiation among populations of two cyprinids *Leuciscus cephalus* and *Rutilus rutilus*, in a large European river. *Canadian Journal of Fisheries and Aquatic Sciences*, **56**, 1659–67.

Larson, G.L. & McIntire, C.D. (1993) Food habits of different phenotypes of threespine stickleback in Paxton Lake, British Columbia. *Transactions of the American Fisheries Society*, **122**, 543–9.

Levings, C.D. & Lauzier, R.B. (1991) Extensive use of the Fraser River basin as winter habitat by juvenile chinook salmon (*Oncorhynchus tshawytscha*). *Canadian Journal of Zoology*, **69**, 1759–67.

Lindström, T. (1970) Habitats of whitefish in some north Swedish lakes at different stages of life history. In: *Biology of Coregonid Fishes*, (eds C.C. Lindsey & C.S. Woods), pp. 461–79. University of Manitoba Press, Winnipeg, Canada.

Lowe-McConnell, R.H. (1987) *Ecological Studies in Tropical Fish Communities*. Cambridge University Press, Cambridge, UK.

McCart, P. (1970) Evidence for the existence of sibling species of pygmy whitefish (*Prosopium coulteri*) in three Alaskan lakes. In: *Biology of Coregonid Fishes*, (eds C.C. Lindsey & C.S. Woods), pp. 81–98. University of Manitoba Press, Winnipeg, Canada.

McClenaghan, L.R., Smith, M.H. & Smith, M.W. (1985) Biochemical genetics of mosquitofish. IV. Changes of allele frequency through time and space. *Evolution*, **39**, 451–60.

McMahon, T.E. & Holtby, L.B. (1992) Behaviour, habitat use and movements of coho salmon (*Oncorhynchus kisutch*) smolts during seaward migration. *Canadian Journal of Fisheries and Aquatic Sciences*, **49**, 1478–85.

Marschall, E.A., Quinn, T.P., Roff, D.A., *et al.* (1998) A framework for understanding Atlantic salmon (*Salmo salar*) life history. *Canadian Journal of Fisheries and Aquatic Sciences*, **55** (Suppl.1), 48–58.

Mitton, J.B. & Lewis, W.M. (1989) Relationships between genetic variability and life-history features of bony fishes. *Evolution*, **43**, 1712–23.

Moser, M.L., Olson, A.F. & Quinn, T.P. (1991) Riverine and estuarine migratory behavior of coho salmon (*Oncorhynchus kisutch*) smolts. *Canadian Journal of Fisheries and Aquatic Sciences*, **48**, 1670–8.

Murray, C.B. & Rosenau, M.L. (1989) Rearing of juvenile chinook salmon in nonnatal tributaries of the lower Fraser River, British Columbia. *Transactions of the American Fisheries Society*, **118**, 284–9.

Nakamoto, R.L. (1994) Characteristics of pools used by adult summer steelhead oversummering in the New River, California. *Transactions of the American Fisheries Society*, **123**, 757–65.

Näslund, I., Degerman, E. & Nordwall, F. (1998) Brown trout (*Salmo trutta*) habitat use and life history in Swedish streams: possible effects of biotic interactions. *Canadian Journal of Fisheries and Aquatic Sciences*, **55**, 1034–42.

Nelson, J.S. (1994) *Fishes of the World*, 3rd edn. John Wiley & Sons, New York.

Nikolsky, G.V. (1963) *The Ecology of Fishes*. Academic Press, London.

Northcote, T.G. (1969) Patterns and mechanisms in the lakeward migratory behaviour of juvenile trout. In: *Symposium on Salmon and Trout in Streams*, (ed. T.G. Northcote), pp. 183–203. H.R. MacMillan Lectures in Fisheries, University of British Columbia, Vancouver.

Northcote, T.G. (1981) Juvenile current response, growth and maturity of above and below waterfall stocks of rainbow trout, *Salmo gairdneri. Journal of Fish Biology*, **19**, 741–51.

Northcote, T.G. (1993) A review of management and enhancement options for the Arctic grayling (*Thymallus arcticus*) with special reference to the Williston Reservoir watershed in British Columbia. *Fisheries Management Report 101*. BC Ministry of Environment, Lands and Parks, Fisheries Branch, Victoria, Canada.

Northcote, T.G. (1995) Comparative biology and management of Arctic and European grayling (Salmonidae, *Thymallus*). *Reviews in Fish Biology and Fisheries*, **5**, 141–94.

Northcote, T.G. (1997) Potamodromy in Salmonidae – living and moving in the fast lane. *North American Journal of Fisheries Management*, **17**, 813–29.

Northcote, T.G. (2000a) An updated review of grayling biology, impacts, and management. *Peace/Williston Fish and Wildlife Compensation Program Report 211*. Prince George, British Columbia, Canada.

Northcote, T.G. (2000b) Ecological interactions among an orestiid (Pisces: Cyprinodontidae) species flock in the littoral zone of Lake Titicaca. In: *Ancient Lakes: Biodiversity, Ecology and Evolution*, (eds A. Rossiter & H. Kawanabe), pp. 399–420. Advances in Ecological Research 31, Academic Press, San Diego.

Northcote, T.G. & Ennis, G.L. (1994) Mountain whitefish biology and habitat use in relation to compensation and improvement possibilities. *Reviews in Fisheries Science*, **2**, 347–71.

Northcote, T.G. & Hartman, G.E. (1988) The biology and significance of stream trout populations living above and below waterfalls. *Polskie Archiwum Hydrobiologii*, **35**, 409–42.

Northcote, T.G. & Kelso, B.W. (1981) Differential response to water current by two homozygous LDH phenotypes of young rainbow trout (*Salmo gairdneri*). *Canadian Journal of Fisheries and Aquatic Sciences*, **38**, 348–52.

Nshombo, M. (1994) Polychromatism of the scale-eater *Perrisodus microlepis* (Cichlidae, Teleostei) in relation to foraging behavior. *Journal of Ethology*, **12**, 141–61.

Olver, C.H. (1991) Lake trout *Salvelinus namaycush*. In: *Trout*, (eds J. Stolz & J. Schnell), pp. 286–99. Stackpole Books, Harrisburg, PA.

Pennisi, E. (2000) Nature steers a predictable course. *Science*, **287**, 207–9.

Quinn, T.P. & Unwin, M.J. (1993) Variation in life history pattern among New Zealand chinook salmon (*Oncorhynchus tshawytscha*) populations. *Canadian Journal of Fisheries and Aquatic Sciences*, **50**, 1414–21.

Riddell, B.E & Leggett, W.C. (1981) Evidence of an adaptive basis for geographic variation in body morphology and time of downstream migration of juvenile Atlantic salmon (*Salmo salar*). *Canadian Journal of Fisheries and Aquatic Sciences*, **38**, 308–20.

Robinson, B.W. & Wilson, D.S. (1994) Character release and displacement in fishes: a neglected literature. *American Naturalist*, 144, 596–627.

Robinson, B.W. & Wilson, D.S. (1996) Genetic variation and phenotypic plasticity in a trophically polymorphic population of pumpkinseed sunfish (*Lepomis gibbosis*). *Evolutionary Ecology*, 10, 631–52.

Robinson, B.W., Wilson, D.S., Margosian, A.S. & Lotito, P.T. (1993) Ecological and morphological differentiation by pumpkinseed sunfish in lakes without bluegill sunfish. *Evolutionary Ecology*, 7, 451–64.

Rossiter, A. & Yamagishi, S. (1997) Intraspecific plasticity in the social system and mating behaviour of a lek-breeding cichlid fish. In: *Fish Communities in Lake Tanganyika*, (eds H. Kawanabe & M. Nagoshi), pp. 195–217. Kyoto University Press, Japan.

Rundle, H.D., Nagel, L., Boughman, J.W. & Schluter, D. (2000) Natural selection and parallel speciation in sympatric sticklebacks. *Science*, 287, 306–8.

Ryan, P.M. (1993) Natural lake use by juveniles: a review of the population dynamics of Atlantic salmon in Newfoundland, Canada. *Canadian Manuscript Report, Fisheries and Aquatic Sciences 2222*, pp. 3–14.

Schluter, D. & McPhail, J.D. (1993) Character displacement and replicate adaptive radiation. *Trends in Ecological Evolution*, 8, 197–200.

Scott, W.B. & Crossman, E.J. (1973) *Freshwater Fishes of Canada*. Bulletin 184, Fisheries Research Board of Canada, Ottawa.

Scrivener, J.C., Brown, T.G., & Anderson, B.C. (1994) Juvenile chinook salmon (*Oncorhynchus tshawytscha*) utilization of Hawks Creek, a small and nonnatal tributary of the upper Fraser River. *Canadian Journal of Fisheries and Aquatic Sciences*, 51, 1139–46.

Scrivener, J.C., Tschaplinski, P.J. & Macdonald, J.S. (1998) An introduction to the ecological complexity of salmonid life history strategies and of forest harvesting impacts in coastal British Columbia. In: *Carnation Creek and Queen Charlotte Islands Fish/Forestry Workshop: Applying 20 Years of Coast Research to Management Solutions*, (eds D.L. Hogan, P.J. Tschaplinski & S. Chatwin), pp. 23–8. British Columbia Ministry of Forests, Victoria, Canada.

Seehausen, O. (2000) Explosive speciation rates and unusual species richness in haplochromine cichlid fishes: effects of sexual selection. In: *Ancient Lakes: Biodiversity, Ecology and Evolution*, (eds A. Rossiter & H. Kawanabe), pp. 237–74. Advances in Ecological Research 31, Academic Press, San Diego.

Shaw, P.W., Carvalho, G.R., Magnurran, A.E. & Segers, B.H. (1994) Factors affecting the distribution of genetic variability in the guppy, *Poecilia reticulata*. *Journal of Fish Biology*, 45, 875–88.

Skúlason, S., Snorrason, S.S., Noakes, D.L.G. & Ferguson, M.M. (1996) Genetic basis of life history variations among sympatric morphs of Arctic char, *Salvelinus alpinus*. *Canadian Journal of Fisheries and Aquatic Sciences*, 53, 1807–13.

Smith, S.B. (1969) Reproductive isolation in summer and winter races of steelhead trout. In: *Symposium on Salmon and Trout in Streams*, (ed. T.G. Northcote), pp. 21–38. H.R. MacMillan Lectures in Fisheries, University of British Columbia, Vancouver.

Smoker, W.W. & Fukushima, M. (1997) Determinants of stream life, spawning efficiency, and spawning habitat in pink salmon in the Auk Lake system, Alaska. *Canadian Journal of Fisheries and Aquatic Sciences*, 54, 96–104.

Stewart, D.C., Smith, G.W. & Youngson, A.F. (2002) Tributary-specific variation in timing of return of adult Atlantic salmon (*Salmo salar*) to fresh water has a genetic component. *Canadian Journal of Fisheries and Aquatic Sciences*, 59, 276–81.

Svärdson, G. (1998) Postglacial dispersal and reticulate evolution of Nordic coregonids. *Nordic Journal of Freshwater Research*, 74, 3–32.

Taylor, E.B. & Bentzen, P. (1993) Evidence for multiple origin and sympatric divergence of trophic ecotypes of smelt (*Osmerus*) in northeastern North America. *Evolution*, 47, 813–32.

Teel, D.J., Milner, G.B., Winns, G.A. & Grant, W.S. (2000) Genetic population structure and origin of life history types in chinook salmon in British Columbia, Canada. *Transactions of the American Fisheries Society*, **129**, 194–209.

Tibbets, C.A. & Dowling, T.E. (1996) Effects of intrinsic and extrinsic factors on population fragmentation in three species of North American minnows (Teleostei: Cyprinidae). *Evolution*, **50**, 1280–92.

Uiblein, F., Jagsch, A., Kössner, G., Weiss, S., Gollmann, P. & Kainz, E. (2000) Untersuchungen zu lokaler Anpassung, Gefahrdung und Schutz der Äsche (*Thymallus thymallus*) in drei Gewässern in Oberösterreich. *Österreichs Fischerei*, **53**, 89–165.

White, H.C. & Huntsman, A.G. (1938) Is local behaviour in salmon heritable? *Journal of the Fisheries Research Board of Canada*, **4**, 1–18.

Wickett, W.P. (1964) An unusually late-spawning British Columbia chum salmon. *Journal of the Fisheries Research Board of Canada*, **21**, 657.

Withler, I.L. (1966) Variability in life history characteristics of steelhead trout (*Salmo gairdneri*) along the Pacific Coast of North America. *Journal of the Fisheries Research Board of Canada*, **23**, 365–93.

Willson, M.F. (1997) Variation in salmonid life histories: patterns and perspectives. *Research Paper PNW-RP-498*. Pacific Northwest Research Station, Forest Service, US Department of Agriculture, Portland, OR.

Woody, C.A., Olsen, J., Reynolds, J. & Bentzen, P. (2000) Temporal variation in phenotypic and genotypic traits in two sockeye salmon populations, Tustumena Lake, Alaska. *Transactions of the American Fisheries Society*, **129**, 1031–43.

Young, K.A. (1999) Environmental correlates of male life history variation among coho salmon populations from two Oregon coastal streams. *Transactions of the American Fisheries Society*, **128**, 1–16.

Chapter 7
Fish migration and passage in forested watersheds

T.G. NORTHCOTE AND S.G. HINCH

Introduction

Fish migration, and the physical, physiological and behavioural necessities to permit passageway through inland waters are all essential to maintenance of many fish populations in forested watersheds. Various conditions associated with deforestation may cause severe difficulties in migration or passage. Migratory behaviour itself has the potential to make fish populations vulnerable to passageway problems (Groot 1982). Evidence supporting these views has been mounting for decades in salmonid fishes in North America and Europe, but now is becoming recognized in freshwater fish populations of South America, the Antipodes and elsewhere (Northcote 1998).

Our purposes here are to (1) define the process of fish migration in inland waters, (2) outline the types and patterns of fish migration with examples from temperate and subtropical or tropical areas, (3) discuss briefly the sensory mechanisms involved in migration, (4) review the physiological capabilities, behavioural tactics and energetics of fish in migration, and (5) discuss the difficulties for fish in surmounting passageway problems, upstream and downstream, resulting from natural and forestry-related obstructions and barriers.

Fish migration defined

There are four essential features of fish migration: (1) individuals cyclically alternate between at least two separated habitats (spawning and feeding) and often three or even more (see Fig. 1 in Northcote 1997); (2) the alternation between habitats occurs with a reasonably predictable sequence; (3) a large fraction of any given population participates in the migratory stages of the life cycle; (4) at some stage the migration occurs as an active, directed movement although it may occur as a passive drift downstream for other stages. The most common migratory phases can be categorized as trophic migrations to feeding habitats, refuge migrations to survival habitats used during periods of severe environmental conditions (e.g. overwintering or oversummering in temperate or arctic regions), and reproductive migrations to spawning habitats. Together they complete the migratory cycle, given elsewhere in more detail (Northcote 1978, 1984, 1991, 1997: Fig. 1).

Types and patterns of migration

Interest in the descriptive aspects of fish migration has produced a plethora of terminology, some of which must be presented before delving into more fundamental features relevant to fish–forestry interactions. Much of it hinges around the origin, destination and direction with respect to water current where the migration occurs. If the migration is between inland waters and the sea then the term applied is *diadromy*, which in turn has two related forms: *anadromy* – 'up-running', involving at one stage a migration from marine waters up rivers or streams to reproduce in freshwater, and *catadromy* – 'down-running', involving at one stage a migration down streams or rivers to reproduce in the sea. If the migratory cycle occurs solely within flowing inland waters, regardless of its direction with respect to current, it is called *potamodromy*. Lake-dwelling fishes which enter tributary streams and rivers to reach spawning habitats are said to make *adfluvial* migrations. Other terms exist but need not be dealt with here.

Lucas & Baras (2001) reviewed migratory behaviour of arctic, subarctic, temperate and tropical freshwater fishes for 44 taxonomic groupings. Some 36 groups of fish, probably representing over 350 species, show well-documented migrations in inland waters (Table 7.1). Over 64 of these species (mainly sharks, rays, mullets and gobioids) are diadromous, over 84 (mainly lampreys, sturgeons, anchovies etc., northern smelts and salmonids) are anadromous, and over 25 (mainly freshwater eels) are catadromous. Potamodromy occurs in over 169 species and no doubt this form of migration is greatly underestimated in minnows, characins and catfishes. Twenty-nine of the 36 groups of fish (81%) have migrant species in temperate inland waters, 15 in subtropical (42%) and 10 (28%) in tropical inland waters. Clearly migratory behaviour seems much more common for temperate inland waters, but detailed information is only available for a small fraction of all freshwater fish, of which there are some 10,000 species (see Table 1.2 in Chapter 1).

Diadromy is said to occur in about 230 species of fish (McDowall 1988), many of which may make only short or brief excursions into estuaries or freshwater. The proportion of freshwater fish species exhibiting diadromy within an area seems to vary with length of coastline relative to land mass and also with latitude (McDowall 1996). For example, in some large insular areas such as New Zealand over half of the freshwater fishes are diadromous and in some such as Newfoundland (Hammar 1987) or Hawaii (Kinzie 1991), all species follow that type of migratory behaviour. Along the Atlantic coast of North America (25–61° N latitude) the number of diadromous species peaks just under 20 (about a third of the freshwater species in the region) at about 45° N. Along its Pacific coast (25–71° N) it peaks at 20 (again nearly a third of the total freshwater species in the region) at just over 60° N, and ranges in New Zealand from nearly 90% of its freshwater fish fauna at 34° S latitude to about 70% at 46° S latitude (McDowall 1996). Diadromy facilitates invasion of new or reinvasion of formerly perturbed inland waters (McDowall 1996), but is also a behavioural feature that can create difficulties for species attempting to gain access to former habitats that become obstructed naturally or anthropogenically (Northcote 1998).

Anadromy is a very common form of migration in salmonids, especially in the salmon, trout and char, but also occurs in whitefish and some grayling populations.

Members of this family of freshwater fish are widely distributed in forested regions of North America and Eurasia (see Scott & Crossman (1973) for a map of their native world distribution). Salmonids have been introduced into forested regions of Central and South America, Africa, India, New Guinea, New Zealand, Australia and some major nearshore as well as smaller offshore oceanic islands. Salmonids commonly migrate into and spawn in forested watersheds. They are the most studied group with respect to fish–forestry interactions, which in many regions of the world have been going on for decades. Furthermore they support important food, commercial and recreational fisheries.

In many salmonids the marine migratory legs of anadromy can be very long – thousands of kilometres are not uncommon. They are usually short and coastal in whitefish and in some grayling populations (Scott & Crossman 1973; Reist & Bond 1988; Bond & Erickson 1992, 1997). In the large (seventh order) glacial Taku River (Canada, Alaska), some of the Pacific salmon (*Oncorhynchus* spp.) juveniles remain in Alaskan lower reaches which then must be carefully considered as rearing habitat for whole-river management purposes (Murphy *et al.* 1997). In anadromous Atlantic salmon (*Salmo salar*) young the freshwater movements increase in distance over the first several years up to the late smolt stage (McCormick *et al.* 1998). Anadromous salmonid populations of the far north, such as the Arctic char, may need freshwater habitat for overwintering so the sea resident period can be no longer than 2 months (Gulseth & Nilssen 2000). The anadromous migratory patterns followed by lacustrine and fluvial forms of Dolly Varden char, as proposed by Armstrong & Morrow (1980), now seem even more complex than previously thought, given that up to 58% of the lacustrine forms overwinter in the sea near Juneau, Alaska, rather than in lakes (Bernard *et al.* 1995). Both anadromous and non-anadromous but still migratory forms of inconnu (*Stenodus leucichthys*) occur in the Mackenzie River system of northern Canada, so these differences have to be accounted for in their management (Howland *et al.* 2000).

Catadromous migrations are much less common than anadromous ones. They occur in 15 species of freshwater eels in inland waters and seas of eastern North America, Europe, northern and southeastern Africa, India, China, Korea, Japan, some of the major islands of southeast Asia, New Zealand and Australia. Catadromy occurs in a few galaxid fishes from southern South America, southernmost Africa, New Zealand, New Caledonia and Australia. It also occurs in some gobies, often the most abundant fish in freshwater of tropical and subtropical oceanic islands (Nelson 1994), and in a few other groups (Table 7.1). Young eels of the elver stage move from marine waters into river systems to freshwater feeding habitats. Later at maturity eels move downstream and out to sea; some species and stocks move long distances offshore. Galaxids and gobies generally move over shorter distances both in freshwater and marine legs of their migrations.

Potamodromy in river-dwelling whitefishes, graylings and other salmonids includes trophic, refuge and reproductive migratory stages (Northcote 1997). The riverine distribution pattern for Arctic grayling, wherein the larger and older fish inhabit upstream reaches, can be explained by differential migrations upstream and downstream between fast and slow growing fish (Hughes 1999). Juvenile downstream migration in

Table 7.1 Type, extent and general location of well-documented migratory behaviour[a] in 36 major groups of fishes frequenting inland waters of the world

| Fish group | Number of species showing migratory behaviour[b] | | | | General location |
	Diadromy[c]	Anadromy	Catadromy	Potamodromy[d]	
Lampreys		9		2	temperate[e]
Sharks, rays	10				subtropical, tropical
Sturgeons		10		3	temperate
Paddlefishes				2	temperate
Gars				2	temperate
Osteoglossids				3	temperate, tropical
Tarpons	1				subtropical
Freshwater eels			15		temperate, subtropical, tropical
Anchovies, shads, herrings, menhaden	1	22		2	temperate, subtropical, tropical
Milkfish	1				tropical, subtropical
Carps, minnows				>33	temperate, subtropical
Suckers				19	temperate
Loaches				2	temperate, subtropical
Characins				>31	subtropical, tropical
Catfishes		1		11	temperate, subtropical, tropical
Pikes, mudminnows	1			4	temperate
Smelts (northern)		7		1	temperate
Noodlefishes		>1			temperate
Smelts, graylings (southern)		1		1	temperate

Group	Anadromous	Amphidromous	Catadromous	Total	Climatic region
Galaxids		>3	>1	>3	temperate
Salmonids[d]	21	1		34	temperate, subtropical
Trout-perches	1			1	temperate
Cods[e]	1			1	temperate
Mullets		1	14	14	temperate, subtropical
Silversides		1		1	temperate, tropical
Sticklebacks	3			3	temperate
Cottids		1?		1?	temperate
Snooks			2	2	temperate, subtropical
Moronid basses	2			2	temperate
Percichthyid perches		1	3	3	temperate
Perches	2?		1	4	temperate
Tigerperches		1	1	2	subtropical, tropical
Jungleperches		>2	1	>2	tropical
Sandperches		1		1	temperate, subtropical
Gobioids[c]	>1	>33	1	>33	subtropical, tropical
Flatfishes		2	2	>3	temperate, subtropical, tropical
Totals	>64	>84	>25	>169	

[a] Information mainly from text in Lucas & Baras (2001); other inputs from Northcote (1997) and McDowall (1988, 1996).

[b] Same species within a group may be recorded in more than one migratory type.

[c] Here includes amphidromous fishes whose larvae migrate to sea soon after hatching; see Lucas & Baras (2001).

[d] Here includes adfluvial migrations from lakes into tributary spawning habitats.

[e] Here and below may include a few species also found in arctic inland waters.

tributaries by adfluvial cutthroat trout to a Utah reservoir occurs mainly at night and shortly after emergence (Knight *et al.* 1999), similar to that of underyearlings for several lake-dwelling but inlet spawning populations of trout (Northcote 1969). Several species of salmonids make regular seasonal use of small tributaries during riverine residence, so their passageway into and out of such important temporary habitat is critical to their production (Hartman & Brown 1987; Bramblett *et al.* 2002). Information on potamodromous migrations, which occur in the many other groups of fishes (Table 7.1), is provided by Lucas & Baras (2001). The important concept of core area use in rivers and streams for seasonal habitat occupation, along with associated movement to and from such areas, is now emerging from radio-tagging studies in several groups of fishes such as sturgeon (Knights *et al.* 2002), bull trout (K. Bray & P. Mylechreest, pers. comm.), brook trout (Curry *et al.* 2002) and Atlantic salmon (Hiscock *et al.* 2002).

In lakes, some salmonids (especially sockeye and kokanee *Oncorhynchus nerka*) undertake extensive and regular diel vertical migrations (Northcote 1967; Narver 1970; Levy 1990, 1991), even under ice in winter (Steinhart & Wurtzbaugh 1999). They occur as a complex compromise to optimize feeding, growth and predator avoidance. These migrations, especially in their near-surface portion, can bring the fish into direct contact with various forestry activities and also their environmental effects (see Chapter 14).

Many non-salmonid fishes make extensive and multidirectional migrations (see Table 7.1). These include anadromy in smelts, clupeid shads (McDowall 1996; Blaber 1997) and sturgeon (Veinott *et al.* 1999; Fox *et al.* 2000), and potamodromy in paddlefish (Lein & De Vries 1998), galaxids (Richardson *et al.* 2001) and cyprinids (Lindsey & Northcote 1963; Tyus 1990; McAda & Kaeding 1991; Osmundson *et al.* 1998). For the latter group, precise reproductive homing has been documented in roach (*Rutilus rutilus*) by L'Abe'e-Lund & Vollestad (1985) and in *Leuciscus cephalus* by LeLouarn *et al.* (1997). Cyprinids in some African lakes have spawning migrations much like trout in temperate lakes (G.F. Hartman, pers. comm.). Schindler (1999) has developed a model to examine the lake onshore–offshore diel migratory strategies of minnows. Catostomids may migrate into lake tributaries to spawn and their young move back to lakes at a young stage (Geen *et al.* 1966). For non-salmonid tropical freshwater fishes, Blaber (1997) has reviewed both anadromous and potamodromous migratory patterns.

Several groups of fish exhibit what has been termed 'partial migration' (Jonsson & Jonsson 1993), i.e. where a particular population can be composed of both migratory and resident individuals, the latter often being dwarfs compared with the migratory forms. This phenomenon is common in at least nine species of salmonids and two species of smelts (see Table 1 in Jonsson & Jonsson 1993).

Mechanisms of fish migration in freshwater

The control of diadromous migration in fishes involves a complex and hierarchical series of mechanisms. This has long been recognized (McCleave *et al.* 1984), as has the importance of olfactory imprinting and other controls in the precision of reproductive homing (Døving *et al.* 1974; Hasler & Scholtz 1983; Smith 1985; Groot *et al.* 1986). Multiple mechanisms are used in the homing phenomenon of migration in rivers (Northcote 1984) and in lakes (Hodgson *et al.* 1998). The olfactory organ of lake salmonids may discriminate among different intensities of various freshwater odours (Sato *et al.* 2000). Olfaction also plays an important role in homing and estuarine migration of the catadromous eel (Barbin 1998). A suite of guidance cues may be integrated through the sensory capabilities used by fish (visual, olfactory and others). These may be regulated by genetic control, by previous experience and learned behaviour as young, and by past as well as present environmental conditions (such as water temperature, light and photoperiod, water currents). Among salmonids, each species may have different strategies of homing migration (Northcote 1978, 1998) and use different sensory abilities affected by hormonal states (Ueda 1998).

Although home stream odours may be imprinted by learning as juveniles and remembered by returning adults several years later, there has been continuing research to identify just when and how such imprinting occurs. Experimental work on coho salmon juveniles suggests that imprinting only occurs during the short period of parr to smolt transformation (Dittman *et al.* 1996). Nevertheless, under wild rearing conditions young salmon can learn site-specific home odours long before reaching the smolt stage. Several species of salmonids that leave freshwater rearing areas as young fry are able to reproductively home with high precision. The field is still under active research by several groups (Barinaga 1999). Because of the importance of 'home stream' olfactory cues, imprinted as young and used by returning adults several years later, it is not difficult to see the implications of changing stream odours by various human activities, including forestry.

Physiological capabilities, behavioural tactics and energetics

Physiological capabilities

Species vary in their physiological capabilities to migrate. Natural selection has resulted in some species or stocks having superior migratory abilities. These capabilities, however, have costs in fecundity, morphology, or other aspects of fitness associated with maintaining them (Crossin 2002).

There are two fundamental types of swimming, sustained and burst. Their magnitude and duration vary among stocks and species in relation to underlying physiological factors and particular environmental features. Sustained swimming is the slowest class of speeds and can usually be maintained indefinitely without causing fatigue. Burst swimming is the fastest class and when elicited usually causes fatigue in less than 1 minute. An intermediate class of speeds, termed 'prolonged', reflects a zone of transition between sustained and burst, and may be the most common one used by upstream migrating individuals (Beamish 1978).

At the physiological level, swimming capabilities are governed by muscular, enzymatic and metabolic factors. Fish swimming musculature is made up of two types, 'red' and 'white' muscle (Gill *et al.* 1989). In red muscle, glucose is converted to energy in the presence of oxygen (aerobic metabolism). This function can go on indefinitely provided that oxygen and glucose are present in the blood. Red musculature is used for day-to-day routine activities (e.g. sustained swimming). White muscle also converts glucose, via glycogen, to energy but it does so without oxygen (anaerobic metabolism), and is only used when blood oxygen is lacking and aerobic swimming is not possible, for instance when fish try to ascend fast stretches of rivers (e.g. burst swimming). However, anaerobic metabolism is much less efficient for conversion of glucose to energy than is aerobic metabolism and its byproduct, lactic acid, can itself impair swimming, leading to fatigue if it cannot be eliminated quickly. In some circumstances, blood lactic acid can accumulate to such high levels during hyperactive swimming that blood acidosis causes mortality (Black 1958). Thus, species or stocks with relatively high proportions of white muscle, and good abilities to eliminate lactic acid and recover from strenuous exercise, may be able to migrate more successfully through fast water. However, high proportions of white muscle will be disadvantageous for steady cruising activities, which may be required for feeding or other behaviour. Trade-offs in relative proportions of muscle types may depend on more than just migratory constraints in a fish's life history (Gill *et al.* 1989).

Maximum sustained speeds vary considerably among sizes and life stages within species, and among taxa. Within a species, larger fish generate greater tail thrust and can attain higher absolute swimming speeds than smaller fish. Yet relative to body size, small fish are better fitted to sustained swimming. For example, 8-cm sockeye salmon (*Oncorhynchus nerka*) have sustained swimming speeds of 6.7 L s^{-1} (lengths per second) but 60-cm sockeye can only sustain 2.1 L s^{-1} (Brett & Glass 1973). Maturation stage can also affect swimming performance, irrespective of body size. For instance, 13-cm Atlantic salmon smolts have a maximum sustained speed of 114 cm s^{-1} whereas that for the same sized parr is 82 cm s^{-1} (Peake *et al.* 2000). Some differences among taxa in maximum sustained speeds can be attributed to differences in body shape and fin position, features that affect their ability to generate power and overcome drag (Webb 1984, 1995). Adult salmonids, with their fusiform bodies and medially positioned pelvic fins, can attain high sustained swimming speeds. These reach 178 cm s^{-1} in sockeye salmon, 138 cm s^{-1} in brown trout (*Salmo trutta*) and 122 cm s^{-1} in brook trout (*Salvelinus fontinalis*). In adult fishes with more laterally compressed body shapes and anteriorly positioned pelvic fins they are lower (e.g. 118 cm s^{-1} in smallmouth bass (*Micropterus dolomieu*) and 67.9 cm s^{-1} in walleye (*Stizostedium vitreum*)) (Brett & Glass 1973; Peake *et al.* 1997, 2000; Bunt *et al.* 1999).

Maximum sustained swimming speeds are governed by water temperature. Fish body temperature is usually the same as surrounding water temperature, and standard metabolism (the energetic costs of respiration) increases exponentially with increasing water temperature. Maximum sustained swimming speeds also increase with increasing temperature, peak at a temperature specific for species and size, and decline with further increases in temperature. Thus, the temperature where sustained speeds are maximal may be different, even for closely related species (Brett 1995).

Temperature also affects prolonged swimming speeds, the transition speeds that involve both aerobic and anaerobic metabolism. Provided that temperatures are within a species' thermal tolerance, warmer temperatures permit higher speeds before fatigue ensues (Brett 1967; Mesa & Olson 1993). Although larger fish swim faster than small fish and may exhibit higher prolonged speeds, in relation to their size, larger fish tend to fatigue at slower speeds than small fish (Brett 1967; Mesa & Olson 1993). Fatigue velocities (c. 15–65 cm s^{-1}) of young flannelmouth suckers (*Catostomus latipennis*) increased with temperature and fish size over a tested range of 10–20°C and 25–115 mm total length (Ward *et al.* 2002).

Maximum burst speeds are periodic high acceleration sprints and last for only a few seconds. Bainbridge (1958) was the first to rigorously examine 'voluntary' burst swimming. He quantified adult sprint speeds for dace (*Leuciscus leuciscus*; 220 cm s^{-1}), trout (*Salmo irideus*; 330 cm s^{-1}) and goldfish (*Carassius auratus*; 88 cm s^{-1}). Relative to their body size, dace and trout can reach burst speeds of 10–12 L s^{-1} and goldfish 8–10 L s^{-1}. Webb (1995) predicted that adult salmonids could sprint at speeds up to 10 L s^{-1}, and juveniles up to 25 L s^{-1}. Adult sockeye salmon migrating upstream have been observed by telemetry to regularly attain speeds of over 9 L s^{-1} and occasionally up to 12 L s^{-1} (Fig. 7.1) (Hinch *et al.* 2002). However, sprint speeds remain inadequately studied for

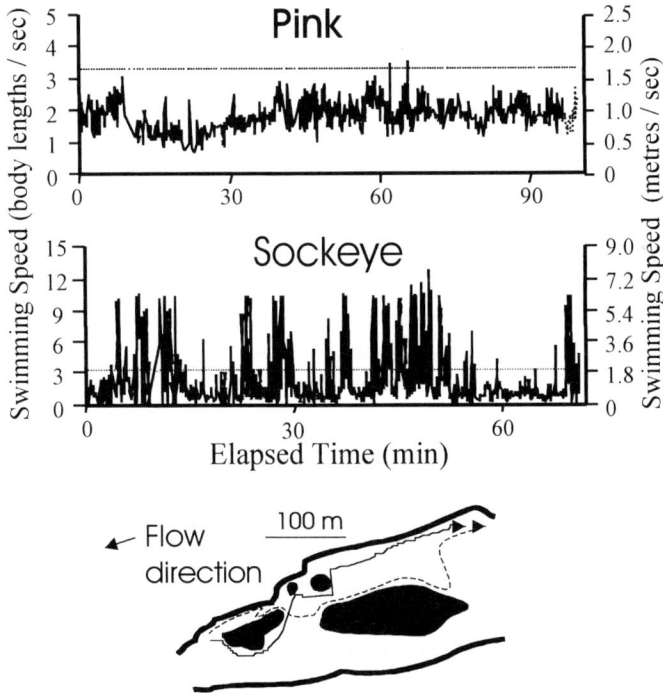

Fig. 7.1 Plots of instantaneous swimming speeds (upper two panels), estimated from EMG transmitters surgically implanted into main swimming muscles, for an adult pink and sockeye salmon to pass through a 400-m reach in the Fraser River canyon (lower panel). Speeds are given in absolute (m sec^{-1}) and relative (body lengths sec^{-1}) measures. The horizontal lines on upper two panels indicate transition zones between low-cost aerobic metabolism and high-cost anaerobic metabolism. This reach contains four islands, white water rapids and multidirectional currents. The lines on the reach diagram (lower panel) represent the upstream trajectory of a pink (broken) and a sockeye (solid) salmon through the reach. Arrowheads indicate direction of fish migration. Additional details in Hinch *et al.* (2002).

most species, despite their importance for understanding migration energetics and passage success. Temperature does not seem to affect maximum burst swimming speeds in fish (Peake *et al.* 2000), but more research is needed.

Behavioural tactics

Physiological capabilities set the stage upon which swimming behaviour acts. This is particularly evident in spawning migrations of salmon, although field studies linking adult fish migration tactics, swimming abilities and energetics have focused on only a few stocks of salmonids. Energy must be conserved for gamete production, development of secondary sexual characteristics, and spawning activities. Where spawning migrations are energetically expensive (e.g. high water temperatures and high water velocities) and with cessation of feeding (as generally occurs in upstream migrating salmonid adults), energy conservation should be under strong natural selection. These adult migrants should be energy-efficient wherever possible. There are several behavioural tactics that migrants could use to conserve energy. The limited field studies suggest that adults may invoke most of these, and switch among them depending on local conditions. Migrants could swim in low speed or reverse-flow current paths (Hinch & Rand 2000) or, where possible, utilize upstream tidal transport (Levy & Cadenhead 1995). In these instances, migrants use sustained speeds and avoid burst swimming. This is a predominant tactic used by adult pink salmon (*O. gorbuscha*) (Fig. 7.1) (Hinch *et al.* 2002).

Paradoxically, under some circumstances, burst swimming can be energetically efficient if alternated with coasting. Such 'burst-then-coast' swimming could be up to 60% more efficient than swimming at sustained speeds (Weihs 1974). Sockeye salmon appear to occasionally invoke this behaviour (Fig. 7.1) (Hinch *et al.* 2002). Another tactic to save energy is to swim in configurations, such as schools, which enable fish to receive locomotory benefits from conspecifics via a reduction in incident water velocity (Weihs 1975).

A final tactic for conserving energy involves steady swimming at sustained speeds, but speeds that are constantly being adjusted to minimize the total energy expended in moving mass through unit distance. Termed swimming at hydrodynamically (Weihs 1973) or metabolically 'optimal speeds' (Ware 1975), this behaviour has recently been documented in migrating salmon, but seems prevalent only at sites where the current speeds encountered are relatively slow (<25 cm s^{-1}) (Hinch & Rand 2000). Analogous studies on adults of other species, or on juvenile migrants, are lacking.

Energetics

Much more is known about the energetics of adult spawning migrations than those of juveniles. This is particularly the case in adult anadromous fish for whom body energy analyses are simplified because they rarely eat and only use reserve energy to fuel their upstream migration. Across stocks and species, typically 50–70% of initial energy is used to complete such migration (Brett 1995; Jonsson *et al.* 1997; Leonard

& McCormick 1999). Energy use is accelerated when migrants encounter elevated temperature, elevated flow rates, or river constrictions. The latter two usually result in higher encountered water velocities (Hinch & Rand 1998; Rand & Hinch 1998). Energy exhaustion is believed to be a factor causing pre-spawning mortality in some stocks (Rand & Hinch 1998; Leonard & McCormick 1999). In iteroparous stocks (spawning several times over lifespan), post-spawning survival rates are negatively related to energetic losses during spawning (Jonsson *et al.* 1997). Bernatchez & Dodson (1987) reviewed energy use studies of both iteroparous and semelparous (spawn only once over lifetime) species and summarized energetic costs of 15 anadromous spawning stocks that migrate upriver distances ranging from 30 to 1200 km. Short distance migrants (<100 km; river lamprey (*Lampetra fluviatilus*), cisco (*Coregonus artedii*), shad (*Alosa sapidissima*) and alewife (*A. pseudoharengus*)) were the least efficient in energy use (all <20 kJ kg^{-1} km^{-1}), whereas long distance migrants (>1000 km; chinook salmon (*O. tshawytscha*), sockeye salmon and chum salmon) were most efficient (all <6 kJ kg^{-1} km^{-1}). This pattern of increasing energy conservation with increasing migratory difficulty has also been observed in a multi-stock examination within one species and one watershed (Fraser River sockeye salmon; Crossin 2002).

In contrast with adults, relatively little is known about energetic costs of juvenile fish migrations. Unlike most adults during spawning migrations, juveniles do feed during their migrations (e.g. Brett 1995). Increasing (or at least not losing) weight is extremely important for maximizing chances of survival by juveniles when entering their new environments. Energetic state is an important determinant of migration initiation in juvenile salmon. A survey of juvenile Atlantic salmon populations revealed that local environmental conditions, specifically water temperature and day length, were good predictors of smoltification (Metcalfe & Thorpe 1990). These conditions reflect changes in growth opportunity and the potential for large energetic gains by migrating to a new environment. For instance, energy availability to young anadromous brown trout was 10% greater than same-aged non-anadromous residents (Jonsson & Jonsson 1998).

However, environmental conditions can be detrimental if anthropogenically altered. For example, when stream temperatures increased in Carnation Creek, British Columbia, following logging, coho (*O. kisutch*) and chum (*O. keta*) salmon emerged several weeks earlier than normal, entering the ocean at a time when coastal survival rates were poor – thus the number of returning adults was negatively affected by the juvenile migratory response to temperature (Hartman & Scrivener 1990). Presumably the energetic costs of juvenile downstream migrations are low as these fish can seemingly move passively, except where lakes or reservoirs are encountered. Indeed, river flow speeds positively correlate with travel times of downstream salmonid smolts (Berggren & Filardo 1993). Yet juveniles are not 'involuntarily' carried downstream. The prolonged swimming speeds of juveniles often exceed the slowest current speeds of the rivers they migrate in (Peake & McKinley 1998). Moreover, juveniles are capable of swimming upstream in rivers with strong currents by using reverse-flow fields near the river banks (McLaughlin & Noakes 1997), similar to the tactic used by adult migrants (see above).

Migration obstructions and barriers

Obstructions (features that slow a migration) and barriers (those that stop a migration) occur both naturally and anthropogenically. Natural fluctuations in turbidity, dissolved oxygen, temperature and discharge can create minor obstructions and cause slight delays in migrations (reviewed by Bjornn & Reiser 1991). However, most stocks have evolved under these varying natural conditions and may not be seriously affected by modest levels of obstruction. Because upstream spawning migrations occur at discrete and precise times each year, more substantial delays could affect proper timing of reproduction, or impair spawning by depleting energy reserves (see above), increasing stress, or increasing susceptibility to disease. Natural barriers, such as waterfalls, rapids or in some cases debris jams, may prevent upstream migratory access, but such barriers are often species- or season-specific (Bjornn & Reiser 1991). There are few natural barriers to downstream migrations. This is particularly true for most juvenile migrations in temperate regions, which are timed to coincide with spring runoff when water temperatures are usually modest and stream flows adequate for passage. Nevertheless, there may be many man-made obstructions even to downstream migration caused by hydroelectric and other types of dams.

All of the natural obstructions listed above can be affected by forestry practices (Chapter 13). For instance, logging can increase stream turbidity, reduce dissolved oxygen, increase stream temperature and reduce apparent summer stream flows and water velocities. Reduction in the latter two could facilitate a spawning migration if it means that energy can be saved because current speeds are reduced. In some cases, streams can become entirely dewatered due to logging-induced landslides and bank erosion, thereby eliminating migration routes and spawning areas. Furthermore, small bridges and road culverts – if partially plugged with debris – can impede fish passage upstream and downstream.

Perhaps the greatest environmental impact of forestry on fish migration is that caused by stream crossings. These can have a variety of effects, but most importantly for migrations, they can create water velocity obstructions or barriers. The scope of these problems can be appreciated when one considers the number of crossings in existence. In British Columbia alone there are over 225,000 stream crossings with 10,000 new ones being installed each year (Harper & Quigley 2000). The most common crossing types used in a forestry context are round or oval corrugated metal pipes (the standard culvert), open/natural bottom structures made of logs, metal or concrete (open box culvert), variants of these with 'baffles' or other forms of surface 'roughness' (Bates & Powers 1998), and bridges. In some regions, slab and ford crossings (roads built through the stream) are also present (Warren & Pardew 1998). Pipe culverts are by far the most common extant crossings, and because they are the cheapest of structures, they are still the ones most frequently being installed today.

Structures that constrict the channel and accelerate water speeds can cause barriers for migrations by creating water velocities that exceed fish maximum speeds or endurance. By scouring out the streambed at the downstream end of the structure thereby 'perching' it with drops to the pool below often well over a metre, access by fish can be prevented. Pipe culverts are notorious for creating these barriers. In recent surveys, 25% of pipe culverts in British Columbia were impassable by salmonids (Harper & Quigley 2000), and in Arkansas, USA, pipe culverts acted as barriers to the movements of nine different fish species (mostly cyprinids, catostomids and fundulids; Warren & Pardew 1998).

An understanding of species- and size-specific prolonged and burst swimming speeds is required as a starting point to predict whether fish can pass upstream through a culvert that is not perched. Indeed, we know a considerable amount about prolonged speeds for several species and size classes (e.g. Jones *et al.* 1974; Scruton *et al.* 1998; Toepfer *et al.* 1999). Moreover, numerous equations have been developed by engineers to understand and predict water velocities and discharges from culverts with differing size and shape characteristics (e.g. Bates *et al.* 1999). Thus, with a basic understanding of hydraulics and swimming performance, it should be possible to design and install culverts of dimensions that will not generate velocities that exceed the swimming endurance of the species and size of concern. As highlighted by the recent culvert surveys described above, this has turned out to be much more difficult than was once believed.

With few recent exceptions (e.g. Belford & Gould 1989), most swimming performance experiments used to predict culvert swimming abilities are conducted in swimming tunnels (tubes, flumes) where fish swim in the light against water that is propelled past them. These apparatuses are designed to have no channel turbulence and no gradient, and are a poor mimic of a long dark corrugated culvert. In addition, fish that choose to not swim in a trial are normally removed from the study. Standard corrugated culverts have considerably more surface roughness, boundary layers and turbulence than swimming tunnels and these features can affect passage success (Bates & Powers 1998). Small fish may be able to utilize boundary layers for passage but turbulence could also overpower them (Bates & Powers 1998). Large-scale river turbulence is believed to cause confusion in adult migrants (Hinch & Rand 1998).

New culvert designs include baffles or bed material embedded into their bottoms. The belief is that these designs aid passage by 'weak-swimming' fish, those with naturally low prolonged or burst swimming speeds, and enable 'strong-swimming' fish to utilize lower swimming speeds and thus conserve energy (reviewed in Poulin & Argent 1997). However, little biological evaluation has been made of these culvert modifications, and in general, we know very little about how turbulence influences fish swimming performance. We also know very little about burst swimming in fish, which they no doubt use frequently to pass through standard pipe culverts. Thus we must be more cautious when applying laboratory-derived swimming criteria to culvert design and selection – the approach that most management agencies have adopted (e.g. Poulin & Argent 1997). These criteria should be set very conservatively. There have been surprisingly few rigorous studies (except see Belford & Gould 1989; Bates & Powers 1998; Warren & Pardew 1998) that have examined swimming abilities of fish through actual or simulated pipe culverts, and many more are needed. Until then, the most 'risk-adverse' approach in culvert choice to minimize chances of creating a migratory obstruction would be to use natural bottomed culverts that span the entire channel width or else bridges (Harper & Quigley 2000).

Concluding comments

Freshwater fish migration, a cyclic alternation between two and often three or more essential habitats – spawning, feeding and survival (under severe conditions) – normally occurs at particular life history stages with regular seasonal and annual timing. It usually involves a large fraction of any given population, although not necessarily all individuals. And the cycle commonly includes energetically demanding and directed movement upstream against current at one stage along with downstream less demanding movement or even drift at another stage. Overall, however, the cycle is key to survival of the species that engage in this remarkable behaviour. Sometimes the complete cycle occurs over thousands of kilometres, but distance is no measure of importance to the populations involved, so migrations of a few kilometres or even a few tens of metres can be just as vital to their long-term survival.

Two groups of freshwater fishes, salmonids and eels, are universally recognized to be phenomenal migrants. Both are abundant and important in forested watersheds over many regions of the world. Even with these groups there is still much to be learned about their migratory behaviour, the many mechanisms used in completing their migratory cycle, their energetic capabilities and tactics, and particularly the effects of forestry practices on their migrations. We are now becoming aware that migration is an important feature in the life history of nearly 40 different groups of freshwater fishes, a number of which occur in arctic, subarctic, temperate, subtropical and tropical areas of the world, often forested or partially so. For many of these fishes we only have rudimentary knowledge of their migrations, surely not enough to effectively prevent their severe loss resulting from forest removal taking place over much of their inland habitats.

The need for more relevant migratory research is great and this will be featured along with other needs in the concluding chapters (Part VIII) of this book. The message must be effectively presented to government agencies locally, nationally and internationally, and to the forest industry. Furthermore the problems and means for their solution also need to be given clearly to relevant professionals, and those in training, in both forestry and fisheries. Together, with appropriate knowledge and funding, they will work out the ways to maintain the many important migratory fish populations in forested watersheds of the world.

Acknowledgements

We are most grateful for the careful reviews at the manuscript stage provided by Drs K.L. Howland, M. Shrimpton and W.M. Tonn. Dr G.F. Hartman gave helpful suggestions in final editorial review.

References

Armstrong, R.H. & Morrow, J.E. (1980) The Dolly Varden charr. In: *Charrs: Salmonid Fishes of the Genus Salvelinus*, (ed. E.K. Balon), pp. 99–140. Dr W. Junk, The Hague, The Netherlands.

Bainbridge, R. (1958) The swimming speed of fish as related to size and to the frequency and amplitude of the tail beat. *Journal of Experimental Biology*, **35**, 109–33.

Barbin, G.P. (1998) The role of olfaction in homing and estuarine migratory behavior of yellow-phase American eels. *Canadian Journal of Fisheries and Aquatic Sciences*, **55**, 564–75.

Barinaga, M. (1999) Salmon follow watery odors home. *Science*, **286**, 705–6.

Bates, K. & Powers, P. (1998) Upstream passage of juvenile coho salmon through roughened culverts. In: *Fish Migration and Fish Bypasses*, (eds M. Jungwirth, S. Schmutz & S. Weiss), pp. 192–202. Blackwell Science, Oxford.

Bates, K., Bernard, R., Heiner, B., Klava, P. & Powers, P. (1999) *Fish Passage Design at Road Culverts: A Design Manual for Fish Passage at Road Crossings*. Washington Department of Fish and Wildlife, Habitat and Lands Program, Environmental Engineering Division, Olympia, Washington.

Beamish, F.W.H. (1978) Swimming capacity. In: *Fish Physiology*, Vol. 7, (eds W.S. Hoar & D.J. Randall), pp. 101–86. Academic Press, New York.

Belford, D.A. & Gould, W.R. (1989) An evaluation of trout passage through six highway culverts in Montana. *North American Journal of Fisheries Management*, **9**, 437–45.

Berggren, T.J. & Filardo, M.J. (1993) An analysis of variables influencing the migration of juvenile salmonids in the Columbia River basin. *North American Journal of Fisheries Management*, **13**, 48–63.

Bernard, D.R., Mepler, K.R., Jones, J.D., Whalen, M.E. & McBride, D.N. (1995) Some tests of the 'migration hypothesis' for anadromous Dolly Varden (southern form). *Transactions of the American Fisheries Society*, **124**, 297–307.

Bernatchez, L. & Dodson, J.J. (1987) Relationship between bioenergetics and behaviour in anadromous fish migrations. *Canadian Journal of Fisheries and Aquatic Sciences*, **44**, 399–407.

Bjornn, T.C. & Reiser, D.W. (1991) Habitat requirements of salmonids in streams. In: *Influences of Forest and Rangeland Management on Salmonid Fishes and their Habitats*, (ed. W.R. Meehan), pp. 83–138. American Fisheries Society Special Publication 19.

Blaber, S.J.M. (1997) *Fish and Fisheries of Tropical Estuaries*. Chapman & Hall, New York.

Black, E.C. (1958) Hyperactivity as a lethal factor in fish. *Journal of the Fisheries Research Board of Canada*, **15**, 573–86.

Bond, W.A. & Erickson, R.V. (1992) Anadromous coregonids of a Canadian arctic estuary. *Polskie Archiwum Hydrobiologii*, **39**, 431–41.

Bond, W.A. & Erickson, R.V. (1997) Coastal migrations of Arctic ciscoes in the eastern Beaufort Sea. *American Fisheries Symposium*, **19**, 155–64.

Bramblett, R.G., Bryant, M.D., Wright, B.E. & White, R.G. (2002) Seasonal use of small tributary and main-stem habitats by juvenile steelhead, coho salmon, and Dolly Varden in a southeastern Alaska drainage basin. *Transactions of the American Fisheries Society*, **131**, 498–506.

Brett, J.R. (1967) Swimming performance of sockeye salmon (*Oncorhynchus nerka*) in relation to fatigue time and temperature. *Journal of the Fisheries Research Board of Canada*, **24**, 1732–41.

Brett, J.R. (1995) Energetics. In: *Physiological Ecology of Pacific Salmon*, (eds C. Groot, L. Margolis & W.C. Clarke), pp. 1–68. UBC Press, Vancouver.

Brett, J.R. & Glass, N.R. (1973) Metabolic rates and critical swimming speeds of sockeye salmon (*Oncorhynchus nerka*) in relation to size and temperature. *Journal of the Fisheries Research Board of Canada*, **30**, 379–87.

Bunt, C.M., Katopodis, C. & McKinley, R.S. (1999) Attraction and passage efficiency of white suckers and smallmouth bass by two Denil fishways. *North American Journal of Fisheries Management*, **19**, 793–801.

Crossin, G. (2002) *Effects of ocean climate and upriver migratory constraints on the bioenergetics, fecundity, and morphology of wild, Fraser River salmon*. MSc thesis, University of British Columbia.

Curry, R.A., Sparks, D. & van de Sande, J. (2002) Spatial and temporal movements of a riverine brook trout population. *Transactions of the American Fisheries Society*, **131**, 551–60.

Dittman, A.H., Quinn, T.P. & Nevitt, G.A. (1996) Timing of imprinting to natural and artificial odors by coho salmon (*Oncorhynchus kisutch*). *Canadian Journal of Fisheries and Aquatic Sciences*, **53**, 434–42.

Døving, K.B., Nordeng, H. & Oakley, B. (1974) Single unit discrimination of fish odours released by char (*Salmo alpinus* L.) populations. *Comparative Biochemistry and Physiology A. Comparative Physiology*, **45A**, 1051–63.

Fox, D.A., Hightower, J.E. & Paruka, F.M. (2000) Gulf sturgeon spawning migration and habitat in the Choctawatchee River system, Alabama-Florida. *Transactions of the American Fisheries Society*, **129**, 811–26.

Geen, G.H., Northcote, T.G., Hartman, G.F. & Lindsey, C.C. (1966) Life histories of two species of catostomid fishes in Sixteenmile Lake, British Columbia, with particular reference to inlet stream spawning. *Journal of the Fisheries Research Board of Canada*, **23**, 1761–88.

Gill, H.S., Weatherley, A.H., Lee, R. & Legere, D. (1989) Histochemical characterization of myotomal muscle of five teleost species. *Journal of Fish Biology*, **34**, 375–86.

Groot, C. (1982) Modifications on a theme – a perspective on migratory behavior of Pacific salmon. In: *Proceedings of the Salmon and Trout Migratory Behavior Symposium*, (eds E.L. Brannon & E.O. Salo), pp. 1–21. University of Washington, Seattle.

Groot, C., Quinn, T.P. & Hara, T.J. (1986) Responses of migrating adult sockeye salmon (*Oncorhynchus nerka*) to population-specific odours. *Canadian Journal of Zoology*, **64**, 926–32.

Gulseth, O.A. & Nilssen, K.J. (2000) The brief period of spring migration, short marine residence, and high return rate of a northern Svalbard population of Arctic char. *Transactions of the American Fisheries Society*, **129**, 782–96.

Hammar, J. (1987) Zoogeographical zonation of fish communities in insular Newfoundland: a preliminary attempt to use the Arctic char population ecology to describe early postglacial colonization interactions. *International Society of Arctic Char Fanatics Information Series*, **4**, 31–8.

Harper, D.J. & Quigley, J.T. (2000) No net loss of fish habitat: an audit of forest road crossings of fish-bearing streams in British Columbia, 1996–1999. *Canadian Technical Report of Fisheries and Aquatic Sciences 2319*.

Hartman, G.F. & Brown, T.G. (1987) Use of small, temporary, floodplain tributaries by juvenile salmonids in a west coast rain-forest drainage basin, Carnation Creek, British Columbia. *Canadian Journal of Fisheries and Aquatic Sciences*, **44**, 262–70.

Hartman, G.F. & Scrivener, J.C. (1990) Impacts of forestry practices on a coastal stream ecosystem, Carnation Creek, British Columbia. *Canadian Bulletin of Fisheries and Aquatic Sciences*, **223**.

Hasler, A.D. & Scholtz, A.T. (1983) *Olfactory Imprinting and Homing in Salmon*. Springer-Verlag, Berlin.

Hinch, S.G. & Rand, P.S. (1998) Swim speeds and energy use of upriver-migrating sockeye salmon (*Oncorhynchus nerka*): role of local environment and fish characteristics. *Canadian Journal of Fisheries and Aquatic Sciences*, **55**, 1821–31.

Hinch, S.G. & Rand, P.S. (2000) Optimal swim speeds and forward assisted propulsion: energy conserving behaviours of up-river migrating salmon. *Canadian Journal of Fisheries and Aquatic Sciences*, **57**, 2470–8.

Hinch, S.G., Standen, E.M., Healey, M.C. & Farrell, A.P. (2002) Swimming patterns and behaviour of upriver migrating adult pink (*Oncorhynchus gorbuscha*) and sockeye (*O. nerka*) salmon as assessed by EMG telemetry in the Fraser River, British Columbia, Canada. *Hydrobiologia*, **483**, 147–60.

Hiscock, M.J., Scruton, D.A., Brown, J.A. & Clarke, K.D. (2002) Winter movement of radio-tagged juvenile Atlantic salmon in Northeast Brook, Newfoundland. *Transactions of the American Fisheries Society*, **131**, 577–81.

Hodgson, J.R., Schindler, D.E. & He, X. (1998) Homing tendency of three piscivorous fishes in a north temperate lake. *Transactions of the American Fisheries Society*, **127**, 1078–81.

Howland, K.L., Tallman, R.F. & Tonn, W.M. (2000) Migration patterns of freshwater and anadromous inconnu in the Mackenzie River system. *Transactions of the American Fisheries Society*, **129**, 41–59.

Hughes, N.F. (1999) Population processes responsible for larger-fish-upstream distribution patterns of Arctic grayling (*Thymallus arcticus*) in interior Alaskan runoff rivers. *Canadian Journal of Fisheries and Aquatic Sciences*, **56**, 2292–9.

Jones, D.R., Kiceniuk, J.W. & Bamford, O.S. (1974) Evaluation of the swimming performance of several fish species from the Mackenzie River. *Journal of the Fisheries Research Board of Canada*, **31**, 1641–7.

Jonsson, B. & Jonsson, N. (1993) Partial migration: niche shift versus sexual maturation in fishes. *Reviews in Fish Biology and Fisheries*, **3**, 348–65.

Jonsson, N. & Jonsson, B. (1998) Body composition and energy allocation in life-history stages of brown trout. *Journal of Fish Biology*, **53**, 1306–16.

Jonsson, N., Jonsson, B. & Hansen, L.P. (1997) Changes in proximate composition and estimates of energetic costs during upstream migration and spawning in Atlantic salmon. *Journal of Animal Ecology*, **66**, 425–36.

Kinzie, R.A. (1991) Hawaiian freshwater ichthyofauna. In: New directions in research, management and conservation of Hawaiian freshwater stream ecosystems. *Proceedings of the 1990 Symposium on Freshwater Stream Biology and Fisheries Management*, pp. 18–39. Division of Aquatic Resources, Department of Land and Natural Resources, Honolulu, Hawaii.

Knight, C.A., Orme, R.W. & Beauchamp, L.P. (1999) Growth, survival and migration patterns of juvenile adfluvial Bonneville cutthroat trout in tributaries of Strawberry Reservoir, Utah. *Transactions of the American Fisheries Society*, **128**, 553–63.

Knights, B.C., Vallazza, J.M., Zigler, S.J. & Dewey, M.R. (2002) Habitat and movement of lake sturgeon in the Upper Mississippi River system, USA. *Transactions of the American Fisheries Society*, **131**, 507–22.

L'Abe'e-Lund, J.H. & Vollestad, A. (1985) Homing precision of roach *Rutilus rutilus* in Lake Arungen, Norway. *Environmental Biology of Fishes*, **13**, 235–9.

Lein, G.M. & De Vries, D.R. (1998) Paddlefish in the Alabama River drainage: population characteristics and the adult spawning migration. *Transactions of the American Fisheries Society*, **127**, 441–54.

LeLouarn, H., Bagliniere, J.L., Marchand, F. & Hamonet, J.M. (1997) Caracteristiques biologiques et ecologiques du chevaine (*Leuciscus cephalus*) dans quelques rivieres de la facade atlantique francaise. *Bulletin Scientifique et Technique Institut National de la Recherche Departement d'Hydrobiologie (France)* no. 29; 22.

Leonard, J.B.K. & McCormick, S.D. (1999) Effects of migration distance on whole-body and tissue-specific energy use in American shad (*Alosa sapidissima*). *Canadian Journal of Fisheries and Aquatic Sciences*, **56**, 1159–71.

Levy, D.A. (1990) Reciprocal diel vertical migration behavior in planktivores and zooplankton in British Columbia lakes. *Canadian Journal of Fisheries and Aquatic Sciences*, **47**, 1755–64.

Levy, D.A. (1991) Acoustic analysis of diel vertical migration behavior of *Mysis relicta* and kokanee (*Oncorhynchus nerka*) within Okanagan Lake, British Columbia. *Canadian Journal of Fisheries and Aquatic Sciences*, **48**, 67–72.

Levy, D.A. & Cadenhead, A.D. (1995) Selective tidal stream transport of adult sockeye salmon (*Oncorhynchus nerka*). *Canadian Journal of Fisheries and Aquatic Sciences*, **52**, 1–12.

Lindsey, C.C. & Northcote, T.G. (1963) Life history of redside shiners, *Richardsonius balteatus*, with particular reference to movements in and out of Sixteenmile Lake streams. *Journal of the Fisheries Research Board of Canada*, **20**, 1001–30.

Lucas, M.C. & Baras, E. (2001) *Migration of Freshwater Fishes*. Blackwell Science, Oxford.

McAda, C.W. & Kaeding, L.R. (1991) Movements of adult Colorado squawfish during the spawning season in the upper Colorado River. *Transactions of the American Fisheries Society*, **120**, 339–45.

McCleave, J.D., Arnold, G.P., Dodson, J.J. & Neill, W.H. (1984) *Mechanisms of Migration in Fishes*. Plenum Press, New York.

McCormick, S.D., Hansen, L.P., Quinn, T.P. & Saunders, R.L. (1998) Movement, migration, and smolting of Atlantic salmon (*Salmo salar*). *Canadian Journal of Fisheries and Aquatic Sciences*, **55** (Suppl. 1), 77–92.

McDowall, R.M. (1988) *Diadromy in Fishes: Migrations between Freshwater and Marine Environments*. Croom Helm, London.

McDowall, R.M. (1996) Diadromy and the assembly and restoration of riverine fish communities: a downstream view. *Canadian Journal of Fisheries and Aquatic Sciences*, **53** (Suppl.1), 219–36.

McLaughlin, R.L. & Noakes, D.L.G. (1997) Going against the flow: an examination of the propulsive movements made by young brook trout in streams. *Canadian Journal of Fisheries and Aquatic Sciences*, **55**, 853–60.

Mesa, M.G. & Olson, T.M. (1993) Prolonged swimming performance of northern squawfish. *Transactions of the American Fisheries Society*, **122**, 1104–10.

Metcalfe, N.B. & Thorpe, J.E. (1990) Determinants of geographical variation in the age of seaward migrating salmon, *Salmo salar*. *Journal of Animal Ecology*, **59**, 135–45.

Murphy, M.L., Koski, K.V., Lorenz, J.M. & Thedinga, J.F. (1997) Downstream migrations of juvenile Pacific salmon (*Oncorhynchus* spp.) in a glacial transboundary river. *Canadian Journal of Fisheries and Aquatic Sciences*, **54**, 2837–46.

Narver, D.W. (1970) Diel vertical movements of under-yearling sockeye salmon and the limnetic zooplankton in Babine Lake, British Columbia. *Journal of the Fisheries Research Board of Canada*, **27**, 281–316.

Nelson, J.S. (1994) *Fishes of the World*, 3rd edn. John Wiley & Sons, New York.

Northcote, T.G. (1967) The relation of movements to production in freshwater fishes. In: *The Biological Basis of Freshwater Fish Production*, (ed. S.D. Gerking), pp. 315–344. Blackwell Science, Oxford.

Northcote, T.G. (1969) Patterns and mechanisms in the lakeward migratory behaviour of juvenile trout. In: *Symposium on Salmon and Trout in Streams*, pp. 183–203. H.R. MacMillan Lectures in Fisheries, University of British Columbia, Vancouver.

Northcote, T.G. (1978) Migratory strategies and production in freshwater fishes. In: *Ecology of Freshwater Fish Production*, (ed. S.D. Gerking), pp. 326–59. Blackwell Science, Oxford.

Northcote, T.G. (1984) Mechanisms of fish migration in rivers. In: *Mechanisms of Migration in Fishes*, (eds J.D. McCleave, G.P. Arnold, J.J. Dodson & W.H. Neill), pp. 317–55. Plenum Press, New York.

Northcote, T.G. (1991) Migration. In: *Trout*, (eds J. Stolz & J. Schnell), pp. 84–95. Stackpole Books, Harrisburg, PA.

Northcote, T.G. (1997) Potamodromy in Salmonidae – living and moving in the fast lane. *North American Journal of Fisheries Management*, **17**, 1029–45.

Northcote, T.G. (1998) Migratory behaviour of fish and its significance to movement through fish passage facilities. In: *Fish Migration and Fish Bypasses*, (eds M. Jungwirth, S. Schmutz & S. Weiss). Blackwell Science, Oxford.

Osmundson, D.B., Ryel, R.J., Tucker, M.E., Burdick, B.D., Elmblad, W.R. & Chart, T.E. (1998) Dispersal patterns of subadult and adult Colorado squawfish in the upper Colorado River. *Transactions of the American Fisheries Society*, **127**, 943–56.

Peake, S. & McKinley, R.S. (1998) A re-evaluation of swimming performance in juvenile salmonids in relation to downstream migration. *Canadian Journal of Fisheries and Aquatic Sciences*, **55**, 682–7.

Peake, S., McKinley, R.S. & Scruton, D.A. (1997) Swimming performance of various freshwater Newfoundland salmonids relative to habitat selection and fishway design. *Journal of Fish Biology*, **42**, 710–23.

Peake, S., McKinley, R.S. & Scruton, D.A. (2000) Swimming performance of walleye (*Stizostedion vitreum*). *Canadian Journal of Zoology*, **78**, 1686–90.

Poulin, V.A. & Argent, H.W. (1997) *Stream Crossing Guidebook for Fish Streams: A Working Draft for 1997/1998*. Prepared for BC Ministry of Forests; Ministry of Employment and Investment; Ministry of Environment, Lands and Parks; Canada Department of Fisheries and Oceans.

Rand, P.S. & Hinch, S.G. (1998) Swim speeds and energy use of river migrating adult sockeye salmon: simulating metabolic power and assessing risk of energy depletion. *Canadian Journal of Fisheries and Aquatic Sciences*, **55**, 1832–41.

Reist, J.D. & Bond, W.A. (1988) Life history characteristics of migratory coregonids of the lower Mackenzie River, Northwest Territories, Canada. *Finnish Fisheries Research*, **9**, 133–44.

Richardson, J., Rowe, D.K. & Smith, J.P. (2001) Effects of turbidity on the migration of juvenile banded kokopu (*Galaxias fasciatus*) in a natural stream. *New Zealand Journal of Marine and Freshwater Research*, **35**, 191–6.

Sato, K., Shoji, T. & Ueda, H. (2000) Olfactory discriminating ability of lacustrine sockeye and masu salmon in various freshwaters. *Zoological Science*, **17**, 313–17.

Schindler, D.E. (1999) Migration strategies of young fishes under temporal constraints: the effect of size-dependent overwinter mortality. *Canadian Journal of Fisheries and Aquatic Research*, **56** (Suppl. 1), 61–70.

Scott, W.B. & Crossman, E.J. (1973) *Freshwater Fishes of Canada, Bulletin 184*. Fisheries Research Board of Canada, Ottawa.

Scruton, D.A., McKinley, R.S., Booth, R.K., Peake, S.J. & Goosney, R.F. (1998) Evaluation of swimming capability and potential velocity barrier problems for fish. Part A. Swimming performance of selected warm and cold water fish species relative to fish passage and fishway design. *CEA Project 9236 G 1014*. Montreal, Quebec.

Smith, R.J.F. (1985) *The Control of Fish Migration*. Springer-Verlag, Berlin.

Steinhart, G.B. & Wurtsbaugh, W.A. (1999) Under-ice diel vertical migrations of *Oncorhynchus nerka* and their zooplankton prey. *Canadian Journal of Fisheries and Aquatic Research*, **56** (Suppl. 1), 152–61.

Toepfer, C.S., Fisher, W.L. & Haubelt, J.A. (1999) Swimming performance of the threatened leopard darter in relation to road culverts. *Transactions of the American Fisheries Society*, **128**, 155–61.

Tyus, H.M. (1990) Potamodromy and reproduction of Colorado squawfish in the Green River Basin, Colorado and Utah. *Transactions of the American Fisheries Society*, **119**, 1035–47.

Ueda, H. (1998) Correlations between homing, migration, and reproduction of chum salmon. *North Pacific Anadromous Fish Commission Bulletin*, **1**, 112–17.

Veinott, G., Northcote, T.G., Rosenau, M. & Evans, R.D. (1999) Concentrations of strontium in the pectoral fin rays of white sturgeon (*Acipenser transmontanus*), by laser ablation sampling-inductively coupled plasma-mass spectrometry as an indicator of marine migrations. *Canadian Journal of Fisheries and Aquatic Sciences*, **56**, 1981–90.

Ward, D.L., Maughan, O.E. & Bonar, S.A. (2002) Effects of temperature, fish length, and exercise on swimming performance of age-0 flannelmouth sucker. *Transactions of the American Fisheries Society*, **131**, 492–7.

Ware, D.M. (1975) Growth, metabolism, and optimal swimming speed of pelagic fish. *Journal of the Fisheries Research Board of Canada,* **32**, 33–41.

Warren, M.L. & Pardew, M.G. (1998) Road crossings as barriers to small-stream fish movement. *Transactions of the American Fisheries Society,* **127**, 637–44.

Webb, P.W. (1984) Form and function in fish swimming. *Scientific American,* **251**, 72–82.

Webb, P.W. (1995) Locomotion. In: *Physiological Ecology of Pacific Salmon,* (eds C. Groot & W.C. Clarke), pp. 71–99. UBC Press, Vancouver.

Weihs, D. (1973) Optimal fish cruising speed. *Nature,* **245**, 48–50.

Weihs, D. (1974) Energetic advantages of burst swimming of fish. *Journal of Theoretical Biology,* **48**, 215–29.

Weihs, D. (1975) Some hydrodynamic aspects of fish schooling. In: *Swimming and Flying in Nature,* Vol. 2, (eds T.Y.T. Wu, C.J. Browkaw & C. Brennen), pp. 703–18. Plenum Press, New York.

Chapter 8
Aspects of fish reproduction and some implications of forestry activities

G.F. HARTMAN AND T.E. MCMAHON

Introduction

The maintenance of fish populations, and the values associated with them, are dependent upon successful reproduction. This activity is crucial to fishes, and the effects of forestry upon it are important. Its full coverage among freshwater fishes would be a huge undertaking. Therefore, this chapter intends to introduce the reader, particularly the non-biologist, to some aspects of reproduction among fishes so that they can appreciate why it may be necessary to apply protective measures during forestry operations.

Although species vary widely in their reproductive requirements, all share basic needs that must be met for successful reproduction. Their young must:

(1) develop and enter the world at an optimal time for feeding;
(2) encounter favourable temperatures and water flows and have sufficient oxygen concentrations for successful development; and
(3) be subjected to minimal predation (Munro 1990a).

Requirements for these conditions, described as 'ultimate factors' (Huntela & Stacey 1990), apply for the great numbers of species of fish in tropical or temperate zones. They are met in an array of reproductive strategies (Breder & Rosen 1966).

Natural selection is based on the three sets of conditions (above) which maximize survival of the offspring. It is also based upon 'proximate factors', i.e. internal rhythms and the external factors that trigger spawning at the optimal time (Huntela & Stacey 1990). These selective processes shape the reproductive strategies of species and, accordingly, their life histories. Despite each species' unique reproductive strategy, it has been possible to describe the reproductive behaviour of groups of fishes (Breder & Rosen 1966; Munro 1990b) and group those that have evolved common strategies into 'reproductive guilds' (Balon 1975, 1981).

Within these broad groupings there are specific elements, such as egg and sperm development, spawning behaviour, incubation and hatching, emergence from the substrate (in some species) and life during the succeeding months, that may be affected by various forestry activities. Forestry effects will be diverse depending upon the kinds of forestry impact and the features of reproduction involved. A concise synthesis of such factors is not simple, given that there are 34 orders and approximately 10,000 species of freshwater fishes worldwide (Chapter 1), and the reproductive biology of most of them is poorly understood. However, a basis for understanding and predicting responses to forestry activities can be achieved by classifying reproductive guilds according to their sensitivity to impacts from forestry operations, i.e. increased erosion, sedimentation, turbidity, and water temperature, decreased shading and simplification of the environment (Austen *et al.* 1994; Scott & Helfman 2001).

In this chapter, we attempt to link knowledge of reproductive requirements and reproductive diversity to forest harvest assessment by:

(1)   describing how key features of egg fertilization and embryonic development may respond to changes in environmental controlling factors such as temperature;
(2)   examining the diversity of reproductive strategies and their relative susceptibility to forestry activities;
(3)   using detailed information about the response of salmonids to forest harvest to illustrate the types of information required to fully assess the effects of forestry activities in other, less studied species or groups;
(4)   providing general guidelines for assessing risk to reproductive success.

Egg and embryo development and hatching

The process of reproduction begins with the sexual maturation of the fish and the formation of reproductive products. 'Reproductive products' are commonly categorized as eggs (unfertilized ova), embryos (fertilized, developing eggs enclosed within the egg membrane) and larvae (hatched embryos). Readers are referred to Hoar & Randall (1988) for extensive treatment of embryonic development processes.

Control of development processes

The timing of development of eggs and sperm, the behaviour leading up to and including spawning, the fertilization process and the development of embryos are all under the control of interacting hormonal systems, endogenous rhythms and seasonal environmental cues. For review of these topics, the reader is referred to some key references (Blaxter 1969; Hoar 1969; Idler *et al.* 1987; Hoar & Randall 1988; Walther & Fyhn 1993; Goetz & Thomas 1995).

Reproductive success among fishes may be affected by environmental conditions that impair the processes of egg development, incubation (embryo development) and hatching. Environmental conditions occurring long before spawning may affect reproductive success because the sequence of hormonal events controlling oocyte development begins, in some cases, many months before ovulation (Sumpter *et al.* 1984). Changes in temperature, photoperiod and food supply may affect liver function, which ultimately governs oocyte and sperm development (Mommsen & Walsh 1988). Therefore, human activities that alter oxygen concentration, temperature or water quality may affect reproduction at its earliest stages.

Embryo development

Immediately following fertilization, the embryo begins a sequence of developmental changes that continue through and beyond hatching (Hoar & Randall 1988; Blaxter 1969, 1988). Among fishes, there is a great variety of patterns of development, involving variability in egg size, size of larvae at hatching, time from fertilization to hatching, size of larvae at the first feeding and timing of critical developmental periods (Blaxter 1988).

During some periods of the development sequence the embryo is particularly sensitive to hypoxia, adverse temperature and physical shock. These sensitivities may vary among species (Blaxter 1988). For example, Balon (1985) illustrated important inter-specific differences in the development of respiratory systems during embryonic development, which are critical in determining responses to hypoxia, and ultimately, embryo survival. Because forestry operations in any part of the world may alter water temperature and oxygen conditions, or create physical disturbance, these impacts and their potential effects on eggs and embryos should be considered during forestry operations.

Critical sensitivity periods

Among all groups of fishes, critical periods during embryonic and early larval development may occur at hatching, first feeding, development of gills, filling of swim-bladder and metamorphosis of larvae (Blaxter 1988).

During embryonic development, requirements and sensitivities to oxygen and temperature vary depending upon the fish species and stage of development. Rombough (1988) summarized lower lethal and no-effect levels of dissolved oxygen for teleost fishes. Increased sensitivities to low oxygen and elevated temperature were reported for salmonids, coregonids (*Coregonus albula*) and cyprinids (*Cyprinus carpio*) at gastrulation and immediately before hatching (Kamler 1992). Some species experience critical periods at the onset of exogenous feeding (Kamler 1992).

Development to hatching of salmon and trout embryos, and presumably those of most other species of fish, is dependent on accumulated temperature units (ATUs, or 'degree days', number of days × average temperature). Although developing eggs require temperatures within specific ranges, time from fertilization to hatching decreases as temperature increases. Blaxter (1988; Fig. 18) illustrated this relationship for 11 species of fish.

The number of ATUs required varies somewhat with the ambient temperature during development (Alderdice & Velsen 1978). Billard & Jensen (1996) illustrated this by developing different mathematical models that predicted required ATUs for salmon and steelhead trout egg incubation. Development times from fertilization to 50% hatch and 50% emergence varied inversely with temperature within a range from 2 to 14°C, among five species of Pacific salmon. Peaks of mortality occurred at the stage when the developing embryo had surrounded the yolk, at the time of formation of eye pigmentation, and at hatching.

Temperature changes may produce effects that are not critical at a specific time, but rather produce important responses that either occur later or occur over an extended period of time. Such effects may not always be negative. Subtle changes in temperature have a marked influence on the time of hatching and the sequence of succeeding life history events. For example, post-logging temperature increases of ~1°C during winter in a coastal stream in British Columbia resulted in advancement of emergence by as much as 6 weeks in coho and chum salmon (Holtby *et al.* 1989). Such changes re-set the timing of critical life history stages with effects that are dependent upon the environment and requirements of the species.

Dissolved oxygen concentration, like temperature, can have direct lethal effects if too low, or can cause indirect effects by altering timing of key life history events such as hatching. The effects of hypoxia on hatching time and survival may vary depending upon the stage at which the stress occurs. Alderdice *et al.* (1958) demonstrated that below a critical minimum, eggs died regardless of stage of development, and at low, but non-lethal oxygen levels, the frequency of deformity occurrence increased greatly. Low oxygen concentrations immediately before hatching caused premature hatching, while reductions at other stages caused a delay in development that was inversely proportional to oxygen concentration (Alderdice *et al.* 1958).

Fish eggs may be sensitive to mechanical shocks that occur during or as a result of forestry operations, e.g. blasting, jet boat operation and shifting gravel. Battle (1944) did early studies on the effects of the mechanical shock caused by dropping teleost eggs. Jensen & Alderdice (1983) developed a mechanical shock device that permitted quantification of shock effect on salmon eggs. They demonstrated three levels of sensitivity in coho salmon eggs, with the highest in the period from 5 to 15 days after fertilization. At the 5–6-day developmental stage, sensitivity to shock was scaled from chinook> coho> steelhead> chum> pink> sockeye (Jensen & Alderdice 1989). Chinook salmon eggs exhibited maximum sensitivity to pressure, from passage of jet boats, at about 9 days of age (Sutherland & Ogle 1975).

Jensen (2002) developed formulae to convert drop height shock into units applicable to such forms of shock as those caused by blasting. Blasting is not permitted near reproducing fish in Canada; however, it is a logging road construction activity that may occur in some parts of the world. In summary, while mechanical shock effects from forestry have not been studied directly, the above literature suggests that they should be considered in the assessment of potential impacts.

Reproductive diversity

A complex of ecosystem features influences the distribution and survival of fish populations. These features include gross characteristics of the areas chosen for reproduction, and they also include the site-specific habitat features. There is an array of patterns of reproduction related to migration routes, spawning habitat features and larval rearing habitat.

Reproduction takes place in lakes and river systems distributed over six continents from polar regions to the equator. The systems used range from a few hundred hectares to thousands of km². Spawning occurs in environments ranging from those with extreme seasonal temperature and light variation, at high latitudes, to those with relatively uniform seasonal temperatures and photoperiods near the equator. Reproduction takes place in stream systems of all sizes. It occurs in systems with and without well-developed estuaries, and those with or without seasonal flood pulses (Fig. 8.1). Given the antiquity of fishes and the diversity of environments they occupy, it is not surprising that a great array of patterns of reproduction have evolved (Breder & Rosen 1966; Balon 1975; Munro 1990b).

Fecundity

Fecundity, the number of eggs in the ovaries of a fish, is a common measure of reproductive potential (Moyle & Cech 1988). The numbers and sizes of eggs produced vary among and within freshwater species. A 100-cm chinook salmon (*Oncorhynchus tshawytscha*) may produce about 13,600 eggs, while an 85-cm common carp (*Cyprinus carpio)* can produce approximately 2.2 million (Scott & Crossman 1973). The number of eggs increases with length within a species, and total egg count can vary >10-fold (Blaxter 1969).

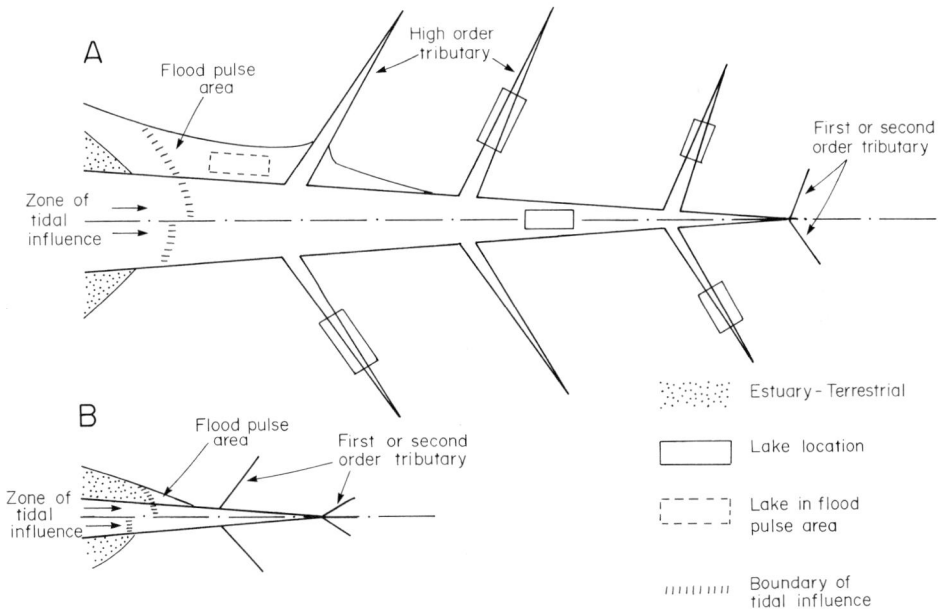

Fig. 8.1 Schematic outlines of large and small drainages, with potential locations of tributaries, lakes and flood-pulse areas indicated. Dashed line up the centre of each divides it to indicate systems with and without flood pulses. Terrestrial estuary area is dotted in, and an upper limit of tidal influence is indicated. Lakes may occur in any part of the drainage including the flood-pulse zone. Systems may occur in a range of conditions, from full seasonal flood pulse to none, and a range of sizes from <10 to millions of square kilometres.

There is also a connection between egg size and number, and degree of care. Species such as the carp scatter their eggs over the bottom or disperse them into open water. The chinook salmon provides some care by burying its eggs. Egg size and number may vary within a genus of fish depending upon degree of care. *Tilapia tholloni*, a substrate brooder, lays 500–3000 eggs, depending on its size. *T. mossambica* and *T. macrocephala*, which are mouth brooders, lay larger eggs and <500 of them (Blaxter 1969).

Spawning movements and spawning location

There is a wide diversity in the migration patterns, distances moved and location of spawning habitats used by freshwater fishes (Fig. 8.2). Such diversity is evident among different species of fishes, even within a family, inhabiting the same watershed (Fig. 8.2). Reproductive activities of species within these families may occur in all parts of the river system. Differences in seasonal timing further complicate the picture. For example, it is common for populations that spawn in small, higher elevation tributaries to migrate and spawn at much different times than their larger conspecifics in big rivers.

Factors that control migration and spawning differ between temperate and tropical areas. Timing of spawning is controlled primarily by rainfall and water levels in the tropics and by temperature and photoperiod in the temperate zones (Munro 1990b).

In Fig. 8.2, we have shown approximate routes taken to and from spawning areas because successful reproduction, including the survival of young fish, may depend upon the habitats passed through and conditions within them. Detailed examination of early rearing strategies of a single species within a small drainage shows an even greater level of complexity in habitat use (Fig. 8.3). Figures 8.2 and 8.3 indicate why it is desirable to know when and where spawning and early life history stages occur, so as to fully address forestry impacts.

Identification of spawning habitat is particularly critical because of the size of areas used. The size of the spawning area used by salmon or trout is very small relative to the size of the rearing area. Pacific salmon populations may use a few kilometres of stream for spawning, but the areas of ocean that they occupy during rearing and growth are vast (Groot & Margolis 1991). A population of large rainbow trout, from Kootenay Lake, occupies a primary spawning area of only 7000 m² (Hartman & Galbraith 1970), but they rear in a total area of approximately 399 km² (Sparrow *et al.* 1964). Similarly, Amazonian catfish (*Brachyplatystoma* and *Pseudoplatystoma*) migrate distances in excess of 3000 km, from the lower Amazon River, to where they spawn in only a few tributaries, e.g. Madeira River (Barthem & Goulding 1997).

In these examples, two relevant points emerge. First, for species whose reproductive habitats occupy only a small part of their entire range, 'small area' effects on the limited spawning habitat may result in 'large-scale' effects on the entire population. Second, for species that migrate over long distances to upstream spawning areas, forestry or other land use activities at small but strategic locations may interrupt movement and preclude spawning.

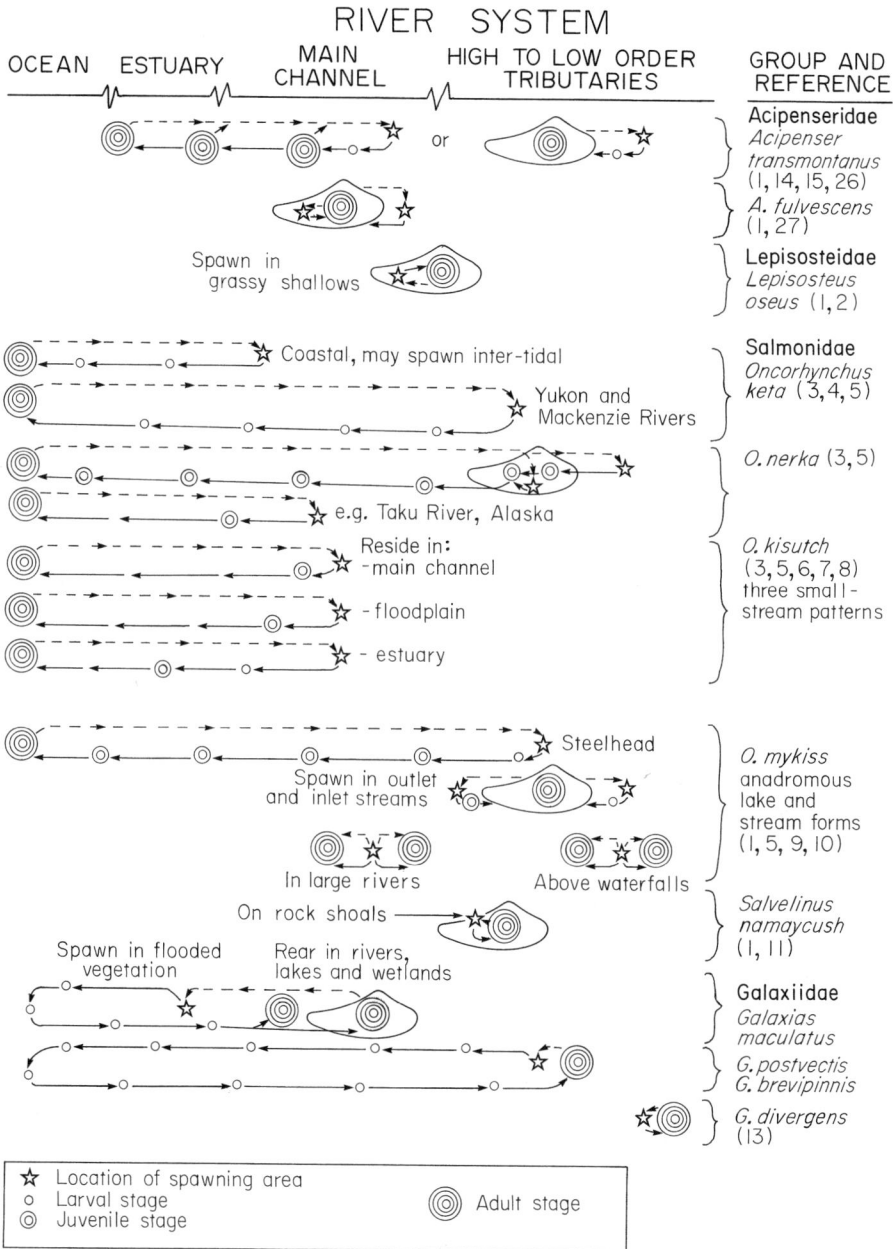

Fig. 8.2 Locations of spawning and rearing areas from ocean to upper parts of river systems. Locations of spawning and rearing areas, and the migration routes that link them, are those that are most common for a species or population type. Lakes may be located in the main stem, river tributaries or floodplain. They may be high in the system (see *Oncorhynchus mykiss*) or they may be low in it. Approximate distance upstream and into lower tributaries is indicated by spawning location. Movement distances indicated in the figure are not to the same scale. The very generalized pattern for characins is based on a description of movements described in Goulding (1980). (*Continued.*)

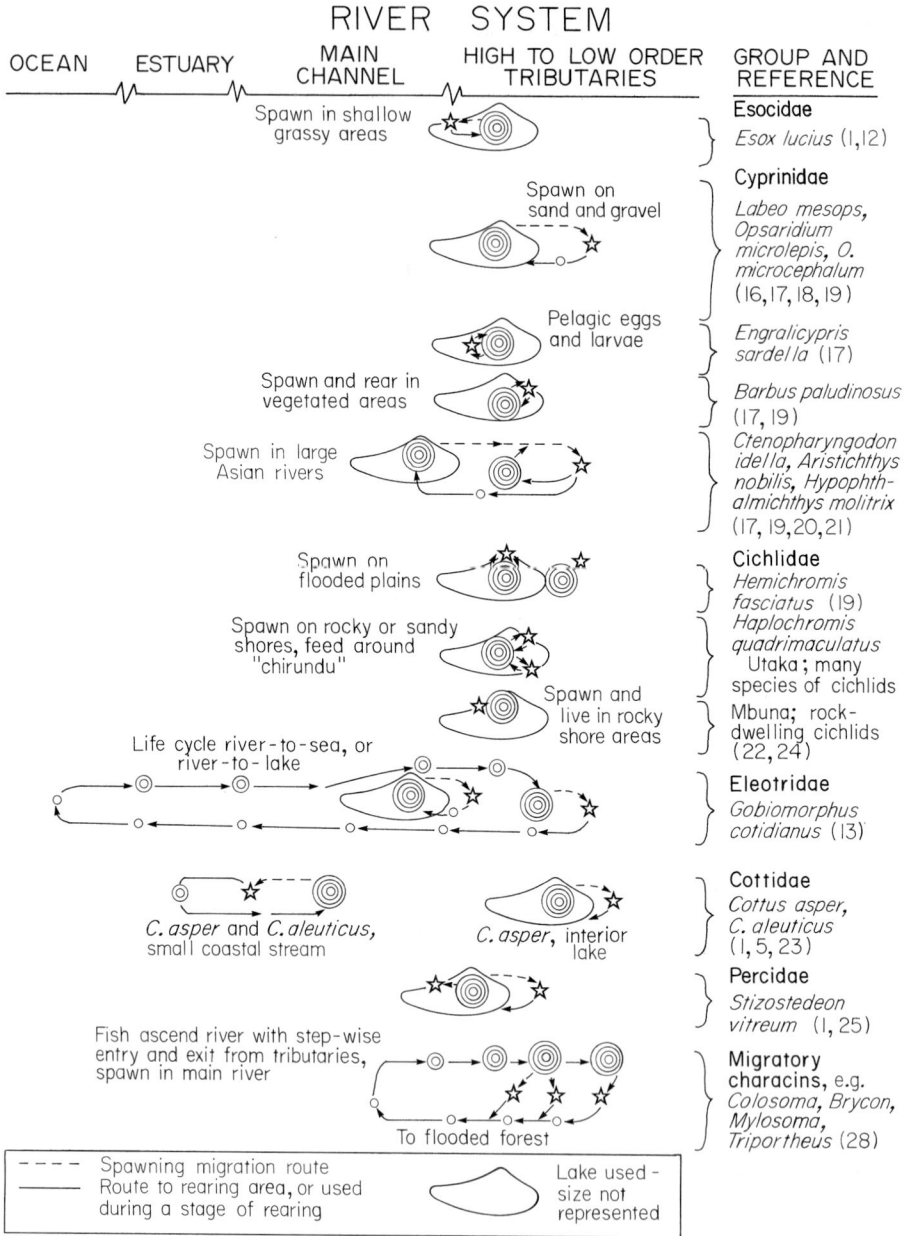

Fig. 8.2 (*Continued.*) References are numbered as follows: 1, Scott & Crossman (1973); 2, Dean (1972); 3, Groot & Margolis (1991); 4, Groot (1989); 5, Carl *et al.* (1967); 6, Hartman & Scrivener (1990); 7, Brown (1985); 8, Tschaplinski (1987); 9, Lindsey *et al.* (1959); 10, Northcote & Hartman (1988); 11, Martin (1957); 12, Fabricius & Gustafson (1958); 13, McDowall (1990); 14, Bruch *et al.* (2001); 15, Kempinger (1988); 16, Breder & Rosen (1966); 17, Fishbase (2001); 18, Tweddle (1983); 19, Munro (1990b); 20, Jhingran & Pullin (1988); 21, Nikol'skii (1963); 22, Fryer & Iles (1972); 23, Ringstad (1982); 24, Ribbink *et al.* (1983); 25, McElman & Balon (1985); 26, Hochleithner & Gessner (1999); 27, Ferguson & Duckworth (1997); 28, Goulding (1980).

There is a wide range of migration patterns and spawning locations within families and, indeed, species of fish (Fig. 8.2). Fish from the small number of families illustrated use small temperate streams, and the upstream and downstream reaches of large rivers. They use flooded estuary vegetation, vegetated, rocky and sandy lakeshore habitat, and flooded lowlands. The white sturgeon, *Acipenser transmontanus*, which resides largely in freshwater, includes populations that enter the estuary or the sea. Its members also occur far inland, in lake–river systems (Fig. 8.2). The chum salmon, *Oncorhynchus keta*, is represented by some populations that spawn in, and just upstream from, the estuaries of small streams (Groot 1989), and by others that spawn in the Yukon and Mackenzie rivers ~ 2000 km from the sea. The rainbow trout, *O. mykiss*, is represented by anadromous populations that spawn in large rivers, by populations that spawn in inlet and outlet streams of the same lake (Lindsey *et al.* 1959), and by those isolated above waterfalls, in small streams (Northcote & Hartman 1988). The coho salmon, *O. kisutch*, which spawns in both large and small streams, exhibits a range of reproduction/growth strategies even within a small drainage (Brown 1985) (Fig. 8.3).

In summary, freshwater fish reproduce in many types of habitat and exhibit a variety of life history patterns (Fig. 8.2). Actual egg deposition may occur on sand, rubble, in or on gravel, on vegetation or in open water. Depending upon the spawning and early rearing biology of the species, erosion, sedimentation, smothering of vegetation or turbidity can have negative effects.

Reproduction: microhabitats and reproductive styles

The guild or reproductive styles concept (Balon 1975, 1981) provides an ecologically relevant classification of spawning sites used by various species of fish. Balon's classification indicates the degree of care that developing larvae will receive, and the specific features of the environment in which development will occur. These classification features provide an indication of the manner in which developing larvae may be affected by a particular land use impact. Balon (1975, 1981) and related papers also include information on three other elements that influence vulnerability to forestry activities during reproduction:

- the nature of the vascular system of the embryo
- the behaviour of the larvae or fry, and
- the type and degree of parental care.

The range of development of vascular systems and relationship to the rearing environments is discussed in Balon (1975, 1985). The structure of the blood circulatory system, behaviour and physiology are all related to the reproductive strategy of the species of fish involved. Rombough (1988) covered the behaviour and physiology of larval fishes in a major review.

There are three major categories of reproductive styles (Balon 1981) which influence vulnerability of the embryos and larvae to land use impacts:

SEASON	MAJOR HABITAT TYPES				
	Off-Channel	Main Channel	Upper Estuary	Estuarine Drainages	Ocean
First Spring		emergence			fry
Summer		90%	10%		
Autumn Redistribution					juveniles
Winter	20%	75%		5%	
Second Spring					smolts age 1+
Second Year					smolts age 2+

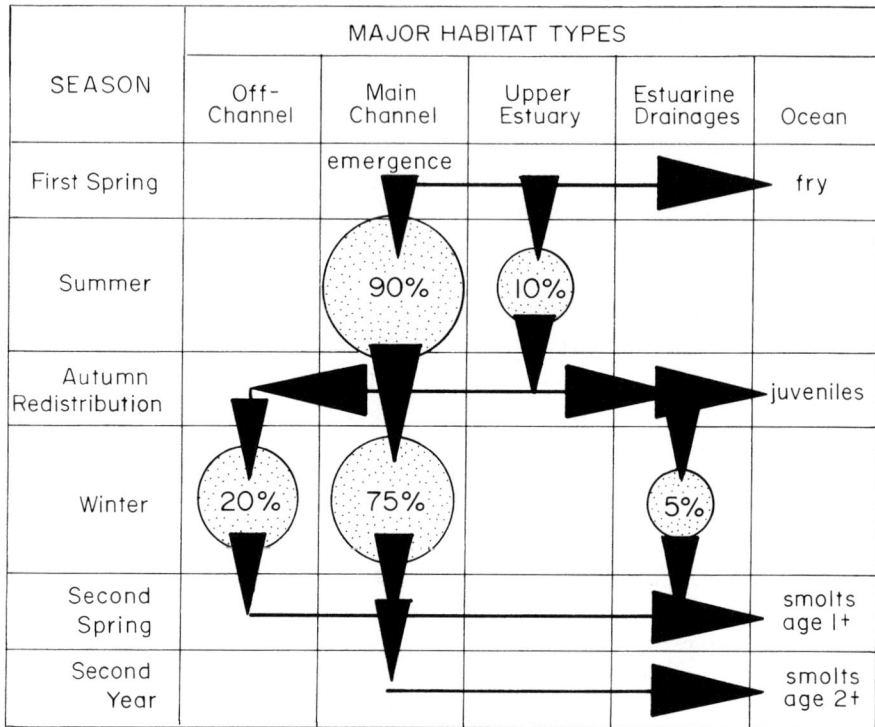

Fig. 8.3 Major habitat types and life history strategies used by coho salmon juveniles in Carnation Creek, British Columbia. Percentages indicate the portion of the population using each habitat type during summer and winter. Redrawn with permission from Scrivener *et al.* (1998).

- non-guarders
- guarders
- bearers.

(See Balon (1975) for a detailed listing of 'reproductive styles' and an explanation of the classification.)

Non-guarders include species that scatter eggs upon open substrate (Fig. 8.4D) or disperse them into open water. Typically, pelagic larvae have weakly developed vascular systems and require clean, oxygenated waters. Non-guarders also include fish that lay eggs in crevices, bury them in the streambed (Fig. 8.4A) or deposit them in the cavities in mussels or sponges. The larvae of non-guarders that lay eggs on or among vegetation may avoid light and hide in locations where oxygen is limited. They have well-developed circulatory systems (Balon 1975).

The guarders include fish that deposit eggs in cavities in the stream bottom (Fig. 8.4B, C) or build nests and guard them. These species may fan the eggs to increase oxygen supply. They also may guard the young for some time after they hatch and become free-swimming. The larvae of some of these forms stick to the plants with adhesive organs, and fan themselves to assist respiration.

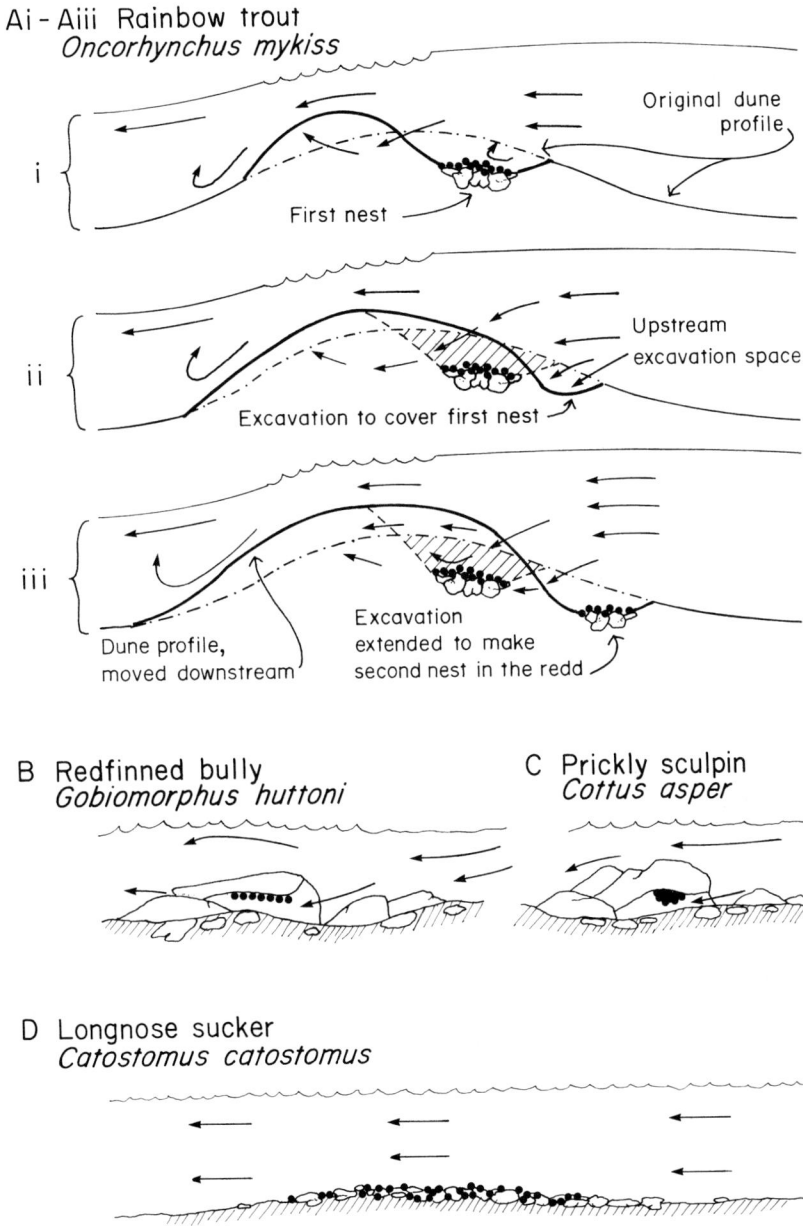

Fig. 8.4 Spawning sites of rainbow trout (Ai–iii), redfinned bully (B), prickly sculpin (C), and longnose sucker (D). Eggs, black dots, are not drawn to scale. Arrows indicate direction of current, and in Ai–Aiii, its entry into the gravel. Sketch Ai shows nest pit and eggs, original dune profile and tail-spill from excavated nest. Aii shows first nest covered. Aiii shows second nest and eggs in the redd, and dune profile extended downstream (information on dune structure from Hartman & Galbraith 1970). (B & C) Nests of redfinned bully and prickly sculpin show eggs attached to the under-surface of rocks, where they are guarded by male fish, and protected from moderate current and sediment. (D) Longnose sucker spawning site. Eggs are not protected from current and sediment. (Note: A redd is a site with one or more nests.)

The bearers include species that either build nests, spawn and brood the young in their mouths (external bearers), or bear live young (internal bearers). These external bearers may clear sediment from the nest site, and are able to move the young to locations of higher oxygen supply.

Tropical freshwater fishes: habitats and vulnerability

Tropical freshwater fishes spawn in a wide array of habitats including lakes, deltas, rivers, small streams, seasonally flooded forests, flooded grasslands and plains, and swamps (Munro 1990b). Often, spawning in tropical fishes is attuned to monsoons and associated floods (Munro 1990b). We have attempted to summarize Asian, African and neotropical freshwater fishes, from within broadly defined size classes, that use the various spawning habitats (Table 8.1). In the case of some species, we were certain that we were interpreting habitat type correctly; in other cases it was more difficult to specify the types of habitat used. For some species, there were spawning records for more than one kind of habitat. Such species were recorded twice in the table.

With regard to sensitivity to disturbance from forestry activities, fishes that spawn in rivers, small streams and flooded land are probably more at risk. These habitats are especially vulnerable to increases in turbidity, sedimentation, temperature change and habitat simplification (Scott & Helfman 2001). We surmise that species that spawn in the lower reaches of large rivers, in which the impacts of upstream tributary events are dampened, are less at risk than those in small, high-energy tributaries located in areas of unstable soils. A predominant number of species of 'large' neotropical and African characoid fishes make use of river, stream and flooded land for spawning. Twelve out of 13 of the 'small' Asian cyprinids reviewed all use streams, flooded land and swamps for spawning.

In contrast, species that spawn in large stable lakes, e.g. cichlids, may be less vulnerable to forestry impacts. The present-day lacustrine species of cichlids were derived from riverine forms that existed before the present-day lakes were formed (Fryer & Iles 1972). The geological creation of large African lakes, where rivers once occurred, has produced strong environmental pressure favouring lake spawning among such species groups as the cichlids. Those species that spawn in vegetated lake shore areas where plants may be affected by turbidity may be more at risk than those that choose open sandy lake shore areas.

Reproduction in salmonids – risks and impacts

The abundant literature on salmonid reproductive behaviour illustrates the many interacting factors that may influence reproductive success by forestry activities. Here, we will review the diversity in reproductive behaviour of this group and discuss specific factors associated with logging that may impair reproductive success. Specific factors we consider are the effects of cover loss, water temperature change, sediment deposition, channel bed erosion and water quality change. Our goal is to indicate the range of variability of potential impacts that occur within a genus or a species of fish,

Table 8.1 Numbers of species of tropical freshwater fish using types of spawning habitat indicated

Taxonomic group, and geographic location and size of fish in group		Spawning habitat					
Superorder, order or family	Geographic location, and where available, size of fish in group	Lake	Delta or estuary	River or stream	Flooded area or plain	Flooded forest	Swamp
Osteoglossomorpha	---, ----	12	5	10	2	1	2
Clupeidae	---, ----	7	2	5	0	0	0
Characoidei	Neotropical, small	0	0	8	0	0	0
	African, small	4	0	3	1	0	0
	Neotropical, large	1	0	12	10	0	0
	African, large	3	0	11	5	0	2
Gymnotoidei	Neotropical, ---	0	0	3	1	0	0
Cyprinoidei	Asian, small	1	0	7	2	0	3
	African, small	4	1	9	1	1	1
	Asian, large	2	0	6	4	1	0
	African, large	2	0	10	4	0	0
Siluriformes	Asian, ------	2	1	7	1	0	0
	Africa and Middle East, -------	12	1	14	5	0	2
	Neotropical, --	0	0	2	1	0	1
Cichlidae	Neotropical, --	13	0	6	4	0	0
	Asian, ------	2	2	0	0	0	0
	African, ------	51	3	13	4	0	6

Information from Munro (1990b). Sizes of fish in groups, where given, are from Munro (1990b).

and thereby demonstrate why detailed local knowledge of reproductive life history is imperative if forestry planning is to be effective.

Overview of salmonid reproductive behaviour

Spawning in salmonids occurs primarily in spring and autumn, depending on species, but may occur during any month of the year. Information on spawning times, for salmonids in Canada, is provided in Groot & Margolis (1991), Scott & Crossman (1973), and for those in British Columbia, in Carl *et al.* (1967). Spawning time tends to be consistent from year to year within populations but varies among populations (Scott 1990).

Rainbow trout (*Oncorhynchus mykiss*), steelhead, an anadromous form of rainbow trout, and the species in the cutthroat trout (*O. clarki*) complex (Behnke 1997) spawn during the spring and early summer. The huchen and taimen, *Hucho hucho* and *H. taimen*, and the lenok *Brachymystax lenok*, spawn during the spring and early summer (Nikol'skii 1963). Brook trout (*Salvelinus fontinalis*) and Arctic char (*S. alpinus*), brown trout (*Salmo trutta*), Atlantic salmon (*S. salar*) and Pacific salmon (*Oncorhynchus* species), spawn during the late summer or autumn (Scott & Crossman 1973; Groot & Margolis 1991).

Spawning behaviour involves nest site selection, courtship, aggression (especially among males), egg release and fertilization, and in many species, covering of the eggs with gravel. During the pre-spawning period, some species of salmonids such as rainbow trout and coho salmon make use of cover (Hartman & Galbraith 1970). During nest construction salmonids clear away fine sediment and create a nest pocket that contains relatively less sediment than the surrounding gravel. The shape of the gravel mound over the nest forces water flow through the gravel around the eggs (Fig. 8.4A).

Courtship, aggression and egg burying behaviour were described for seven species of Pacific salmon (Groot & Margolis 1991), for chum salmon and rainbow trout (Tautz & Groot 1975), for rainbow trout (Hartman 1969, 1970), for bull trout (Leggett 1969), and for Arctic char (Fabricius 1953; Fabricius & Gustafson 1954). Visual, tactile and olfactory cues are all important during spawning. Females select the nest sites and determine bottom features using visual and tactile cues (Fabricius & Gustafson 1954; Breder & Rosen 1966; McCart 1969; Hartman 1970). Males court the females with a series of movements and body contacts (Leggett 1969; Tautz & Groot 1975; Newcombe & Hartman 1980). Males respond to olfactory cues from females. These cues attract them to the females and determine timing of behavioural stages during spawning (Newcombe & Hartman 1973; Liley 1982; Scott *et al.* 1994). Water quality conditions that impair visual or olfactory responses potentially affect spawning success.

Cover loss

Many species of salmonids rely on the cover of deep pools or large woody debris as resting or staging areas during their extensive upstream spawning migrations. For example, rainbow trout in large streams occupy deep pools prior to spawning. They may escape into such habitat if disturbed during spawning or move back to them during the first 2 days of spawning, during which activity is predominantly nocturnal (Hartman 1969). Forest removal, channel widening and aggradation may fill in such habitats (Hartman & Miles 2001). In small streams, the same fluvial changes may result in loss of large woody debris cover, adjacent to spawning sites, for species such as coho salmon. In the Pacific Northwest, summer runs of salmonids (fish enter the stream in the summer and remain there for months until spawning) are the most at risk of extirpation. This risk is due to the loss of deep pools and cover, or to excessive warming in the main-stem rivers where they reside before moving upstream into tributaries to spawn (Nicholas & Hankin 1988; Nehlsen *et al.* 1991).

Water temperature changes

Changes in water temperature can have profound effects on salmonid reproductive activities, at all stages from adult migration to seaward movement of fry. Beschta *et al.* (1987) provided a good overview of forestry effects on stream temperature, and examined effects of temperature change on the survival of developing salmon embryos.

Effects of temperature changes on reproduction may begin at the stage of spawning migration. Elevated temperatures during migration interrupt movement, increase metabolic rate and stress levels, and reduce energy reserves – all of which affect reproductive success. Water temperatures may affect gonad development and the percentage of a population that spawns. Elevated temperatures during spawning reduce sperm motility and impair fertilization success (Macdonald *et al.* 1998).

Following spawning, incubation, development and emergence may be affected by elevated temperature. In an earlier part of this chapter we touched on experimental work on the effect of water temperature on incubation rates. In the stream, temperature changes have direct effects that occur during the incubation period. However, subsequent or indirect effects may extend well beyond that period. In a long-term study on the western coast of British Columbia, Holtby (1988) showed that small (<1°C) logging-related temperature increases, during winter, caused early emergence of coho fry. This resulted in a linked series of changes that extended from egg deposition (through incubation and post-hatch life in the creek) to adult return. The effects on coho salmon were not all negative (Hartman & Scrivener 1990). However, changes in winter temperature caused chum salmon, a species whose fry do not remain in the stream for an extended period, to go to sea earlier than usual, at a time when conditions for development and survival were less than optimal (Hartman & Scrivener 1990).

Sockeye salmon, with a primary distribution from 46° to 65° N latitude, spawn over a wide range of conditions in coastal and interior waters. An important part of their juvenile life history occurs in lakes, where they may be less vulnerable to direct forestry impacts unless such activities increase turbidity and reduce plankton production. In addition, sockeye exhibit reproductive adaptations for life histories occurring within different geographic areas. Interior populations, with eggs incubating at relatively low mean temperatures, spawn early (August) whereas coastal populations, with eggs incubating at higher temperatures, spawn late (November). The timing of spawning and the rate of incubation are adapted to produce young that emerge and enter lakes under optimal feeding conditions (Brannon 1987). Local stream temperature changes could upset geographically tuned development timing. Extremely low stream temperatures in the interior of British Columbia can cause the formation of ice within the streambed, and force sockeye larvae to move to avoid freezing.

Streamside clearing along streams in the northern interior of British Columbia affected sockeye salmon through a different process than that which occurred with coho salmon in a small coastal stream. In streams in the Takla Lake area, most of the larval development occurs between August and November. Hypothetically, an increase of 1.5°C, as a potential result of streamside clearing, would advance the development rate such that larvae would emerge and enter the lake out of synchrony with the timing of spring zooplankton production (Macdonald *et al.* 1998).

Scour and re-deposition, and fine sediment impacts

All salmonids require clean gravel for reproductive success. Egg-to-fry survival is inversely related to the amount of sediment (<0.85 mm) in the gravel (Everest *et al.* 1987). During nest construction, salmonids clear away fine sediment and create a nest pocket (Fig. 8.4A) that contains less sediment than the surrounding gravel (Peterson & Quinn 1996). Because the shape of the mound over the nest forces water flow through the gravel around the eggs (Chapman 1988), excess suspended sediment, if present, would enter the egg pocket with the intra-gravel flow. This material could smother the developing embryos or restrict movement and emergence of the larvae.

Salmonids in the Pacific Northwest have evolved under conditions of storms, floods, streambed scour and deposition, and freezing. Features of their reproductive biology such as relatively large egg numbers, burying of eggs, seasonal timing of spawning and mobility of larvae adapt them to such conditions. Although scour and deposition conditions occur naturally, forestry activities can increase their frequency and severity regardless of the region of the world in which they occur. Road construction and logging in streamside and up-slope areas, singly or together, can increase sediment loading (see Chapter 13) (Hartman & Scrivener 1990; Scott & Helfman 2001). These activities alter both streambed stability and gravel quality, and hence egg-to-fry survival.

Hydrological regime, stream channel type and streambed scour have been shown to constrain the distribution of salmonids and influence their spawning time. The depths at which eggs are buried may represent an adaptation to depths of scour during incubation (Montgomery *et al.* 1999). Scour accounted for a part of the mortality of coho salmon eggs in a small western coast, British Columbia stream (Holtby & Healey 1986). Under flood flows of 48.8 m³s⁻¹ in this 10 km² drainage, modelled scour depths were 98 cm (Haschenburger 1998). Extreme scour conditions will affect reproduction in stream-spawning salmonids and other taxa of fish that bury their eggs or lay them in cavities in the streambed.

Salmonid larvae (alevins), in the gravel, exhibit behaviour that can partially combat the effects of accumulated sediment. 'Ventilation swimming' (Bams 1969) may keep the immediate intra-gravel space around the fish free of sediment. Alevins may 'cough' to back-flush sediment from the gills, and in extreme situations they may exit from the gravel and re-enter it elsewhere (Bams 1969).

Sediment may also block interstitial spaces in the gravel, restricting movement and 'entombing' salmonid alevins at the time of emergence. Entombment was considered to be the major cause of alevin mortality, and the size and survival of chum fry was cor-related with composition of the top layer of gravel in a western coast stream in British Columbia (Scrivener & Brownlee 1989). However, Crisp (1993) showed that 25% of brown trout alevins, and 40% of those of Atlantic salmon, could emerge through up to 8 cm of sand.

Reproduction in non-salmonids

Risks and impacts

The information on reproductive biology and related habitat use is not developed as well for any other group as it is for salmonids. Consequently, the nature of our discus-sion about other groups is much more speculative. Extrapolation of information from one region to another involves some risk. However, information about salmonids may be extrapolated most usefully to other species that:

- spawn in streams
- are temperature-sensitive
- hide eggs in cavities or scatter them within the surface of a gravel bottom
- reqire sediment-free, well-oxygenated water.

River and stream spawning

Breder & Rosen (1966) listed 27 families with representatives that spawn in rivers or streams. This number appears large as it includes families with representatives that also spawn in lakes and ponds. The number of obligate stream spawners is lower. How-ever, diverse groups of North American families include stream spawners that may be subject to physical disturbances in the same way as salmonids. Cyprinids such as the horny head chub (*Nocomis biguttatus*), the river chub (*N. micropogon*) and the creek chub (*Semotilus atromaculatus*) spawn in nests in streams. Members of the sucker family (*Catostomus catostomus*, *C. commersoni* and *Moxostoma anisurum*) scatter their eggs in streams (Scott & Crossman 1973). Some species of sculpins (Fig. 8.4C) and darters spawn in cavities among the stream bottom stones. In the family Percidae, the walleye scatters its eggs on the stream bottom, whereas females of several species of darters burrow into the bottom so as to bury their eggs (Scott & Crossman 1973). At least three species of bully (family Eleotridae) in New Zealand (McDowall 1990) spawn in cavities (Fig. 8.4B) in the stream bottom.

Eggs deposited in cavities or buried by fish in the foregoing species would presum-ably be affected negatively by streambed erosion of more than a few centimetres. Those deposited in crevices in the gravel, by species such as suckers or walleye, would also be impacted by erosion or damaged by deposited sediment. In Appalachia, highland fish faunas suffered following forest removal and accompanying stream sedimentation, and the subsequent increase in embeddedness of bottom stones. Scott & Helfman (2001) showed that homogenization of habitat caused a decrease in relative abundance of species of darters, sculpins and cyprinids that spawned in crevices among the rocks. Furthermore, an invasion of 'generalist' species that were more sediment tolerant occurred with such 'homogenization'.

Lake spawning

In a relative sense, impacts from forestry create less risk for lake-spawning species than for those that spawn in streams and rivers. For example, although sedimentation of reproductive habitats, following logging, increased in boreal lakes, its effects on lake trout, whitefish, sucker and yellow perch reproduction were not clearly demonstrable (Gunn & Sein 2000; St Onge & Magnan 2000). The greatest risk from increased turbidity and sedimentation may be for species that utilize vegetation or bare substrate for spawning. Increased turbidity from forestry operations may also reduce reproductive success by impairing phytoplankton/zooplankton production or altering community composition.

In comparison to that available for salmonids, there is a scarcity of information about forestry impacts on tropical species in both lakes and large rivers. We speculate that there may be important effects of suspended sediment on developing larvae from such species as the grass carp (*Ctenopharyngodon idella*), the silver carp (*Hypophthalmichthys molitrix*) and the bighead carp (*Aristichthys nobilis*), as they drift downstream in the large Asian rivers in which they occur. There may be counterbalancing effects of increased turbidity in reducing risk of predation versus decreasing ability to see food (Newcombe 2001). Removal of flooded forests will create very serious negative effects for those species that leave main channels in flood-pulse rivers, such as the Amazon, to feed in the flooded areas. Elevated water temperatures may compound such trophic effects. Increased turbidity and deposited sediment, arising from upstream logging, may cause negative effects by reducing or destroying lake bottom vegetation used by fish that spawn on plants.

Factors influencing reproductive success

This section attempts to summarize major categories of factors, and moderating conditions associated with them, that determine risk to fish during reproduction. Reproductive success among fish populations may be influenced by completely different sets of factors (Table 8.2). Some of these operate at a 'macro-scale' through effects such as climate, geology, and the area, size and energy of streams used. Some operate at the level of the population or individual fish through the behaviour and physiology of the fish. At the level of human society, public interest and quality of science and administration that underlie forestry planning also determine the success of populations of

Table 8.2 Major categories of influence and condition ranges within them that may affect reproductive success among fish within watershed system

Category	Condition range and risk direction
Climate and geology	• Climate determines degree of temperature increase with deforestation: risk potential increases from coastal cool to interior warm/dry zones, and from north to south • Climate and geology determine the range of peak flows in rain or rain-on-snow zone: increasing scour potential with increasing stream power and peak flow potential • Geology and soils determine risk of mass failures and landslides that cause negative effects of sedimentation on one hand, and creation of complex habitat sections on the other
Size and nature of watershed	• Drainage area relative to cutting area: increasing risk with percentage of drainage area cut. Smallest tributaries affected most because all or a large part of one may be included within a single cut-block • Hydrology and channel features may determine the optimal season for reproduction in drainage basins of different types • Number of lakes in drainage: increasing stability in reproductive environment with increasing lake number
Biology of fish	• Duration of incubation time: risk increases with incubation time • Parental care: risk decreases with burial of eggs and guarding/fanning in gravel or rocky bottom streams • Fish undertaking longer migrations may be at greater risk of some form of habitat impact along the route
Biology of embryo	• Requirements for high oxygen, relatively stable temperature and disturbance-free incubation environment: risk increases with species sensitive to hypoxia • Requirement for relatively stable temperature and absence of physical shock: risk increases with physical disturbance of the environment • Duration of high sensitivity periods during embryo development: risk increases with duration of period of any type of sensitivity
Status of science and quality of management	• Amount and quality of ecological knowledge: risk increases for reproductive success and other life history stages with decreasing quality and amount of science • Transfer of knowledge: risk increases with decreasing amount and quality of information transfer and exchange • Quality of forestry planning and actual implementation of plans: risk increases with decreasing planning quality and with the gap between planning and actual implementation

spawning fish (Table 8.2). These diverse categories of factors and ranges of conditions within them provide a setting within which impacts on reproduction may be considered in any geographic region.

Summary

In this chapter we have attempted to outline some features of egg development, fertilization, and embryonic and larval development. We have also described some of the many reproductive strategies of fishes to indicate their diversity. We have used information on salmonids to illustrate how forestry activities might affect their reproduction. Because the reproductive requirements of many tropical freshwater species are not well known, we have provided a framework that might be used to guide people, in various regions, when assessing important features of fish reproduction. Knowledge about such features will guide fish biologists and foresters in understanding the potential effects of forestry in other regions. Important issues include:

- How fish eggs and embryos develop, because this is a critical period for all fish.
- How stage-specific, critical sensitivities to low oxygen, elevated temperature and physical disturbance may affect eggs and embryos.
- Attention to the work of Balon (1975, 1981), because that research links spawning habitat use, egg characteristics, embryological development of circulatory systems and behaviour of alevins.
- Consideration of a small number of examples that illustrate the diversity of combinations of movement to spawning areas, spawning sites used and movement of young fish away from them (Table 8.1, Fig. 8.2).
- Review of aspects of spawning biology of salmonids to illustrate diversity within that group, and the complex ways that forestry-related activities affect reproductive success.
- Consideration of Tables 8.1 and 8.2, showing that there are some watershed environments in which reproductive activities may be more vulnerable to road construction, forest removal, yarding, etc. within them than they are in others.
- Illustration of how reproductive success among fishes in any part of the world is influenced by a suite of macro-scale factors (Table 8.2).

There are vital aspects of the reproductive biology of fishes that should be known so that fish biologists and foresters might plan forestry operations and anticipate impacts from them. The following requirements are more than a blind request for 'more research':

- identification of life history patterns
- identification of spawning areas and environmental features in them
- determination of spawning time
- description of characteristics of eggs

- determination of features of egg and larval developmental physiology, incubation time and oxygen requirements, and
- determination of life history requirements for embryos and larvae.

There is a need to have this kind of information and mesh it with knowledge of the physical effects of forestry activities (see Chapter 13). Scott & Helfman (2001) offer a relevant quote, 'Because of the multi-disciplinary nature of any conservation-oriented investigation, studies of basic biology, ecology, and behavior must be encouraged and supported'.

Research in one geographical region of the world can provide guidance for understanding and managing forestry impacts in other areas (see Chapter 13). However, within any region, the working framework must include research that provides specific local management-related information within a context of understanding of the watershed processes. This foundation plus local habitat experience, and an appreciation of fish resource values, provides a framework for protecting fish during the vital stages of reproduction as well as all other stages.

Acknowledgements

We gratefully acknowledge the considerable help from G. Miller and G. Pattern (Pacific Biological Station Library), and A. Thompson, W. Hartman and Charles Scrivener. Drs D. Alderdice and E. Groot carefully reviewed the manuscript and offered many valuable suggestions. We appreciate all such help.

References

Alderdice, D.F & Velsen, F.P. (1978) Relation between temperature and incubation time for eggs of chinook salmon (*Oncorhynchus tshawytscha*). *Journal of the Fisheries Research Board of Canada*, **35**, 69–75.

Alderdice, D.F., Wickett, W.P. & Brett, J.R. (1958) Some effects of temporary exposure to low dissolved oxygen levels on Pacific salmon eggs. *Journal of the Fisheries Research Board of Canada*, **15**, 229–50.

Austen, D.J., Bayley, P.B. & Menzel, B.W. (1994) Importance of the guild concept to fisheries research and management. *Fisheries*, **19**, 12–20.

Balon, E.K. (1975) Reproductive guilds of fishes: a proposal and definition. *Journal of the Fisheries Research Board of Canada*, **32**, 821–64.

Balon, E.K. (1981) Additions and amendments to the classification of reproductive styles in fishes. *Environmental Biology of Fishes*, **6**, 377–90.

Balon, E.K. (1985) *Early Life Histories of Fishes: New Developments, Ecological and Evolutionary Perspectives*. Dr W. Junk Publishers, Dordrecht, the Netherlands.

Bams, R.A. (1969) Adaptations of sockeye salmon associated with incubation in stream gravels. In: *Symposium on Salmon and Trout in Streams*, (ed. T.G. Northcote), pp. 71–99. H.R. Macmillan Lecture Series in Fisheries, Institute of Fisheries, University of British Columbia, Vancouver.

Barthem, R. & Goulding, M. (1997) *The Catfish Connection: Ecology, Migration, and Conservation of Amazonian Predators*. Columbia University Press, New York.

Battle, H.I. (1944) Effects of dropping on the subsequent hatching of teleostean ova. *Journal of the Fisheries Research Board of Canada*, 6, 252–5.

Behnke, R.J. (1997) Evolution, systematics, and structure of *Oncorhynchus clarki clarki*. In: *Sea-run Cutthroat: Biology, Management, and Future Conservation*, (eds J.D. Hall, P.A. Bisson & R.E. Gresswell), pp. 3–6. Oregon Chapter of the American Fisheries Society, Corvallis, OR.

Beschta, R.L., Bilby, R.E., Brown, G.W., Holtby, L.B. & Hofstra, T.D. (1987) Stream temperature and aquatic habitat: fisheries and forestry interactions. In: *Streamside Management: Forestry and Fishery Interactions*, (eds E.O. Salo & T.W. Cundy), pp. 191–232. Contribution No. 57. College of Forest Resources, and Institute of Forest Resources, University of Washington, Seattle, WA.

Billard, R. & Jensen, J.O. (1996) Gamete removal, fertilization and incubation. In: *Principles of Salmonid Culture: Developments in Aquaculture and Fisheries Science*, Vol. 29, (eds W. Pennell & B.A. Barton), pp. 291–364. Elsevier, Amsterdam.

Blaxter, J.H. (1969) Development: eggs and larvae. In: *Fish Physiology, Vol. III, Reproduction and Growth Bioluminescence, Pigments and Poisons*, (eds W.S. Hoar & D.J. Randall), pp. 177–252. Academic Press, New York.

Blaxter, J.H. (1988) Pattern and variety in development. In: *Fish Physiology, Vol. XI, The Physiology of Developing Fish, Part A, Eggs and Larvae*, (eds W.S. Hoar & D.J. Randall), pp. 1–58. Academic Press, San Diego, CA.

Brannon, E.L. (1987) Mechanisms stabilizing salmonid fry emergence timing. In: *Sockeye Salmon (Oncorhynchus nerka): Population Biology and Future Management*, (eds H.D. Smith, L. Margolis & C.C. Wood), pp. 120–4. Canadian Special Publication of Fisheries and Aquatic Sciences 96.

Breder, C.M. & Rosen, D.E. (1966) *Modes of Reproduction in Fishes*. Natural History Press, Garden City, New York.

Brown, T.G. (1985) *The role of abandoned stream channels as over-wintering habitat for juvenile salmonids*. MSc thesis, University of British Columbia.

Bruch, R.M., Dick, T.A. & Choudhury, A. (2001) *A Field Guide for Identification of Stages of Gonad Development in Lake Sturgeon, (Acipenser fulvescens Rafinesque), With Notes on Lake Sturgeon Reproductive Biology and Management Implications*. Wisconsin Department of Natural Resources, Fisheries Management.

Carl, G.C., Clemens, W.A. & Lindsey, C.C. (1967) *The Fresh-water Fishes of British Columbia*. British Columbia Museum Publication, Handbook No. 5. Queen's Printer, British Columbia.

Chapman, D.W. (1988) Critical review of variables used to define effects of fines in redds of large salmonids. *Transactions of the American Fisheries Society*, 117, 1–21.

Crisp, T.D. (1993) The ability of U.K. salmonid alevins to emerge through a sand layer. *Journal of Fish Biology*, 43, 656–8.

Dean, E.L. (1972) *Reproductive biology of the longnose gar (Lepisosteus osseus L.) in Lake St. Clair*. MSc thesis, University of Guelph, Ontario.

Everest, F.H., Beschta, R.L., Scrivener, J.C., Koski, K.V., Sedell, J.R. & Cederholm, C.J. (1987) Fine sediment and salmonid production: a paradox. In: *Streamside Management: Forestry and Fishery Interactions*, (eds E.O. Salo & T.W. Cundy), pp. 98–142. Contribution No. 57. College of Forest Resources, and Institute of Forest Resources, University of Washington, Seattle, WA.

Fabricius, E. (1953) Aquarium observations on the spawning behaviour of the Char, *Salmo alpinus*. *Institute of Freshwater Research, Drottningholm Report*, 34, 14–48.

Fabricius, E. & Gustafson, K-J. (1954) Further aquarium observations on the spawning behaviour of the Char, *Salmo alpinus* L. *Institute of Freshwater Research, Drottningholm Report*, 35, 58–104.

Fabricius, E. & Gustafson, K-J. (1958) Some new observations on the spawning behaviour of the pike, *Esox lucius* L. *Institute of Freshwater Research, Drottningholm Report*, **39**, 23–54.

Ferguson, M.M. & Duckworth, G.A. (1997) The status and distribution of lake sturgeon, *Acipenser fulvescens,* in the Canadian provinces of Manitoba, Ontario and Quebec: a genetic perspective. In: *Sturgeon Biodiversity and Conservation,* (eds V.J. Birstein, J.R. Waldman & W.E. Bemis), pp. 299–309. Kluwer Academic Publishers, Dordrecht, the Netherlands.

Fishbase (2001) Species summary http://www.fishbase.org

Fryer, G. & Iles, T.D. (1972) *The Cichlid Fishes of the Great Lakes of Africa: Their Biology and Evolution.* Oliver & Boyd, Edinburgh.

Goetz, F.W. & Thomas, P. (1995) Reproductive physiology of fish. *Proceedings of the Fifth International Symposium on the Reproductive Physiology of Fish.* University of Texas, Marine Science Institute, Austin, TX.

Goulding, M. (1980) *The Fishes and the Forest: Explorations in Amazonian Natural History.* University of California Press, Berkeley, CA.

Groot, C. & Margolis, L. (1991) *Pacific Salmon: Life Histories.* UBC Press, University of British Columbia, Vancouver.

Groot, E.P. (1989) *Intertidal spawning of chum salmon: saltwater tolerance of early life stages to actual and simulated intertidal conditions.* MSc thesis, University of British Columbia.

Gunn, J.M. & Sein, R. (2000) Effects of forestry roads on reproductive habitat and exploitation of lake trout *(Salvelinus namaycush)* in three experimental lakes. *Canadian Journal of Fisheries and Aquatic Sciences* **57** (Suppl. 2), 97–104.

Hartman, G.F. (1969) Reproductive biology of the Gerrard stock rainbow trout. In: *Symposium on Salmon and Trout in Streams,* (ed. T.G. Northcote), pp. 53–67. H.R. MacMillan Lecture Series in Fisheries, Institute of Fisheries, University of British Columbia, Vancouver.

Hartman, G.F. (1970) Nest digging behavior of rainbow trout (*Salmo gairdneri*). *Canadian Journal of Zoology*, **48**, 1458–64.

Hartman, G.F. & Galbraith, D.M. (1970) *The Reproductive Environment of the Gerrard Stock Rainbow Trout.* Province of British Columbia, Ministry of Recreation and Conservation, Fisheries Management Publication No. 15.

Hartman, G. & Miles, M. (2001) Assessment of techniques for rainbow trout transplanting and habitat management in British Columbia. *Canadian Manuscript Report of Fisheries and Aquatic Sciences No. 2562.*

Hartman, G.F. & Scrivener, J.C. (1990) Impacts of forestry practices on a coastal stream ecosystem, Carnation Creek, British Columbia. *Canadian Bulletin of Fisheries and Aquatic Sciences 223.*

Haschenburger, J.K. (1998) Channel scour and fill in coastal streams. In: *Carnation Creek and Queen Charlotte Islands Fish/Forestry Workshop: Applying 20 Years of Coastal Research to Management Solutions,* (eds D.L. Hogan, P.J. Tschaplinski & S. Chatwin), pp. 109–17. BC Ministry of Forests, Land Management Handbook No. 41.

Hoar, W.S. (1969) Reproduction. In: *Fish Physiology, Vol. III, Reproduction and Growth Bioluminescence, Pigments, and Poisons,* (eds W.S. Hoar & D.J. Randall), pp. 1–72. Academic Press, New York.

Hoar, W.S. & Randall, D.J. (1988) *Fish Physiology: The Physiology of Developing Fish, Vol. XI, Part A, Eggs and Larvae.* Academic Press, San Diego, CA.

Hochleithner, M. & Gessner, J. (1999) *The Sturgeons and Paddlefishes (Acipenseriformes) of the World: Biology and Aquaculture.* AquaTech Publications, Kitzbuehel, Austria.

Holtby, L.B. (1988) Effects of logging on stream temperatures in Carnation Creek, British Columbia, and associated impacts on the coho salmon (*Oncorhynchus kisutch*). *Canadian Journal of Fisheries and Aquatic Sciences,* **45**, 502–15.

Holtby, L.B. & Healey, M.C. (1986) Selection for adult size in female coho salmon (*Oncorhynchus kisutch*). *Canadian Journal of Fisheries and Aquatic Sciences,* **43**, 1946–59.

Holtby, L.B., McMahon, T.E. & Scrivener, J.C. (1989) Stream temperature and inter-annual variability in the emigration timing of coho salmon (*Oncorhynchus kisutch*) smolts and fry and chum salmon (*O. keta*) fry from Carnation Creek, British Columbia. *Canadian Journal of Fisheries and Aquatic Sciences,* **46**, 1396–405.

Huntela, A. & Stacey, N.E. (1990) Cyprinidae. In: *Reproductive Seasonality in Teleost Fishes: Environmental Influences,* (eds A.D. Munro, A.P. Scott & T.J. Lam), pp. 53–77. CRC Press, Boca Raton, FL.

Idler, D.R., Crim, L.W. & Walsh, J.M. (1987) *Reproductive Physiology of Fish 1987.* Marine Sciences Research Laboratory, Memorial University of Newfoundland, St John's, Newfoundland.

Jensen, J.O. (2002) New mechanical shock velocity units in support of criteria for protection of salmonid eggs from blasting or seismic disturbance. In: *Incubation of Fish: Biology and Techniques, International Congress on Biology of Fish.* 21–26 July 2002, University of British Columbia, Vancouver.

Jensen, J.O. & Alderdice, D.F. (1983) Changes in mechanical shock sensitivity of coho salmon (*Oncorhynchus kisutch*) eggs during incubation. *Aquaculture,* **32**, 303–12.

Jensen J.O. & Alderdice, D.F. (1989) Comparison of mechanical shock sensitivity of eggs of five Pacific salmon (*Oncorhynchus*) species and steelhead trout (*Salmo gairdneri*). *Aquaculture,* **78**, 163–81.

Jhingran, V.S. & Pullin, R.S. (1988) *A Hatchery Manual for the Common, Chinese and Indian Major Carps.* International Center for Living Aquatic Resources Management, ICLARM Contribution 252.

Kamler, E. (1992) *Early Life History of Fish: An Energetics Approach.* Chapman & Hall, London.

Kempinger, J.J. (1988) Spawning and early life history of lake sturgeon in the Lake Winnebago system, Wisconsin. In: *11th Annual Larval Fish Conference, American Fisheries Society Symposium 5,* (ed. R.D. Hoyt), pp. 110–22. American Fisheries Society, Bethesda, MD.

Leggett, J.W. (1969) *The reproductive biology of the Dolly Varden char Salvelinus malma Walbaum.* MSc thesis, University of Victoria, BC.

Liley, N.R. (1982) Chemical communication in fish. *Canadian Journal of Fisheries and Aquatic Sciences,* **39**, 22–35.

Lindsey, C.C., Northcote, T.G. & Hartman, G.F. (1959) Homing of rainbow trout to inlet and outlet spawning streams at Loon Lake, British Columbia. *Journal of the Fisheries Research Board of Canada,* **16**, 695–719.

McCart, P. (1969) Digging behaviour of *Oncorhynchus nerka* spawning in streams at Babine Lake, British Columbia. In: *Symposium on Salmon and Trout in Streams,* (ed. T.G. Northcote), pp. 39–51. H.R. MacMillan Lecture Series in Fisheries, Institute of Fisheries, University of British Columbia, Vancouver.

Macdonald, J.S., Scrivener, J.C., Patterson, D.A. & Dixon-Warren, A. (1998) Temperatures in aquatic habitats: the impacts of forest harvesting and the biological consequences to sockeye salmon incubation habitats in the interior of B.C. In: *Fish-Forest Conference: Land Management Practices Affecting Aquatic Ecosystems,* (eds M.K. Brewin & D.M. Monita), pp. 313–24. Canadian Forest Service, Northern Forest Centre, Information Report Nor-X-356.

McDowall, R.M. (1990) *New Zealand Freshwater Fishes: A Natural History and Guide.* Heinemann Reed, MAF Publishing Group, Birkenhead, Auckland, New Zealand.

McElman, J.F. & Balon, E.K. (1985) Early ontogeny of walleye, *Stizostedeon vitreum,* with steps of saltatory development. In: *Early Life Histories of Fishes: New Developmental, Ecological and Evolutionary Perspectives,* (ed. E.K. Balon), pp. 92–131. Dr W. Junk Publishers, Dordrecht, the Netherlands.

Martin, N.V. (1957) Reproduction of lake trout in Algonquin Park, Ontario. *Transactions of the American Fisheries Society* **86**, 231–44.

Mommsen, T.P. & Walsh, P.J. (1988) Vitellogenesis and oocyte assembly. In: *Fish Physiology, Vol. XI, The Physiology of Developing Fish, Part A, Eggs and Larvae*, (eds W.S. Hoar & D.J. Randall), pp. 347–406. Academic Press, San Diego, CA.

Montgomery, D.R., Beamer, E.M., Pess, G.R. & Quinn, T.P. (1999) Channel type and salmonid spawning distribution and abundance. *Canadian Journal of Fisheries and Aquatic Sciences*, **56**, 377–87.

Moyle, P.B. & Cech, J.J. (1988) *Fishes: An Introduction to Ichthyology*. Prentice Hall, Englewood Cliffs, NJ.

Munro, A.D. (1990a) General introduction. In: *Reproductive Seasonality in Teleosts: Environmental Influences*, (eds A.D. Munro, A.P. Scott & T.J. Lam), pp. 1–11. CRC Press, Boca Raton, FL.

Munro, A.D. (1990b) Tropical freshwater fish. In: *Reproductive Seasonality in Teleosts: Environmental Influences*, (eds A.D. Munro, A.P. Scott & T.J. Lam), pp. 145–239. CRC Press, Boca Raton, FL.

Nehlsen, W., Williams, J.E. & Lichatowich, J.A. (1991) Pacific salmon at the crossroads: stocks at risk from California, Oregon, Idaho, and Washington. *Fisheries*, **16**, 4–21.

Newcombe, C. P. (2001) *Impact Assessment Model for Clear Water Fishes Exposed to Conditions of Reduced Water Clarity*. Ministry of Water, Land and Air Protection, Habitat Protection Branch, Victoria, BC.

Newcombe, C. & Hartman, G. (1973) Some chemical signals in the spawning behavior of rainbow trout (*Salmo gairdneri*). *Journal of the Fisheries Research Board of Canada*, **30**, 995–7.

Newcombe, C.P. & Hartman, G.F. (1980) Visual signals in the spawning behaviour of rainbow trout. *Canadian Journal of Zoology*, **58**, 1751–7.

Nicholas, J.W. & Hankin, D.G. (1988) Chinook salmon populations in Oregon coastal river basins: description of life histories and assessment of recent trends in run strengths. *Fishery Information Report 88–1*. Oregon Department of Fish and Wildlife.

Nikol'skii, G.V. (1963) *Special Ichthyology*. Israel Program for Scientific Translations Ltd, printed in Jerusalem by S. Monson.

Northcote, T.G. & Hartman, G.F. (1988) The biology and significance of stream trout populations (*Salmo spp.*) living above and below waterfalls. *Polskie Archiwum Hydrobiologii*, **35**, 409–42.

Peterson, N.P. & Quinn, T.P. (1996) Persistence of egg pocket architecture in redds of chum salmon, *Oncorhynchus keta*. *Environmental Biology of Fishes*, **46**, 243–53.

Ribbink, A.J., Marsh, B.A., Marsh, A.C., Ribbink, A.C. & Sharp, B.J. (1983) A preliminary survey of the cichlid fishes of rocky habitats in Lake Malawi. *South African Journal of Zoology* **18**, 149–310.

Ringstad, N. (1982) Carnation Creek watershed project freshwater sculpins: genus *Cottus* a review. In: *Proceedings of the Carnation Creek Workshop: A Ten-year Review*, (ed. G.F. Hartman), pp. 219–39. Pacific Biological Station, Nanaimo, BC.

Rombough, P.J. (1988) Respiratory gas exchange, aerobic metabolism, and effects of hypoxia during early life. In: *Fish Physiology, Vol. XI, The Physiology of Developing Fish, Part A, Eggs and Larvae,* (eds W.S. Hoar & D.J. Randall), pp. 59–161. Academic Press, San Diego, CA.

Scott, A.P. (1990) Salmonids. In: *Reproductive Seasonality in Teleosts: Environmental Influences*, (eds A.D. Munro, A.P. Scott & T.J. Lam), pp. 33–51. CRC Press, Boca Raton, FL.

Scott, A.P., Liley, N.R. & Vermeirssen, E.L. (1994) Urine of reproductively mature rainbow trout, *Oncorhynchus mykiss* (Walbaum), contains a priming pheromone which enhances plasma levels of sex steroids and gonadotropin 11 in males. *Journal of Fisheries Biology*, **44**, 131–47.

Scott, M.C. & Helfman, G.S. (2001) Native invasions, homogenization, and the mismeasure of integrity of fish assemblages. *Fisheries*, **26**, 6–15.

Scott, W.B. & Crossman, E.J. (1973) *Freshwater Fishes of Canada*. Fisheries Research Board of Canada, Bulletin 184.

Scrivener, J.C. & Brownlee, M.J. (1989) Effects of forest harvesting on spawning gravel and incubation survival of chum (*Oncorhynchus keta*) and coho salmon (*O. kisutch*) in Carnation Creek, British Columbia. *Canadian Journal of Fisheries and Aquatic Sciences*, **46**, 681–96.

Scrivener, J.C., Tschaplinski, P.J. & Macdonald, J.S. (1998) An introduction to the ecological complexity of salmonid life history strategies and of forest harvesting impacts in British Columbia. In: *Carnation Creek and Queen Charlotte Islands Fish/Forestry Workshop: Applying 20 Years of Coastal Research to Management Solutions*, (eds D.L. Hogan, P.J. Tschaplinski & S. Chatwin), pp. 23–8. British Columbia Ministry of Forests, Research Program, Land Management Handbook No. 41.

Sparrow, R.A., Larkin, P.A. & Rutherglen, R.A. (1964) Successful introduction of *Mysis relicta* Lovén into Kootenay Lake, British Columbia. *Journal of the Fisheries Research Board of Canada*, **21**, 1325–7.

St Onge, I. & Magnan, P. (2000) Impact of logging and natural fire on fish communities in Laurentian Shield lakes. *Canadian Journal of Fisheries and Aquatic Sciences*, **57** (Suppl 2), 165–74.

Sumpter, J.P., Scott, A.P., Baynes, S.M. & Witthames, P.R. (1984) Early stages in the reproductive cycle of virgin female rainbow trout (*Salmo gairdneri* Richardson). *Aquaculture, **43**, 235–42.

Sutherland, A.J. & Ogle, D.G. (1975) Effect of jet boats on salmon eggs. *New Zealand Journal of Marine and Freshwater Research*, **9**, 273–82.

Tautz, A.F. & Groot, C. (1975) Spawning behavior of chum salmon (*Oncorhynchus keta*) and rainbow trout (*Salmo gairdneri*). *Journal of the Fisheries Research Board of Canada*, **32**, 633–42.

Tschaplinski, P.J. (1987) *Comparative ecology of stream-dwelling and estuarine juvenile coho salmon (Oncorhynchus kisutch) in Carnation Creek, Vancouver Island, British Columbia*. PhD thesis, University of Victoria, BC.

Tweddle, D. (1983) Breeding behaviour of the Mpasa, *Opsaridium microlepis* (Gunther) (Pisces; Cyprinidae), in Lake Malawi. *Journal of the Limnological Society of South Africa*, **9**, 23–8.

Walther, B.T. & Fyhn, H.J. (1993) *Physiological and Biochemical Aspects of Fish Development*. University of Bergen, Norway.

Chapter 9
Foraging ecology: from the fish to the forest

M. KARAGOSIAN AND N.H. RINGLER

Introduction

The diversity of structures, physiology and behaviours associated with the acquisition of energy by fish is bewildering. This reflects the number of known species and their presence and persistence in virtually every aquatic habitat on earth (Moyle & Cech 2000). The physiochemical and biological differences in these habitats are reflected energetically by the many forms of food available to fish. Persistence of fish in these diverse environments is primary evidence that the fish have found ways to exploit these food resources and successfully convert them to reproducing offspring. Food intake is linked with several proximate measures of fitness, particularly rapid growth to a large size. Fish quickly graduate from sizes that make them vulnerable to gape-limited predators (e.g. Werner *et al*. 1983; Werner & Hall 1988; Turner & Mittelbach 1990; Sogard 1997) and successfully reach reproductive maturity if an adequate energetic intake is attained. At maturity the quality of eggs and young, parental care, some secondary sexual characters and behaviours, and overall reproductive success are directly responsive to ration size and foraging success (Stanley & Wooton 1986; Ridgway & Shuter 1994; Fletcher & Wooton 1995; Wiegmann *et al*. 1997; Ali & Wooton 1999; Smith & Wooton 1999).

Like most vertebrates (e.g. Holling 1959), fishes are concerned with two major issues while foraging: (1) finding food and (2) getting that food into the gut. Early investigations into the foraging behaviour of fish – Ivlev (1961) and Rashevsky (1959) – placed these two issues on each end of a continuum of constraints on foraging success as a function of the density of food in the environment. When the density of food in a given patch of environment is low, finding that food should occupy most of the activity of the fish and any effort invested into searching should result in increased foraging success. At higher food densities less time is needed for search, so the fish is constrained by how fast it can get that food into its mouth and move on to the next food item. A discovery–handling continuum provides us with a framework to mechanistically analyse the foraging ecology of fish in light of their amazing diversity. This is the organizing principle of the first two sections of this chapter. Selected components (and associated behaviours and structures) of the discovery and handling process in fish will be introduced, and the nature of the alteration of these components by ecological factors will be discussed. The third section of this chapter examines the ecology of foraging fish in selected aquatic systems (trophic environments). We emphasize those systems that are intimately connected energetically with the surrounding forested terrestrial ecosystem, with aspects of that connection influenced by the foraging of the fish there. It is hoped that the first part of the chapter will provide an individually based framework in which to think about how fish exploit these trophic environments.

Discovery

The subcomponents of the discovery process represent those behaviours that fish employ to search for food, detect it, discriminate it as food, and then decide to eat. These behaviours and associated morphologies represent adaptations that contribute to maximization of energetic intake and minimization of the time spent foraging (Ware 1982). Behavioural and morphological aspects of the searching process for fish discussed here include foraging mode, body plans and swimming speeds. Components of the detection and discrimination processes discussed are visual acuity, the ecology of the reactive distance and other behavioural measures of detection, and reactive volumes. The relation of these items to the selection process is considered.

Searching

Foraging mode

The tactics employed by a fish when searching for food can be placed along a continuum ranging from an ambush, sit-and-wait tactic to a pure cruising one (Fig. 9.1; Webb 1984; O'Brien *et al*. 1990). These tactics are correlated with the general physical structure and the available prey of the fish's environment (Webb 1984). Cruising fish forage on widely spaced individuals or patches of prey organisms in environments with little physical impediment to sustained swimming. An ambush tactic can be expected in a structurally complex environment with a high local availability of prey items. Characterization of each tactic is made by examining locomotor patterns (O'Brien *et al*. 1989), visualized as distance travelled/time. A cruising mode is indicated by a large distance travelled in a short period of time, while an ambush mode is characterized as a short distance travelled in a comparatively large period of time. In several species of fish an intermediate, saltatory mode has been recognized (O'Brien & Evans 1991; Karagosian 2001). A saltatory pattern is comprised of intermediate runs separated by frequent pauses that are hypothesized to facilitate search and repositioning while monitoring the environment for predators (O'Brien *et al*. 1989).

Body forms

There are morphological correlates in fish of the various foraging modes (Fig. 9.1). A cruising predatory fish must be able to sustain swimming for long periods of time, either moving through vast distances in the pelagia of large lakes or the ocean, or holding position in flowing water. A fish adopting an ambush foraging strategy, on the other hand, would need to generate a large amount of power for fast starts and accelerations over short time periods. Generalizations can be made about the body plans (shape, fin

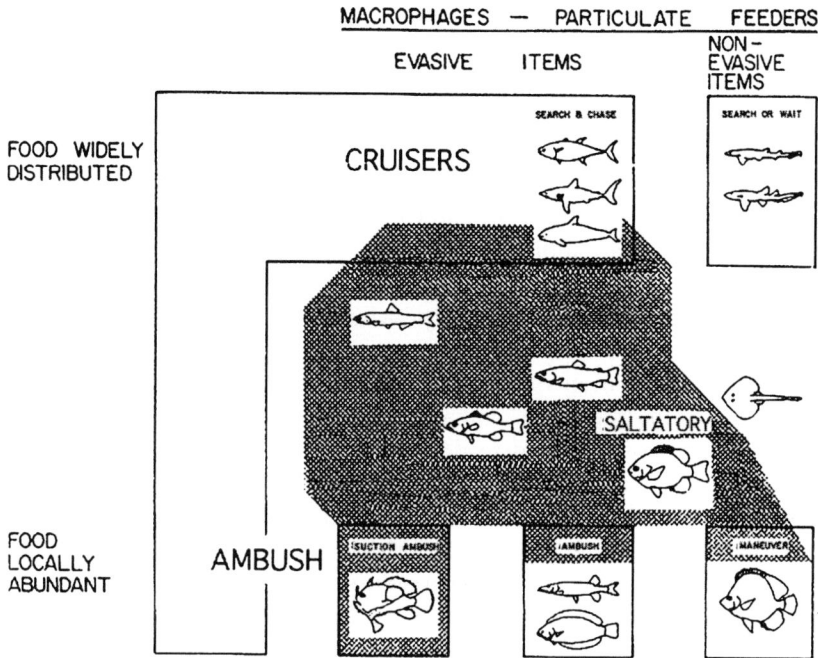

Fig. 9.1 The relationship between fish foraging mode, body form and properties of prey. Specialized forms are contained within blocks, intermediate and generalist forms are contained within the shaded region. Modified from Webb (1984).

size and placement, etc.) appropriate to these specialized strategies (Fig. 9.1) (Webb 1975, 1984). There are, however, innumerable potential prey that have intermediate distributions, the exploitation of which would be severely limited by specialization in body form. Many fish are thus locomotor generalists (Webb 1984), exhibiting body plans intermediate between those of cruising and ambush specialists (Fig. 9.1). These fish will exhibit more saltatory locomotor patterns and be able to handle a more biologically and physically variable environment (O'Brien *et al.* 1990).

Optimal swimming speeds

The adjustment of swimming speed adds flexibility to the body form–search strategy relationship, particularly for fish engaged in cruising or saltatory search. This flexibility enables the fish to match its foraging strategy with the ecological properties of its food and preserve the inherent efficiency of the strategy. Condition-specific optimal speeds of foraging can be identified that provide for a maximal rate of food intake and/or minimal energetic and temporal expenditures. These optimal speeds while foraging are demonstrably different from the optimal speed for non-foraging movement, with regard to both absolute value and the criteria upon which the optimality is based. Ware (1975, 1978) showed that an optimal speed for non-foraging swimming maximized the ratio of the distance travelled to energy used, while the optimal foraging swimming speed maximized net energy gain. Unique optimal swimming speeds exist for different densities of prey (Ware 1978) as well as prey types (O'Brien *et al.* 1989; but see Feldman & Savitz 1999).

Detection and response

The processes of detecting a potential prey item and discriminating it from other objects in the environment are additional subcomponents of discovery, and are intimately associated with and dependent upon the search strategy employed. The effectiveness of a given search strategy entails efficient positioning and repositioning of the sensory apparatus and the temporal and spatial geometry of its perceptual limits. Detection and response utilize a variety of sensory and central neural systems (Ringler 1994), either singly or in some integrated combination. Here we consider the utilization of one sensory modality, vision, for the detection and discrimination of a prey item or items. Vision is certainly one of the most dominant sensory modalities for foraging in fish, and by far the most studied with regard to its contribution to foraging behaviour. Many commercially and recreationally valuable fish species are particulate feeders, be they piscivores or planktivores, and thus vision is the predominant method of finding food. Recognition of the importance of fish predation as a form of 'top-down' regulation of productivity in many aquatic systems (Kerfoot & Sih 1987; Lazzaro 1987; Northcote 1988; Gerking 1994) has stimulated a great deal of research into the behavioural aspects of visual feeding behaviour over the past five decades.

Measuring visual acuity

Quantification of visual fish foraging behaviour usually begins with measures of visual acuity and range. The size and position of the lens of a fish eye determines the size and distance of a particle that can be properly focused on the retina. The lens is fixed in shape and is the only portion of the fish eye having a refractive index different from water. The normal 'relaxed' position of the lens within the eye allows for resolution of nearfield objects while more distant objects are resolved by moving the lens (accommodation), often in a dynamic interplay with displacement of the fish toward the object (Trevarthen 1968). The degree and direction of accommodation varies among fish species and is correlated with foraging behaviour, prey type and the degree of dominance of vision in feeding (Sivak 1973).

Visual acuity increases with fish size (Browman *et al.* 1990; Walton *et al.* 1994; Wanzenbock *et al.* 1996). A measure of visual acuity in fish is the minimum separable angle (MSA), the smallest visual angle a detected prey object subtends on the fish retina. The MSA is a function of lens diameter and cone density, so it is inversely proportional to fish size. The MSA can be measured behaviourally via the reactive distance of fish to prey of various sizes (Hairston *et al.* 1982). Reactive distance is a fundamental behavioural measure of visual acuity and is the basis for the geometric determination of visual fields and volumes that are essential in describing the process of discovery and detection.

Reactive distance

The reactive distance of a fish to its prey is defined as that distance from the fish a particle is recognized as a potential prey item by the fish and an attack is commenced (Gerking 1994). Reactive distance is measured as the distance between the prey and a point either on the snout or between the eyes of the fish when the fish oriented towards the prey and began to accelerate towards it. We can view reactive distance as the outward manifestation of an integration of sensory, cognitive and digestive (satiative), and motivational (Colgan 1973) processes which is subsequently modified by prevailing ecological conditions and features of the prey organisms.

The general pattern which emerges is that reactive distance is a sigmoidal function of light intensity (e.g. Confer *et al.* 1978). At a fixed light level, various properties of the prey will determine the reactive distance of a fish. The most influential of these properties are revealed by studies which have observed, in the mixed zooplankton prey assemblages of many lakes and ponds, that certain prey forms or types are more susceptible to differential predation and possible elimination by planktivorous fish. Among the first to document such phenomena were Hrbacek (1962) and Brooks & Dodson (1965), who found that in lakes inhabited by large populations of planktivorous fish the larger-bodied zooplankton tended to be eliminated or reduced numerically, and smaller body forms dominated the prey assemblage. This is in contrast to geographically and physically similar lakes without a significant planktivore, where large-bodied forms dominated. A subsequent 'size efficiency hypothesis' was posed (Brooks & Dodson 1965) which generalized the above findings and stimulated a great deal of research into the mechanisms involved in differential predation based on body size. Reactive distance is an increasing function of prey body size for planktivorous fish. Several studies have shown a linear relationship (Confer & Blades 1975; Schmidt & O'Brien 1982; Wright & O'Brien 1982, 1984; Link 1998), while others have revealed a decelerating, asymptotic increase in reactive distance as a function of prey size (Ware 1973; Werner & Hall 1974; Confer *et al.* 1978; Luecke & O'Brien 1981; Dunbrack & Dill 1983).

Other physical features of potential prey items may override size as the chief determinant of vulnerability to predation. Zooplankton of higher relative contrast, in the form of pigmentation and colour differences, are more susceptible to fish predation than those of lower relative contrast (Zaret 1972; Zaret & Kerfoot 1975), because fish have greater reactive distances to higher contrast forms (Wright & O'Brien 1982). Similarly, fish have greater reactive distances to moving prey items than stationary ones (Ware 1973; Wright & O'Brien 1982, 1984). Lazzaro (1987), O'Brien (1987) and Gerking (1994) have reviewed the effects of light and prey features such as size, contrast and movement on the reactive distance of planktivorous fish.

Reactive distance is also a function of the current neural and systemic state of the fish. Reactive distance increases with experience (Ware 1971) and is inversely proportional to the level of satiation (Confer *et al.* 1978).

Turbidity and the reactive distance

Increased turbidity in aquatic systems is a common result of forestry activities (Tebo 1955; Brown & Krygier 1971; Steedman & France 2000) and can have great consequences for the fish communities present in these systems by impairing visually based foraging (Utne-Palm 2002). Increases in suspended particulates can impair growth of fish via diminished foraging performance, altered activity levels, and stress (Berg & Northcote 1985; McLeay *et al.* 1987; Benfield & Minello 1996; but see Boehlert & Morgan 1985; Gregory & Northcote 1993; Utne-Palm, 2002). Reactive distance is a decreasing, non-linear function of turbidity level (Vinyard & O'Brien 1976; Wright & O'Brien 1984; Barrett *et al.* 1992; Gregory & Northcote 1993; Benfield & Minello 1996; Sweka & Hartman 2001).

Alternative behavioural measures

The particular foraging mode used by a fish may necessitate alternative measures of visual perception and performance. Cruising fish predominantly search for and detect prey at the periphery of their visual volume, while saltatory searchers scan their entire visual volume (O'Brien *et al.* 1990). Mean reactive distance of cruising fish will then likely reflect the maximum distance at which a particular prey item can be detected, while the mean reactive distance for saltatory searchers will underestimate this value. To estimate the maximum distance at which a particular prey can be detected by a saltatory fish, O'Brien *et al.* (1990) suggested using the value of mean reactive distance which bounds 90% of the observations in the mean reactive distance distribution, the maximum location distance (MLD). For bluegill sunfish and threespine sticklebacks, MLD was greater than the mean reactive distance (Walton *et al.* 1994, 1997; Karagosian 2001).

Detailed examination of the prey capture behaviour of drift-feeding fish highlights an additional subtlety in interpreting the information contained within reactive distance. Drift-feeding fish usually hold a position close to the substrate and make forays away from this holding position to capture drifting food (e.g. Kalleberg 1958; Hartman 1963; Ringler 1979a, 1985). This holding position is in a flow refuge (or low flows) adjacent to higher or unobstructed flows which carry drifting food (Wankowski 1981), and represents a compromise between expending energy holding position and foraging in areas of a high drift density or rate (Fausch 1984). As the drifting food is in motion relative to the fish, the distance the fish must travel to capture the prey item is often different from the reactive distance (Wankowski 1981), and dependent upon water velocity (Wankowski & Thorpe 1979; Hughes & Dill 1990). This distance has been termed the capture or attack distance (Dunbrack & Dill 1983; Flore *et al.* 2000) and is measured as the distance between the positions of the fish's snout when it reacted to and consumed the prey. Attack or capture distance, for a given prey size, fish size and water velocity, is generally less than the reactive distance (Dunbrack & Dill 1983; Metcalfe *et al.* 1986; O'Brien & Showalter 1993; but see Flore *et al.* 2000) and inversely proportional to water velocity (Godin & Rangeley 1989).

Visual fields and volumes

Quantification of reactive distance, MLD, or attack (capture) distance is essential to the construction of search areas and volumes. Search areas and volumes govern encounter rates with prey and are useful models of the discovery process. There are several methods of determining these areas and volumes that are dependent upon the foraging mode of the fish under study.

For fish which search while stationary, such as saltatory searchers, the search volume is roughly spherical with a radius equal to the reactive distance or the MLD (Luecke & O'Brien 1981; Wright & O'Brien 1984; O'Brien *et al.* 1989; Walton *et al.* 1994). Reactive distances and capture success are greatest in the forward-directed hemispheres, often slightly above or to the side of the fish's head (Luecke & O'Brien 1981; Schmidt & O'Brien 1982; Wright & O'Brien 1984). This led to O'Brien *et al.* (1989) calculating the search volume of white crappie, which employs a saltatory search strategy, as a wedge-shaped region that is a function of the MLD (Fig. 9.2A).

The reactive volumes of fish that utilize a cruising strategy or drift feed must take into consideration movement of the fish or the flow past the fish, respectively. Reactive distance measured in each of three dimensions about a cruising fish is greatest in the forward-directed region (Confer *et al.* 1978; Dunbrack & Dill 1984; Link 1998). In addition, there is more of a contribution from laterally and upwardly directed reactive distances in these fish than stationary searchers, so the shape of these areas can be quite complex (Confer *et al.* 1978). To compute the volume searched for cruising fish, search area is multiplied by the distance moved by the fish (e.g. Eggers 1977). There are several rapid methods of estimating search volume for a cruising fish (see Link 1998).

The reactive volume of a drift-feeding fish is computed by determining the product of the search area and the water velocity. Fausch (1984), drawing upon qualitative features of Wankoswki's (1981) observations of drift-feeding Atlantic salmon, hypothesized that position-holding drift-feeding fish (coho salmon, brook and brown trout) foraged in an area that was one-quarter of a semicircular area with a radius equal to two body lengths. Hughes & Dill (1990), studying drift-feeding Arctic grayling, provided a refinement (Fig. 9.2B) by incorporating maximum capture distance (MCD, see above), which is an increasing function of reactive distance and maximum sustainable swimming speed and a decreasing function of water velocity. As a result the search area, and thus the drift-feeding search volume, is a decreasing function of water velocity (Hughes & Dill 1990; O'Brien & Showalter 1993) but is larger in flowing water than standing water (Flore *et al.* 2000).

Lowered light intensity shrinks the reactive volume (Luecke & O'Brien 1981; Schmidt & O'Brien 1982; Wright & O'Brien 1984) while larger prey sizes will increase the reactive volume for saltatory and cruising fish (Schmidt & O'Brien 1982; Wright & O'Brien 1984; Link 1998). The search volume of drift-feeding juvenile nase is a linear function of fish size (Flore *et al.* 2000).

A.

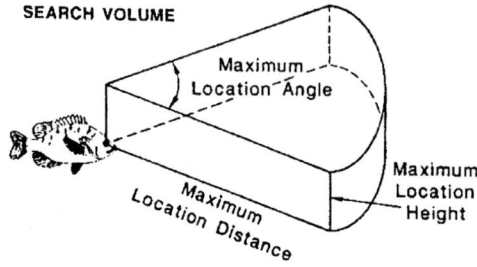

B.

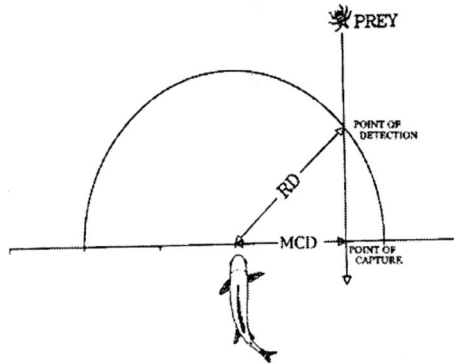

Fig. 9.2 Mode-specific foraging areas and volumes. (A) Search volume for a fish engaged in saltatory search. Modified from O'Brien *et al.* (1989). (B) Drift-feeding area model incorporating attack distance (AD). RD = reactive distance. Modified from Hughes and Dill (1990).

Reactive volumes and prey selection

Because reactive and search volumes describe the geometry of the space monitored by fish, constrained by sensory limits and ecological factors, they represent a 'snapshot' of environmental information. Of the array of prey items in the environment, then, the fish will be visually aware of only those within its search volume. These prey items represent the 'effective' prey density from which to select prey (Werner & Hall 1974), distinct from the overall environmental density of food. Many ecologists have investigated the nature of the selection process when fish are confronted with multiple prey types (different species, body forms, body sizes, movement patterns, etc.) within their search volume (see reviews by Hyatt 1979; Ringler 1979b, 1994; Lazzaro 1987; Gerking 1994). The various theories and predictive models of selection (Werner & Hall 1974; Confer & Blades 1975; O'Brien *et al*. 1976; Gibson 1980; Eggers 1982; Wetterer & Bishop 1985; Bannon & Ringler 1986) are direct conceptual descendants of the pioneering fieldwork that established a prey-size basis for prey selection by planktivorous fish (Hrbacek 1962; Brooks & Dodson 1965). The predictive models generally hold up well under given experimental and observational environmental conditions.

Handling

'Handling' is a term encompassing those activities, separate from search, to which an organism must devote time and energy in order to capture a prey item (Holling 1959). For a fish this term encompasses pursuit, capture, retention and processing (O'Brien *et al*., 1989, 1990). Each must be sequentially completed successfully for the successful ingestion of food. Handling and searching were historically viewed as distinct phenomena, where engaging in one precluded performing the other (Holling 1959). Studies of fish that utilize a saltatory search mode, however, indicate that searching can take place during the sequence of events that make up handling (Visser & Reinders 1981; Karagosian 2001), blurring the temporal distinction between the two. Their separation is useful, however, in a qualitative discussion. Like the discovery process, handling is dependent upon features of the prey species, and there are morphological correlates. These correlates are readily observable in jaws, pharyngeal teeth and gill rakers, and help to maximize the intake of energy and minimize the time spent per prey individual.

Jaws

The evolution of jaws in fish has been one of increasing flexibility (Lauder 1982). Articulation of the mandible and separation of the premaxilla from the skull have resulted in one of the most important morphological features and evolutionary steps of fish, protrusible jaws (Alexander 1967; Lauder 1982; Gerking 1994). The protrusible mouth allows the fish to accelerate the volume of water surrounding a prey item toward the mouth, greatly increasing the probability of capture for small (relative to the size of the fish) prey items. Variations on this theme in many groups of fish have produced some remarkable and highly specific morphological correlations with some prey items. There are specialized jaw structures for scraping, biting, picking, crushing, grabbing, etc., often within (intra- and inter-species) groups of fish that enable coexistence via food partitioning (Luczkovich *et al*. 1995; Barlow 2000).

Teeth

There is a close match between the form of the pharyngeal teeth and prey type in fish (Wootton 1992; Bond 1996). This match serves to increase the probability of success in capture and retention and/or the efficient extraction of energy in the processing step.

There are both inter-species and intra-species examples of these different adaptations (Fig. 9.3A). For example, piscivores possess canid teeth which pierce and hold highly mobile, soft-bodied prey. This facilitates successful manipulation and positioning of the fish prey for swallowing (Juanes & Conover 1994; L'Abee-Lund *et al.* 1996). Molluscivores, insectivores, and fish that feed on other aquatic macroinvertebrates, on the other hand, are typified by molariform teeth (Wootton 1992) enabling the fish to crush hard shells and chitinous cuticles to expose the energy-rich soft internal parts of the prey. There are many other diverse pharyngeal specializations of fish that facilitate the extraction of energy via the shredding, scraping, piercing and crushing of prey (Wootton 1992; Bond 1996).

Gill rakers

Fish that feed on zooplankton or suspended particulate algae (suspension feeders) (Moyle & Cech 2000) have reduced pharyngeal teeth and rely on other structures such as specialized collecting surfaces (Sanderson *et al.* 1991) and packed long, fine, sieve-like gill rakers (Jobling 1995) to strain and collect small particles (Fig. 9.3B). Piscivores and other macrophagous fish have more stout, widely spaced and stronger gill rakers that reduce the risk of blockage or the potential for injury (Fig. 9.3B).

Handling time

A components view of handling as a sequential process suggests several points at which diverse biological and physical phenomena can act to modify handling time. Fish and prey size are the predominant influences on the time spent by the fish on each prey item (Werner 1974; Bindoo & Aravindan 1992), with handling time inversely proportional to fish size (gape width) for a given size prey. This is the basis for diet expansion as a fish grows (Gerking 1994). The time devoted to handling individual prey items is also influenced by motivational processes, with handling time increasing as satiation is approached (Ware 1972; Werner 1974; Bindoo & Aravindan 1992; Karagosian 2001).

Trophic environments

Forestry practices can impact all lotic and lentic aquatic systems within a watershed (see Chapter 1). The unperturbed, or natural, constancy and/or variability of these systems have shaped the evolution of fish species, populations and communities that are suited to partition and exploit the various energy sources found in each. The ecology of the energy sources, and the constraints that must be dealt with by fish in exploiting them, make up the trophic environment of the system under consideration. In this section, the trophic environment of two major types of aquatic systems, temperate low order lotic environments and tropical forested floodplains, will be discussed. These two types of systems are emphasized for two reasons. First, the influence of the forest on the trophic environment of these systems is physically and biologically direct. The forest provides physical structure, maintains habitat heterogeneity and supports the

A.

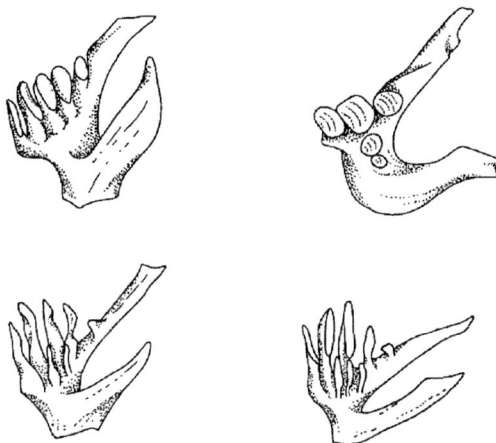

B.

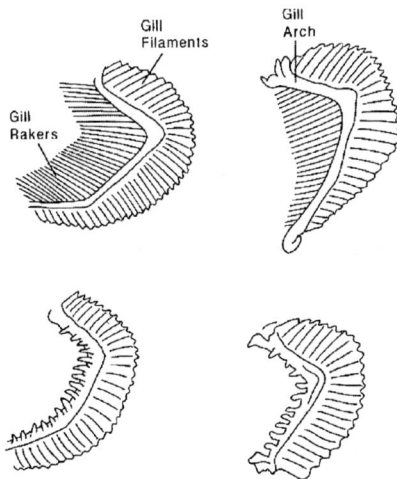

Fig. 9.3 Morphological correlation of pharyngeal teeth and gill rakers with preferred prey types. (A) Pharyngeal teeth of various cyprinid species. Top figures: molluscivores and plant eaters; bottom figures: insectivore (insect larvae) and crustacean eater. (B) Gill rakers from planktivores (top figures), an insectivore (bottom left), and a piscivore (bottom right). From Jobling (1995).

production of fish with direct energetic subsidies and/or by driving the production of lower trophic levels. Second, there is clear evidence of a significant energetic reciprocity between these aquatic systems and the nearby forest that is mediated by the feeding habits of fish.

Low order lotic systems

Streams have by far been the most extensively studied aquatic systems with regard to forestry impacts (see Chapters 1 and 13), given their ubiquity within forested watersheds and their importance as habitat for the natural production of commercially and recreationally valuable salmonids. The effects of altering the surrounding terrestrial habitat are particularly acute for these lower order systems because of (1) their small size (particularly wetted width and depth) and (2) their dependence upon allochthonous material and energy sources to support the production of lower trophic levels. The small wetted width of the stream relative to the area of the canopy of the bordering forest results in the blockage of large amounts of solar energy, leading to low levels of primary production. As a result, the primary production of the surrounding terrestrial environment is a necessary subsidy to secondary production of the lower trophic levels of streams (Minshall 1967; Vannote *et al.* 1980; Wallace *et al.* 1997) and thus to the production of fish.

Foraging in low order forested streams

Flowing water necessitates drift-feeding for many fish, while the constrained cross-sectional area of streams puts a premium on adequate feeding positions. As a result, intra-species competition for foraging space is a major structuring force of the fish communities in these systems, and territorial behaviour is the chief mechanism by which this structure is maintained. Territory size is an increasing, decelerating function of body size in juvenile salmonids (Grant & Kramer 1990) that is density-independent (Elliot 1990; but see Bohlin *et al.* 1994) and inversely proportional to food (drift) density (Grant *et al.* 1998) and water velocity (Kalleberg 1958). Territory size is thus a reliable predictor of the density of juvenile salmonids in small streams (Grant & Kramer 1990). Because of its relationship with body size, territorial adjustment during growth is one of the chief density-dependent mechanisms of self-thinning in stream populations (Chapman 1966; Elliot 1990; Bohlin *et al.* 1994; Grant *et al.* 1998). For a given stream section the mosaic of salmonid territories reflects a dominance hierarchy because territories are heterogeneous with regard to profitability, with the most dominant fish occupying the best positions (Kalleberg 1958; Chapman & Bjornn 1969; Bachman 1984; Elliot 1990). Ritualistic displaying and physical contact are the behavioural methods by which territories are acquired and defended (Keenleyside & Yamamoto 1962) and the dominance hierarchies maintained. The frequency and extent of this costly defence can be ameliorated by increases in habitat heterogeneity, which enhances not only the number of refuges available for drift-feeders, but also the degree of visual isolation between neighbouring territory holders (Kalleberg 1958).

Energetic subsidization

Habitat heterogeneity and its influence on stream fish-feeding territories, and thus density and production (Bardonnet & Bagliniere 2000), is a product of channel characteristics. In small streams, geomorphology and large woody material from the surrounding forest interact dynamically to produce the physical character of a stream channel (see Chapters 1 and 3). Allochthonous input of large woody debris (LWD) thus can directly influence the quality and number of feeding positions in a stream. In addition, LWD is habitat for many stream macroinvertebrates, which use this material as refuge or food; thus the presence of LWD can increase the density of many organisms in the stream (see Chapters 1 and 3). An increase in prey density translates into increased prey drift, thereby elevating the quality of existing feeding territories.

Aside from providing structural elements that indirectly benefit resident fish, the surrounding forest supplies direct energetic subsidies in the form of terrestrial arthropods (Mason & MacDonald 1982; Garman & Moring 1993). These items can represent the most significant source of energy throughout the year and be crucial to the persistence and production of stream fish (Nakano *et al.* 1999a). In a second order stream rainbow trout selectively fed on terrestrially derived drift, as these prey items were larger than benthically derived larval and adult prey (Nakano *et al.* 1999b). Active selection and preference of abundant and energetically superior terrestrially derived arthropods by fish can significantly reduce fish predation pressure on benthic invertebrates, an effect that can have ramifications for the production of more distal trophic groups in streams (Nakano *et al.* 1999a).

Because terrestrial arthropods are a necessary energetic subsidy in low order streams, alteration of riparian and adjacent forested areas can have significant energetic consequences for stream fish. Unaltered grassland and forested riparian areas contributed significantly more arthropod biomass to the drift of several New Zealand streams than did altered (introduced non-native vegetation and grazed areas) riparian areas (Edwards & Huryn 1996). Similarly, more structurally complex and dense riparian areas contributed more to the drift of several Alaskan streams (Wipfli 1997), where terrestrially derived foods are a significant late-season energy source for several salmonids. Experimentally blocking the input of terrestrial invertebrates into reaches of a headwater stream led to a greater reliance of fish on stream benthos during those seasons when terrestrially derived drift was usually the most important and superior energy source for resident salmonids (Nakano *et al.* 1999b).

Energetic reciprocity

Energetic subsidization is not a one-way street in low order lentic systems, as the aquatic habitat can provide significant energetic and nutritive inputs that enhance the production of different levels of nearby terrestrial food webs. Several of these inputs are mediated directly or indirectly by fish foraging (Table 9.1). Spawning salmon (*Onchorhynchus* spp.) bring downstream or marine-derived nutrients and energy to upstream reaches of forested streams (e.g. Rand *et al.* 1992; Kline *et al.* 1993). Biomass accumulated while these semelparous fish were feeding and growing in the ocean or

Table 9.1 Some examples of energetic reciprocities between streams and floodplain environments and their nearby terrestrial ecosystems

	Terrestrial–aquatic	Aquatic–terrestrial
Streams	Energetic subsidies	Decrease export of energy
	• Directly to fish	• Depress emergence of aquatic
	• Indirectly via lower trophic levels	insect adults
	Physical subsidies	Import of marine/downstream
	• Large woody debris	nutrients
	• Directly provides habitat/station	• Carcass decomposition
	for fish	• Carcass scavenging
	• Directly provides habitat for	• Predation on spawners
	lower trophic levels	
	• Indirectly alters fish habitat by	
	modifying channel	
	characteristics	
Flooded forest	Energetic subsidies	Alter plant distribution
	• Directly to fish	• Seed predation
		• Seed dispersal
		Import of nutrients
		• Fish defecation, decomposition
		• Predation on fish

in large freshwater lotic systems can be transferred to adjacent terrestrial systems via carcass decomposition (Johnson & Ringler 1979) and scavenging and predation on spawners by terrestrial vertebrates (Willson *et al.* 1998; Hilderbrand *et al.* 1999a; Restani *et al.* 2000; Helfield & Naiman 2001). These nutrients represent an important contribution to both terrestrial plant and vertebrate productivity (Hilderbrand *et al.* 1999a, 1999b). More indirectly, fish predation can depress the export of emerging adults of aquatic invertebrates which can represent a significant contribution to the yearly energy budget of the insectivores of nearby terrestrial systems (Jackson & Fisher 1986; Nakano & Murakami 2001). Because reduction of terrestrial input into the drift can shift salmonid predation pressure from terrestrially derived drift to benthic organisms (Nakano *et al.* 1999b), this reciprocity is dynamic and under the influence of multiple interacting controlling factors.

Forested floodplains and flooded forests

One of the most intimate relationships between fish, their aquatic habitat and the surrounding forested region can be found in river systems with forested floodplains. Some of the most conspicuous examples are the flooded forests of the Amazon and its tributaries. Levels of primary production in a great portion of the waters of the Amazon and its tributaries are low due to a chronic shortage of mineral salts and the presence of humic acids (Gottsberger 1978; Goulding 1980). Thus, the input of allochthonous energy sources is a year-round necessity for the trophic function of these systems and the support of fish production. Terrestrially derived foods such as leaves, arthropods, flowers and monkey faeces are consumed year-round by adult fish (Goulding 1980). In the Amazon watershed there is a close temporal match between the phenologies of the plants of the flooded forest and the fish (particularly the characins of the genera *Mylossoma*, *Colossoma* and *Brycon*) that eat them (Gottsberger 1978; Goulding 1980). Parallel evolutionary modification of plant and fish life histories in this system is evidenced by complementary reproductive and feeding adaptations. The onset of the high water season triggers both fruit production and fish movement upstream into tributaries and then into the flooded forest. Once in the flooded forest there is a bounty of seeds, fruits and nuts that the fish exploit, and these items dominate the diet of these fish.

Energetic subsidization and foraging in the forest

The energy budget of an adult fish in a given year is predominantly balanced by foraging for 5 or 6 months during this wet season within the flooded forest on fruits, nuts and seeds. This reliance is so pronounced that the large characins can be considered terrestrial herbivores (Horn 1997). Aggregations of fish are often found underneath the branches of the plants producing these items, attracted by the sound of falling nuts and fruits. These fish may even be anticipating fruit fall after recognizing, possibly by the submerged trunks or buttressed roots, a given tree as one that could potentially provide food (Gottsberger 1978; Goulding 1980), suggesting a great capability for learning on the part of these fish in such a complex environment. During this flooding period fat stores are created which sustain both activity through the dry season, and gonad production and spawning movements in the wet season (Goulding 1980). As a result, very high production rates and large standing crops of fish are maintained in these nutrient-poor waters (Gottsberger 1978).

Nuts and seeds without a fleshy matrix dominated the diet of large characins foraging in the flooded forest of an Amazon tributary (Goulding 1980). The massive mandibles, strong jaw muscles, and multicuspid molariform and incisive teeth of these fish are suited for crushing and chopping these very hard, high protein items that are unavailable to most other fish because of their inability to process hard seeds. This processing destroys the seeds, so large characins represent significant sources of seed mortality. Seeds embedded in fleshy fruits, however, realized greater post-consumptive viability (as measured by germination success of recovered seeds; Gottsberger 1978; Goulding 1980) because of their smaller size and the tendency of fish to swallow many of them whole when eating pieces of fruit. The level of destruction of seeds by fish is also dependent upon fish size and species. Larger *Colossoma* spp. generally require less mechanical processing of material before swallowing, allowing more seeds to pass unharmed. Catfish (*Doradid* spp.) lack the dentition powerful enough to crush seeds, so those seeds were generally still viable after catfish consumption (Kubitzki & Zuburski 1994; Souza-Stevaux *et al.* 1994). Within the characins, differential mortality of seeds is realized at the generic level. Goulding (1980) found that rubber tree seeds were always crushed by *Colossoma* before ingesting them, while some rubber tree seeds ingested by *Brycon* spp. were spared. This is due to the large broad molars of *Colossoma*, used to crush and ingest all portions of the seed. *Brycon*, on the other hand, with its peg-like teeth often cracked the hard outer shell (which has little nutritional value) and swallowed the intact kernel. Differences in dentition among the characins also lead to differential choice of seed type, enabling them to effectively partition the food base of the flooded forest and separate themselves ecologically (Goulding 1980). This wet season partitioning (coupled with the unique ability to create fat stores that essentially allows these fish to stop feeding in the dry season) makes these fish exceptions to the general rule in the trophic biology of tropical fishes, that the greatest amount of segregation and feeding specialization occurs during the dry season when food is limiting and competition is greatest (Lowe-McConnell 1987).

Reciprocity

These interacting phenomena of selection by fish and the differential mortality imposed on the various forms of plant seeds in the flooded forest highlight the reciprocity that exists between the fish and plants of the flooded forest (Table 9.1). The plants provide the vast majority of the energy necessary to balance the yearly energy budget of adult fish, while the fish, although significant seed predators, are dispersal vectors for several species of plants. Gottsberger (1978) notes that most of the species of flooded forest plants which are dispersed by fish are primitive angiosperms, suggesting that ichthychory is a primitive form of dispersal that has been maintained in the Amazon because the fish in these nutrient-poor waters depend so heavily on allochthonous material.

There are several ecological issues that must be addressed for fish to be successful dispersal agents. Broadly, these issues are: (1) the level of mortality of seeds in the gut; (2) the quantity and frequency of intact seeds in the gut; and (3) the probability of the seeds reaching a favourable germination spot. The degree of seed mortality is related to the degree of processing and form of seed (i.e. naked seed or nut versus fruit) as discussed above, as well as the length of time in the gut and the resilience of the unbroken protective sheath or pericarp. Issue 2 is often met, as we saw above, given the selection and the consumption of large quantities of the product of a single plant while aggregating below it (Goulding 1980). Issue 3 is met via the movement of the fish back into the tributaries and main channel after foraging in the flooded forest, often proceeding upstream, and bringing the seeds with them (Kubitzki & Ziburski 1994; Souza-Stevaux *et al.* 1994; Horn 1997).

Consideration of these issues in relation to other possible modes of dispersal of the seeds of these rainforest plants highlights the importance of fish in these systems, revealing the extent to which these regions are highly integrated systems (Kubitzki & Ziburski 1994). Aside from fish, water flow and other vertebrates (e.g. monkeys, turtles and birds) have been suggested as dispersal agents (Goulding 1980; Kubitzki & Ziburski 1994). Among the vertebrates Kubitzki & Ziburski (1994) rank fish first among the seed dispersal animals in the rainforest by virtue of the quantity and frequency of ingestion of the various forms of seeds and the upstream movement by the fish associated with spawning. This ranking is irrespective of the fact that fish are also the top ranked seed predator (Goulding 1980). This importance is growing even more as deforestation for lumber and creation of agricultural space reduces the habitats available for the terrestrial vertebrates, hence reducing their abundance and potential as dispersal agents.

Summary and concluding remarks

On a gross level, discovery and handling are the main functional elements in energy acquisition. Visually based foraging in fish, particularly the planktivores, is by far the most studied aspect of fish feeding ecology and lends itself well to functional generalizations. Similarly, consideration of a fish's environment as an energetically dynamic system, of which the individual fish and the surrounding terrestrial system are integral elements, suggests some functional interpretations. Two such energetic phenomena, mediated and influenced in part by the feeding of fish, are subsidization and reciprocity. Low order streams and forested floodplains are two types of ecosystems where the action of these phenomena is striking and most susceptible to perturbation by forestry practices.

The assessment of the impact of forestry practices on fish production necessitates an understanding of the ecology of the energy acquisition process on several levels. In this chapter we considered aspects of two levels, individual fish behaviour and the environment in which it is found. A complete assessment would include energetic consideration of ecological physiology (e.g. digestive and stress physiology), population ecology (e.g. life histories, stock recruitment, predator–prey interactions) and seasonality in biological and physiochemical elements of the environment. Integrative approaches utilizing multi-level information, such as bioenergetic and individual-based population modelling, are being constantly refined. Their application to fish production–forestry issues could hold much promise in gaining insight into environmental issues and generating hypotheses for explanatory and, ultimately, rehabilitative and preventive action.

References

Alexander, R.M. (1967) *Functional Design in Fishes*. Hutchinson, London.

Ali, M. & Wooton, R.J. (1999) Effect of variable food levels on reproductive performance of breeding female three-spined sticklebacks. *Journal of Fish Biology*, 55, 1040–53.

Bachman, R.A. (1984) Foraging behaviour of free-ranging wild and hatchery brown trout in a stream. *Transactions of the American Fisheries Society*, 113, 1–32.

Bannon, E. & Ringler, N.H. (1986) Optimal prey size for stream resident brown trout (*Salmo trutta*): tests of predictive models. *Canadian Journal of Zoology*, 64, 704–13.

Bardonnet, A. & Bagliniere, J-L. (2000) Freshwater habitat of Atlantic salmon (*Salmo salar*). *Canadian Journal of Fisheries and Aquatic Sciences*, 57, 497–506.

Barlow, G.W. (2000) *The Cichlid Fishes*. Perseus, Cambridge, MA.

Barrett, J.C., Grossman, G.D. & Rosenfeld, J. (1992) Turbidity-induced changes in the reactive distance of rainbow trout. *Transactions of the American Fisheries Society*, 121, 437–43.

Benfield, M.C. & Minello, T.J. (1996) Relative effects of turbidity and light intensity on reactive distance and feeding of an estuarine fish. *Environmental Biology of Fishes*, 46, 211–16.

Berg, L. & Northcote, T.G. (1985) Changes in territorial, gill-flaring, and feeding behaviour in juvenile coho salmon (*Oncorhynchus kisutch*) following short-term pulses of suspended sediment. *Canadian Journal of Fisheries and Aquatic Sciences*, **42**, 1410–17.

Bindoo, M. & Aravindan, C.M. (1992) Influence of size and level of satiation on prey handling time in *Channa striata* (Bloch). *Journal of Fish Biology*, **40**, 497–502.

Boehlert, G.W. & Morgan, J.B. (1985) Turbidity enhances feeding abilities of larval Pacific herring, *Clupea harengus pallasi*. *Hydrobiologia*, **123**, 161–70.

Bohlin, T., Dellefors, C., Faremo, U. & Johlander, A. (1994) The energetic equivalence hypothesis and the relation between population density and body size in stream-living salmonids. *The American Naturalist*, **143**, 478–93.

Bond, C.E. (1996) *Biology of Fishes*, 2nd edn. Saunders, New York.

Brooks, J.L. & Dodson, S.I. (1965) Predation, body size, and composition of plankton. *Science*, **150**, 28–35.

Browman, H.I., Gordon, W.C., Evans, B.I. & O'Brien, W.J. (1990) Correlation between histological and behavioral measures of visual acuity in a zooplanktivorous fish, the white crappie (*Pomoxis annularis*). *Brain Behavior and Evolution*, **35**, 85–97.

Brown, G.W. & Krygier, J.T. (1971) Clear-cut logging and sediment production in the Oregon Coast Range. *Water Research*, **7**, 1189–98.

Chapman, D.W. (1966) Food and space as regulators of salmonid populations in streams. *The American Naturalist*, **100**, 345–57.

Chapman, D.W. & Bjornn, T.C. (1969) Distribution of salmonids in streams with special reference to food and feeding. In: *Symposium on Salmon and Trout in Streams*, (ed. T.G. Northcote), pp. 153–76. University of British Columbia, Vancouver.

Colgan, P. (1973) Motivational analysis of fish feeding. *Behaviour*, **45**, 38–66.

Confer, J.L. & Blades, P.I. (1975) Omnivorous zooplankton and planktivorous fish. *Limnology and Oceanography*, **20**, 571–9.

Confer, J.L., Howick, G.L., Corzette, M.H., Kramer, S.L., Fitzgibbon, S. & Landesberg, R. (1978) Visual predation by planktivores. *Oikos*, **31**, 27–37.

Dunbrack, R.L. & Dill, L.M. (1983) A model of size-dependent surface feeding in a stream-dwelling salmonid. *Environmental Biology of Fishes*, **8**, 203–16.

Dunbrack, R.L. & Dill, L.M. (1984) Three-dimensional prey reaction field of juvenile coho salmon (*Oncorhynchus kisutch*). *Canadian Journal of Fisheries and Aquatic Sciences*, **41**, 1176–82.

Edwards, E.D. & Huryn, A.D. (1996) Effect of riparian land use on contributions of terrestrial invertebrates to streams. *Hydrobiologia*, **337**, 151–9.

Eggers, D.M. (1977) The nature of prey selection by planktivorous fish. *Ecology*, **58**, 46–59.

Eggers, D.M. (1982) Planktivore preference by prey size. *Ecology*, **63**, 381–90.

Elliot, J.M. (1990) Mechanisms reponsible for population regulation in young migratory trout, *Salmo trutta*. III. The role of territorial behaviour. *Journal of Animal Ecology*, **59**, 803–18.

Fausch, K.D. (1984) Profitable stream positions for salmonids: relating specific growth rate to net energy gain. *Canadian Journal of Zoology*, **62**, 441–51.

Feldman, A.C. & Savitz, J. (1999) Influence of prey behaviour on selective predation by lake trout (*Salvelinus namaycush*) under laboratory conditions. *Journal of Freshwater Ecology*, **14**, 399–406.

Fletcher, D.A. & Wooton, R.J. (1995) A hierarchical response to differences in ration size in the reproductive performance of female three-spined sticklebacks. *Journal of Fish Biology*, **46**, 657–68.

Flore, L., Reckendorfer, W. & Keckels, H. (2000) Reaction field, capture field, and search volume of 0+ nase (*Chondrostoma nasus*): effects of body size and water velocity. *Canadian Journal of Fisheries and Aquatic Sciences*, **57**, 342–50.

Garman, G.C. & Moring, J.R. (1993) Diet and annual production of two boreal river fishes following clearcut logging. *Environmental Biology of Fishes*, **36**, 301–11.

Gerking, S.D. (1994) *Feeding Ecology of Fish*. Academic Press, San Diego, CA.

Gibson, R.M. (1980) Optimal prey-size selection by three-spined sticklebacks (*Gasterosteus aculeatus*): a test of the apparent size hypothesis. *Zeitschrift für Tierpsychologie*, **52**, 291–307.

Godin, J.-G.J. & Rangeley, R.W. (1989) Living in the fast lane: effects of cost of locomotion on foraging behaviour in juvenile Atlantic salmon. *Animal Behavior*, **37**, 943–54.

Gottsberger, G. (1978) Seed dispersal by fish in the inundated regions of Humaita, Amazonia. *Biotropica*, **10**, 170–83.

Goulding, M. (1980) *The Fishes and the Forest: Explorations in Amazonian Natural History*. University of California, Berkeley, CA.

Grant, J.W.A. & Kramer, D.L. (1990) Territory size as a predictor of the upper limit to population density of juvenile salmonids in streams. *Canadian Journal of Fisheries and Aquatic Sciences*, **47**, 1724–37.

Grant, J.W.A., Steingrimsson, S.O., Keeley, E.R. & Cunjack, R.A. (1998) Implications of territory size for the measurement and prediction of salmonid abundance in streams. *Canadian Journal of Fisheries and Aquatic Sciences*, **55** (Suppl. 1), 181–90.

Gregory, R.S. & Northcote, T.G. (1993) Surface, planktonic, and benthic foraging by juvenile chinook salmon (*Oncorhyncus kisutch*) in turbid laboratory conditions. *Canadian Journal of Fisheries and Aquatic Sciences*, **50**, 233–40.

Hairston, N.G., Jr., Li, K.T. & Easter, S.S. (1982) Fish vision and the detection of planktonic prey. *Science*, **218**, 1240–2.

Hartman, G.F. (1963) Observations on behaviour of juvenile brown trout in a stream aquarium during winter and spring. *Journal of the Fisheries Research Board of Canada*, **20**, 769–87.

Helfield, J.M. & Naiman, R.J. (2001) Effects of salmon-derived nitrogen on riparian forest growth and implications for stream productivity. *Ecology*, **82**, 2403–9.

Hilderbrand, G.V., Hanley, T.A., Robbins, C.T. & Schwartz, C.C. (1999a) Role of brown bears (*Ursus arctos*) in the flow of marine nitrogen into a terrestrial ecosystem. *Oecologia*, **121**, 546–50.

Hilderbrand, G.V., Schwarz, C.C., Robbins, C.T., *et al.* (1999b) The importance of meat, particularly salmon, to body size, population productivity, and conservation of North American brown bears. *Canadian Journal of Zoology*, **77**, 132–8.

Holling, C.S. (1959) Some characteristics of simple types of predation and parasitism. *Canadian Entomologist*, **91**, 385–98.

Horn, M.H. (1997) Evidence for dispersal of fig seeds by the fruit-eating fish *Brycon guatemalensis* Regan in a Costa Rican tropical rain forest. *Oecologia*, **109**, 259–64.

Hrbacek, J. (1962) Species composition and the amount of zooplankton in relation to the fish stock. *Rozpravy Ceskoslovenske Akademie Ved Rada Matematick Pvirodnich Ved*, **72**, 1–116.

Hughes, N.F. & Dill, L.M. (1990) Position choice by drift-feeding salmonids: model and test for Arctic grayling (*Thymallus arcticus*) in subarctic mountain streams, interior Alaska. *Canadian Journal of Fisheries and Aquatic Sciences*, **47**, 2039–48.

Hyatt, K.D. (1979) Feeding strategy. In: *Fish Physiology*, vol. VIII. Academic Press, New York.

Ivlev, V.S. (1961) *Experimental Ecology of the Feeding of Fishes*. Yale University Press, New Haven, CT.

Jackson, J.K. & Fisher, S.G. (1986) Secondary production, emergence, and export of aquatic insects of a Sonoran desert stream. *Ecology*, **67**, 629–38.

Jobling, M. (1995) *Environmental Biology of Fishes*. Chapman & Hall, London.

Johnson, J.H. & Ringler, N.H. (1979) The occurrence of blowfly larvae (Diptera: Calliphoridae) on salmon carcasses and their utilization by juvenile salmon and trout. *Great Lakes Entomologist*, **12**, 137–9.

Juanes. F. & Conover, D.O. (1994) Piscivory and prey size selection in young-of-the-year bluefish: predator preference or size-dependent capture success? *Marine Ecology Progress Series*, **114**, 59–69.

Kalleberg, H. (1958) Observations in a stream tank of territoriality and competition in juvenile salmon and trout (*Salmo salar* L. and *S. trutta* L.). *Report of the Institute of Freshwater Research Drottningholm*, **39**, 55–98.

Karagosian, M. (2001) *The influence of prey density and environmental complexity on the foraging behaviour of the threespine stickleback* (Gasterosteus aculeatus L.). PhD dissertation, State University of New York College of Environmental Science and Forestry, Syracuse, NY.

Keenleyside, M.H.A. & Yamamoto, F.T. (1962) Territorial behaviour of juvenile Atlantic salmon (*Salmo salar* L.). *Behaviour*, **19**, 139–69.

Kerfoot, W.C. & Sih, A. (1987) *Predation: Direct and Indirect Impacts on Aquatic Communities*. University Press of New England, NH.

Kline, T.C., Goering, J.J., Matheson, O.A., Poe, P.H., Parker, P.L. & Scalan, R.S. (1993) Recycling of elements transported upstream by runs of Pacific salmon: $d^{15}N$ and $d^{13}C$ evidence in the Kvichak River watershed, Bristol Bay, Southwestern Alaska. *Canadian Journal of Fisheries and Aquatic Sciences*, **50**, 2350–65.

Kubitzki, K. & Ziburski, A. (1994) Seed dispersal in flood plain forests of Amazonia. *Biotropica*, **26**, 30–43.

L'Abee-Lund, J.H., Aass, P. & Saegrov, H. (1996) Prey orientation in piscivorous brown trout. *Journal of Fish Biology*, **48**, 871–7.

Lauder, G.V. (1982) Patterns of evolution in the feeding mechanism of Actinopterygian fishes. *American Zoologist*, **22**, 275–85.

Lazzaro, X. (1987) A review of planktivorous fishes: their evolution, feeding behaviours, selectivities, and impacts. *Hydrobiologia*, **146**, 97–167.

Link, J. (1998) Dynamics of lake herring (*Coregonus artedi*) reactive volume for different crustacean zooplankton. *Hydrobiologia*, **368**, 101–10.

Lowe-McConnell, R.H. (1987) *Ecological Studies in Tropical Fish Communities*. Cambridge University Press, Cambridge.

Luczkovich, J.J., Motta, P.J., Norton, S.F. & Liem, K.F. (1995) *Ecomorphology of Fishes*. Kluwer, Norwell, MA.

Luecke, C. & O'Brien, W.J. (1981) Prey location volume of a planktivorous fish: a new measure of prey vulnerability. *Canadian Journal of Fisheries and Aquatic Sciences*, **38**, 1264–70.

McLeay, D.J., Birtwell, I.K., Hartman, G.F. & Ennis, G.L. (1987) Responses of Arctic grayling (*Thymallus arcticus*) to acute and prolonged exposure to Yukon Placer mining sediment. *Canadian Journal of Fisheries and Aquatic Sciences*, **44**, 658–73.

Mason, C.F. & McDonald, S.M. (1982) The input of terrestrial invertebrates from tree canopies to a stream. *Freshwater Biology*, **12**, 305–11.

Metcalfe, N.B., Huntingford, F.A. & Thorpe, J.E. (1986) Seasonal changes in feeding motivation of juvenile Atlantic salmon (*Salmo salar*). *Canadian Journal of Zoology*, **64**, 2439–46.

Minshall, G.W. (1967) Role of allochthonous detritus in the trophic structure of a woodland springbrook community. *Ecology*, **48**, 139–49.

Moyle, P.B. & Cech, J.J., Jr (2000) *Fishes, An Introduction to Ichthyology*, 4th edn. Prentice Hall, NJ.

Nakano, S. & Murakami, M. (2001) Reciprocal subsidies: dynamic interdependence between terrestrial and aquatic food webs. *Proceedings of the National Academy of Sciences USA*, **98**, 166–70.

Nakano, S., Kawaguchi, Y., Taniguchi, Y., *et al*. (1999a) Selective foraging on terrestrial invertebrates by rainbow trout in a forested headwater stream in northern Japan. *Ecological Research*, **14**, 351–60.

Nakano, S., Miyasaki, H. & Kuhara, N. (1999b) Terrestrial-aquatic linkages: riparian arthropod inputs alter trophic cascades in a stream food web. *Ecology*, **80**, 2435–41.

Northcote, T.G. (1988) Fish in the structure and function of freshwater ecosystems: a 'top-down' view. *Canadian Journal of Fisheries and Aquatic Sciences*, **45**, 361–79.

O'Brien, W.J. (1987) Planktivory by freshwater fish: thrust and parry in the pelagia. In: *Predation: Direct and Indirect Impacts on Aquatic Communities*, (eds W.C. Kerfoot & A. Sih), pp.3–16. New England Press, Hanover, NH.

O'Brien, W.J. & Evans, B.I. (1991) Saltatory search behaviour in five species of planktivorous fish. *Verhandlungen Internationale Vereinigung Limnologie*, **24**, 2371–6.

O'Brien, W.J. & Showalter, J.J. (1993) Effects of current velocity and suspended debris on the drift feeding of Arctic grayling. *Transactions of the American Fisheries Society*, **122**, 609–15.

O'Brien, W.J., Slade, N.A. & Vineyeard, G.L. (1976) Apparent size as the determinant of prey selection by bluegill sunfish. *Ecology*, **57**, 1304–10.

O'Brien, W.J., Evans, B.I. & Browman, H.I. (1989) Flexible search tactics and efficient foraging in saltatory searching animals. *Oecologia*, **80**, 100–10.

O'Brien, W.J., Browman, H.I. & Evans, B.I. (1990) Search strategies of foraging animals. *American Scientist*, **78**, 152–60.

Rand, P.S., Hall, C.A.S., McDowell, W.H., Ringler, N.H. & Kennen, J.G. (1992) Factors limiting primary productivity in Lake Ontario tributaries receiving salmon migrations. *Canadian Journal of Fisheries and Aquatic Sciences*, **49**, 2377–85.

Rashevsky, N. (1959) Some remarks on the mathematical theory of nutrition of fishes. *Bulletin of Mathematical Biophysics*, **21**, 161–83.

Restani, M., Harmata, A.R. & Madden, E.M. (2000) Numerical and functional responses of migrant bald eagles exploiting a seasonally concentrated food source. *Condor*, **102**, 561–8.

Ridgway, M.S. & Shuter, B.J. (1994) The effects of supplemental food on reproduction in parental male smallmouth bass. *Environmental Biology of Fishes*, **39**, 201–7.

Ringler, N.H. (1979a) Selective predation by drift feeding brown trout (*Salmo trutta*). *Journal of the Fisheries Research Board of Canada*, **36**, 392–403.

Ringler, N.H. (1979b) Prey selection by benthic feeders. In: *Predator-Prey Systems in Fishery Management*, (eds R. Stroud & H. Clepper), pp. 219–29. Sport Fishing Institute, Washington, DC.

Ringler, N.H. (1985) Individual and temporal variation in prey switching by brown trout, *Salmo trutta*, to invertebrate drift. *Copeia*, **4**, 918–26.

Ringler, N.H. (1994) Fish foraging: adaptations and patterns. In: *Advances in Fish Biology and Fisheries*, Vol. 1, (eds H.R. Singh & S.Z. Qasim), pp. 33–59. Vedams Books International, Delhi, India.

Sanderson, S.L., Cech, J.J., Jr & Patterson, M.R. (1991) Fluid dynamics in suspension feeding blackfish. *Science*, **251**, 1346–8.

Schmidt, D. & O'Brien, W.J. (1982) Planktivorous feeding ecology of Arctic grayling (*Thymallus arcticus*). *Canadian Journal of Fisheries and Aquatic Sciences*, **39**, 475–82.

Sivak, J.G. (1973) Interrelation of feeding behaviour and accommodative lens movements in some species of North American freshwater fishes. *Journal of the Fisheries Research Board of Canada*, **30**, 1141–6.

Smith, C. & Wooton, R.J. (1999) Parental energy expenditure of the male three-spined stickleback. *Journal of Fish Biology*, **54**, 1132–6.

Sogard, S.M. (1997) Size-selective mortality in the juvenile stage of teleost fishes: a review. *Bulletin of Marine Science*, **60**, 1129–57.

Souza-Stevaux, M.C.D., Negrelle, R.R.B. & Citadini-Zanette, V. (1994) Seed dispersal by the fish *Pterodorus granulosus* in the Parana River basin, Brazil. *Journal of Tropical Biology*, 10, 621–6.

Stanley, B.V. & Wooton, R.J. (1986) Effects of ration and male density on the territoriality and nest building of male three-spined sticklebacks (*Gasterosteus aculeatus*). *Animal Behavior*, 34, 527–35.

Steedman, R.J. & France, R.L. (2000) Origin and transport of aeolian sediments from new clearcuts into boreal lakes, northwestern Ontario, Canada. *Water, Air, and Soil Pollution*, 122, 139–52.

Sweka, J.A. & Hartman, K.J. (2001) Influence of turbidity on brook trout reactive distance and foraging success. *Transactions of the American Fisheries Society*, 130, 138–46.

Tebo, L.B. (1955) Effects of siltation resulting from improper logging on the bottom fauna of a small trout stream in the southern Appalachians. *Progressive Fish-Culturist*, 17, 64–70.

Trevarthen, C. (1968) Vision in fish: the origins of the visual frame for action in vertebrates. In: *The Central Nervous System and Fish Behaviour*, (ed. D. Ingle), pp. 61–94. University of Chicago Press, Chicago, IL.

Turner, A.M. & Mittelbach, G.G. (1990) Predator avoidance and community structure: interactions among piscivores, planktivores, and plankton. *Ecology*, 71, 2241–54.

Utne-Palm, A.C. (2002) Visual feeding of fish in a turbid environment: physical and behavioral aspects. *Marine and Freshwater Behaviour and Physiology*, 35, 111–28.

Vannote, R.L., Minshall, G.W., Cummins, K.W., Sedell, J.R. & Cushing, C.E. (1980) The river continuum concept. *Canadian Journal of Fisheries and Aquatic Sciences*, 37, 130–7.

Vinyard, G.L. & O'Brien, W.J. (1976) Effects of light and turbidity on the reactive distance of bluegill (*Lepomis macrochirus*). *Journal of the Fisheries Research Board of Canada*, 33, 2845–9.

Visser, M. & Reinders, L.J. (1981) Waiting time as a new component in functional response models. *Netherlands Journal of Zoology*, 31, 315–28.

Wallace, J.B., Eggert, S.L., Meyer, J.L. & Webster, J.R. (1997) Multiple trophic levels of a forest stream linked to terrestrial litter inputs. *Science*, 277, 102–4.

Walton, W.E., Easter, S.S., Jr, Malinoski, C. & Hairston, N.G., Jr (1994) Size-related change in the visual resolution of sunfish (*Lepomis* sp.). *Canadian Journal of Fisheries and Aquatic Sciences*, 51, 2017–26.

Walton, W.E., Emiley, J.A. & Hairston, N.G., Jr (1997) Effect of prey size on the estimation of behavioural visual resolution of bluegill (*Lepomis macrochirus*). *Canadian Journal of Fisheries and Aquatic Sciences*, 54, 2502–8.

Wankowski, J.W.J. (1981) Behavioural aspects of predation by juvenile atlantic salmon (*Salmo salar* L.) on particulate, drifting prey. *Animal Behavior*, 29, 557–71.

Wankowski, J.W.J. & Thorpe, J.E. (1979) Spatial distribution and feeding in Atlantic salmon (*Salmo salar*) juveniles. *Journal of Fish Biology*, 14, 239–47.

Wanzenbock, J., Zaunreiter, M., Wahl, C.M. & Noakes, D.L.G. (1996) Comparison of behavioural and morphological measures of visual resolution during ontogeny of roach (*Rutilis rutilis*) and yellow perch (*Perca flavescens*). *Canadian Journal of Fisheries and Aquatic Sciences*, 53, 1506–12.

Ware, D.M. (1971) Predation by rainbow trout (*Salmo gairdneri*): the effect of experience. *Journal of the Fisheries Research Board of Canada*, 28, 1847–52.

Ware, D.M. (1972) Predation by rainbow trout (*Salmo gairdneri*): the influence of hunger, prey density, and prey size. *Journal of the Fisheries Research Board of Canada*, 29, 1193–201.

Ware, D.M. (1973) Risk of epibenthic prey to predation by rainbow trout (*Salmo gairdneri*). *Journal of the Fisheries Research Board of Canada*, 30, 787–97.

Ware, D.M. (1975) Growth, metabolism, and optimal swimming speed of a pelagic fish. *Journal of the Fisheries Research Board of Canada*, 32, 33–41.

Ware, D.M. (1978) Bioenergetics of pelagic fish: theoretical change in swimming speed and ration with body size. *Journal of the Fisheries Research Board of Canada*, 35, 220–8.

Ware, D.M. (1982) Power and evolutionary fitness of teleosts. *Canadian Journal of Fisheries and Aquatic Sciences*, **39**, 3–13.

Webb, P.W. (1975) Hydrodynamics and energetics of fish propulsion. *Bulletin of the Fisheries Research Board of Canada*, **190**, 1–158.

Webb, P.W. (1984) Body form, locomotion, and foraging in aquatic vertebrates. *American Zoologist*, **24**, 107–20.

Werner, E.E. (1974) The fish size, prey size, handling time relation in several sunfishes and some implications. *Journal of the Fisheries Research Board of Canada*, **31**, 1531–6.

Werner, E.E. & Hall, D.J. (1974) Optimal foraging and the size selection of prey by the bluegill sunfish (*Lepomis macrochirus*). *Ecology*, **55**, 1042–52.

Werner, E.E. & Hall, D.J. (1988) Ontogenetic habitat shifts in bluegill: the foraging rate-predation risk trade-off. *Ecology*, **69**, 1352–66.

Werner, E.E., Gilliam, J.F., Hall, D.J. & Mittelbach, G.G. (1983) An experimental test of the effects of predation risk on habitat use in fish. *Ecology*, **64**, 1540–8.

Wetterer, J.K. & Bishop, C.J. (1985) Planktivore prey selection: the reactive field volume model vs. the apparent size model. *Ecology*, **66**, 457–64.

Wiegmann, D.D., Baylis, J.R. & Hoff, M.H. (1997) Male fitness, body size and timing of reproduction in smallmouth bass, *Micropterus dolomieui*. *Ecology*, **78**, 111–28.

Willson, M.F., Gende, S.M. & Marston, B.H. (1998) Fishes and the forest: expanding perspectives on fish-wildlife interactions. *Bioscience*, **48**, 455–62.

Wipfli, M.S. (1997) Terrestrial invertebrates as salmonid prey and nitrogen sources in streams: contrasting old-growth and young-growth riparian forests in southeastern Alaska, U.S.A. *Canadian Journal of Fisheries and Aquatic Sciences*, **54**, 1259–69.

Wootton, R.J. (1992) *Fish Ecology*. Chapman & Hall, New York.

Wright, D.I. & O'Brien, W.J. (1982) Differential location of Chaoborus larvae and Daphnia by fish: the importance of motion and visible size. *American Midland Naturalist*, **108**, 68–73.

Wright, D.I. & O'Brien, W.J. (1984) The development and field test of a tactical model of the planktivorous feeding ecology of white crappie (*Pomoxis annularis*). *Ecological Monographs*, **54**, 65–98.

Zaret, T.M. (1972) Predators, invisible prey, and the nature of polymorphism in the cladocera (class crustacea). *Limnology and Oceanography*, **17**, 171–84.

Zaret, T.M. & Kerfoot, W.C. (1975) Fish predation on *Bosmina longirostris*: body-size selection versus visibility selection. *Ecology*, **56**, 232–7.

Part IV
Forestry Activities

Chapter 10
Forest harvest and transportation

P. SCHIESS AND F. KROGSTAD

Introduction

Timber is harvested to produce economic returns from forest lands, but it also impacts the environment. Forestry operations are often viewed as a trade-off between economic benefits and environmental impacts, in which any additional environmental protection is seen as reducing economic returns. In exploring the economic and environmental costs of forest harvest and roading, however, it is common to find that options for improving the economics can often improve the environmental impacts as well. An understanding of the operational considerations of logging and roading is the first step for understanding current forest practices, and for identifying options for improving economic and environmental returns from the forest.

Every logging operation consists of the following phases:

(1)   Felling – the trees are cut down.
(*)   Processing – trees are cut into logs and limbs are removed.
(2)   Yarding – logs are moved from the stump to the landing.
(3)   Loading – logs are loaded for transport to the mill.
(4)   Hauling – transport to the mill (usually by truck).
       *Processing can occur at different phases, depending on the harvest system used.

While there are many economic impacts of forest harvesting (from the chainsaw manufacturer to the paper consumer) in this chapter we will consider only the costs of those who pay the loggers, reap the benefits, and thus decide how the forest will be managed. Once a harvest unit has been delineated and trees to be harvested have been identified, then the total value of the resulting logs can be determined from current mill prices, and every other aspect of the harvest operation is just a cost.

return = value – felling – processing – yarding – loading – hauling

Each of these costs includes not just the cost of labour and fuel, but the depreciation and risk entailed in owning the equipment. These costs and benefits of logging are both economic and environmental. A thinning operation that improves habitat value also involves yarding (with its direct costs) and roads (with their sediment and habitat fragmentation costs).

Harvest operations can be divided between primary (stump-to-truck) transportation and haul (secondary transport) to the mill (Conway 1982). Primary transportation moves low log volumes slowly (from 1 tonne in a cable thinning operation to 10 tonnes in a forwarder moving at walking speeds). Secondary transportation on smooth road surfaces and straight road alignments moves large volumes at high speeds (25–35 tonnes at 30–100 km/h).

Timber harvest – stump-to-truck

The conversion of a stand of trees into truckloads of logs at the mill is an industrial process. Specialized equipment and workers are assigned to different tasks and must coordinate their outputs and timing to avoid slowdowns and bottlenecks in the process. The following sections outline options for felling the trees: delimbing, bucking into logs, yarding to the road and loading onto trucks.

Felling and processing

Felling is the first step in converting trees into lumber, paper, and other products (Staaf & Wiksten 1984). Trees are felled either manually with a chainsaw or mechanically with a feller-buncher or harvester (Fig. 10.1). The felled tree can either be processed

Fig. 10.1 A feller-buncher (top left) fells trees into bunches (bottom left). A harvester (top right) fells and then delimbs and bucks the trees in its path, creating a slash mat that protects the soil. Logs are dropped in bundles or bunches to both sides of the yarding corridor (bottom right).

(delimbed and bucked into logs) there in the woods, or processed at the landing. Whether manual or mechanized, limbing and bucking can be time-consuming. The lengths into which the logs are bucked, however, determine the price received at the mill and the profitability of the whole operation. Mills that turn logs into products such as lumber and plywood of specific sizes need logs at least that large. When bucking the downed tree into logs, the logger must identify the best possible combination of log sizes, while avoiding breaks and other log defects. As much as 40% of the possible value can be lost in felling and processing (Conway 1982).

Manual felling is the preferred method in large timber and on steep terrain inaccessible to mechanized equipment. Manual felling is also dangerous, resulting in high levels of injuries and even fatalities (Workers' Compensation Board 1999), resulting in insurance rates as high as 60% of the faller's wage. Manual thinning in dense stands is difficult, since felled trees often are caught by the residual trees, causing unsafe conditions and delays.

The high cost of manual felling facilitated the move towards mechanized felling and processing (limbing and bucking). Feller-bunchers can drop timber into bunches (Fig. 10.1) that can be more easily gathered for yarding than the scattered pieces left by manual felling. Mechanized felling operations are limited by tree size and terrain. Mechanized felling operations are the preferred method in timber stands with tree diameters typically <50 cm. Most operational machines are limited to slopes <40%, although in Europe harvesters are being used on slopes as steep as 50–55% (Heinimann *et al.* 2001).

Harvesters fell the trees, then limb and buck them into logs. Processing the trees in front of the equipment creates a mat of slash, which protects the soil from compaction as the harvester and then the forwarder move over it. The slash is usually sufficiently mixed with soil so the fire hazard is not increased. Harvesters are used primarily in thinning. The short (<8 m) logs produced by the harvesters can be efficiently handled in dense residual stands, improving production and reducing damage to residual trees – crucial considerations in thinning operations.

Yarding

There is a wide range of equipment systems used to transport the logs to the landing. For simplicity, we can group these systems according to whether the yarding equipment drives out to the stump (ground-based systems) or stays at the landing (cable-based systems) or flies to the log (helicopter yarding).

Ground-based yarding

Ground-based systems (Fig. 10.2) drive to the downed trees or logs and bring them back to the landing. Where topography, soils and piece size permit, ground-based yarding is usually the most effective and flexible option for getting logs to the landing. Horses are one of the oldest ground-based systems, and still fill a limited niche. The advent of wheeled or tracked vehicles dramatically reduced yarding costs by increasing yarding

Fig. 10.2 Ground-based equipment travels to the stump and brings logs/trees back to the landing. A wheeled out tracked grapple skidder (top left) can quickly grab bunches created by a feller-buncher, although a cable winch with chokers can replace the grapple in manually felled timber. A clam-bunk skidder (top right) can assemble and skid a much larger turn. A forwarder (bottom left) can collect a large turn from logs that have been felled and processed by a harvester. A shovel or log loader (bottom right) swings trees/logs towards the landing.

speed and load size. Lifting the front end (skidding) or the entire log (forwarding) or swinging the log (shovel yarding) not only reduce the soil disturbance and erosion but also reduce the fuel and time costs of overcoming friction and hang-ups.

Yarding production rates (volume per hour) and production costs (cost per volume) are affected by stand density (trees per area), log sizes, yarding distances, payload and operator skill. The skidder or forwarder drives to the felled trees or logs, and collects these pieces to form a turn, which it brings back to the landing. The combination of high stand density and large piece size allows for fast accumulations of a turn. Low piece densities, such as in a partial cut, require more time to accumulate a comparable load, which can have an adverse effect on production rates and cost. Production rates and costs are also affected by skidding speed and payload brought to the landing. Wheeled skidders carry low payloads, but at 8–10 km/h are typically the fastest. The slow moving clam-bunk and forwarder carry a much larger payload. The wheeled skidders need to make two-to-four trips to equal the forwarding capacity of a forwarder or a clam-bunk, which can be important on sensitive soils where the number of trips over a skid trail may have to be limited to avoid site degradation. If the retention spacing of a thinning is large enough, skidders or forwarders can avoid damaging the retained trees, although the retained trees can still be at risk from soil compaction and root damage.

Ground-based systems are limited by soils (wheels/tracks on cohesive/frictional soils), obstacles (e.g. stumps, fallen logs, boulders) and terrain surface roughness (Malmberg 1989). Ground-based machinery typically is restricted to slopes <30–40% but can be limited to even lower slopes by any combination of soil conditions, obstacles and ground roughness. Heavy or prolonged rain may further reduce or even prohibit the use of ground-based equipment. Mechanized felling-processing equipment is less affected by soil and slope conditions than skidders or forwarding equipment, which must cross this ground repeatedly.

Ground-based yarding on unfavourable soils and slopes can also cause soil compaction and erosion, although slash from harvesters can reduce these impacts. In the past, soil impacts were not fully recognized, and skidder trails disturbed much of the harvest area. More recently, the use of designated skid trails has resulted in significantly less ground disturbance. This disturbance is concentrated towards the landings, since relatively few logs come from the far side of the harvest area but all logs must travel to the landing. If the roads (and landings) are located near streams, then sediment eroded from these disturbed areas is more likely to be delivered to the streams than if the road network is not near the stream network.

Cable-based yarding

Instead of moving the yarder to the log, cable systems move the log to the yarder. By stringing cables out into the woods and pulling the logs to the landing, cable systems overcome the topographic, soil and speed limitations of ground-based logging systems. However, cable systems need considerable time to set up and to move to different parts of the harvested area.

Highlead

Early cable systems consisted of a steam engine turning a drum around which a cable was wound, with the end of the cable pulled to the log by horse. A second haulback drum and a block (pulley) in the woods were soon added to haul the mainline back out (Samset 1985). To minimize log drag and hang-ups, some vertical lift is provided by passing the mainline through a pulley in a spar tree or tower, held upright by guyline cables. The main and a haulback line form a loop, connected by the butt rigging, to which several short cables called chokers are attached, that are wrapped around logs by the choker setter crew. This highlead system is the simplest of cable systems (Fig. 10.3), with a crew of 5–8.

Highlead systems have no ability to reach sideways (lateral reach) beyond the length of the chokers, and must change cable locations frequently. Passing through two or more pulleys (tail and corner block) at the far side of the harvest, the main and haulback lines form a triangle or fan-shaped setting that allows rapid shifting in cable location. Once all the logs have been yarded from along the cable's path, a new block is added along the back of the loop, and the loop is shortened by releasing it from the current

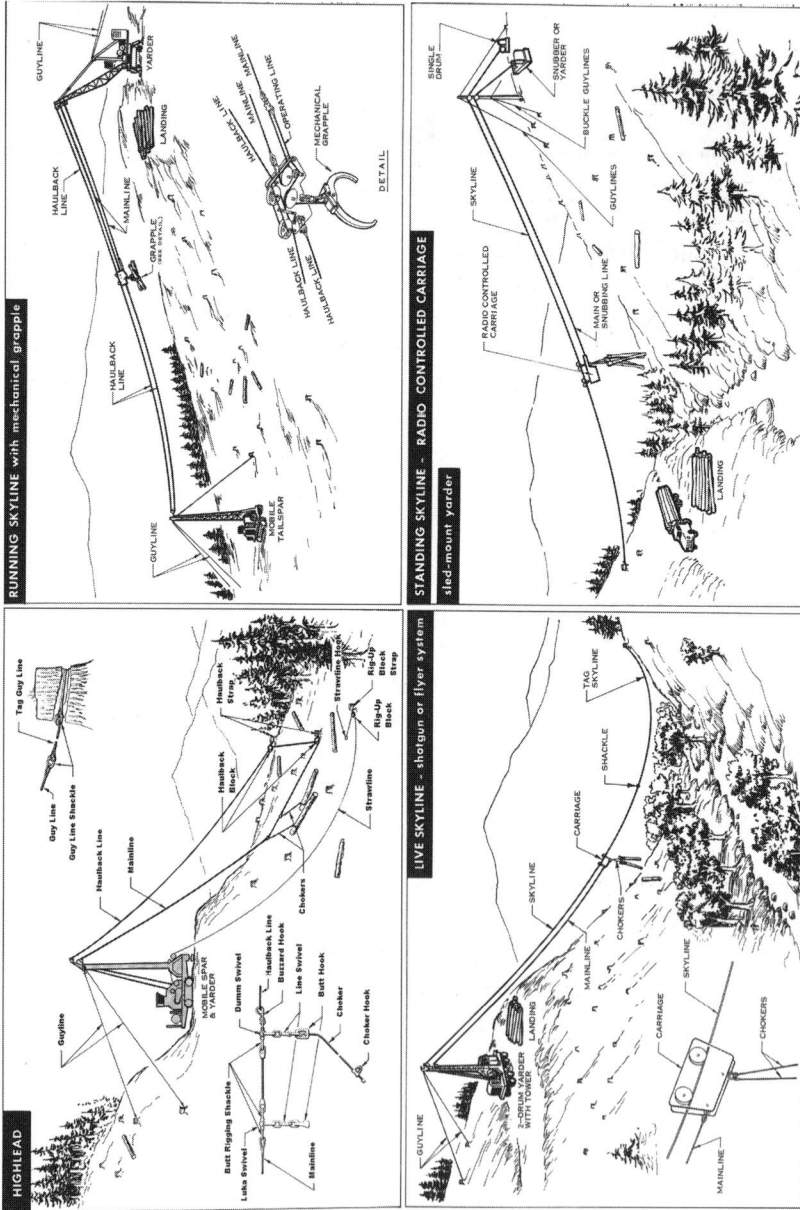

Fig. 10.3 Cable yarders remain stationary while pulling logs to the landing. A highlead system (top left) provides limited vertical lift to the logs being pulled to the landing. A running skyline (bottom left) provides vertical lift by maintaining cable tension. The grapple carriage shown provides no lateral reach. A live skyline (top right) can lower the carriage to the ground or can be raised to provide lift. A standing skyline (bottom right) uses a radio-controlled carriage containing a winch with a dropline (Studier and Binkley 1974).

block and letting it snap to this next corner block. As logging progresses, the cable sweeps across the harvest area, successively producing a radial pattern of cable roads. Standing trees would get in the way, so this process is only appropriate in clear-cuts. The highlead system has limited ability to provide vertical lift. Available vertical lift, log control and production decrease rapidly away from the tower, so highlead yarding distances are usually <300 metres (Studier & Binkley 1974).

Highlead systems are not appropriate for most variable-retention silviculture systems, and are unsuited to dispersed retention such as thinning, although some pie-shaped retention aggregates at the end of corridors may be feasible. Soil disturbance and erosion from dragging logs is typically small compared with skidder yarding and is concentrated near the landing, so it is important that the landing is not located near the stream network (Megahan 1980).

Some vertical lift can be obtained by braking the haulback drum while pulling in on the mainline drum. Tension can be maintained more efficiently by mechanically interlocking the drums so that one pulls in while the other lets out. Replacing the butt rigging with a carriage that rides on the haulback line produces a running skyline (Fig. 10.3). Set-up or rigging time can be improved by replacing the tower with a self-propelled crane and reducing the number of guylines to two. The mobility of the crane allows for rapid shifts in yarder location. When used with carriages with lateral reach, running skylines can be used for variable retention silviculture, especially if aggregated or in a heavy partial cut. They are not appropriate in thinning because the moving lines damage the residual trees.

Adding a grapple attachment (Fig. 10.3) to the running skyline carriage allows grapple yarding. This remotely operated mechanical grapple eliminates the inherently dangerous job of manually looping chokers around the logs. A mobile backspar (often an excavator) is used to increase lift and to move the whole operation more quickly from one corridor to the next. The grapple yarder has no lateral reach, so these yarding corridors are parallel and narrowly spaced. Where slopes are too steep (>40–45%) to serve as a landing, logs can be landed at the roadside or on the road itself, in which case roads should be wider than normal (MacDonald 1999). Grapple yarding is usually limited to about 150–200 metres, preferably less. The combination of short yarding distances, fast line speeds, minimal hook times and deployment in areas with large log dimensions can yield high production rates of 400–500 m³ per shift. Lacking lateral reach, grapple yarders are not appropriate systems for most retention silviculture, especially if the retained trees are dispersed.

True skylines

Vertical lift can also be provided by adding a carriage that runs on a dedicated skyline suspended between the yarder and the backspar, such as a tree or stump (Fig. 10.3). These systems are usually employed in steep terrain, where road access is difficult and therefore longer yarding distances are required. When logging uphill, gravity is sufficient to return the carriage to the woods and no haulback line is needed. When yarding down to the yarder, skylines require a haulback, requiring additional rigging or set-up time.

Simple carriages with chokers directly attached access only the wood in the yarding corridor. Carriages with chokers attached to a dropline can be pulled laterally to yard 10–30 metres on each side of the skyline corridor. The dropline can be attached to the mainline (mechanical slackpulling), or attached to a small, radio-controlled winch in the carriage.

If both ends of the skyline cable are anchored, then the skyline cannot be lowered or raised, and a dropline is usually needed to reach the wood. This standing skyline system is typically used where full suspension along the full length of the yarding corridor is required or desired, such as to move logs from one side of a stream to the other, or for yarding long distances up to 1000–1500 metres. Where topography allows, the system can even carry logs over the tops of standing trees. Topographic limitations can be overcome by hanging the skyline over intermediate supports.

Alternately, the skyline can be raised and lowered by running one end of the skyline to a drum on the yarder to create a live skyline or slackline system (Fig. 10.3). Where chokers are attached directly to the carriage, the skyline must be lowered to the choker setters, but this configuration has no lateral excursion capability. Where chokers are attached to a dropline they can be pulled laterally beyond the yarding corridor.

The weight of the logs, carriage and cable causes the cable to sag towards the ground. The more it can be allowed to sag, the more weight it can carry. This turn weight can often be increased with a tall tower and often by elevating the skyline tail hold in a tree or up the opposite side of the valley. Concave topography provides near ideal conditions for skyline operations whereas convex topography is very difficult to log at any distance without intermediate supports. Skyline systems usually yard logs with one end off the ground (partial suspension), although full suspension (logs are lifted entirely off the ground) is also possible. The ability to provide partial or full suspension improves log control and reduces hang-ups, thus improving production and reducing ground disturbance.

Silvicultural systems that require a significant number of retained trees (e.g. thinnings, partial cuts) require a cable system that allows for both lateral reach and partial or full load suspension. Highlead systems, allowing neither, are appropriate only for clear-cut operations. Grapple yarding systems can provide partial or full load suspension but lack lateral reach capabilities, so they are suitable only for clear-cuts. Their lateral reach and load suspension make standing skyline the cable system of choice for thinning or high levels of tree retention.

Helicopter yarding

Where soils, topography or other issues inhibit cable- or ground-based yarding, the high cost of helicopter yarding can be justified. Helicopter operations usually have very high operating costs (US\$16,000–28,000 per shift) and production rates (up to 1500 m^3/shift) compared with ground-based or cable-based systems. These costs necessitate running the helicopters at near-capacity payloads of 2000–10,000 kg (Studier & Neal 1993). Helicopter logging also requires a large and highly skilled crew, and substantial planning and coordination to maintain efficiency. Fuel must be kept on site, logs must always be ready for lifting, log destination points kept clear, and fallers must ensure that log weights do not exceed the helicopter's payload capacity. Maximum yarding distances should not exceed 2000 metres (a flight time of 2 minutes), with level to downward flight paths so gravity can assist the loaded flight. Helicopters may use roadsides or water bodies as landings. Helicopters are more sensitive to wind and fog than are cable-based systems. Wind speeds above 50 km/h will usually ground a helicopter operation.

Loading – the interface between primary and secondary transport

Operations at the landing can be just as varied as the yarding operations that moved the logs there. Logs are commonly loaded onto trucks with an excavator-type loader. It would be costly to have trucks waiting at the landing, so the landing also serves as a log storage site. Different log species and sizes are often used in different mills, so logs are often sorted by destination. Whole trees can be delimbed and bucked at the landing by mechanized processors. Small pieces can be debarked and chipped at the landing.

The shape and size of a landing depends on whether logs/trees are yarded to a single landing by stationary towers or distributed along the road by mobile yarders or ground-based systems. Yarding and loading can be more flexible and efficient when the wood is not all yarded to a single location. In steep topography it can be difficult to find or create a horizontal space large enough for landing operations. Log storage space can fill quickly, necessitating a loader on-site during yarding. Log processing and storage is thus restricted in such confined landings (Schuh & Kellog 1988). Swing yarders that move from one yarding corridor to the next can stack the small volume from each corridor at roadside. This wood can be loaded after the yarding has been completed.

The environmental impact of sediment eroded from landings depends largely on their location. It is more likely to reach vulnerable streams if roads and landings are located near streams. The excavated soils and logging debris from a cable landing on steep slopes can provide a significant source of subsequent landsliding. Helicopter yarding operations can require very large flat landings and this entails significant soil disturbance.

Harvest systems

With the variety of felling, processing, yarding and loading options available, the number of possible combinations of these components is very large. Topography, timber and soils tend to favour some combinations over others, however, and the requirements of each component tend to make it work better with some components than others. Where soils, topography and log size permit, labour costs can be reduced by using mechanized systems that drag (skidders), carry (forwarders), or swing (shovel yarders) the logs to the landing. When topography or soils inhibit ground operations, cables strung from towers can drag or fly the log to the landing. When neither ground nor cable systems will work effectively, more expensive helicopter systems can fly logs to the landing. When yarding distances are large, more time is spent driving to and from the woods, thus favouring the larger loads of forwarders and clam-bunks, and the vertical lift afforded by skyline systems. When the road density is high and yarding distances are short, inbound and outbound speeds are less important than the time needed to pick up and drop a load (favouring skidders and shovel yarders) and change cable corridors (favouring highlead and grapple systems).

In retention harvests and thinnings, harvest systems are constrained both by what is cut and what is left. Harvested logs must be pulled out without causing excessive damage to the retained trees. Highlead and grapple yarder systems cannot manoeuvre logs through the standing trees into the cable corridor. Ground-based and some cable systems (skylines with lateral reach) can move through the thinned stand, but have difficulties on side slopes, with logs swinging downhill and damaging the retained timber. The size and number of trees removed can further constrain the harvest. There is usually a negative relationship between harvested volume and harvest unit costs ($/m^3). Lower timber volumes can render unprofitable all but the most efficient harvest system, necessitating careful consideration of where the conversion from tree to logs will take place.

In thinning operations, the small logs and narrowly spaced residual trees tend to favor shortwood methods, in which trees are felled, delimbed and bucked in the harvest area, then carried by forwarder to the landings. Tops and branches can be left scattered in the cut-over area or be concentrated along skidding trails for equipment to travel over. The resulting pulpwood and sawlogs (usually 4–6 m) allow for better handling, resulting in less damage to residual trees and smaller landing sizes.

When not prohibited by landing and log size, yarding and processing costs can be reduced by yarding the whole tree (branches and tops attached) to the landing. Whole trees are processed at the landing, or transported as whole trees to a central processing yard or mill. Processing at the landing may involve bucking to logs and/or chipping of the tops or the whole tree (typical for early thinning material). Limbs, tops, and sometimes bark, accumulate at the landing and may require disposal. The whole tree approach reduces the slash left on the site, which may reduce fire risks, but the removal of branches, needles and leaves from the site can reduce available nutrients. The additional processing equipment and residue accumulation necessitates a large landing. This whole tree approach lends itself to highly mechanized operations, requiring high capital costs but few people, with a feller-buncher followed by grapple skidder or clambunk skidder, and processor equipment at the landing.

In mature large diameter timber stands, trees are felled, limbed and bucked at the stump with a chainsaw to log-lengths (6–15 m) or topped to tree-length (top diameter of 7–10 cm). The large log-lengths are yarded by cable systems or simple ground-based equipment such as grapple skidders. Tree-length methods may require additional processing at the landing (some delimbing and bucking) and have larger landing requirements than shortwood methods.

Roads and haul – secondary transport

Forest roads exist to provide cost-effective transportation of people and equipment to the forest, and forest products from the forest to mills. In this section we focus on haul traffic, but road use for forest administration and recreation follow similar patterns, with similar economic and environmental costs. The total cost of a road is the sum of the cost to build it, plus the costs to maintain it, plus the costs to drive on it:

total = construction + maintenance + traffic

For example, consider the costs for three alternate road options (Table 10.1). The total annual cost is the sum of the road construction cost (divided by the amortization period), the annual maintenance cost and the annual trucking cost. Note that in this example, the low standard option has the lowest construction costs, the high standard has the lowest trucking costs, but the total cost is minimized by the moderate standard road. This approach can also be used to evaluate alternative routes, such as a longer route on a gentle favourable grade versus a shorter route with a steeper grade.

If the annual haul volume in Table 10.1 were larger, the haul costs would be proportionally higher for each option, and the total cost would be lowest for the high design standard. Similarly, if the haul volume were lower, then haul cost would be proportionally smaller, the construction costs begin to dominate and the low standard road would have the lowest total cost. Traffic volume on little-used roads does not justify high construction costs, since the resulting improvements in haul time (or transport cost savings) do not offset the increased construction costs.

While the road network covers the entire landscape being accessed, the traffic is concentrated on just a few roads (Fig. 10.4). Most of the roads in a road network carry little traffic, but most of the traffic to and from the woods is concentrated along a few heavily used roads. Traffic volume on the many, little-used roads does not warrant high construction costs since the resulting improvements in haul time (transport cost) do not offset the increased construction costs. In order to minimize the total cost of a road network, high transport costs are allowed along less used roads so as to minimize the construction costs, while high construction costs are allowed along the heavily used roads so as to minimize traffic costs. In general, the heavily travelled roads will tend

Table 10.1   Costs* for three alternate road options

| | Design standard | | |
	High	Moderate	Low
Construction	$40,000	$22,000	$15,000
Depreciation	$1,600	$880	$600
Maintenance	$300	$400	$500
Traffic	$2,500	$3,000	$3,500
Total cost	$4,400	$4,280	$4,600

*The total annual cost of a kilometre of road is the sum of the 25-year depreciated construction costs, the annual maintenance and the cost of hauling 10 million cubic metres of material over it. With increasing design standard, construction costs increase but maintenance and traffic costs drop, yielding a total cost that is minimized by the moderate road standard for this moderate haul volume.

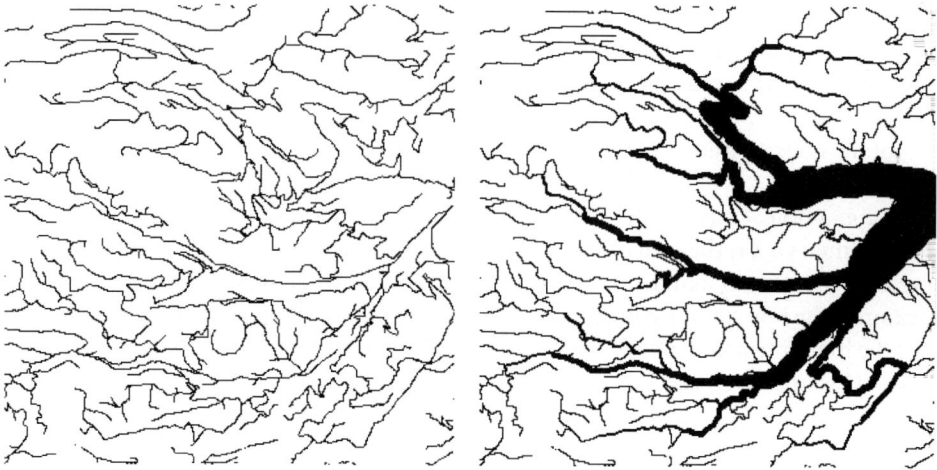

Fig. 10.4 The road network (left) is spread across a landscape (in central Washington State), but haul traffic (right) is concentrated on a relatively small fraction of these roads. Haul traffic (line thickness) determines potential environmental impact and optimal construction and maintenance levels.

to be straight, flat and wide, with good surfacing, while the less travelled roads are narrower, steeper and less straight, with little or no surfacing. The various trade-offs are summarized according to traffic service levels in Table 10.2.

Trucking costs

When moving to and from the woods, time is money, and forest roads exist to reduce this transportation cost. Travel time is the inverse of travel speed, so increasing speed reduces costs. For example, at US $70/h (the typical cost for truck and driver), increasing travel speed from 10 to 20 km/h on a 0.5-km segment of road reduces haul costs by $3.50 per round trip. Much like non-forest roads, travel speed is reduced by sharp turns, steep grades and limited sight distances. Along narrow forest roads, travel speed can be further limited by road width, turnout spacing and road surface roughness.

The environmental costs of forest road traffic are similar to the economic costs. Steep grades, sharp turns, rutted and potholed road surface not only slow traffic, but also increase the erosion of road surface materials. For example, a road segment with asphalt surfacing provides a smooth running surface allowing for higher haul speeds and little or no sediment generation. A native surfaced road (with no surface improvement) is less expensive to build but will generate more sediment, reduce haul speeds and increase trucking costs.

The haul cost and road erosion are not evenly or randomly distributed across the landscape, but instead are concentrated onto a few heavily used roads (Fig. 10.4). Economic and environmental costs can be minimized by building these few heavily used roads to a high standard. The many little-used roads contribute comparatively little to total trucking costs and road erosion. Trucks travel only a short distance on these low

Table 10.2 Traffic service levels control the design, construction and use of forest roads (USDA Forest Service 1982)

	A	B	C	D
Flow	Free-flowing with adequate parking facilities	Congested during heavy traffic such as during peak logging or recreation	Interrupted by limited passing facilities, or slowed by the road condition	Flow is slow or may be blocked by an activity. Two-way traffic is difficult and may require backing to pass
Volumes	Uncontrolled; will accommodate the expected traffic volumes	Occasionally controlled during heavy use periods	Erratic; frequently controlled as the capacity is reached	Intermittent and usually controlled. Volume is limited to that associated with the single purpose
Vehicle types	Mixed; includes the critical vehicle and all vehicles normally found on public roads	Mixed; includes the critical vehicle and all vehicles normally found on public roads	Controlled mix; accommodates all vehicle types including the critical vehicle. Some use may be controlled to vehicle types	Single use; not designed for mixed traffic. Some vehicles may not be able to negotiate. Concurrent use traffic is restricted
Critical vehicle	Clearances are adequate to allow free travel; overload permits are required	Traffic controls needed where clearances are marginal; overload permits are required	Special provisions may be needed. Some vehicles will have difficulty negotiating some segments	Some vehicles may not be able to negotiate. Loads may have to be off-loaded and walked in
Safety	Safety features are a part of the design	High priority in design; some protection is accomplished by traffic management	Most protection is provided by management	The need for protection is minimized by low speeds and strict traffic controls

Table 10.2 (*Continued.*)

	A	B	C	D
Traffic management	Normally limited to regulatory, warning and guide signs and permits	Employed to reduce traffic volume and conflicts	Traffic controls are frequently needed during periods of high use by the dominant resource activity	Used to discourage or prohibit traffic other than that associated with the single purpose
User costs	Minimize; transportation efficiency is important	Generally higher than 'A' because of slower speeds and increased delays	Not important; efficiency of travel may be traded for lower construction cost	Not considered
Alignment	Design speed is the predominant factor within feasible topographic limitations	Influenced more strongly by topography than by speed and efficiency	Generally dictated by topographic features and environmental factors. Design speeds are generally low	Dictated by topography, environmental factors and the design and critical vehicle limitations. Speed is not important
Road surface	Stable and smooth with little or no dust, considering normal season of use	Stable for the predominant traffic for the normal use season. Periodic dust control for heavy use or environmental reasons. Smoothness is commensurate with the design speed	May not be stable under all traffic or weather conditions during normal use season. Surface rutting, roughness and dust may be present, but controlled for environmental or investment protection	Rough and irregular. Travel with low clearance vehicle is difficult. Stable during dry conditions. Rutting and dust controlled only for soil and water protection

volume roads before entering higher volume roads, which make up most of their trip to the mill. These many little-used roads can thus be built to low standards (Tables 10.1 and 10.2) with relatively little increase in haul cost and sediment impacts.

Construction

The construction of a road involves converting the natural topography into a road structure with specific design elements (Fig. 10.5), each serving a particular function in maintaining the road and the access it provides to the forest. In constructing a forest road, the trees, organic material and topsoil are removed first. The topography must then be excavated and moved to match the intended road alignment. Drainage structures and surfacing layers may then be added to minimize future road damage and thus reduce future maintenance and environmental costs.

The organic material is sometimes incorporated into the road structure. Over time this material will decay, weakening the road structure and sometimes causing landsliding. These problems can be avoided by first removing any vegetation (clearing) and scraping away the remaining organic materials such as stumps and duff layer (grubbing). Depending on vegetation and side slopes, clearing and grubbing can comprise 10–20% of the total road construction cost.

A level surface is constructed by excavating, moving and sometimes compacting soil. This earthwork can be done either by multiple passes of a bulldozer or by hydraulic excavator (Fig. 10.6). Several earthwork strategies have been developed to minimize these economic and environmental costs. Excavated material can rapidly be pushed down-slope with a bulldozer blade, but this uncompacted side cast material, resting on organic material, can become saturated with water and unstable. Excavating down to a level surface (full bench) and hauling away the excavated material (end haul) can become expensive if there are no stable dumping locations nearby. On steeper slopes, excavators allow the precise placement of material, reducing excavation requirements and improving fill construction, and are generally superior to bulldozer in both cost and quality of construction (FAO 1989).

The width that must be cleared and the volume that must be excavated increase with the gradient of the pre-existing side slope. This additional earthwork increases both

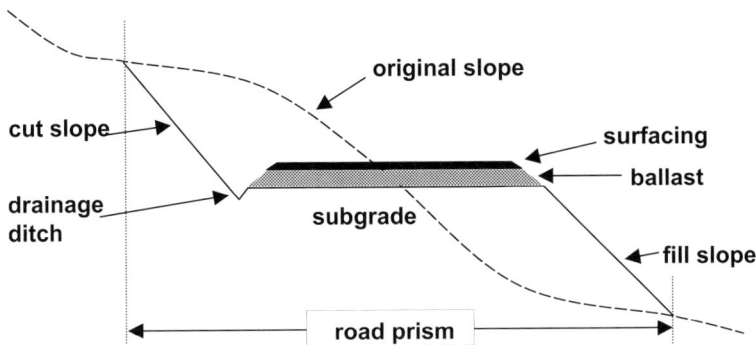

Fig. 10.5 Road structural terms.

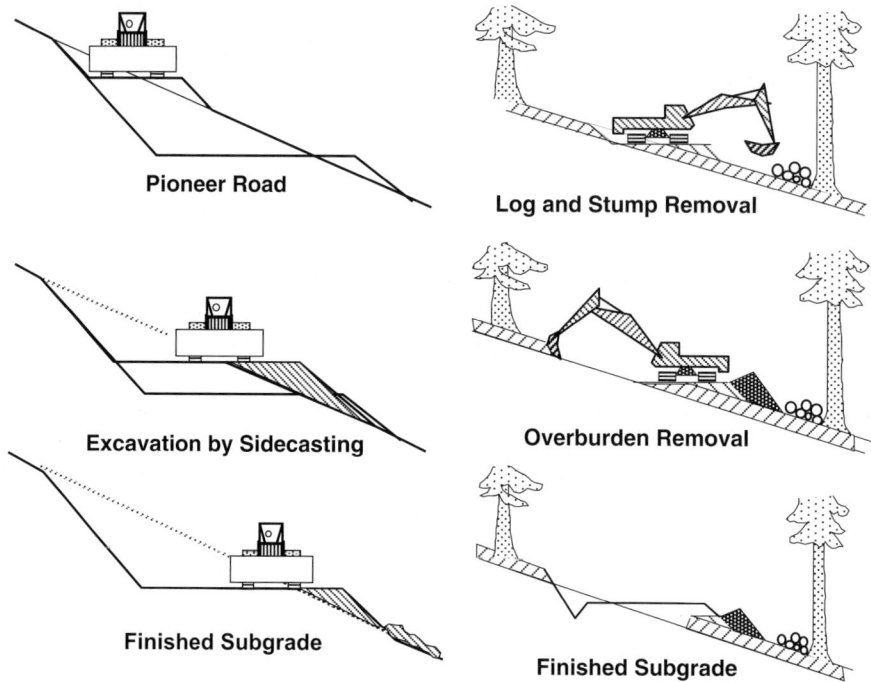

Fig. 10.6 A bulldozer (left) can rapidly construct a road prism by repeated passes in which soil is pushed (sidecast) downhill. Excavators (right) proceed more slowly, but their greater control of placed material can dramatically reduce the earthwork and environmental impacts on steeper slopes.

the construction cost and the potential for environmental impacts. The resulting cut and fill slopes are steeper than the pre-existing topography. The more soil that is side-cast and the more the uphill soil is undercut, the greater the risk of future landsliding. The cut and fill slopes, devoid of a protective organic layer, will also erode much more rapidly than the natural forest soils. Some of this erosion can be reduced if the cut and fill slopes can be revegetated after the road has been completed.

In some cases, the cleared and flattened soil can serve as the running surface of the road. Such native surface roads are common when traffic is to be restricted to the dry season. When these roads are wet, they tend to be vulnerable to traffic damage, such as deep ruts and high erosion rates. Adding an armouring layer of rock or gravel (ballast or base course) over the native soil can dramatically reduce traffic damage and sediment production (Kochenderfer & Helvey 1984) and maintenance requirements. Over time, traffic will push this armouring layer into the underlying soil (subgrade) which is then exposed to erosion and rutting. The rate at which the rock (ballast) layer is pushed into the subgrade is a function of traffic volume, the vehicle weight, the area over which this load is spread and the strength of the underlying soil (subgrade). Local pressure can be reduced by increasing the number of axles on each truck and/or decreasing tyre pressure, but the most common way to spread load is to thicken the ballast layer. The strength of the subgrade can be increased by compaction, but subgrade soil will rapidly lose strength if it gets saturated. A surfacing layer of finer materials can be laid atop the ballast and graded outwards to shed rainwater and reduce infiltration to the subgrade. The ballast and surfacing is typically the major cost item of a road and can range from 30% to 60% of the total road construction cost.

The subgrade can also become saturated by surface and subsurface water from the surrounding forest. This water can be routed away from the road through a network of ditches and culverts. Unfortunately, this network can deliver sediment-laden runoff from the road surface and cut slope wherever it crosses the stream network. Sediment delivery to the stream network can be reduced by the judicious placement of cross-drains and culverts to divert ditch water onto the forest floor where it can get filtered and trapped by the organic layer before it reaches a stream (Wiest 1998).

The surest way to avoid road–stream interactions, however, is to not build roads near streams. This may at first seem difficult because stream networks (like road networks) cross the entire landscape, seemingly necessitating numerous road–stream crossings. Valley bottoms are frequently the flattest and lowest gradient parts of the landscape, and many existing road alignments run parallel to or even directly up stream channels. Ridges form another low gradient network, however, which covers the entire landscape, never crossing streams, and being midway between streams maximizes road–stream separation. Locating roads along this ridge network can minimize road–stream crossings, and the resulting landings will concentrate their soil disturbance away from the stream network.

Maintenance

Once built, a road will begin to be damaged by traffic and other causes. Vehicles travelling over the surface impose both compressive and shearing forces that can rut and otherwise damage the road, slowing traffic and eroding sediment. This road damage can be reduced by periodic maintenance. Potholes, rutting and washboarding can be eliminated by minor grading. The drainage network and cross-drains must also be periodically cleared of vegetation, sediment and debris, which can slow or divert runoff and cause saturation and instability of the road prism. Maintenance costs can add up over time but they are often not considered in the initial planning of a road. While reducing erosion from road damage, maintenance is itself a source of soil disturbance and erosion, which must be weighed against its environmental benefits.

If a road will not be used for several years, maintenance might be discontinued, but if the road is not blocked, traffic can continue to cause damage. Even if it is blocked, natural hydrologic, geomorphic and vegetative processes will continue, leaving the drainage network vulnerable to sediment and debris accumulation, and the erosion, saturation and landsliding they can cause. The economically cheapest approach might be to ignore this road damage and repair it only when reactivating the road, but the sediment delivered to the stream from the resulting drainage failures and landsliding may be environmentally unacceptable.

It may be cheaper to deconstruct much of the road instead of maintaining the drainage network during years of road inactivity. Deconstruction usually involves removal of cross-drains and stream culverts that might plug and fail. Water bars can also be built across the road to route water off the road and dissipate its erosive force. Such alterations can be easily reversed when the road is reopened.

Instead of this partial deconstruction and then reconstruction, it may turn out to be better to totally eliminate the existing road, and build a replacement in a better location. Existing roads were built with technologies and management objectives that may no longer be valid. Stream-proximate road alignments and unstable fill slopes can be difficult and costly to 'fix'. The cost of ballast/surfacing and drainage requirements may account for up to 80% of a new road construction, but a road following a ridge alignment may have so little chance of delivering sediment to the stream that no ballast is needed to prevent environmental impacts. This approach of building a whole new road is commonly overlooked in favour of repairing and upgrading existing or even overgrown road alignments.

If a road will only be used for a single dry season (and immediately deconstructed), then it need not be constructed to the drainage and slope stability standards of a permanent road. This approach can reduce both construction costs and environmental impact. If cut and fill slopes do not need to withstand wet season saturation, then they can be made steeper, reducing earthwork and clearing requirements. Logs and stumps can be used to stabilize and steepen the toe of the fill (a practice that should not be used in the construction of permanent roads). The subgrade is unlikely to become saturated during the dry season, so the road surface can be out-sloped to eliminate a roadside ditch and to further reduce road width. Where the native soil is strong enough, the use of ballasting and surfacing material can also be avoided, eliminating the cost of acquiring, hauling and placing it, and the environmental costs of leaving or removing it. Where soils are soft and/or saturated, a good surfacing can be made by chipping the unmerchantable wood and debris, which can then be spread across the site when the road is eliminated. If on the other hand the temporary road is built to the standards of a permanent road, it will be wider, and more costly, with much more exposed soil. It will take longer to build and unbuild, so it will be more likely to deliver sediment during the wet season, assuming it is removed.

Road and harvest planning – bringing it all together

Harvest operations are about getting timber from the stump to the mill at the lowest economic and environmental cost (Table 10.3). As outlined in the previous sections, selection of the appropriate system is based on a number of variables, including:

- topography (slope steepness and variability)
- soil (saturation, composition, sensitivity to disturbance)
- silvicultural system (clear-cut, thinning, retention level/pattern)
- timber characteristics (tree size, volume, density)
- potential road access constraints
- equipment characteristics and performance (production and cost)
- processing (limbing and bucking) location
- stream and wetland distribution
- mill and market requirements.

Table 10.3 Total system costs, stump-to-truck, production rates and production costs for six harvest systems (silvicultural system, tree size and yarding distance affect production rates and costs)

Type of system and system crew	Total monthly owning and operating costs (1000 US$)	Economical external yarding distance limit (m)	Daily production range (truckloads)[1]	Production cost (US$/m³)
Ground-based, mechanized system 1 feller-buncher 1 Shovel/log loader 1 delimber 1 log loader 4-man crew	40–44	100–160	15–20[3]	8–12
Ground-based, semi-mechanized system 2 wheeled skidders 1 log loader 3-man crew + 3 fallers	38–44	250–300	11–16[2] 5–8[4]	11–15 21–40
Cut-to-length system 1 harvester 1 forwarder 2-man crew	35–40	250–300	4–7[4]	20–36
Standing skyline – small tower 1 log loader 5-man crew + 3 fallers	42–48	250–300	2–5[4]	32–100
Live skyline – large tower 1 log loader 7-man crew + 3–4 fallers	64–67	400–600	12–15[2] 7–12[3]	13–20 25–40
Helicopter, medium size 2 log loaders 14-man crew + 6–8 fallers	370–420	1500–2000	20–24[2] 11–15[3]	56–64 105–133

[1]A truckload is based on 22 tonnes or about 23-25 m³ based on wood density.
[2]Production rates based on clear-cut operations, average tree size 3 m³.
[3]Production rates based on clear-cut operations, average tree size 1 m³.
[4]Production based on thinning operations, average tree size 0.5 m³.

In the preceding sections, harvest systems and roads were discussed in isolation. However, the design of seemingly unrelated operations, such as a road segment from point A to point B, or the delineation of a particular harvest setting, should not be considered separately. The fact that harvesting costs tend to increase with increasing yarding distance might suggest a goal of reducing yarding distances so as to reduce harvest costs (stump-to-truck). On the other hand, shorter yarding distances usually require more roads, increasing road costs. This contradiction suggests that it is not the harvest or road costs that must be minimized, but rather their sum. The task is to find the optimal yarding distance (the point at which the total road/harvest cost is minimized) for a given system and use it as a guideline in planning harvest boundaries and road access options. The planning process involves finding the optimal mix of harvest system and road network for an area and time-frame beyond any single operation (Cullen & Schiess 1992).

When determining an optimal yarding distance, all costs must be considered. Unfortunately, yarding distances are often optimized to minimize just yarding and road construction alone, without consideration of maintenance and environmental costs. This has commonly resulted in high road densities serving cost-efficient short yarding systems. Over time the environmental costs (landslides and other erosion) of these high road densities have necessitated economically costly maintenance and/or decommissioning activities. If these costs had been included in the original planning, then a longer optimal yarding distance would have been selected, with a less dense road network.

For any given system, road–stream separation may be the most effective tool for reducing sediment delivery to the streams. Sediment is produced on almost all forest roads, landings and yarding operations, but its delivery to streams is a function of the distance and routing to a stream. The further the sediment has to flow across the forest floor, the more it can be filtered, and the less likely it is to be delivered to the stream network. Roadside ditches and culverts (that deliver to stream crossings) short-circuit this filtering, so road alignment should avoid streams wherever possible. Road–stream proximity can be avoided by noting that the stream network is the topographic opposite of the ridge network; never crossing ridges and always maximizing its distance from the ridge network. A network of primary and secondary roads following ridge networks (and crossing the stream network only rarely) would minimize sediment delivery to the stream network (Krogstad & Schiess 2000). Shifting from a riparian-based road network to a ridge-based road network will necessitate building new roads and might even increase road density. In shifting to a ridge-based network, however, new road construction may actually improve water quality by routing traffic over roads that deliver less sediment to the stream. Such a solution is only possible in the context of comprehensive harvest and transportation planning at the landscape or watershed level.

References

Conway, S. (1982) *Logging Practices, Principles of Timber Harvesting Systems.* Miller Freeman Publications, San Francisco, CA.

Cullen, J. & Schiess, P. (1992) Integrated computer aided timber harvest planning. In: *Planning and Implementing Future Forest Operations. Proceedings, International Mountain Logging & 8th PNW Skyline Symposium*, (eds P. Schiess & J. Sessions), pp. 22–33. University of Washington, Seattle.

FAO (1989) *Watershed Management Field Manual: Road Design and Construction in Sensitive Watersheds*. Food and Agriculture Organization of the United Nations, Rome.

Heinimann, H.R., Stampfer, K., Loschek, J. & Caminada, L. (2001) Perspectives on central European cable yarding systems. In: *2001 A Forest Engineering Odyssey. Proceedings, International Mountain Logging 11th PNW Skyline Symposium*, (eds P. Schiess & F. Krogstad), pp. 268–79. University of Washington, Seattle.

Kochenderfer, J.N. & Helvey, J.D. (1984) Soil losses from a minimum-standard truck road constructed in the Appalachians. In: *Mountain Logging Symposium*, (eds P. Peters & J. Luchok), pp. 215–75. West Virginia University, Morgantown, WV.

Krogstad, F. & Schiess, P. (2000) Haul routing: an overlooked factor in environmentally driven road decommissioning. In: *Technologies for New Millennium Forestry: Proceedings of the 23rd Annual Meeting of the Council on Forest Engineering*, Kelowna, British Columbia.

MacDonald, A.J. (1999) *Harvesting Systems and Equipment in British Columbia*. Crown Publications, Victoria.

Malmberg, C.E. (1989) *The Off-Road Vehicle*. Technical Association of the Pulp and Paper Industry, Atlanta.

Megahan, W.F. (1980) Nonpoint source pollution from forestry activities in the western United States: results of recent research and research needs. In: *US Forestry and Water Quality: What Course in the 80s?, Proceedings of the Water Pollution Control Federation Seminar*, pp. 92–151. Richmond.

Samset, I. (1985) *Winch and Cable Systems*. Martinus Nijhoff/Dr. W. Junk, Dordrecht, the Netherlands.

Schuh, D. & Kellog, L. (1988) Mechanized delimbing at a cable landing. In: *International Mountain Logging and Seventh Pacific Northwest Skyline Symposium*, (ed. J. Sessions), pp. 112–20. Oregon State University, Corvallis, OR.

Staaf, K.A.G. & Wiksten, N.A. (1984) *Tree Harvesting Techniques*. Martinus Nijhoff/Dr.W. Junk, Dordrecht, the Netherlands.

Studier, D.D. & Binkley, V.W. (1974) *Cable Logging Systems*. Oregon State University, Corvallis, OR.

Studier, D.D. & Neal, J. (1993) *Helicopter Logging: A Guide for Timbersale Preparation*. US Forest Service, PNW Region, Portland, OR.

USDA Forest Service (1982) *Transportation Engineering Handbook*.

Wiest, R.L. (1998) *A Landowner's Guide to Building Forest Access Roads*. USDA Forest Service, NE Area, Radnor.

Workers' Compensation Board (1999) *Cable Yarding Systems Handbook*. Workers' Compensation Board of British Columbia, Vancouver.

Chapter 11
Silviculture

J.E. BARKER

Introduction

Silviculture involves intervention in natural processes, altering them in ways that significantly influence the structure and function of forested ecosystems.

Traditional silvicultural approaches consist of documented experiences, codified into a sequence of treatments. Results typically have been evaluated in terms of regeneration success and wood volumes, but during the past few decades there has been pressure to explicitly include non-timber values in the development of silvicultural systems. Generalized, recipe-type silvicultural systems are becoming less acceptable as value systems become more complex and the numbers of interacting factors to be considered increase. Alternative silvicultural treatments are under considerable pressure to demonstrate their anticipated effects in a quantitative manner, not only for timber but also for numerous other attributes. Complex models are necessary to achieve this, and their development will depend on the availability of a designed range of alternative systems, maintained and monitored for significant time periods, for calibration. These are scarce and the complexity of their structure is beyond the scope of this chapter. However, currently there are tools available that assist silviculturalists in the analysis and application of various practices individually and their impact principally on timber yield. These will be discussed briefly in the following sections.

Background

Silviculture as a formal discipline originated in the 1700s in response to public concern regarding the degraded condition of European forests. Unstructured use of the forest for firewood, charcoal and litter removal had led to a crisis situation and local wood shortages. Those public concerns about degradation that shaped silvicultural practices 300 years ago are still in force today and are having a marked influence on silvicultural practices. Silviculture is thus a blend of science, experience and public acceptability.

In days of early settlement, forests were often viewed as an impediment to other uses such as agriculture, mining or settlement. This is still the case where population pressures are leading to conversion of forest land to other uses. In such situations, there is little interest in silviculture. However, as the forest is diminished and the values flowing from it are threatened, silviculture begins to develop and flourish.

Baker (1950) discusses the tension between economic and biological aspects of silviculture and the need to obtain maximum benefit for minimum cost. Because of the complexity of forested ecosystems, it is often difficult to predict treatment impacts. Also, social goals change and are hard to define over the lengthy time period of a forest rotation. Consequently, silviculture is situation-specific, and must be flexible to adapt to changes in both the biological values involved and the socio-economic environment in which it exists.

In North America, the definitions of forestry and forest management are changing in response to the changing values of society. Forest management, according to the 1958 definition (SAF 1958): 'was the application of business methods and technical forestry principles to the operation of a forest property'. Forest management was viewed as being concerned with the application of business methods and technical forestry principles while management for other values was recognized as separate forms of management. The unfortunate consequence of this fragmentation (or specialization) was the implication that forest management differed from watershed management or wildlife management and was controlled by economics at the expense of environmental externalities. A consequence of this reductionism has been the creation of a number of alternative specialties, and the perception that forestry is narrowly focused on timber production. Now however, the revised forestry dictionary of terms (Helms 1998) does not categorize management at all. The definition of silviculture has been changed to explicitly recognize the importance of non-timber values. Rather than simply '… producing and tending a forest …' (SAF 1958), the definition of silviculture now includes 'meet(ing) the diverse needs and values of landowners and society on a sustainable basis' (Helms 1998).

Thus sustainable forest management, as opposed to sustained yield management, now explicitly emphasizes the inclusive nature of forestry, which shapes both timber and non-timber resource values such as water, aesthetics, biodiversity, habitat, etc. in ways that will help meet objectives for all uses, whereas formerly it was an implicit assumption that this would occur if sustained yield was the goal.

Basic concepts

Silvics

Silvics is the study of how different species of trees reproduce, grow, and develop in relation to their environment. Taxonomy, geographic distribution, climatic factors, soils, and topography all form a part of silvics, as well as associated vegetation and competition, pests and disease factors. How individual species respond to these factors at a very basic level is a function of their physiology. Burns & Honkala (1990) have summarized a wealth of silvics information on individual tree species. Such information underlies the development of biologically feasible silviculture.

Range

All species occupy a geographic range, which is characterized by a range of climatic, topographic, and soil conditions. Within this range, depending on the species and its evolutionary history, patterns of genetic variation can develop resulting in a species having significantly different silvical characteristics in different portions of its range. As a result, generalizations must be treated cautiously.

Light and shade tolerance

Silviculturalists control light through manipulation of stand density. In order to interpret and manage stand responses to light, an understanding of the relative shade tolerance of the different species is needed, particularly for regeneration. The relative ability of trees to grow and develop in the shade of competitors is a working definition of shade tolerance. Baker (1950) has categorized species as to their relative shade tolerance. Relative, as opposed to absolute, refers to a species' behaviour in comparison with other species. Shade-intolerant trees require open conditions for regeneration and early growth. Where they occur in existing forests, and are represented as intermediate or suppressed crown classes (see Fig. 11.1) they are likely to be slow-growing, non-vigorous trees; candidates for death and exclusion from the stand. On the other hand, shade-tolerant trees, even if currently represented as intermediate or suppressed, can be expected to grow and represent future crop trees, since they maintain the capacity for growth and development in shaded conditions. As a consequence, the crown classification for a tree will have a different physiological connotation for different species. Douglas fir (*Pseudotsuga menziesii*), for example, can be regenerated in shade in the drier parts of its range but is highly shade-intolerant in the wetter coastal Pacific Northwest.

Fig. 11.1 Crown classes: D, dominant; C, co-dominant; I, intermediate; S, suppressed.

Reproduction

A substantial portion of a silviculturalist's job involves regeneration. Some species such as lodgepole pine (*Pinus contorta*) will produce seed at very young ages (5–10 years) while others such as the true firs (*Abies* spp.) may take 20–30 years before seed production begins. Deciduous species such as red alder (*Alnus rubra*) may reach sexual maturity between 3 and 5 years. Once sexual maturity is reached, periodicity (number of years between significant seed crops) becomes a factor of importance if natural regeneration is to be expected. Some species such as Douglas fir produce medium to large seed crops every 5–7 years whereas red alder produces crops every 1–3 years. Such differences impact the species composition of a regenerating stand and create challenges in meeting silvicultural goals where undesirable species have higher and more frequent seed production. For example, the presence of species which are prolific seeders in riparian leave zones (red alder is a good example) can result in reforestation problems on adjacent cut-blocks where the species is not wanted.

Seed dissemination differs among species, but most commercial conifers in North America rely on wind dispersal. The degree to which natural seeding is effective on large cut-blocks is limited as a majority of seed falls within about two tree lengths of the parent source. However, depending on species tolerance, sufficient advance regeneration may be present which is released on harvest.

Some species such as broadleaf maple (*Acer macrophyllum*), redwood (*Sequoia sempervirens*), red alder, aspen and poplar (*Populus* spp.) regenerate vegetatively from stump sprouts or root suckering as well as through seed.

Ecological classification

Landform, topographic position and soils in conjunction with climatic factors are reflected in patterns of vegetation. Recurring, similar communities can be defined and identified as spatially explicit units in terms of these factors (Hills 1952; Krajina 1965; Daubenmire 1968; Pojar *et al.* 1987; Meidinger & Pojar 1991). Such classifications allow silviculturalists to identify acceptable species for reforestation, their growth potential, associated vegetation and possible weed competition, stand health risks and many other important considerations that go into adoption of a silvicultural system.

Stand structure

The basic silvicultural unit of treatment is the forest stand or an aggregation of vegetation, sufficiently uniform in species, structure, age, productivity or other characteristics, that it can be identified as unique. The structural components of a stand are termed cohorts or groups of trees – essentially the same age – originating as a unit from the same disturbance event. A stand may contain a single cohort or multiple cohorts depending on stand history and the type of silvicultural system adopted.

Within cohorts, trees differentiate into crown or social classes during stand development that characterize within-stand vertical structure. These classes serve to help identify the competitive status of a tree within a stand. Figure 11.1 illustrates the four crown classes that are commonly used.

Structure can also be characterized by stand tables (frequency distributions based on stand diameters) and stock tables (distribution of volume by diameter classes).

Horizontal structure (termed 'clumpiness' or 'patchiness') is difficult to quantify and can refer to the arrangement of individual trees within a stand, a cohort or among cohorts within stands. Gaps or areas between clumps play an important role in creating environments for regeneration and influence stand volume development in complex ways. Such structure forms an important aspect of habitat value.

These measures are useful tools for prescribing and monitoring the effectiveness of silvicultural treatments.

Stand development

Stands progress through a number of developmental stages, each presenting its own unique opportunities.

Establishment stage

During the establishment phase, young seedlings are free from competition with other trees within their cohort. However, treatment to reduce competition with other species, such as brush species, is often required. Delaying or neglecting such treatment can result in either mortality or a long-term reduction in the growth potential of the crop species, particularly shade-intolerant species, which can lose crown and require several years to recover after release.

Nutritional demands are high as new crown material is being produced, so addition of fertilizers may be useful, provided that competing vegetation is not the primary beneficiary of the added nutrients. Combining weed control with fertilization may be advantageous for this reason.

Exclusion stage

At this stage, established seedlings have formed a closed canopy. Individual trees compete with one another, slowing their rate of diameter growth. Trees begin to express their bole quality, particularly in the bottom logs, which comprise close to 50% of the volume and an even higher proportion of the tree's value. At this stage a pre-commercial thinning (PCT) may be undertaken to favour crop trees with well-formed boles, free from kinks, forks, and other stem defects, and increase their diameter growth by concentrating the productive capacity of the site on the remaining trees. However, excessive branch sizes can develop if thinning is too heavy. Green branch pruning at this stage may be undertaken to improve stem quality.

As this stage progresses, trees segregate into social classes and, in the absence of PCT, suppression leading to mortality will begin to thin the stand. Lower branches die as they become shaded and the crown length of individual trees stabilizes. Because of the low light levels beneath the closing canopy, understory vegetation is usually sparse.

Commercial thinning may be used to salvage volume losses from mortality, promote greater diameter increment on selected stems and perhaps stimulate the development of understory vegetation for habitat purposes. Opening the stand at this point in its development may lead to blow-down problems, particularly if the stand has previously been very dense.

Mature stage

As the stand ages further, tree height growth slows, crowns begin to flatten and rate of mortality stabilizes at a fairly low level. Light penetration to the forest floor increases and shade-tolerant understory species begin to develop, leading to a more multi-storied canopy structure and a more varied habitat. Silvicultural interventions at this stage are generally directed towards regeneration, as stands have a reduced capacity to respond to thinning and crown release.

Silvicultural systems

A silvicultural system is comprised of a number of activities, applied sequentially to forest stands to achieve defined goals. Once deemed the prerogative of the forest manager, these goals are now more frequently developed through some form of consultative planning process involving forest managers and non-timber resource managers as well as an interested public. Once the goals are agreed upon, silvicultural prescriptions are developed that produce the appropriate stand conditions. These prescriptions specify the treatments to be applied, when they are to be applied, and a procedure for monitoring and adjustment. The cycle of harvest, renewal, growth and tending is specified but the methods used will probably change as a better understanding of natural processes develops and social values change.

The terms silviculture and management at one time tended to be interchangeable. Today however, silviculture refers to the actual hands-on application of cultural techniques in forest stands. Forest management, on the other hand, now deals with a much broader range of activities.

Classification of silvicultural systems

Classification of silvicultural systems is treated in detail by a number of authors (Troup 1952; Spurr 1956; Matthews 1989). Given the very large number of forest types and treatments and objectives that are possible, a classification approach using broad general principles is most appropriate.

One simple classification is based on age structure. Even-aged and uneven-aged structures represent the two extremes in such a classification.

A second classification is based on the type of propagule used to regenerate the forest stand. Thus, regeneration may be from seed (or rooted cuttings (stecklings) deployed as if they were seedlings) or from coppicing techniques where the new forest arises as vegetative sprouting from stumps or root suckering of the previous stand. These are termed high forest and low forest systems respectively.

In addition, silvicultural systems can be further classified by:

- structure remaining after harvest
- purpose for which the structure is being maintained
- length of time that structure is maintained.

There is a continuum of possible structures between a single cohort, even-aged stand arising from a single, complete harvest, i.e. a clear-cut with no residual structure, through a range of variants where residual stems are left following harvest. These residual stems may be left for periods of time sufficient to provide adequate seed fall, shelter, or both, before their removal. In other cases, the residual structures may be maintained throughout the next rotation, to provide structural diversity. Finally, multi-storied, uneven-aged stands, consisting of several cohorts of different ages created and maintained through successive harvest entries, represent the other end of the continuum.

Even-aged silvicultural systems

Even-aged systems are generally applicable to shade-intolerant species which require full-light conditions for their regeneration and development. However, species classified as shade-tolerant may also be managed in this fashion, particularly where climatic conditions are not extreme. Even-aged stands are characterized by diameter distributions resembling a bell-shaped or normal distribution and a limited variation in tree age that is not related to tree size. Even-aged stands can be produced using a range of harvest systems.

Clear-cutting

Trees are removed either all at one time or over a small fraction of a rotation and a new cohort is established. A clear-cut can be defined as an area where over 50% of the area is free from the influence of standing trees. It is generally assumed that the effects of an adjacent area extend approximately 1.5–2 times the height of the edge trees. The actual size of edge effect depends on the aspect of the edge, and the slope of the terrain.

Seed tree

Seed tree harvest systems involve leaving either individual trees or small patches distributed across a cut-block to provide a seed source for establishing natural regeneration. The trees left standing (leave trees) should be healthy and of good form to ensure good genetic value of offspring and must also be wind firm, capable of standing long enough to provide the seed. In view of the periodicity of seed crops, this could be as long as 5–10 years. Following establishment of a second crop, the seed trees are removed to allow the new cohort to develop freely. This removal should take place while the young trees are small and supple to avoid excessive damage.

Shelterwood

Regeneration of some species requires amelioration of the climatic extremes experienced in an open environment in order to regenerate and survive. To provide such protection, some trees are left standing. The number of trees left is variable depending on the shade tolerance of the species to be regenerated and the degree of shelter needed.

Classical shelterwood systems involve three harvest entries, although two usually prove satisfactory. First, an optional preparatory cut may be carried out. This step is needed when the stand is dense, with small crowns and there is a concern about stability following harvest. The aim is to create crown volume and wind stability on the residual trees. The time necessary to develop stable, large crowns will be a function of the characteristics of the particular stand, its developmental history and the risk of damaging winds.

Following the preparatory cut, a seed cut is carried out to provide sufficient light for the establishment and early growth of the new stand. Regeneration usually is obtained from natural seed fall but planting is also an option. Once satisfactory regeneration has been achieved, a final removal cut occurs, which removes all the overstory trees and leaves the new cohort to develop freely. Depending on the spatial organization of these cuts, they may be classed as strip or group shelterwoods. Strip shelterwoods are often organized to progress into the damaging wind direction so that cut edge exposure is reduced. In either case, cutting is designed to maintain a residual stand influence over the regeneration until it is well established.

Uneven-aged systems

Uneven-aged systems are most applicable to species tending towards the shade-tolerant end of the spectrum. Harvesting in such stands is carried out in a cyclic fashion, with entries removing trees either in small patches or as individual trees at some defined interval, termed a cutting cycle (often 10–20 years), and leaving behind a forest structure which consists of at least three distinct age cohorts. The overall stand table for an uneven-aged stand resembles a reverse 'J' shape with larger numbers of trees in the smaller diameter classes and fewer in the larger diameter classes. This overall structure may be comprised of a number of smaller even-aged cohorts, distributed through the stand. The residual structure must be sufficient to maintain stand productivity while ensuring that conditions favourable to regeneration are met.

The pattern of removal may be varied. A group selection method removes groups of trees where the size of the group is such that there is continued influence of the surrounding trees over the area harvested. In cases where individual trees are removed, scattered throughout the stand, the approach is termed a uniform system.

There has been continuing debate as to whether uneven-aged forest structures are superior to even-aged structures in terms of ecological stability, social acceptability and productivity. In recent years, silvicultural methods that mimic natural disturbance patterns have gained currency as a means of addressing these issues. Natural uneven-aged systems, by the fact that they exist, and have done so for millennia, demonstrate their long-term stability so that silvicultural systems patterned on them can also be expected to be stable. Regardless of the truth behind such suppositions, uneven-aged approaches must now meet the combined constraints imposed by biological, as well as social and economic, factors before the adoption of this particular form of silvicultural system can be successful.

Uneven-aged silvicultural systems are often (but not necessarily always) designed on the basis of patterns observed in natural forests. Silvicultural activities are directed to creating and maintaining such patterns. This involves harvesting so as to ensure adequate regeneration plus control of stem exclusion rates over time, within all co-horts, in a way that ensures a sustainable overall stand structure. Serious sustainability problems can arise if owners seek to maximize short-term financial gain by removing only high-valued trees and leave the poorer material behind, creating a green illusion. Care is therefore necessary to define the appropriate measures necessary to produce a sustainable structure.

To attain such a structure, silviculture in uneven-aged stands uses a defined diameter class distribution, a maximum size of crop tree and stand basal area as a means of reaching and maintaining sustainability. Diameter class distributions that typify naturally occurring forest structures for the species under consideration are frequently, although not necessarily, used as a model for sustainable structure. The actual stand structure being managed is then modified through a series of cuts so that the stand's diameter distribution comes to resemble this desired distribution, usually a reverse 'J' shape, and basal area stocking is maintained at a level where full site occupancy is achieved (see below) while regeneration is facilitated. (The basal area of a stand is the sum of the cross-sectional areas of boles measured at 1.3 m above the ground.)

This is called a 'Bdq' approach to creating and maintaining a stand structure and can include both intermediate and final harvest. It involves controlling the stand basal area (B), the maximum diameter desired (d) and the rate of stem exclusion, q, which defines the slope of the reverse 'J' shaped diameter distribution.

Careful attention to basal area control is essential, as retaining too much basal area will preclude adequate regeneration and unduly impact growth of smaller stems, while removing too much will reduce total stand volume growth and could promote the establishment of undesirable weed species. The growth rates in the different diameter classes must be understood to ensure that the stems retained are capable of producing the necessary aggregate growth (Schutz 1997).

The interval between successive interventions to produce and maintain the desired structure is called the cutting cycle. The cutting cycle is determined by the site productivity, the rate of growth and the limits on basal area stocking, as well as the amount of timber necessary for an economic harvest. Achieving a sustainable harvest requires a balance between these factors. For example, to achieve low visual impact, a low harvest removal per entry leaving high residual stocking levels may be prescribed. However, this alternative must be accompanied by more frequent entries, more frequent site disturbance and road use if the defined harvest level is to be maintained. Such considerations play a role in riparian management where a trade-off between frequency of disturbance and the impact of the individual disturbance must be attained.

European experience with the Plenterwald system has been summarized by Spathelf *et al.* (1997), Kenk (1997) and Schutz (1997). They conclude that while selection forests may have aesthetic appeal, they are highly artificial systems requiring high levels of management expertise, good record-keeping, suitable markets, dense road systems and careful harvesting methods. To achieve success requires long-term commitment and continuity of effort, which can be a problem where the political and economic climates fluctuate, ownership changes or the political landscape is altered.

Retention systems

Harvest patterns that leave portions of the original stand structure behind are termed retention systems and can be described by the gap distributions or clumping pattern of trees that are retained. Mitchell & Beese (2002) outline the essential elements of retention systems. Group retention refers to cases where the retained trees are in patches or small groups, whereas dispersed retention refers to a situation where individual trees are left standing throughout the block. Depending on the scale involved, a combination of both group and dispersed patterns may be present, leading to the term variable retention. Retention systems have evolved to deal with a number of concerns. One purpose of retention is the provision of 'lifeboat habitats' designed to provide sufficient habitat structure and connectedness within the landscape for species present prior to harvest. Another purpose is to maintain a forested landscape for aesthetic reasons. The principal difference between retention systems and traditional silvicultural systems is that the residual structure is not being left to obtain regeneration but to retain some desired structural elements from the previous stand. Subcategories of retention systems may be defined based on residual stand density (Mitchell & Beese 2002). Retention is applicable to both even- and uneven-aged systems. Individual, large, long-lived trees or groups of trees could be maintained in stands being regenerated and managed as uneven-aged stands to provide structural elements, such as large snags or relic species, not present in the normal uneven-aged system.

Regeneration

Regeneration is an essential step in sustainable silviculture. At the regeneration stage, many future options are defined. Such things as species composition, growing space and genetic quality can be controlled most easily at this stage.

Either natural or artificial methods or a combination of the two are possible. Achieving successful regeneration by any of these approaches depends on site conditions and other factors such as whether the existing seed source or existing seedling advanced growth represent the species best suited both ecologically and economically to the situation.

The regeneration period is a sensitive period in stand development where tolerance of adverse environmental effects, such as frost or drought, is not high. Timing can also play an important role. In British Columbia for example, on rich alluvial sites, a delay of 1–2 years in establishing a conifer crop such as Sitka spruce (*Picea sitchensis*) can result in dense thickets of salmon berry (*Rubus spectabilis*), greatly reduced timber yields and/or a need for expensive silvicultural rehabilitation treatments. It is important to understand the potential limiting factors that will be encountered on a site in order to ameliorate their effects during the regeneration phase, either through site preparation or weed control activities. An understanding of environmental and vegetation changes that follow the site preparation activity is desirable since some treatments may produce undesirable side effects. For example, burning may result in increased erosion and loss of organic matter or the invasion of undesirable competing vegetation.

Site preparation

Site preparation aims at creating conditions favourable for seedling establishment and survival. This involves modifying numerous factors such as soil disturbance, slash loading and competing vegetation to create a local environment suited to the species desired.

In situations where forest productivity is high, and rotations are short, intensive site preparation, weed control and fertilization are frequently combined to ensure that maximum growth response can be obtained from planted, high genetic gain seedlings. In some situations where water tables are close to the surface, some form of ploughing or bedding may be used to improve soil moisture conditions. (Bedding is the creation of a continuous raised mound of earth.)

In colder climates, raised seedbeds may be created by mounding equipment or by ploughing, which improves soil temperature through cold air drainage and increased exposure to incoming solar radiation.

Site preparation increases the number of available planting spots, which improves the spatial occupancy of the site and usually results in lower planting costs because of better access. In more extensive silvicultural situations where costs of planting cannot be justified, and natural regeneration from seed is the goal, scarification to provide a more receptive seedbed is beneficial.

Controlled burning

Controlled use of fire is a widely practised form of site preparation following harvest. It also has a place in the management of natural areas where fire has been excluded as a natural phenomenon.

When applied in conjunction with modern fire management technology, quite favourable results can be achieved without damage to adjacent values while removing accumulations of slash, knocking back competing vegetation and providing access to planting spots. Since much of the nutrition on a site is contained in the organic layers, it is important to control burn intensity. Spring burning, when organic layers are still moist but surface debris is dry, is a preferred approach that minimizes such impacts.

During recent years in British Columbia, forestry practices have increased the structural complexity of the landscape with the creation of residual structures such as riparian zones or wildlife tree patches in or immediately adjacent to cut-blocks. The costs of burning have undergone dramatic increases, as these non-timber values must be protected from damage. Other concerns related to air quality and CO_2 emissions are factors impacting on the use of prescribed burning.

Thus a decision to burn requires a site-specific evaluation of the costs and benefits of the practice, both in terms of the site itself as well as any other values likely to be impacted.

Mechanical site preparation

Spot cultivation, discing, disc trenching, ploughing, as well as bedding, are common mechanical approaches. Various kinds of machinery are used depending on terrain, availability and end purpose. On soft ground, for example, low ground pressure vehicles are needed. Tracked and wheeled primary movers such as skidders or bulldozers frequently are used to tow a variety of implements. Chopping rollers, sharkfin barrels, root rakes, ploughs, mulchers, tined buckets, anchor chains and ripping tines are all part of the array of tools available.

The aims of mechanical preparation are varied and depend on the local conditions. For example, mechanical means may be used in conjunction with fire where the fuels are manipulated prior to the burn to enhance burnability. In southeastern USA, bedding ploughs are commonly used, where water tables are near the surface, to provide raised planting sites. (A bedding plough has two coulters (discs) mounted a bed-width apart on a single axle.) In the boreal forest, disc trenchers are a means of creating raised ground to provide warmer, better-drained conditions for either seeding or planting. In New Zealand, ripping tines mounted on powerful primary movers are used to disrupt impervious layers to improve rooting access and improve wind stability in the developing crop. In coastal areas of British Columbia, large residual stumps and heavy volumes of debris, on or within the soil profile, can make continuous methods such as ploughing or trenching impossible. Excavator-type machines with powered scarifying heads or rakes, capable of moving around such obstructions, are then used.

Chemical

Chemicals can be used to reduce competing vegetation on the site or may be used to desiccate the vegetation to improve subsequent burning. Public concern over the use of chemicals has meant that this use is highly regulated and subject to intensive evaluation before application.

Administrative approaches to controlling the use of herbicides vary globally. The approach used in British Columbia is presented here as an example of a working system.

The Canadian federal government is responsible for ensuring the health and safety aspects of the chemicals for use. This responsibility is met through a registration process. Following a detailed testing and evaluation protocol, a Pest Control Product number (PCP number) is issued. The label issued with the product outlines the conditions for use. In the case of British Columbia, only four chemicals are registered for site preparation use: glyphosate (Vision®), 24D (amine and ester forms), triclopyr (Release®) and hexazinone (Velpar®). Of these, the most commonly used is glyphosate.

Given a registered herbicide, the British Columbian provincial government assumes responsibility for ensuring that application procedures are appropriate and meet environmental and safety concerns. In order to achieve this, a permitting process exists. Detailed pest management plans are required which define the area involved along with application details such as chemical used, timing and rate of application, signage as well as designation of non-treated areas such as buffers around water bodies. These applications are advertised and if there are no substantive objections, the Province then issues a permit, which is subject to appeal. Applicators must be licensed or under the supervision of a person with a licence.

In deciding which herbicide is to be used, factors such as target vegetation, potential for movement, length of time of residual effects and effects on non-target species need to be considered. For example, glyphosate is a broad-spectrum herbicide while 24D is selective with low impact on grasses. Both have relatively short residual lifetimes while that of triclopyr is longer. Hexazinone has the capability of moving through mineral soils and thus may pose concerns in riparian zone contamination.

Details of the British Columbian requirements and a detailed discussion of chemical properties and target vegetation can be found in a comprehensive field handbook (Otchere-Boateng 2002).

Application methods can be grouped broadly into spraying or injection techniques. Spraying is carried out in a number of ways such as aerially, ground hoses operating from a pumper truck or through use of backpack sprayers. Each has its own advantages, which relate to the terrain, the sensitivity of the area and the need to control drift. For site preparation purposes, there is no need for selectivity unless advance regeneration is present. Injection techniques often involve simple tools such as a hatchet and oil can. Injection lances are available which drive capsules containing herbicide into the plant to be killed. These techniques allow close control over the boundary of the treatment area.

Ground-based manual applications require great care to avoid undue applicator exposure, and use of protective gear is of major importance.

These methods help ensure that adverse effects are avoided. For example, results of a large-scale, ecosystem-based study including the use of aerially sprayed glyphosate over an aquatic habitat in the Carnation Creek watershed on Vancouver Island are summarized by Reynolds *et al.* (1993). Impacts on salmonids and other aquatic organisms were judged as short-term and small enough to be thought acceptable. The herbicide degraded rapidly and, after a year's time, the chemical had either disappeared or was bound up in inactive forms in the organic sediments.

Natural regeneration

If a suitable amount of advance regeneration is present prior to harvest, it may be conserved to provide the new crop. If seed fall is required, cut-block size may be adjusted to fit seed flight distances or, alternatively, seed parents may be left standing on the block. Seedbed condition is another factor that plays an important role. Most species benefit from some sort of ground disturbance which results in a seedbed composed of mixed organic and mineral components. The extent to which disturbance occurs and the distribution pattern of the disturbance on the ground can have a large influence on the survival and growth of the new crop. Where disturbance from harvest is inadequate, mechanical scarifiers have been developed and can be used to improve the quantity and quality of the seedbed if burning is undesirable.

Artificial regeneration

Because natural regeneration has a probabilistic component which may lead to the need for expensive follow-up weed control, incomplete site occupancy and less desirable species reaching unacceptable proportions in the stand, there has been an emphasis on planting as a means of regeneration in many managed forests. This approach also permits introduction of genetic gain from tree improvement programmes.

Successful planting requires substantial planning and scheduling. Seed must be obtained that is genetically and ecologically appropriate for the site, orders for seedlings must be placed ahead for delivery on time and the harvested site must be prepared and ready to receive the trees.

It is important that seedlings be appropriately conditioned to survive through the handling, shipping and establishment period. This may involve cold storage for fairly extensive periods of time between lifting from the nursery and planting in the field, which makes seedling physiological conditioning important to avoid cold damage and desiccation. Pre-lifting conditioning such as blackout shading, root pruning or droughting is commonly used to induce a hardened or dormant condition which improves survival.

Planting is best done immediately prior to breaking of spring dormancy so that the seedlings are entering a favourable period for both root and shoot growth. Trees planted in the fall are entering a period of lower temperatures and slower growth and must endure several months of adverse conditions, which limits the build-up of photosynthetic reserves and the establishment of adequate root development.

Seedlings may be lifted and cold stored well before spring planting where immediate planting is not feasible due to factors such as site availability or physiological condition of the seedlings. Seasonal variation in inherent patterns of root growth capacity and frost hardiness show that lifting of seedlings in the nursery is best done after a period of dormancy has been experienced (Dunsworth 1988; Simpson 1990) but before seedlings begin to lose dormancy in preparation for growth resumption in the spring.

Mishandling of seedlings during lifting, storage and planting reduces survival. Warm temperatures and high humidity conditions in storage leads to mould and depletion of carbohydrate reserves. High temperatures in storage containers exposed to direct sun can kill seedlings. Root exposure and desiccation during planting must be minimal. Such effects are cumulative and care is needed throughout all stages of activity.

Careful selection of appropriate micro-sites for the planted seedlings is essential. Depressions or areas of rotten wood or exposed subsoil material should be avoided. Although it is advisable to ensure that roots have contact with mineral soil, there are situations where substantial organic horizons may preclude this. If these organic horizons remain moist, survival and growth will be acceptable.

Planting methods must be appropriate for the rooting habit of the seedlings being planted. For container stock, narrow shovels or dibbles may be appropriate and provide for high planting rates. However, for bare root trees, a wider shovel or a mattock is preferable, since they can be used to dig a hole large enough to accommodate the roots. If the planting hole is too shallow, roots are forced into a horizontal position, leading to a permanent distortion known as 'hockey sticking' or 'J' roots. Survival may be adequate in such situations, but instability becomes a problem as the trees grow.

Vegetation management

Following establishment, young trees undergo a period where there is often significant competition from undesirable species (weeds). If competition is severe enough, trees either die or endure an extended period of slow growth from which they never recover. Vegetation management aims at ensuring that trees reach what may be called a 'free to grow' condition without experiencing growth check or mortality. Vegetation management integrates a number of approaches – biological, mechanical and chemical – to achieve the objective of a free growing crop. For example, large planting stock may be used, allowing reductions in the amount of weed control required. Application of fertilizer may further enhance early growth but this requires weed control. Figure 11.2 illustrates the interaction responses that result from using different combinations of treatments.

Vegetation management requires careful planning to assess the nature of the problem and identify the appropriate tools for use in terms of their cost-effectiveness. Such activities are greatly facilitated by use of ecological classification (Pojar *et al.*1987). It is also necessary that treatments are actually carried out – on time with quality control

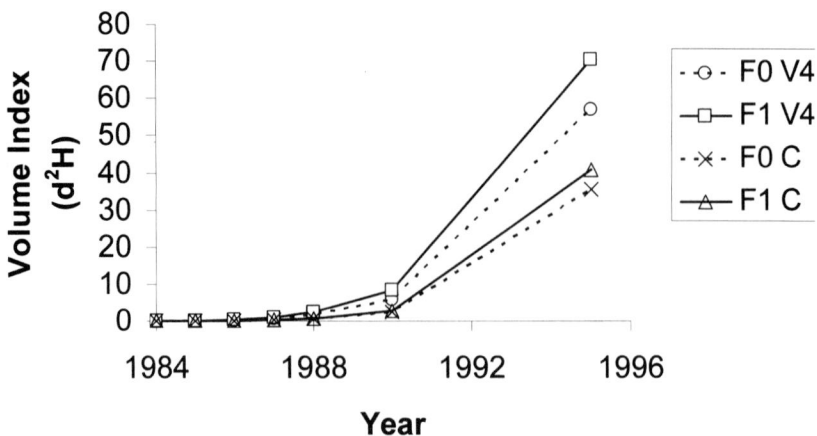

Fig. 11.2 Response to fertilization with and without weed control; Sitka spruce 11 years from planting. F0, unfertilized; F1, fertilized; V4, hexazinone applied for 4 years; C, no weed control. (Source: author's unpublished data.)

and a system for monitoring and assessing results. Further details on integrated vegetation management can be found in Comeau *et al.* (1999).

Thinning

Following establishment, natural patterns of mortality through inter-tree competition reduce the number of live trees, usually from thousands to hundreds. The rate at which this reduction occurs varies in relation to the number, size and species of trees. Thinning represents active intervention into this natural process at a number of intermediate ages between establishment and final harvest to achieve a variety of goals.

Thinning does not have regeneration as a goal. Rather, it aims at redistributing the growth potential of a site onto fewer, larger stems and recovering volumes that would otherwise be lost to mortality.

Thinning classification

One approach to thinning follows natural stand replacement patterns where smaller trees die and their volume is lost from the stand. Thinning smaller trees from a stand (thinning from below or low thinning) follows this philosophy. Historically, low thinning is categorized into a series of grades from A to E representing increasingly heavy interventions. The A and B grades are generally too light for today's markets and are of historical interest only. More commonly, C, D and E grades, which involve successively higher removals from the co-dominant and dominant crown classes, are used.

A different approach involves removing only those trees competing directly and strongly with the crop trees. This form of thinning (thinning from above or crown thinning) leaves the smaller trees and removes the larger competitors. This approach aims at stimulating the growth of the residual crop trees rather than salvaging low valued mortality.

A third approach is systematic thinning whereby trees are removed in some arbitrary fashion without reference to tree quality. This approach is applicable in very dense young stands where mechanical mowing approaches can be applied or where tree quality is uniform and there is little to gain from a selection process.

A fourth method is called selection thinning where the largest trees are removed. This approach is most applicable where there are large 'wolf' trees or large individuals of an undesirable species present that are having an adverse effect on stand development.

Low thinning and crown thinning are the most commonly used methods.

Thinning intensity

The prescription of how many trees to remove, not simply the class of tree to be removed, requires an understanding of the quantitative relationship between stand growth and stand density. Trade-offs between the associated habitat and timber values of different stand structures can then be evaluated in a meaningful, quantitative way.

Langsaeter (1941) postulated a relationship between basal area stocking and total stand growth that has been widely adopted as a basis for assessing the impacts of stand density on growth. This relationship hypothesizes four zones of stand density: zone A, free growth; zone B, incipient competition; zone C, full site occupancy and zone D, overstocked.

Studies by Møller (1947) showed that litter fall volumes per hectare remained rather constant, as did bole increment, over a wide range of site qualities and stand densities, suggesting that tree crowns grow to occupy a finite volume of space following crown closure and that this volume can be filled by either a smaller number of large trees or a larger number of small trees without compromising the actual amount of foliage produced and the resulting site productivity. Quantification of the relationships between site occupancy and growth rates is thus essential to the projection of alternative stand density treatments. Efforts to do this have been concentrated in the B to D Langsaeter zones or the range between a maximum stocking density and the minimal stocking where crown closure begins to occur (Fig. 11.3).

Studies by Reineke (1933) demonstrated the existence of a maximum density beyond which plants of a given dimension cannot maintain themselves. This concept of a limiting maximum density line provides an upper limit (zone D) against which various stocking alternatives can be compared.

At the other end of the stand density scale, an approach that identifies the minimum number of open-grown crowns capable of fully covering an area was proposed by Krajicek *et al.* (1961). By determining crown widths and crown areas of open-grown trees, and assuming a crown shape function (circles or hexagons) one can estimate the number of open-grown trees that would fit on an area. This can be taken to approximate the point at which inter-tree competition begins or the boundary between zone A and zone B of Langsaeter.

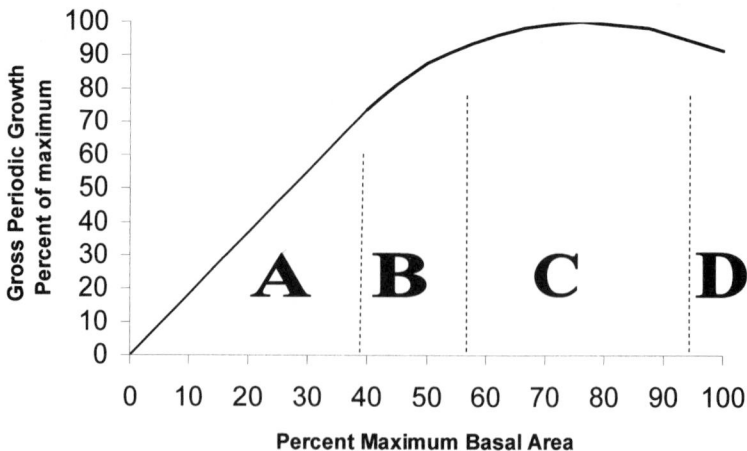

Fig. 11.3 Langsaeter relationship. Zone A: growth increases linearly with increase in stand density. Zone B: growth rate increase slows with increased stand density as competition begins to occur. Zone C: growth rate is insensitive to change in stand density. Zone D: growth declines with increased stand density. (100% maximal basal area corresponds to the limiting density relationship identified by Reineke 1933.)

These principles and their relationship with crop size parameters are most clearly portrayed visually as stand density management diagrams (SDMDs) (Drew & Flewelling 1977, 1979).

SDMDs (Fig. 11.4) define a response surface that facilitates projection of stand growth and development through time as a function of stand density, tree height and tree size, allowing visualization of the growth and site occupancy implications of a wide range of stand tending options. SDMDs have been produced for a number of coniferous species (Drew & Flewelling 1979; Farnden 1996; Newton 1997) using either long-term growth plot data or simulations from growth models (Mitchell 1975; Mitchell & Cameron 1985).

Pruning and wood quality

Stand density also influences the physical properties of wood. At wider spacing, branch size increases, reducing strength properties and straightness in sawn products. Tree crowns are deeper, delaying the production of knot-free wood on the lower bole.

As stands develop, and crowns close, the lower branches die. Species with resinous branches growing in low humidity conditions may require several decades to shed their branches. As a result, long rotations are required to produce any significant amounts of clear wood from most temperate coniferous species if reliance is placed on natural branch shedding.

If structural grades of wood are desired, then the presence of knots is unimportant provided that knot size is kept within the limits of the grade rules.

In addition to quality reductions arising from branches, the intrinsic properties of wood formed in the region of the green crown (juvenile wood) are undesirable. Such things as low basic density and dimensional instability cause degrade, not only from warping and checking but also because of low intrinsic strength properties arising from cellular properties such as micro-fibril orientation.

One way of managing these undesirable traits is to actively control crown dimensions either by undertaking mechanical pruning or by keeping trees closely spaced and crowns small. To obtain large diameter second-growth trees with low amounts of juvenile wood, either green pruning or long rotations are required. There are thus competing pressures on stand management practices – wide spacing to increase tree diameter growth rates and reduce rotation lengths, and wood quality pressures favouring closer spacing, smaller branches and denser wood.

To produce clear sawn timber or veneer, it is necessary to produce an adequate sheath of knot-free wood around a small central core of knotty wood. Prompt pruning involving green branch removal will achieve this provided that the amount of green crown removal is not excessive. Delaying pruning until the branches die lengthens rotations since the knotty core is large. Delay also results in dead, bark-encased knots and does not take full advantage of the vigorous growth exhibited by young trees. Achieving a balance between growth and quality involves understanding how crown removals influence growth in both height and diameter. If too much crown is removed, both bole and height growth are reduced. Removal of between 25 and 40% of the

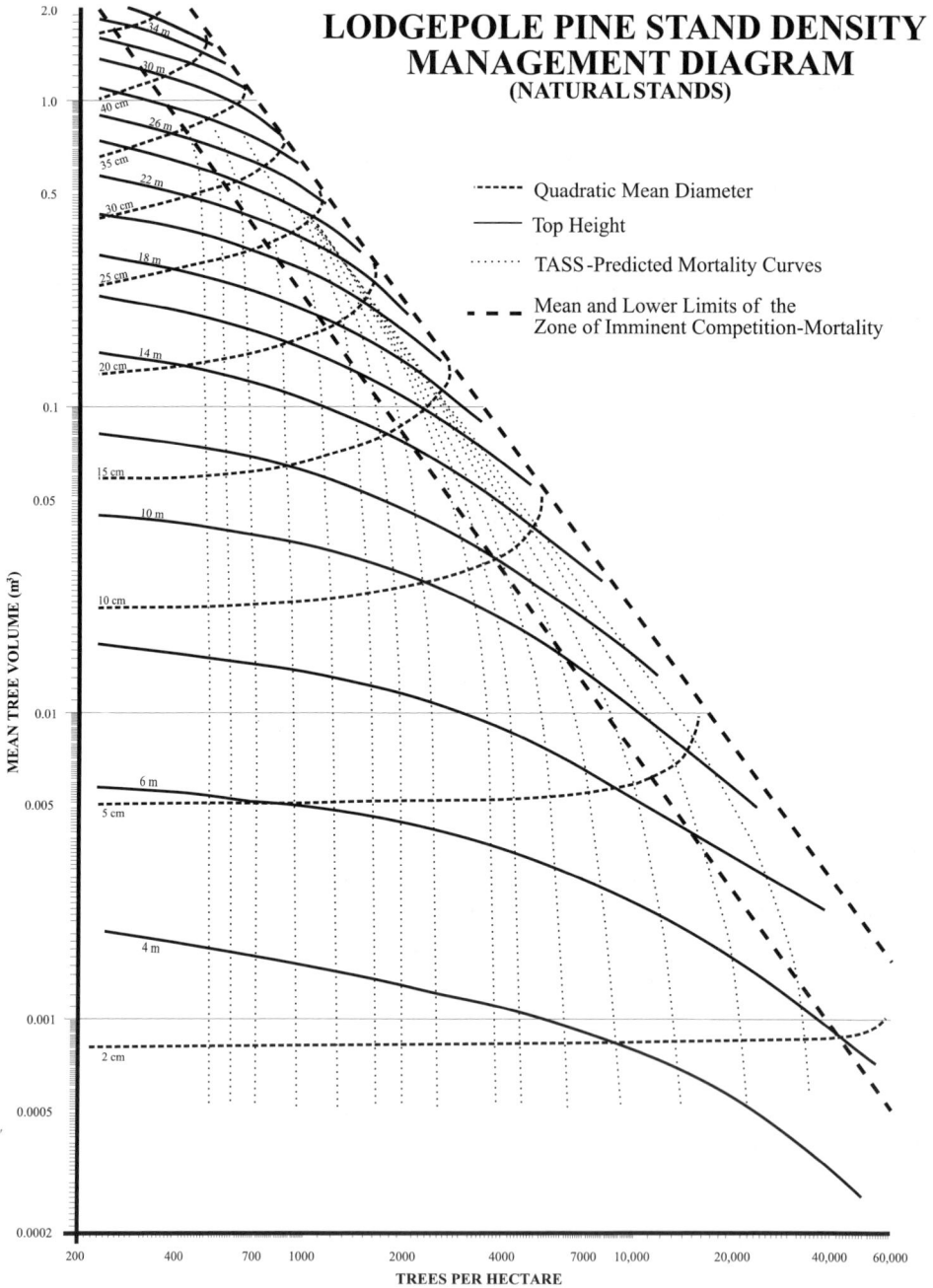

Fig. 11.4 A stand density management diagram. Beginning with any initial number of trees, a stand's development is followed up the mortality lines. Age is indicated by the intersection with the height lines. At any point, the number of stems may be reduced by thinning to a different mortality line. Crop parameters such as stem volume, stand volume and average diameter are available.

green crown length has minimal impacts (Møller 1960; Staebler 1964; DeMontigny & Stearns-Smith 2001).

Pruning may be done in stages (lifts) to control the size of the knotty core with two to three lift pruning being common in fast-growing species such as radiata pine.

The knotty core size is also influenced by the pruning tool used and the quality of work done. If cuts are made too far from the bole, the knotty core will be larger because of the branch stubs, which must heal.

Pruning should be carried out at an early age if optimal economic benefits are to be realized. Since the production of clear sawn lumber generally requires that a sheath of pruned bole wood approximately 12–15 cm thick be produced and because of limitations on the rate of crown removal, knotty cores of between 13 and 20 cm can be expected; final crop diameters must be at least 45–50 cm to achieve the optimal benefits from pruning.

Diameter growth following pruning should be rapid so that a merchantable-sized bole is reached quickly and the compounded pruning costs are minimized. This implies that thinning should accompany pruning. However, log grading rules and pricing policies that place restrictions on ring width can limit the ability to achieve rapid growth rates.

Other factors influence pruning. Species producing epicormic branches, such as western red cedar and Sitka spruce, must be pruned with care. Exposure of the bole to direct sunlight can cause initiation of these whisker-like branches that arise from dormant buds in the bark. In some instances, sudden exposure of the bole results in cambial damage (sunscald) which can cause serious defect or top death.

Hardwood species often have a habit of producing heavy top branching, which limits the length of bole produced. A judicious pruning to remove large side branches can often pay off in substantially improved bole length and form.

Where fire danger is significant, pruning of lower branches can reduce the ladder effect where fire goes from the ground to the tree crowns via the flammable branches.

Pruning is expensive and must be used in appropriate situations. Species which are valued for clear, decorative, knot-free properties and which receive a high premium should be favoured. Species used for structural purposes, especially where the wood is not visible, do not warrant pruning unless branch size is excessive.

Fertilization

As pressures for increased timber production from a reduced managed land base intensify, the attractiveness of investment in forest fertilization increases as a means of either maintaining or increasing harvest levels. Unlike agriculture, forestry operates over extended periods of time and returns to such investments have to be carefully analysed. Several questions need to be answered:

- Which elements are lacking?
- What is the expected actual stand growth response?
- How much and in what form should the element(s) be added?

- What application method will be effective?
- Will there be impacts on water quality?
- What is the impact of stand growth response at the forest level?

The initial stage often involves field personnel noting some observable foliar symptoms such as chlorosis or bronzing. Symptoms typical of various nutrient deficiencies have been identified and published by numerous authors (Leaf 1968; Ballard & Carter 1985; van den Driessche 1989) for a number of tree species. These observations can be supplemented by determining foliar nutrient concentrations and comparing them to published deficiency levels. It is important that within-tree sampling position and season are standardized.

In other situations, there may be no overt signs of deficiency, in which case only direct testing for a response will provide the necessary information.

Having identified that a deficiency is likely to exist, the question then arises as to whether a response to fertilization can be reliably expected and what dosage would need to be applied to obtain an optimal response. The answer to this question can be complicated if addition of one element induces a deficiency in another (van den Driessche & Ponsford 1995). It is desirable to determine the range of sites over which results are predictable. Species such as Douglas fir are quite predictable based on the extensive work by the Pacific Northwest Fertilizer Cooperative, whereas with western hemlock, results are more variable and site-specific. Positive responses in boreal pine species have been well documented (Krause *et al.* 1987).

In the absence of any information on responses, a prudent step is to establish some form of quantitative response trials that assess a range of dosages and nutrient elements, both singly and as mixtures over a range of sites. Responses are best detected using field techniques, which vary from plot techniques to foliar analyses. An approach known as vector analysis (Weetman & Algar 1976; Timmer & Morrow 1984) is a useful, quick approach. Responses to a range of treatments on small plots are evaluated based on sampling a fixed number of needles (often 100) before treatment and one growing season after treatment. The method is best suited to growth determinant species although some interpretations are possible with non-determinant species (Weetman *et al.* 1989).

Upon determining that a response can be expected, larger trials are needed to obtain volume responses on a per hectare basis to provide meaningful management information for assessment of the impacts on forest level harvests. There are indications that growth measurement at breast height may not give an accurate measure of response, as fertilization may shift the form of the tree bole away from the normal shape (Mitchell & Kellogg 1972; Snowdon *et al.* 1981).

The most commonly deficient elements in forestry are nitrogen (N) and phosphorus (P). Rates of application vary but are commonly between about 150 and 250 kg/ha (elemental) for N with responses lasting between 5 and 7 years. For P, rates between 50 and 100 kg/ha are common with responses lasting up to 20 years. In some cases, mixtures of N + P can give substantial responses (Weetman *et al.* 1989; Prescott & Weetman 1994).

Application techniques depend on a number of factors including terrain, stand age and environmental concerns.

Where terrain is easy, and access is possible, mechanical spreading techniques are possible. Such approaches may involve banding or side dressing as well as broadcasting. In more difficult terrain, helicopters or fixed wing application are favoured. Helicopters are favoured where placement is an issue such as riparian avoidance. Uniformity of application is achieved through the use of electronic navigational tools to control flight paths and a swath overlap between one-third and one-half of the swath width.

The formulation of the added nutrient can be important for a number of reasons. For example, urea at 46% N would be preferred to ammonium nitrate at 28% N since less bulk is required to deliver the same dosage. However, volatilization losses may be more severe with urea, particularly if weather conditions are such that the granules of urea (prills) do not get rapidly incorporated into the acidic soil environment.

A second consideration involves solubility. Application of rock phosphate delivers a highly insoluble form of P, which delivers for an extended time period, whereas superphosphate is highly soluble and much more immediately available. The dynamics of solubility may thus influence the nature of the response over time and influence loss from the site.

From an investment point of view, application of fertilization in conjunction with thinning within 20 years of harvest is often a preferred tactic that permits an early return on the investment. However, early rotation applications may be justified in situations where responses are large and prolonged resulting from shifts induced in the mineral cycling conditions on the site (Weetman & Prescott 2001).

Landscape analysis

Silvicultural activities are planned and implemented at the stand level but may have additional effects at the forest level. Sustainable harvest planning takes into account all stands, their performance and their availability for harvest. In forests which have an irregular age class structure, timing of availability can have a substantial effect on harvest level. Consequently, silvicultural treatments which impact the age at which a stand becomes merchantable can have an immediate impact on the harvest level for the forest before the individual treated stand matures. This is called an 'allowable cut effect' (ACE), which results from a reduction in the length of time that the existing older merchantable resource must be rationed before the young crop becomes merchantable. In forests with a more balanced age class structure, such ACE effects are minor to non-existent.

Consequently, a decision as to whether to undertake a particular stand treatment should consider the impact not only on the performance of the stand but also on the forest as a whole.

Spatial factors such as viewscape quality, habitat distribution and rate of watershed recovery, which constrain harvests, can also be influenced by silviculture. For example, in watersheds, crown cover must be adequate to manage water flow. Watershed recovery following harvest depends on the rate at which crown cover is re-established.

Prompt replanting with large planting stock at close spacing combined with early fertilization can speed recovery of canopy closure and reduce the impact of the spatial constraint on harvest levels, as existing merchantable stands are freed from the imposed constraints, allowing an immediate increase in harvest.

Forest level analyses are complex and require sophisticated computerized tools that allow the integration of stand level factors into an overall forest level picture. Multi-resource inventory data provide an essential component if inter-related factors are to be integrated to help define a desired future forest condition. Since the increased harvest levels are predicated on anticipated patterns of stand development, monitoring systems must be in place to ensure that the treatments are, in fact, carried out and that responses are in line with those predicted.

Silviculture, in the modern sense of sustainable forest management, can thus be seen to form an integral part of a complex management system, which draws upon a wide range of information for its successful implementation.

References

Baker, F.S. (1950) *Principles of Silviculture.* McGraw-Hill, New York.

Ballard, T.M. & Carter, R.E. (1985) Evaluating forest stand nutrient status. *Land Management Report # 20.* British Columbia Ministry of Forests.

Burns, R.M. & Honkala, B.H. (1990) Silvics of North America: 1. Conifers; 2. Hardwoods. *Agriculture Handbook No. 654.* US Department of Agriculture, Forest Service, Washington, DC.

Comeau, P.G., Biring, B.S. & Harper, G. (1999) Conifer response to brushing treatments: a summary of British Columbia Data. *British Columbia Ministry of Forests Extension Note 41*, December 1999.

Daubenmire, R.F. (1968) *Plant Communities: A Textbook of Plant Synecology.* Harper & Row, New York.

De Montigny, L. & Stearns-Smith, S. (2001) Pruning density and severity in coastal western hemlock: 4-year results. *Extension Note Number 51.* British Columbia Ministry of Forests, Research Branch, Victoria, BC.

Drew, T.J. & Flewelling, J.W. (1977) Some recent Japanese theories of yield-density relationships and their application to Monterey pine plantations. *Forest Science*, 23, 517–34.

Drew, T.J. & Flewelling, J.W. (1979) Stand density management: an alternative approach and its application to Douglas-fir plantations. *Forest Science*, 25, 518–32.

Dunsworth, B.G. (1988) Impact of lift date and storage on field performance for Douglas-fir and western hemlock. *General Technical Report No. RM-167.* Rocky Mountain Forest and Range Experiment Station, USDA Forest Service.

Farnden, C. (1996) Stand density management diagrams for lodgepole pine, white spruce and interior Douglas-fir. *Information Report BC-X-360.* Canadian Forest Service, Victoria, BC.

Helms, J.A. (1998) *The Dictionary of Forestry.* Society of American Foresters, Bethesda, MD.

Hills, G.A. (1952) The classification and evaluation of site for forestry. *Report Number 24.* Research Branch, Ontario Department of Lands and Forests.

Kenk, G. (1997) The uneven-aged silvicultural systems in the coniferous dominated forests of the Black Forest. In: *Proceedings of the IUFRO Interdisciplinary Uneven-aged Management Symposium*, (ed. W.H. Emmingham), pp. 1–20. Oregon State University.

Krajicek, J.E., Brinkman, K.A. & Gingrich, S.F. (1961) Crown competition – a measure of density. *Forest Science*, 7, 35–42.

Krajina, V.J. (1965) Biogeoclimatic zones and biogeocoenoses of British Columbia. *Ecology of Western North America*, 1, 1–17.

Krause, H.H., Weetman, G.F., Koller, E. & Veilleux, J.M. (1987) Interprovincial forest fertilization program 1968–1983. *Information Report DPC-X-21*. Canadian Forestry Service, Ottawa.

Langsaeter, A. (1941) Om tynning I enaldret gran-og furuskog. [On thinning in even-aged pine, spruce and fir stands.] *Meddel f. d. Norske Skogforsøksvesen*, 8, 131–216.

Leaf, A.L. (1968) K, Mg and S deficiencies in forest trees. In: *Forest Fertilization – Theory and Practice, Proceedings of a Symposium on Forest Fertilization*. Gainsville, FL, 1967, pp. 88–122. Tennessee Valley Authority, National Fertilizer Development Center, Muscle Shoals, AL.

Matthews, J.D. (1989) *Silvicultural Systems*. Oxford University Press, Oxford.

Meidinger, D. & Pojar, J. (1991) Ecosystems of British Columbia. *Special Report Number 6*. Research Branch, BC Ministry of Forests, Victoria, BC.

Mitchell, K.J. (1975) Dynamics and simulated yield of Douglas-Fir. *Forest Science Monographs*, 17, 1–39.

Mitchell, K.J. & Cameron, I.R. (1985) Managed stand yield tables for coastal Douglas-fir: initial density and pre-commercial thinning. *Land Management Report No. 31*. Ministry of Forests, British Columbia.

Mitchell, K.J. & Kellogg, R.M. (1972) Distribution of area increment over the bole of fertilized Douglas-Fir. *Canadian Journal of Forest Research*, 2, 95–7.

Mitchell, S.J. & Beese, W.J. (2002) The retention system: reconciling variable retention with the principles of silvicultural systems. *Forestry Chronicle*, 78, 397–403.

Møller, C.M. (1947) The effect of thinning, age and site on foliage, increment and loss of dry matter. *Journal of Forestry*, 45, 393–404.

Møller, C.M. (1960) The influence of pruning on the growth of conifers. *Forestry*, 33, 37–53.

Newton, P.F. (1997) Stand density management diagrams: review of their development and utility in stand-level management planning. *Forest Ecology & Management*, 98, 251–65.

Otchere-Boateng. J. (2002) *Herbicide Field Handbook*. *Handbook Number 6* (revised). Forest Practices Branch, British Columbia Ministry of Forests, Victoria, BC.

Prescott, C.E. & Weetman, G.F. (1994) *Salal Cedar Hemlock Integrated Research Program: A Synthesis*. Faculty of Forestry, University of British Columbia, Vancouver, BC.

Pojar, J., Klinka, K. & Meidinger, D. (1987) Biogeoclimatic ecosystem classification in British Columbia. *Forest Ecology & Management*, 22, 119–54.

Reineke, L.H. (1933) Perfecting a stand-density index for even-aged forests. *Journal of Agricultural Research*, 46, 627–38.

Reynolds, P.E., Scrivener, J.C., Holtby, L.B. & Kingsbury, P.D. (1993) Review and synthesis of Carnation Creek herbicide research. *Forestry Chronicle*, 69, 323–30.

SAF (1958) *Forest Terminology – A Glossary of Technical Terms Used in Forestry*. Society of American Foresters, Washington, DC.

Schutz, J.-P. (1997) The Swiss experience: more than 100 years of experience with a single tree selection management system in mountainous mixed forests of spruce, fir and beech from an empirically developed utilization in small scale private forest to an elaborate and original concept of silviculture. In: *Proceedings of the IUFRO Interdisciplinary Uneven-aged Management Symposium*, (ed. W.H. Emmingham), pp. 21–34. Oregon State University.

Simpson, D.G. (1990) Long nights and moisture stress affect Douglas-fir seedling growth, dormancy and root growth potential. *Canada-BC FRDA Research Report Number 151*.

Snowdon, P., Waring, H.D. & Woollons, R.C. (1981) Effect of fertilizer and weed control on stem form and average taper in plantation-grown pines. *Australian Forest Research*, 11, 209–21.

Spathelf, P., Spiecker, H. & Rogers, R. (1997) What can we learn from selection forests, 'Plentarwald', in the Black Forest? In: *Proceedings of the IUFRO Interdisciplinary Uneven-aged Management Symposium*, (ed. W.H. Emmingham), pp. 145–65. Oregon State University.

Spurr, S.H. (1956) German silvicultural systems. *Forest Science*, **2**, 75–80.

Staebler, G.R. (1964) Height and diameter growth for four years following pruning of Douglas-fir. *Journal of Forestry*, **62**, 406.

Timmer, V.R. & Morrow, L.D. (1984) Predicting fertilizer growth response and nutrient status of jack pine by foliar diagnosis. In: *Forest Soils and Treatment Impacts, Proceedings of the 6th North American Forest Soils Conference*, (ed. E.L. Stone), pp. 335–51. Knoxville, TN, June 1983.

Troup, R.S. (1952) *Silvicultural Systems*, 2nd edn. Clarendon Press, Oxford.

van den Driessche, R. (1989) Nutrient deficiency symptoms in container-grown Douglas-fir and white spruce seedlings. *FRDA Report No. 100*. Forestry Canada/BC Ministry of Forests, Canada.

van den Driessche, R. & Ponsford, D. (1995) Nitrogen induced potassium deficiency in white spruce (*Picea glauca*) and Engelmann spruce (*Picea engelmannii*) seedlings. *Canadian Journal of Forest Research*, **25**, 1445–54.

Weetman, G.F. & Algar, D. (1976) Selection cutting in over-mature Spruce-Fir stands in Quebec. *Canadian Journal of Forest Research*, **6**, 69–77.

Weetman, G.F. & Prescott, C. (2001) The structure, functioning and management of old-growth cedar-hemlock-fir forests on Vancouver Island, British Columbia. In: *The Forests Handbook, Vol. 2: Applying Forest Science for Sustainable Management*, (ed. J. Evans), pp. 275–87. Blackwell Science, Oxford.

Weetman, G.F., Fournier, R.M., Barker, J., Schnorbus-Panozzo, E. & Germain, A. (1989) Foliar analysis and response of fertilized chlorotic western hemlock and western red cedar reproduction on salal-dominated cedar/hemlock cutovers on Vancouver Island. *Canadian Journal of Forest Research*, **19**, 1512–20.

Chapter 12
Manufacturing processes and their impact on effluent discharges

N. MCCUBBIN, E.R. HALL AND K.J. HALL

Introduction

Scope of chapter

This chapter describes pulp and paper manufacturing processes, with an emphasis on those aspects of the technology that have significant impact on effluent discharge characteristics. Processes described are those in use at the opening of the twenty-first century, with some discussion of historical processes and probable future developments.

A perspective on the scale of the industry worldwide is presented, including both modern 'industrial' mills and artisanal operations.

This text is intended for readers with little past exposure to the industry. Recent comprehensive textbooks on the manufacturing processes include Kocurek (1986–1989) and Gullichsen *et al.* (2000), both multi-volume series. Smook (1988) describes the industry well in one volume, while Dence & Reeve (1996) focus on pulp bleaching. These all include information on environmental issues, although their emphasis is on production of pulp and paper.

The environmental regulatory agencies in the European Union have developed a useful report (IPPC 2000) on environmental protection technology, which also includes data on a number of low effluent mills.

Overview of processes

The end product of papermaking is normally paper or board for communications, packaging or hygienic purposes, as well as much smaller quantities for diverse uses such as air filters, cigarette paper, tea bags, etc.

Regardless of the final product, the first stage is to produce pulp: a matrix of fibres with all unacceptable bark, dirt, lignin and other contaminants removed. The pulping operations may or may not be on the same site as the papermaking activities. Pulping (with any associated bleaching) normally has a much greater impact on effluent characteristics than the manufacture of paper itself.

The production capacity of pulp mills varies from 1 tonne/day to 6000 tonnes/day. The following discussion focuses on installations producing several hundred tonnes per day or more.

Wood is the raw material for most pulp mills, but before it is used the bark is removed, generating solid waste. Up to 15% of the raw wood is removed in this way, and can be a significant source of toxic substances in the mill effluent.

Pulping operations consist essentially of separating the useful fibre from the raw material, and cleaning it to the extent necessary for the final product specifications. The mass yield of useful fibre can be from about 35% to 99% of the raw material, so that the quantity of waste products can range from approximately 2000 kg/tonne product to under 10 kg/tonne. In some processes, chemicals are added, which further increase the quantity of waste material by up to about 50%.

Most mills that operate with process yield towards the lower end of the above-mentioned range use the kraft process, where from 90% to 99% of the waste is recovered and burned, almost always in an environmentally sound manner.

The pulp fibre is the principal raw material for papermaking, but up to about 25% by weight of the finished paper is added, mostly in the form of inert fillers and coatings.

In both the pulp mill and in the papermaking operations, the fibres are handled as a dilute slurry in water, containing between 0.5% and 15% dry matter ('consistency' is the normal industry term for the fraction of fibre in the pulp suspension). Since various process steps require different consistencies, water and fibre are separated and recombined at several points in the manufacturing process, leading to losses of fine fibre, and incidental contamination of the water used.

Both the pulping and papermaking processes can use up to about 200 m³ water per tonne product (m³/t), but most mills use under 100 m³/t, and some, only a few cubic metres/tonne. Since there is a loss of several cubic metres/tonne of product due to evaporation, a small proportion of mills operate without any discharge of liquid effluents. These are generally mills manufacturing products such as unbleached board or roofing materials that can accept significant quantities of dirt. Typical effluent discharges from the mainstream products (office paper, kraft pulp, newsprint) are 50–100 m³/t, while a few mills of these types discharge under 20 m³/t pulp.

Process classifications

The wood fibres may be separated either mechanically or chemically. Current chemical pulping processes are kraft, sulphite and soda, in declining order of application. The other processes are conventionally considered to be mechanical, but in recent years the distinction between sulphite and mechanical has become less clear.

The principal processes in pulp and papermaking are described briefly in Table 12.1.

Typical characteristics of the pulping processes of most interest to this report are summarized in Table 12.2. This tends to understate the complex and diverse industrial processes involved. The reader is cautioned that there will always be exceptional situations where data will fall outside the ranges given.

Most of the pulping processes are used in mills where paper production is integrated with the pulping operation. The papermaking operation will generally not generate great amounts of BOD (biochemical oxygen demand), but will generate a significant amount of wastewater. The discharges are mixed, so that while the above-mentioned

Table 12.1 Principal pulp and papermaking unit operations

Process	Description
Debarking	Removes bark and any frozen dirt from tree. When wet debarking was employed in the past, this was a major source of water pollution, but today many mills use dry debarking systems, or avoid debarking by purchasing wood chips from sawmills
Chipping	The debarked wood is cut into chips. Today, many of the chips used are from sawmill waste. Chipping is usually undertaken at the sawmills, and generates no liquid effluents, although dust can be an issue, close to the mill
Kraft pulping	Wood chips are reacted ('cooked' in industry terminology) with sodium hydroxide and sodium sulphide to separate the fibres. The mixture of organics which are not bound to the fibre, along with water and the spent pulping chemicals, is known as black liquor. Approximately 1700 kg black liquor solids are produced per tonne pulp produced. From 97% to 99.5% of this black liquor is recovered by washing the pulp, depending on mill equipment. The remainder of the black liquor is lost to the wastewater treatment plant or is carried with the pulp to the bleaching, or papermaking and boardmaking departments
Sulphite pulping	Similar to the kraft process, except that the pulping chemicals are acidic and, in practice, the efficiency of recovery of the non-fibrous organic substances is lower than for kraft. This process was dominant in the past, but is largely supplanted by the kraft process today
Thermomechanical pulping (TMP)	Wood chips are broken into fibres in refiners, by the application of considerable amounts of electric energy
Chemi-thermomechanical pulping (CTMP)	Wood chips are mechanically broken into fibres, as in the TMP process, but with the addition of some sodium sulphite

Table 12.1 (*Continued.*)

Process	Description
Groundwood	Wood is broken into fibres by holding logs against a grindstone, under water. Groundwood has largely been replaced by TMP and CMP
Recycled paper pulping	Recovered paper and board is broken into fibres by agitation in water
Deinking	In some recycled pulp mills, ink is removed, and the pulp may be bleached to produce fibres suitable for manufacture of tissue or communications papers
Papermaking	The pulp is processed mechanically, perhaps blended with other pulps, then converted to paper or board. Fillers, coating and other chemicals may be added. This operation generates wastewater which carries some lost fibre, but the quantities discharged to receiving waters are generally trivial in today's mills equipped with effluent treatment systems
Kraft recovery (of spent cooking chemicals and non-fibrous organics)	The black liquor recovered from kraft pulping is concentrated by evaporation, and burned to produce steam and to recover the inorganic pulping chemicals as 'smelt'. The recovered smelt is converted to cooking liquor for reuse in the pulping operation. Effluents generated include evaporator condensates as well as planned and unplanned discharges of black liquor. The black liquor discharges can be a significant contributor to the total mill pollutant discharge
Sulphite recovery	Spent cooking liquors, including the non-fibrous organics, are concentrated and burned, to recover some of the inorganic cooking chemicals as well as heat energy

Several variations of the above processes exist, many intermediate between two of the main classifications shown.

Table 12.2 Characteristics of pulping processes and effluents, prior to treatment

Type of pulp	Yield	BOD kg/tonne	Effluent flow m³/t	Principal pulping chemicals	Chemical feed (total inorganic) kg/tonne
Mechano-sulphite spectrum of pulping processes					
**Low-yield sulphite	45–55%	40–100	100–300	Na, Mg, SO$_2$	450
*High-yield sulphite	55–75%	140–250	50+	Na, SO$_2$	200–400
Ultra high-yield sulphite	80–90%	50–100	50+	Na, SO$_2$	150
*BCTMP	80–90%	50–80	0–25	Na, SO$_2$, H$_2$O$_2$	
CTMP	88–92%	40–60	8–25	Na, SO$_2$	50
TMP	90–95%	25–50	8–20	None	0
Pressurized groundwood	93–96%	11–20	8–20	None	0
Traditional groundwood	93–96%	10–15	50	None	0
Miscellaneous pulping processes					
**Kraft (unbleached)	43–58%	15–30	0–100	Na$_2$S, NaOH	50 (makeup)
*Kraft (bleached)	40–55%	20–50	20–200	Na$_2$S, NaOH	100
**Dissolving pulps	30–45%	50–250	40–200	Na$_2$S, NaOH	50–100
*Semichemical (NSSC)	70–80%	80–120	0–100	Na$_2$CO$_3$, SO$_2$ (perhaps)	150
Deinking	50–90%	10–50	2–100	HOOH, NaOH, NaSiO$_2$	
Recycled (without deinking)	90–99%	5–20	10–100	None	None

The above values are typical for the industry. A few mills operate beyond the ranges indicated.
*Indicates that much of the unused material is recovered in some mills.
**Indicates that much of the unused material is recovered in most mills.

data on BOD generated indicate the total BOD in the effluent, the flows of wastewater will generally be much higher. The BOD discharge is characteristic of the process, whereas wastewater flows are quite dependent on equipment selected and operating practices, so can vary widely.

Historical perspective

The pulp and paper industry has existed for hundreds of years, and by the nineteenth century was a significant polluter of watercourses. Except for a few isolated cases, there was little attempt to reduce effluent discharges until the 1950s, when a number of mills in North America and Europe installed effluent treatment systems, generally using the classic primary (sedimentation) and secondary (biological) processes, similar to those used for municipal sewage. A considerable body of scientific literature has emerged, with several hundred thousand papers having been published that relate to environmental aspects of the pulp and paper industry.

Since the 1960s, there has been a steady advance in the implementation of effluent treatment and improvement of the manufacturing processes. At the beginning of the twenty-first century some pollutants, such as mercury and zinc, had been eliminated, and discharge of others reduced dramatically. There is no simple way to quantify the improvements, since the reductions in discharges of different pollutions vary from minor to essentially 100%. Discharges of the widely known bulk parameters, such as BOD, suspended solids and AOX (adsorbable organic halogen) have generally been reduced by 90–99% over the period. The changes in manufacturing processes that have been implemented have caused further improvements in effluent characteristics. The use of persistent toxicants to control biological growth within the manufacturing process has been virtually eliminated, and the extensive replacement of wet debarking methods by dry systems has reduced the quantity of the many natural insecticides, fungicides and other toxic substances in bark that can leach into the effluents.

The result of these developments in the industry is that any literature published before about 1995 on pulp industry effluents and their environmental impacts must be interpreted with considerable caution, as the characteristics of the industry effluents have changed significantly since that time, so that the problems that remain differ significantly from those of the past.

Worldwide pulp and paper manufacture

The pulp and paper industry has a significant impact on the environment and economy of a number of countries. Most countries have at least a few paper mills, but many have no pulp manufacturing. The USA is the largest producer in the world, while Canada, Japan, China, Sweden, Finland, Brazil and Chile are all major players.

Industrial scale mills

For the purposes of discussion, 'industrial-scale' refers mostly to pulp mills producing over 100 tonnes/day, or paper mills producing over about 50 tonnes/day. The vast majority of the published literature refers to such installations, and this chapter refers to this class of mill, unless otherwise noted. There are about 500 industrial-scale pulp mills, and several thousand industrial-scale paper mills in the world.

Artisanal mills

Mills producing under about 100 tonnes/day are generally quite primitive, and can discharge relatively large quantities of pollutants. In the 1980s there were many thousand such mills in the world, but a high proportion have been closed down, and most of the remainder will probably disappear in the near future.

Production capacity

Quantities produced vary widely from one country to another, with a high proportion being concentrated in the top 10 countries, as shown in Table 12.3.

China is unusual, in that there are about 5000 mills in the country, but the average production is very low. The vast majority of Chinese mills are of artisanal scale, and

Table 12.3 Worldwide pulp and paper production

Pulp			Paper and board		
Country	Number of mills	Production (million tonne/year)	Country	Number of mills	Production (million tonne/year)
USA	183	57	USA	512	85
Canada	48	26	Japan	473	32
China	5000	17	China	4750	31
Finland	43	12	Canada	102	21
Sweden	45	12	Germany	192	18
Japan	44	11	Finland	46	14
Brazil	69	7	Sweden	48	11
Russia	35	6	France	131	10
Indonesia	14	4	S. Korea	123	9
Chile	11	3	Italy	202	9
Total top ten	5500	156		6600	240
Total world	6000	188		8900	323
Total excluding China	500	171		4100	292
Top 10 as fraction	92%	83%		74%	74%

Source: Paperloop (2001). Some values rounded. Numbers of mills reported in other sources vary slightly. Notice that most of the mills in China are on an artisanal scale, producing only a few tonnes/day.

are quite primitive, and highly polluting relative to their size. These are being rapidly replaced by modern, industrial-scale operations. Until the late twentieth century, there were also many tiny, artisanal mills in some other Asian countries, but most have been shut down.

Of the 500 or so industrial-scale pulp mills in the world, roughly 400 use the kraft process, including almost 300 that produce bleached pulp. These mills normally have a capacity between 500 and 3000 tonnes/day (180,000 to 1,000,000 tonnes/year).

Process yield

The wood fed to a pulp mill contains approximately 50% cellulose fibres, 20–30% lignin, 5% extractives, and about 25% hemicelluloses. The actual values depend on the wood species used.

For higher grades of paper, such as printing and tissue, only the fibre and hemicellulose is useful, while some paper grades, such as newsprint, can use much of the lignin also.

Chemical pulping methods remove most of the lignin, while mechanical processes retain the lignin in the fibres. All the material that is not retained in the product has relatively high BOD and COD (chemical oxygen demand), and some of it is toxic to fish. Most chemical pulp mills recover and burn a high proportion of the unwanted organic matter (normally with only trivial atmospheric emissions), while this is rare in mechanical pulp mills.

The yield of marketable fibre and the efficiency of any recovery systems are key determinants of the characteristics of effluent discharged by any pulp mill. Typical values for the principal processes are shown in Table 12.2.

Recycled fibre fed to a pulping operation can vary from high-grade waste requiring minimal processing, to highly contaminated, post-consumer wastes. The yield depends on the raw material, as well as the manufacturing process.

The environmental significance of the substance discharged varies widely from one process to another, but it is generally true that 'less is better' for any one process. In all cases, the efficiency of the wastewater treatment process (WWTP) has a major impact on effluent characteristics.

Wood preparation

Many mills today use wood chips from sawmills as the primary raw material. Others receive wood in the form of logs ('roundwood') so have to remove the bark and cut the wood into chips. In stone groundwood mills, which are uncommon nowadays, the logs are debarked, but not chipped.

Debarking

Bark contains a variety of natural insecticides, biocides and toxic substances that the tree produces for self-protection in the forest.

Bark and the associated sand, gravel and other dirt are removed from logs mechanically. In the past, most mills used wet processes, generating effluent contaminated with resin acids and other organic substances leached from the bark that was quite toxic, unless treated biologically.

Modern mills use dry debarking systems, so there is no liquid effluent generated. In older mills, dry systems have been steadily replacing the wet processes over the past 30 years.

Leachate from outdoor bark storage is toxic, and is routed to biological treatment systems in most mills today.

There has been a steady tendency over the past 20 years for mills to purchase wood in the form of chips produced at sawmills from the waste from lumber manufacture. The sawmills all debark the wood by dry mechanical equipment, so that the question of effluent does not arise.

Some mills in cold climates have log de-icing showers and dry debarking. Such systems normally generate very little effluent.

Chipping

Except for mills using the stone groundwood pulping processes, the debarked logs are cut into chips before being fed to the pulping operation. This process does not generate any effluent.

It is common to wash the chips in very hot water in mills using the thermomechanical pulping process (TMP). This generates a small effluent flow, contaminated mostly with suspended solids that are readily removed in primary treatment. Minor quantities of resin acids and similar substances may be extracted from the wood, and discharged with the effluent.

Kraft pulping

Kraft is the most widely used process for converting wood into pulp, and is also the most complex.

The distinguishing characteristic of this process is that the wood chips are broken down to individual fibres by reaction ('cooking' is the normal industry terminology) in strong alkali at elevated temperature. Overall process yield is around 40%, so that approximately 1500 kg waste product is generated per tonne pulp produced, in the form of dissolved organic material. Much of this is recovered and burned to produce energy and to recover the cooking chemicals, normally with trivial atmospheric emission. The key factor affecting the generation of liquid effluents is the efficiency of this recovery.

A significant proportion of the kraft pulp produced is bleached, generally by processes that involve discarding all unwanted material. Typically the quantity of organic matter rejected in bleaching is between 30 and 100 kg/tonne product, mixed with up to about 150 kg/tonne inorganic chemicals, mostly sodium salts.

A typical kraft pulp mill, including a bleach plant, is shown in Fig. 12.1, and is described below. This produces pulp at about 12% consistency, which would either be dried for sale or pumped to an adjacent paper mill.

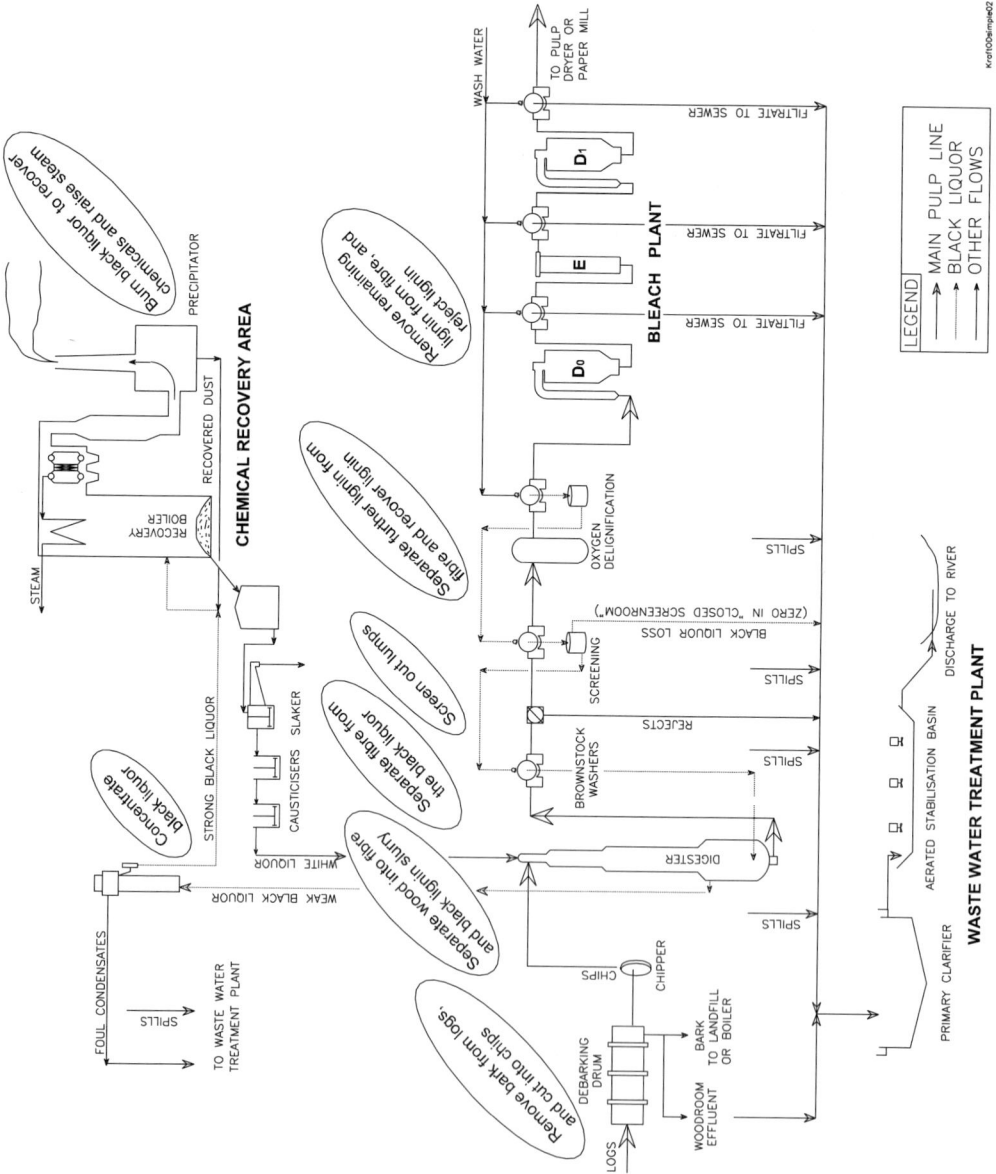

Fig. 12.1 Typical kraft mill with closed screen room and oxygen delignification system.

Fibre separation

Digesters and cooking

As shown in Fig. 12.1, wood chips are mixed in the digester with white liquor (sodium hydroxide and sodium sulphide solution) and heated with steam to 'cook' them. The process is continued until the wood is delignified as far as possible, without destroying or weakening the cellulose fibre. The final lignin content is defined by a standard analysis, which gives a 'kappa number', proportional to the lignin concentration. Softwood pulps for bleaching are conventionally cooked to about 30 kappa, while hardwoods are cooked to about 20 kappa.

Today, extended cooking technology is often used to reduce the kappa number to about 70–95% of the traditional values. Many mills have modified older digesters to profit from the extended cooking technology.

The pulp leaves the digester as a very dark slush, at very roughly 10% consistency, carried by the black liquor.

Black liquor exerts a high BOD (approximately 35% of its weight), high COD (roughly 1 kg/kg black liquor solids), is very toxic to fish, and is highly coloured.

The digester system is essentially a closed process, so that it has no direct contribution to mill effluents, except for any leaks in the system due to failures of piping or equipment.

Coarse screening (knotting)

After leaving the digester, the pulp passes through coarse screens, normally known as knotters, to remove large wood particles that have not been sufficiently broken down to separate fibres. The fraction of wood rejected in this way can be reduced to almost nil by modern chip screening systems, which are becoming more common practice.

The rejected knots normally carry black liquor, so if they are disposed of to landfill or the mill sewer, they can be a significant contribution to the contamination of the mill effluent discharge.

The knotting equipment is omitted from Fig. 12.1, in the interests of simplicity.

Brown stock washing

The pulp is separated from the black liquor in a series of countercurrent brown stock washers, by passing water through a mat of pulp fibres to displace the black liquor.

Figure 12.1 depicts the traditional vacuum drum type of washer widely used in older mills. The equipment is physically open, and subject to frequent overflows, if not well maintained and operated. This can be a major source of pollutant discharge.

A number of more effective washing devices have been introduced since 1980, and are steadily replacing the above-mentioned drum washers. These include twin roll presses, multi-stage belt washers, and various proprietary designs such as the DD® and Compaction-Baffle® washers. These generally wash more efficiently, and most are physically closed and pressurized so that overflows do not occur.

The most efficient brown stock washing systems recover 99.5% of the black liquor, whereas some older systems were only about 95% efficient, and thus caused the discharge of 10 times more black liquor solids than the best systems.

The recovered black liquor is routed to the evaporators for eventual recovery of the heat and chemical values.

Brown stock screening

After washing, the pulp passes through fine screens, known normally as 'brown stock screens' to remove fibre bundles and shives (very small uncooked chips).

This process requires that the pulp be diluted from approximately 12% consistency to below 2.5% consistency, and then rethickened. In the past, much of the water from the rethickening stage (known as the 'brown stock decker') was discharged to the mill sewer, carrying residual black liquor solids. Normally known as 'unbleached white water' (UBWW), this waste stream was often identified as a major contributor to mill effluent. It is essentially diluted black liquor, with its undesirable characteristics.

A screening system as described above is known as an 'open screen room'.

The screens themselves can be either 'atmospheric' or 'pressurized'. In the former, the pulp slurry is exposed to the atmosphere, and mixes with air, incidentally to the screening process. This creates foam, which frequently overflows, contributing to the mill effluent. The foam is rich in kraft soap and resin acids, so is more toxic than the black liquor. The LC_{50} of soap and foam has been reported to be in the parts-per-million range. (The LC_{50} is the concentration of an effluent in clean water that will kill 50% of a sample of test organisms in defined laboratory conditions, in a defined time. The time period is typically 48 or 96 hours.)

In principle, it is quite simple to adjust water flows so that there is no excess unbleached white water, creating a 'closed screen room', but this was very difficult in practice until pressure screens were introduced. Today, an increasing number of mills have been successful in closing the screening operation, thus eliminating the discharge of unbleached white water.

After screening, pulp is normally further delignified by reaction with oxygen and alkali, but may simply be stored at around 12% consistency prior to bleaching.

Oxygen delignification

Oxygen delignification (OD) is an extension of the kraft or sulphite cooking process that delignifies the wood further, while avoiding the strength loss that would occur if the cooking process were prolonged; refer to Fig. 12.1 for a typical installation.

In OD, the pulp is delignified by reaction with oxygen under pressure in strongly alkaline conditions. In almost all mills, the soluble organic material separated from the pulp fibres is recovered and recycled to the mill's chemical recovery system, where it is incinerated. The recovered material is quite similar to black liquor, except that the organic material is oxidized to some extent.

The kappa number is a measure of the lignin content of the pulp, so reducing the kappa number with OD results in a corresponding reduction in the material that will eventually be removed from the pulp in the downstream bleaching operations, and discharged to the mill's effluent treatment plant. The cost of bleaching chemicals is approximately proportional to the kappa number of the pulp entering the bleach plant.

An oxygen stage will allow most bleached kraft mills to reduce bleach plant BOD discharges by approximately 65% and colour by 70%. Discharges of organochlorines will be reduced by approximately 35–65%.

A very small proportion of mills have OD systems as described above, but do not recover the waste organics. Further, a few North American mills have adopted the 'mini-OD' process, which is essentially a low capital cost, low efficiency variation of conventional OD, where the organics removed are discharged. Both these variations of the OD process reduce organochlorine discharge, but tend to increase effluent colour.

Oxygen delignification is being used to produce a variety of pulp grades, and there are over 200 oxygen delignification systems operating in the 400 kraft mills around the world, with several more planned or under construction. Essentially all new fibrelines and those rebuilt since 1980 outside North America incorporate oxygen delignification. Approximately half the new or rebuilt fibrelines in North America since 1985 have included OD.

Black liquor solids carried into the bleaching process

Whether oxygen delignification is used or not, the quantity of free black liquor solids (i.e. solids that could theoretically be removed by washing with water) carried with the pulp into the bleaching process has a significant impact on the characteristics of the effluent from the bleach plant.

While most of the recent research on pulp washing has focused on COD in the liquor travelling with the pulp as the key criterion of performance, most mills in North America use the traditional saltcake content as the principal parameter. A pulp stream that carries 1 kg/tonne sodium sulphate will carry about 1 kg of COD. Actual values vary by ±50% from this rough guideline.

Typically, mills report saltcake losses between 4 and 40 kg/tonne pulp. It is technically feasible to limit the quantity of organics carried forward to the subsequent bleaching process to under 5 kg COD per tonne pulp, and this is a common design basis for new installations.

Conventional practice is to install two post oxygen washers, and recirculate the filtrate as shown in Fig. 12.1. However, if the filtrate from the second post oxygen washer is sewered directly, and fresh water used as shower water on the first washer, then the flow of organics into the first bleaching stage would be reduced by about half, at the cost of some increase in mill effluent BOD, and reduction in bleach chemical cost. This is practised in some mills, particularly those under more pressure to reduce AOX discharges than to improve overall effluent characteristics.

Recovery of black liquor

Overview and environmental significance

The existence of the recovery cycle sets the kraft pulping industry apart from most other subsectors of the paper industry, and from many other industries. This integral part of the process has the capability of converting almost all the pollutants formed during manufacture into energy, without generating any significant air pollution or other environmental problems.

The chemical recovery system in a kraft mill collects the non-fibrous residual material from the pulp mill, known as 'black liquor', and burns it in a specialized reactor (the recovery boiler) to produce steam and a byproduct (smelt) which consists of sodium salts. The smelt is dissolved in water and reacted with calcium oxide to convert the principal sodium salt present (sodium carbonate) to sodium hydroxide. The product (white liquor) is the principal chemical used in the digester.

A flowsheet for the complete cycle is shown in Fig. 12.1, and the principal chemical cycles are shown in Fig. 12.2.

Whereas pulping technology has changed substantially since 1985, and bleaching processes have advanced dramatically over the same period, there have been no major developments in the chemical recovery processes in practical use. There have, of course, been evolutionary developments, resulting in major reductions in atmospheric emissions and somewhat increased capacity for older equipment.

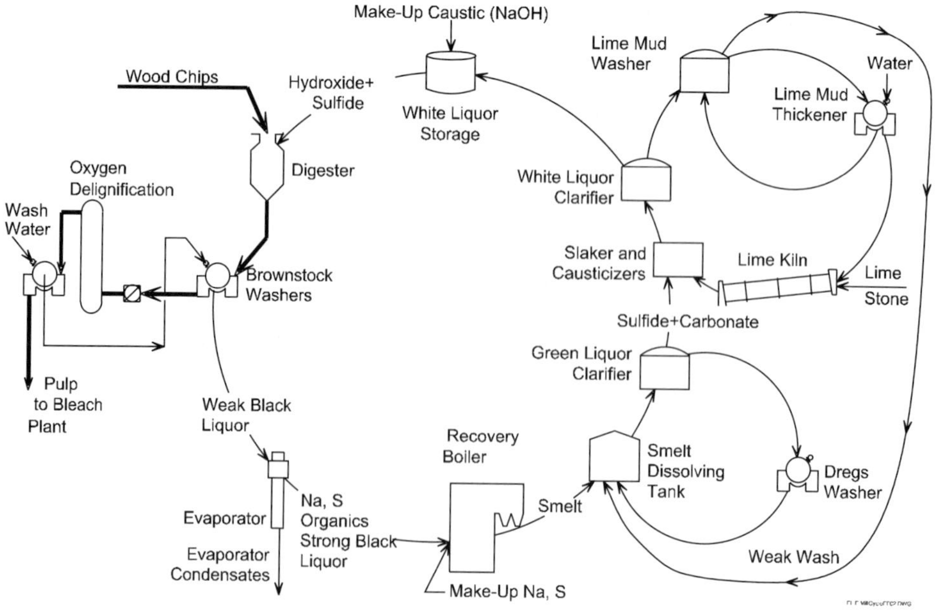

Fig. 12.2 Kraft mill chemical cycle.

In principle, the recovery cycle is closed with respect to aqueous emissions, except for the evaporator condensates. The recovery system processes approximately 1600 kg of black liquor solids per tonne pulp, so that accidental losses of even a small fraction of the material recycled can have a major impact on effluent characteristics. To illustrate: the loss of 1% of this stream would exceed the planned discharge from a well designed and operated pulp mill, for an equal period of time.

Recovery chemical cycle

The overall recovery cycle is shown in Fig. 12.1. If a mill does not have an oxygen delignification system, the cycle is similar, except that the wash water is applied to the brown stock washers, and the pulp discharged from them is routed directly to the bleach plant.

The weak black liquor recovered from the pulp in the brown stock washers is an aqueous solution, typically between 10% and 18% concentration. It has to be concentrated by evaporation to at least 63% dry matter before being burned in the recovery boiler. The recent design trend is to burn the liquor at higher concentrations, with at least one mill in Finland routinely firing liquor at over 80% concentration. When the liquor firing concentration is raised, the capacity of the boiler to burn liquor is increased by up to about 15%.

The evaporator condensates are a mixture of methanol, volatile sulphides and terpenes, often contaminated with black liquor. They are quite toxic to waterborne organisms, but if accidental black liquor contamination is avoided, the remaining toxic substances are readily biodegradable, so may have little or no effect on the toxicity of biologically treated kraft mill effluent.

The green liquor and white liquor, as well as the diluted forms such as 'weak wash', are very toxic to fish, primarily because of the sulphide content. However, this is very effectively oxidized to sodium sulphate in biological treatment systems, so is not normally a contributor to mill effluent toxicity.

Bleaching chemical pulps

Both kraft and low-yield sulphite pulps require bleaching if they are to be used for manufacture of communications grades of paper, or most forms of tissue. Unbleached kraft is used primarily for packaging grades. The bleaching process is similar for both types of pulp.

After brown stock washing (and, if used, oxygen delignification) the pulp is still brown, with a colour not unlike a brown paper grocery bag. The pulp is bleached in slurry form, by reaction with various chemicals, generally in alternating acid and alkaline stages, to remove almost all of the remaining lignin and produce white pulp.

The principal unit operations used in bleaching pulp are summarized in Table 12.4, along with the standard abbreviations for each.

Table 12.4   Kraft and sulphite bleaching operations

Abbreviation	Unit operation
O	Oxygen delignification (actually a pre-bleaching process). Treatment of pulp with elemental oxygen, in alkaline conditions. Sometimes called oxygen bleaching
C	Chlorination stage, where pulp is treated with gaseous, molecular chlorine, primarily to chlorinate the residual lignin, so that it can later be solubilized
E	Caustic extraction. Dissolution of reaction products with sodium hydroxide
Eo	As 'E' above, with the addition of about 5 kg/tonne elemental oxygen
Ep	As 'E' above, with the addition of about 3 kg/tonne hydrogen peroxide
Eop	Combination of Eo and Ep. Becoming very common practice
Z	Treatment of pulp with ozone, under acid conditions
D	Reaction with chlorine dioxide, applied as an aqueous solution
C/D	Chlorination stage with chlorine dioxide. Use of this process is declining
Cd	Chlorination stage with ClO_2 addition after Cl_2. Use of this process is declining
Dc	Sequential addition of ClO_2 followed by chlorine. Use of this process is declining
H	Reaction with hypochlorite (normally sodium). Use of this process is declining
P	Reaction with hydrogen peroxide (as the principal reagent)
Q	Chelating agents such as EDTA
Y	Reaction with hydrosulphite (also known as dithionite)
W	Wash stage (indicated only where a washer would not normally be expected, e.g. WW where two-stage washing is installed)
N	No-wash (indicated only where a washer would normally be expected)

Subscripts are frequently used to indicate percentage substitution of molecular chlorine where relevant, for example C_{d70} would imply 70% substitution of molecular chlorine with chlorine dioxide. Subscripts (0, 1, 2) are often used to identify multiple implementations of the same chemical reaction. Most mills use chlorine dioxide in the C stage, and the abbreviation 'C' is often loosely used to refer to a C/D stage.

Three to six bleaching stages are required to bleach pulp, depending on the end product, wood species and process economics. Common sequences include DEopD, DEoDED, DEopDND, DEopD and DcEoDED.

Where the only chlorine-based chemical used for bleaching is chlorine dioxide, the process is defined by convention as being 'elemental chlorine-free' (ECF).

Bulk delignification

The lignin that was not washed out of the pulp in the brown stock washing systems is first solubilized with chlorine and chlorine dioxide, then the chlorinated and oxidized products are extracted with sodium hydroxide (caustic). After each reaction, the pulp is washed and the filtrate is discarded. These stages are identified as D_0 and E, in Fig. 12.1.

Traditionally, the first bleaching stage was known as the 'chlorination stage', or the 'C'-stage. Today, most mills have discontinued the use of chlorine, so the first stage is widely known as the 'D_0' stage ('D-zero').

These bulk delignification stages remove from 2% to 6% of the pulp entering the bleach plant. Most of the pollutants discharged from bleaching are formed therein, and discharged with the washer filtrates. The kappa number of the pulp is between 2 and 4 after most bulk delignification stages. The wide range in fraction of pulp removed is due to varying input kappa values, which in turn depend on wood species and upstream processing. Since the organic material removed from the pulp in bleaching is normally discharged from the mill, the kappa number of the pulp entering the bleach plant has a major impact on the effluent characteristics.

Brightening

After the above treatment, the pulp is light brown, and it is bleached white in the brightening stages, commonly using the sequence chlorine dioxide-caustic extraction-chlorine dioxide ($D_1E_2D_2$) or perhaps simply in one stage (D_1), as shown in Fig. 12.1.

Until the last few years, sodium hypochlorite (hypo) was used widely, and was the principal source of chloroform emissions from bleaching.

The quantities of chemical used in the brightening stages are low, relative to those in the bulk delignification stages, and the formation of pollutants is also low.

Ozone delignification in bleaching

The greatest level of pre-bleaching delignification normally attainable by the use of modern cooking and oxygen delignification corresponds to approximately 10 kappa number. It is possible to reduce this to approximately 5 kappa by reaction with ozone under acid conditions. If this is done, then the demand for chlorine-based chemicals is reduced, thus reducing formation of AOX, and facilitating recovery of the organic discharge.

A few mills use a ZD or DZ sequence, where there is no washing between the two reaction vessels.

Mechano-sulphite spectrum of pulping

Apart from the dominant kraft process, most of the pulp produced in the world from wood is manufactured by a family of processes that can best be considered as a continuous spectrum from low-yield sulphite through semichemical pulp, chemi-thermomechanical pulp (CTMP) and thermomechanical pulp (TMP), to the traditional groundwood pulping process. Both chemical and mechanical pulps are included in this spectrum. The term mechano-sulphite has been adopted herein for these pulps, because there is no other generally accepted terminology.

It is conventional to consider those pulps in this spectrum with yields of over 80% to be mechanical pulp, while those of lower yield are categorized as chemical pulp. Some characteristics of the spectrum from groundwood through to low-yield sulphite pulp are shown in Table 12.2.

Groundwood pulping

Also known as stone groundwood (SGW), this was the earliest form of mechanical pulping used commercially, but has largely been replaced by thermomechanical and similar processes, or by recycled fibre.

In this process, logs are forced into contact with a revolving grindstone in the presence of water to reduce the wood to a macerated fibrous condition. The water applied cools, cleans and lubricates the stone and conveys the pulp away from the stone. A few percent of the organics in the wood are solubilized, causing the water carrying the pulp to exert a BOD when discharged, and to be toxic; refer to Table 12.2 for typical values.

Virtually all groundwood mills are integrated with paper mills producing newsprint and similar grades. Most groundwood mills still in operation are old, and discharge large flows of dilute effluent. Flows of 100 m³/t are not uncommon.

A few modern groundwood mills have been built in the USA and Scandinavia, where the grinder operates under pressure, and the water system is designed to minimize flow, concentrating the effluent. Mitchell & Tappio (1990) reported on a Finnish newsprint machine integrated with a modern pressure groundwood mill where the total effluent flow was slightly under 9 m³/t product.

Thermomechanical pulping

Thermomechanical pulping (TMP) separates the wood fibres in a device known as a refiner, where the chips pass between two serrated plates, one (or both) of which are rotating. This process requires almost twice as much electrical energy as the above-mentioned groundwood process, but the mechanical properties of the pulp are substantially better. BOD and toxicity are increased, as indicated in Table 12.2.

Sulphite pulping

Sulphite pulping was the most widely used chemical pulping process in the early days of the industry. Most former sulphite mills have been converted to kraft over the past 50 years. Refer to Kocurek (1986–1989: Vol. 4) for detailed process descriptions.

In sulphite pulping the fibres are separated by the action of sulphur dioxide and a metallic base, under pressure and at an elevated temperature. The traditional sulphite process is now generally known as low-yield sulphite. The following hybrid pulping processes have been developed from the traditional process:

- chemi-mechanical pulping (CMP)
- chemi-thermomechanical pulping (CTMP)
- high-yield sulphite (HYS)
- ultra high-yield sulphite (UHYS)
- neutral sulphite semichemical (NSSC).

The classic sulphite process is similar to kraft, in that the wood fibres are separated chemically, comparable wastes are generated, and most of the unwanted organic matter is recovered and incinerated. Traditionally, sulphite mills operated without any recovery of cooking wastes, but that is relatively rare today. Recovery is generally less efficient in sulphite mills than kraft, for a number of technical and economic reasons, and the evaporation stage releases considerably larger quantities of low molecular weight organic material.

The sulphite cooking reaction uses an acidic solution of one or more forms of the sulphites of calcium, ammonium, sodium or magnesium. Calcium was the original base, but has largely been replaced by the soluble bases, since there is no accepted process for the recovery of calcium-based spent cooking liquors.

The process and equipment used for the manufacture and bleaching of sulphite pulp are similar to those for kraft, with various differences dictated by the process chemistry. The recovery processes for spent sulphite liquor (the sulphite analogue of kraft's 'black liquor') are quite different from kraft.

Chemi-thermomechanical pulping

A number of TMP mills react sodium sulphite with the pulp immediately before refining, to soften the chips, reduce energy consumption and modify pulp characteristics. This process is known as CTMP. Chemical charge varies, and can result in pulping yield from levels below that of the TMP process, down to about 80%. The organic content of the effluent is proportional to the fraction of the raw wood that is not converted to pulp. Unless recovery of these organics is practised (which is rare) the BOD and COD can be relatively high. Bleached CTMP mills at Meadow Lake and Chetwynd, in Canada, recover all organics, and operate without any liquid effluent discharge.

Bleaching mechano-sulphite pulps

The low-yield sulphite pulps are commonly bleached by oxygen and chlorine-based processes very similar to those used for kraft pulps.

The higher yield pulps are often bleached by oxidative chemicals, principally hydrogen peroxide and sodium hydrosulphite. Chlorine-based bleaching is ineffective for pulps with a process yield above about 50%. Zinc hydrosulphite was used to bleach mechanical pulps in the past, but is widely prohibited today.

Where the improvement in brightness desired is only a few ISO (International Organization for Standardization) brightness units, the process is generally referred to as 'brightening'.

In all cases, the bleaching reagent is added to the pulp at controlled temperature and pH conditions. The bleaching systems may be quite powerful, and attain 80 or more ISO brightness units. Such processes release considerable quantities of dissolved organics, so the mill effluent BOD and COD are high, as shown in Table 12.2. The organic substances released from the pulp fibres are washed out and discharged.

In operations that are not equipped with pulp washers, the organics released in the bleaching reactions travel with the pulp to the paper machines. Most of the dissolved material is discharged from the paper manufacturing operation, along with the excess white water.

Semichemical pulp

This type of pulp is often known as NSSC (neutral sulphite semichemical), but the technological advances have made the inclusion of the term 'neutral sulphite' in the name somewhat obsolete. Semichemical pulp is almost always used on-site for the production of corrugating medium in an integrated mill. The demand for this type of pulp has largely been replaced by recycled fibre. The process has disappeared from Western Europe, and use is declining in the rest of the world.

Delignification occurs by cooking wood chips in a pressure vessel, the digester, with (traditionally) sodium sulphite. The cooked pulp is discharged (blown) to a tank at atmospheric pressure, and pulping is completed using refiners. Recovery of the spent cooking liquor is sometimes practised. While desirable environmentally, the economics of recovery are poor, which has accelerated the trend toward abandoning the process in favour of recycled fibre.

As indicated in Table 12.2, the raw BOD for this type of mill is high, so biological treatment is relatively expensive. The effluent toxicity is relatively high, with LC_{50} typically under 10%, upstream of effluent treatment.

Effluent generation by mechano-sulphite mills

It is clear from consideration of the mass balance of a mechano-sulphite mill that the quantity of organic material released from the wood is related to the pulping yield by a relationship of the form:

organic loss % = 100 × (1 – yield)

where yield is expressed as a fraction, loss is expressed as kg/tonne pulp.

One can expect the BOD, COD and toxicity of the effluent to rise as the organic loss rises. Where bleaching is practised, yield evaluations must consider the losses in bleaching as well as in pulping.

Discharge prevention at source

Apart from replacing the more polluting processes with more modern, less polluting ones, there are relatively few opportunities for significant improvements in effluent quality by in-plant process modifications for the mechano-sulphite mills. The principal opportunity is to reduce the effluent flow to lower operating costs and/or improve the effectiveness of external treatment systems.

Pulp from recovered paper

Recovered paper refers to paper that is considered waste by the user, and which is recovered for processing into useful products. While the term 'waste paper' is widely used by the public, the industry prefers the terms 'recovered paper' or 'secondary fibre'.

In this context, 'pulping' refers to the use of recovered paper as raw material for paper manufacturing, and as a replacement or partial replacement for virgin wood pulp. There has always been some reuse of secondary fibre, and the rate of its use rose rapidly through the 1990s. Current rates of increase are well below those of the 1990s, as the practical limits of paper recovery are reached.

Generally, paper can be produced from recovered fibre less expensively, using less energy, and with less effluent discharge than from virgin wood fibre. However, site-specific design and operating factors vary widely.

Simple pulping of recovered fibre

Recovered paper is reduced to a pulp slurry by vigorous agitation in a tank full of water, generally known as a pulper. Contaminants are removed mechanically by screens and centrifugal cleaners. The use of non-deinked waste papers is generally restricted to production of various packaging grades, or some building products.

Wastes produced consist of heavy trash, which is collected in relatively dry form, and fine granular material along with small fibres, which are lost with the excess white water flows. The latter also contain a variety of soluble materials that are washed from the pulp, and which will exert a BOD, and may be toxic to fish. As indicated in Table 12.2, the BOD discharge is generally low.

Where the raw material consists primarily of old corrugated containers and similar material, a significant amount of dissolved organic material can be extracted from the raw material, and discharged with the wastewater.

In this type of recovered fibre process, it is often technically feasible to close the water cycle completely, and operate with zero effluent (except for the heavy trash). This practice is becoming common, particularly in Western Europe.

De-inking

Some recovered waste paper is used in production of newsprint, tissue, and other paper grades of relatively high quality, which necessitates the removal of ink and virtually all other impurities. This cleaning process is generally known as de-inking. After pulping of the waste paper and removal of gross contamination, surfactants are added to the slurry to separate ink from the fibres. The objective is to disperse the ink in very small particles that have minimal affinity for fibres, so the two can be separated, either by washing or by dissolved air flotation.

Washing pulp in a de-inking mill is very different from the washing processes used in kraft and sulphite mills to recover soluble solids, because de-inking washers are designed to avoid forming a mat of fibres, so that the small ink particles and non-fibrous filler material will not be retained. This results in loss of most of the fillers in the waste stock, as well as significant quantities of the fine fibres, and generates relatively large effluent flows, in the order of 50–100 m^3/t.

Newsprint grade de-inking operations will discharge approximately 20 kg BOD/ tonne product, whether the wash or flotation de-inking process is used. The effluent BOD can be reduced to about 10 mg/L or less by biological treatment, which corresponds to under 0.1 kg/tonne for a modern flotation de-inking mill.

Other pulping processes

The foregoing discussion of pulp and papermaking technology addresses only the mainstream processes that are used to manufacture the majority of the world's pulp and paper. A wide variety of other processes exist, but are used in only one or a few mills. Some of these are mentioned briefly below, but space precludes extensive discussion.

Non-wood fibre sources

Most of the paper in the world is manufactured from wood fibres, but some mills use non-wood annual plants as raw material. Generally, they can be pulped by one or more of the processes described above for wood pulp, with modifications to suit the particular raw material. A variety of technical and economic problems prevent efficient recovery of the non-fibrous organic wastes, so the effluent discharges are often highly polluting by today's standards. Most non-wood mills are small, which mitigates environmental damage, but also inhibits the economic introduction of advanced pollution prevention technology.

Dissolving pulps

A dozen or so wood pulp mills produce chemically pure cellulose, for uses where the fibrous nature is irrelevant, such as for manufacture of food additives, viscosity control agents, explosives and textiles. These use the kraft or sulphite processes, with additional process stages to remove hemicelluloses and other components that are not desirable for specific end uses.

Effluent discharges are generally similar to paper-grade kraft and sulphite mills, and vary widely depending on the degree of pollution prevention practised. Some mills are at the top end of the world range of effluent discharges, while others are at the bottom.

Soda pulping

A few mills operate the 'soda process', which is quite like kraft pulping, but uses sulphur. With respect to effluent characteristic and discharge control, it is quite similar to the kraft process.

Solvent pulping

Many researchers have proposed the use of organic solvents to separate fibres, instead of the conventional kraft and mechano-sulphite processes. Such processes are normally asserted to be environmentally cleaner than kraft, but analysis of the processes does not support the assertion. Since none are in full-scale operation, there is a lack of data on their environmental performance on an industrial scale.

Papermaking

Paper mills may be integrated with one or more pulp mills, receiving their pulp as a slurry in water, by pipeline. Alternatively, they may be 'non-integrated', meaning that they purchase all fibre required from pulp mills, normally in dry form. It is common practice for integrated paper mills to use a mixture of purchased fibre, and pulp produced on-site.

The following general discussion on paper machines applies to integrated as well as to non-integrated mills. However, there is one important difference between these two generic types. In an integrated mill the pulp stock arrives at the paper machine at relatively low consistency so that the paper machines will always have an excess of process water. It is possible to return the excess process water (white water) to the pulp mill, at least to some extent. The mill effluent flow and rate of discharge of many contaminants is dependent on how successful the mill is in resolving the associated problems.

The preceding discussion has emphasized the relationship between process yield and effluent characteristics, with respect to pulp mills. This is not a useful concept for paper mills, since it is common practice to add up to about 25% non-fibrous material to the fibrous raw material. This material is mostly calcium carbonate and/or clay, so is relatively inert with respect to environmental considerations. Smaller quantities of substances such as latex, starch, rosin, dyes and bactericides are also added, which can have some effect on the effluents.

Stock preparation

The various kinds of pulp and any additives are blended and then passed through screens and centrifugal cleaners to remove physical contaminants. The main stock stream or a side stream may pass through refiners to modify the fibre properties before feeding it to the paper machine.

The BOD discharge is generally negligible, but can be several kg/tonne paper if there are significant starch, or coating, leaks, other losses due to equipment weakness or operator error. Good housekeeping is the prime control measure. The stock preparation area will not normally contribute to toxicity of the effluent to fish, unless toxic substances are added.

The only other environmental effect of the stock preparation system is the discharge of very small quantities (<1 kg/tonne) of dirt and several kg/tonne of fibre. This material will be removed from the effluent by the primary clarifier, and may contribute to the mill's total solid waste flow.

Due to the extensive recycling of water in paper mills, particularly those integrated with pulp mills, the above-mentioned BOD and suspended solids may exit the mill at a point quite far removed from the stock preparation department.

Paper machines

The term 'paper machine' is used for equipment that converts the low consistency stock from the stock preparation system into a roll of dry paper, ready for shipment or cutting into sheets.

In the traditional Fourdrinier-type paper machine, the stock from the headbox is first dewatered by being deposited on to a horizontal wire mesh conveyor belt, called the former. The stock is fed to the former at a consistency of about 0.3–1%, and leaves at about 20% consistency. It is then pressed between rotating rolls to raise the solids content to about 40%, and finally dried by passing it over steam-heated rolls. Any dissolved material (including oxygen-demanding or toxic substances) contained in the water passing into the dryer will either be destroyed or volatilized by the heat of the dryer, or will become a constituent of the paper.

In an increasing number of modern paper machines, the forming section uses two wire conveyor belts, and the sheet of paper is formed between them. There is a wide variety of these 'twin-wire' formers. While the differences between them have a major impact on paper properties and the grades of paper that can be manufactured on any one machine, they do not make much difference in discharges of waste to the environment. However, conversion of an older single-wire paper machine to twin-wire can affect suspended solids discharges indirectly, because the consistency and volume of white water normally rises. This can result in increased discharges to sewer if the white water system is not upgraded simultaneously.

Water balance for paper machines

The paper-forming process requires removal of 100–200 m^3 of water per tonne of product. The water contains significant quantities of fibres, and is known as white water. Most of it is recycled to the stock preparation system, while some must be discharged to sewer or recycled to other parts of the mill. Management of the white water system is the most environmentally significant item in controlling effluents from paper machines. When coated paper is manufactured, starch, latex and other coating materials can be present, in addition to fibres, and may have an appreciable BOD and COD, but are not generally considered to be toxic to fish.

Except for paper machines incorporating a coater, very little COD or toxicity is generated in the paper machine area, so that the content of such substances in the effluent is dependent on the quantity carried into the machine with the pulp. As these pollutants are mostly soluble, the split between effluent and the paper sheet will be approximately in proportion to the ratio of flows of water into the dryer (normally about 1.5 m³/t paper) and the total effluent flow (varying from a few m³/t paper to over 100 m³/t).

When manufacturing a product in which some dirt is acceptable (roofing felt, insulating board, some packaging, etc.), it is possible to reduce water use to be equal to the mill's evaporative losses, so that there is no liquid effluent, and all contaminants leave the mill with the paper. This is accomplished by some 100 mills today. Success is also contingent on the ability of the product to accept the dissolved substances that enter the mill, mostly with the raw fibre.

Effluent treatment

External treatment is utilized to reduce effluent discharge, as a supplement or alternative to discharge control at source. Treatment may involve a range of physical-chemical and biological measures.

- Primary treatment involves removal of suspended solids, normally by sedimentation.
- Secondary treatment removes soluble organic materials, primarily affecting BOD and toxicity for the discharged stream.
- Tertiary treatment refers to additional processes, perhaps designed to remove specific contaminants from secondary-treated effluent.

Most external effluent treatment processes rely upon, at least to some extent, concentrating the pollutants into a side stream, normally a sludge. Many require addition of chemicals, which may result in additional sludge formation. There is no single process that is environmentally best, even if cost is ignored.

Most mills today operate primary and secondary treatment systems, but few use tertiary treatment. The principal secondary processes used are activated sludge treatment (AST) and aerated stabilization basins (ASB).

Activated sludge treatment

Activated sludge treatment is a widely used biological process normally applied to primary-treated effluents. The principle is to create the conditions for a high concentration of micro-organisms to grow on the soluble materials in the effluent. The micro-organisms are later separated from the treated effluent by sedimentation and recycled in the process as a sludge. Excess sludge is concentrated and then incinerated or landfilled. AST can achieve lower BOD and TSS (total suspended solids) discharges than ASB, but generates significant quantities of waste sludge, which require chemical addition and are difficult to dispose of in an environmentally satisfactory manner. Biodegradable organics will be largely destroyed or mineralized. The non-biodegradable substances and the heavy metals removed from the wastewater are stored in the waste sludge. Energy and chemical requirements for sludge handling make this process more expensive per unit weight of BOD removed than an ASB system.

Aerated stabilization basin

Aerated stabilization basin treatment is a biological process widely applied to primary-treated pulp and paper industry effluents. The principle is based on the growth of low concentrations of micro-organisms on the soluble materials in the effluent without the sludge recycle that is characteristic of the above-mentioned AST process. When the micro-organisms die, the sludge is used as food by other micro-organisms, and thus the BOD is digested. The successful operation of an ASB system involves the control of the non-digestible sludge so that minimal quantities of TSS and BOD are discharged in the final effluent without the dredging of accumulated sludge being necessary.

Tertiary treatment

The two technologies potentially applicable in the pulp and paper industry are chemically assisted coagulation and granular filtration. Either can reduce effluent BOD from a well-operated AST system by in the order of 33%, and will reduce TSS and phosphorus discharges by about 50%. The attainable reductions in heavy metals are site-specific.

Performance attainable by biological treatment systems

The actual performance of biological treatment systems varies widely from mill to mill, and is often inferior to that indicated in textbooks. Reasons for this range from sloppy operation to real technical difficulties that have not yet been completely resolved, despite about 100 years of experience in building and operating such systems. In all the mills that the writers are aware of with excellent performance of their biological treatment systems, considerable technical and personnel resources are committed to their operation.

There is a minimal practical attainable concentration of the parameters commonly used for defining effluent quality, as shown in Table 12.5 (McCubbin *et al.* 1992). This is based on reported operations for systems treating pulp and paper industry wastes.

As mills reduce water consumption, the concentrations of pollutants in the untreated effluents may rise, but this can normally be countered by efficient treatment plant operation, so that mills with low effluent volumes can also achieve low discharges of most pollutants.

Table 12.5 Attainable concentrations for TSS, BOD, P and N in ASB and AST treatment

	Aerated stabilization basin				Activated sludge treatment			
	LTA	AVG_{95}	AVG_{30}	MAX_{95}	LTA	AVG_{95}	AVG_{30}	MAX_{95}
TSS	46	45	55	70	21	20	30	40
BOD	21	20	30	50	11	10	15	25
Total phosphorus	0.8	0.8	1.0	1.5	0.8	0.8	1.0	2.0
Total Kjeldahl nitrogen	9	9	10	18	10	10	12	20
NH_3-nitrogen	1.7	1.5	2.0	4.5	0.9	0.8	2.0	6.0
NO_3-nitrogen	0.7	0.5	1.2	2.0	0.7	0.5	0.8	1.0

Data are shown as mg/L. LTA, long-term average; AVG_{30}, 30 days rolling average; AVG_{95}, annual 95th percentile average; MAX_{95}, maximum day, 95th percentile.

References

Dence, C. & Reeve, D. (1996) *Pulp Bleaching – Principles and Practices*. TAPPI Press, Atlanta, GA.

Gullichsen, J. & Paulapuro, H. (2000) *Papermaking Science and Technology*, (series of 19 books). Fapet Oy, Helsinki, in cooperation with TAPPI Press and the Finnish Paper Engineers Association.

IPPC (2000) *Integrated Pollution Prevention and Control*. Reference Document on Best Available Techniques in the Pulp and Paper Industry. Available from World Trade Center, Isla de la Cartuja s/n, E-41092 Seville, Spain. Internet: http://eippcb.jrc.es

Kocurek, M.J. (1986–1989). *Pulp and Paper Manufacture*, 5 volumes. Series of textbooks published by the Joint Textbook Committee of the Paper Industry. PAPTAC, Montreal.

McCubbin, N., Edde, H., Barnes, E., Folke, J., Bergman, E. & Owen, D. (1992) *Best Available Technology for the Ontario Pulp and Paper Industry*. Ontario Ministry of the Environment, Water Resources Branch.

Mitchell, G. & Tapio, M. (1990) PGW/CPGW Mill Report Low Effluent Load, Reduced Steam Contamination. *Pulp and Paper*, **June** 1990.

Paperloop (2001) *Annual Review 2001*. Report published by Paperloop.com (assoc. with Miller Freeman), San Francisco, CA.

Smook, G.A. (1988) *Handbook for Pulp and Paper Technologist*. Joint Textbook Committee, TAPPI Press, Atlanta, GA.

Part V
Forestry Effects on Aquatic Systems and Fishes

Chapter 13
Effects of forest management activities on watershed processes

G.F. HARTMAN

Introduction

This chapter reviews effects of forestry activities on watershed ecosystems and fishes within them. The basic characteristics of streams are determined by climatically influenced hydrological circumstances, geological conditions and the forest systems that arise under such conditions (Chapter 3). The interplay among forestry activities, hydrologic and geomorphic processes, forests and associated fish habitat are at the foundation of forestry–fish interactions.

This chapter examines five related topics:

- Hydrologic and geomorphic processes and their primary roles in stream ecosystem function.
- Effects of individual forest management activities (road construction and use, felling and yarding, and silvicultural work).
- Interconnections among impact processes, and how complexes of processes affect stream ecosystems in time and space.
- Extrapolation of information about forestry effects on fish from one region to another, and limitations of such extrapolation.
- Generalizations about effects of forestry impacts applicable to forestry/fish research and management programmes.

Hydrological geomorphological processes

For reviews, the reader is referred to Hewlett (1982), forest hydrology; Church & Eaton (2001), hydrological effects of forest harvest in the Pacific Northwest; Sullivan *et al.* (1987), links between forests, channel conditions and fishes; Jones & Mulholland (2000), interactions between groundwater and surface water, groundwater chemistry and biotic conditions. Literature on terrain stability, slides and gully assessment pertaining to western Canada includes Rood (1984), Hogan (1986), Krag *et al.* (1986), Rollerson (1992) and Hogan *et al.* (1998a). Geomorphic hazards in managed forests in Asia, Europe and North America are dealt with in Rice (1991).

Hydrologic processes

In the hydrological cycle, evaporated water moves from the surface of lakes, streams and oceans to the atmosphere where it condenses, falls to the earth surface and enters the soil and vegetation, from which some of it re-evaporates. Soil water that is not evaporated either moves to the streams, lakes and oceans, or is taken up by vegetation and transpired to the atmosphere. The main features of this hydrological cycle are illustrated in Hetherington (1987). Hydrologic processes within a forested ecosystem, elaborated somewhat in Fig. 13.1, are part of this cycle. Water may reach the ground

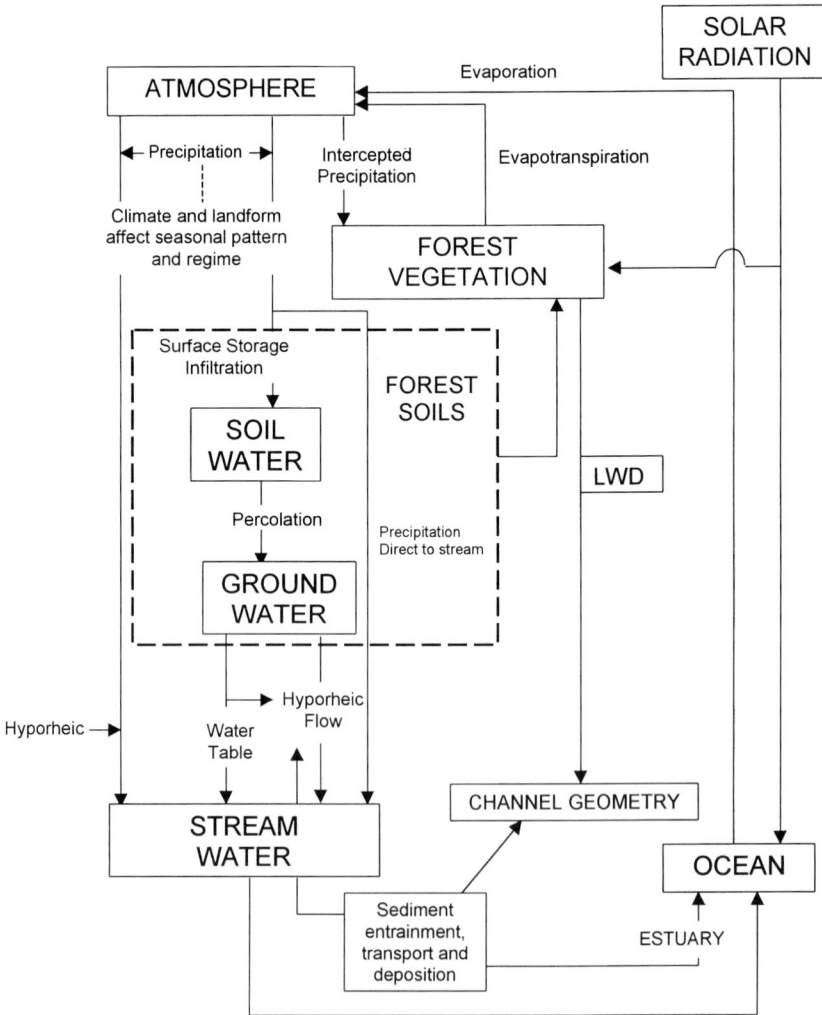

Fig. 13.1 Generalized model of the hydrological subsystem in a forest–stream ecosystem. Connections to woody debris movement and channel geometry are indicated to emphasize the role of hydrologic processes.

from sources other than rain. In coastal zones, fog is intercepted by trees and drips to the soil (Hetherington 1987). The rates of the above processes are influenced by climate, geography, forest type and forest removal.

Precipitation reaching the ground may flow overland, enter the soil and become subsurface flow that reaches a stream, or moves as groundwater directly to the sea or a lake. Groundwater or stream water that enters the zone below the streambed and moves from there to the stream and back is referred to as hyporheic flow. Groundwater is distinguished from hyporheic flow by entering or leaving the stream only once (Harvey & Wagner 2000). Groundwater and hyporheic flow are indicated in Fig. 13.1 because of their important roles in stream ecological processes (Jones & Mulholland 2000).

Hyporheic flow exchange brings stream water into contact with geochemically and microbially active sediment. Carbon and oxygen carried into the hyporheic environment support metabolic processes in the extensive fauna there. This fauna contributes to food sources of small invertebrates and larval fish in the stream (Hakenkamp & Palmer 2000).

Figure 13.1 also indicates linkages between hydrological processes, woody debris and sediment movement that are critical in determining channel geometry. The processes indicated by connecting arrows in Fig. 13.1 may occur in any region where there is forest vegetation. However, their rates can be altered by forest removal, and other activities that disturb forest soils.

Geomorphic processes

Geomorphic processes, linked to hydrologic systems, determine key features in stream channels. Many geomorphic processes begin on the valley slopes and move woody debris, sediment and litter down-slope into the stream channel (Fig. 13.2). These processes are influenced by the movement of water which links most physical events within a watershed. The processes can occur naturally; however, forest management activities may alter their rates and severity (see points A to F, Fig. 13.2, and Table 13.1, effects on fish).

Mass movement, sheet erosion and bank erosion (Fig. 13.2) are caused by different processes. Sheet erosion results from accumulated rain-splash and dislodgement of soil particles. The rate of such erosion is a function of hill-slope, vegetation cover, soil type, length of slope and rainfall regime (Reid 1993).

Mass movement processes affect the entire surface profile of the soil and occur as four different processes (Swanston 1971, 1991):

- creep – slow down-slope movement of the soil mantle
- debris torrents – fast movement of water-charged soil alluvium and vegetation down stream or ephemeral channels
- debris avalanches – rapid, shallow mass movement from the hill-slope, commonly containing a high concentration of water, and
- earth-flow – slow down-slope movement of material, planar to irregular in form and 3–20 cm thick.

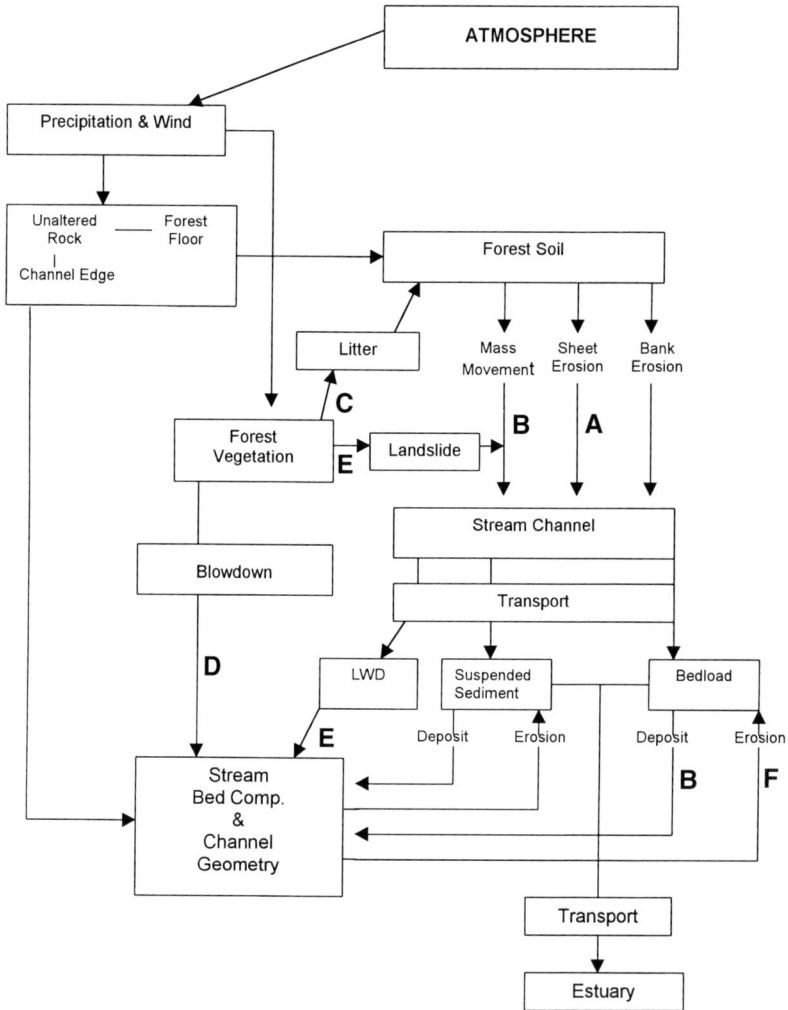

Fig. 13.2 Schematic diagram of geomorphic processes where forestry activities may have watershed effects. (A) Erosion of side-cast material along roads. (B) Forestry-induced increases in landslide frequency – increased delivery to channel. (C) Reduced litter input with clearing. (D) Reduced large woody debris input from riparian zone with clearing. (E) Formation of debris jams from landslides (debris torrents), natural or logging-induced. (F, B) Increased erosion and deposition with streamside logging and bank breakdown.

Geology, soil mechanics and soil moisture affect slope stability and distribution of tree roots. The latter are important because they increase the shear strength of soils. The age of the root systems modifies root strength and hence shear strength of the soil. Forest removal is, for this reason, an important cause of loss of soil strength and mass soil movements. Root strength decreases for 5–10 years after cutting and re-planting, and then begins to increase again (Abe & Ziemer 1991).

Table 13.1 Influences of timber harvest on physical characteristics of stream environments, potential changes in habitat quality and resultant consequences for salmonid growth and survival

Forest practice	Potential change in physical stream environment	Potential change in quality of salmonid habitat	Potential consequences for salmonid growth and survival
Road construction and use	Increased sediment to the stream from: sheet erosion of side-cast material, landslides and sediment wash-in from road surfaces during use	Temporary increases in suspended sediment and turbidity, and increase in fine sediment in streambed gravel	Effects on fish behaviour and feeding efficiency, and effects on incubating eggs
	Loss of riparian forest where roads approach or cross the stream	Loss of large woody material, and local loss of shading with consequent effects on cover and water temperature	Effects on survival during periods when cover is needed. Positive effects on growth possible in sections of cold streams
Timber harvest from streamside areas	Increased incident solar radiation	Increased stream temperature; higher light levels; increased autotrophic production	Reduced growth efficiency; increased susceptibility to disease; increased food production; changes in growth rate and age at smolting
	Decreased supply of large woody debris	Reduced cover; loss of pool habitat; reduced protection from peak flows; reduced storage of gravel and organic matter; loss of hydraulic complexity	Increased vulnerability to predation; lower winter survival; reduced carrying capacity; less spawning gravel; reduced food production; loss of species diversity
	Short-term loss of leaf litter input	Alteration of insect food supply	Potential alteration of invertebrate fauna and fish feeding
	Addition of logging slash (needles, bark, branches)	Short-term increase in dissolved oxygen demand; increased amount of fine particulate organic matter; increased cover	Reduced spawning success; short-term increase in food production; increased survival of juveniles

Table 13.1 (*Continued.*)

Forest practice	Potential change in physical stream environment	Potential change in quality of salmonid habitat	Potential consequences for salmonid growth and survival
	Erosion of stream banks	Loss of cover along edge of channel; increased stream width, reduced depth	Increased vulnerability to predation; increased carrying capacity for age -0 fish, but reduced carrying capacity for age -1 and older fish
		Increased fine sediment in spawning gravels and food production areas	Reduced spawning success; reduced food supply
Timber harvest from hill-slopes; forest roads	Altered stream flow regime	Short-term increase in stream flows during summer	Short-term increase in survival
		Increased severity of some peak flow events	Embryo mortality caused by bed-load movement
	Soil exposure, gouging by yarding; accelerated surface erosion and mass wasting	Increased fine sediment in stream gravels	Reduced spawning success; reduced food abundance; loss of winter hiding space
		Increased supply of coarse sediment	Increased or decreased rearing capacity
		Increased frequency of debris torrents; loss of instream cover in the torrent track; improved cover in some debris jams	Blockage to migrations; reduced survival in the torrent track; improved winter habitat in some torrent deposits
	Increased nutrient runoff	Elevated nutrient levels in streams	Increased food production

Activity		
Scarification and slash burning (preparation of soil for reforestation)	Increased nutrient runoff	Short-term elevation of nutrient levels in streams	Temporary increase in food production
	Inputs of fine organic and inorganic matter	Increased fine sediment in spawning gravels and food production areas; short-term increase in dissolved oxygen demand	Reduced spawning success
Fertilization	Input of nitrogen as urea, nitrate or ammonium	Potential short-term enhancement of periphyton production and stream insect growth	Low toxicity to fish and potential increase in food availability
		Potential temperature increase	Stress on fish if temperature increase is large
Herbicide application	Loss of riparian vegetation in small stream over-spray areas	Loss of leaf litter input and effects on heterotrophic production	Decrease in food availability
		Increase in insect drift in some applications	Short-term increase in food availability and medium-term reduction in standing crop of food organisms
		Deposition of herbicide in the water – water quality change	Range of toxic effects from acute to sublethal to fish. 96-hr LC_{50} range from 1.0 mg/L to 670, and one extreme of 30,000 mg/L. Many low toxicity compounds in use
Insecticide application		Water toxicity increase; loss or reduction of aquatic insect fauna.	Range of toxic effects from acute to sublethal. Most high toxicity to fish: 96-hr LC_{50} range from 0.0014 to 1000 mg/L. Effects on fish behaviour at low concentrations
		Some extreme toxicity to invertebrates: 96-hr LC_{50} from 0.00015 to 3.0 mg/L	Potential effects on availability of fish food organisms, and hence fish growth

Based on Hicks *et al.* (1991b) with information added to include silvicultural activities.

Other landscape attributes such as slope angle, position on the slope, and slope morphology and curvature influence the probability of failure at a particular location (Rollerson 1992).

Debris torrents and avalanches tend to be episodic in occurrence. From 1810 to 1991, almost 85% of the sediment and woody debris delivered to streams in the Queen Charlotte Islands (British Columbia) was generated in seven large events (Hogan *et al.* 1998b). At least four of these were coincident with rainfall of 45 cm or more in a 24-hour period. In mountainous old-growth watersheds, debris avalanches at junctions with small tributaries provide the majority of woody debris in the channels (Maser & Sedell 1994). This process of wood input, plus that caused by blow-down, plays an important role in determining channel and fish habitat structure (see Fig. 13.2E & D).

Bank loss may not be considered a mass movement, but it can affect water quality in streams. It may occur steadily from erosion, or episodically as bank collapse. In Carnation Creek, a small western coast British Columbian stream, bank erosion along 50–75-m logged reaches was as much as three times higher than in reaches with leave strips (Toews & Moore 1982). Bank erosion sets many channel changes in motion, including channel widening, streambed aggradation and sediment release, with downstream impacts on streambed composition (Hartman & Scrivener 1990).

These elements of hydrologic and geomorphic processes, briefly touched upon, are at the foundation of stream channel conditions that are influenced by logging.

Forestry activities: effects on fish

General

Forestry operations within a watershed can simultaneously affect several different conditions in the system (Reid 1993). However, this chapter will consider single activities and their effects first because planning elements and regulations, within a watershed, tend to be designed to ameliorate the effects of single activities. The 'single' activities fall within one of three major categories of forest management:

- road construction and use
- timber harvest from streamside and up-slope areas
- silviculture.

These activities create potential changes in the physical stream environment affecting the quality of salmonid habitat and life history activities of the fish themselves (Table 13.1). The above three categories of activity may affect four different regimes or conditions within the stream itself (Hartman & Scrivener 1990):

- temperature
- hydrological
- water quality and nutrient flux
- channel geometry.

The movement of water into and through a watershed (Fig. 13.1) carries nutrients through the drainage, and provides the energy that influences sediment and large woody debris movement and deposition (Fig. 13.2). Forestry activities change the scale or frequency of elements in such processes (see connecting arrows in Fig. 13.1 and arrows and capital letters in Fig. 13.2). Forest removal may produce further important changes by altering the temperature regime within the stream. Different activities produce different sets of impacts. Logging effects begin with road construction.

Road construction and operation

Road construction and use have many effects on a stream system; e.g. changing stream hydrographs, altering sediment deposition, changing infiltration rates, intercepting and diverting subsurface flow, and increasing peak flows by such alteration of the functional drainage network (Furniss *et al.* 1991). For a full review of effects of roads on forest ecosystems see Gucinski *et al.* (2001).

During the life of a road, different effects arise. Construction work may release sediment from newly exposed surfaces, road use can result in surface sediment being washed into streams, and old, unused roads may collapse causing mass movement of soil. The significance of these effects depends upon soil and terrain features, road density and the size of the area influenced.

Scales of construction

The total kilometres of road constructed and the intensity of use are related to the size of the area logged and the scale of development of the industry. They provide one indication of the overall scale of forestry activities. Because of differences in terrain, period of logging and method of harvest, the relationship between total area cut and road length may vary within areas. Total lenghts of forestry roads for British Columbia, Washington State (Department of Natural Resource Forests lands) and Montana plus Northern Idaho National Forests are 303,600 km, 11,437 km and 104,911 km, respectively. Within these jurisdictions, road densities in logged areas range from 1 to 5 km/km^2 (Table 13.2).

Road density, averaged over the total forest area of British Columbia, appears to be low. However, in 80 of 144 groups of watersheds in the southern half of the province, they ranged from 0.5 to >2 km/km^2, within watershed groupings where 10 to >30% of the riparian areas had been logged (Anonymous 2001). The number of stream crossings per unit of area, an important consideration regarding road density, is determined by both road and stream density. The seriousness of effects of road crossings is related to mechanisms of habitat impairment, e.g. migration blockage, sediment production, and slope failure. Numbers of impassable culverts per km^2, and the length of habitat from which spawning and rearing fish are excluded, are more meaningful indicators of effects than simple road density (Gucinski *et al.* 2001).

Table 13.2 Areas of forest, controlling jurisdiction, harvest areas and rates, and road construction information for some selected jurisdictions in North America (it has not been possible to obtain information that is consistent for time interval and jurisdictional area)

Province or state and controlling jurisdiction	Forested area (ha × 1000)	Average annual cut (ha × 1000)	Total area cut (ha × 1000)	Total km of road	Road density km/km²	References
Alberta	38,200	34.0 (1966–1998)	1,123.2	–	–	Canadian Council of Forest Ministers (2001);
(Environmental Protection and Natural Resources Services)	34,964[a]	27.5[a]	908.0[a] (1966–1998)		2–5 for 48% of area	Hayduke & Associates (1997)
British Columbia (Ministry of Forests)	60,600	204	ca 5,300 (1975–2000)	303,600 (to 2000)	1.0 Central and South 1.5 Vancouver Island	Anonymous 2001; Canadian Council of Forest Ministers (2001)
Alaska (Tongass National Forest)	6,714	3.48 (1950–1996)	160.6 (1950–1996)	–	–	Bryant & Everest (1998); Anonymous (2000a)
Washington State Dept Natl Res.	850	8.8 (1950–1986)	–	11,437	2.17	Washington State Dept Natl Res., general agency information; C.J. Cederholm (pers. comm.);
US Dept Agriculture, Forest Service; Bureau of Land Management, State, Private, and Native, WA	–	148.2 (1950–1986)	13,547 (1950–1986)	–	–	Larsen (1989)

Oregon US Dept Agriculture, Forest Service;	11,331	278.6 (1988–1999)	3,343	–	2.1[b]	Anonymous (1993); Oregon Dept of Forestry website
Bureau of Land Management, State, Native and private OR		335.7 (1976–2000)				
Montana-Idaho, US Dept Agriculture, Forest Service	10,283	34.8	1,777 (1950–2000)	104,911	1.02	US Forest Service, Region 1, Transportation Dept, (pers.comm.)
Canada (10 provinces)	327,390[c] 234,500[d] 119,000[e]	–	–	–	–	Canadian Council of Forest Ministers (2001)
US Dept of Agriculture, Forest Service – National Forests total	75,948 50,857 (with roads)	–	–	–	–	Anonymous (2000a)

[a]Boreal Forest Natural Region, [b]range of the spotted owl, [c]provincial total, [d]commercial forest, [e]managed forest.

Effects on a watershed system

Road construction and operation have an array of watershed impacts (Table 13.1 and Fig. 13.2). Right-of-way clearing for roads results in loss of riparian vegetation at crossings and where road locations approach the stream. To the extent that clearing of riparian trees occurs, there will be more light reaching the stream with consequent effects on water temperature, litter input and primary production. There will also be a reduction in the long-term potential for large wood input and a possible increase in stream bank erosion (Fig. 13.2).

Side-cast material may contribute to the overall production of sediment within a watershed by increasing the area for sheet erosion (Fig. 13.2). In the Clearwater River system (Washington State) side-cast material produced 3 tons of sediment for each km^2 (Cederholm & Reid 1987) with particle size of <2 mm. The amount and grain size of eroded material reflected both soil type and construction method. In the Carnation Creek watershed, roads adjacent to the stream were constructed primarily by excavating the road grade and end-hauling the material away. Short-term problems occurred during periods of heavy rain at four of eight sampling sites (Toews & Brownlee 1981). However, in the long term, side-cast material did not contribute to sediment loading in the stream (Hartman & Scrivener 1990).

Up-slope roads that are poorly planned increase the frequency of landslides in steep or unstable terrain. Such movements influence large woody debris input, sediment dynamics and channel geometry (Fig. 13.2). Slope angle, location, soil type, proximity to bedrock and subsurface water content all affect the probability of road-related landslides (Swanston 1971, 1991). Such risk factors apply in any part of the world.

Road fills, side-cast material on steep slopes, wood-bearing side-cast material, debris, under-sized culverts, water rerouted to unstable road sections and poor maintenance all contribute to up-slope road failures and landslides. The frequency of erosion gullies on steep slopes, below roads, can be dramatically increased. Road-related slides may reach lakes, or end up in 'storage' locations adjacent to streams. Complexes of roads and clear-cuts may result in multiple input locations of slides and surface erosion.

Slide frequency can be increased by poor road construction. Frequency of road-related failures was 30–300 times higher (Furniss *et al.* 1991), and 42 and 133 times higher than expected on unroaded slopes (Rood 1984). Volumes of sediment from landslides and debris flows were 186 ton/km^2/yr in the Clearwater River basin. Of this, 57 ton/km^2/yr were composed of material <2 mm, and intra-gravel sediment (>0.85 mm) increased significantly after >3% of the basin area was covered in roads.

Erosion of road surfaces, when used during periods of rain, is a source of fine sediment input in some watersheds. Its production from road surfaces is a function of lithology and soils (Duncan & Ward 1985), and intensity of road use (Cederholm & Reid 1987).

The percentages of roaded area in a basin and fines in streambed gravel are positively related (Cederholm & Reid 1987). Sediment concentration increased with culvert discharge, and heavily used road sections contributed about 130 times as much sediment as unused sections. Paved roads yielded <1% as much as a heavily used section (Reid

& Dunne 1984). In the Clearwater River basin, road surface erosion contributed 500 ton/km road/yr (material <2 mm) during heavy use, 64 ton/km of road, during temporary non-use and 40 tons/km/yr, during moderate use (Cederholm & Reid 1987).

Such sediment input may affect fish eggs and alevins directly (Everest *et al.* 1987), or depending upon concentration and duration, it will affect fish behaviour and survival (Newcombe & Jensen 1996). Road-generated sediment production may occur in drainages in any part of the world where there are logging roads, rain and fine-grain road material.

Roads may intercept natural flow routes, change flow and soil moisture patterns, compact soil and increase surface runoff. Road surfaces produce more excess water and the ditches by them act as first order stream channels sending water more quickly to the adjacent stream (Wemple 1994; Murphy 1995).

Soil channels (macropores) exist in most forest soils (Whipkey 1967; Cheng *et al.* 1975). They provide pathways of low resistance to subsurface storm flow. Interception and redirection of such flow by road cut-banks affect runoff timing. In the Carnation Creek drainage such interception and redirection following road construction changed down-slope groundwater levels, and rerouted water into ditches (Hetherington 1998; Wigmosta & Perkins 2000).

Information on the effects of roads on flow regimes is conflicting and the relative importance of the effects of roads versus clearing is not always evident (Wright *et al.* 1990). Roads have been shown to increase, decrease, or have no effect on peak discharges (Jones & Grant 1996). Bowling & Lettenmaier (2000) found that the effects of roads plus forest harvest were equal or slightly larger than the sum of the two. Roads increase peak flows after they cover 12% of a basin (Rothacher 1973; Harr *et al.* 1975). Road surfaces covering 5% (Ziemer 1981) and 8% (Rothacher 1970) had no significant effect on flow parameters. Jones & Grant (1996) reported somewhat different results in Oregon. There was a higher than expected, but not significant, number of events in which peak flows were increased and 'begin times' were advanced in a study of storm events in a 100-ha watershed (6% area in roads, but no forest removal).

The layout of logging operations within watersheds frequently requires stream crossings. Culverts are the most common water crossings but they offer risks for fish. Installation and presence may cause high water velocities, excessive drops, altered stream hydraulics, erosion, sedimentation and channel change, all of which have directly related impacts on fish (Toews & Brownlee 1981).

Falling and yarding

Falling and yarding affect both up-slope and riparian areas, but their effects are different. Clearing of the riparian zone has great potential to affect stream processes because of proximity to the channel.

In considering forestry impacts, the riparian zone must be viewed as a three-dimensional zone with height, width and length along the stream (Gregory *et al.* 1991). The riparian forest has effects on the soil and stream water chemistry, shading and temperature, litter input, production of woody debris and terrestrial insect 'fall-in'. The hyporheic zone may extend latterly under, and affect riparian vegetation development.

The amount and chemical nature of water reaching the stream channel may be altered in locations where it passes through the root zones of either or both, up-slope and riparian trees (Fig. 13.3).

Changes that occur in the riparian zone affect the network of trophic processes that, with the structural environment, sustain fish populations. Clearing of the riparian zone may alter the scale or rate of processes indicated at A, B, D and E, in Fig. 13.4. Changes in the volume of woody debris may further alter the retention of salmon carcasses and subsequent nutrient input and food for macroinvertebrates that they contribute (Cederholm *et al.* 2000).

Effects on hydrological processes

Forest removal, particularly on valley slopes, has critical effects on hydrological regimes and processes related to them (Hetherington 1987). Forest removal changes water yield, storm flow conditions, peak flow responses, snowmelt rates and summer low flow levels.

Water yield, the total volume of surface runoff from a watershed, represents the precipitation that falls, and is not lost through evapotranspiration or groundwater flow (Church & Eaton 2001). Studies have shown fairly consistently that forest removal

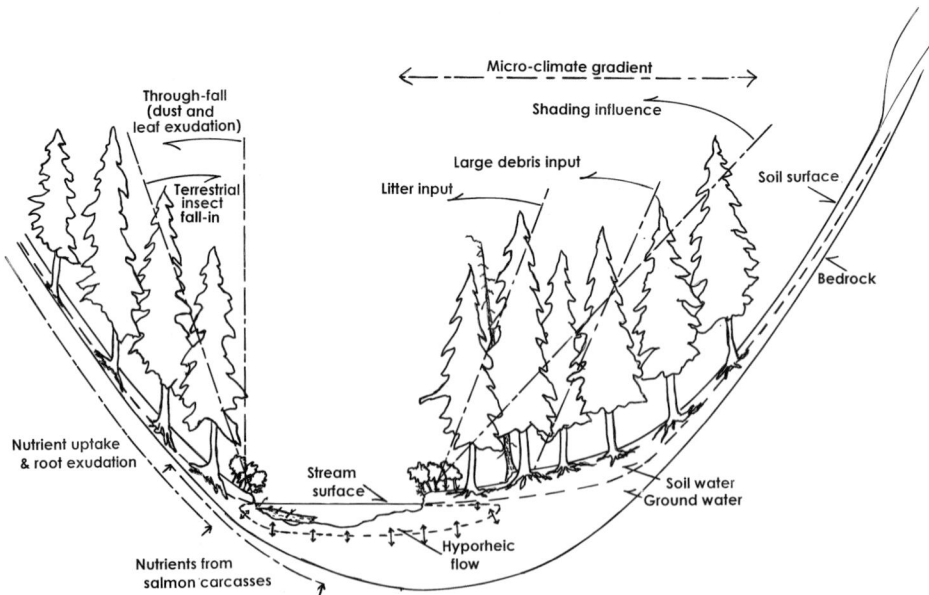

Fig. 13.3 Hypothetical two-dimensional cross-section of a stream channel and riparian zone indicating shading and micro-climate effects of forest, zones from which large woody debris, litter and terrestrial insects arrive, soil, ground and hyporheic water connections to the stream, and nutrient sources. The third dimension of the riparian zone extends up and down the stream on an axis along which stream size, litter input effects, relative amount of shading and role of riparian woody debris changes.

increases water yield by decreasing evapotranspiration, and that water yield tends to rise with increase in forest removal (Keppeler & Ziemer 1990). Experimental work conducted in western Oregon compared effects of clear-cutting (including roads), and patch-cutting with controls. Slash in both treated areas was burned after logging. Annual water yields increased by 40% in the clear-cut area, and 8–16% in the patch-cut area (Hicks *et al.* 1991a).

Water yield increases, following forest removal, differ among different forest types. Removal of 15% of coniferous forest cover increased water yield by 40 mm (measured as water depth over the entire drainage). It increased by 25 mm with 10% removal of a hardwood forest and by 10 mm with 10% removal of shrub and brush cover (Bosch & Hewlett 1981).

Changes in water yield following forest removal are greatest in areas of high mean annual precipitation. There may, however, be exceptions to this generalization. Water yield from Carnation Creek where precipitation was high, was lower than that predicted (Bosch & Hewlett 1981). This was possibly due to the thin soils and rapid drainage in the basin (Hetherington 1987). Harr (1983) found a small decrease in water yield after forest removal in a small watershed in Oregon. The removal of forest eliminated 'fog drip', which contributed to water yield. During winter, in areas where forests trap wind-blown snow, forest removal may eliminate the trapping effect and also cause a reduction in total runoff from the area (Hetherington 1987).

The duration of water yield increases differs depending upon forestry treatments. Hicks *et al.* (1991a) reported that water yield in the patch-cut area remained elevated for about 20 years after the treatment, and might remain higher than the control for about 50 years. Most of such increased water yield occurred during the wettest months.

The species of trees used in reforestation may affect the pattern of change in water yield following clearing. In Australia, water yield rose following clearing, but fell below pre-harvest levels in sites that were re-forested with eucalypt species (Chapter 30).

Increases in water yield during periods of summer low flow may be advantageous to fish. In western North America, fish habitat space may be limited by low flows during late summer. Hetherington (1987) reported the following increases in August–September flows after fire or logging:

- 10 to 36% after wildfire in British Columbia
- 133 to 318% after forest removal in New Brunswick (Canada)
- 78% increase in Carnation Creek, British Columbia
- a doubling of yield from a clear-cut plot in Saskatchewan (Canada).

However, stream-flow increases during the summer low flow period may not be evident if the flow runs subsurface through an aggraded streambed following logging.

The effect of logging on peak flows is of special interest because such events are often the most important in determining litter movement, nutrient export, sediment transport, debris movement and re-structuring of fish habitat. The effects of clear-cutting and road building (individually or combined) on storm flows have been studied extensively. The evidence on the effects of peak flow responses to forest harvest is mixed (Jones & Grant 1996, 2001; Thomas & Megahan 1998, 2001).

The effects of clear-cutting on peak flows may vary depending on the type of storm event (snowmelt, rain-on-snow, or rain), on the location of roads and cut-blocks and on the characteristics of the basin (Church & Eaton 2001). In many cases road building plus forest removal affects peak flows. Peak flow increases were noted after road construction plus 82% forest removal (Harr *et al.* 1975). Shelterwood cutting and patch-cutting resulted in peak flow increases of 50% and 11% respectively (Harr *et al.* 1979). Hetherington (1982) found that peak flows increased by about 20% following logging (90% cleared) and road building. He attributed part of the increase to each activity.

In a study of six watersheds of 6200–63,700 ha and three of 60–100 ha, Jones & Grant (1996) found an increase in peak discharges, advanced 'begin-times' and delayed peak flow times in the small drainages. In the large basins, they showed that peak discharges were significantly related to cumulative area harvested and recent rates of cut. Peak discharges increased by 0.014 m^3/sec/km^2 for each 1% difference in area cut.

The largest increase in rainfall-induced peak flow, in cleared areas, may occur during the first autumn rains. Soils in cleared areas contain relatively higher water content than soils in forested areas, so modest amounts of rainfall produce peak flows that are elevated above those from comparable forested areas. During winter, when soils are saturated, the proportion of rainfall that becomes storm flow can be as high as 90%, and the incremental storage capacity of the forest floor is no longer a factor in the peak size of storm flow (Church & Eaton 2001).

In coastal zones of western North America, largest peak flow events occur during periods when relatively warm rain falls into a snow-laden forest canopy. Such storms are called 'rain-on-snow events'. The main incremental effect is caused by advection of warm, moist air passing through the forest. This sustains rapid melting of snow by using the energy released during the condensation process (Church & Eaton 2001). About six times more energy is released during condensation of 1 g of water than during its freezing. This accelerated snowmelt does not occur to the same degree in clear-cuts, and annual peak flows may be reduced by clearing. Forests may, therefore, regulate snowmelt rates, and the distribution and scale of clear-cuts may determine how peak flows will respond. It may be possible to desynchronize snowmelt with clear-cuts, in a basin, and regulate peak flows (Hetherington 1987).

In cold, dry regions, snowfall on trees does not lead to the same peak flow conditions as in coastal areas. It may be removed from the trees by wind or sublimation, reducing the total snow burden and the subsequent runoff (Hetherington 1987).

Soil disturbance and sediment input

Forest harvesting can influence sediment transfer (mobilization, transport and deposition) by changing the hydrological conditions in the basin. It may also influence sediment movement by making more of it available for transfer by exposing slope soils, altering slope stability, damaging stream banks and depositing debris in stream channels and gullies (Church & Eaton 2001).

Falling and yarding in both up-slope and streamside areas influence soil stability, sediment mobilization and input to streams in several ways. Up-slope falling and yarding cause soil exposure, compaction and gouging. Surface erosion potential in cleared areas is directly related to the amount of bare, compacted soil that is exposed to rainfall and runoff, and is added to that caused by roads (Chamberlin *et al.* 1991). Soil disturbance may be as low as 2% in unlogged areas, and as much as 16% in cleared and yarded areas. As much as 80% of the disturbance that occurs in cleared and yarded areas can be caused by gouging and deposition (Smith & Wass 1982). Such disturbance causes soil compaction, water rerouting and increased stream sedimentation. It may also cause reduced root and soil strength, and increased risk of slope failure.

Falling and yarding in the riparian zone may cause sedimentation in a stream through different mechanisms to those that occur after up-slope logging. In Carnation Creek, this activity caused loading of small debris, erosion and collapse of stream banks, and release of gravel, sand and fine sediment. This resulted in deposition of sediment, streambed aggradation and channel widening in the immediate vicinity, and deposition of finer material in the lower reaches and estuary. The latter reduced egg-to-fry survival of young salmon (Hartman & Scrivener 1990).

The different sources, input routes, timing and particle size of sediment from streamside and up-slope falling and yarding produce an array of biological effects on invertebrate faunas and fish (Table 13.1, and Everest *et al.* 1987).

Effects on input of woody debris

The important role of large wood in stream channels is multifaceted, and the implications of its loss are indicated in Table 13.1 and in Chapter 3. The primary reason for some repeat coverage is to emphasize that up-slope and streamside cutting may have different effects on the entry routes, availability, timing of input and stability of woody debris.

In mountainous watersheds, wood is deposited by debris avalanches at the junctions with larger channels (Maser & Sedell 1994). These episodic events, often centuries apart in unlogged watersheds, form complex hot spots for biological activity and fish production.

Non-episodic input of large wood from the riparian zone is caused by tree blow-down, erosion of root support, disease or insect infestation (Maser & Sedell 1994). Streamside cutting stops such input processes. In addition, it may affect the duration of time that large woody material remains in the channel by initiating a sequence of disturbance events:

- weakening of stream banks
- addition of small, unstable woody material
- reduction of large wood volume and stability
- widening and aggradation of the channel.

Effects on input of leaf litter

The removal of riparian forest affects the composition and timing of leaf litter input. As it breaks down, this material provides the foundation of heterotrophic production in a stream ecosystem. It provides food directly for insects and also forms the substrate upon which bacterial layers develop to provide further insect food. In addition, nutrient release from decomposing litter contributes to production of stream algae (Fig. 13.4). Forestry-related changes in the timing and type of litter input from different species of vegetation may alter the relative importance of heterotrophic and autotrophic production, timing of production, and species composition of invertebrate faunas.

Effects on water chemistry

The flux of nutrients and water across the boundaries of an ecosystem is critical for the function and maintenance of the natural system. Biological structure and diversity,

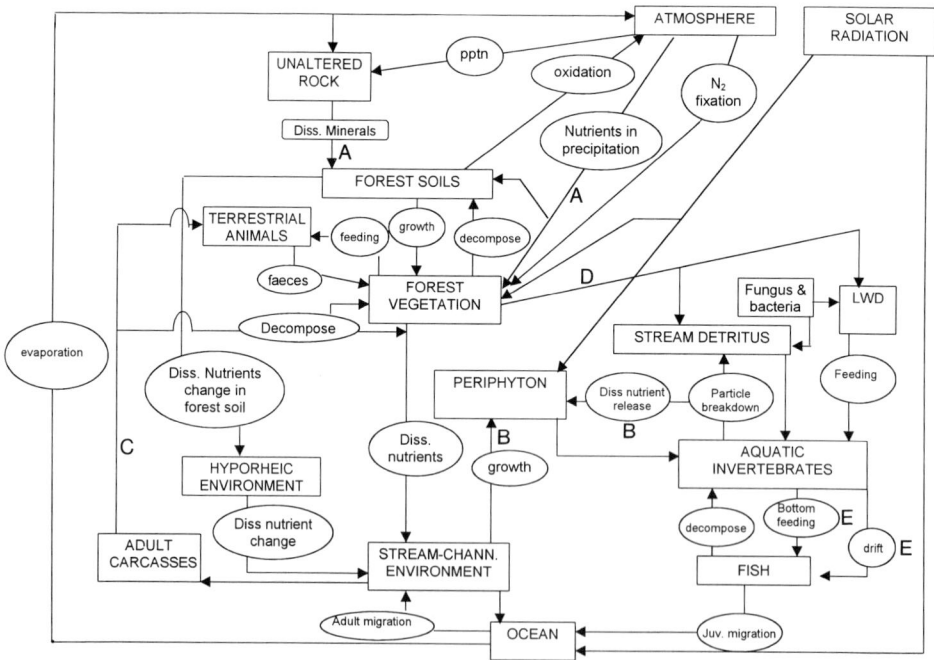

Fig. 13.4 Schematic diagram indicating nutrient pathways, autotrophic processes, heterotrophic processes and delivery of marine-derived nutrients to the forest–stream ecosystem. Nutrients in precipitation and throughfall, and from weathering of rock. (A, B) Nutrient release from litter processing. (B) Enhanced periphyton production with riparian tree removal. (C) Delivery of marine-derived nutrients to stream ecosystem affected by over-fishing and habitat-related population change. (D) Loss of riparian trees causing change in stream detritus input and reduction of large wood input affecting heterotrophic production and substrate for insect production. (E) Delivery of invertebrate food organisms to fish affected by balance in autotrophic and heterotrophic production.

geological conditions, climate and season affect this flux of water and movement of chemicals within a system (Likens *et al.* 1977). Nutrients, within a stream ecosystem, originate from unaltered rock, atmospheric inputs and decomposition of forest materials and fish carcasses. Changing the amounts and routing of nutrient inputs has the potential to alter the relative importance of production processes in the stream environments (Fig. 13.4), and to affect fish production (Table 13.1).

Conditions in a clear-cut forest and associated nutrient fluxes change on both long and short timescales. Bormann & Likens (1979) proposed a sequence of responses, following clear-cutting, that extend for over a century and, ultimately, move the system back to pre-disturbance conditions. Gregory *et al.* (1987) suggested a pattern of nutrient change that extends over 100 years following logging.

Chemical conditions along the hydrological pathway from upland to the stream, and within it, are altered by both deforestation (Likens *et al.* 1970; Bormann & Likens 1979; Scrivener 1982; Dahlgren 1998), and afforestation (Hornung & Reynolds 1995). Results of studies on nutrient concentration patterns following clear-cutting are not consistent if comparisons are made among results from Scrivener (1982) for Carnation Creek, Bormann & Likens (1979) for Hubbard Brook, and Dahlgren (1998) for Caspar Creek. Concentrations of most ions increase following clear-cutting; however, the longevity and intensity of responses vary depending upon geography, climate and drainage conditions (Likens *et al.* 1970; Dahlgren & Driscoll 1994). Post-logging nutrient pulses may be short-lived, occur during winter freshets, and provide little improvement for insect and fish production (Scrivener 1982).

Radiant energy and water temperature

The removal of riparian forest has the potential to increase the total radiant energy spectrum reaching the stream surface. Changes in two major components, long wave radiation and ultraviolet, may have significant but different effects on the stream ecosystem.

Forest canopy screens out 95% of the total available light that may reach a stream in dense coastal forests, and up to 43% in moderate canopy densities (Stockner & Shortreed 1976). Loss of such screening has effects on a stream ecosystem that vary according to the size of the stream and the riparian forest characteristics. The removal of dense riparian forest cover from a first or second order stream will have greater impact than the removal of sparse riparian forest from a seventh order stream. The light-energy effects of streamside clear-cutting differ with geographic locations and vary directly with amount of stream bank canopy removed and inversely with stream size (Beschta *et al.* 1987).

Water temperature in streams is one of the most important factors in controlling aquatic processes. Increase in radiant energy entering the stream is considered to be the major cause of temperature change following logging (Brown 1969, 1980). The most severe effects of riparian zone forest loss, for salmonids, occur in the southern portions of their ranges where populations may be constrained by temperature conditions even when there are no logging activities (Beschta *et al.* 1987). Summer maximum stream temperatures were increased up to 12.5°C in Oregon (Moring 1975), and up to

10.5°C in eastern North America (Beschta *et al.* 1987). Elevated stream temperature during summer was directly detrimental to fish in Alsea River tributaries in Oregon (Moring 1975).

Effects of logging were different in Carnation Creek. Stream temperature during summer increased only ~ 2.5°C, following logging in this British Columbian stream. However, such modest increases had indirect, but significant effects upon coho salmon by altering the timing of life history events. Modest increases in water temperature during winter decreased the time required for egg incubation, shortened the time to fry emergence, and permitted a longer growing season for the young fish. Modest temperature increases during summer increased growth rates, and decreased the age at which young coho salmon moved to sea (Holtby 1988; Holtby & Scrivener 1989; Hartman & Scrivener 1990). Temperature increases during spring caused earlier movement to the ocean, and reduced marine survival.

Stream temperatures usually decrease, during winter, after logging, but results are not consistent. They were from 0.5°C lower to 0.9°C higher following logging beside streams in Pennsylvania (Beschta *et al.* 1987), and about 1.0°C higher in a coastal British Columbian stream (Holtby 1988).

A summary of studies in six areas in eastern North America and five from the Pacific Northwest of North America showed that changes in water temperatures affect other conditions (Beschta *et al.* 1987). Alteration of thermal conditions in a stream may affect oxygen concentration and utilization, metabolism and development of stream organisms (Beschta *et al.*1987), fish swimming performance, metabolic efficiency and growth. Depending upon the degree of change, it may cause long-term stress and disease-related mortality or short-term lethal effects (Brett 1995).

Because forest removal affects both light and temperature conditions in a stream, it is difficult to separate the effects of these simultaneous changes on production. Primary influence on algal production has been related to light condition in some studies and to nutrient or temperature conditions in others (Chapter 3). Stream temperature increases led to higher growth rates in some insect species (Cummins *et al.* 1973) and accelerated development and emergence of some. In some instances, temperature increases led to decreased production and diversity of stream insect faunas (Beschta *et al.* 1987).

Ultraviolet (UV) light penetrating a stream may affect the biota in it. Canopy removal may influence production positively by increasing light availability in some spectral components, but it may influence it negatively by increasing intensity of others. UV light can reduce photosynthesis and growth of benthic diatoms (Bothwell *et al.* 1994; Watkins *et al.* 2001) and zooplankton (Williamson *et al.* 2001). Bothwell *et al.* (1994) demonstrated that while UV radiation (wavelengths 320–400 nm) reduced diatom growth, it concurrently inhibited chironomid numbers more than it reduced diatom growth. Grazing was decreased and there was an increase in algal volume in spite of reduced growth under UV light.

The manner in which UV light may affect a section of stream within a clear-cut is modified by several factors. UV light attenuation is controlled by dissolved organic carbon (Williamson *et al.* 2001). Its effect within a clear-cut may decrease along the stream if there is an increase in dissolved organic carbon. Canopy removal and increased UV may produce differential effects on periphyton production and insect grazing that are confounding.

Post-logging silviculture treatments

Silvicultural methods and reasons for use of various cutting patterns, herbicide use and fertilizer application are explained in Chapter 11. The effects of silviculture have been separated from other forestry activities in this chapter because they occur at different stages of forestry operations and include some different impacts.

Slash burning and mechanical site preparation (scarification)

Slash burning, once carried out extensively in Pacific Northwest parts of North America, has declined. Mechanical site preparation has increased (Anonymous 2000b). However, burning of slash or the forest as a whole still occurs on a large scale in other parts of the world (Cochrane *et al.* 1999; Goldammer 1999; Nepstad *et al.* 1999; Wuethrich 2000). Regardless of the reason for burning, it has important potential effects on nutrient fluxes in the drainage areas involved.

Following scarification or slash burning, as well as falling and yarding, there is usually a short period of increase in the flux of dissolved material through the stream system (Fredriksen 1971; Brown *et al.* 1973; Moring 1975; Scrivener 1982; Feller & Kimmins 1984). It is followed by a decline, over a period of up to 8 years, to pre-disturbance, or lower, levels (Feller & Kimmins 1984).

Concentration patterns of nitrate and phosphate are of particular interest because they are key nutrients in autotrophic production and are important constituents for litter processing in heterotrophic production (Fig. 13.4, and Chapter 3). The concentration of nitrate is characteristically high following clear-cutting, mechanical site preparation and slash burning (Fredriksen 1971; Moring 1975; Scrivener 1982; Feller & Kimmins 1984). However, in watersheds of coastal British Columbia, where phosphate concentrations are naturally low, increases may be below detectable levels (Scrivener 1982; Feller & Kimmins 1984).

Vegetation control – herbicides

Different types of forest management chemicals have been used for four silvicultural purposes: controlling non-commercial vegetation, retarding fire, controlling insects and accelerating commercial tree growth (Norris *et al.* 1991).

During the past few decades 11 different herbicides have been used (Norris *et al.* 1991). In British Columbia, sprays containing glyphosate have been used for the past two decades over about 25,000 ha/yr. However, since 1991 the use of stem girdling and injection have assumed prominence and risen to about 45,000 ha/year (Anonymous 2000c).

Compendiums of references and abstracts and toxicological assessments were prepared for glyphosate (Giesy *et al.* 2000; Sullivan & Sullivan 2000), and for triclopyr herbicide (Kidd & Mihajlovich 1998). The toxicity of the herbicides to fish varies widely (96-hour LC_{50} values from 0.035 to 310 mg/L) depending upon the herbicide and the species of fish studied (Norris *et al.* 1991).

Glyphosate was applied by helicopter in the Carnation Creek watershed for alder and shrub control (Reynolds 1989). The treatment increased water conductivity by 20% and nitrate concentration by 300% (Scrivener 1989). Caged fish exhibited some 'qualitative' evidence of stress and uncaged fish moved temporarily out of the tributary (Holtby & Baillie 1989).

Activities such as herbicide application, that reduce stream shading, shift trophic processes toward more periphyton production and potentially, a different invertebrate fauna. Vegetation control and consequent effects on stream temperature may alter the rates at which biotic processes occur in the stream. Changes in streamside vegetation, caused by herbicide application, potentially alter the species composition of leaf litter input, the rates of litter conditioning and utilization by insects (Table 13.1). Effects of herbicide application on vegetation, shading, water temperature and litter fall may be similar regardless of region, but modified by climate, stream condition and the nature of the vegetation involved.

Fertilizers

Fertilizers are used in many regions of the world, with N and P compounds applied at concentrations from 90 to 740 kg/ha (Shepard 1999). Urea is the most common chemical used, and most of the fertilizer that goes into streams enters directly within the first 48 hours. It can have a positive effect if stream productivity is nitrate-limited. Peak concentrations of nitrate and ammonia occur with freshets and are not usually toxic to fish (Norris *et al.* 1991; Shepard 1999).

Pesticides

There is a long history and an extensive literature on the efforts to control forest insects with different types of pesticides. Norris *et al.* (1991) provided a full review of the topic. The annual use of 71,000 kg of forest pesticides in USA is about 0.05% of the annual agricultural application.

References on the impact and fate of forest pesticides in aquatic systems may be found in Anonymous (2000b). A large number of insecticides, herbicides and fungicides were used until 1978 (Brown 1978; Norris *et al.* 1991). They had different modes of action, toxicity levels, effects on invertebrates, fish, birds and mammals, and fates within streams and the forest floor (Brown 1978; Garner & Harvey 1984; Wit 1985). Pesticide use has been partially replaced by use of viral controls and pheromone traps, methods that are less harmful to fish (Garner & Harvey 1984).

Complex impacts and process responses

Forestry impacts: single and cumulative

In any forestry operation, a combination of two or several of the effects listed in Table 13.1 can occur. There may be different combinations of impacts in different logging situations. Furthermore, cumulative impacts may result from forestry operations plus other land use activities (Reid 1993).

Two or more different forestry activities may cause the same type of impact. For example, additive sediment increases may occur from road construction and use, hill-slope erosion from falling and yarding, and bank erosion from streamside clearing (e.g. Cederholm *et al.* 1981). Although such sediment production may not occur in the same way, the effects on fish habitat may be the same.

Alternately, there can be more than one impact from one particular forestry activity (Table 13.1), e.g. riparian clearing may cause increase in sediment release, loss of woody debris and temperature change. Such impacts may occur either as a linked sequence or concurrently. The effects of impacts on fish may be immediate or delayed until some stage later in the life history.

Changes in temperature regime can be accompanied by a decrease in stream channel complexity and cover, a decline in the quality of the reproductive habitat and changes in trophic processes. In Carnation Creek the sequence of impacts caused by change in the temperature regime was accompanied by, and inter-connected with, changes in the hydrologic system, stream channel processes and trophic system (Hartman & Scrivener 1990). Such complex responses in a logged or partially logged forest–stream ecosystem are more difficult to understand and their consequences are more difficult to predict than those arising from a single activity.

Impacts may be evident at the site where the causal activity took place, or they may occur at another location in the system. Activities that occur in the upstream reaches of a watershed frequently affect fish habitat and fish, in a sequence of events, further downstream. The creation of landslides in the upper part of a stream system affects sediment deposition and scour, and channel structure and debris distribution in the lower reaches. The removal of riparian trees in the upper part of the drainage can affect water temperature, ion concentrations and litter-based trophic processes in the lower parts. The duration of such effects may not be similar and they may affect different subsystems indicated in Fig. 13.4. Managers must understand enough about stream processes to be able to anticipate the effects of forestry activities on the whole system, not just on the site where the activities take place.

Sequences of effects may follow each other through more than one life history stage with effects on survival showing up latently. An example of such effects involved changes in stream temperature in which incubation time of eggs was shortened, growth rates of fish in one or more seasons changed, and the timing of smolt transformation was accelerated (Holtby & Scrivener 1989). These changes produced survival effects that were both positive and negative, but the specific effect depended upon the life stage and species involved.

Discussion has focused on some of the more complex and inter-connected processes that may be caused by forestry activities. Results, briefly discussed above, indicate that logging in complex ecosystems, such as those found in the tropics, will produce complex sequences of impacts that may be difficult to interpret or predict. This poses the question, 'How much can information generated in one place be applied in another?'

Transfer of information and understanding

Predicting effects of single activities

The listing of the changes in quality of salmonid habitat and some impacts on growth and survival (Table 13.1) provides a framework for extrapolation of understanding from a single forestry activity. Many of the impacts listed in the second column of the table, '*Potential change in physical stream environment*', will occur predictably in other regions of the world, as described in Table 13.1. The severity of effect of each activity will be modified by the quality of forestry planning. If we consider the third column of influences, '*Potential change in quality of salmonid habitat*', predicting effects from each activity will be more problematic. Such effects will depend upon the life history characteristics of fish, the biology of their food organisms and the severity of forestry treatments. These may vary greatly from one region to another. Attempts to predict effects on fish (the fourth column), '*Potential consequences for salmonid growth and survival*', from each activity may be yet more difficult because growth and survival are influenced by many factors in a system. The complexity of influences, in Table 13.1, increases from left to right across the table and with this increase, extrapolation becomes more difficult. In this regard, Hartman & Hicks (1992) attempted to summarize the transferability of information, e.g. that summarized in Table 13.1, from North America to New Zealand. Using a subjective 'response similarity' rating, the average transferability decreased from 2.1 to 1.6 to 0.9 from column two to three to four, respectively.

Thus far, consideration of predictability of effects has been limited to one activity and one physical change resulting from it; however, a forestry operation will include several activities.

Predicting effects from multiple activities

It will be more difficult to predict effects of the combined array of logging activities because responses will reflect the cumulative effects of all changes in the physical environment, the consequent changes of the quality of habitat, and the biotic responses to them. Because of this, transferability of information or predicting effects of multiple activities from one region to another will be considered in a different context.

Naiman *et al.* (1992) proposed a comprehensive approach to understanding the array of stream types that included criteria for ecologically 'healthy' watersheds. They proposed a hierarchy of controlling components and factors within them that influenced others within the ecosystem. The hierarchy also provides a framework to help explain what ecosystem-level of forestry impact information may be extrapolated most meaningfully from one region to another.

The hierarchy among controlling factors was established on the basis of the spheres of influence of five components that had decreasing influence on the level of control within the system (Naiman *et al.* 1992):

(1)   basin geomorphology
(2)   hydrologic pattern
(3)   water quality
(4)   riparian forest characteristics
(5)   habitat characteristics.

The predominant control is from basin geomorphology and hydrology downward although there may be some feedback from lower components, e.g. (4) and (5). Because of the increasing complexity of influences from (1) to (5), it may be harder to predict forestry effects on habitat characteristics and fish than on geomorphic and hydrologic responses. Although it is difficult to use scientific results in one region to predict forestry effects in another, there are some generalizations that should be kept in mind by scientists and managers alike.

Generalizations

Some generalizations that have emerged from research in North America and should be kept in mind by anyone interpreting research results and then trying to extrapolate them to another system are:

- Forestry activities have effects on fish that may vary depending upon species and life history stage.
- Effects appear and are manifested on different timescales.
- Elevation, aspect and basin configuration will have important effects on the nature of impacts even for stream systems in the same geographic region.
- Effects from single forestry activities may be simple. However, overall impacts will most likely occur from combinations of activities, and as such their effects on fish will be complex and difficult to predict.
- Effects of short-term climate patterns and fishing may confound those arising from forestry activities.
- There is a critical need to have first-hand, site-specific knowledge of any system under consideration. It is necessary to understand the processes that occur in them in order to interpret forestry effects.

Many of the basic physical effects of forestry activities on watershed conditions will be the same regardless of region. However, the way these are reflected among populations of fishes will not necessarily be the same. The tolerances of the various species will be different to those of salmonids. The responses of populations will be different depending upon whether they are in the middle or at the fringe of their ranges.

Information and ideas provided herein should provide both directly applicable information and guidance for planning studies and carrying out management elsewhere. Forestry impacts can be understood and planned for. Three elements are essential:

- research on local watersheds
- site-specific information on areas to be logged, and
- experience on the part of forestry and fish habitat managers.

However, these important elements are of limited value if they are not applied within a broader context of sound governance, where there is a genuine commitment to sustainable management of forests and associated fish resources.

Acknowledgements

I am very grateful for the help from G. Miller and G. Pattern of the DFO Library at the Pacific Biological Station, Nanaimo. The following people helped me in many ways: Connie Bresnahan, Mason Bryant, Jeff Cederholm, Jim Hall, Ward Hartman, Barbara Knight (Weyerhaeuser), David Mayhood, Charles Scrivener, and Ann Thompson. My grateful thanks to them.

References

Abe, K. & Ziemer, R.R. (1991) Effect of tree roots on shallow-seated landslides. In: *Proceedings of the IUFRO Technical Session on Geomorphic Hazards in Managed Forests, General Technical Report PSW-130,* (tech. coordination R.M. Rice), pp. 11–20. USDA Pacific Southwest Research Station.
Anonymous (1993) *Draft Supplemental Environmental Impact Statement on Management of Habitat for Late-Successional and Old-Growth Forest Related Species within the Range of the Northern Spotted Owl.* United States Department of Agriculture, Forest Service.
Anonymous (2000a) *Land Areas of the National Forest System.* United States Department of Agriculture, Forest Service, FS-383.
Anonymous (2000b) *Just the Facts: A Review of Silvicultural and other Forestry Statistics.* British Columbia Ministry of Forests.
Anonymous (2000c) *Pesticide Use and Vegetation Management: Vol. 1: Impacts of Pesticides on Wildlife, Wildlife Habitat and Vegetation Diversity.* Silviculture Practices Branch, BC Ministry of Forests.
Anonymous (2001) *Environmental Trends in British Columbia 2000.* British Columbia Ministry of Environment, Lands and Parks.
Beschta, R.L., Bilby, R.E., Brown, G.W., Holtby, L.B. & Hofstra, T.D. (1987) Stream temperature and aquatic habitat: fisheries and forestry interactions. In: *Streamside Management: Forestry and Fishery Interactions,* (eds E.O. Salo & T.W. Cundy), pp. 191–232. Contribution No. 57. College of Forest Resources, University of Washington, University of Washington Institute of Forest Resources, Seattle, WA.

Bormann, F. & Likens, G.E. (1979) *Pattern and Process in a Forested Ecosystem.* Springer-Verlag, New York.

Bosch, J.M. & Hewlett, J.D. (1981) A review of catchment experiments to determine the effect of vegetation changes on water yield and evapotranspiration. *Journal of Hydrology*, 55, 3–23.

Bothwell, M.L., Sherbot, D.M. & Pollock, C.M. (1994) Ecosystem response to solar ultraviolet-B radiation: influence of trophic-level interactions. *Science*, 265, 97–100.

Bowling, L.C. & Lettenmaier, D.P. (2000) The effects of forest roads and harvest on catchment hydrology in a mountainous maritime environment. In: *Land Use and Watersheds: Human Influence on Hydrology and Geomorphology in Urban and Forest Areas*, (eds M.S. Wigmosta & S.J. Burgess), pp. 145–64. American Geophysical Union.

Brett, J.R. (1995) Energetics. In: *Physiological Ecology of Pacific Salmon*, (eds C. Groot, L. Margolis & W.C. Clarke), pp. 3–68. UBC Press, Vancouver.

Brown, A.W. (1978) *Ecology of Pesticides.* John Wiley & Sons, New York.

Brown, G.W. (1969) Predicting temperatures of small streams. *Water Resources Research*, 5, 68–75.

Brown, G.W. (1980) *Forestry and Water Quality.* Oregon State University Bookstores, Corvallis, OR.

Brown, G.W., Gahler, E.R. & Marston, R.B. (1973) Nutrient losses after clear-cut logging and slash burning in the Oregon Coast Range. *Water Resources Research*, 9, 1450–3.

Bryant, M.D. & Everest, F.H. (1998) Management and condition of watersheds in southeast Alaska: the persistence of anadromous salmon. *Northwest Science*, 72, 249–67.

Canadian Council of Forest Ministers (2001) *Compendium of Canadian Forestry Statistics.* <http://nfdp.ccfm.org>

Cederholm, C.J. & Reid, L.M. (1987) Impact of forest management on coho salmon (*Oncorhynchus kisutch*) populations of the Clearwater River, Washington: a project summary. In: *Streamside Management: Forestry and Fishery Interaction*, (eds E.O. Salo & T.W. Cundy), pp. 373–98. Contribution No. 57. College of Forest Resources, University of Washington, University of Washington Institute of Forest Resources, Seattle, WA.

Cederholm, C.J., Reid, L.M. & Salo, E.O. (1981) Cumulative effects of logging and road sediment on salmonid populations in the Clearwater River, Jefferson County, Washington. In: *Proceedings from the Conference: Salmon-spawning Gravel: A Renewable Resource in the Pacific Northwest?* pp. 38–74. State of Washington Water Research Center Report 39. Washington State University, Pullman, WA.

Cederholm, C.J., Johnson, D.H., Bilby, R.E., *et al.* (2000) *Pacific Salmon and Wildlife: Ecological Contexts, Relationships, and Implications for Management.* Special Edition Technical Report. Washington Department of Natural Resources, Olympia, WA.

Chamberlin, T.W., Harr, R.D. & Everest, F.H. (1991) Timber harvesting, silviculture, and watershed processes. In: *Influences of Forest and Rangeland Management on Salmonid Fishes and their Habitats*, (ed. W.R. Meehan), pp. 181–205. American Fisheries Society Special Publication 19, Bethesda, MD.

Cheng, J.D., Black, T.A. & Willington, R.P. (1975) The generation of storm flow from small forested watersheds in the Coast mountains of southwestern British Columbia. *International Association of Scientific Hydrology, Publication* 117, 542–51.

Church, M. & Eaton, B. (2001) Hydrological effects of forest harvest in the Pacific Northwest. *Technical Report to Central Coast Land Resource Management Planning Table.* University of British Columbia, Vancouver.

Cochrane, M.A., Alencar, A., Schulze, M.D., *et al.* (1999) Positive feedbacks in the fire dynamic of closed canopy tropical forests. *Science*, 284, 1832–5.

Cummins, K.W., Petersen, R.C., Howard, F.O., Wuycheck, J.C. & Holt, V.I. (1973) The utilization of leaf litter by stream detritivores. *Ecology*, 54, 336–45.

Dahlgren, R.A. (1998) Effects of forest harvest on stream-water quality and nitrogen cycling in the Caspar Creek watershed. *General Technical Report GSW-GTR-168-Web*. USDA Forest Service, Pacific Southwest Research Station.

Dahlgren, R.A. & Driscoll, C.T. (1994) The effects of whole-tree clear-cutting on soil processes at Hubbard Brook experimental forest, New Hampshire, USA. *Plant and Soil*, **158**, 239–62.

Duncan, S.H. & Ward, J.W. (1985) The influence of watershed geology and forest roads on the composition of salmon spawning gravel. *Northwest Science*, **59**, 204–12.

Everest, F.H., Beschta, R.L., Scrivener, J.C., Koski, KV., Sedell, J.R. & Cederholm, C.J. (1987) Fine sediment and salmonid production: a paradox. In: *Streamside Management: Forestry and Fishery Interactions*, (eds E.O. Salo & T.W. Cundy), pp. 98–142. Contribution No. 57. College of Forest Resources, University of Washington, University of Washington Institute of Forest Resources, Seattle, WA.

Feller, M.C. & Kimmins, J.P. (1984) Effects of clearcutting and slash burning on streamwater chemistry and watershed nutrient budgets in southwestern British Columbia. *Water Resources Research*, **20**, 29–40.

Fredriksen, R.L. (1971) Comparative chemical water quality – natural and disturbed streams following logging and slash burning. In: *Forest Land Uses and Stream Environment*, (eds J.T. Krygier & J.D. Hall), pp. 125–137. Oregon State University.

Furniss, M.J., Roelofs, T.D. & Yee, C.S. (1991) Road construction and maintenance. In: *Influences of Forest and Rangeland Management on Salmonid Fishes and their Habitats*, (ed. W.R. Meehan), pp. 297–323. American Fisheries Society Special Publication 19, Bethesda, MD.

Garner, W.Y & Harvey, J. (1984) *Chemicals and Biological Controls in Forestry*. American Chemical Society, Washington, DC.

Giesy, J.P., Dobson, S. & Soloman, K.R. (2000) Ecotoxicological risk assessment for Roundup herbicide. *Reviews of Environmental Contaminant Toxicology*, **167**, 35–120.

Goldammer, J.G. (1999) Forests on fire. *Science*, **284**, 1782–3.

Gregory, S.V., Lamberti, G.A., Erman, D.C., Koski, KV., Murphy, M.L. & Sedell, J.R. (1987) Influence of forest practices on production. In: *Streamside Management: Forestry and Fishery Interactions*, (eds E.O. Salo & T.W. Cundy), pp. 233–55. Contribution No. 57. College of Forest Resources, University of Washington, Seattle, WA, and University of Washington, Institute of Forest Resources.

Gregory, S.V., Swanson, F.J., McKee, W.A. & Cummins, K.W. (1991) An ecosystem perspective of riparian zones: focus on links between land and water. *BioScience*, **41**, 540–51.

Gucinski, H., Furniss, M.J., Ziemer, R.R. & Brookes, M.H. (2001) Forest roads: a synthesis of scientific information. *General Technical Report PNW-GTR-509*. USDA Forest Service.

Hakenkamp, C.C & Palmer, M.A. (2000) The ecology of hyporheic meiofauna. In: *Streams and Ground Waters*, (eds J.B. Jones & P.J. Mulholland), pp. 307–36. Academic Press, San Diego, CA.

Harr, R.D. (1983) Potential for augmenting water yield through forest practices in western Washington and western Oregon. *Water Resources Research*, **19**, 383–92.

Harr, R.D., Harper, W.C., Krygier, J.T. & Hsieh, F.S. (1975) Changes in storm hydrographs after road building and clear-cutting in the Oregon Coast Range. *Water Resources Research*, **11**, 436–44.

Harr, R.D., Fredricksen, R.L. & Rothacher, J. (1979) Changes in streamflow following timber harvest in southwestern Oregon. *Research Paper PNW-249*. United States Department of Agriculture, Forest Service, Pacific Northwest Forest and Range Experiment Station.

Hartman, G.F. & Hicks, B.J. (1992) Application of fisheries/forestry research. In: *Proceedings of the Fisheries/Forestry Conference, 27–28 February 1990, Christchurch* (eds J.W. Hayes & S.F. Davis). *New Zealand Freshwater Fisheries Report* **136**, 78–86.

Hartman, G.F. & Scrivener, J.C. (1990) Impacts of forestry practices on a coastal stream eco-system, Carnation Creek, British Columbia. *Canadian Bulletin of Fisheries and Aquatic Sciences*, **223**.

Harvey, J.W. & Wagner, B.J. (2000) Quantifying hydrologic interactions between streams and their subsurface hyporheic zones. In: *Streams and Ground Waters*, (eds J.B. Jones & P.J. Mulholland), pp. 3–44. Academic Press, San Diego, CA.

Hayduke & Associates Ltd (1997) *Boreal Forest Natural Region Linear Disturbance Inventory*. Alberta Environmental Protection Natural Resources Services, Recreation & Protected Areas Division.

Hetherington, E.D. (1982) A first look at logging effects on the hydrologic regime of Carnation Creek experimental watershed. In: *Proceedings of the Carnation Creek Workshop, A 10 year Review*, (ed. G.F. Hartman), pp. 45–63. Pacific Biological Station, Nanaimo, BC.

Hetherington, E.D. (1987) The importance of forests in the hydrological regime. In: *Canadian Aquatic Resources*, (eds M.C. Healey & R.R. Wallace), pp. 179–211. *Canadian Bulletin of Fisheries and Aquatic Sciences,* **215**.

Hetherington, E.D. (1998) Watershed hydrology. In: *Carnation Creek and Queen Charlotte Islands Fish/Forestry Workshop: Applying 20 Years of Coast Research to Management Solutions*, (eds D.L. Hogan, P.J. Tschaplinski & S. Chatwin), pp. 33–40. Land Management Handbook, No. 41. British Columbia Ministry of Forests, Research Branch, Victoria, BC.

Hewlett, J.D. (1982) *Principles of Forest Hydrology*. University of Georgia Press, Athens, GA.

Hicks, B.J., Beschta, R.L. & Harr, R.D. (1991a) Long-term changes in streamflow following logging in western Oregon and associated fisheries implications. *Water Resources Bulletin*, **27**, 217–26.

Hicks, B.J., Hall, J.D., Bisson, P.A. & Sedell, J.R. (1991b) Responses of salmonids to habitat changes. In: *Influences of Forest and Rangeland Management on Salmonid Fishes and their Habitats*, (ed. W.R. Meehan), pp. 483–518. American Fisheries Society Special Publication 19.

Hogan, D.L. (1986) Channel morphology of unlogged, logged, and debris torrented streams in the Queen Charlotte Islands. *Land Management Report Number 49*. BC Ministry of Forests and Land.

Hogan, D.L., Tschaplinski, P.J. & Chatwin, S. (1998a) *Carnation Creek and Queen Charlotte Islands Fish/Forestry Workshop: Applying 20 Years of Coastal Research to Management Solutions*. Land Management Handbook 41. British Columbia Ministry of Forests Research Program.

Hogan, D.L., Bird, S.A. & Rice, S. (1998b) Stream channel morphology and recovery process. In: *Carnation Creek and Queen Charlotte Islands Fish/Forestry Workshop: Applying 20 Years of Coastal Research to Management Solutions*, (eds D.L. Hogan, P.J. Tschaplinski & S. Chatwin), pp. 77–96. Land Management Handbook 41. British Columbia Ministry of Forests Research Program.

Holtby, L.B. (1988) Effects of logging on stream temperatures in Carnation Creek, British Columbia, and associated impacts on coho salmon (*Oncorhynchus kisutch*). *Canadian Journal of Fisheries and Aquatic Sciences,* **45**, 502–15.

Holtby, L.B. & Baillie, S.J. (1989) Effects of the herbicide Roundup® (Glyphosate) on coho salmon fingerlings in an over-sprayed tributary of Carnation Creek, British Columbia. In: *Proceedings of the Carnation Creek Herbicide Workshop*, (ed. P.E. Reynolds), pp. 273–85. Forest Pest Management Institute, Forestry Canada, Sault Ste Marie, Ontario.

Holtby, L.B. & Scrivener, J.C. (1989) Observed and simulated effects of climatic variability, clear-cut logging, and fishing on the numbers of chum salmon (*Oncorhynchus keta*) and coho salmon (*O. kisutch*) returning to Carnation Creek, British Columbia. In: *Proceedings of the National Workshop on Effects of Habitat Alteration on Salmonid Stocks*, (eds C.D. Levings, L.B. Holtby & M.A. Henderson), pp. 62–81. *Canadian Special Publication Fisheries and Aquatic Sciences* **105**.

Hornung, M. & Reynolds, B. (1995) The effects of natural and anthropogenic environmental changes on ecosystem processes at the catchment scale. *TREE*, **10**, 443–9.

Jones, J.A. & Grant, G.E. (1996) Peak flow responses to clear-cutting and roads in small and large basins, western Cascades, Oregon. *Water Resources Research*, **32**, 959–74.

Jones, J.A. & Grant, G.E. (2001) Comment on 'Peak flow responses to clear-cutting and roads in small and large basins, western Cascades, Oregon: A second opinion' by R.B. Thomas and W.F. Megahan. *Water Resources Research*, **37**, 175–8.

Jones, J.B. & Mulholland, P.J. (2000) *Streams and Ground Waters*. Academic Press, San Diego, CA.

Keppeler, E.T. & Ziemer, R.R. (1990) Logging effects on streamflow: water yield and summer low flows at Caspar Creek in northwestern California. *Water Resources Research*, **26**, 1669–79.

Kidd, F.A. & Mihajlovich, M. (1998) *Triclopyr Herbicide: A Bibliography of Technical References of Non-target Field and Laboratory Results*. MarCon International Inc. and Incremental Forest Technologies Ltd.

Krag, R.K., Sauder, E.K. & Wellburn, G.V. (1986) A forest engineering analysis of the landslides in logged areas on the Queen Charlotte Islands, British Columbia. *Land Management Report Number 43*. BC Ministry of Forests and Lands.

Larsen, D.N. (1989) *1986 Washington Timber Harvest Report*. Washington State Department of Natural Resources; Analysis and Planning.

Likens, G.E., Bormann, F.H., Johnson, N.M., Fisher, D.W. & Pierce, R.S. (1970) Effects of forest cutting and herbicide treatment on nutrient budgets in the Hubbard Brook watershed-ecosystem. *Ecological Monographs*, **40**, 23–47.

Likens, G.E., Bormann, F.H., Pierce, R.S., Eaton, J.S. & Johnson, N.M. (1977) *Bio-Geo-Chemistry of a Forested Ecosystem*. Springer-Verlag, New York.

Maser, C. & Sedell, J.R. (1994) *From the Forest to the Sea*. St Lucie Press, Delray Beach, FL.

Moring, J.R. (1975) The Alsea watershed study: effects of logging on the aquatic resources of three headwater streams of the Alsea River, Oregon. Part II – Changes in environmental conditions. *Fishery Research Report Number 9*. Oregon Department of Fish and Wildlife, Corvallis, OR.

Murphy, M.L. (1995) Forestry impacts on freshwater habitat of anadromous salmonids in the Pacific Northwest and Alaska – requirements for protection and restoration. BC Ministry of Forests and Lands. *Decision Analysis Series No. 7*. NOAA National Marine Fisheries Service, Alaska Fisheries Science Center, Auke Bay Laboratory, NOAA Coastal Ocean Program.

Naiman, R.J., Beechie, T.J., Benda, L.E., *et al.* (1992) Fundamental elements of ecologically healthy watersheds in the Pacific Northwest Coastal Ecoregion. In: *Watershed Management: Balancing Sustainability and Environmental Change*, (ed. R.J. Naiman), pp. 127–88. Springer-Verlag, New York.

Nepstad, D.C., Veríssimo, A., Alencar, A., *et al.* (1999) Large-scale impoverishment of Amazonian forests by logging and fire. *Nature*, **398**, 505–8.

Newcombe, C.P. & Jensen, J.O. (1996) Channel suspended sediment and fisheries: a synthesis for quantitative assessment of risk and impact. *North American Journal of Fisheries Management*, **16**, 693–727.

Norris, L.A., Lorz, H.W. & Gregory, S.V. (1991) Forest chemicals. In: *Influences of Forest and Rangeland Management on Salmonid Fishes and their Habitats*, (ed. W.R. Meehan), pp. 207–96. American Fisheries Society Special Publication Number 19.

Reid, L.M. (1993) Research and cumulative watershed effects. *General Technical Report PSW-GTR-141*. USDA Forest Service, Pacific Southwest Research Station.

Reid, L.M. & Dunne, T. (1984) Sediment production from forest road surfaces. *Water Resources Research*, **20**, 1753–61.

Reynolds, P.E. (1989) *Proceedings of the Carnation Creek Herbicide Workshop*. Forest Pest Management Institute, Forestry Canada, Sault Ste Marie, Ontario.

Rice, R.M. (1991) Proceedings of the IUFRO technical session on geomorphic hazards in managed forests. *General Technical Report PSW-GTR-130.* USDA Forest Service.

Rollerson, T.P. (1992) Relationships between landslide attributes and landslide frequencies after logging: Skidegate Plateau, Queen Charlotte Islands. *Land Management Report Number 76.* BC Ministry of Forests.

Rood, K.M. (1984) An aerial photograph inventory of the frequency and yield of mass wasting on the Queen Charlotte Islands, British Columbia. *Land Management Report Number 34.* BC Ministry of Forests.

Rothacher, J. (1970) Increases in water yield following clear-cut logging in the Pacific Northwest. *Water Resources Research,* 6, 653–8.

Rothacher, J. (1973) Does harvest in west slope Douglas-fir increase peak flow in small forest streams? *Research Paper PNW-163.* USDA Forest Service.

Scrivener, J.C. (1982) Logging impacts on the concentration pattern of dissolved ions in Carnation Creek, British Columbia. In: *Proceedings of the Carnation Creek Workshop: A Ten-year Review,* (ed. G.F. Hartman), pp. 64–80. Pacific Biological Station, Nanaimo, BC.

Scrivener, J.C. (1989) Comparative changes in concentration of dissolved ions in the stream following logging, slash burning, and herbicide application (glyphosate) at Carnation Creek, British Columbia. In: *Proceedings of the Carnation Creek Herbicide Workshop,* (ed. P.E. Reynolds), pp. 197–211. Forest Pest Management Institute, Forestry Canada, Sault Ste Marie, Canada.

Shepard, J. (1999) Water quality effects of forest fertilization. *Technical Bulletin No. 782.* National Council for Air and Stream Improvement (NCASI).

Smith, R.B. & Wass, E.F. (1982) Ground-surface characteristics and vegetative cover associated with logging and broadcast burning. In: *Proceedings of the Carnation Creek Workshop: A Ten-year Review,* (ed. G.F. Hartman), pp. 100–9. Pacific Biological Station, Nanaimo, BC.

Stockner, J.G & Shortreed, K.R. (1976) Autotrophic production in Carnation Creek, a coastal rainforest stream on Vancouver Island, British Columbia. *Journal of the Fisheries Research Board of Canada,* 33, 1553–63.

Sullivan, D.S. & Sullivan, T.P. (2000) *Non-Target Impacts of the Herbicide Glyphosate: A Compendium of References and Abstracts,* 5th edn. Information Report, Applied Mammal Research Institute, Summerland, BC.

Sullivan, K., Lisle, T.E., Dolloff, C.A., Grant, G.E. & Reid, L.M. (1987) Stream channels: the link between forests and fishes. In: *Streamside Management: Forestry and Fishery Interactions,* (eds E.O. Salo & T.W. Cundy), pp. 39–97. Contribution No. 57. College of Forest Resources, Institute of Forest Resources, University of Washington, Seattle, WA.

Swanston, D.N. (1971) Principal mass movement processes influenced by logging, road building, and fire. In: *Forest Land Uses and Stream Environment,* (eds J.T. Krygier & J.D. Hall), pp. 29–39. Oregon State University.

Swanston, D.N. (1991) Natural processes. In: *Influences of Forest and Rangeland Management on Salmonid Fishes and their Habitats,* (ed. W.R. Meehan), pp. 139–179. American Fisheries Society Special Publication 19, Bethesda, MD.

Thomas, R.B. & Megahan, W.F. (1998) Peak flow responses to clear-cutting and roads in small and large basins, western Cascades, Oregon: a second opinion. *Water Resources Research,* 34, 3393–403.

Thomas, R.B. & Megahan, W.F. (2001) Reply (to Jones, J.A. and Grant, G.E. 2001. *op. cit.*). *Water Resources Research,* 37, 181–3.

Toews, D.A & Brownlee, M.J. (1981) *A Handbook for Fish Habitat Protection on Forest Lands in British Columbia.* Land Use Unit, Habitat Protection Division, Department of Fisheries and Oceans, Canada.

Toews, D.A. & Moore, M.K. (1982) The effects of three streamside logging treatments on organic debris and channel morphology of Carnation Creek. In: *Proceedings of the Carnation Creek Workshop, A 10 Year Review,* (ed. G. Hartman), pp. 129–53. Pacific Biological Station, Nanaimo, BC.

Watkins, E.M., Schindler, D.W., Turner, M.A. & Findlay, D. (2001) Effects of solar ultraviolet radiation on epilithic metabolism, and nutrient and community composition in a clear-water boreal lake. *Canadian Journal of Fisheries and Aquatic Sciences*, **58**, 2059–70.

Wemple, B. (1994) *Hydrologic integration of forest roads with stream networks in two basins, western Cascades, Oregon*. MS thesis, Oregon State University.

Whipkey, R.Z. (1967) Storm runoff from forested catchments by subsurface routes. *International Association of Scientific Hydrology, Publication* **85**, 773–9.

Wigmosta, M.S. & Perkins, W.A. (2000) Simulating the effects of forest roads on watershed hydrology. In: *Land Use and Watersheds: Human Influence on Hydrology and Geomorphology in Urban and Forest Areas*, (eds M.S. Wigmosta & S.J. Burgess), pp. 127–43. American Geophysical Union.

Williamson, C.E., Olson, O.G., Lott, S.E., Walker, N.D., Engstrom, D.R. & Hargreaves, B.R. (2001) Ultraviolet radiation and zooplankton community structure following deglaciation in Glacier Bay, Alaska. *Ecology*, **82**, 1748–60.

Wit, J.M. (1985) *Chemistry, Biochemistry, and Toxicology of Pesticides*. Cooperative Extension Services, Oregon State University, Corvallis, OR.

Wright, K.A., Sendek, K.H., Rice, R.M. & Thomas, R.B. (1990) Logging effects on streamflow: storm runoff at Caspar Creek in Northwestern California. *Water Resources Research*, **26**, 1557–667.

Wuethrich, B. (2000) Combined insults spell trouble for rainforests. *Science*, **289**, 35–7.

Ziemer, R.R. (1981) Stream flow response to road building and partial cutting in small streams of northern California. *Water Resources Research*, **17**, 907–17.

Chapter 14
Effects of forestry on the limnology and fishes of lakes

T.G. NORTHCOTE, M. RASK AND J. LEGGETT

Introduction

Effects of the wide range of different forestry practices on lakes are said to be 'grossly understudied' in comparison with such research on streams (Laird & Cumming 2001; Laird *et al.* 2001). This is especially surprising when for decades in much of the limnological literature flowing waters were largely ignored. Wetzel (2001) only included rivers in the third edition of his comprehensive text on limnology. In part, the lack of attention given to effects of forestry on lakes may have arisen because many of those on streams were so obvious and those on moderate to large sized lakes far less so, being slower to respond or buffered to some degree by dilution in the larger water volumes involved. Furthermore, the response of lakes to at least some of the effects of deforestation may not resemble those of streams, despite the suggestion to the contrary by Carignan & Steedman (2000).

Lake effects of forestry are now being addressed with ever-increasing interest, especially in the last few years, with lakes being included in the sequence of recent workshops, conferences and journal supplements on the subject (Kelso & Wooley 1996; Brewin & Monita 1998; Carignan & Steedman 2000), in addition to an outburst of separate journal publications. A preliminary literature review of forest harvesting impacts on lake ecosystems (Miller *et al.* 1997) draws heavily on inferences from stream effects, and specifically for lakes includes only that by Rask *et al.* (1993) for a lake in Finland, the study by France & Peters (1995) of four lakes in Ontario, Canada, a report on water chemistry changes of a lake in coastal British Columbia (Nordin 1996), and another in central British Columbia by Parkinson *et al.* (1977). Diagrams in the review by Miller *et al.* (1997) showing logging influences on lakes via stream inputs and those at or near lake edges provide useful overviews of theoretical effects and response directions. Magnitude of effects from three physical and three chemical factors on watershed inputs via streams and near-lakeshore logging both resolve into an increase in the magnitude of impact with area harvested and a decrease in magnitude of impact with increasing lake volume.

We consider the effects of forestry on lakes with a very broad perspective – spatially including those arising from their airsheds, their watersheds, and internally (shorelines, surfaces, midwaters, benthic waters and bottoms); temporally (seasonally, annually, over decades and even centuries); and functionally (giving attention to effects on the many physical, chemical and biological processes involved in lakes – see Chapter

4). Effects on temperate and boreal lakes are now being especially well examined, but some tropical lake information is included.

Given the paucity of long-term studies on effects of forestry practices on lakes and their fish communities, it may be possible to gain insights from wildfire defor-estation, realizing that their mechanisms are different, as well as their spatial extent and frequency of occurrence. Therefore impacts on water quality and other effects may differ (Carignan & Steedman 2000). In addition, lake paleolimnological studies provide a means of examining effects over a timescale not available for streams and permit long-term coverage of deforestation effects over decadal to millennial periods; see for example Warwick (1980), Blais *et al.* (1998), Paterson *et al.* (1998), Keller *et al.* (2001). Effects on fish from forestry have long been recognized in North America (Knight 1907a, 1907b; Elson *et al.* 1972) and this form of anthropogenic disturbance may cause perturbations requiring much more recovery time than those arising from natural disturbances (Rapport & Whitford 1999). Nevertheless there are few detailed studies that document effects of forestry practices on fish in lakes. Therefore we also try to include forestry–lake–fish interactions by combining existing knowledge on the hydrological, physical-chemical and biological responses of lakes to forestry with ecological features of common fish species.

Airshed-borne effects

Although airborne acidic fallout is well known for its effects on lakes (see Chapter 4), the potential of nutrient input coming to lakes from forestry activities via rainfall or particulate ash dryfall has rarely been evaluated. That coming from sawmill waste burners and slash burning was considered in the nutrient income to Kootenay Lake, Canada and was judged to be small in relation to water-borne drainage basin sources – agricultural fertilizers, domestic waste outfalls, non-forestry industrial operation losses (Northcote 1973).

Decreased airborne litterfall through riparian deforestation can affect temperate lake metabolism (France & Peters 1995). A large portion of major ionic solutes in rain of the Central Amazon Basin appeared to originate within the rainforest. It was considered unlikely that this would remain unchanged with extensive cutting and dis-turbance of that large forest (Lesack & Melack 1991). Nevertheless subsequent solute mass balance studies (Lesack & Melack 1996) revealed that the chemical composition of rain seemed to be derived from a constant source of oceanic and rainforest aerosols, not significantly influenced by local or regional biomass burning (Williams *et al.* 1997). Andreae & Merlet (2001) provide a global review of emissions from biomass burning. In a comparison of rainfall versus throughfall in an archipelago of the Negro River, Brazil, Filoso *et al.* (1999) concluded that the latter was an important nutrient source to a floodplain lake.

Watershed-borne effects

In forested catchments the influence of forests on both the hydrology and the physical and chemical characteristics of lakes is of essential importance. Forests effectively prevent or restrict hydrological extremes such as floods, erosion and eutrophication of surface waters (Kenttämies & Saukkonen 1996). Forestry practices that affect the hydrology and material balances of the soil, such as cutting, drainage, fertilization and soil scarification will, correspondingly, affect the water bodies.

Watershed-borne effects resulting from forestry on the limnology of lakes and their fish communities may be large or small, and long-term or short-term. Effects depend on the relative size differences of lakes and their watersheds, and lake positioning within them (see Chapter 4). One of the more obvious and immediate effects of forestry activities coming from road construction and clear-cutting in lake watersheds is that of increased fine sediment loading. Such was the case for Tumuch Lake (only 7.3% of watershed logged) in central British Columbia, Canada, where the high density of logging roads on moderately to highly erodable silty loam soils doubled its sediment loading compared with that of an unlogged reference lake (Parkinson *et al.* 1977). A recent study in north coastal to central British Columbia found that the largest increases in sediment yield to lakes occurred in heavily forest-harvested and roaded catchments (Schiefer *et al.* 2001). Clarke *et al.* (1998) showed that there was a significant increase in fine sediment input to the Copper Lake system of Newfoundland, Canada, mainly from logging road crossings of its tributaries. In northwestern Montana, USA, timber harvest over the last 150 years increased mass sedimentation rate in three lakes up to 14 times the rate at which it occurred before European settlement (Spencer & Schelske 1998). Logging road construction was the largest contributor, although natural land disturbances (floods, wildfires) also were involved in the increase.

Land deforestation starting mainly with agricultural clearing in the large Lake Simcoe system, Ontario, Canada (lake and catchment areas 722 and 3572 km^2, respectively) brought about a sharply accelerating erosional load of total phosphorus to the lake from a natural background level of just over 30 metric tonnes per year to over triple that by the mid-1950s, decreasing since then to about 65 by the 1990s (Evans *et al.* 1996). In the Great Lake (Canada) basin, marshes in forested landscapes were clear and nutrient-poor while those in deforested agricultural areas were turbid and nutrient-rich (Crosbie & Chow-Fraser 1999). Recently much attention has been given to forestry-related effects on water quality of Canadian boreal lakes. In most cases forest harvesting caused significant nutrient increases in both N and P as well as other components such as dissolved organic carbon (Carignan *et al.* 2000; Devito *et al.* 2000; Lamontagne *et al.* 2000; Prepas *et al.* 2001). Effects were usually directly proportional to the land area harvested divided by lake volume or area. Lakes with the largest catchment and thus the shortest residence time had the strongest nutrient response to forest harvesting. There was no evidence that 20-, 100- or 200-m wide buffer strips influenced lake response (Devito *et al.* 2000). Rask *et al.* (1993) also found higher total phosphorus and nitrogen concentrations in a Finnish lake after catchment deforestation. With water renewal times of about 10 years, three small (30 ha) boreal forest lakes showed only minor water quality changes by the third year after moderate to intensive watershed logging (Steedman 2000). Contrary to the hypothesis of Evans *et al.* (2000), mean daily phosphorus export from logged and unlogged subcatchments of a small boreal lake were not much different.

Specific in-lake effects

Physical-chemical

Specific impacts of forestry treatments on lakes are usually seen more easily in headwaters where relatively larger proportions of catchments of small lakes may be affected by forestry impacts. Furthermore, the catchments of large lakes are often subject to other human activities such as those of industry or agriculture, that may mask the effects of forestry. However, in sparsely populated areas of northeastern Finland, eutrophication and decreased oxygen concentrations in the hypolimnia of some large lakes (nearly 100 km^2 in surface area) have been attributed to large-scale forestry ditching and clear-fellings (Sandman *et al.* 1994).

Deforestation around lake margins can affect several internal physical-chemical characteristics. In a small Ontario lake, thermocline deepening probably occurred as a result of wildfire deforestation of its catchment (Schindler *et al.* 1990). Logging increased mixing depth through increased wind action on a small Finnish lake (Rask *et al.* 1993). Removal of protective riparian trees through wind blow-down and wildfires deepened the thermoclines of several lakes in Ontario. Among 63 other Ontario lakes, those with riparian trees removed by clear-cutting or wildfires had thermocline depths 2 m deeper per unit fetch than those with intact lakeshore tree cover (France 1997). In northwestern Ontario, shoreline logging did not significantly increase average littoral temperatures in two boreal lakes, but increases were higher in the littoral area of clear-cut shorelines than they were along undisturbed shorelines or shorelines with buffer strips (Steedman *et al.* 2001). Clear-cut logging around three small (30–40 ha) Ontario lakes produced only a minor increase (5% or less) in their mid-lake wind speeds, but no change in spring or autumn circulation efficiency or stratification duration (Steedman & Kushneriuk 2000).

Log storage in semi-enclosed embayments in large lakes can produce severe depression of dissolved oxygen (down to 2 mg/L or less). This depression is the result of elevated biochemical oxygen demand from gelatinous leachate coatings that develop on freshly cut floating logs (Levy *et al.* 1989). Hypolimnetic oxygen levels in Lake Simcoe, Ontario, decreased from about 4.5 mg/L in 1975 to 2.0 mg/L by 1993, driven in large part by high land erosional and increasing point source loading of total phosphorus (Evans *et al.* 1996). Cutting, scarification, fertilization and drainage of lake catchments increase the leaching of phosphorus and nitrogen. This may result in eutrophication of lakes if over 10–30% of catchments are affected (Kenttämies & Saukkonen 1996). Leaching of nutrients has been shown to continue for 5–15 years after such forestry activities (Rask *et al.* 1993; Holopainen & Huttunen 1998).

Primary production

Forestry activities have the potential to affect lake waters in several key ways that could alter primary production coming from phytoplanktonic, macrophytic and periphytic sources (see definitions in Chapter 4).

Comparison of 7 logged, 9 burnt and 16 reference watershed lakes treatments in Quebec, Canada (Planas *et al.* 2000) showed that phytoplanktonic algal biomass increased up to three-fold in the perturbed lakes. The extent of algal increase was proportional to the area of the catchment perturbed divided by the lake surface area. Mean primary production of phytoplankton in a Finnish lake was three- to six-fold higher after watershed clear-cutting (Kenttämies 1988; Rask *et al.* 1993), along with an increase in phytoplanktonic biomass from 10 beforehand to >20 mg m^{-3} afterwards. In that lake and also in another study on forestry impacts on small Finnish lakes, slight blooms of blue-green algae were detected (Rask *et al.* 1998). A comparison with the total number of heavy algal blooms in lakes in Finland during 1982–1994 has revealed that 7% (90 cases out of 1300) probably originated from forestry practices. Blue-green algae dominated the blooms, indicating a good supply of phosphates. An important green alga causing mass blooms was *Gonyostomum semen*, which favours brown-water lakes with a fairly high phosphorus content (Kenttämies *et al.* 1995).

Long-term assessments of forest harvesting effects on some elements of lake limnology can be made by paleolimnological studies. In a small (8.1 ha) bog lake in Michigan, USA, Scully *et al.* (2000) followed pre- and post-clear-cut logging (ca. 1870–1890) changes in mixing regime, along with those on autotrophic and phytoplanktonic communities. The deepwater populations of anaerobic photosynthetic bacteria were eliminated for over 100 years by forest harvesting, metalimnetic chrysophyte phytoplankters were reduced in abundance, but epilimnetic phytoplanktonic groups (blue-greens, chlorophytes and cryptophytes) were unaffected. In a study on coastal and central interior lakes in British Columbia, small but significant changes in diatom abundance and species composition were found between clear-cut and reference lakes (Laird & Cumming 2001; Laird *et al.* 2001). Similar observations have been made in several Finnish lakes of different size (Simola 1983; Turkia *et al.* 1998).

Zooplankton

In log storage embayments of Babine Lake, British Columbia, crustacean zooplankton abundance, an important food source for juvenile sockeye salmon, decreased significantly compared with that in a nearby reference site (Power & Northcote 1991). In the first year after catchment logging of nine boreal Quebec lakes, the small rotifer-sized zooplankton (100–200 µm) increased 59% in biomass, whereas the larger calanoid copepods decreased 43% in biomass, but by the second and third year after logging there were no significant differences for any of the limnoplanktonic size fractions among logged and unlogged watershed lakes (Patoine *et al.* 2000). In a Swedish study, forestry-related draining of peatland in the catchment of two lakes did not affect the species composition of zooplankton. However, the cladoceran biomass in the lake closer to the drained peatland decreased clearly, apparently due to increased amounts of inedible particles in the water. In the other lake downstream and further from the drained area, zooplankton biomasses slightly increased (Bergquist *et al.* 1984). In small lakes in northeastern Finland, the amounts of crustacean zooplankton increased after catchment forestry treatments (Rask *et al.* 1998). These variable results suggest that the responses of zooplankton to catchment forestry are not consistent. Results may depend on both the suitability of the forestry-induced changes to the food requirements of the zooplankters and on the predatory effects of macroinvertebrates and fish at the same time.

Benthic invertebrates

Log storage areas in lakes and those affected by lakeshore sawmills may show marked changes in their benthic invertebrate communities. In the southern interior of British Columbia, benthic invertebrate abundance near a log storage and sawmill area at Monte Lake was significantly higher (mainly oligochaetes and gammarids) than that in an unaffected area, but benthic diversity was very much higher in the latter (T.G. Northcote, unpublished data). Benthic insect larvae (mainly dipterans, ephemeropterans and trichopterans) were greatly reduced in a Babine Lake embayment used for log storage compared with those in reference sites (Levy *et al.* 1989). Sawmill bark and debris accumulation was thought to have had a negative effect on Nipigon Bay littoral substrate (Kelso & Cullis 1996). In contrast to the gradual accumulation and periodic shifting of small and fine woody debris, found in the littoral zones of lakes around and near log storage and sawmill operations, the coarse woody debris associated with these areas may remain in place for centuries. This coarse and more stable material may provide important habitat for floral and faunal communities, including macrobenthic invertebrates and fish (Guyette & Cole 1999).

The effects of catchment forestry on zoobenthos of lakes may be due to changes in the bottom substrate after increased sedimentation, or to changes in oxygen conditions. In a Swedish lake, the biomasses of bottom fauna increased, mainly due to increased amounts of chaoborid and chironomid larvae (Simonsson 1987). Similar observations were also made in a Finnish lake study (Rask *et al.* 1998); there apparently a result of increased availability of allochthonous particulate organic matter for food of chironomid larvae and small mussels. However, in running waters, negative effects of forestry-induced inorganic sedimentation on zoobenthos are well known (Vuori *et al.* 1998).

Fish

Impacts on fish in lakes from catchment forestry effects can occur if essential changes in their habitat take place. These may include changes in riparian vegetation, littoral zone woody debris structure, sediment inputs, light conditions, oxygen conditions and supply of primary nutrients. The effects may be direct, for example degradation of spawning grounds, or indirect as a result of increased biological productivity. Short-term suspended sediment pulses can have negative effects on feeding and territorial behaviour of juvenile salmonids in streams (Berg & Northcote 1985), and could have similar effects in lake littoral zones.

In certain coastal areas of Finland and Sweden, forestry-related ditching of sulphide- and aluminium-containing soils has resulted in fish kills due to toxic effects of acidity and aluminium (Ramberg 1976; Hudd *et al.* 1996). Usually such observations have been made in running or Baltic coastal waters, but at least one lake case is known from Sweden (Vallín 1953). In a fish status survey on effects of acidic precipitation in the mid-1980s (Rask & Tuunainen 1990), observations were made on decreases and disappearances of roach (*Rutilus rutilus*), an acid-sensitive cyprinid. In some of the lakes, roach populations obviously were affected also by ditching of catchment peatlands for forestry purposes, which then resulted in high loads of humic acidic substances to the lakes. Increased iron concentrations due to forestry practices together with low pH may be toxic to fish (Peuranen *et al.* 1994).

The fish community structure of lakes may differ from region to region and be differentially sensitive to catchment forestry. For example, in a comparison of fish communities in small lakes of Finland and Wisconsin, USA (Tonn *et al.* 1990), it was shown that fish assemblages of Finnish lakes are composed of few species with wide environmental tolerances and little inter-specific regulation. In contrast, in Wisconsin lakes, the number of species tends to be higher and the role of inter-specific interactions, especially predation, is more important in structuring the fish communities. In such lakes, forestry impacts might be more severely expressed on fish.

The Tumuch Lake study (Parkinson *et al.* 1977) in central British Columbia found few measurable effects on fish growth rates, but only a small part (7.3%) of its catchment was logged. Furthermore, at higher trophic levels such as that occupied by fish, other biological interactions often control population dynamics, making it difficult to fully assess effects of catchment perturbations (Rask *et al.* 1993). Lakeshore spawning and incubation sites of brook trout (*Salvelinus fontinalis*) were affected by groundwater flow coming from at least 20 m into the land catchment. Nevertheless a 90-m wide shoreline buffer zone from logging gave protection to only 9–55% of the required recharge area (Curry & Devito 1998). Through an experimental simulation of reduced lake trout reproductive habitat, the results of Gunn & Sein (2000) suggest that this species can tolerate considerable spawning habitat loss but that the great increase in angler access via logging roads may result in their excessive sport-fishing exploitation. In a Quebec study of the effects of deforestation (9 lake watersheds logged, 9 burnt, 20 reference), St-Onge & Magnan (2000) could find no significant differences in the abundant fishes (eight different species) among the three lake groups, although proportions of small yellow perch and white sucker were significantly lower in populations of impacted lakes (logged and wildfire pooled).

In small headwaters lakes in Fennoscania the fish fauna is usually very simple, and usually consists of one to five species. The most common species, European perch (*Perca fluviatilis*), northern pike (*Esox lucius*) and roach are all capable of completing their life cycle in shallow waters and varying environmental conditions. Therefore, no direct effects of catchment forestry on them were recorded even though in studied sites in Finland (Rask *et al.* 1993, 1998, 2000) and Sweden (Bergquist *et al.* 1984) changes were recorded in water quality, primary productivity, zooplankton and zoobenthos.

Indirectly, increased densities of planktonic crustaceans were seen as an increased proportion in the diet of small (<15 cm) perch in two of the four target lakes of a Finnish study (Rask *et al.* 2000). Further, a slight increase in the growth of 0+ perch was recorded in the same lakes after the forestry treatments. These changes were interpreted to be consequences of increased biological productivity of the lakes. The competitive relationships of perch and roach are well studied (Persson 1983, 1986, but see also Horppila *et al.* 2000) and oligotrophic conditions are more favourable for perch whereas roach tend to be superior in eutrophic lakes (Svärdson 1976). Thus one can suppose that forestry-induced eutrophication of lakes favours roach.

Most obvious direct effects of habitat changes on fish have taken place in some lakes inhabited by fish species such as the vendace (*Coregonus albula*) and burbot (*Lota lota*). Both are cold water species that need hard bottom substrates for spawning. In an 11-km² lake in eastern Finland, annual vendace catches used to be 5000–6000 kg in the early 1960s but are now only a few kilograms. In the time between, the catchment of the lake has been subject to intensive forestry including clear-felling, soil scarification, ploughing, ditching and fertilizing. There are almost no other sources of sediment loading to the lake, so the forestry-related increase in sedimentation of spawning grounds and worsening of oxygen conditions next to the bottom were given as the reasons for loss of the vendace population (Tossavainen 1997).

In Sweden, industrialization beginning in the twelfth century around its four largest lakes demanded, among other things, development of arable land and use of large forests so these human impacts could well have affected their fish stocks and diversity that early (Degerman *et al.* 2001).

Mercury cycling

Extensive logging of lake catchments may disrupt the normal cycling of mercury. Garcia & Carignan (1999) reported significantly higher methyl mercury in zooplankton in lakes with recently logged watersheds (135 ng g⁻¹ dry wt) than in lakes with undisturbed watersheds (112 ng g⁻¹ dry wt). Furthermore, northern pike had significantly higher mercury concentrations (3.4 µg g⁻¹) 1–2 years after watershed logging than did those in reference watershed lakes (1.9 µg g⁻¹) (Garcia & Carignan 2000).

The occurrence of high amounts of mercury in the forested catchments of remote areas, mostly due to air pollutants, and the enrichment of methyl mercury in aquatic ecosystems after forestry practices, also is a severe environmental problem in northern Europe, both for Sweden (Lindqvist *et al.* 1991) and Finland (Verta 1990). However, the mercury concentrations in muscle tissue of European perch showed a slightly decreasing trend after catchment forestry treatments (Rask *et al.* 2000). This was interpreted as being due to an overall decrease in mercury levels after reduced atmospheric deposition (Munthe *et al.* 1995). On the other hand, there are recent records indicating that despite the decreased mercury precipitation, levels in northern pike muscle tissue after catchment forestry practices are at similar levels to those in the 1980s and 1990s (Porvari *et al.* 2003).

Forestry chemicals in lakes

A wide range of chemicals may be applied to forested, deforested and reforested landscapes with an equally wide range in objectives and responses, some of them inadvertent. Norris *et al.* (1991) cover many of these, mainly with respect to streams. Other aspects are given in Chapters 11 and 12. Wood sugars may be leached from fresh logging slash around lakeshores. Such leachate could increase biochemical oxygen demand and consequentally decrease dissolved oxygen. It could also stimulate excessive periphytic algal growth. Toxic compounds may leach from fresh logs of some tree species such as the western red cedar (*Thuja plicata*). This leachate can affect aquatic organisms (Peters *et al.* 1976).

A forest insecticide, tebufenoxide, seems to have variable effects on zooplankton communities in lakeshore enclosures (Kreutzweiser & Thomas 1995; Kreutzweiser *et al.* 1998), the first study showing significant decreases in cladocerans and the later one not, even at three times the expected environmental concentration. The broad-spectrum biocide pentachlorophenol (PCP), used largely as a wood preservative, may find its way into lakes. Samis *et al.* (1993) show that 10-day exposure to PCP at a concentration of 173 µg/L significantly reduces bluegill (*Lepomis macrochirus*) growth by 75% and their food intake by 29%. Even at concentrations of 48 µg/L their growth rate was reduced by 29%. Similar growth depressant effects of PCP have been recorded for sockeye salmon (Webb & Brett 1973), rainbow trout (Hodson & Blunt 1981), several species of cyprinids (Ward & Parrish 1980; Holcombe *et al.* 1982; Borgmann & Ralph 1986), a species of cichlid (Kreuger *et al.* 1968) and a species of catfish (Belliyappa & Reddy 1986).

Effects of forestry on fishing: what is the public reaction?

A mail survey on recreational fisheries in Finland was directed to 20,000 people in 1991. The sample was random but regionally stratified in order to ensure sufficient participation from all parts of the country (Lappalainen *et al.* 1994). A list of 13 environmental changes was given and the people were asked to assess the degree of negative impacts of each change on their fishing waters. Forestry practices, especially ditching, were ranked second among the 13 environmental impacts by people that fished in lakes; 9% of them considered forest and peatland ditching a severe harm to their fishing waters. For agriculture, judged the most damaging factor, the figure was 14% (Lappalainen 1995). Regionally the results were logical: the negative experiences of ditching were reported most numerously in the regions where ditching had been most intensive. Also, other observations made by the people, like muddy odour or taste of fish, rapid fouling of fishing gear and increased brown colour of water were thought to be typical consequences of forestry activities.

Lakeshore management zones and forestry practices

The importance of lakeshores is becoming more recognized by foresters, fish biologists and fisheries managers. Lakeshore management areas have been defined in many ways but are recognized for their importance by most government agencies in North America and Europe. The objectives of lakeshore management areas are to prevent rutting, erosion compaction and damage to lakeshore soils associated with logging activity; to provide shade, litter fall, insects and a supply of organic debris to the lake; and to filter up-slope runoff and stabilize the lake shore. Furthermore, such areas provide habitat for forest-dwelling species associated with the lakeshore, reduce and restrict user access, and present a visual barrier that helps maintain quality of recreational experience from the lake as well as its shoreline. Forestry operations within a lakeshore management area can vary from no harvesting, to selective cutting or small clear-cut blocks, depending on the lake value and timber type surrounding the lake.

In British Columbia, Canada, as directed by the Forest Practices Code, a lake classification must be provided by the District Forest Manager to the harvesting licensee for all lakes over 5 ha in surface area before timber harvesting can occur. On large lakes over 1000 ha, different portions of the lakeshore may be assigned different classifications to reflect differences in lakeshore values. The classification is given to the lakeshore management zone which is part of the Lakeshore Management Area, comprised of riparian and lakeshore management zones (Fig. 14.1). The riparian reserve zone is measured from the high water mark of the lake or, if a lake is bordered by shrubs and other wetland vegetation, from the outer edge of the wetland vegetation. The distance of the riparian management zone is set by regulation at 10 m and allows for no harvesting except under special circumstances, which must be agreed upon by both the District Forest Manager and a designated environmental official. The distance of the lakeshore management zone can vary depending on the lake value but in most cases the default is 200 m. Lake classification guidebooks have been developed in British Columbia to address the harvesting of forests adjacent to lakes. In some Forest Districts, lake classification teams, involving biologists, foresters, planners, licensees, recreationalists, First Nations (descendants of those peoples that came from Siberia millennia ago) and public groups and users, are formed to assist the District Forest Manager in providing guidance for harvesting within the lakeshore management zone.

In Oregon, USA, there is no riparian management area required for lakes smaller than 0.2 ha that do not support fish. Larger lakes require a riparian management area of 30 m up-slope from the high water mark. Protection is given to live tree retention, soil and hydrologic functions, understory vegetation, and snag and down-wood retention.

The standards in the province of Alberta, Canada, do not permit disturbance of trees, or road and landing construction, within 100 m of the high water level for lakes even where there is little or no recreational, waterfowl or sport-fishing potential. On lakes exceeding 4 ha in area and with recreational, waterfowl or sport-fishing potential, there is to be no disturbance or removal of timber within 100 m of the high water level, nor is there to be any landing construction within 200 m of this level without written approval of the Forest Superintendent.

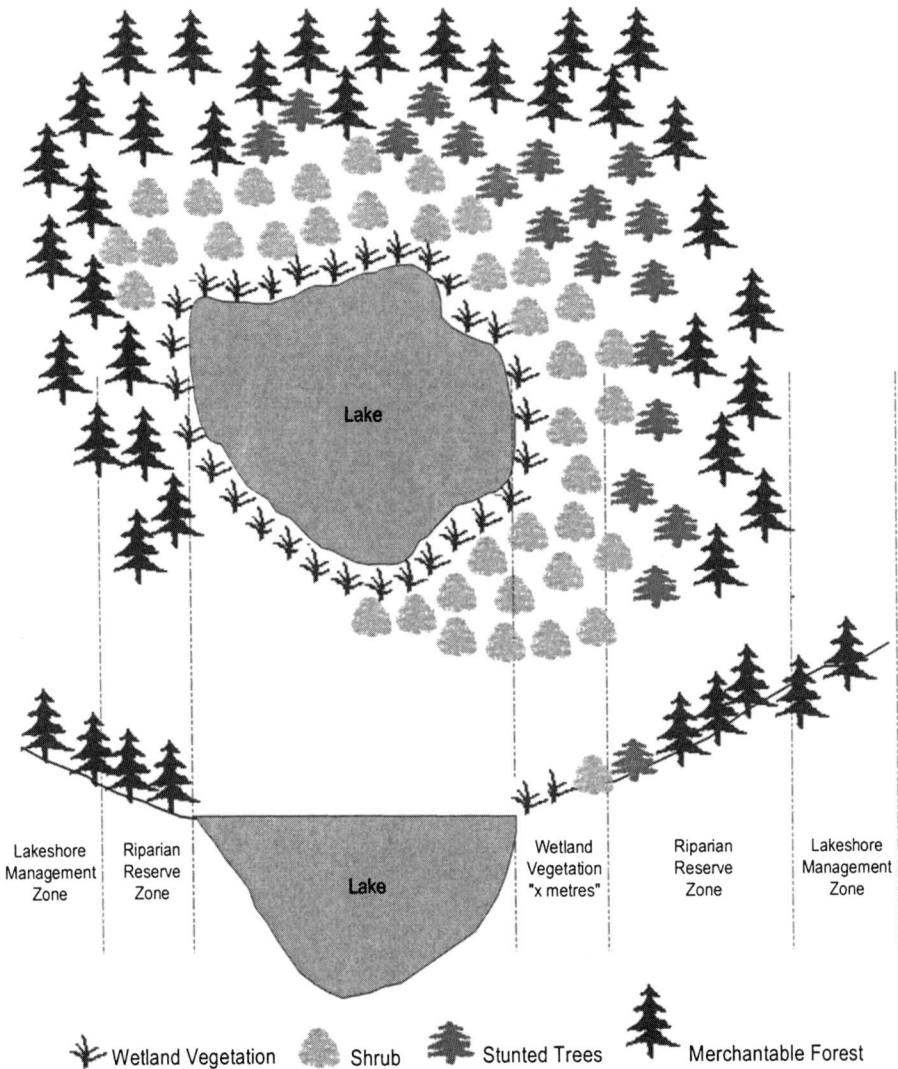

Fig. 14.1 Diagrammatic representation (not to scale) of a lakeshore management area in British Columbia, Canada.

In Nova Scotia, Canada, forestry operators must establish a special management zone of at least 20 m in width along the lakeshore. Where land adjoining the lake has a steep slope the width is to be increased 1 m for each additional 2% of slope above 20% up to a maximum width of 60 m. Forestry operators must ensure that the understory vegetation and non-commercial trees within 20 m of the lake edge are retained to the fullest extent possible, and no activity is to be conducted within 20 m that would result in sediment being deposited in the watercourse.

In Finland, the guidelines for good practice in forestry and water protection (Hän-ninen *et al.* 1998) give two goals: (1) to maintain the high quality of waters, and (2) to maintain the biodiversity of aquatic ecosystems. These goals can be achieved by a series of measures including: decreasing nutrient loads, preventing erosion and load of solid substances, preventing fertilizers and biocides from getting into waters, preventing oil discharges from forestry machinery, treating groundwater areas with special care, protecting small waters, taking key biotopes of aquatic systems into account and taking care of landscape aspects. Furthermore, the guidelines give examples of managing with buffer strips, constructing forest roads, preventing loads of solid substances from ditches with ditch breaks, sedimentation pools, filtering fields, sedimentation basins, and their different combinations.

Closing comment

Despite the paucity of previous long-term studies on forestry effects on lakes, it is evident that the situation is now rapidly changing with a strong series of works, especially in boreal North America and Fennoscania. Unfortunately there seem to be few coordinated or large-scale studies of forestry-driven effects on lakes in tropical regions. There, in many ways, the effects of forestry practices will be confounded with those coming from agricultural-driven deforestation and from hydroelectric impoundment.

Acknowledgements

We are grateful to Drs C. Araujo Lima and J.M. Melack for helpful information and references. Drs K. Kenttämies (Finland), D.J. McQueen, G.F. Hartman and J.C. Lyons (Canada), gave helpful advice.

References

Andreae, M.O. & Merlet, P. (2001) Emission of trace gases and aerosols from biomass burning. *Global Biogeochemical Cycles*, **15**, 1–13.

Belliyappa, G. & Reddy, S.R. (1986) Growth and conversion efficiency of the catfish *Heteropneustes fossilis* exposed to sublethal concentrations of sodium pentachlorophenate. *Polish Archives of Hydrobiology*, **33**, 121–8.

Berg, L. & Northcote, T.G. (1985) Changes in territoriality, gill flaring, and feeding behaviour in juvenile coho salmon (*Oncorhynchus kisutch*) following short-term pulses of suspended sediment. *Canadian Journal of Fisheries and Aquatic Sciences*, **42**, 1410–17.

Bergquist, B., Lundin, L. & Andersson, A. (1984) *Effects of Peatland Drainage on Hydrology and Hydrobiology. The Basin Siksjöbäcken.* University of Uppsala, Institute of Limnology. [In Swedish, English summary.]

Blais, J.M., France, R.L., Kimpe, L.E. & Cornett, R.J. (1998) Climatic changes in northwestern Ontario have had a greater effect on erosion and sediment accumulation than logging and fire: evidence from ^{210}Pb chronology in lake sediments. *Biogeochemistry*, **43**, 235–52.

Borgmann, U. & Ralph, K.M. (1986) Effects of cadmium, 2,4-dichlorophenol, and pentachlorophenol on feeding, growth, and particle size-conversion efficiency of white sucker (*Cato-*

stomus commersoni) larvae and young common shiners (*Notropis cornutus*). *Archives of Environmental Contamination and Toxicology*, **15**, 473–80.

Brewin, M.K. & Monita, D.M.A. (1998) *Forest-Fish Conference: Land Management Practices Affecting Aquatic Ecosystems*. Abstract. Natural Resources Canada, Canadian Forest Service, Edmonton, Alberta.

Carignan, R. & Steedman, R.J. (2000) Impacts of major watershed perturbations on aquatic ecosystems. *Canadian Journal of Fisheries and Aquatic Sciences*, **57** (Suppl. 2), 1–4.

Carignan, R., D'Arcy, P. & Lamontagne, S. (2000) Comparative impacts of fire and forest harvesting on water quality in Boreal Shield lakes. *Canadian Journal of Fisheries and Aquatic Sciences*, **57** (Suppl. 2), 105–17.

Clarke, K.D., Scruton, D.A. & McCarthy, J.H. (1998) The effect of logging and road construction on fine sediment yield in streams of the Copper Lake watershed, Newfoundland, Canada: initial observations. In: *Forest-Fish Conference: Land Management Practices Affecting Aquatic Ecosystems*, (eds M.K. Brewin & D.M.A. Monita), pp. 353–60. *Information Report NOR-X-356*. Canadian Forest Service, Edmonton, Alberta.

Crosbie, B. & Chow-Fraser, P. (1999) Percentage land use in the watershed determines the water quality and sediment quality of 22 marshes in the Great Lakes basin. *Canadian Journal of Fisheries and Aquatic Sciences*, **56**, 1781–91.

Curry, R.A. & Devito, K.J. (1998) Hydrologeology of brook trout (*Salvelinus fontinalis*) spawning and incubation habitats: implications for forestry and land use development. In: *Forest-Fish Conference: Land Management Practices Affecting Aquatic Ecosystems*, (eds M.K. Brewin & D.M.A. Monita), pp. 77–82. *Information Report NOR-X-356*. Canadian Forest Service, Edmonton, Alberta.

Degerman, E., Hammar, J., Nyberg, P. & Svärdson, G. (2001) Human impact on fish diversity in the four largest lakes of Sweden. *Ambio*, **30**, 522–8.

Devito, K.J., Creed, I.F., Rothwell, R.L. & Prepas, E.E. (2000) Landscape controls on phosphorus loading to boreal lakes: implications for the potential impacts of forest harvesting. *Canadian Journal of Fisheries and Aquatic Sciences*, **57** (Suppl. 2), 1977–84.

Elson, P.F., Saunders, J.W. & Zitko, V. (1972) Impact of forest-based industries on freshwater-dependent fish resources in New Brunswick. *Technical Report of the Fisheries Research Board of Canada*, **326**.

Evans, D.O., Nicholls, K.H., Allan, Y.C. & McMurtry, M.J. (1996) Historical land use, phosphorus loading, and loss of fish habitat in Lake Simcoe, Canada. *Canadian Journal of Fisheries and Aquatic Sciences*, **53** (Suppl. 1), 194–218.

Evans, J.E., Prepas, E.E., Devito, K.J. & Kotak, B.G. (2000) Phosphorus dynamics in shallow subsurface waters in an uncut and cut subcatchment of a lake on the Boreal Plain. *Canadian Journal of Fisheries and Aquatic Sciences*, **57** (Suppl. 2), 60–72.

Filoso, S., Williams, M.R. & Melack, J.M. (1999) Composition and deposition of throughfall in a flooded forest archipelago (Nero River, Brazil). *Biogeochemistry*, **45**, 169–95.

France, R. (1997) Land-water linkages: influences of riparian deforestation on lake thermocline depth and possible consequences for cold stenotherms. *Canadian Journal of Fisheries and Aquatic Sciences*, **54**, 1299–305.

France, R.L. & Peters, R. (1995) Predictive model of the effects on lake metabolism of decreased airborne litterfall through riparian deforestation. *Conservation Biology*, **9**, 1578–86.

Garcia, E. & Carignan, R. (1999) Impact of wildfire and clear-cutting in the boreal forest on methyl mercury in zooplankton. *Canadian Journal of Fisheries and Aquatic Sciences*, **56**, 339–46.

Garcia, E. & Carignan, R. (2000) Mercury concentrations in northern pike (*Esox lucius*) from boreal lakes with logged, burned, or undisturbed catchments. *Canadian Journal of Fisheries and Aquatic Sciences*, **57** (Suppl. 2), 129–35.

Gunn, J.M. & Sein, R. (2000) Effects of forestry roads on reproductive habitat and exploitation of lake trout (*Salvelinus namaycush*) in three experimental lakes. *Canadian Journal of Fisheries and Aquatic Sciences*, **57** (Suppl. 2), 97–104.

Guyette, R.P. & Cole, W.G. (1999) Age characteristics of coarse woody debris (*Pinus strobus*) in a lake littoral zone. *Canadian Journal of Fisheries and Aquatic Sciences*, **56**, 496–505.

Hänninen, E., Kärhä, S. & Salpakivi-Salomaa, P. (1998) *Forestry and Water Protection*. Metsäteho, Helsinki. [In Finnish.]

Hodson, P.V. & Blunt, B.R. (1981) Temperature-induced changes in pentachlorophenol chronic toxicity to early stages of rainbow trout. *Aquatic Toxicology*, **1**, 113–27.

Holcombe, G.W., Phipps, G.L. & Fiaudt, J.T. (1982) Effects of phenol, 2, 4-dimethylphenol, 2,4-dichlorophenol and pentachlorophenol on embryo, larvae and early-juvenile fathead minnows. *Archives of Environmental Contamination and Toxicology*, **11**, 73–8.

Holopainen, A.-L. & Huttunen, P. (1998) Impact of forestry practices on ecology of algal communities in small brooks in the Nurmes experimental area, Finland. *Boreal Environmental Research*, **3**, 63–73.

Horppila, J., Ruuhijärvi, J., Rask, M., Karppinen, C., Nyberg, K. & Olin, M. (2000) Seasonal changes in the food composition and relative abundance of perch and roach – a comparison between littoral and pelagic zones of a large lake. *Journal of Fish Biology*, **56**, 51–72.

Hudd, R., Kjellman, J. & Urho, L. (1996) The increase of coincidence in relative year-class strengths of coastal perch (*Perca fluviatilis* L.) stocks in the Baltic Sea. *Annals Zoologica Fennici*, **33**, 383–7.

Keller, W., Dixit, S.S. & Heneberry, J. (2001) Calcium declines in northeastern Ontario lakes. *Canadian Journal of Fisheries and Aquatic Sciences*, **58**, 2011–20.

Kelso, J.R.M. & Cullis, K.I. (1996) The linkage among ecosystem perturbations, remediation, and success of the Nipigon Bay fishery. *Canadian Journal of Fisheries and Aquatic Sciences*, **53** (Suppl. 1), 67–78.

Kelso, J.R.M. & Wooley, C. (1996) Introduction to the International Workshop on the Science and Management for Habitat Conservation and Restoration Strategies (HabCARES). *Canadian Journal of Fisheries and Aquatic Sciences*, **53** (Suppl. 1), 1–2.

Kenttämies, K. (1988) The effects of modern boreal forestry practices on waters. In: *Proceedings of the XX Nordic Hydrological Conference, NHP Report 44*, (ed. J. Kajander), pp. 142–62.

Kenttämies, K. & Saukkonen, S. (1996) *Forestry and Waters*. Summary of the joint project on the effects of forestry on waters and their abatement (METVE). Publications of the Finnish Ministry of Agriculture and Forestry 4/1996. [In Finnish, English summary.]

Kenttämies, K., Lepistö, L. & Vilhunnen, O. (1995) The impact of forestry on the harmful algal blooms in Finnish lakes. *Finnish Environment*, **2**, 229–39. [In Finnish.]

Knight, A.P. (1907a) A further report upon effects of sawdust on fish life. *Contributions to Canadian Biology*, **1902–5**, 37–54.

Knight, A.P. (1907b) Sawdust and fish life. Final Report. *Contributions to Canadian Biology*, **1902–5**, 111–20.

Kreutzweiser, D.P. & Thomas, D.R. (1995) Effects of a new moult-inducing insecticide, tebufenozide, on zooplankton communities in lake enclosures. *Ecotoxicology*, **4**, 307–28.

Kreutzweiser, D.P., Dunn, J.M., Thompson, D.G., Pollard, H.G. & Faber, M.J. (1998) Zooplankton responses to a novel forest insecticide, tebufenozide (RH-5992), in littoral lake enclosures. *Canadian Journal of Fisheries and Aquatic Sciences*, **55**, 639–48.

Krueger, H.M., Saddler, J.B., Chapman, G.A., Tinsley, I.J. & Lowry, R.R. (1968) Bioenergetics, exercise, and fatty acids of fish. *American Zoologist*, **8**, 119–29.

Laird, K. & Cumming, B. (2001) A regional paleolimnological assessment of the impact of clear-cutting on lakes from the central interior of British Columbia. *Canadian Journal of Fisheries and Aquatic Sciences*, **58**, 492–505.

Laird, K., Cumming, B. & Nordin, R. (2001) A regional paleolimnological assessment of the impact of clear-cutting on lakes from the west coast of Vancouver Island, British Columbia. *Canadian Journal of Fisheries and Aquatic Sciences*, **58**, 479–91.

Lamontagne, S., Carignan, R., D'Arcy, P., Prairie, Y.T. & Paré, D. (2000) Element export in runoff from eastern Canadian Boreal Shield drainage basins following forest harvesting and wildfires. *Canadian Journal of Fisheries and Aquatic Sciences*, 57 (Suppl. 2), 118–28.

Lappalainen, A. (1995) Observations of recreational fishermen and fish farmers on the forestry-induced effects on surface waters. *Finnish Environment*, 2, 329–34.

Lappalainen, A., Hildén, M. & Leinonen, K. (1994) Acidification and recreational fisheries in Finland: a mail survey of potential impacts. *Environmental Management*, 18, 831–40.

Lesack, L.F.W. & Melack, J.M. (1991) The deposition, composition, and potential sources of major ionic solutes in rain of the Central Amazon Basin. *Water Resources Research*, 27, 2953–77.

Lesack, L.F.W. & Melack, J.M. (1996) Mass balance of major solutes in a rainforest catchment in the Central Amazon: implications for nutrient budgets in tropical rainforests. *Biogeochemistry*, 32, 115–42.

Levy, D.A., Northcote, T.G., Hall, K.J. & Yesaki, I. (1989) Juvenile salmonid responses to log storage in littoral habitats of the Fraser River estuary and Babine Lake. In: *Proceedings of the National Workshop on Effects of Habitat Alteration on Salmonid Stocks*, (eds C.D. Levings, L.B. Holtby & M.A. Henderson). *Canadian Special Publication Fisheries and Aquatic Sciences*, 105, 82–91.

Lindqvist, O., Johansson, K., Aastrup, M. *et al.* (1991) Mercury in Swedish environment – recent research on causes, consequences and corrective methods. *Water Air Soil Pollution*, 55, 1–261.

Miller, L.B., McQueen, D.J. & Chapman, L. (1997) Impacts of forest harvesting on lake ecosystems: a preliminary literature review. *BC Environment Wildlife Bulletin* B-84, Victoria, Canada.

Munthe, J., Hultberg, H. & Iverfeldt, Å. (1995) Mechanisms of deposition of methyl mercury and mercury to coniferous forests. *Water Air Soil Pollution*, 80, 363–71.

Nordin, R. (1996) Watershed ion export and lake water chemistry changes in a British Columbia lake after watershed logging. Draft unpublished research report. BC Ministry of Environment, Lands and Parks, Victoria, Canada.

Norris, L.A., Lorz, H.W. & Gregory, S.V. (1991) Forest chemicals. In: *Influences of Forest and Rangeland Management on Salmonid Fishes and their Habitats*, (ed. W.R. Meehan), pp. 207–96. American Fisheries Society Special Publication 19, Bethesda, MD.

Northcote, T.G. (1973) Some impacts of man on Kootenay Lake and its salmonoids. *Great Lakes Fishery Commission Technical Report 25*, Ann Arbor, MI.

Parkinson, E.A., Slaney, P.A. & Halsey, T.G. (1977) Some effects of forest harvesting on a central interior lake in British Columbia. BC Ministry of Recreation and Conservation, Fisheries Management Publication No. 17.

Paterson, A.M., Cumming, B.F., Smol, J.P., Blais, J.M. & France, R.L. (1998) Assessment of the effects of logging, forest fires and drought on lakes in northwestern Ontario: a 30-year paleolimnological perspective. *Canadian Journal of Forest Research*, 28, 1546–56.

Patoine, A., Pinel-Alloul, B., Prepas, E.E. & Carignan, R. (2000) Do logging and forest fires influence zooplankton biomass in Canadian Boreal Shield lakes? *Canadian Journal of Fisheries and Aquatic Sciences*, 57 (Suppl. 2), 155–64.

Peters, G.B., Dawson, H.J., Hrutfiord, B.F. & Whitney, R.R. (1976) Aqueous leachate from western red cedar: effects on some aquatic organisms. *Journal of the Fisheries Research Board of Canada*, 33, 2703–9.

Persson, L. (1983) Effects of intra- and interspecific competition on dynamics and size structure of a perch (*Perca fluviatilis*) and a roach (*Rutilus rutilus*) population. *Oikos*, 41, 126–32.

Persson, L. (1986) Temperature induced shift in foraging ability in two fish species, roach (*Rutilus rutilus*) and perch (*Perca fluviatilis*): implications for coexistence in poikilotherms. *Journal of Animal Ecology*, 55, 829–39.

Peuranen, S., Vuorinen, P.J., Vuorinen, M. & Hollender, A. (1994) The effects of iron, humic acids and low pH on the gills and physiology of brown trout (*Salmo trutta*). *Annals Zoologica Fennici*, **31**, 389–96.

Planas, D., Desrosiers, M., Groulx, S-R., Paquet, S. & Carignan, R. (2000) Pelagic and benthic algal responses in eastern Canadian Boreal Shield lakes following harvesting and wildfires. *Canadian Journal of Fisheries and Aquatic Sciences*, **57** (Suppl. 2), 136–45.

Porvari, P., Verta, M., Munthe, J. & Haapanen, M. (2003) Forestry practices increase mercury and methyl mercury output from boreal forest catchments. *Environmental Science Technology*, **37**, 2389–93.

Power, E.A. & Northcote, T.G. (1991) Effects of log storage on the food supply and diet of juvenile sockeye salmon. *North American Journal of Fisheries Management*, **11**, 413–23.

Prepas, E.E., Pinel-Alloul, B., Planas, D., Míthot, G., Paquet, S. & Reedyk, S. (2001) Forest harvest impacts on water quality and aquatic biota on the Boreal Plain: introduction to the TROLS lake program. *Canadian Journal of Fisheries and Aquatic Sciences*, **58**, 421–36.

Ramberg, L. (1976) Effects of forestry operations on aquatic ecosystems. In: *Man and Boreal Forest*, (ed. C.O. Tamm), pp. 143–9. *Ecological Bulletin (Stockholm) 21.*

Rapport, D.J. & Whitford, W.G. (1999) How ecosystems respond to stress. *BioScience*, **49**, 193–203.

Rask, M. & Tuunainen, P. (1990) Acid-induced changes in fish populations of small Finnish lakes. In: *Acidification in Finland*, (eds P. Kauppi, K. Kenttämies & P. Anttila), pp. 911–27. Springer-Verlag, Berlin.

Rask, M., Arvola, L. & Salonen, K. (1993) Effects of catchment deforestation and burning on the limnology of a small forest lake in southern Finland. *Verhandlungen Internationale Vereinigung für theoretische und angewandte Limnologie*, **25**, 525–8.

Rask, M. Nyberg, K., Markkanen, S.-L. & Ojala, A. (1998) Forestry in catchments: effects on water quality, plankton, zoobenthos and fish in small lakes. *Boreal Environmental Research*, **3**, 75–86.

Rask, M., Nyberg, K., Karppinen, C., *et al.* (2000) Responses of perch (*Perca fluviatilis*) populations to catchment forest clearcutting and soil scarification in small lakes in eastern Finland. *Verhandlungen Internationale Vereinigung für theoretische und angewandte Limnologie*, **27**, 1122–6.

Samis, A.J.W., Colgan, P.W. & Johansen, P.H. (1993) Pentachlorophenol and reduced food intake of bluegill. *Transactions of the American Fisheries Society*, **122**, 1156–60.

Sandman, O., Turkia, J. & Huttunen, P. (1994) *Long-term Effects of Forestry Practices on Large Lakes; A Sedimentary Study of the Lakes Änättijärvi and Lentua in Kuhmo, Finland.* Publications of the Water and Environment Administration – Series A 179, pp. 3–56. [In Finnish, English summary.]

Schiefer, E., Reid, K., Burt, A. & Luce, J. (2001) Assessing natural sedimentation patterns and impacts of land use on sediment yield: a lake-sediment based approach. In: *Watershed Assessment in the Southern Interior of British Columbia: Workshop Proceedings*, (eds D.A.A. Toews & S. Chatwin), pp. 209–36. BC Ministry of Forests, Victoria, Canada.

Schindler, D.W., Beatty, K.G., Fee, E.J., *et al.* (1990) Effects of climatic warming on lakes in the central boreal forest. *Science*, **250**, 967–70.

Scully, N.M., Leavitt, P.R. & Carpenter, S.R. (2000) Century-long effects of forest harvest on the physical structure and autotrophic community of a small temperate lake. *Canadian Journal of Fisheries and Aquatic Sciences*, **57** (Suppl. 2), 50–9.

Simola, H. (1983) Limnological effects of peatland drainage and fertilization as reflected in the varved sediment of a deep lake. *Hydrobiologia*, **106**, 43–57.

Simonsson, P. (1987) Environmental effects of draining wetland and forest. Final report from a group of projects. *Naturvardsverket, Report 3270.*

Spencer, C.N. & Schelske, C.L. (1998) Impact of timber harvest on sediment deposition in surface waters in northwest Montana over the last 150 years: a paleolimnological study. In: *Forest-Fish Conference: Land Management Practices Affecting Aquatic Ecosystems*, (eds

M.K. Brewin & D.M.A. Monita), pp. 187–201. *Information Report NOR-X-356.* Canadian Forest Service, Edmonton, Alberta.

Steedman, R.J. (2000) Effects of experimental clearcut logging on water quality in three small boreal forest lake trout (*Salvelinus namaycush*) lakes. *Canadian Journal of Fisheries and Aquatic Sciences*, **57** (Suppl. 2), 92–6.

Steedman, R.J. & Kushneriuk, R.S. (2000) Effects of experimental clearcut logging on thermal stratification, dissolved oxygen, and lake trout (*Salvelinus namaycush*) habitat volume in three small boreal forest lakes. *Canadian Journal of Fisheries and Aquatic Sciences*, **57** (Suppl. 2), 82–91.

Steedman, R., Kushneriuk, R. & France, R. (2001) Littoral water temperature response to experimental shoreline logging around boreal forest lakes. *Canadian Journal of Fisheries and Aquatic Sciences*, **58**, 1638–47.

St-Onge, I. & Magnan, P. (2000) Impact of logging and natural fires on fish communities of Laurentian Shield lakes. *Canadian Journal of Fisheries and Aquatic Sciences*, **57** (Suppl. 2), 165–74.

Svärdson, G. (1976) Interspecific population dominance in fish communities of Scandinavian lakes. *Report of the Institute of Freshwater Research, Drottningholm*, **55**, 144–71.

Tonn, W.M., Magnuson, J.J., Rask, M. & Toivonen, J. (1990) Intercontinental comparison of small-lake fish assemblages: the balance between local and regional processes. *American Naturalist*, **136**, 345–75.

Tossavainen, T. (1997) Environmental management plan for Lake Kuohattijärvi, Nurmes, eastern Finland. *Technical Report 14/1997.* Regional Environment Center on North Carelia. [In Finnish.]

Turkia, J., Sandman, O. & Huttunen, P. (1998) Paleolimnological evidence of forestry practices disturbing small lakes in Finland. *Boreal Environmental Research*, **3**, 45–61.

Vallín, S. (1953) Zwei azidotropen Seen im Kustgebiet von Nordschweden. *Report of the Institute of Freshwater Research Drottninholm*, **34**, 167–89.

Verta, M, (1990) Mercury in Finnish forest lakes and reservoirs: anthropogenic contribution to the load and accumulation in fish. *Publications of the Water and Environment Research Institute, Finland*, **6**.

Vuori, K.-M., Joensuu, I., Latvala, J., Jutila, E. & Ahvonen, A. (1998) Forest drainage: a threat to benthic biodiversity of boreal headwater streams? *Aquatic Conservation: Marine and Freshwater Ecosystems*, **8**, 745–59.

Ward, G.S. & Parrish, P.R. (1980) Evaluation of early life stage toxicity tests with embryos and juveniles of sheepshead minnows. In: *Aquatic Toxicology*, (eds J.G. Eaton, R. Parrish & A.C. Hendricks), pp. 243–47. American Society for Testing Materials, Philadelphia.

Warwick, W.F. (1980) Paleolimnology of the Bay of Quinte, Lake Ontario: 2800 years of cultural influence. *Bulletin of Canadian Fisheries and Aquatic Sciences*, **206**.

Webb, P.W. & Brett, J.R. (1973) Effects of sublethal concentrations of sodium pentachlorophenate on growth rate, conversion efficiency, and swimming performance in underyearling sockeye salmon (*Oncorhynchus nerka*). *Journal of the Fisheries Research Board of Canada*, **30**, 499–507.

Wetzel, R.G. (2001) *Limnology*, 3rd edn. Academic Press, San Diego, CA.

Williams, M.R., Fisher, T.R. & Melack, J.M. (1997) Chemical composition of rain in the central Amazon, Brazil. *Atmospheric Environment*, **31**, 207–17.

Chapter 15
Effects of forestry on estuarine ecosystems supporting fishes

C.D. LEVINGS AND T.G. NORTHCOTE

Introduction

Harvesting, handling, processing and shipping of wood are among the oldest types of industrial activities and, given that humans tend to concentrate at river mouths, it is not surprising that estuarine ecosystems have been modified by forestry in a variety of direct and indirect ways. Temperate forests in estuaries needed to be cut down to build wharves at the world's ancient estuarine ports. Mangrove forests were removed to build harbours in tropical and subtropical regions, changing the configuration of the estuarine ecosystem permanently. A few studies have tried to examine ecosystem effects of loss of the entire floodplain forest in an estuary. Using an energy flow model, Wolff *et al.* (2000) suggested that logging of mangroves in a tropical estuary would result in decreased energy flow through mangrove detritus with possible effects on mangrove crabs (*Ucides cordatus*), a harvested species.

Most studies on the effects of forestry in estuaries have been conducted on specific topics and with a relatively narrow focus. Transport of logs to or between estuaries has physico-chemical effects on fish habitat that have been documented in a number of studies, especially on the west coast of North America where estuaries are used for rearing by juvenile salmonids. Effects of logging in the catchment basin can also propagate into the estuary and lower river; for example, through changes in the sediment transport regime. Input of detrital material from forests in the catchment (e.g. fluvial particulate organic carbon (FPOC) from leaves) may also be decreased by tree harvesting, although FPOC is probably held in an intact basin by retention structures such as large cobble and woody debris (Naiman & Sibert 1978). Water quality impacts from wood-processing industries began with the industrial revolution when the pulp and paper mills disposed their effluents in estuaries. Several major research programmes on the biological effects of pulp mills on fish and ecosystems in estuaries started in the late 1940s in North America and Europe and led to major advances in effluent treatment technology in the past two decades. However, the problem is still relatively unstudied in the tropics and subtropics.

In this chapter we provide a focused review on effects of forestry on fish and their supporting habitats in estuaries. A number of review articles or books have been published on separate problems such as log storage and the role of wood in estuarine ecosystems (Sedell *et al.* 1991; Maser & Sedell 1994), and there is an extensive literature on effects of wood-processing industries, especially pulp mills (e.g. Servos *et al.* 1994; see also Chapter 16, this volume). However, this chapter is an integrated overview of an array of problems. Shortfalls in knowledge and recommended research are also given. Possibly because the importance of estuaries for fish and wildlife has only been appreciated in the past few decades (see Chapter 5), most of our knowledge of forestry effects on estuarine ecosystems is of recent origin. Many of the data are from temperate estuaries, but wherever possible we have tried to bring in information from subtropical and tropical systems.

Freshwater entering estuaries from upstream

Discharge

Effects on discharge and water yield from logging in the catchment basin have been documented in studies (e.g. Swank *et al.* 2001) and these effects propagate downstream into the estuary. As an example, increased flows into the estuary would likely result in reduced penetration of the salt wedge upstream into the lower river. Upriver deforestation effects on discharge into South African estuaries are noted in Allanson & Baird (1999). No doubt these have occurred in other heavily logged river systems throughout the world, although they may have been masked in many of the large rivers along the Pacific coast of North America and in those of South America and elsewhere in tropical estuaries (Blaber 1997). The broad-leaved climax forests of Hong Kong watersheds were cleared over several millennia of human impact and the steep topography of the region and its climatic seasonality promote a highly pulsed discharge down to the estuarine reaches of its streams (Dudgeon 1996).

Sediments

Upstream sources of sediment are controlling factors for accretion or erosion in the estuary and excess sedimentation there can result from forest harvesting in the catchment basin. The magnitude of the estuarine effect depends on several correlated variables such as the natural sediment load of the river, climate, soils in the basin, degree and duration of deforestation, headwater elevation and distance from the sediment source to the estuary. In an estuary in New Zealand, apparent sedimentation rates increased from 0.1 to 0.3 mm y^{-1} in pre- and post-Polynesian times to 11 mm y^{-1} after commercial logging began in the river basin (Sheffield *et al.* 1995). The major polluter of tropical estuaries and adjacent critical habitat such as coral reefs is mud, which in parts of Australia has been mobilized by removal of coastal mangrove forests. In this situation transport of the sediment downriver from the catchment basin was not involved (Wolanski & Spagnol 2000). In Carnation Creek, BC, Canada, movement of gravel and sand from logged headwaters led to sediment build-up in the estuary and hence changes in its capacity to rear coho (*Oncorhynchus kisutch*) salmon fry (Tschaplinksi 1987). At the Skokomish River delta in Puget Sound, WA, USA, the combined effects of logging and water withdrawal led to the loss of 15–19% of highly productive low intertidal surface area and a 17% loss of eelgrass (*Zostera marina*) habitat. This was because of a loss of sediment transport capacity in the lower river and estuary due to water diversion and increased sediment deposition from forest harvesting (Jay & Simenstad 1996).

Pollutants

The transport of organic pollutants into the estuarine environment from wood-processing industries far upstream has been documented in the Fraser River system, Canada. As an example, Richardson & Levings (1996) found 2,3,7,8-tetrachloro-dibenzofurans (2.5–6.5 pg/g wet mass) in aquatic insects at a sample site in the lower river, just above tidal influence. The likely sources of these contaminants were pulp mills several hundred kilometres upstream. Dioxin and furans from starry flounder (*Platichthys stellatus*) and peamouth (*Mylocheilus caurinus*) caught in the lower estuary proper likely originated from both the upstream pulp mills and local industrial activity, including that related to forestry (Raymond *et al.* 2001).

Log transport to, in and through estuaries

Logs are transported into estuaries for storage, transhipment or processing. They are moved to the estuary by a variety of methods: river flow, transported by rail or truck and then dumped into the estuary, towed into the estuary by tugboats, dumped from barges, or more recently, dropped directly into the estuary by helicopter. Each of the transport modes has somewhat different ecological impacts.

River flow

In the late nineteenth century, log 'drives' were used in Washington and Oregon to move wood from the catchment basins of rivers to sawmills or processing areas in estuaries. In some instances this required temporary river dams, which stored floating logs. When water behind the dam was released to bring logs to the estuary, the resulting flush of channel sediments filled estuarine channels (Sedell *et al.* 1991) that would have been important fish habitat. The inadvertent drift of large volumes of waste wood and the river transport of logs from the catchment basin into the estuary of the Fraser River resulted in smothering of estuarine marsh plants (sedges, rushes). Clean-up programmes have been conducted by conservation groups to restore productivity (http://www.riverworks.org/field.htm). However a moderate amount of wood, especially stumps, may be important for progradation of marsh plants onto estuarine tidal mud flats as the wood creates gouges in the sediment where roots and rhizomes can establish plants (Maser & Sedell 1994).

Rail and truck

One of the direct consequences of rail and truck transport is loss of critical wetland habitat by physical changes in the estuary to create platforms for putting logs into the water. In the Kokish estuary in British Columbia, for example, dykes and railway embankments built by the forest industry reduced river flows to sectors of the estuary, leading to changes in the estuarine vegetation. The plant community shifted toward salt-tolerant species such as *Salicornia virginica* and away from brackish species such as *Carex lyngbyei* (Campbell & Bradfield 1988). Direct loss of a variety of coastal habitats such as rockweed (*Fucus* spp.) and eelgrass can also occur at sites where ramps are constructed (Fig. 15.1a) to slide logs from trucks and railcars into water. During this process, bark and wood debris is lost and can smother intertidal and shallow subtidal benthic habitats (Fig. 15.1b). Jackson (1986) reported bark deposits as thick as 60 cm

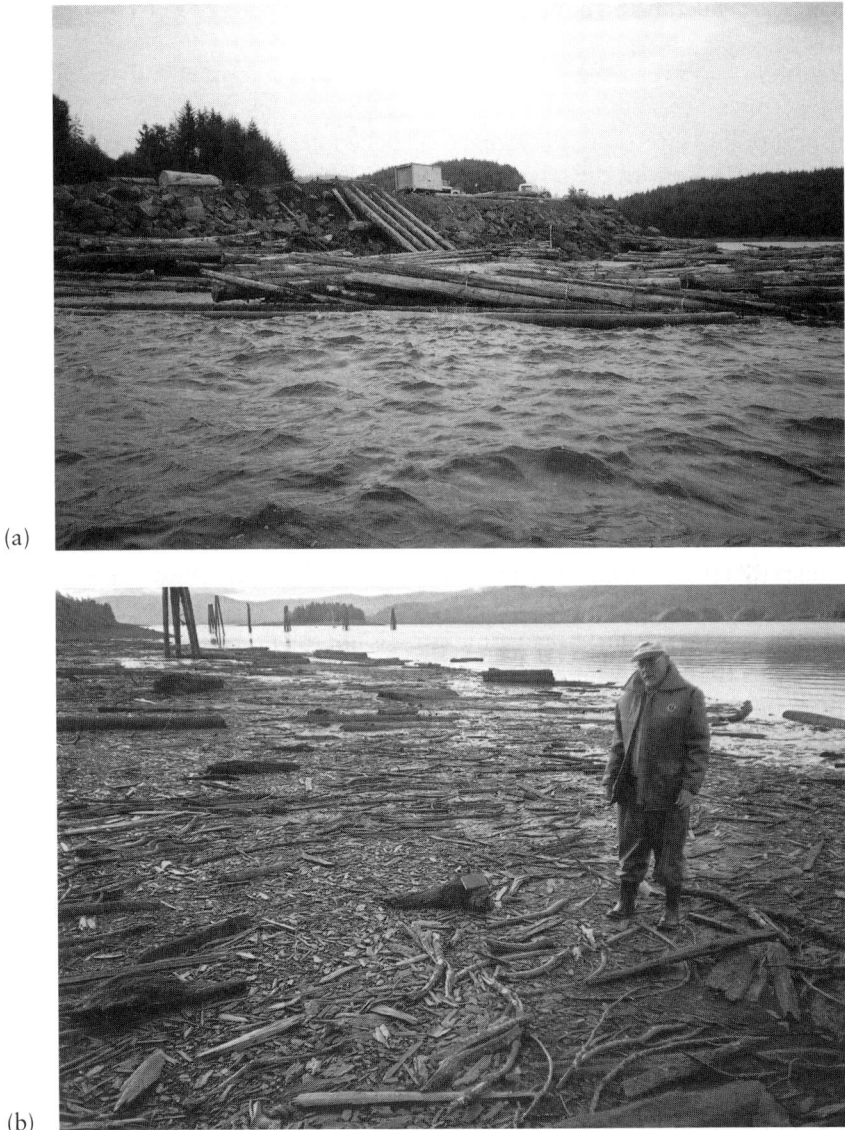

(a)

(b)

Fig. 15.1 (a) Ramps used to slide logs from trucks into an estuary on Skidegate Channel, Queen Charlotte Islands, BC, Canada. (b) Accumulation of wood fragments and debris at the base of a log dump ramp used for several decades at Bearskin Bay, Queen Charlotte Islands, BC, Canada. (Photographs taken in August 2000, courtesy of Ray Sjolund, Department of Fisheries and Oceans.)

at such sites in southeast Alaska. In such extreme conditions infaunal invertebrates can be virtually eliminated and, in less severe instances, deposit-feeding invertebrates (e.g. the polychaete *Pectinaria granulata*) almost totally replaced suspension feeders such as bivalves (e.g. *Macoma truncata*). Jackson (1986) and Conlan & Ellis (1979) found that as little as 2.5 cm of bark deposit excluded several species of molluscs and poly-chaetes. Survival and condition of *Protothaca staminea* and *Mytilus edulis* decreased with increasing depth and duration of bark coverage (Freese & O'Clair 1987).

Helicopter

Transport of logs directly to the estuary via helicopter is now common in British Co-lumbia and Alaska with the advent of selective logging. Single logs or small bundles are dropped into water from an altitude of a few tens of metres. Usually the logs are dropped vertically into the water and if they are of sufficient length may disrupt benthic habitats on the bottom of the estuary. Some debarking and debris loss is also likely. Although concern has been raised about such ecological impact, there has been no study of effects, to our knowledge.

Log and wood product storage in estuaries

Log rafts are stored in estuaries for variable amounts of time, sometimes up to several months, awaiting processing in sawmills or pulp mills (Faris & Vaughan 1985). It has been suggested that in some situations storage in estuaries is necessary to prevent dam-age to the wood from boring organisms (e.g. the crustacean *Limnoria tripunctata* and the mollusc *Teredo navalis*) because the latter is not tolerant of brackish or freshwater (Anonymous 1980). Depending on local circumstances, the logs could be stored above high tide on the adjacent shoreline, or in the intertidal zone, or moored over deeper water. There are both chemical and physical effects of these alternatives.

Leaching from stored and waste wood

Buchanan *et al.* (1976) investigated the toxicity of extracts from the bark of Sitka spruce and western hemlock. In bioassays with pink salmon fry (*O. gorbuscha*), hem-lock extract was found to be lethal to 50% of test fish at 100–120 mg L^{-1} and Sitka spruce extract at much lower levels (50–56 mg L^{-1}). Sitka spruce bark was more toxic than hemlock in bioassays with adult and larval pink shrimp (*Pandalus borealis*) and larval Dungeness crab (*Cancer magister*). Aqueous leachate from western red cedar is known to affect salmonids (Peters *et al.* 1976) and its logs are often stored in estuarine areas. Some of the toxic organics present in bark leachates may be compounds such as dehydroabietic acid, abietic acid and pimaric acid. These compounds were found in relatively low concentrations in sediments characterized by 30-year-old bark in sedi-ments at an abandoned log port in New Zealand (Healey *et al.* 1999).

Land sorting and storage

A dry land sort is usually a gravel or paved log storage area, sometimes several hectares in area, located immediately above the high tide boundary in the estuary. Construction of these facilities can result in loss of important estuarine riparian habitat (Brownlee *et al.* 1984). In the Pacific Northwest, logs are usually brought to the 'sort' by truck, then separated by species and size using forklifts or front-end loaders. They are then slid down ramps into the estuary and made into rafts for towing to sawmills or pulp mills. Debarking and loss of wood waste into the estuary is likely. Runoff from a dry land sort or other wood processing operations such as sawmill can include metals and other toxic chemicals (Bailey *et al.* 1999).

Intertidal storage

Grounding of log rafts on the estuarine sediments has direct and significant impacts because the sediments are compacted at low tide by the weight of the logs and mixing is reduced because of lower wave action behind the rafts. Although not specifically investigated, there are likely two major impacts on seagrasses such as eelgrass: shading and repeated grounding with the rise and fall of the tide. In addition, the competency of the estuary to transport sediment will be reduced because of decreased current action, leading to fine sediment deposition and inhibition of interstitial water circulation, with accompanying reduction in dissolved oxygen levels. These changes were observed in the Nanaimo River estuary, BC, by Sibert & Harpham (1979), who investigated effects on meiofauna, especially harpacticoid copepods. Although they did not find reduced populations of harpacticoids under the log rafts, there was a change in the dominant species, with only the smaller interstitial species (*Enhydrosoma uniarticulatum*) found there. Effects on epibenthic harpacticoids such as *Harpacticus uniremis*, a species important in chum salmon fry (*O. keta*) diets, were equivocal. In a related study at the Nanaimo estuary, McGreer *et al.* (1984) investigated recovery of part of an intertidal habitat after removal of log rafts. Benthic communities at the location where logs were removed and a reference site were similar after 13 months. Results of a similar study with macrobenthos at the Snohomish River estuary in Washington, USA, were very different. Smith (1977) showed that recovery of invertebrate communities after removal of logs occurred within 1 or 2 months, with recolonization slower in winter. We could not find quantitative documentation for recovery of seagrasses or marsh plants from log rafting effects (but cf. Plate 4h,i). However, it is likely that several years are required, judging from transplant and colonization data from a habitat restoration experiment at the Campbell River estuary (Brownlee *et al.* 1984; Dawe *et al.* 2000).

Decomposition of bark and wood debris falling off logs stored in poorly flushed estuaries leads to anaerobic, sulphide-rich habitats where bacterial mats (e.g. *Begiattoa*) develop, replacing the natural algal-based food web. Organic content of sediments from wood waste at the Snohomish estuary ranged from 6.9% to 9.9%, significantly higher than in unaffected mud flats (Smith 1977). Decomposition effects can propagate into higher trophic levels such as gammarid amphipods. At a relatively poorly flushed sector of the Squamish River estuary in BC, Stanhope & Levings (1985) investigated the secondary production of the amphipod *Eogammarus confervicolus*, important as a food item for juvenile salmon. Secondary production was reduced from approximately 21 g dry weight $m^{-2} y^{-1}$ at a reference area to 6 g dry weight $m^{-2} y^{-1}$ at a part of the estuary affected by log storage. A similar pattern was found for productivity of an intertidal isopod (*Gnorimosphaeroma insulare*) in the same area (Stanhope *et al.* 1987). In parts of estuaries swept by tidal and river currents, effects can be less severe. For example Levy *et al.* (1982, 1989), working at the mouth of the Fraser River, BC, found that chinook salmon fry were more abundant in a log storage area relative to a reference marsh. The abundance of their amphipod prey species (*E. confervicolus*, *Corophium* spp.) was higher in the log storage area, perhaps because the wood debris provided cover for these crustaceans. Feeding habits of the chinook salmon fry were different in the log storage site – fewer insects and more mysids were consumed in comparison with the reference marsh nearby.

Estuarine habitats can also be disrupted and eroded by propeller action and scouring of towboats moving logs around in the estuary at high tide. On one occasion at the Nanaimo River estuary, a plume of mud 30 m in length was observed in the wash of a towboat. A strong odour of hydrogen sulphide was detected, indicating that the propeller wash was suspending material from below the redox potential discontinuity in the sediments (Sibert & Harpham 1979).

Log debris problems may be reduced through proper procedures. Means for controlling sawlog debris in the lower Fraser River, BC, down to its estuary are reviewed by Fairbain & Peterson (1975).

Deep water storage

Very large volumes of wood in log rafts are often stored outside the immediate river mouth area, anchored to the steep fjord shoreline in deep estuaries. Bark and wood debris from the anchored logs sink to the bottom, disrupting sediment habitats and biota. Depending on wood species, depth, temperature and other factors, most of the wood and bark decomposes. If bottom currents are weak in the deep areas, anaerobic conditions with hydrogen sulphide and low dissolved oxygen develop. Benthic communities are often impaired under these circumstances (McGreer *et al.* 1985) and in severe situations invertebrates are totally excluded. In some cases the more refractory components of wood can resist decomposition and will remain intact for decades without imposing a significant biochemical oxygen demand (BOD). Wooden planks and chips from a sawmill abandoned about 50 years were intact at a British Columbia inlet (Fournier & Levings 1982) and diverse invertebrate communities characterized the sediments in the area.

The loss of particular species in an azoic zone under a heavily used log storage area, and a species shift toward tolerant species in the moderately polluted zone on its perimeter, conforms to the well-known response pattern from hypoxia and sulphide toxicity described by Pearson & Rosenberg (1978). One of the early detailed studies on the ecological effects of wood waste was a comparative photographic survey of macrobenthos at log storage and reference sites conducted using SCUBA by McDaniel (1975) in Howe Sound, BC. Numbers of filter feeders were reduced and numbers of deposit-feeding organisms were increased at sites heavily impacted by wood waste from log storage (McDaniel 1975). Conlan & Ellis (1979) obtained similar results using quantitative grab sampling in a British Columbia fjord. More recently, studies have been initiated in deeper water, beyond SCUBA range, where remote observation techniques are required to augment grab sampling. Williamson *et al.* (2000) used side scan sonar to detect sunken logs in Douglas Channel. This technique can only be used to map the distribution of wood waste if particles are >25 cm, so side scan sonar has limitations for mapping fine particulate material such as decomposing bark or wood 'mulch'. Anderson *et al.* (2002) showed that digital acoustic classification methods (QTC Vies; Series IV) were effective for mapping log waste in coastal Newfoundland. This method might be useful for mapping wood mulch and small particles of wood. Kirkpatrick *et al.* (1998) used a manned submersible to give excellent faunal data at a deep water (40–70 m) log transfer facility in Alaska. Of 91 taxa observed during their study, most (69 species) were found on rocky, bark-free habitat; significantly reduced species richness was found in all bark-dominated habitats.

Data on the effects of wood waste on the autecology of benthic organisms are scarce in the literature. Benthic organisms such as crabs and shrimp avoid wood in sediment relative to natural substrates, as shown in the behavioural experiments conducted by Chang & Levings (1976). O'Clair & Freese (1988) found that fecundity and number of ovigerous Dungeness crabs were reduced at an embayment affected by wood waste in Alaska. No published papers could be found on effects of wood waste on bottom-dwelling fish. However, it is expected that species such as lingcod (*Ophiodon elongatus*), which require intact and shallow rocky bottom habitat for nesting (Low & Beamish 1978), would be affected by sunken wood debris.

Barging

Log barges enable transportation of large volumes of wood from remote coastal areas to storage or processing areas without the need for rafting and towing the logs. While the logs are being lifted by crane onto or off barges, bark and wood splinters are lost into the estuary. Some barges are self-tipping, which launches the logs into the water at a considerable speed, resulting in further debarking, splintering and possibly gouging the estuary floor if the water is shallow. To our knowledge the ecological impact of this has not been investigated.

Barge shipment of wood byproducts such as chips used in pulping can also lead to wood in estuarine benthic habitats when the chips blow or spill off the barges into the water. The wood particles eventually sink to the bottom and in some inlets this has led to build-up of chips on benthic habitats (e.g. Thornbrough Channel in Howe Sound, BC; McDaniel *et al.* 1978). Effects can also arise in the pelagic environment from floating wood. Wood particles in water from chip barges and log rafts can also affect fish gills by causing lesions that reduce respiratory capacity and can even cause direct trauma by piercing the gill filaments (Magor 1988). Black & Miller (1993) found that salmon being reared in an aquaculture operation near a pulp mill consumed woody debris and bark, decreasing survival and causing tissue damage to fish stomachs.

Wood waste dredged from around sawmills and barged to ocean dumpsites for disposal can also impair benthic habitats. In Alberni Inlet, BC, Levings *et al.* (1985) found a shift in polychaete feeding types toward subsurface deposit feeders such as *Capitella* spp. at the dump site, and away from more sensitive species such carnivores and filter feeders. Other hardy opportunistic species living on rather than in the wood waste are found in large densities at the dump sites relative to natural muddy sediment habitats. An example is the squat lobster (*Munida quadrispina*) found in thousands per trawl (McDaniel *et al.* 1978) at dumpsites in Howe Sound. It is possible that this species may be using the waste wood as cover.

Pulp mills in estuaries

Numerous pulp and paper mills around the world are located in estuaries. Even in the early 1900s, effects of pulp mill pollution on fish life in lower reaches of eastern Canadian waters were being noted (Knight 1901). Effects of the mills on estuarine ecosystems, including tropical areas (Blaber 1997), have focused on (1) disruption of the sediment ecosystem owing to excess sedimentation of waste fibre or suspended solids; (2) effects on water quality, especially toxicity of the effluent and reduction of dissolved oxygen owing to the biochemical oxygen demand of the waste; (3) tainting and bioaccumulation of organic and inorganic toxic materials (Servos *et al.* 1994).

Suspended solids

Older groundwood pulp and paper mills were inefficient, sometimes losing 10–25 t per day of waste wood to the estuary (Pearson 1971; Pearson & Rosenberg 1976), causing development of a fibre blanket on the bottom which smothered benthic organisms and also created high dissolved oxygen (DO) demand in the water column. Severe reductions in shallow water benthic productivity and diversity were documented in numerous studies around the world (e.g. Leppakoski 1975; Pearson & Rosenberg 1976; Fournier & Levings 1982). Modern bleached kraft mills lose much less fibre (e.g. Australia 4.2 kg per net t production or 2100 kg per day for a 500 t mill in 1997; http://www.ppmfa.com.au/environment/). If natural sedimentation is sufficient, fibre beds that are not deepening with fresh waste can be covered or 'capped' with natural sediment. Using images analysed from the Sediment Vertical Profiling System, SAIC (1999) found that infaunal organisms had colonized 6–8 cm of silt above an extensive fibre bed in a Washington harbour. The depth of redox potential discontinuity layer, which provides an index of the 'health' of the sediment ecosystem, also indicated normal functioning above the fibre bed.

Water quality

Before the introduction of modern treatment facilities such as biobasins for effluent treatment to reduce BOD, pulp mill waste effluent disposed of in estuaries caused very low DO levels in water masses. The effluent was also highly toxic to almost all organisms. Tully's (1949) early investigation of reduced DO from pulp mill effluent entering the Somass River estuary in British Columbia is a classic documentation of oceanographic aspects of this problem. Effects on young salmon were reviewed by Alderdice & Brett (1957). Effluent disposal in estuaries resulted in fish kills, severely degraded receiving water and effects at all levels of the estuarine food web (Coloedy & Wells 1992). Entire embayments were characterized by low DO, especially in the bottom waters of highly stratified estuaries and fjords (Waldichuk 1962).

Laboratory experiments showed that fish mortality and sublethal effects of pulp mill effluent were typically caused by combinations of factors, especially low DO, temperature and toxicants such as terpenes present in the wood (Davis 1973; Howard 1975; Rogers *et al.* 1992). More recent studies have focused on biochemical methods to measure the response of detoxification enzymes in fish (e.g. EROD – ethoxy resorufin O-deethylase). An example is EROD induction in mummichog (*Fundulus heteroclitus*) from organic contaminants in the effluents (e.g. Couillard & Nellis 1999). Another emerging area of concern is the effect of bleached kraft pulp mill effluent as an endocrine disrupter for estuarine fish reproduction (e.g. Dube & MacLatchy 2000). Such effluent can modify the reproductive period of fish in estuaries (Leblanc *et al.* 1997), or cause avoidance behaviour in other estuarine fishes (Lewis & Livingston 1977).

Effects on intertidal organisms (e.g. oysters *Crassostrea gigas*) can occur when effluent is dispersed over beaches, as was documented in early studies in Spain (Niell & Buela 1976) and British Columbia (Quayle 1969). In the former study, lignosulphate residues in the beach sediments were negatively correlated with the biomass of algae (two species of *Fucus*, as well as *Ascophyllum nodosum*, and *Pelvetia canaliculata*). Parker & Sibert (1976) working in Alberni Inlet, BC, concluded that low DO was not necessarily related to BOD, but to reduced photosynthetic activity because of the darkening effect of pulp mill effluent colour. Stockner & Costella (1976) conducted laboratory experiments with phytoplankton and concluded that high concentrations of effluent toxicants also could be involved. Sibert and Brown (1975) investigated heterotrophic conversion of metabolizable organic carbon from pulp mill effluent and concluded that 70% of the daily input into a British Columbia inlet was converted to particulate organic carbon.

Effluent contaminants and tainting

Zinc used in pulp mills as a brightener for paper products became a problem in British Columbia when it was found to bioconcentrate in oysters. In 1973, the industry switched to boron, which was non-toxic to the oysters and did not bioaccumulate (Thompson *et al.* 1976). However, zinc contamination is still a problem in some parts of the world, such as in South America where paper mills in Sepetiba Bay and Basin in Brazil released 2.4 t of zinc per year to the estuary (Barcellos & Lacerda 1994). In the mid-1980s, fisheries for bream (*Abramis brama*) and pike (*Esox lucius*) in a Finnish estuary were affected by tainting ('off flavour') from pulp mill effluent and from mercury used in slimicides (Paavilainen *et al.* 1985). In the 1980s, dioxin contamination from pulp mills in British Columbia caused the closure of several major commercial and recreational fisheries for Dungeness crab in coastal waters near an estuary (Nassichuk

1992). The fisheries were closed because consumption of the crabs was a significant health risk to humans. Macdonald *et al.* (1992) showed that the appearance of a dioxin compound (2,3,7,8-TCDF) in sediment cores from the Strait of Georgia correlated temporally with the use of chlorine to bleach pulp. Pulp mills have now significantly reduced chlorine use, so dioxin levels in the crabs are now lower and most fisheries are reopened. However, some areas are still closed (Levings *et al.* 2003).

Fungicides and wood preservatives

A variety of metal and organochlorine compounds have been used in or near estuaries to prevent wood rot or infestation by fungus (Carey & Hart 1988; Macdonald *et al.* 1992). Most of these have been found to be toxic to fish and other estuarine organisms. Wood treated with the preservative chromated copper arsenate (CCA) was found to affect estuarine benthic infauna by reducing species richness and diversity in sediments adjacent to treated wood structures. CCA also affected the development and growth of mysids. Wood that had been in the estuary was less toxic, indicating the CCA leached out over time (Weis & Weis 1996). Creosote is another wood preservative that is widely used on structures such as docks and pilings in estuaries. This compound was found to have significant effects on embryonic herring (*Clupea harengus*) in laboratory experiments (Vines *et al.* 2000). The hatching rate of embryos exposed to creosote was 90% lower than control embryos and 72.4% lower than embryos exposed to untreated wood, and the LC_{50} for hatching success was 0.05 mg/L. Kruznyski & Birtwell (1994) investigated the sublethal effects of an antisapstain fungicide (2-(thiocyanomethyl-thio) benzothiazole – TCMTB) using a predation bioassay approach. Chinook salmon smolts that had been exposed to the toxicant were consumed by predatory rockfish (*Sebastes flavidus*) in preference to the control group by a factor of 5.5:1.

Concluding comments

The interaction between changes in water quality and physical changes in the catchment basin and estuary might be a fruitful area for research, perhaps using ecosystem models. Results of these analyses would be directly usable for managers involved in integrated management of the coastal zone. In areas where deep estuaries such as fjords are used as log storage areas, work is needed on methods and impact assessment on deep water fish habitat and ecosystems. Priority areas for research on water quality relating to the pulp and paper industry include research on biomarkers, organic micropollutants and endocrine-disrupting chemicals. Although the impacts of individual forestry activities on estuarine fish habitats have been documented, there is an urgent need for methods and analyses that consider their cumulative effects. Most of the literature on effects of forest harvesting on estuarine ecosystems originates from the west coast of North America but because of differences in fish communities, forest types, geology, climate and forestry practices it is unlikely that results can be extrapolated to other regions without local calibration. Therefore there is a definite requirement for detailed research studies elsewhere in the world. A first priority might be tropical and subtropical estuaries where information appears to be particularly lacking.

Acknowledgements

K.J. Hall, G.F. Hartman and J.D. Pringle read the manuscript and provided numerous useful comments.

References

Alderdice, D.F. & Brett, J.R. (1957) Some effects of kraft mill effluent on young Pacific salmon. *Journal of the Fisheries Research Board of Canada*, **14**, 783–95.

Allanson, B.R. & Baird, D. (1999) *Estuaries of South Africa*. Cambridge University Press, Cambridge, UK.

Anderson, J.T., Gregory, R.S. & Collins, W.T. (2002) Acoustic classification of marine habitats in coastal Newfoundland. *Journal of Marine Science*, **59**, 156–67.

Anonymous (1980) *Nanaimo Estuary Fish Habitat and Log Management Task Force; Report of the Log Management Subcommittee to the Steering Committee*. Victoria, BC, Canada.

Bailey, H.C., Elphick, J.R., Potter, A. & Zak, B. (1999) Zinc toxicity in stormwater runoff from sawmills in British Columbia. *Water Research*, **33**, 2721–5.

Barcellos, C. & Lacerda, L.D. (1994) Cadmium and zinc source assessment in the Septetiba Bay and Basin. *Environmental Monitoring Assessment*, **29**, 183–99.

Blaber, S.J.M. (1997) *Fish and Fisheries of Tropical Estuaries*. Chapman & Hall, New York.

Black, E.A. & Miller, H.J. (1993) The effect of consumption of woody debris on marine cage-reared salmon. *Journal of Applied Ichthyology*, **9**, 41–8.

Brownlee, M.J., Mattice, E.R. & Levings, C.D. (1984) The Campbell River estuary: a report on the design, construction, and preliminary follow-up study findings of the intertidal marsh islands created for the purposes of estuarine rehabilitation. *Canadian Manuscript Report Fisheries and Aquatic Science*, **1789**.

Buchanan, D.V., Tate, P.S. & Moring, J.R. (1976) Acute toxicities of spruce and hemlock bark extracts to some estuarine organisms in southeastern Alaska. *Journal of the Fisheries Research Board of Canada*, **33**, 1188–92.

Campbell, A. & Bradfield, G.E. (1988) Short-term vegetation change after dyke breaching at the Kokish Marsh, northeastern Vancouver Island. *Northwest Science*, **31**, 153–62.

Carey, J.H. & Hart, J.H. (1988) Sources of chlorophenolic compounds to the Fraser River estuary. *Water Pollution Research Journal of Canada*, **23**, 55–68.

Chang, B.C. & Levings, C.D. (1976) Laboratory experiments on the effects of ocean dumping on benthic invertebrates. *Technical Report Fisheries Research Board of Canada*, **637**.

Coloedy, A.G. & Wells, P.G. (1992) Effects of pulp and paper mill effluents on estuarine and marine ecosystems in Canada: a review. *Journal of Aquatic Ecosystem Health*, **1**, 201–26.

Conlan, K.E. & Ellis, D.V. (1979) Effects of wood waste on sand-bed benthos. *Marine Pollution Bulletin*, **10**, 262–7.

Couillard, C.M. & Nellis, P. (1999) Organochlorine contaminants in mummichog (*Fundulus heteroclitus*) living downstream from a bleached-kraft pulp mill in the Miramichi Estuary, New Brunswick, Canada. *Environmental Toxicology & Chemistry*, **18**, 2545–56.

Davis, J.C. (1973) Sublethal effects of bleached kraft pulp mill effluent on respiration and circulation in sockeye salmon (*Oncorhynchus nerka*). *Journal of the Fisheries Research Board of Canada*, **30**, 369–77.

Dawe, N.K., Bradfield, G.E., Boyd, W.S., Tretheway, D.E.C. & Zolbrod, A.N. (2000) Marsh creation in a northern Pacific estuary: is thirteen years of monitoring vegetation dynamics enough? *Conservation Ecology,* **4,** 12–36.

Dube, M.G. & MacLatchy, D.L. (2000) Endocrine responses of *Fundulus heteroclitus* to effluent from a bleached-kraft pulp mill before and after installation of reverse osmosis treatment of a waste stream. *Environmental Toxicology & Chemistry,* **19,** 2788–96.

Dudgeon, D. (1996) Anthropogenic influences on Hong Kong streams. *GeoJournal,* **40,** 53–61.

Fairbain, B. & Peterson, K. (1975) Controlling sawlog debris in the lower Fraser River. *Technical Report No. 5.* Westwater Research Centre, University of British Columbia.

Faris, T.L. & Vaughan, K.D. (1985) Log transfer and storage facilities in Southeast Alaska: a review. *General Technical Report PNW-174.* US Department of Agriculture, Forest Service, Pacific Northwest Range Experiment Station, Portland, OR.

Fournier, J.A. & Levings, C.D. (1982) Polychaetes recorded near two pulp mills on the northern coast of British Columbia: a preliminary taxonomic and ecological account. *Syllogeus National Museum of Canada,* **40,** 1–91.

Freese, J.L. & O'Clair, C.E. (1987) Reduced survival and condition of *Protothalca staminea* and *Mytilus edulis* buried by decomposing bark. *Marine Environmental Research,* **23,** 49–64.

Healey, T.R., Wilkins, A.L. & Leipe, T. (1999) Extractives from a coniferous bark dump in coastal estuarine sediments. *Journal of Coastal Research,* **13,** 293–6.

Howard, T.E. (1975) Swimming performance of juvenile coho salmon (*Oncorhynchus kisutch*) exposed to bleached kraft pulpmill effluent. *Journal of the Fisheries Research Board of Canada,* **32,** 789–93.

Jackson, R.G. (1986) Effects of bark accumulation on benthic infauna at a log transfer facility in Southeast Alaska. *Marine Pollution Bulletin,* **17,** 258–62.

Jay, D.A. & Simenstad, C.A. (1996) Downstream effects of a water withdrawal in a small, high-gradient basin: erosion and deposition on the Skokomish River delta. *Estuaries,* **19,** 501–16.

Kirkpatrick, B., Shirley, T.C. & O'Clair, C.E. (1998) Deep-water bark accumulation and benthos richness at log transfer and storage facilities. *Alaska Fishery Research Bulletin,* **5,** 103–15.

Knight, A.P. (1901) The effects of polluted waters on fish life. *Contributions to Canadian Biology,* **1901,** 9–18.

Kruzynski, G.M. & Birtwell, I.K. (1994) A predation bioassay to quantify the ecological significance of sublethal responses of juvenile chinook salmon (*Oncorhynchus tshawytscha*) to the antisapstain TCMTB. *Canadian Journal of Fisheries and Aquatic Sciences,* **51,** 1780–90.

Leblanc, J., Couillard, C.M. & Brethes, J.-C. (1997) Modifications of the reproductive period in mummichog (*Fundulus heteroclitus*) living downstream from a bleached kraft pulp mill in the Miramichi Estuary, New Brunswick, Canada. *Canadian Journal of Fisheries and Aquatic Sciences,* **54,** 2564–73.

Leppakoski, E. (1975) Assessment of degree of pollution on the basis of macrobenthos in marine and brackish-water environments. *Acta Academiae Aboensis, Series B35,* **2.**

Levings, C.D., Anderson, E.P. & O'Connell, C.W. (1985) Biological effects of dredged material disposal in Alberni Inlet. In: *Wastes in the Ocean, Vol. 6, Nearshore Waste Disposal,* (eds B. Ketchum, J.M. Capuzzo, W.V. Burt, I.W. Duedall & P.K. Park), pp. 132–55. John Wiley & Sons, New York.

Levings, C.D., Stein, J., Stehr, C. & Samis, S. (2003) Introduction to the PICES Practical Workshop: objectives, overview of the study area, and projects conducted by the participants. *Marine Environmental Research,* **57,** 3–18.

Levy, D.A., Northcote, T.G. & Barr, R.M. (1982) Effects of estuarine log storage on juvenile salmon. *Technical Report 26.* University of British Columbia, Westwater Research Centre.

Levy, D.A., Northcote, T.G., Hall, K.J. & Yesaki, I. (1989) Juvenile salmon responses to log storage in littoral habitats of the Fraser River estuary and Babine Lake, British Columbia. In: *Proceedings of the National Workshop on Effects of Habitat Alteration on Salmonid*

Stocks, (eds C.D. Levings, L.B. Holtby & M.A. Henderson), pp. 82–91. *Canadian Special Publication of Fisheries and Aquatic Science*, **105**. Ottawa.

Lewis, F.G. & Livingston, R.J. (1977) Avoidance of bleached kraft pulpmill effluent by pinfish (*Lagodon rhomboides*) and gulf killifish (*Fundulus grandis*). *Journal of the Fisheries Research Board of Canada*, **34**, 568–70.

Low, C.J. & Beamish, R.J. (1978) A study of the nesting behavior of lingcod (*Ophiodon elongatus*) in the Strait of Georgia, British Columbia. *Technical Report of Fisheries and Marine Service, Canada*, **843**.

Macdonald, R.W., Cretney, W.J., Crewe, N. & Paton, D. (1992) A history of octachloro-dibenzofuran-p-dioxin, 2,3,7,8-tetrachlorodibenzofuran, and 3,3′,4,4′-tetrachlorobiphenyl contamination in Howe Sound, British Columbia. *Environmental Science and Technology*, **26**, 1544–50.

McDaniel, N.G. (1975) A survey of the benthic macroinvertebrate fauna and solid pollutants in Howe Sound. *Technical Report Fisheries Research Board of Canada*, **385**.

McDaniel, N.G., Levings, C.D., Goyette, D. & Brothers, D. (1978) Otter trawl catches at disrupted and intact habitats in Howe Sound, Jervis Inlet, and Bute Inlet British Columbia, August 1976 to December 1977. *Fisheries and Environment Canada, Fisheries and Marine Service Data Report*, **92**.

McGreer, E.R., Moore, D.M. & Sibert, J.R. (1984). Study of the recovery of intertidal benthos after removal of log booms, Nanaimo River estuary, British Columbia. *Canadian Technical Report Fisheries and Aquatic Sciences*, **1246**.

McGreer, E.R., Munday, D.R. & Waldichuk, M. (1985) Effects of wood waste for ocean disposal on the recruitment of marine macrobenthic communities. *Canadian Technical Report Fisheries and Aquatic Sciences*, **1398**.

Magor, B.G. (1988) Gill histopathology of juvenile *Oncorhynchus kisutch* exposed to suspended wood debris. *Canadian Journal of Zoology*, **66**, 2164–9.

Maser, C. & Sedell, J. (1994) *From the Forest to the Sea: The Ecology of Wood in Streams, Rivers, Estuaries, and Oceans*. St Lucie Press, Delray Beach, FL.

Naiman, R.J. & Sibert, J. (1978) Transport of nutrients and carbon from the Nanaimo River to its estuary. *Limnology and Oceanography*, **23**, 1183–93.

Nassichuk, M.D. (1992) Dioxin mediated shellfish closures in Howe Sound. In: *Proceedings of the Howe Sound Environmental Workshop*, (eds C.D. Levings, R.B. Turner & B. Ricketts), pp. 215–28, *Canadian Technical Report Fisheries and Aquatic Sciences*, **1879**.

Niell, F.X. & Buela, J. (1976) Incidencia de vertidos industriales en la estructura de poblaciones intermareales I. Distribucion y abundancia de Fucaceas caracteristicas. *Investigacion Pesquera*, **40**, 137–49.

O'Clair, C.E. & Freese, J.L. (1988) Reproductive condition of Dungeness crabs, *Cancer magister*, at or near log transfer facilities in southeastern Alaska. *Marine Environmental Research*, **26**, 57–81.

Paavilainen, K., Langi, A. & Tana, J. (1985) Effect of pulp and paper mill effluents on a fishery in the Gulf of Finland. *Finnish Fisheries Research*, **6**, 81–91.

Parker, R.R. & Sibert, J. (1976) Responses of phytoplankton to renewed solar radiation in a stratified inlet. *Water Research*, **10**, 123–8.

Pearson, T.H. (1971) The benthic ecology of Loch Linne and Loch Eil, a sea-loch system on the west coast of Scotland. III. The effect on the benthic fauna of the introduction of pulpmill effluent. *Journal Experimental Marine Biology and Ecology*, **6**, 211–33.

Pearson, T.H. & Rosenberg, R. (1976) A comparative study of the effects on the marine environment of wastes from cellulose industries in Scotland and Sweden. *Ambio*, **5**, 77–9.

Pearson, T.H. & Rosenberg, R. (1978) Macrobenthic succession in relation to organic enrichment and pollution in the marine environment. *Marine Biology Annual Review*, **16**, 229–311.

Peters, G.B., Dawson, H.J., Hrutfiord, B.F. & Whitney, R.R. (1976) Aqueous leachate from western red cedar: effects on some aquatic organisms. *Journal of the Fisheries Research Board of Canada*, **33**, 2703–9.

Raymond, B.A., Shaw, D.P., Kim, K., *et al.* (2001) Fraser River action plan resident fish contaminant and health assessment. *Fraser River Action Plan Report 1998–20*. Canada Department of Environment.

Richardson, J.S. & Levings, C.D. (1996) Chlorinated organic contaminants in benthic organisms of the lower Fraser River, British Columbia. *Water Pollution Research Research Journal of Canada*, **31**, 153–62.

Rogers, I.H., Macdonald, J.S. & Sadar, M. (1992) Uptake of selected organochlorine contaminants in fishes resident in the Fraser River estuary, Vancouver, British Columbia. *Water Pollution Research Journal of Canada*, **27**, 733–49.

Quayle, D.B. (1969) Pacific oyster culture in British Columbia. *Fisheries Research Board of Canada, Bulletin*, **169**.

SAIC (Science Applications International Corporation) (1999) *Port Angeles Harbor Wood Waste Study*. Prepared for Washington State Department of Ecology, Bothell, WA, USA.

Sedell, J.R., Leone, F.N. & Duval, W.S. (1991) Water transportation and storage of logs. In: *Influences of Forest and Rangeland Management on Salmonid Fishes and their Habitat*. American Fisheries Society Special Publication 19.

Servos, M.R., Munkittrick, K.R., Carey, J.H. & Van Der Kraak, G.J. (1994) *Environmental Fate and Effects of Pulp and Paper Mill Effluents*. St Lucie Press, Delray Beach, FL.

Sheffield, A.T., Healey, T.R. & McGlone, M.S. (1995) Infilling rates of a steepland catchment estuary, Whangamata, New Zealand. *Journal of Coastal Research*, **11**, 1294–308.

Sibert, J. & Brown, T.J. (1975) Characteristics and potential significance of heterotrophic activity in a polluted fjord estuary. *Journal of Experimental Marine Biology and Ecology* **19**, 97–104.

Sibert, J.R. & Harpham, V.J. (1979) Effects of intertidal log storage on the meiofauna and interstitial environment of the Nanaimo River delta. *Technical Report Fisheries Research Board of Canada*, **883**.

Smith, J.E. (1977) A baseline study of invertebrates and the environmental impact of intertidal log rafting on the Snohomish River delta. *Final Report March 1977*. Washington Cooperative Fishery Research Unit, College of Fisheries, University of Washington, Seattle.

Stanhope, M.J. & Levings, C.D. (1985) Growth and production of *Eogammarus confervicolus* (Amphipoda: Anisogammaridae) at a log storage site and in areas of undisturbed habitat within the Squamish estuary, British Columbia. *Canadian Journal of Fisheries and Aquatic Sciences*, **42**, 1733–40.

Stanhope, M.J., Powell, D.W. & Hartwick, E.B. (1987) Population characteristics of the estuarine isopod *Gnorimosphaeroma insulare* in three contrasting habitats: sedge marsh, algal bed, and wood debris. *Canadian Journal of Zoology*, **65**, 2097–104.

Stockner, J.G. & Costella, A. (1976) Marine phytoplankton growth in high concentrations of pulpmill effluent. *Journal of the Fisheries Research Board of Canada*, **33**, 2758–65.

Swank, W.T., Vose, J.M. & Elliott, K.J. (2001) Long-term hydrologic and water quality responses following commercial clearcutting of mixed hardwoods on a southern Appalachian catchment. *Forest Ecology and Management*, **143**, 163–78.

Thompson, J.A.J., Davis, J.C. & Drew, R.E. (1976) Toxicity, uptake, and survey studies of boron in marine environments. *Water Research*, **10**, 869–75.

Tschaplinski, P. (1987) The use of estuaries by juvenile coho salmon. In: *Proceedings of the Workshop: Applying 15 Years of Carnation Creek Results*, (ed. T.W. Chamberlin), pp. 123–42. Pacific Biological Station, Nanaimo, BC.

Tully, J.P. (1949) Oceanography and prediction of pulp mill pollution in Alberni Inlet. *Bulletin of the Biological Board of Canada*, **83**.

Vines, C.A., Robbins, T., Griffin, F.J. & Cherr, G.N. (2000) The effects of diffusible creosote-derived compounds on development in Pacific herring (*Clupea pallasi*). *Aquatic Toxicology*, **51**, 225–39.

Waldichuk, M. (1962) Some water pollution problems connected with disposal of pulp mill wastes. *Canadian Fish Culturist*, **31**, 3–34.

Weis, J.S. & Weis, P. (1996) The effects of using wood treated with chromated copper arsenate in shallow-water environments: a review. *Estuaries*, **19**, 306–10.

Williamson, C.J., Levings, C.D., Macdonald, J.S., White, E., Kopeck, K. & Pendray, T. (2000) A preliminary assessment of wood debris at four log dumps on Douglas Channel, British Columbia. *Canadian Manuscript Report Fisheries and Aquatic Science*, **2539.**

Wolanski, E. & Spagnol, S. (2000) Pollution by mud of Great Barrier Reef coastal waters. *Journal of Coastal Research*, **16**, 1151–6.

Wolff, M., Koch, V. & Isaac, V. (2000) A trophic flow model of the Caete Mangrove Estuary (North Brazil) with considerations for the sustainable use of its resources. *Estuarine Coastal and Shelf Science*, **50**, 789–803.

Chapter 16
Environmental effects of effluents from pulp and paper mills

K. R. MUNKITTRICK

Introduction

Effluents are discharged from pulp and paper mills into a wide variety of surface waters, and it is often difficult to generalize responses in organisms exposed downstream due to large differences in receiving environments, responses or species. Pulp production in Canada in 2001 was almost 30 million tonnes (FPAC 2002a), and the industry is one of Canada's largest employers. Aside from the production of domestic sewage, wastewater output from the pulp and paper industry exceeds all other in Canada. Environmental concerns about the effects of pulp and paper operations peaked in the late 1980s, and the Canadian pulp and paper industry has spent more than $6 billion on environmentally upgrading their processes and waste treatment systems since 1990 (FPAC 2002b); similar changes have been taking place in most of the industrialized nations. There are more than 130 pulp or paper mills across Canada (Fig. 16.1), with a wide variety of process and treatment types, receiving environments and dilution potential at the sites (Environment Canada 2002). There are also significant large pulp and paper industries in Sweden, Finland and the USA, and smaller industries in New Zealand and Australia.

The industry has been active for more than 100 years, and impacts, concerns and issues have been evolving continuously. Environmental concerns arising in the mid-1980s prompted the development of a series of international conferences on the environmental effects of pulp mills, with conferences conducted in Sweden in 1991 (Södergren 1992), Canada in 1994 (Servos *et al.* 1996), New Zealand in 1997 (Anonymous 1997) and Finland in 2000 (Ruoppa *et al.* 2000). These conference proceedings review much of the available recent information.

This chapter will discuss some of the recent studies on the responses of fish populations to pulp mill effluent exposure conducted worldwide. Pulp and paper mills require large amounts of wood fibre, electricity, water and labour. Consequently, pulp mills are seldom present in isolation but are co-located near sewage discharges, power facilities, wood sources and water sources. Furthermore, wood fibre supplies were often historically transported to facilities by floatation (in some locations until the 1990s), and there are often considerable fibre and bark deposits upstream and downstream of many mill facilities, as well as near coastal facilities. Impacts on fish associated with pulp and paper wastewaters being discharged into streams, rivers, lakes, estuaries and open coastline will be reviewed. The potential roles of historical fibre deposits, confounding discharges, and land-based activities associated with related activities are

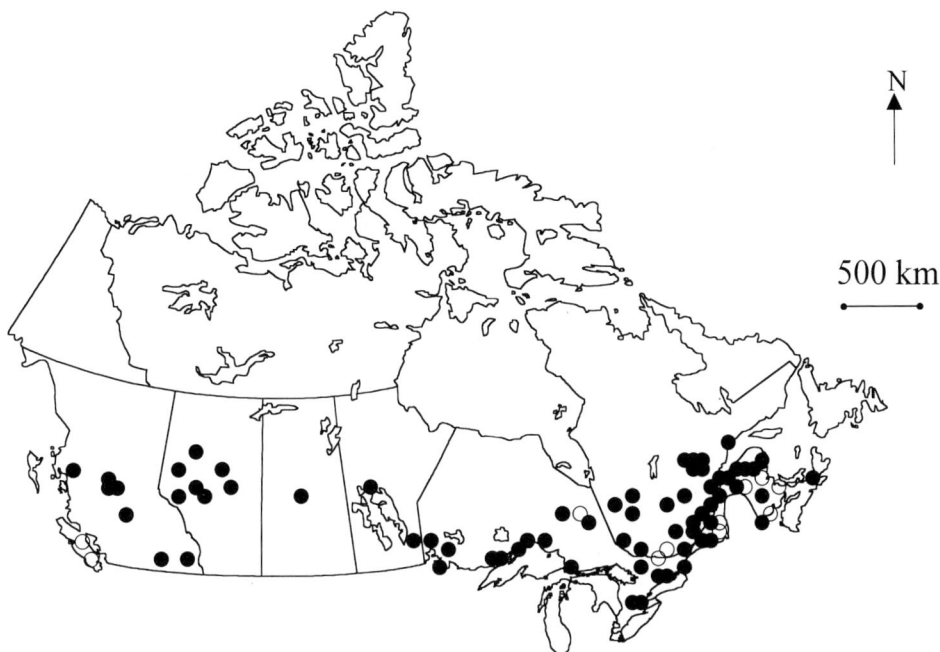

Fig. 16.1 Locations of sites where Environmental Effects Monitoring programs conducted fish studies during cycle 2 of the pulp mill monitoring program (1996–2000). The open circles represent ongoing research studies.

not directly considered. It is clear that this approach of fragmentation of issues will disappear as we move towards trying to deal with ecosystem impacts in a holistic and sustainable manner.

The massive investments made by the industry during recent modernization efforts have greatly improved environmental quality at many sites. This improvement has changed the impacts and responses, the concerns, and the responsible chemicals from those of papers published prior to the mid-1980s. The improving environmental situation means that present-day impacts are often reduced from the dramatic changes seen 40 years ago to subtle changes, often within the magnitude of the responses seen due to natural changes in environmental conditions. The process of trying to trace down the origin of impacts is complex, and further complicated by the lack of consensus about the presence of, or seriousness of, impacts seen in receiving environments. There is an intensive, ongoing, environmental monitoring programme near Canadian pulp and paper mills that is generating additional, valuable data for helping to understand the magnitude of changes near pulp and paper operations (Environment Canada 2002). This review will emphasize studies involving potential impacts of pulp mill effluents on fish populations. Extensive studies have also been conducted on benthic communities (Environment Canada 2002), sublethal toxicity (Scroggins *et al.* 2002) and chemical exposures and bioaccumulation, but they will not be dealt with here.

The pulp and paper industry

The purpose of pulp production is to remove, digest or dissolve most of the components of thousands of tonnes of trees to extract and purify the cellulose fibres used to make paper (more details can be found in Dubé & Munkittrick in press). The goal of pulping is to extract the maximum yield of wood fibre with the minimum degradation in quality. The extraction and digestion processes result in the concentration and/or generation of waste material that has to be disposed of. Some mills have a variety of process types, use different tree species in different lines or different seasons, and have different markets and customers that require a change or shift in processes during different production cycles to yield different qualities of fibres. The shifting wood quality, process type, operational efficiencies and waste treatment qualities constantly change the quality and composition of waste streams. The marked differences between mills in processes, strategies, wood sources and product requirements further increase the complexity of generalizing between mills.

Pulping process types include kraft (or sulphate), sulphite, groundwood, thermomechanical, chemi-thermomechanical and deinking processes. There are numerous configurations of mills, and combinations of processes, including mills that use multiple processing lines with different technologies. In general, trees are debarked or chipped, digested, washed, bleached, washed and dried to yield the fibre. Significant waste streams are generated during digestion (condensates, black liquor), bleaching (bleach effluent) and washing (brown stock wastes). There have been many recent changes in technologies within the pulp mills, including:

- movements towards dry debarking to reduce waste generation
- movements towards recovery of energy losses through recycling of waste streams
- development of oxygen delignification to decrease bleaching requirements
- alterations in bleaching strategies to reduce formation of chlorinated organic wastes by switching from elemental chlorine that was common in the 1970s, to chlorine dioxide (elemental chlorine-free or ECF bleaching).

Many countries, however, still use primitive processes and waste treatment strategies, depending on their market needs and environmental requirements. Pulp mill effluents contain hundreds or thousands of different chemicals, many of which remain unidentified. Compounds of concern have included organochlorine compounds (chlorinated phenols, catechols, guaiacols, dioxins, furans and resin and fatty acids), and extractive compounds (resin and fatty acids, phytosterols and phenols).

Waste treatment has evolved from primary treatment, with screening and settling of solids in clarifiers or settling basins (summarized in Dubé & Munkittrick in press), to secondary treatment. Secondary treatment usually consists of aerated stabilization basins that hold the waste for periods of time, or activated sludge treatment systems that aim to maximize biological activity by maintaining aerobic conditions and optimal nutrient concentrations in the basins. The goal of secondary treatment is to reduce biological oxygen demand, but the process also reduces suspended solids, acute lethality and effluent temperatures. Many pulp mills have also reduced waste discharges by treating, reducing or recirculating individual waste lines. There are several consequences to reductions in water use, including reduced spills within the mill, reduced effluent flows, increased retention time in waste treatment systems and increased concentrations of remaining chemicals.

Although waste treatment and process changes have dramatically reduced waste loadings, there is a law of diminishing returns, where further decreases in chemical loadings cost progressively more and more money. Environmental targets are needed to document when waste reductions have led to sustainability. Furthermore, the obvious benefits of reductions in water use are compromised by end-of-pipe effluent acute toxicity regulations that limit chemical concentrations in effluent and ignore loadings. These issues will continue to influence environmental impacts in the near future.

History of environmental concerns

Prior to the 1950s, little attention was given to the potential for effluents to impair environmental quality, but since 1950, environmental concerns can be summarized into three historical stages (reviewed more fully in Dubé & Munkittrick in press). From the 1950s to late 1970s, environmental concerns were predominantly associated with oxygen demand, suspended solids, nutrient loading and acute lethality to fish in receiving waters. In the 1970s and early 1980s, the focus of pulp and paper effects assessment was on trying to identify the chemicals responsible for acute toxicity of pulp and paper effluents; especially in terms of the roles of resin and fatty acids and chlorinated phenolics (reviewed in Dubé & Munkittrick in press).

Many of these compounds are relatively easily dealt with in secondary treatment systems and, in the 1980s, attention shifted to the potential environmental effects of chlorinated organic chemicals originating in pulp mills. The discovery of persistent chlorinated organic compounds (dioxin and furan congeners) bioaccumulating in aquatic biota led to regulations restricting their discharge. The concerns over the need for secondary waste treatment were quickly followed first by concerns about chlorinated organics, and then by the potential chronic environmental effects of exposure to pulp mill effluents.

There is considerable controversy about whether impacts are associated with effluents from modernized pulp mills with good quality control and waste treatment. Prior to the 1970s, impacts were dramatic, obvious and extensive at many sites due to the absence of significant waste treatment. As waste treatment and pulping processes have become more environmentally friendly, impacts have been reduced. It is common now for fish to be able to survive for months in 100% effluent, and for sublethal receiving environment changes to be related to subtle changes in growth rates, maturity and reproductive development. In the late 1990s to present, the assessment of pulp and paper effluent effects on fish has focused on fish reproduction and the discharge of chemical compounds that have the potential to disrupt normal reproductive processes. The reproductive effects of exposure to pulp mill effluents are not consistent between mills, studies, seasons or species and the basis for the inconsistencies is unknown (Dubé & Munkittrick in press).

Effects of effluent exposures

Studies conducted in Sweden in the early 1980s provided some of the first evidence that effluents from some pulp mills were capable of inducing toxic responses in fish at very low concentrations in the receiving environment (reviewed in Södergren 1992; Munkittrick *et al.* 1997, 2003). The changes in growth, biochemistry and deformities were dose-dependent and the area over which the effluents exerted health effects was considered to be large (>8–10 km from the pulp mill; >1000-fold dilution of the effluent). Subsequent Swedish studies, conducted after modernization of bleaching processes, showed that reproductive effects persisted (i.e. delayed sexual maturation, smaller embryos) (Sandström 1996; Larsson *et al.* 1997).

These studies were followed by a series of confirmation studies in Canada (reviewed in Munkittrick *et al.* 1998). Recent studies have been undertaken in Finland, the USA, New Zealand, Australia, Portugal, Taiwan and Chile (Table 16.1). A variety of monitoring designs have been used to assess the effects of pulp mill effluents on aquatic ecosystems including field assessments, caging studies, artificial stream or mesocosm studies, and laboratory toxicity tests (reviewed in Dubé & Munkittrick in press).

The original studies in Sweden were widely and heavily criticized (reviewed in Munkittrick *et al.* 2003), because of the absence of supporting data from North American receiving waters. The reviews suggested that the study sites were unique and that such effects would not occur near well-operated mills. Over the next few years, supporting data were obtained (Table 16.1) in Canada, the USA, Finland and New Zealand (reviewed in Munkittrick & Sandström 2003) confirming that effluent from some discharges was capable of affecting fish performance at much lower concentrations than previously suspected.

Impacts on the onset of spawning or time to maturation, the gonadal size or numbers of eggs produced, and secondary sex characteristics of fish have been seen in the laboratory and in the field (reviewed in Munkittrick *et al.* 1998). The impacts appear to operate through changes in the ability of fish to control their production of steroid hormones (reviewed in Munkittrick *et al.* 1997, 1998), and it is certain that the effluent from some pulp mills is capable of impacting reproductive performance of fish at low concentrations in the receiving environment. Some field and laboratory studies have shown that pulp mill effluents discharged from mills with modernized processes and secondary effluent treatment can affect fish reproduction (reviewed in Munkittrick *et al.* 1997, 2003). However, other studies have failed to observe reproductive effects in fish at environmentally relevant effluent concentrations at other sites (e.g. Swanson 1994). The factors determining which mills will or will not demonstrate responses are unknown, but may be associated with the specific receiving environment for the discharge and the site-specific factors associated with exposure and sensitivity of local fish species.

Interest was focused on the potential for physiological indicators to identify pulp mill effects in fish (Munkittrick & Sandström 2003). Induction of liver detoxification enzymes (mixed function oxygenase or MFO), for example, received much attention as a physiological indicator of exposure of fish to pulp mill effluent (see references in Martel *et al.* 1997). Similar attention became focused on the potential impacts on reproductive steroid hormones (see review in McMaster *et al.* 1996). Laboratory exposures of fish commonly emphasized the physiological endpoints (Table 16.2). Laboratory exposures have examined responses in terms of MFO induction, changes in sex steroids, stress responses and immune responses after exposures for various durations (Table 16.2). A variety of species has been used, but mill testing has predominantly been with kraft mill effluent. In general, the results of exposures have been supportive of field studies, demonstrating MFO induction (i.e. Martel *et al.* 1997), steroid hormone depression (e.g. McMaster *et al.* 1996; McCarthy *et al.* 2003) and changes in physiological parameters (Table 16.2).

Artificial streams have also been used to assess the effects of pulp mill effluents on fish (e.g. Dubé *et al.* 2002). The emphasis in Canada has largely been on developing the technology for use in Environmental Effects Monitoring Programs (Dubé *et al.* 2002), especially in cases where the receiving environments are difficult to sample or interpretation is complicated by confounding discharges. The exposure periods are generally much longer for mesocosm studies than for laboratory studies (Table 16.3), and many of the endpoints are whole organism integrative measures.

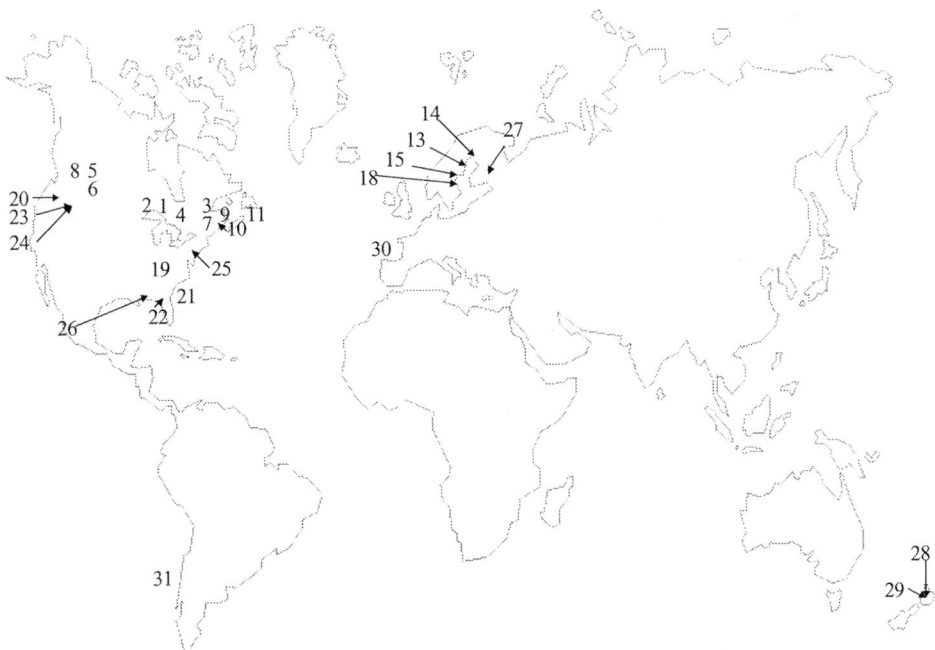

Fig. 16.2 Locations of research studies on pulp mill impacts on fish. Numbered sites correspond to those located in Table 16.1.

Table 16.1 Summary of studies of pulp mill effects on fish

Location		Species		Mill type	Source or reviewed in
Canada	1. Jackfish Bay ON	White sucker	*Catostomus commersoni*	Bleached kraft	Munkittrick *et al.* 1998
		Longnose sucker	*Catostomus catostomus*	Bleached kraft	Munkittrick *et al.* 1998
		Lake whitefish	*Coregonus clupeaformis*		
	2. Ontario (8+ mills)	White sucker	*Catostomus commersoni*	Various	Munkittrick *et al.* 1998
	3. St Maurice River, QC			Bleached kraft	Bussières *et al.* 1998
	4. Moose River basin			Various	Munkittrick *et al.* 1998
					Janz *et al.* 2001
	5. Wapiti/Smoky River, AL	Trout perch	*Percopsis omiscomaycus*	Various	Gibbons *et al.* 1998b
		Longnose sucker	*Catostomus catostomus*	Bleached kraft	Swanson 1994
		Mountain whitefish	*Prospium williamsoni*		
		Longnose sucker	*Catostomus catostomus*		McMaster *et al.* 2000
		Trout-perch	*Percopsis omiscomaycus*		
	6. Athabasca River, AL	Spoonhead	*Cottus ricei*	Bleached kraft	Gibbons *et al.* 1998a
		Sculpin			
	7. St Francois River, QC	White sucker	*Catostomus commersoni*	Various	Kovacs *et al.* 2002
		Smallmouth bass	*Micropterus dolomieu*		
		Tesselated darter	*Etheostoma olmstedi*		
	8. Fraser River	Chinook salmon	*Oncorhynchus tshawytscha*	Various	Wilson *et al.* 2000
	9. Miramichi River, NB	Mummichog	*Fundulus heteroclitus*	Various	LeBlanc *et al.* 1997
					Fournier *et al.* 1998
	10. New Brunswick and Nova Scotia (4 mills)	Tomcod	*Microgadus tomcod*	Various	Williams *et al.* 1998
		Tomcod	*Microgadus tomcod*	Various	Couillard *et al.* 1999
	11. Port Hamon, Birchy Cove, NF	Winter flounder	*Pleuronectes americanus*		Khan & Hooper 2000
	12. Various (65 mills)	Various		Various	Environment Canada 1997
Sweden	13. Norrsundet	Eurasian perch	*Perca fluviatilis*	Bleached kraft	Sandström 1996
					Ericson & Larsson 2000

Country	Location	Common name	Species	Effluent type	Reference
	14. Husum			Bleached kraft	Sandström 1996
	15. Karlsborg			Bleached kraft	
	16. Munksund			Unbleached kraft	
	17. Lövholmen			Unbleached kraft	
	18. Mönsterås	Eelpout	*Zoarces viviparus*	Bleached kraft	Larsson *et al.* 2000
	19. Pigeon River, TN	Redbreast sunfish	*Lepomis auritus*	Bleached kraft	Adams *et al.* 1992 (in Munkittrick *et al.* 1998)
USA	20. Columbia River, WA	White sturgeon	*Acipenser transmontanus*	Bleached kraft + non-pulp mill sources	Foster *et al.* 2001
	21. St Johns, FL	Largemouth bass	*Micropterus salmoides*	Bleached kraft	Sepúlveda *et al.* 2002
	22. Fenholloway River	Mosquitofish	*Gambusia holbrooki*	Bleached kraft	Parks *et al.* 2001
	23. MacKenzie River, OR	Various	Various	Bleached and unbleached	Hall *et al.* 2000
	24. Willamette River, OR	Various	Various		
	25. Codorus Creek, PA				
	26. Leaf River, MS				
	Nor defined	Bluegill sunfish	*Lepomis macrochirus*	Bleached kraft	D'Surney *et al.* 2000
	Not defined	Blue catfish	*Ictalurus furcatus*	Unbleached kraft	Felder *et al.* 1998
Finland	27. Lake Saimaa	Perch	*Perca fluviatilis*	CTMP and kraft mill	Karels & Oikari 2000
	28. Tarawera River	Rainbow trout	*Oncorhynchus mykiss*		Donald 1997
New Zealand	29. Waikato River	Rainbow trout	*Oncorhynchus mykiss*		Richardson 1997
		Goldfish	*Carassius auratus*		
		Common bully	*Gobiomorphus cotidianus*		
Portugal	30. Vouga River	Eel (caged)	*Anguilla anguilla*	Bleached kraft	Pacheco & Santos 1999; Ferreira *et al.* 2002
Chile	31. Biobio River		*Mugil cephalus*	Bleached kraft	Gaete *et al.* 2000
			Eleginoops maclovinus		Barra *et al.* 2001

Table 16.2   Summary of laboratory fish studies with pulp mill effluents

Species	Species (Latin)	Endpoints	Duration	Mill type	Author
Fathead minnows	*Pimephales promelas*	Growth, reproduction	Up to 180 days	Bleached kraft	Robinson 1994
					Kovacs *et al.* 1997
			Up to 200 days	Bleached kraft	Borton *et al.* 2000a
			90–120 days	Bleached sulphite	Parrott *et al.* 2000c
Rainbow trout	*Oncorhynchus mykiss*	MFO induction	4 days	Various	Marrel and Kovacs 1997
		Physiological, biochemical, growth	4.5 months	Bleached kraft	Mattson *et al.* 2001b
		Vitellogenin, steroids, MFOs	21 days	Bleached kraft	Tremblay & Van Der Kraak 1999
	Channa punctatus	Immune endpoints	15–90 days	Paper mill	Fatima *et al.* 2001
	Heteropneustes fossilis	Haematology	90 days	Paper mill	Ahmad *et al.* 1998
Bluegill sunfish	*Lepomis macrochirus*	Haematology	56 days	Bleached kraft	D'Surney *et al.* 2000
		Organ size and biochemical	30 days	Unbleached kraft	Felder *et al.* 1998
Chinook salmon	*Oncorhynchus tshawytscha*	MFO, DNA adducts	28 days	Bleached kraft	Wilson *et al.* 2001
Goldfish	*Carassius auratus*	Sex steroids	Up to 20 days	Various	McCarthy *et al.* 2003
					Parrott *et al.* 2000b
Largemouth bass	*Micropterus salmoides*	Messenger RNA	7 days	Bleached kraft	Denslow *et al.* 2000
European whitefish	*Coregonus lavaretus*	Stress responses	3–6 weeks	Bleached kraft	Lappivaara 2001
		Vitellogenin gene, physiological parameters	30 days	Bleached kraft	Soimasuo *et al.* 1998b

Roach	*Rutilus rutilus*	Immune system, physiological changes	21 days	Bleached kraft	Aaltonen et al. 2000a
Tilapia	*Oreochromis mossambicus*	MFO	3–7 days	Bleached kraft	Chen et al. 2001
Eel	*Anguilla anguilla*	EROD, stress parameters	Up to 3 s, up to 188 hours	Bleached kraft / Bleached kraft	Pacheco & Santos 1999 / Santos & Pacheco 1996
Common jollytail	*Galaxis maculates*	Whole organism, MFO, histology	Months	Eucalypt-based pulp mill	Woodworth et al. 1997
Mosquitofish	*Gambusia affinis*	External characteristics	21 days	Not stated	Ellis et al. 2000
Exposure to specific compounds					
Rainbow trout	*Oncorhynchus mykiss*	MFOs, developmental abnormalities	4–32 days	Retene	Billiard et al. 1997 / Fragoso et al. 2000 / Oikari et al. 2002
Zebrafish	*Danio rerio*	Developmental abnormalities / Steroid levels, vitellogenin	14 days / 21 days	Retene / β-Sitosterol	Billiard et al. 1997 / Tremblay & Van Der Kraak 1998, 1999
Brown trout	*Salmo trutta lacustris*	Reproduction and offspring survival	4.5 months	Wood sterols	Lehtinen et al. 1999
Eelpout	*Zoarces viviparus*	Reproduction and offspring survival	8+ months	Wood sterols	Mattson et al. 2001a

Table 16.3 Long-term mesocosm studies with pulp mill effluents

Species	Endpoints	Duration	Mill type	Author
Brown trout *Salmo trutta*	Steroid levels, egg hatchability	4 months	Mechanical	Johnsen *et al.* 2000
	Physiological, biochemical	8 weeks	Thermomechanical	Johnsen *et al.* 1998
Rainbow trout *Oncorhynchus mykiss*	Growth, gonad development, steroids, vitellogenin	2 months	Bleached kraft/TMP	van den Heuvel *et al.* 2002
	Whole organism, biochemical	6–9 months	Bleached kraft	Hall *et al.* 1997
Largemouth bass, Bluegill sunfish, Golden shiners *Micropterus salmoides, Lepomis macrochirus* *Notemigonus crysoleucas*	Production, whole organism, steroids MFOs	510–610 days	Bleached kraft	Borton *et al.* 2000b
European whitefish *Coregonus lavaretus*	Stress responses, vitellogenin gene	3–6 weeks	Bleached kraft	Lappivaara 2001 Mellanen *et al.* 1999
Mummichog *Fundulus heteroclitus*	Whole organism and sex steroids	7–57 days	Bleached kraft	Dubé & MacLatchy 2000, 2001 Dubé *et al.* 2002
	Whole organism and sex steroids	30–57 days	Thermomechanical	Dubé *et al.* 2002
Stickleback *Gasterosteus aculeatus*	Length, weight	Up to 126 days, 16 months	Bleached magnephite pulp Bleached kraft	Lehtinen 1997 Tana *et al.* 1997

There is a wide variety of concerns about the studies on fish ranging from the suitability of reference sites to potential confounding factors (Kovacs *et al.* 1997), to the ecological relevance of changes that are seen (Munkittrick *et al.* 2000; Kovacs *et al.* 2002). Problems encountered during environmental monitoring of fish exposed to pulp mill effluents have been reviewed (Environment Canada 1997; Munkittrick *et al.* 2002). In many field studies, the concentration and duration of exposure to effluent is difficult to quantify because of complex effluent dispersion in the water column, mobility of fish in and out of the effluent plume, and the unavoidable presence of other effluent discharges. In these situations, caging studies and mesocosm approaches have been developed to isolate the potential impacts of effluents while simulating receiving environment conditions (Tables 16.3 and 16.4).

Caging studies have been used in Finland for studying pulp mill impacts for more than a decade, and have been used more recently in New Zealand and Portugal (Table 16.4). Caging studies in Canada have emphasized responses in sex steroids or characteristics of the MFO induction associated with exposure. Biochemical responses to exposure can occur within days of exposure (e.g. Parrott *et al.* 1999; Hewitt *et al.* 2002), but effects appear to be transient after exposure ceases (Munkittrick *et al.* 1999). Caging exposures are generally not favoured in Canadian studies (Environment Canada 1997), and stress responses in wild fish put into cages can occur quickly (Jardine *et al.* 1996), often altering the interpretation of some of the reproductive endpoints.

Environmental Effects Monitoring Programs

There is an enormous amount of information being generated in Canada through the Environmental Effects Monitoring (EEM) Program. While EEM programs have been developed in both Sweden (Sandström *et al.* 1997; Swedish EPA 1997) and Australia (Keough & Mapstone 1997), neither program is mandatory for all operating mills. The Canadian program was implemented in 1992 as part of a set of revised pulp and paper effluent regulations (Walker *et al.* 2002). The regulatory package emphasized the elimination of acute toxicity in whole effluent and the elimination of persistent organochlorine compounds, but included a requirement for mill discharging effluents to conduct an EEM program on a cyclical basis. The objective of the EEM program is to assess the adequacy of national regulations for protecting fish, fish habitat and the use of fisheries resources. The EEM program was designed with requirements to consider effluent toxicity, benthic community structure, fish populations, fish consumption and fish contamination. Methodologies for EEM are available in extensive Technical Guidance Documents and interpretation documents (see www.ec.gc.ca/eem). More recent guidance produced for the new metal mining program is also available (Ribey *et al.* 2002).

The fish population requirements are related to measurements of life history characteristics. The first cycle of studies was completed in April 1996 and >115 studies reported results (fish studies reviewed in Environment Canada 1997). Expert working groups were established to review the studies, and to make recommendations for changes to the regulations to allow more site-specificity and flexibility in the second

Table 16.4 Summary of caging studies looking at responses of fish to pulp and paper mill effluents

Species		Endpoints	Duration	Exposure	Author
White sucker	*Catostomus commersoni*	MFO	Up to 14 days	Bleached kraft	Munkittrick *et al.* 1999
Goldfish	*Carassius auratus*	Steroids	Up to 16 days	Bleached kraft	McMaster *et al.* 1996
Crucian carp	*Carassius carassius*	Organ size, histology, steroids, vitellogenin	Living in secondary treatment pond	Bleached kraft	Kukkonen *et al.* 1999
Eel	*Anguilla anguilla*	MFO and biochemical	3 days	Bleached kraft	Pacheco & Santos 1999
Whitefish	*Coregonus lavaretus*	MFO, bile conjugates, vitellogenin, steroids	1 month	Bleached kraft	Karels *et al.* 1999 Lappivaara *et al.* 2002
		Physiological stress responses	30 days	Various	Mellanen *et al.* 1999
		Vitellogenin gene	30 days	Various	Soimasuo *et al.* 1998a
		Physiological parameters	30 days	Various	
Largemouth bass	*Micropterus salmoides*	Whole organism, steroids, vitellogenin	28–56 days	Various	Sepúlveda *et al.* 2001
Long-finned and short-finned eels	*Anguilla diefenbachii*, *A. australia*	EROD, biochemical, whole organism	32–108 days	Various	Jones *et al.* 1997

cycle (Environment Canada 1997). Cycle 2 pulp and paper EEM reported in April of 2000, and results have recently been summarized (Environment Canada 2002). The fish studies were summarized by Munkittrick *et al.* (2002) and Courtenay *et al.* (2002), and included data from 65 different sites. Three distinct response patterns were visible for fish, ranging from eutrophication to food limitation, but the national trend demonstrated was metabolic disruption (Munkittrick *et al.* 2002). Metaboloic disruption consists of reduced gonad sizes, with increased condition, liver size and size-at-age (Environment Canada 2002).

Attempts to identify the causes of impacts associated with pulp mill effluents

Once sublethal biological effects were found in fish in the late 1980s, it was widely assumed that identification of the responsible compounds would come quite quickly. This has not happened for a variety of reasons, including:

- laboratory testing depends on surrogate species and on biochemical testing
- effluents vary in chemical content on a daily (or more frequent) basis, and the stability of the unknown chemicals in transport to laboratories is unknown
- laboratory testing of pure compounds has unknown relevance with regard to the potential bioavailability, environmental fate, uptake and metabolism of those compounds from complex effluent mixtures
- biochemical lesions have not been directly associated with the responses of concern, and responses can be inconsistent.

Responses to exposures to different types of mill and treatment processes have been compared in field studies or laboratory studies, but responses have been inconsistent and impossible to generalize (reviewed in Kovacs *et al.* 1997; Munkittrick *et al.* 1998). Studies have also attempted to correlate responses with mill processes, isolate specific waste streams, chemicals or chemical fractions associated with responses, and have tested specific suspect compounds (reviewed in Dubé & Munkittrick in press).

Various approaches have been taken to try and identify or isolate the responsible chemicals, including fractionation of the effluent (Hewitt *et al.* 1996; Burnison *et al.* 1999), the reproduction of responses in small-scale models (Coakley *et al.* 2001), and the definition of the characteristics of the responsible chemicals (Munkittrick *et al.* 1999). These approaches have sometimes focused on isolated biochemical responses such as the induction of liver detoxification enzymes (MFOs) (Martel & Kovacs 1997; Schnell *et al.* 2000), or the ability to produce sex steroid hormones (McMaster *et al.* 1996; McCarthy *et al.* 2003; Parrott *et al.* 1999). Some experiments have moved towards allowing the fish to accumulate the chemicals during short-term effluent exposures, and fractionating the liver-associated contaminants to track responses. The accumulated chemicals can be separated into different chemical fractions and responses can be followed in terms of potential for MFO induction (Parrott *et al.* 2000a), or responses of oestrogen receptors, androgen receptors and sex steroid-binding proteins (Hewitt *et al.* 2000, 2002). These studies have shown similar patterns in the bioavailability and solubility of chemicals functioning as ligands for sex steroid receptors at a bleached kraft mill (Hewitt *et al.* 2000) and more recently at a bleached sulphite/groundwood mill (Hewitt *et al.* in press), where effects on fish reproduction have been observed.

Studies have also approached the issue by the isolation of individual waste streams (e.g. Martel *et al.* 1997; Parrott *et al.* 1999, 2000a) or by following the responses of waste streams to process or treatment changes (Dubé & MacLatchy 2000, 2001). If the responsible process could be isolated, it may be possible to alter the process or the treatment, or to selectively treat the specific line to reduce the consequences of the chemicals without needing to know the identity of the chemicals or their characteristics.

It is clear from exposures that there are multiple compounds that can be associated with responses in different effluents and in different waste streams, and that the responses may not be consistent between mills. It is not clear if the various biochemical lesions are associated with the same chemicals, or whether they are responsible for the whole organism consequences of exposure.

Chemicals playing undefined roles in effects

During studies in the early 1990s, it became apparent that:

- it would be more difficult than originally assumed to identify the responsible chemicals
- there may be more than a single chemical or group of chemicals responsible for impacts, and
- the responsible chemicals may be previously unidentified chemicals (reviewed in Dubé & Munkittrick in press).

Although several compounds have been identified which can be associated with some of the responses, there are concerns about the bioavailability, environmental fate, uptake and metabolism that are poorly understood for most of the compounds. Furthermore, there are also inconsistencies with responses that reduce the level of confidence that any of the compounds are directly related to all of the reproductive responses in wild fish (Van Der Kraak *et al.* 1998).

The observation that reproductive effects occurred at mills with different bleaching technologies suggested that the responsible compounds might be natural wood compounds as opposed to cooking or bleaching chemicals (reviewed in Dubé & Munkittrick in press). Some tannins, natural wood compounds which form the natural defence mechanism of a tree to ward off insect infestations, and some plant sterols are present in pulp mill effluents. Recent evidence indicates that several unidentified chemicals related to lignin degradation products are associated with steroid reductions in chemical recovery condensates at a bleached kraft mill, and work is ongoing to characterize these compounds (Hewitt *et al.* 2002). We do not know any more about these natural compounds now than we knew about chlorophenols in the early 1990s, and conclusive evidence about their potential role in impacts will not come quickly. Laboratory testing is dependent on surrogate species, and relies heavily on biochemical testing. There are concerns that biochemical changes may not occur until reproduction is already disrupted, or may not be directly related to reproductive impairment. Furthermore, the capacity of single chemicals to impact reproductive performance has unknown relevance to the capability of those compounds to exert influence within an effluent composed of hundreds of compounds.

It is clear that compounds associated with the responses can originate from a variety of sources, including:

(1)   The bleaching process: the presence of dioxins and furans in effluents in the late 1980s and 1990s was widely believed to be responsible for some of the effects (reviewed in Munkittrick *et al*. 1997). The disappearance of some biochemical responses after a short-term maintenance shutdown suggested that these compounds were not playing a direct role in the biochemical impacts (Munkittrick *et al*. 1998).

(2)   Weak black liquor carryover with the pulp stock is probably one source, as it contains many compounds associated with some of the responses (i.e. phytosterols, resin and fatty acids) (reviewed in Dubé & MacLatchy 2000, 2001).

(3)   Condensates also appear to be a source of compounds causing MFO induction (Martel *et al*. 1997), acute toxicity, sublethal toxicity and sex steroid depressions in fish (Dubé & MacLatchy 2000, 2001). Dubé & MacLatchy (2000) observed that condensates may be a more important source causing acute toxicity of a final pulp mill effluent on fish than other sources (weak black liquor carryover), although these results are mill-specific. At the mill studies, the causative substance(s) in condensates were low molecular weight, neutral compounds volatilized into the condensates through weak black liquor evaporation and removed by reverse osmosis with molecular weight cut-off of 60 Da. In these studies, phenolics were identified as a major compound class removed by reverse osmosis (Dubé & MacLatchy 2001). Bleaching may play a role in modifying these phenolic compounds, affecting their biological activity (Hewitt *et al*. 1996).

(4)   Secondary treatment (Dubé *et al*. 2002): some steroid responses appear at some mills to be worse after secondary treatment. Hewitt *et al*. (2001) conducted a survey of effluents before and after treatment from a variety of pulping process types and found differences in the ability of effluents to interact with sex steroid receptors. There was no correlation with process type, wood furnish or treatment with the ability of effluent extracts to compete for fish sex steroid receptors; activity after treatment was reduced at some mills but increased at other sites, with no apparent pattern.

The compounds responsible for the reproductive impacts will be very difficult to identify from a suite of potential compounds with demonstrated activity for at least some of the responses. The examination of dioxin's roles took several years, since it could cause the responses and was present in effluents (reviewed in Munkittrick *et al*. 1998). It will be difficult to compare single chemical exposures and whole effluent exposures, and the methods followed by PAPRICAN in their TMP fractionations provide a useful model (Martel *et al.* 1997). However, ecological assessments need to play an important role in verification of decisions. The use of field studies at sites where dioxin levels have dramatically decreased has been most useful in the assessment (Munkittrick *et al.* 1997, 2000). Since process changes are occurring for a variety of reasons, and responses are improving at most well-studied sites, there is the chance that newer processes and better operation will reduce reproductive impacts before the responsible chemicals are identified. However, total recovery has not yet been seen in any field studies (Munkittrick *et al.* 1997, 2003).

Conclusions

Various studies have documented the potential of effluents from some pulp and paper mills to affect the growth, condition and reproductive performance of fish in their discharge zones. The 'risk' seems to be lower when effluents are diluted to <1% in the receiving area, but differences between mills, receiving environments and species prevent any generalizations. There is a lot of controversy surrounding the existence and importance of impacts.

The ultimate goal of environmental assessment and monitoring is to attain scientifically based information on environmental health and transfer that information into a process of environmental management. With respect to the pulp and paper sector, it has been difficult to get consensus among multi-stakeholder groups on definitions of what constitutes an effect, when is a response an impact, and when does an impact become damage. For example, after 12 years of studies conducted on the potential impacts of pulp mill effluents at Jackfish Bay, Lake Superior (Munkittrick *et al.* 1997), there is no consensus on whether impacts exist. The controversy rests on whether the delayed maturity and altered gonadal sizes observed at this site represent 'impacts'.

The failure to achieve consensus on the existence or importance of changes, coupled with the improvements in habitat and partial recovery in general health and reproductive function in fish seen at several sites, has led to a widely held perception that the issue has been solved. In terms of the reproductive impacts, there is still some doubt as to whether chlorine was involved in the original responses, whether secondary treatment changed the level of impact, and whether the responsible compounds are known (see Munkittrick & Sandström 2003). In terms of ecological assessment, too much emphasis is being placed on potential confusion associated with natural variability, reference site variability, confounding factors and ecological relevance (see Munkittrick *et al.* 2003). Taken together, there are several reasons why caution should still be exercised as to whether this issue has been 'solved':

(1) The sites with the longest running studies maintain reproductive impacts after considerable modernization of mills (Larsson *et al.* 1997; Munkittrick *et al.* 1997) and impacts have been seen at mills with modernized bleaching (Munkittrick *et al.* 2000).

(2) New sites with new impacts are being reported (Larsson *et al.* 2000) and new studies in Portugal, Chile and the USA are reporting impacts (Table 16.1).

(3)   Reductions in effluent toxicity associated with improved treatment allow fish to inhabit areas of higher effluent concentrations, thereby increasing their relative exposure.

(4)   Reductions in the effluent concentrations of some compounds previously associated with biological responses may allow other compounds to exert impacts which were previously prevented by biological availability or transformation, or were otherwise masked (i.e. changes in secondary sex characteristics at Jackfish Bay; Munkittrick *et al.* 2003).

(5)   New studies are suggesting that some effluents may exert new or more potent effects after secondary treatment.

Ecological assessments can be dramatically affected by interpretation difficulties and personal biases related to differences in the interpretation of changes (reviewed in Munkittrick & Sandström 2003); these include:

•   Biochemical differences would have to be persistent year-round and under all conditions to be important, and must be linked to higher levels of organization to be useful.

•   Organismal and suborganismal changes which do not impact the population and community level are not ecologically relevant.

•   Population level changes which do not impact abundance are not important.

•   Abundance changes are not important if populations do not become rare or extirpated.

One of the major factors limiting progress into interpretation of effects is the misconception that all impacts are adverse, and therefore are unacceptable. A fish may be impacted at the biochemical level without impacting survival, and individual survival can be impacted without damaging the population. One area of research into improving ecological field assessments focuses on data interpretation and determining the ecological relevance of changes (not necessarily the statistical significance of changes). Munkittrick *et al.* (2000) have developed an effects-based assessment approach to assess the cumulative effects of anthropogenic stressors (including pulp and paper mill effluent) on fish. In this approach, a method suggested to assess whether changes observed in fish are ecologically relevant is to measure the magnitude of the change relative to natural variability measured at reference sites.

Once the scientifically based information is attained for a particular aquatic system, it should be incorporated into a decision-making framework that also incorporates other stakeholder issues of economic and sociological importance. In Sweden, decisions on environmental quality objectives may be based on ethical concerns and public perceptions and not strictly on ecological relevance (Swedish EPA 1997). It is the responsibility of the scientific process to objectively quantify and provide information on the types of changes, the distribution of changes, the frequency and duration of changes, and the relevance of changes to other levels of organization so that this information can be used to make more informed decisions.

The question of ecological relevance may not always be directly relevant to the question of defining unacceptable changes. In some cases, changes that are not ecologically relevant, such as tainting or undesirable external lesions, may be deemed unacceptable by stakeholder negotiation. In other cases, changes in sexual maturity or fecundity, which are ecologically relevant, may be deemed acceptable when examined within the context of sociological and economic factors that play a role in management decisions. The priorities for ecological assessment have to be on defining impacts, understanding the mechanisms, and defining the factors limiting the performance of a particular system.

Acknowledgements

The author is supported by a Canada Research Chair Fellowship and an NSERC operating grant.

References

Aaltonen, T.M., Jokinen, E.I., Lappivaara, J., *et al.* (2000a) Effect of primary and secondary-treated bleached kraft mill effluents on the immune system and physiological parameters of roach. *Aquatic Toxicology*, **51**, 55–67.

Aaltonen, T.M., Jokinen, E.I., Salo, H.M., *et al.* (2000b) Modulation of immune parameters of roach, *Rutilus rutilus*, exposed to untreated ECF and TCF bleached pulp effluents. *Aquatic Toxicology*, **47**, 277–89.

Ahmad, I., Fatima, M., Athar, M., *et al.* (1998) Responses of circulating fish phagocytes to paper mill effluent exposure. *Bulletin of Environmental Contamination and Toxicology*, **61**, 746–53.

Anonymous (1997) *Proceedings of the 3rd International Conference on the Environmental Fate and Effects of Pulp and Paper Effluents*, 10–13 November 1997, Rotorua, New Zealand. Forest Research Institute, Rotorua, New Zealand.

Barra, R., Notarianni, V. & Gentili, G. (2001) Biochemical biomarker responses and chlorinated compounds in the fish *Leuciscus cephalus* along a contaminant gradient in a polluted river. *Bulletin of Environmental Contamination and Toxicology*, **66**, 582–90.

Billiard, S.M., Querbach, K. & Hodson, P.V. (1997) Toxicity of retene to early life stages of two freshwater fish species. In: *Proceedings of the 3rd International Conference on the Environmental Fate and Effects of Pulp and Paper Effluents*, 10–13 November 1997, Rotorua, New Zealand. pp. 362–9. Forest Research Institute, Rotorau, New Zealand.

Borton, D., Streblow, W., Bousquet, T. & Cook, D. (2000a) Fathead minnow (*Pimephales promelas*) reproduction during multi-generation life cycle tests with kraft mill effluents. In: *Proceedings of the 4th International Conference on Environmental Impacts of the Pulp and Paper Industry*, 12–15 June 2000. pp. 52–157. Finnish Environment Institute Report 417.

Borton, D.L., Bradley, W.K., van Veld, P. & Bousquet, T. (2000b) Survival, growth, reproduction and bioindicator responses of fish exposed in experimental streams to biological treated effluent from a O+ECF kraft mill. In: *Proceedings of the 4th International Conference on Environmental Impacts of the Pulp and Paper Industry*, 12–15 June 2000. pp. 219–25. Finnish Environment Institute Report 417.

Burnison, B.K., Comba, M.E., Carey, J.H., *et al.* (1999) Isolation and tentative identification of compound in bleached kraft mill effluent capable of causing mixed function oxygenase induction in fish. *Environmental Toxicology and Chemistry*, 18, 2882–7.

Bussières, D., Gagnon, M.M., Dodson, J. & Hodson, P.V. (1998) Does annual variation in growth and sexual maturation of white sucker (*Catostomus commersoni*) confound comparison between pulp mill contaminated and reference river? *Canadian Journal of Fisheries and Aquatic Sciences*, 55, 1068–77.

Chen, C.C., Liu, M.C., Shih, M.L., *et al.* (2001) Microsomal monooxygenase activity in tilapia (*Oreochromis mossambicus*) exposed to a bleached kraft pulp mill effluent using different exposure systems. *Chemosphere*, 45, 581–8.

Coakley, J., Hodson, P.V., van Heinengen, A. & Cross, T. (2001) MFO induction in fish by filtrates from chlorine dioxide bleaching of wood pulp. *Water Research*, 35, 921–8.

Couillard, C.M., Williams, P.J., Courtenay, S.C. & Rawn, G.P. (1999) Histopathological evaluation of Atlantic tomcod (*Microgadus tomcod*) collected at estuarine sites receiving pulp and paper mill effluent. *Aquatic Toxicology*, 44, 263–78.

Courtenay, S.C., Munkittrick, K.R., Dupuis, H.M.C., *et al.* (2002) Quantifying impact of pulp mill effluent on fish in Canadian marine and estuarine environments: problems and progress. *Water Quality Research Journal of Canada*, 37, 79–99.

Denslow, N.D., Saunders, V., Chow, M.M., *et al.* (2000) The estrogen mimicking potential of effluent from a CEHD type paper mill in Central Florida. In: *Proceedings of the 4th International Conference on Environmental Impacts of the Pulp and Paper Industry*, 12–15 June 2000. pp. 227–31. Finnish Environmental Institute Report 417.

Donald, R. (1997) Status and health of rainbow trout (*Oncorhynchus mykiss*) in the Tarawere River, New Zealand. In: *Proceedings of the 3rd International Conference on the Environmental Fate and Effects of Pulp and Paper Effluents*, 10–13 November 1997, Rotorua, New Zealand. pp. 299–309.

D'Surney, S.J., Eddy, L.P., Felder, D.P., *et al.* (2000) Assessment of the impact of a bleached kraft mill effluent on a south-central USA river. *Environmental Toxicology*, 15, 28–39.

Dubé, M.G. & MacLatchy, D.L. (2000) Endocrine responses of *Fundulus heteroclitus* to effluent from a bleached kraft pulp mill before and after installation of reverse osmosis treatment of waste stream. *Environmental Toxicology and Chemistry*, 19, 2788–96.

Dubé, M.G. & MacLatchy, D.L. (2001) Identification and treatment of a waste stream at a bleached kraft pulp mill that depresses a sex steroid in the mummichog (*Fundulus heteroclitus*). *Environmental Toxicology and Chemistry*, 20, 985–95.

Dubé, M.G. & Munkittrick, K.R. (in press) Case study: pulp and paper mill impacts. In: *The Toxicology of Fishes*, (eds R. Di Giulio & D. Hinton). Taylor & Francis, New York.

Dubé, M., MacLatchy, D., Culp, J., *et al.* (2002) Utility of mobile, field-based artificial streams for assessing effects of pulp mill effluents on fish in the Canadian environmental effects monitoring (EEM) program. *Journal of Aquatic Ecosystem Stress & Recovery*, 9, 85–102.

Ellis, R.J., van den Heuvel, M.R., Stuthridge, T.R., *et al.* (2000) Masculinization of adult female mosquitofish (*Gambusia affinis*) exposed to pulp and paper mill wastewaters. In: *Proceedings of the 4th International Conference on the Environmental Impacts of the Pulp and Paper Industry*, 12–15 June 2000. p. 226. Finnish Environment Institute Report 417.

Environment Canada (1997) *Fish Survey Expert Working Group: Recommendations from Cycle 1 review.* EEM/1997/6.

Environment Canada (2002) *Review of Environmental Effects Monitoring Studies, Cycle 2 Results.* Environment Canada, Hull, QC.

Ericson, G. & Larsson, A. (2000) DNA adducts in perch (*Perca fluviatilis*) living in coastal waters polluted with bleached kraft pulp mill effluents. *Ecotoxicology and Environmental Safety*, 46, 167–73.

Fatima, M., Ahamad, I., Siddiqui, R. & Raisuddin, S. (2001) Pulp and paper mill effluent induced immunotoxicity in freshwater fish *Channa punctatus* Bloch. *Archives of Environmental Contamination and Toxicology*, 40, 271–6.

Felder, D.P., D'Surney, S.J., Rodgers, J.H. & Deardorff, T.L. (1998) A comprehensive environmental assessment of a receiving aquatic system near an unbleached kraft mill. *Ecotoxicology*, 7, 313–24.

Ferreira, R.C.F., Graca, M.A.S., Craveiro, S., *et al.* (2002) Integrated environmental assessment of BKME discharged to a Mediterranean river. *Water Quality Research Journal of Canada*, 37, 181–93.

Foster, E.P., Fitzpatrick, M.S., Feist, G.W., *et al.* (2001) Plasma androgen correlation, EROD induction, reduced condition factors and the occurrence of organochlorine pollutants in reproductively immature white sturgeon (*Acipenser transmontanus*) from the Columbia River, USA. *Archives of Environmental Contamination and Toxicology*, 41, 182–91.

Fournier, M., Lacroix, A., Voccia, I. & Brousseau, P. (1998) Phagocytic and metabolic activities of macrophages from mummichog naturally exposed to pulp mill effluents in the Miramichi River. *Ecotoxicology and Environmental Safety B*, 40, 177–83.

FPAC (Forest Products Association of Canada) (2002a) *2001 Annual Review.* (www.cppa.prg/cgi-bin/view pddf.cqi/FPAC2001.pdf)

FPAC (Forest Products Association of Canada) (2002b) Pulp and paper operations in Canada. (www.open.doors.cppa.ca/english/info/work/.htm)

Fragoso, N., Parrott, J.L., Hahn, M.E. & Hodson, P.V. (2000) Chronic retene exposure causes sustained induction of Cyp1A activity and protein in rainbow trout (*Oncorhynchus mykiss*). *Environmental Toxicology and Chemistry*, 17, 2347–53.

Gaete, H., Larrain, A., Bay-Schmith, E., *et al.* (2000) Ecotoxicological assessment of two pulp mill effluents, BioBio River Basin, Chile. *Bulletin of Environmental Contamination and Toxicology*, 65, 183–9.

Gibbons, W.N., Munkittrick, K.R. & Taylor, W.D. (1998a) Monitoring aquatic environments receiving industrial effluents using small fish species. 1. Response of spoonhead sculpin (*Cottus ricei*) downstream of a bleached kraft pulp mill. *Environmental Toxicology and Chemistry*, 17, 2238–45.

Gibbons, W.N., Munkittrick, K.R., McMaster, M.E. & Taylor, W.D. (1998b) Monitoring aquatic environments receiving industrial effluents using small fish species. 2. Comparison between responses of trout-perch (*Percopsis omiscomaycus*) and white sucker (*Catostomus commersoni*) downstream of a pulp mill. *Environmental Toxicology and Chemistry*, 17, 2238–45.

Hall, T.J., Haley, R.K., LaFleur, L.E., *et al.* (1997) Experimental stream responses to biologically treated bleached kraft mill effluent before and after mill process changes to increased chlorine dioxide substitution and oxygen delignification. In: *Proceedings of the 3rd International Conference on the Environmental Fate and Effects of Pulp and Paper Effluents*, 10–13 November 1997, Rotorua, New Zealand. pp. 343–52.

Hall, T., Dudley., Fisher, R. & Borton, D. (2000) Monitoring parameters and examples of initial waster quality, effluent quality and biological characterization for four long-term receiving water study locations in the US. In: *Proceedings of the 4th International Conference on the Environmental Impacts of the Pulp and Paper Industry*, 12–15 June 2000. pp. 247–52. Finnish Environment Institute Report 417.

Hewitt, L.M., Carey, J.H., Munkittrick, K.R. & Dixon, D.G. (1996) Examination of bleached kraft mill effluent fractions for potential inducers of mixed function oxygenase activity in rainbow trout. In: *Environmental Fate and Effects of Pulp and Paper Mill Effluents*, (eds M.R. Servos, K.R. Munkittrick, J.H. Carey & G. Van Der Kraak). pp. 79–94. St Lucie Press, DelRay Beach, FL.

Hewitt, L.M., Parrott, J.L., Wells, K.L., *et al.* (2000). Characteristics of ligands for the Ah receptor and sex steroid receptors in hepatic tissues of fish exposed to bleached kraft mill effluent. *Environmental Science and Technology*, 34, 4327–34.

Hewitt, L.M., Parrott, J.L., Marlatt, V., *et al.* (2001) Effluent and bioavailability studies of bioactive chemicals released from pulp mills of different technologies in Canada. Presented at *4th Annual Meeting of the Society of Environmental Toxicology and Chemistry, Latin America*, Buenos Aires, Argentina, 22–25 October 2001.

Hewitt, L.M., Smyth, S.A., Dubé, M.G., *et al.* (2002) Isolation of compounds from bleached kraft mill chemical recovery condensates associated with reduced levels of circulating testosterone in mummichog *(Fundulus heteroclitus)*. *Environmental Toxicology and Chemistry*, **21**, 1359–67.

Hewitt, L.M., Pryce, A.C., Parrott, J.L., *et al.* (in press) Bioavailability of ligands for sex steroid receptors and the Ah receptor to fish exposed to treated effluent from a bleached sulphite/ groundwood pulp mill. *Environmental Toxicology and Chemistry*.

Holm, S., Sepúlveda, M.S., Quinn, B.P., *et. al.* (2000) Endocrine disruption in largemouth bass and exposure to papermill effluents: a reconnaissance study of the lower St. Johns River basin. In: *Proceedings of the 4th International Conference on the Environmental Impacts of the Pulp and Paper Industry*, 12–15 June 2000. pp. 172–7. Finnish Environment Institute Report 417.

Janz, D.M., McMaster, M.E., Weber, L.P., *et al.* (2001) Recovery of ovary size follicle cell apoptosis, and HSP70 expression in fish exposed to bleached pulp mill effluent. *Canadian Journal of Fisheries and Aquatic Sciences*, **58**, 620–5.

Jardine, J.J., Van Der Kraak, G.J. & Munkittrick, K.R. (1996) Capture and confinement stress in white sucker exposed to bleached kraft pulp mill effluent. *Ecotoxicology and Environmental Safety*, **33**, 287–98.

Johnsen, K., Tana, J., Lehtinen, K.-J., *et al.* (1998) Experimental field exposure of brown trout to river water receiving effluent from an integrated newsprint mill. *Ecotoxicology and Environmental Safety*, **40**, 184–93.

Johnsen, K., Grotell, C., Tana, J. & Carlberg, G.E. (2000) Impact of mechanical pulp mill effluent on egg hatchability of brown trout. *Bulletin of Environmental Contamination and Toxicology*, **64**, 973–9.

Jones, P., Hannah, D., Huser, B., *et al.* (1997) The induction of EROD activity in fish exposed to New Zealand pulp and paper mill effluents. In: *Proceedings of the 3rd International Conference on the Environmental Fate and Effects of Pulp and Paper Effluents*, 10–13 November 1997, Rotorua, New Zealand. pp. 333–42.

Karels, A. & Oikari, A. (2000) Effects of pulp and paper effluents on the reproductive and physiological status of perch *(Perca fluviatilis* L.) and roach *(Rutilus rutilus* L.) during the spawning period. *Annals Zoologica Fennici*, **37**, 65–77.

Karels, A., Soimasuo, M. & Oikari, A. (1999) Effects of pulp and paper effluents on reproduction, bile conjugates and liver MFO (mixed function oxygenase) activity in fish at Southern Lake Saimaa, Finland. *Water Science and Technology*, **40**, 109–14.

Karels, A., Markkula, E. & Oikari, A. (2000) Reproductive, biochemical, physiological and population responses in perch *(Perca fluviatilis* L.) and roach *(Rutilus rutilus* L.) downstream of two elemental chlorine-free pulp and paper mills. *Environmental Toxicology and Chemistry*, **20**, 1517–27.

Keough, M.J. & Mapstone, B.D. (1997) Designing environmental monitoring for pulp mills in Australia. *Water Science and Technology*, **35**, 397–404.

Khan, R.A. & Hooper, R.G. (2000) Decontamination of winter flounder *(Pleuronectes americanus)* following chronic exposure to effluent from a pulp and paper mill. *Archives of Environmental Contamination and Toxicology*, **38**, 197–201.

Khan, R.A. & Payne, J.F. (2000) Some factors influencing EROD activity in winter flounder *(Pleuronectes americanus)* exposed to effluent from a pulp and paper mill. *Chemosphere*, **46**, 235–9.

Kovacs, T.G., Gibbons, J.S., Martel, P.H. & Voss, R.H. (1997) Perspective on the potential of pulp and paper effluents to affect the reproductive capacity of fish: a review of Canadian field studies. *Journal of Toxicology and Environmental Health*, **51**, 305–52.

Kovacs, T.G., Martel, P.H. & Voss, R.H. (2002) Assessing the biological status of fish in a river receiving pulp and paper mill effluents. *Environmental Pollution*, **118**, 123–40.

Kukkonen, J.V.K., Punta, E., Koponen, P., *et al.* (1999) Biomarker responses by Crucian carp (*Carassius carassius*) living in a pond of secondary treated pulp mill effluent. *Water Science and Technology*, **40**, 123–30.

Lappivaara, A. (2001) Effects of acute handling stress on whitefish (*Coregonus lavaretus*) after prolonged exposure to biologically treated and untreated bleached kraft mill effluent. *Archives of Environmental Contamination and Toxicology*, **41**, 55–64.

Lappivaara, A, Mikkonen, J. & Soimasuo, M. (2002) Attenuated carbohydrate and gill Na+, K+ATPase stress responses in white fish caged near bleached kraft mill discharges. *Ecotoxicology and Environmental Safety*, **51**, 5–11.

Larsson, D.G.J., Förlin, L., Lindesjöö, E. & Sandström, O. (1997) Monitoring of individual organism responses in fish populations exposed to pulp mill effluents. Presentation at *Proceedings of the 3rd International Conference on the Environmental Fate and Effects of Pulp and Paper Effluents*, 10–13 November 1997, Rotorua, New Zealand.

Larsson, D.G.J., Hällman, H. & Förlin, L. (2000) More male fish embryos near a pulp mill. *Environmental Toxicology and Chemistry*, **19**, 2911–17.

Leblanc, J., Couillard, C.M. & Brêthes, J.-C.F. (1997) Modifications of the reproductive period in mummichog (*Fundulus heteroclitus*) living downstream from a bleached kraft pulp mill in the Miramichi estuary, New Brunswick, Canada. *Canadian Journal of Fisheries and Aquatic Sciences*, **54**, 2564–73.

Lehtinen, K.-J. (1997) Function and structural variability in aquatic ecosystems and the causal relationship with accumulated energy, nutrients and pulp mill effluents. In: *Proceedings of the 3rd International Conference on the Environmental Fate and Effects of Pulp and Paper Effluents*, 10–13 November 1997, Rotorua, New Zealand. pp. 388–97.

Lehtinen, K.-J., Mattsson, K., Tana, J., *et al.* (1999) Effects of wood-related sterols on the reproduction, egg survival and offspring of brown trout (*Salmo trutta lacustris* L.). *Ecotoxicology and Environmental Safety*, **42**, 40–9.

McCarthy, L.H., Munkittrick, K.R., Blunt, B.R., *et al.* (2003) Assessment of steroid levels in fish exposed to pulp mill effluents. In: *Environmental impacts of pulp and paper waste streams* (eds T. Stuthridge, M. van den Heuvel, N. Marvin, A. Slade & J. Clifford), pp. 342–351. Proceedings from the 3rd International Conference on Environmental Fate and Effects of Pulp and Paper Mill Effluents, Roturia, New Zealand.

McMaster, M.E., Munkittrick, K.R., Van Der Kraak, G.J., *et al.* (1996) Detection of steroid hormone disruptions associated with pulp mill effluent using artificial exposures to goldfish. In: *Environmental Fate and Effects of Pulp and Paper Mill Effluents*, (eds M.R. Servos, K.R. Munkittrick, J.H. Carey & G. Van Der Kraak). pp. 425–37. St Lucie Press, DelRay Beach, FL.

McMaster, M.E., Hewitt, L.M., Parrott, J., *et al.* (2000) Detailed endocrine assessments in wild fish downstream of pulp and paper mills: application to the Northern Rivers Ecosystem Initiative Endocrine Program in Northern Alberta, Canada. In: *Proceedings of the 4th International Conference on the Environmental Impacts of the Pulp and Paper Industry*, 12–15 June 2000, (eds M. Ruoppa, J. Paasivirta, K.-J. Lehtinen & S. Ruonala). pp. 222–37. Finnish Environment Institute Report 417.

Martel, P.H. & Kovacs, T. (1997) A comparison of the potential of primary- and secondary-treated pulp and pulp mill effluents to induce mixed function oxygenase (MFO) activity in fish. *Water Research*, **31**, 1482–8.

Martel, P.H., Kovacs, T.G., O'Connor, B.I. & Voss, R.H. (1997) The source and identity of compounds in a thermomechanical pulp mill effluent inducing hepatic mixed function oxygenase (MFO) activity in fish. *Environmental Toxicology and Chemistry*, **16**, 2375–83.

Mattsson, K., Lehtinen, K.-J., Tana, J., *et al.* (2001a) Effects of wood-related sterols on the offspring of viviparous blenny (*Zoarces viviparus*). *Ecotoxicology and Environmental Safety*, **49**, 144–54.

Mattsson, K., Lehtinen, K.-J., Tana, J., *et al.* (2001b) Effects of pulp mill effluents and restricted diet on growth and physiology of rainbow trout (*Oncorhynchus mykiss*). *Ecotoxicology and Environmental Safety*, **49**, 144–54.

Mellanen, P., Soimasu, M., Holmbom, B., *et al.* (1999) Expression of the vitellogenin gene in the liver of juvenile whitefish (*Coregonus lavaretus* L. *s.l.*) exposed to effluents from pulp and paper mills. *Ecotoxicology and Environmental Safety*, **43**, 133–7.

Mishra, A., Pandey, G.C. & Pandey, A.C. (2000) Dilution-dependent biochemical effects of paper mill effluent in the freshwater Asian walking catfish *Clarias batrachus* Linnaeus. *Journal of Advances in Zoology*, **21**, 65–6.

Munkittrick, K.R. & Sandström, O. (2003) Ecological assessments of pulp mill impacts: issues, concerns, myths and research needs. In: *Environmental impacts of pulp and paper waste streams.* (eds T. Stuthridge, M. van den Heuvel, N. Marvin, A. Slade & J. Clifford), pp. 352–362. Proceedings from the 3rd International Conference on Environmental Fate and Effects of Pulp and Paper Mill Effluents, Roturua, New Zealand.

Munkittrick, K.R., Servos, M.R., Carey, J.H. & Van Der Kraak, G.J. (1997) Environmental impacts of pulp and paper wastewater: evidence for a reduction in environmental effects at North American pulp mills since 1992. *Water Science and Technology*, **35**, 329–38.

Munkittrick, K.R., McMaster, M.E., McCarthy, L.H., *et al.* (1998) An overview of recent studies on the potential of pulp mill effluents to impact reproductive function in fish. *Journal of Toxicology and Environmental Health, Part B*, **1**, 101–25.

Munkittrick, K.R., Servos, M.R., Gorman, K., *et al.* (1999) Characteristics of EROD induction associated with exposure to pulp mill effluent. In: *Impact Assessment of Hazardous Aquatic Contaminants: Concepts and Approaches,* (ed. S.S. Rao). pp. 79–97. Lewis Publishers, Boca Raton, FL.

Munkittrick, K.R., McMaster, M., Van Der Kraak, G., *et al.* (2000) *Development of Methods for Effects-Based Cumulative Effects Assessment Using Fish Populations: Moose River Project.* SETAC Press, Pensacola, FL.

Munkittrick, K.R., McGeachy, S.A., McMaster, M.E. & Courtenay, S.C. (2002) Overview of freshwater fish studies from the pulp and paper Environmental Effects Monitoring program. *Water Quality Research Journal of Canada*, **37**, 49–77.

Munkittrick, K.R., Sandström, O., Larsson, Å., *et al.* (2003) A reassessment of the original reviews of Norrsundet and Jackfish Bay field studies. In: *Environmental impacts of pulp and paper waste streams.* (eds T. Stuthridge, M. van den Heuvel, N. Marvin, A. Slade & J. Clifford), pp. 459–477. Proceedings from the 3rd International Conference on Environmental Fate and Effects of Pulp and Paper Mill Effluents, Roturua, New Zealand.

Oikari, A., Fragoso, N., Leppänen, H., *et al.* (2002) Bioavailability to juvenile rainbow trout (*Oncorhynchus mykiss*) of retene and other mixed-function oxygenase-active compounds from sediments. *Environmental Toxicology and Chemistry*, **21**, 121–8.

Pacheco, M. & Santos, M.A. (1999) Biochemical and genotoxic responses of adult eel (*Anguilla anguilla*) to resin acids and pulp mill effluent: laboratory and field experiments. *Ecotoxicology and Environmental Safety*, **42**, 81–93.

Parks, L.G., Lambright, C.S., Orlando, E.F., *et al.* (2001) Masculinization of female mosquitofish in kraft mill effluent-contaminated Fenholloway River water is associated with androgen receptor agonist activity. *Toxicological Science*, **62**, 257–67.

Parrott, J.L., Jardine, J.J., Blunt, B.R., *et al.* (1999) Comparing biological responses to mill process changes: a study of steroid concentrations in goldfish exposed to effluent and waste streams from a Canadian bleached sulphite mill. *Water Science Technology*, **40**, 115–21.

Parrott J., van den Heuvel, M.R., Hewitt, L.M., *et al.* (2000a) Isolation of MFO inducers from tissues of white suckers caged in bleached kraft mill effluent. *Chemosphere*, **41**, 1083–9.

Parrott, J.L., Jardine, J.J., Blunt, B.R., *et al.* (2000b) Comparing biological responses to mill process changes: a study of steroid concentrations in goldfish exposed to effluent and waste streams from Canadian pulp mills. In: *Proceedings of the 4th International Conference on the Environmental Impacts of the Pulp and Paper Industry*, 12–15 June 2000. pp. 145–51. Finnish Environment Institute Report 417.

Parrott J.L., Wood, C.S., Boutot, P., *et al.* (2000c) Fathead minnow long term growth/reproduction tests to assess final effluent from a bleached sulphite mill. In: *Proceedings of the 4th International Conference on Environmental Impacts of the Pulp and Paper Industry*, 12–15 June 2000. pp. 207–12. Finnish Environment Institute Report 417.

Ribey, S.C., Munkittrick, K.R., Courtenay, S., *et al.* (2002) Development of a monitoring design for examining effect in wild fish associated with discharges from metal mines. *Water Quality Research Journal of Canada*, 37, 229–49.

Richardson, J. (1997) Use of a fish health profile for monitoring pulp and paper mill effluents. In: *Proceedings of the 3rd International Conference on the Environmental Fate and Effects of Pulp and Paper Effluents*, 10–13 November 1997, Rotorua, New Zealand. pp. 474–9.

Robinson, R.D. (1994) *Evaluation and development of laboratory protocols for predicting the chronic toxicity of pulp mill effluents to fish*. PhD thesis, University of Guelph, ND.

Ruoppa, M., Paasivirta, J., Lehtinen, K.-J. & Ruonala, S. (2000) *Proceedings of the 4th International Conference on the Environmental Impacts of the Pulp and Paper Industry*. Finnish Environment Institute Report 417.

Sandström, O. (1996) *In situ* assessments of the impacts of pulp mill effluents on fish. In: *Environmental Fate and Effects of Pulp and Paper Mill Effluents*, (eds M.R. Servos, K.R. Munkittrick, J.H. Carey & G. Van Der Kraak). pp. 449–57. St Lucie Press, DelRay Beach, FL.

Sandström, O., Förlin, L., Grahn, O., *et al.* (1997) *Environmental Impact of Pulp and Paper Mill Effluents: A Strategy for Future Environmental Risk Assessments*. Swedish Environmental Protection Agency Report 4785.

Santos, M.A. & Pacheco, M. (1996) *Anguilla anguilla* L. stress biomarkers recovery in clean water and secondary-treated pulp mill effluent. *Ecotoxicology and Environmental Safety*, 35, 96–100.

Schnell, A., Hodson, P.V., Steel, P., *et al.* (2000) Enhanced biological treatment of bleached kraft mill effluents – II. Reduction of mixed function oxygenase (MFO) induction in fish. *Water Research*, 34, 501–9.

Scroggins, R.P., Miller, J.A., Borgmann, A.I. & Sprague, J.B. (2002) Sublethal toxicity findings by the pulp and paper industry for cycles 1 and 2 of the Environmental Effects Monitoring program. *Water Quality Research Journal of Canada*, 37, 21–48.

Sepúlveda, M.S., Ruessler, D.S., Denslow, N.D., *et al.* (2001) Assessment of reproductive effects in largemouth bass (*Micropterus salmoides*) exposed to bleached/unbleached kraft mill effluents. *Archives of Environmental Contamination and Toxicology*, 41, 475–82.

Sepúlveda, M.S., Johnson, W.E., Higman, J., *et al.* (2002) An evaluation of biomarkers of reproductive function and potential contaminant effects in Florida largemouth bass (*Micropterus salmoides floridanus*) sampled from the Saint Johns River. *Science of the Total Environment*, 289, 133–44.

Servos, M.R., Munkittrick, K.R., Carey, J.H. & Van Der Kraak, G.J. (1996) *Environmental Fate and Effects of Pulp and Paper Mill Effluents*. St Lucie Press, DelRay Beach, FL.

Södergren, A. (1992) *Environmental Fate and Effects of Bleached Pulp Mill Effluents*, Swedish Environmental Protection Agency Report 4031.

Soimasuo, M.R., Karels, A.E., Leppänen, H., *et al.* (1998a) Biomarker responses in whitefish (*Coregonus lavaretus* L. *s.l.*) experimentally exposed in a large lake receiving effluents from pulp and paper industry. *Archives of Environmental Contamination and Toxicology*, 34, 69–80.

Soimasuo, M.R., Lappivaara, J. & Oikari, A.O.J. (1998b) Confirmation of *in situ* exposure of fish to secondary treated bleached-kraft mill effluent using a laboratory simulation. *Environmental Toxicology and Chemistry*, 17, 1371–9.

Stanko J.P., McNatt, H.B. & Angus, R.A. (2001) An investigation of the reproductive physiology of female mosquitofish *Gambusia holbrooki* exposed to pulp mill effluent. *Toxicology*, 164 (Suppl), 210.

Swanson, S.M. (1994) *Wapiti-Smokey River Ecosystem Study*. Final Report. Weyerhauser Canada, Grande Prairie, AL.

Swedish EPA (Environmental Protection Agency) (1997) *Environmental Impacts of Pulp and Paper Mill Effluents: A Strategy for Future Environmental Risk Assessments*. Report 4785.

Tana, J., Lehtinen, K.-J., Mattson, K. & Engström, C. (1997) Effects in mesocosms exposed to bleach plant effluents from ECF and TCF kraft pulp production. In: *Proceedings of the 3rd International Conference on the Environmental Fate and Effects of Pulp and Paper Effluents*, 10–13 November 1997, Rotorua, New Zealand. pp. 423–31.

Tremblay, L.T. & Van Der Kraak, G. (1998) Use of a series of homologous *in vitro* and *in vivo* assays to evaluate the endocrine modulating actions of β-sitosterol in rainbow trout. *Aquatic Toxicology*, **43**, 149–62.

Tremblay, L.T. & Van Der Kraak, G. (1999) Comparison of the effects of the phytosterol β-sitosterol and pulp and paper mill effluents on sexually immature rainbow trout. *Environmental Toxicology and Chemistry*, **18**, 329–36.

van den Heuvel, M.R., Ellis, R.J., Tremblay, L.A. & Stuthridge, T.R. (2002) Exposure of reproductively maturing rainbow trout to a New Zealand pulp and paper mill effluent. *Ecotoxicology and Environmental Safety*, **51**, 65–75.

Van Der Kraak, G., Munkittrick, K.R., McMaster, M.E. & MacLatchy, D.L. (1998) A comparison of bleached kraft pulp mill effluent, 17β-estradiol and β-sitosterol effects on reproductive function in fish. In: *Principles and Processes for Evaluating Endocrine Disruption in Wildlife*, (eds R.J. Kendall, R.L. Dickerson, W.A. Suk & J.P. Giesy). pp. 249–65. SETAC Press, Pensacola, FL.

Walker, S.L., Hedley, K. & Porter, E. (2002) Pulp and paper environmental effects monitoring in Canada: an overview. *Water Quality Research Journal of Canada*, **37**, 7–19.

Williams, P.J., Courtenay, S.C. & Wilson, C.E. (1998) Annual sex steroid profiles and effects of gender and season on cytochrome P450 mRNA levels in Atlantic tomcod (*Microgadus tomcod*). *Environmental Toxicology and Chemistry*, **17**, 1582–8.

Wilson, J.Y., Addison, R.F., Martin, D., *et al.* (2000) Cytochrome P450 1A and related measurements in juvenile Chinook salmon (*Oncorhynchus tshawytscha*) from the Fraser River. *Canadian Journal of Fisheries and Aquatic Sciences*, **57**, 405–13.

Wilson, J.Y., Kruzynski, G.M. & Addison, R.F. (2001) Experimental exposure of juvenile Chinook salmon (*Oncorhynchus tshawytscha*) to bleached kraft pulp mill effluent: hepatic CYP1A induction is correlated with DNA adducts but not with organochlorine residues. *Aquatic Toxicology*, **53**, 49–63.

Woodworth, J.G., Munday, B.L. & Campin, D. (1997) Evaluation of biomarkers for exposure of fish to eucalypt-based pulp mill effluent and for determination of exposure routes. In: *Proceedings of the 3rd International Conference on the Environmental Fate and Effects of Pulp and Paper Effluents*, 10–13 November 1997, Rotorua, New Zealand, pp. 310–21.

Plate 1 Largescale forestry clearcuts in British Columbia (BC), Canada (a–d), and **one of several alternative cutting practices** (e, f). (a) Atleo River watershed, Vancouver Island; (b) central Vancouver Island watershed; (c) Misery Creek watershed, south coastal BC; (d) central BC lake watersheds; (e, f) variable retention cuts in a Vancouver Island stream and lake watershed, respectively. Sources: (a) G. Hutching, Forest & Bird, Wellington, New Zealand; (b, e, f) G.F. Hartman; (c) M. Miles; (d) T.G. Northcote.

Plate 2 (a–c) Debris torrents follow clearcut logging on steep watersheds in coastal BC; (d–f) attempts to cable in stable large wood to provide gravel retention and instream fish cover in debris torrented streams. Other poor logging practices around streams cause water quality problems for fish: (g) falling across streams; (h) yarding across streams; (i) leaving fresh, fine slash. Sources: (a-c, g,i) T.G. Northcote; (d–f) H. Klassen; (h) J.D. Williams.

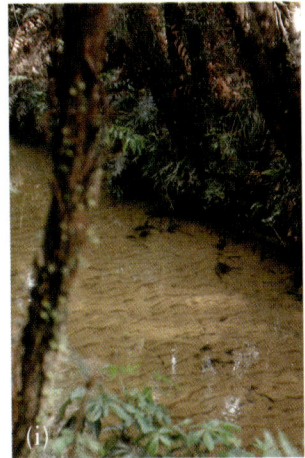

Plate 3 Logging roads and stream sedimentation. Threat to a trophy size rainbow trout fishery in Kootenay Lake, BC. (a) Poorly designed logging road to a small clearcut above the major trout spawning area, Trout Lake outlet; (b) failed logging road feeds sediment to a tributary creek; (c) sediment enters spawning ground and spreads downstream. **Logging road through old lacustrine sediment deposits, central BC.** (d) Logging road feeds sediment downslope to nearby stream; (e) high sediment load carried downstream over salmonid spawning area; (f) severe deposition on stream bottom. **Poor logging road construction and drainage.** (g) sidecast debris into a Yangze River tributary, China; (h) poor road drainage erodes soils into a New Zealand stream; (i) resulting in severe bottom sedimentation. Sources: (a–c) R.A. Rutherglen; (d–f) T.G. Halsey & P.A. Slaney; (g–i) T.G. Northcote.

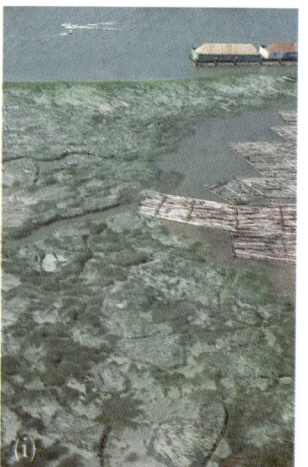

Plate 4 Aquatic dumping, storage, and transport of logs. Lakeshore log dump: (a) Kootenay Lake, BC; lakeshore log storage, (b) Babine Lake, BC; (c) Mazurian lake, Poland. **Splash dam water releases and log drives**: (d) sudden water releases from a 1920s dam in a New Zealand river to float large logs downstream still shows long-term effect of reduced instream gravel and fish cover; (e) log drives occurred on a BC river up to mid 1960s, and (f) still do in some central China rivers. **Estuarine log storage**: (g) large log booms stored in estuarine reaches of Fraser River, BC; (h) log stranding by dewatering at low tide removes productive aquatic vegetation from tidal channels; (i) where such storage stopped, marsh vegetation regenerates. Sources: (a) C. Grubmeyer; (b–d, f–i) T.G. Northcote; (e) G.D. Taylor.

Part VI
North American Fish–Forestry Interactions

Chapter 17
Fish–forestry interactions in Oregon, Washington and Alaska, USA

J.D. HALL, C.J. CEDERHOLM, M.L. MURPHY AND K.V. KOSKI

Perhaps nowhere in the world are the potential conflicts between forestry and fishery interests more intense than on the northwest coast of North America. The watersheds of the region are home not only to some of the continent's most productive commercial forests, but also to streams that are the spawning and nursery areas for ecologically important and economically valuable runs of anadromous salmonids. The many ecological linkages between the forests and the stream habitats that support the fish, combined with their susceptibility to logging (Hicks *et al.* 1991), bring the primary harvesters and consumers of these two resources into conflict.

Disputes between users of these two valuable resources have been adversarial since the two industries began harvesting in the mid-nineteenth century. In Oregon and Washington, the forest products industry has historically been the largest manufacturing sector, greatly exceeding the fishing industry in dollar volume. In contrast, the Alaska salmon fishery provides a larger fraction of that state's income than does forestry. In the past 10–15 years the economic value and importance of both industries have declined in Oregon and Washington. Principal causes have been restrictions on logging on federal lands stemming from requirements of the Endangered Species Act (ESA), and substantial reductions in salmon harvests owing to reductions in fish abundance that have led to some of the ESA listings. In Alaska, salmon fisheries have remained relatively healthy, but demand for timber has fallen. The reductions in harvest imposed on both industries have intensified the conflicts.

That salmonid habitat was influenced by forest practices has always been apparent, but more recent evidence shows that runs of anadromous salmonids can influence productivity of the riparian forest. By transporting nutrients from the ocean to the spawning rivers and their tributaries, salmon are an agent of nutrient renewal in streams and adjacent riparian zones (Cederholm *et al.* 2001). This biological process serves to underscore the subtlety of interactions between forests and fish.

In this chapter we explore the historical roots of the conflicts between forestry and fisheries in this diverse region, describe past and current forest practices, review the scientific evidence of effects of forest practices on the fish species and their habitats, and make recommendations for further improvements in management.

Description of the geographic region

Pacific Northwest

The primary forested regions of Oregon and Washington are in the western third of both states. The most productive forests occupy a series of coastal mountain ranges (the Olympic Mountains to the north, the Coast Range, and the Klamath Mountains to the south) and the Cascade Range to the east. Further east is a high plain, the Columbia Plateau, with many interspersed ranges and less productive forests. About 60% of the region's forestlands are in public ownership.

The coastal mountains consist primarily of uplifted marine sedimentary formations, with scattered intrusions of basalt. The Cascade Range is formed predominantly from basalts. Much of northern Washington was covered with the continental ice sheet during the Pleistocene, and other scattered areas were affected by alpine glaciers. In general, the slopes are steep, and erosion is a prominent feature of the stream and river valleys (Jackson & Kimerling 1993).

The climate of the Pacific Northwest is diverse, with generally moderate temperatures and substantial precipitation. Precipitation shows a strong west–east gradient, influenced mainly by the mountain ranges. Annual precipitation ranges from >300 cm on the western edge of the coastal mountains to <25 cm in the Columbia Plateau. Runoff from the streams in the Coast Ranges and foothills of the Cascades is rain-dominated. Peak flows occur during winter, followed by low flows during summer. The streams of the higher Cascades and eastern mountains show a snowmelt pattern of runoff, with maximum flows during spring.

The major river system of the Pacific Northwest is the Columbia, which forms much of the boundary between Oregon and Washington and extends into British Columbia. However, this river has been so modified by dams (there are more than 50 major hydroelectric facilities on the main stem and tributaries) that forest influences on the river and its tributaries are overshadowed by effects of these dams. Many coastal river basins provide substantial habitat for valued salmonid fishes, and are particularly susceptible to forestry practices. Estuaries along the Pacific Coast provide habitat for anadromous salmonids moving to and from the ocean. Estuaries in Oregon are of limited extent, except for the Columbia River, but Washington's are more extensive, particularly in Puget Sound.

The most commercially important tree species in the Pacific Northwest is Douglas-fir (*Pseudotsuga menziesii*), which occupies nearly two-thirds of all forested land west of the Cascade crest. Other important timber species in this region include western hemlock (*Tsuga heterophylla*) and western redcedar (*Thuja plicata*). Sitka spruce (*Picea sitchensis*) and western hemlock occur in a small zone on the coast, from northern Washington into northern Oregon. The principal timber species east of the Cascades are ponderosa pine (*Pinus ponderosa*), the interior form of Douglas-fir, and lodgepole pine (*Pinus contorta*) (Jackson & Kimerling 1993). Because Douglas-fir is intolerant of shade, it is usually harvested in clear-cuts. This forest practice, with its attendant logging and transportation systems, can have a substantial negative influence on the quality of stream habitat.

The principal freshwater fish species of interest in this region are anadromous salmonids. Major commercial species include chinook (*Oncorhynchus tshawytscha*), coho (*O. kisutch*) and chum (*O. keta*) salmon. Those most sought by recreational anglers include chinook and coho salmon, steelhead (*O. mykiss*) and cutthroat trout (*O. clarki*). There are about 30 other fish species in streams that are subject to timber harvest.

Alaska

The review of fish–forestry interactions in Alaska will focus on Southeast Alaska, where most of the logging and research on fish habitat have been done. The Alexander Archipelago, which makes up Southeast Alaska, is unique in both physiography and richness of fish resources. It is nearly 650 km long and 200 km wide, with extensive sheltered estuaries and a shoreline of over 16,000 km. The climate is maritime, with moderate temperatures ranging from about 7° to 18°C in summer and from about –10° to 5°C in winter. The dominant features of the weather are rain and wind. Annual precipitation averages >250 cm, and peak rainfall occurs from September through November.

The topography of Southeast Alaska is characterized by a massive glaciated mountain range (1800–3000 m in elevation), the crest of which forms the Alaska–Yukon–British Columbia boundary. This mountainous crest is irregularly broken by several deep glaciated valleys with large river systems having headwaters in the interior of the Yukon Territory and British Columbia. Elevations of forested areas extend up to about 900 m in the southern sections of the forest and up to 750 m farther north. Glaciers are the dominant active geological feature, and much of the area has only recently been freed from glacial ice. Over 5400 catalogued salmon streams in the region contribute millions of salmonids to commercial, sport and native subsistence fisheries annually.

Southeast Alaska has the largest contiguous stands of old-growth forest remaining in the USA. These forests are characterized by uneven-aged stands of western hemlock and Sitka spruce, with scattered stands of western redcedar and Alaska cedar (*Chamaecyparis nootkatensis*). Windthrow has been the main element of successional change. The most productive stands of timber are located along tidewater and in valleys adjacent to streams. The large old-growth trees are key elements in the structure and function of stream channels and riparian zones of Southeast Alaska.

Five species of Pacific salmon are present in the region. Pink salmon (*O. gorbuscha*) are the most abundant and occur throughout Southeast Alaska. Chum and coho salmon are also widely distributed. Chinook salmon are mainly present in the larger mainland river systems, and sockeye salmon (*O. nerka*) are typically in mainland rivers with lakes. Three other native salmonids – steelhead, Dolly Varden (*Salvelinus malma*) and cutthroat trout – occur throughout the region. A few species of cottids (*Cottus* spp.) comprise the majority of the remaining freshwater fish fauna.

General features of forestry practices

Historical background

Pacific Northwest

Conflicts between users of forest and fish resources have been common throughout the history of both industries on the west coast of North America. The first sawmills in Oregon and Washington were established around 1840. They were often sited on streams and rivers to provide for easy transport of logs and for disposal of waste. Mill dams were built that blocked migration of anadromous fish, and decomposing sawdust and wood waste took a heavy toll on fish life (Lichatowich 1999).

Transport by water was the principal means of moving logs to mills until well into the 1900s. Log drives scoured stream banks and channels, destroying much of the structure that provided fish habitat. 'Splash dams' were built on streams too small to allow logs to move at normal flows. These dams stored water until volume was sufficient to sluice logs downstream to a mill, and were most damaging in the Northwest, where logs were large and stream gradients steep. They were widespread – over 300 on coastal streams in Oregon and Washington and on Columbia River tributaries (Sedell *et al.* 1991). Their operation continued into the 1950s. As an agent of lasting habitat change in streams and small rivers, the splash dam had no equal.

The advent of heavy-duty logging trucks in the 1940s and 1950s allowed access to steeper, previously inaccessible terrain. However, roads built on steep hillsides often failed, sending tons of sediment into stream channels below. Improperly located or inadequately designed culverts often blocked migration of anadromous fish.

Early logging disregarded streams altogether – logs were regularly felled into stream channels and yarded through streams. Stream channels and adjacent floodplains were often used as roads or skid trails, and logging debris was left where it lay. Early clear-cuts harvested extensive areas at one time, often devastating streams and stream banks. Riparian buffer protection along streams was virtually unknown until the 1960s, and developed slowly from that time.

Alaska

Most timber harvest in Alaska comes from coastal old-growth forests of western hemlock and Sitka spruce. The first consumptive demand for this timber came with Russian colonization in the 1700s. In the early 1800s a foundry was built near Sitka, and local stands of spruce and hemlock were cut for charcoal. The first Alaskan sawmill was built near Sitka in 1833. During World Wars I and II, high quality spruce was sought for construction of military aircraft. Before 1950, much of the logging was done with tractor skidding or A-frame cable logging near mouths of streams or along beaches (Harris & Farr 1974).

The current timber industry developed during the 1950s in the Tongass National Forest, which stretches the entire length of Southeast Alaska and covers about 75% of its land area. The US Forest Service signed 50-year contracts with two timber companies. In return for building and operating pulp mills in Ketchikan and Sitka, the companies were guaranteed long-term access to more than 13 billion board feet of timber (~70 million m^3) on a non-competitive basis (AFSEEE 1996). In the last several decades, harvest has increased on private lands in Southeast Alaska coastal forests. Currently most of the forest resources on land received by Native Corporations in Southeast Alaska in settlement of aboriginal treaty claims (about 220,000 ha) have been clear-cut. Less than half of the original, high volume old-growth forests that once existed in Southeast Alaska remain today (AFSEEE 1996). Alaska does have extensive boreal forests of spruce, birch and poplar in the interior, but these species are typically too small to be of commercial value.

Current forest practices

Oregon

In the 1960s the federal forest management agencies began to consider and implement protection for riparian zones. The first forest practice act in North America that specifically addressed practices in riparian zones on state and private lands became effective in Oregon in 1972. Federal forestry agencies agreed to meet or exceed state standards. These initial efforts focused on preserving some shading over streams for temperature control and on minimizing disturbance to stream banks and channels. Over the years, rules have become more restrictive and comprehensive. In 1987 Oregon's rules were revised to specify numbers and sizes of conifers to be retained in riparian management areas, recognizing the importance of riparian forests in providing for large wood as habitat in streams. The rules underwent major revision in 1994. Today the rules also include specifications for water quality, width of buffer zones, size and spacing of clear-cuts, and restrictions on other forest practices that may affect streams and their fish populations.

The 1994 revision of the Forest Practice Rules allows active management in the Riparian Management Area (RMA), except for a 6-m no-cut buffer immediately adjacent to the stream channel. Activity may include harvest, if the goals for desired future stand condition are met. There are separate criteria for small, medium and large streams, and for fish-bearing, non-fish-bearing and domestic-supply streams without fish (Young 2000). Standards were established for conifer basal area for each class. If the pre-harvest riparian basal area exceeds this standard, the operator can harvest down to the standard. In addition, the owner can place large wood in the stream and receive a basal area credit that may allow additional harvest in the RMA.

The Oregon Department of Forestry maintains a monitoring programme that evaluates how well the rules are protecting riparian functions and providing for desired future conditions. A recent report from this monitoring effort (Dent 2001) found that the rules were generally effective in maintaining potential sources of recruitment of large wood for large streams, but not for small streams. They were considered generally successful in protecting shade on large streams, but less so on medium and less still on small streams. Recommendations have been made to increase the target for basal area and the width of no-cut buffers on small streams.

Current specifications for riparian management on federal lands in Oregon and Washington come from several sources, depending on location (Gregory 1997). In general, federal guidelines are more protective than state forest practice rules.

Washington

In 1974 Washington State established a Forest Practices Act designed to maintain a viable forest products industry and to afford protection to fisheries, water quality and other natural resources. Its first rules were adopted in 1976, and are administered on both state and private forest lands by the state Department of Natural Resources (DNR).

In 1986 a group representing the Native American tribes, some state natural resource agencies, the timber industry, landowners and environmental interests met to negotiate an agreement upon which to base new, more protective forest practices rules. The process resulted in the Timber, Fish, and Wildlife Agreement (TFW) in 1987.

The State Forest Practices Board established a process of adaptive management to monitor and assess implementation of forest practice rules and to achieve desired resource objectives. A Cooperative Monitoring Evaluation and Research Committee was established to develop protocols and standards governing the adaptive management process, and to oversee research projects designed to facilitate the regulatory process.

Watershed analysis was developed in 1992 to address the problem of cumulative effects, those changes to the environment caused by two or more forest practices. The method identifies resource conflicts and hill-slope erosion hazards, allowing areas of resource vulnerability to be documented. Watershed analysis is combined with adaptive management, a process that allows feedback of research, monitoring and evaluation to provide for periodic improvements to the Forest Practice Rules.

In 1996 the state was faced with an imminent listing of several salmon species under the Endangered Species Act, as well as information indicating that riparian protection for public resources was inadequate. Consequently, the TFW participants and representatives from federal agencies and Washington counties agreed to negotiate and to submit to the Forest Practices Board a proposal that would incorporate additional protection. Although the environmental caucus withdrew from negotiations, the remaining caucus group ultimately wrote the Forests and Fish Report (April 1999). In May 2001, directed by state legislation, the Forest Practices Board adopted the Forests and Fish Report recommendations as new rules.

Habitat Conservation Plans (HCPs) are another major tool that has been used in Washington to minimize conflicts caused by listings or potential listing under the ESA. Their purpose is to allow timber harvest on non-federal lands that provide habitat for federally listed species, while minimizing and mitigating adverse effects on these species. One of the primary advantages of an HCP is that landowners can achieve some flexibility in managing for federally listed species on their lands, and can achieve a reasonable level of certainty in how they manage their land in the face of changing, and often increasingly restrictive, regulations.

The DNR prepared a multi-species HCP, adopted in September 1997, to cover 650,000 ha of forested state trust land. The HCP conserves habitat for the Northern Spotted Owl (*Strix occidentalis*), the Marbled Murrelet (*Brachyramphus marmoratus*), and several other listed species in Washington, including runs of several salmonids. One of the provisions of the HCP is a riparian buffer applied to both sides of all perennial streams. The width of the buffer on fish-bearing streams is approximately equal to the site-potential height of trees at 100 years, or 30 m, whichever is greater. A buffer 30 m wide is applied to both sides of perennial non-fish streams. Buffers for seasonal non-fish streams are currently under study.

Alaska

Timber management in coastal Alaska forests has been based on even-aged silviculture on a 100-year rotation (Harris & Farr 1974). Trees are typically harvested by clear-cutting. Natural regeneration is usually relied on for restocking. Size of contiguous clear-cuts has ranged from a few hectares to over 800, averaging about 25 ha on the Tongass National Forest during the peak of timber harvest. Logs are usually yarded by highlead systems, but helicopter yarding is sometimes used where highlead yarding could damage habitat. With little road access on the many islands, logs are usually hauled to log transfer facilities (LTFs), where they are transferred to saltwater for transport in rafts to mills or export sites.

The principal concern about LTFs is that bark and debris from logs accumulate on the bottom substrate and smother marine habitat and invertebrates (Faris & Vaughn 1985). By the mid-1990s, 123 marine log transfer and storage sites had been constructed in Southeast Alaska (AFSEEE 1996). Guidelines for siting and operating LTFs have been used since 1986 to minimize impacts. The guidelines require monitoring for bark accumulation to a water depth of 20 m, although bark deposits are also found in deeper water.

Based on research findings (Murphy *et al.* 1986), the National Marine Fisheries Service issued a policy in 1988 calling for minimum 30-m no-cut buffer zones along both stream banks of anadromous fish streams. This policy became the focus for debate as Congress and the Alaska Legislature considered forest practices on federal, state and private land.

The US Congress passed the Tongass Timber Reform Act in 1990, modifying timber harvest practices in the Tongass National Forest. The Act required no-cut buffer zones at least 30 m wide on each side of all streams with anadromous fish and on tributaries with only resident fish. Buffer zones were not required along streams without fish, but Best Management Practices (BMPs – regulations controlling manner of harvest) establish requirements to protect downstream habitat. In 1994 Congress directed the Forest Service to report on effectiveness of habitat protection. Their report (USDA 1995) concluded that increased protection of headwaters, increased buffers on floodplains and alluvial channels, and analysis of cumulative impacts were necessary to protect fish habitat. This Habitat Assessment drove revision of the Tongass National Forest Plan in 1997 and provides direction for managing riparian areas, as does the *Aquatic Habitat Management Handbook* published by the US Forest Service in 2001.

The Alaska Legislature reformed forest practices on state and private lands in 1990. On state lands, timber harvest is prohibited within 30 m of anadromous fish streams. On private lands, timber harvest is prohibited within 22 m of anadromous fish streams not incised in bedrock. Along anadromous fish streams incised in bedrock, timber harvest within 30 m or to the slope break must follow BMPs. Along small tributaries without anadromous fish, harvest within 15 m or to slope break must follow BMPs.

Although forest practice rules throughout the Pacific Northwest and Alaska have been continually updated and strengthened (Ellefson *et al.* 1997), they continue to be the subject of vigorous debate (Young 2000). The process of evaluating and revising these rules has become more and more litigious. As this chapter went to press, the adequacy of forest practice rules was being challenged by legal actions in both Oregon and Washington.

The scientific basis of riparian and watershed management

The requirements of salmonids for cold water, clean gravel, large wood and other riparian habitat features make them particularly susceptible to habitat alterations from forestry activities. Recent emphasis on management of riparian zones as a means to protect salmonid habitat has its rationale in two principal sources: (1) significantly increased understanding of the structure and function of riparian zones, particularly in relation to habitat for salmonids, and (2) research studies designed to quantify changes in habitat and salmonid populations influenced by logging practices. These two lines of evidence have generally validated the protections provided for riparian corridors by forest practice rules.

Riparian structure and function

Many studies in the past 10–20 years have provided substantially improved knowledge of the structure and function of riparian zones in the Pacific Northwest (Gregory *et al.* 1991; Naiman *et al.* 2000). This work has contributed to improved forest management practices, thus reducing impacts of timber harvest on salmonid habitat.

Owing to earlier logging practices that often filled streams with logging debris, fish managers in the 1950s through 1970s often viewed large wood negatively and supported its removal, principally to allow passage for spawning anadromous fish. During that period tremendous volumes of wood (both logging-related and natural accumulations) were removed from streams in the Northwest. In the past 20 years, large wood has come to be recognized as a vital element that provides structure to the stream system and habitat for salmonids. Large wood and intact riparian zones provide many benefits for salmonids, including stabilizing stream banks, moderating stream temperature increases, forming pools and instream cover, increasing habitat complexity, regulating and trapping sediment, and retaining salmon carcasses for improved nutrient retention (Cederholm *et al.* 1989). Riparian management practices have evolved toward protecting and improving the natural processes that ensure input of large wood to streams over the entire cycle of timber harvest.

There has been a similar evolution in the way in which other elements of the riparian system are viewed. Some of these include a new perspective on the role of episodic natural disturbance in maintaining habitat for salmonids (Reeves *et al.* 1995), and new understanding of the key role played by anadromous salmonids in the nutrient dynamics of both stream and riparian ecosystems (Naiman *et al.* 2000; Cederholm *et al.* 2001).

Research findings on influences of logging practices on salmonids and their habitats

Oregon

Alsea Watershed Study
The Alsea Watershed Study (AWS), established in Oregon in 1957, was the first comprehensive long-term analysis of the effects of forest harvesting on water quality and salmonid resources in coniferous forests of western North America. The study was designed for a 15-year period, including 7 years pretreatment, 1 year of logging and 7 years of post-logging evaluation. Several agencies and organizations cooperated to allow for the broad scope of the work. Three adjacent watersheds in the headwaters of the Alsea River basin were chosen for study (Fig. 17.1). They ranged in area from 70 to 300 ha. One was left undisturbed as a control, one was completely clear-cut with no riparian protection, and the third was cut in three patches with buffers protecting the stream. Principal salmonids in these watersheds were coho salmon and cutthroat trout (Hall & Lantz 1969). Other fishes were the reticulate sculpin (*Cottus perplexus*) and two species of lamprey (*Lampetra* spp.).

Research began in 1958, roads were built in 1965 into the two watersheds to be harvested, and logging occurred in 1966. Post-logging evaluation continued through 1973. The data on physical features collected during the study included stream flow, suspended sediment, stream temperature and quality of the spawning gravel (principally dissolved oxygen and percentage of fine sediment). Biological investigations included enumeration of upstream and downstream migration of all salmonids, periodic population estimation of juvenile salmonids in the three streams, and estimation of survival of coho salmon from egg deposition to emergence. This latter study provided the first quantitative demonstration in a naturally spawning population of the inverse relationship between the amount of fine sediment in spawning gravel and success of emergence (Koski 1966).

The complete clear-cut of Needle Branch, a practice common at the time, resulted in substantial changes in habitat conditions for the salmonids. In the year after logging, maximum stream temperature reached 30°C, compared with the previous maximum of 16°C (Hall & Lantz 1969). Substantial diel variation and influx of cooler groundwater allowed many juvenile salmonids to survive by limiting their exposure to the highest temperatures, although they probably experienced significant stress. Rapid revegetation of the riparian zone moderated the high temperatures; by 1973 stream temperatures had returned to pre-logging levels (Moring & Lantz 1975). The riparian

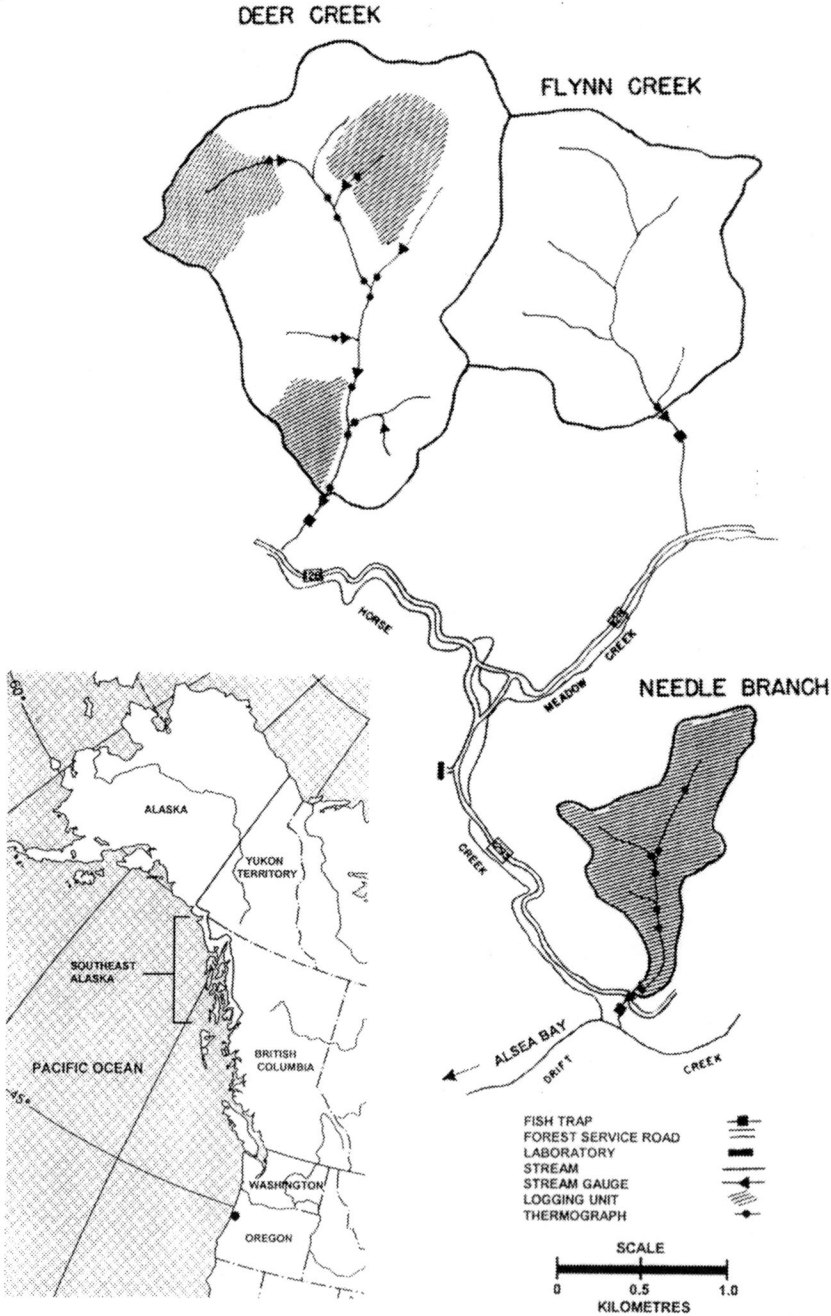

Fig. 17.1 Map of the Alsea Watershed Study showing layout of the logging units in Deer Creek (patch-cut with buffers) and Needle Branch (complete clear-cut, no buffers). Also shown are locations of stream gauges and thermographs. Dot on inset map shows location on the Oregon Coast.

buffers left in the patch-cut watershed protected the stream from significant increase in solar radiation; there was only a minor increase in temperature in this stream.

The substantial loading of fine organic debris, coupled with increased temperature, reduced dissolved oxygen in the surface water of the clear-cut stream to levels near zero in some reaches of the stream during the summer of logging. Dissolved oxygen returned to near normal after removal of slash from the stream in the autumn. Intrusion of organic debris and fine sediment into spawning gravels caused a longer-lasting reduction of dissolved oxygen in intra-gravel water.

Two indices of survival-to-emergence suggested that the reduced quality of spawning gravel adversely affected emergence of coho salmon fry. This added mortality reduced by about 50% the average number of fry that left the watershed soon after emergence. For a few years the downstream migration of coho salmon smolts occurred earlier in the year, but the overall abundance of downstream migrant salmon smolts was not significantly reduced (Hall *et al.* 1987).

A major change was observed in abundance of juvenile cutthroat trout in the clear-cut watershed. Numbers of fish resident in the stream in late summer were reduced to about 40% of pre-logging abundance for the period 1966–1973 (Hall *et al.* 1987; Hall in press).

Substantial year-to-year variation in measures of salmonid populations and their habitat made conclusions difficult in this study. However, there was at least one clear result: the protection to the stream provided by buffer strips in the patch-cut watershed was of great value. Only minimal changes occurred in the stream habitat, and the fish populations in this stream showed no significant change after logging.

Collection of data was discontinued for 15 years after completion of the original AWS, but limited observations in the three watersheds were resumed in 1988. The most complete biological data are annual estimates of salmonid populations in late summer. These estimates show that total numbers of cutthroat trout in the clear-cut stream had generally recovered to pre-logging values, but this recovery had come only in age-0 fish. From 1989 through 1996, numbers of trout age 1 and older were even further reduced, averaging only 20% of their pre-logging abundance. Additionally, year-to-year variability in abundance of juvenile trout increased dramatically after clear-cutting, especially during 1989–1996, when the coefficient of variation reached 3–4 times the pre-logging average (Gregory *et al.* in press).

Other studies in the region

Several studies in the Oregon Cascade Range showed that stream productivity for salmonids could increase in small watersheds harvested without streamside protection, at least for a few years. This was particularly true in high gradient streams where changes in sediment load and temperature were minimal, and where large boulders provided structural habitat when large wood was removed. Comparison of 24 stream reaches showed that biomass of cutthroat trout was highest in clear-cut streams, intermediate in old-growth, and lowest in highly shaded second-growth plantations (Murphy & Hall 1981). There were indications that over the longer term, production might decrease to levels below those in old growth, especially in low gradient systems where sediment accumulation could be substantial. Additional work in the region (Murphy *et al.* 1981; Hawkins *et al.* 1983) found generally similar results. The mechanisms resulting in higher initial trout biomass in clear-cut streams appeared to be greater food abundance and more effective foraging by trout in streams where sunlight was increased at the stream surface (Wilzbach *et al.* 1986). More recent work, part of the Long Term Ecological Research programme at the H.J. Andrews Experimental Forest in the central Cascade Range, has centred on effects of logging and road building on peak stream flow (Jones & Grant 1996) and on the role of large wood as salmonid habitat and in the dynamics of stream ecosystems (Gregory *et al.* 1991).

Connolly and Hall (1999) sampled cutthroat trout populations in 16 headwater streams in the Coast Range, above the extent of other salmonids. The streams were distributed among basins that had been logged 20–30 and 40–60 years previously and a group undisturbed since fires of 125–150 years ago. There were no clear differences in trout abundance related to logging history, but trout biomass was strongly related to the abundance of large wood. They concluded that removal of large wood, where it occurred after earlier logging operations, might hinder recovery of trout over the long term. A possible explanation for the lack of a clear effect of logging on cutthroat trout in these watersheds, in contrast to other coastal results, is the absence of competition from other salmonids, particularly coho salmon (Reeves *et al.* 1997; Gregory *et al.* in press).

Hicks and Hall (2003) estimated salmonid abundance in 30 km of channel in 10 Oregon Coast Range streams of similar basin area. The basins were selected to include two rock types, basalt and sandstone, and a range of timber harvest within each rock type. They found no clear effect of timber harvest on salmonid abundance, but substantial differences in species composition and abundance between rock types. The results emphasize the importance of considering bedrock geology in studies of effects of logging.

The influence of logging and other land use practices on streams east of the Cascade Range has received relatively little study. One significant exception resulted from fortuitous recovery of original data from habitat surveys conducted by the US Bureau of Fisheries in the Columbia River basin between 1934 and 1945. Resurveys using the original protocols showed that substantial changes in pool structure had been caused by land management activity (McIntosh *et al.* 2000). It was not possible to separate the influence of timber harvest from other uses such as livestock grazing and mining, but the evidence is convincing that salmonid habitats have been degraded in basins where land management has been intense.

Washington

Studies of the effects of logging in Washington largely grew out of investigations on pink and chum salmon carried out in Alaska during the 1950s and 1960s (McNeil & Ahnell 1964; McNeil 1966). This work focused on the adverse effects of fine sediments (sand and silt) in spawning gravels. It showed that increases in the percentage of fines could significantly lower embryo and fry survival by reducing dissolved oxygen in intra-gravel water and by preventing fry emergence.

These results prompted initiation in 1966 of a major controlled experiment on the effects of spawning gravel composition on survival-to-emergence of chum salmon at Big Beef Creek in Puget Sound. Koski (1975) found that survival decreased 1.3% for each 1% increment in sand within the spawning gravel. A decrease in fry survival after emergence was directly related to low dissolved oxygen and high percentages of fine sediment in the gravel.

The Clearwater River studies

In 1971 a series of landslides related to logging roads in the steep upper Clearwater River basin led to studies on the effects of fine sediment on salmon spawning and rearing habitats (Cederholm *et al.* 1981; Cederholm & Reid 1987). The Clearwater River drains a 390-km² basin and is the largest tributary of the Queets River, located on the north coast of the Olympic Peninsula. Logging road landslides were found to be a major source of sediment <0.85 mm diameter, and some of this material was retained in spawning gravels downstream. High gradient streams were affected by sediment for only 2 or 3 years, as gravels were presumably flushed clean of sediments during winter storms. However, sediment was retained longer in low gradient channels. Reid and Dunne (1984) determined that erosion on the surface of in-use gravel roads was a significant source of fine sediment, and that road-related landslides increased erosion 2.5 to 4.1 times over road surface erosion. Further studies in the Clearwater River and adjoining river systems confirmed that the percentage of a basin logged and the density of logging roads had a significant relationship to the amount of fine sediments in downstream spawning gravel (Rittmueller 1986).

The Clearwater basin was the site of a number of other studies relating logging and roading practices to fish habitat. Tagart (1984) found that survival of coho salmon fry to emergence in naturally spawned redds was inversely correlated with the percentage of fine sediments in spawning gravel. Edie (1975) found a negative relationship between fine sediment in the substrate and abundance of young trout and a species of sculpin. However, no conclusions could be reached on the overall impact of logging on summer rearing populations of salmonids because standing stocks (g/m²) of rearing juvenile coho salmon were low and highly variable, suggesting that low spawning escapements of coho salmon were limiting population size.

In a controlled field experiment of canopy removal on Bear Creek, a small tributary of the Bogachiel River about 30 km north of the Clearwater basin, researchers found that cool temperatures moderated the effects of an increased food supply that resulted from light-induced enhancement of production at lower trophic levels (June 1981; Martin *et al.* 1981). This study concluded that a streamside management zone was effective for fish protection when it resulted in retaining sufficient large wood to maintain the physical integrity of the stream channel.

Coincident with the studies related to sedimentation, it became evident that healthy salmon habitat was defined by much more than just clean spawning gravels. This meant that more information was needed about such factors as seasonal movements of juvenile salmonids and the function of riparian forests as sources of large woody debris.

In their studies of the impacts of debris removal practices on salmonid rearing habitats and populations, Lestelle and Cederholm (1984) found that stream clearance had little or no immediate effect on resident cutthroat trout numbers and biomass. Large reductions occurred over the first winter, but populations returned to pretreatment levels within 1 year. Subsequent monitoring (10–20 years later) of these same study streams revealed that the proportion of older-aged trout had declined significantly (Cederholm, unpublished data). The decline of older trout appeared to be related to the loss of deep pools that had been formed by large wood.

Grette (1985) studied long-term trends in abundance of large wood in streams and changes in juvenile salmonid rearing habitat. Large wood from old growth was more abundant in unlogged streams than in young, middle-aged, or old second-growth streams. Densities of older-aged juvenile steelhead and cutthroat trout correlated positively with area of pool cover formed by large wood in summer. Densities of coho salmon fry were not correlated with area of cover at summer low flow, but fry numbers in winter were closely related to the amount of wood.

Investigation into the life history of Clearwater River salmonids revealed that during initial fall storms many juvenile coho salmon, steelhead and cutthroat trout left the main river and moved into small tributaries located along the lower floodplain (Cederholm & Scarlett 1982; Peterson 1982). Some coho salmon, for example, spent the winter in tributaries 25–30 km down river from their summer rearing locations. These results pointed out the importance to migrating juvenile salmon of maintaining year-round access to tributaries.

Cederholm and Peterson (1985) determined that woody debris was important for retaining spawned-out salmon carcasses, making them more available as food for a myriad of fish and wildlife consumers (Cederholm *et al.* 1989, 2001). Logging practices and stream clean-out had depleted woody debris in some study streams, reducing potential for carcass retention.

Cederholm and Reid (1987) summarized the Clearwater River studies and concluded that forestry-related mortality in the Clearwater basin was primarily caused by increased sediment load and by alterations in the riparian environment that reduced woody debris and denied access to rearing tributaries in winter. It also was evident that Clearwater River coho salmon were over-fished and subject to the cumulative effects of sport and commercial harvest and habitat loss.

Research studies in other locales

Studies of juvenile cutthroat trout, steelhead and coho salmon in western Washington streams flowing through old-growth forests and areas recently clear-cut found that total salmonid biomasses averaged 1.5 times greater in clear-cut than in adjacent unlogged sections (Bisson & Sedell 1984). However, habitat changes that resulted from timber harvesting and debris removal from the channels apparently favoured age-0 trout to the detriment of juvenile salmon and older trout.

A study in streams draining old-growth, clear-cut and second-growth forests in southwestern Washington found that the amount of large wood decreased as stream size increased in the three stand types, and was greatest at old-growth sites (Bilby &

Ward 1991). In large streams (>10 m wide) the volume of individual pieces was larger at old-growth sites than at clear-cut and second-growth sites, but this difference was not evident in smaller streams. Large streams in old-growth sites had more pools associated with large wood than did streams in the other stand types.

In another study of channel morphology and woody debris in logged and unlogged basins of western Washington, timber harvest did not affect the number of pieces of wood within stream channels, but the size of individual pieces was smaller in harvested basins (Ralph *et al.* 1994). In harvested segments, debris was located toward channel margins and was less likely to provide instream cover during low flow periods.

Studies by Coffin and Harr (1992) suggested that timber harvest can increase release of water from snowpacks during many rain-on-snow conditions by reducing snow interception and increasing heat transfer to the snow. O'Connor and Harr (1994) concluded that fluvial sediment transport in steep, low order channels is large enough to make the bedload storage function of large wood a significant control on bedload yield. Their results point to the importance of maintaining the sediment storage function of large wood in steep drainages.

Alaska

Studies on effects of logging in Southeast Alaska were initiated in 1949 by the US Forest Service at Hollis, on Prince of Wales Island. These initial studies were designed to evaluate effects on productivity of pink and chum salmon, which were the largest salmon fisheries in Southeast Alaska. Several streams near Hollis were studied, and logging was initiated in 1952. In 1956 the Fisheries Research Institute, University of Washington, began a study of effects of logging at Hollis. This research contributed to understanding environmental factors influencing growth, development and mortality of salmon embryos and alevins (McNeil & Ahnell 1964; McNeil 1966). The knowledge gained at Hollis was later used throughout the Pacific Northwest and became the foundation for future research on effects of logging.

Effects of timber harvest on spawning habitat
Reduced dissolved oxygen caused by sedimentation of spawning gravel was a principal cause of egg-to-fry mortality (McNeil & Ahnell 1964). Mortality of deposited eggs caused by low dissolved oxygen in three streams near Hollis was 60–90%. Construction and use of logging roads were usually the greatest sediment sources associated with timber harvest. Sediment delivery to streams in two Chichagof Island watersheds during road construction was about 100 m³ per km of road per year, compared with 0.2 m³ per km of road per year from lightly used roads after construction (Swanston *et al.* 1990). Monitoring, however, indicated that Best Management Practices limited sediment inputs from road construction to levels within the natural range (Paustian 1987). Regular road use can cause chronic sediment input to streams because the gravel road surface may break down with repeated heavy wheel loads under wet conditions. Sediment produced by log hauling is estimated at 50 m³ per km of road per year (Swanston *et al.* 1990).

Compared with roads, tree felling and yarding away from stream banks usually produce negligible sediment because coarse soils with high permeability and rapid revegetation help limit surface erosion after felling and yarding. Sediment production from landslides, however, can be increased by clear-cutting because increased snow accumulation in clear-cuts increases down-slope weight (USDA 1995).

Although effects of timber harvest on sedimentation are difficult to assess because of natural variation in sediment dynamics (Paustian 1987), early studies near Hollis found evidence of increased fine sediment in spawning gravel after timber harvest (Bishop & Stevens 1964). McNeil and Ahnell (1964) reported two to four times more fine sediment in the Harris River during timber harvest.

Effects of timber harvest on stream flow could affect spawning success. Higher peak flows are considered detrimental because bedload movement can kill incubating embryos. Bartos (1993) concluded that peak flows increased after timber harvest in three watersheds with metamorphic geology, but decreased in two watersheds with abundant limestone. Base stream flow in Staney Creek, Prince of Wales Island, increased significantly after about 20% of the watershed was harvested.

Effects of timber harvest on rearing habitat

From the mid-1960s through the 1970s, timber harvest on National Forest lands in Southeast Alaska expanded rapidly, reaching a peak of 550 million board feet (~3 million m³) annually. Watersheds of hundreds of salmon streams were logged, causing concern about effects on species that depended on streams for extended periods of rearing, such as coho salmon and steelhead. The focus of research then shifted to effects of logging on habitat of rearing salmonids. A key development in this period was a new spirit of cooperation between forestry and fishery agencies, with the formation of the Alaska Working Group on Cooperative Forestry–Fisheries Research. The Group identified and prioritized research, facilitated logistical support, and coordinated studies by the participants (Gibbons *et al.* 1987).

Assessing the effects of timber harvest on rearing habitat is more complicated than assessing effects on spawning habitat because of the extended time juvenile salmonids are exposed to freshwater mortality factors before migrating to sea. During 1976–1981, studies focused on the role of old-growth forests in providing rearing habitat and on the natural variation in salmonid abundance and habitat in pristine systems. These studies on Porcupine Creek, Etolin Island, provided a baseline for comparison with logged areas and included research on periphyton, benthos, invertebrate drift, salmonid food habits, seasonal habitat utilization and yield of salmon smolts and adults (Koski 1984). Smolt yield, although difficult to assess, is a key variable in determining effects of timber harvest on species limited by rearing habitat (Koski *et al.* 1984).

Increased temperature caused by timber harvest is usually less of a problem in Southeast Alaska than in other areas because of cooler and cloudier weather. However, temperature can become a problem in some situations. Studies on four drainages showed that timber harvest increased maximum summer temperature up to 5°C (Meehan *et al.* 1969). Extensive studies found that temperature in small Southeast Alaska streams was increased by clear-cutting but did not reach lethal levels (Tyler & Gibbons

1973). Sublethal effects, however, could be important. Because of increased temperature in spring, coho salmon fry could emerge 37 days earlier in clear-cut reaches and 22 days earlier in reaches with buffer zones than in old-growth reaches (Thedinga *et al.* 1989). Some Southeast Alaska streams (wide, shallow, low gradient streams with lake or muskeg sources) are 'temperature-sensitive', and removal of shade canopy could increase temperature to undesirable levels (Gibbons *et al.* 1987).

Removing the forest canopy can increase food availability by increasing aquatic primary production. Where food is limiting, summer density of coho salmon fry tends to be higher in clear-cut than old-growth areas (Murphy *et al.* 1986). Probably more important than the short-term increases in production after clear-cutting are the potential long-term decreases in production caused by increased shading by a dense second-growth canopy in later successional stages (Bjornn *et al.* 1992).

Although timber harvest tends to increase fry abundance in summer by opening the canopy, this positive effect can be nullified by reduced winter habitat (Fig. 17.2). Mortality of rearing salmonids in winter is often substantial (Murphy *et al.* 1986), and juvenile salmonids in Alaska spend more winters in fresh water than do those in more southerly areas. Although summer density of coho salmon fry was greater in both clear-cut streams and those with buffer zones than it was in old-growth streams, pre-smolts in late winter were less abundant in clear-cuts than in old growth, whereas buffered streams maintained the highest pre-smolt density (Thedinga *et al.* 1989). The disadvantage in clear-cuts was a reduction in pools and LWD; the advantage in buffered reaches was a combination of both enhanced food abundance because of more open canopy in summer and increased LWD cover in winter (Murphy *et al.* 1986). Thus, winter habitat is frequently a bottleneck in freshwater production of salmon smolts, and clear-cutting without buffers had its most detrimental effects at this point (Koski *et al.* 1984).

Highest densities of juvenile salmonids are often associated with LWD and pool habitat (Murphy *et al.* 1986). Loss of wood reduces available habitat for juvenile salmonids (Dolloff 1986; Elliott 1986). LWD not only provides cover directly, but also

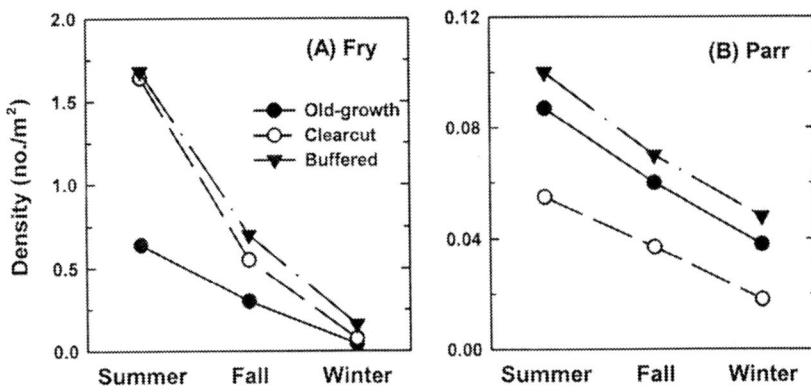

Fig. 17.2 Decline in density of (A) coho salmon fry (age 0) and (B) parr (age 1 and 2) from August 1982 to March 1983 in old-growth, clear-cut and buffered reaches of streams in Southeast Alaska (after Koski *et al.* 1984).

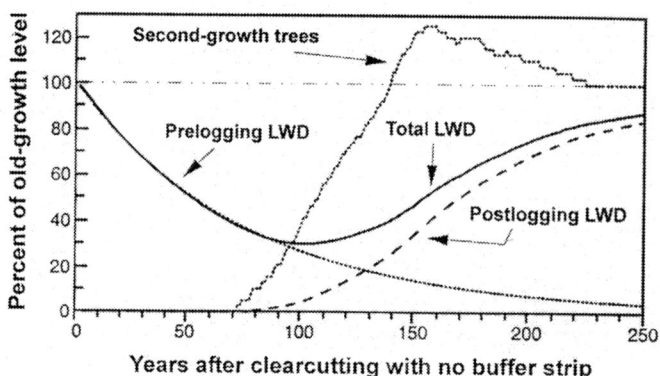

Fig. 17.3 A model of the changes in key pieces of large wood (LWD >60-cm diameter) in a small stream in Southeast Alaska after clear-cut logging without a riparian buffer zone (after Murphy & Koski 1989).

forms 80–90% of pools in valley bottom streams (Heifetz *et al.* 1986) and helps maintain water levels during low flow periods (Lisle 1986). Logging slash is less stable and less effective in forming fish habitat than is large woody debris from natural sources and, when slash moves, it can destabilize natural LWD accumulations downstream (Bryant 1980). In autumn, juvenile steelhead may leave clear-cut reaches where LWD has been depleted and move into old-growth and buffered reaches where LWD is abundant (Johnson *et al.* 1986).

Buffer zones along streams are important to maintain LWD and other riparian functions. Blow-down of buffer zones, once considered damaging to habitat, can enhance winter cover for juvenile salmonids by providing additional LWD, particularly large trees with attached rootwads (Murphy *et al.* 1986). One study in Southeast Alaska showed that 99% of LWD came from within 30 m of the stream channel, through stream undercutting, windthrow, mortality, landslides and beaver activity (Murphy & Koski 1989). Models based on the depletion rate of LWD (Murphy & Koski 1989) indicated that clear-cutting without buffer zones would lead to a long-term reduction in LWD (Fig. 17.3), whereas 30-m buffer zones would maintain LWD in the stream.

Recommendations for improvement of forest practices

In a thoughtful analysis of challenges facing resource managers, Reeves *et al.* (1995) proposed a new paradigm as a template for forest management in the Pacific Northwest. Our recommendations for improvements in harvest practices are based largely on their analysis. They observe that large-scale disturbance has been a pervasive condition of the Pacific Northwest landscape for thousands of years, and that anadromous salmonids have adapted to these dynamic conditions. The underlying basis of their management proposal is a recognition that landscape disturbance caused by timber harvest differs in several important ways from the natural disturbance regime.

- The legacy of natural landslides includes both sediment and large wood, whereas that from clear-cuts includes mostly sediment.
- The interval between natural disturbances has been considerably longer than those between timber harvests.
- The area in early successional stages because of recent wildfires is estimated to have been substantially smaller in historic times than currently exists as a result of timber harvest.
- Natural fires tended to be large and concentrated, whereas the current regime of timber harvest is widely dispersed in small patches.

The outcome of the present regime of disturbance from timber harvest has been a reduction in the complexity of stream habitat. The Reeves proposal suggests altering timber harvest and its accompanying disturbance to more closely mimic the character, timing and spatial scale of the natural disturbance regime to which fish populations in the region have adapted, thereby regaining some of the lost habitat complexity. Their specific management recommendations include:

- increased riparian protection in headwall first and second order tributaries so that the legacy of hill-slope failures will include more large wood;
- longer intervals between timber harvests, depending on the natural disturbance regime and the time needed for favourable habitat conditions to develop;
- concentrated rather than dispersed management that more closely mimics the pattern generated by natural disturbance.

The specifics of their proposal are geared to the ecological and physical conditions of the Oregon Coast Range, where natural disturbance primarily occurred in the form of large catastrophic wildfires and subsequent landslides. However, the principles should be applicable over a wider area, with consideration for prevailing natural disturbance regimes and local conditions. For the most part, these management concepts require large tracts in a single ownership and are thus most applicable to public lands managed by the US Forest Service and Bureau of Land Management. However, some aspects may be useful in providing further protection to fish habitat on private lands.

These new concepts should be incorporated with many of the current management recommendations and could complement the programme of permanently designated reserves in the Northwest Forest Plan. Important as they are, permanent reserves cannot be relied upon to remain static over time, owing to the dynamic nature of natural disturbance. For example, current modelling suggests that only 40–60% of salmonid habitat in third and fourth order forested streams in the sandstone region of the Oregon Coast Range would have been considered in 'good' condition at any one time during presettlement (G. Reeves, PNW Research Station, Corvallis, Oregon, pers. comm.). Thus it is important that other measures be in place to protect fish habitat, including many of those recommended in Murphy (1995): buffer strips, Best Management Practices, watershed analysis and restoration. Replacement of inadequate road culverts is currently a priority, particularly in the Pacific Northwest.

Incorporation of this new paradigm will require that both forest and fish managers adopt new ways of thinking that include longer time-frames, consideration at a landscape scale, and a recognition that disturbance is not necessarily damaging to habitat in the long term (Reeves *et al.* 1995).

A recent review (Fausch *et al.* 2002) further emphasizes the importance of a landscape perspective for fisheries research and management. The authors argue persuasively that both research and management have often been carried out at a spatial scale too small to provide effective information and appropriate management protection for the many species of stream fishes that have declined in abundance.

Many stocks of anadromous salmonids, particularly those in the Pacific Northwest, have been classified as threatened or endangered. It seems clear that further protection is required if these stocks are to thrive. Effects of forest management are only one of many causes for their decline, but any improvement in current forest practices will be a step toward recovery of these valuable resources.

Acknowledgements

We appreciate the assistance of N.P. Peterson, P.A. Bisson and G.H. Reeves, whose reviews of earlier drafts were helpful in improving the presentation.

References

AFSEEE (1996) *Tongass in Transition: Blueprint for a Sustainable Future*. Association of Forest Service Employees for Environmental Ethics, Eugene, OR.

Bartos, L. (1993) Stream discharge related to basin geometry and geology, before and after logging. In: *Proceedings of Watershed '91*, (ed. T. Brock), pp. 29–32. R10-MB-217. USDA Forest Service, Juneau, AK.

Bilby, R.E. & Ward, J.W. (1991) Characteristics and function of large woody debris in streams draining old-growth, clear-cut, and second-growth forests in southwestern Washington. *Canadian Journal of Fisheries and Aquatic Sciences*, **48**, 2499–508.

Bishop, D.M. & Stevens, M.E. (1964) Landslides on logged areas in southeast Alaska. *Research Paper NOR-1*. USDA Forest Service.

Bisson, P.A. & Sedell, J.R. (1984) Salmonid populations in streams in clearcut vs. old-growth forests of western Washington. In: *Fish and Wildlife Relationships in Old-Growth Forests, Proceedings of a Symposium*, (eds W.R. Meehan, T.R. Merrell Jr & T.A. Hanley), pp. 121–9. American Institute of Fishery Research Biologists.

Bjornn, T.C., Brusven, M.A., Hetrick, N.J., Keith, R.M. & Meehan, W.R. (1992) Effects of canopy alterations in second-growth forest riparian zones on bioenergetic processes and responses of juvenile salmonids to cover in small southeast Alaska streams. *Technical Report 92–7*. Idaho Cooperative Fish and Wildlife Research Unit, Moscow, ID.

Bryant, M.D. (1980) Evolution of large, organic debris after timber harvest: Maybeso Creek, 1949 to 1978. *General Technical Report, PNW-101*. USDA Forest Service.

Cederholm, C.J. & Peterson, N.P. (1985) The retention of coho salmon (*Oncorhynchus kisutch*) carcasses by organic debris in small streams. *Canadian Journal of Fisheries and Aquatic Sciences*, **42**, 1222–5.

Cederholm, C.J. & Reid, L.M. (1987) Impact of forest management on coho salmon (*Oncorhynchus kisutch*) populations of the Clearwater River, Washington: a project summary. In:

Streamside Management: Forestry and Fishery Interactions, (eds E.O. Salo & T. Cundy), pp. 373–98. Contribution No. 57. University of Washington, College of Forest Resources, Seattle.

Cederholm, C.J. & Scarlett, W.J. (1982) Seasonal immigrations of juvenile salmonids into four small tributaries of the Clearwater River, Washington, 1977–1981. In: *Proceedings of the Salmon and Trout Migratory Behavior Symposium*, (eds E.L. Brannon & E.O. Salo), pp. 98–110. School of Fisheries, University of Washington, Seattle.

Cederholm, C.J., Reid, L.M. & Salo E.O. (1981) Cumulative effects of logging road sediment on salmonid populations in the Clearwater River, Jefferson County, Washington. In: *Proceedings from the Conference–Salmon Spawning Gravel: A Renewable Resource in the Pacific Northwest?* Report No. 39, pp 38–74. Water Research Center, Washington State University, Pullman.

Cederholm, C.J., Houston, D.B., Cole, D.L. & Scarlett, W.J. (1989) Fate of coho salmon (*Oncorhynchus kisutch*) carcasses in spawning streams. *Canadian Journal of Fisheries and Aquatic Sciences*, **46**, 1347–55.

Cederholm, C.J., Johnson, D.H., Bilby, R.E., *et al.* (2001) Pacific salmon and wildlife: ecological contexts, relationships, and implications for management. In: *Wildlife-Habitat Relationships in Oregon and Washington*, (Managing Directors: D.H. Johnson & T.A. O'Neil), pp. 628–84. Oregon State University Press, Corvallis, OR.

Coffin, B.A. & Harr, R.D. (1992) Effects of forest cover on volume of water delivery to soil during rain-on-snow. TFW-SH1–92–001. Washington Department of Natural Resources Timber/Fish/Wildlife Program.

Connolly, P.J. & Hall, J.D. (1999) Biomass of coastal cutthroat trout in unlogged and previously clear-cut basins in the central Coast Range of Oregon. *Transactions of the American Fisheries Society*, **128**, 890–9.

Dent, L. (2001) Harvest effects on riparian function and structure under current Oregon Forest Practice Rules. *Technical Report 12*. Oregon Department of Forestry, Salem.

Dolloff, C.A. (1986) Effects of stream cleaning on juvenile coho salmon and Dolly Varden in southeast Alaska. *Transactions of the American Fisheries Society*, **115**, 743–55.

Edie, B.G. (1975) *A census of juvenile salmonids in the Clearwater River Basin, Jefferson County, Washington, in relation to logging*. MS thesis, University of Washington, Seattle.

Ellefson, P.V., Cheng, A.S. & Moulton, R.J. (1997) State forest practice regulatory programs: an approach to implementing ecosystem management on private forest lands in the United States. *Environmental Management*, **21**, 421–32.

Elliott, S.T. (1986) Reduction of a Dolly Varden population and macrobenthos after removal of logging debris. *Transactions of the American Fisheries Society*, **115**, 392–400.

Faris, T.L. & Vaughan, K.D. (1985) Log transfer and storage facilities in southeast Alaska: a review. *General Technical Report PNW-174*. USDA Forest Service.

Fausch, K.D., Torgerson, C.E., Baxter, C.V. & Li, H.W. (2002) Landscapes to riverscapes: bridging the gap between research and conservation of stream fishes. *BioScience*, **52**, 483–98.

Gibbons, D.R., Meehan, W.R., Koski, KV. & Merrell, T.R., Jr (1987) History of fisheries and forestry interactions in southeastern Alaska. In: *Streamside Management: Forestry and Fishery Interactions*, (eds E.O. Salo & T.W. Cundy), pp. 297–329. Contribution No. 57. University of Washington, College of Forest Resources, Seattle.

Gregory, S.V. (1997) Riparian management in the 21st century. In: *Creating a Forestry for the 21st Century: the Science of Ecosystem Management*, (eds K.A. Kohm & J.F. Franklin), pp. 69–85. Island Press, Washington, DC.

Gregory, S.V., Swanson, F.J., McKee, W.A. & Cummins, K.W. (1991) An ecosystem perspective of riparian zones. *BioScience*, **41**, 540–51.

Gregory, S.V., Schwartz, J.S., Hall, J.D., Wildman, R.C. & Bisson, P.A. (in press) Long-term trends in habitat and salmonid populations in the Alsea Basin. In: *The Alsea Watershed: Hydrological and Biological Responses to Temperate Coniferous Forest Practices*, (ed. J.R. Stednick). Springer-Verlag, New York.

Grette, G. (1985) *The role of large organic debris in juvenile salmonid habitat in small streams in second-growth and unlogged forests.* MS thesis, University of Washington, Seattle.

Hall, J.D. (in press) Salmonid populations and habitat. In: *The Alsea Watershed: Hydrological and Biological Responses to Temperate Coniferous Forest Practices,* (ed. J.R. Stednick). Springer-Verlag, New York.

Hall, J.D. & Lantz, R.L. (1969) Effects of logging on the habitat of coho salmon and cutthroat trout in coastal streams. In: *Symposium on Salmon and Trout in Streams,* (ed. T. G. North-cote), pp. 355–75. H. R. MacMillan Lectures in Fisheries, University of British Columbia, Vancouver.

Hall, J.D., Brown, G.W. &. Lantz, R.L. (1987) The Alsea Watershed Study: a retrospective. In: *Proceedings of a Symposium: Streamside Management – Fishery and Forestry Interactions,* (eds E.O. Salo & T.W. Cundy), pp. 399–416. College of Forest Resources, University of Washington, Seattle.

Harris, A.S. & Farr, W.A. (1974) The forest ecosytem of southeast Alaska: 7. Forest ecology and timber management.*General Technical Report PNW 25.* USDA Forest Service.

Hawkins, C.P., Murphy, M.L., Anderson, N.H. & Wilzbach, M.A. (1983) Density of fish and salamanders in relation to riparian canopy and physical habitat in streams of the northwestern United States. *Canadian Journal of Fisheries and Aquatic Sciences,* 40, 1173–85.

Heifetz, J., Murphy, M.L. & Koski, KV. (1986) Effects of logging on winter habitat of juvenile salmonids in Alaskan streams. *North American Journal of Fisheries Management,* 6, 52–8.

Hicks, B.J. & Hall, J.D. (2003) Rock type and channel gradient structure salmonid populations in the Oregon Coast Range. *Transactions of the American Fisheries Society,* 132, 468–82.

Hicks, B.J., Hall, J.D., Bisson, P.A. & Sedell, J.R. (1991) Responses of salmonids to habitat changes. In: *Influences of Forest and Rangeland Management on Salmonid Fishes and their Habitats,* (ed. W.R. Meehan). *American Fisheries Society Special Publication,* 19, 483–518.

Jackson, P.L. & Kimerling, A.J. (1993) *Atlas of the Pacific Northwest,* 8th edn. Oregon State University Press, Corvallis, OR.

Johnson, S.W., Heifetz, J. & Koski, KV. (1986) Effects of logging on the abundance and seasonal distribution of juvenile steelhead in some southeastern Alaska streams. *North American Journal of Fisheries Management,* 6, 532–7.

Jones, J.A. & Grant, G.E. (1996) Peak flow responses to clear-cutting and roads in small and large basins, western Cascades, Oregon. *Water Resources Research,* 32, 959–74.

June, J.A. (1981) *Life history and habitat utilization of cutthroat trout (*Salmo clarki*) in a headwater stream on the Olympic Peninsula, Washington.* MS thesis, University of Washington, Seattle.

Koski, KV. (1966) *The survival of coho salmon (*Oncorhynchus kisutch*) from egg deposition to emergence in three Oregon streams.* MS thesis. Oregon State University.

Koski, KV. (1975) *The survival and fitness of two stocks of chum salmon (*Oncorhynchus keta*) from egg deposition to emergence in a controlled-stream environment at Big Beef Creek.* PhD thesis, University of Washington, Seattle.

Koski, KV. (1984) A stream ecosystem in an old-growth forest in southeast Alaska: Part I. Description and characteristics of Porcupine Creek, Etolin Island. In: *Fish and Wildlife Relationships in Old-Growth Forests, Proceedings of a Symposium,* (eds W.R. Meehan, T.R. Merrell, Jr & T.A. Hanley), pp. 47–55. American Institute of Fishery Research Biologists.

Koski, KV., Heifetz, J., Johnson, S., Murphy, M. & Thedinga, J. (1984) Evaluation of buffer strips for protection of salmonid rearing habitat and implications for enhancement. In: *Proceedings: Pacific Northwest Stream Habitat Management Workshop,* (ed. T.J. Hassler), pp. 138–55. American Fisheries Society, Humboldt State University, Arcata, CA.

Lestelle, L.C. & Cederholm, C.J. (1984) Short-term effects of organic debris removal on resident cutthroat trout. In: *Fish and Wildlife Relationships in Old-Growth Forests, Proceedings of a Symposium*, (eds W.R. Meehan, T.R. Merrell Jr & T.A. Hanley), pp. 131–40. American Institute of Fishery Research Biologists.

Lichatowich, J. (1999) *Salmon Without Rivers: A History of the Pacific Salmon Crisis*. Island Press, Washington, DC.

Lisle, T.E. (1986) Effects of woody debris on anadromous salmonid habitat, Prince of Wales Island, southeast Alaska. *North American Journal of Fisheries Management*, 6, 538–50.

McIntosh, B.A., Sedell, J.R., Thurow, R.F., Clarke, S.E. & Chandler, G.L. (2000). Historical changes in pool habitats in the Columbia River Basin. *Ecological Applications* 10, 1478–96.

McNeil, W.J. (1966) Effects of the spawning bed environment on reproduction of pink and chum salmon. US Fish and Wildlife Service. *Fishery Bulletin*, 65, 495–523.

McNeil, W.J. & Ahnell, W.H. (1964) Success of pink salmon spawning relative to size of spawning bed materials. US Fish and Wildlife Service. *Special Scientific Report – Fisheries*, 469.

Martin, D.J., Salo, E.O., White, S.T., June, J.A., Foris, W.J. & Lucchetti, G.L. (1981) *The Impact of Managed Streamside Timber Removal on Cutthroat Trout and the Stream Ecosystem. Part 1 – A Summary*. FRI-UW-8107. University of Washington College of Fisheries, Fisheries Research Institute, Seattle.

Meehan, W.R., Farr, W.A., Bishop, D.M. & Patric, J.H. (1969) Some effects of clearcutting on salmon habitat of two southeast Alaska streams. *Research Paper, PNW-82*. USDA Forest Service.

Moring, J.R. & Lantz, R.L. (1975) The Alsea Watershed Study: effects of logging on the aquatic resources of three headwater streams of the Alsea River, Oregon. Oregon Department of Fish and Wildlife. *Fishery Research Report*, 9.

Murphy, M.L. (1995) Forestry impacts on freshwater habitat of anadromous salmonids in the Pacific Northwest and Alaska: requirements for protection and restoration. *NOAA Coastal Ocean Program Decision Analysis Series No. 7*. NOAA Coastal Ocean Office, Silver Spring, MD.

Murphy, M.L. & Hall, J.D. (1981) Varied effects of clear-cut logging on predators and their habitat in small streams of the Cascade Mountains, Oregon. *Canadian Journal of Fisheries and Aquatic Sciences*, 38, 137–45.

Murphy, M.L. & Koski, KV. (1989) Input and depletion of woody debris in Alaska streams and implications for streamside management. *North American Journal of Fisheries Management*, 9, 427–36.

Murphy, M.L., Hawkins, C.P. & Anderson, N.H. (1981) Effects of canopy modification and accumulated sediment on stream communities. *Transactions of the American Fisheries Society*, 110, 469–78.

Murphy, M.L., Heifetz, J., Johnson, S.W., Koski, KV. & Thedinga, J.F. (1986) Effects of clear-cut logging with and without buffer strips on juvenile salmonids in Alaskan streams. *Canadian Journal of Fisheries and Aquatic Sciences*, 43, 1521–33.

Naiman, R.J., Bilby, R.E. & Bisson, P.A. (2000) Riparian ecology and management in the Pacific Coastal Rain Forest. *BioScience*, 50, 996–1011.

O'Connor, M. & Harr, R.D. (1994) *Bedload Transport and Large Organic Debris in Steep Mountain Streams in Forested Watersheds on the Olympic Peninsula, Washington*. TFW-SH7-94-001. Washington Department of Natural Resources Timber/Fish/Wildlife Program.

Paustian, S.J. (1987) Monitoring nonpoint source discharge of sediment from timber harvesting activities in two southeast Alaska watersheds. In: *Water Quality in the Great Land: Alaska's Challenge*, (ed. R.G. Huntsinger), pp. 153–68. University of Alaska, Fairbanks, IWR-109. American Water Resources Association.

Peterson, N.P. (1982) Immigration of juvenile coho salmon (*Oncorhynchus kisutch*) into riverine ponds. *Canadian Journal of Fisheries and Aquatic Sciences*, 39, 1308–10.

Ralph, S.C., Poole, G.C., Conquest, L.L. & Naiman, R.J. (1994) Stream channel morphology and woody debris in logged and unlogged basins of western Washington. *Canadian Journal of Fisheries and Aquatic Sciences,* **51,** 37–51.

Reeves, G.H., Benda, L.E., Burnett, K.M., Bisson, P.A. & Sedell, J.R. (1995) A disturbance-based ecosystem approach to maintaining and restoring freshwater habitats of evolutionarily significant units of anadromous salmonids in the Pacific Northwest. *American Fisheries Society Symposium,* **17,** 334–49.

Reeves, G.H., Hall, J.D. & Gregory, S.V. (1997) The impact of land management activities on coastal cutthroat trout and their freshwater habitats. In: *Sea-Run Cutthroat Trout: Biology, Management, and Future Conservation,* (eds J.D. Hall, P.A. Bisson & R.E. Gresswell), pp. 138–44. Oregon Chapter, American Fisheries Society, Corvallis, OR.

Reid, L.M. & Dunne, T. (1984) Sediment production from forest road surfaces. *Water Resources Research,* **20,** 1753–61.

Rittmueller, J.F. (1986) *Effects of logging roads on the composition of spawning gravel in streams of the west slope Olympic Mountains, Washington.* MS thesis, Western Washington University, Bellingham.

Sedell, J.R., Leone, F.N. & Duval, W.S. (1991) Water transportation and storage of logs. In: *Influences of Forest and Rangeland Management on Salmonid Fishes and their Habitats,* (ed. W.R. Meehan). *American Fisheries Society Special Publication* **19,** 325–68.

Swanston, D.N., Webb, T.M., Bartos, L, Meehan, W.R., Sheehy, T. & Puffer, A. (1990) The hydrology and soils submodel. In: *SAMM: A Prototype Southeast Alaska Multiresource Model,* (eds R.D. Fight, L.D. Garrett & D.L. Weyermann), pp. 28–45. *General Technical Report PNW-GTR-255.* USDA Forest Service.

Tagart, J.V. (1984) Coho salmon survival from egg deposition to fry emergence. In: *Proceedings of the Olympic Wild Fish Conference,* (eds J.M. Walton & D.B. Houston), pp. 173–81. Peninsula College, Port Angeles, WA.

Thedinga, J.F., Murphy, M.L., Heifetz, J., Koski, KV. & Johnson, S.W. (1989) Effects of logging on size and age composition of juvenile coho salmon (*Oncorhynchus kisutch*) and density of presmolts in southeast Alaska streams. *Canadian Journal of Fisheries and Aquatic Sciences,* **46,** 1383–91.

Tyler, R.W. & Gibbons, D.R. (1973) *Observations of the Effects of Logging on Salmon-Producing Tributaries of the Staney Creek Watershed and the Thorne River Watershed and of Logging in the Sitka District.* Fisheries Research Institute, University of Washington, Seattle.

USDA (1995) Report to Congress. *Synthesis: Anadromous Fish Habitat Assessment.* R10-MB-279. USDA Forest Service Alaska Region, Pacific Northwest Research Station.

Wilzbach, M.A., Cummins, K.W. & Hall, J.D. (1986) Influence of habitat manipulations on interactions between cutthroat trout and invertebrate drift. *Ecology,* **67,** 898–911.

Young, K.A. (2000) Riparian zone management in the Pacific Northwest: who's cutting what? *Environmental Management,* **26,** 131–44.

Chapter 18

Fish–forestry interaction research in coastal British Columbia – the Carnation Creek and Queen Charlotte Islands studies

P.J. TSCHAPLINSKI, D.L. HOGAN AND G.F. HARTMAN

Introduction

Two major investigations of the effects of forestry practices on watersheds and fish have been conducted in the coastal region of British Columbia (BC), Canada, within the past three decades. The Carnation Creek Experimental Watershed Project, on the southwest coast of Vancouver Island, was initiated in 1970, and the Fish–Forestry Interaction Program (FFIP) on the Queen Charlotte Islands (QCI) was begun in 1981 and concluded by 1994. The Carnation Creek project is a long-term case study now in its 32nd year. The FFIP was a synoptic study covering 30 watersheds. This review summarizes and compares these two projects which used different study designs to examine forestry effects on channel morphology, fish habitat and fish populations.

The Carnation Creek project was initiated by researchers to learn about biological and physical processes operating within a coastal watershed, and to study the effects of forestry practices upon these processes (Lewis 1998; Tschaplinski 2000). In contrast, the FFIP study was implemented under the direction of senior government managers after large landslides occurred in 1979 following logging on unstable hillslopes, and after confrontations between government and industry, and widespread public criticism of forestry practices (Lewis 1998). Both studies have made significant contributions to our knowledge of how watersheds function in northwestern North America and how forestry activities may affect these functions. The two projects were based on different but complementary research approaches, which have enabled the results, taken together, to have significantly more application to forest practices than the results of either study taken individually. Both studies have played critical roles in the development of forest management regulations, guidelines and practices in BC and elsewhere in the Pacific Northwest.

The Pacific coastal region of British Columbia

The coastal region of BC includes all offshore islands such as Vancouver Island and the Queen Charlotte Islands, and most of the area covered by the Coast Mountains on the continental mainland (Fig. 18.1). This topographically diverse region includes coastal lowlands and broad alluvial valley bottoms associated with major rivers, but

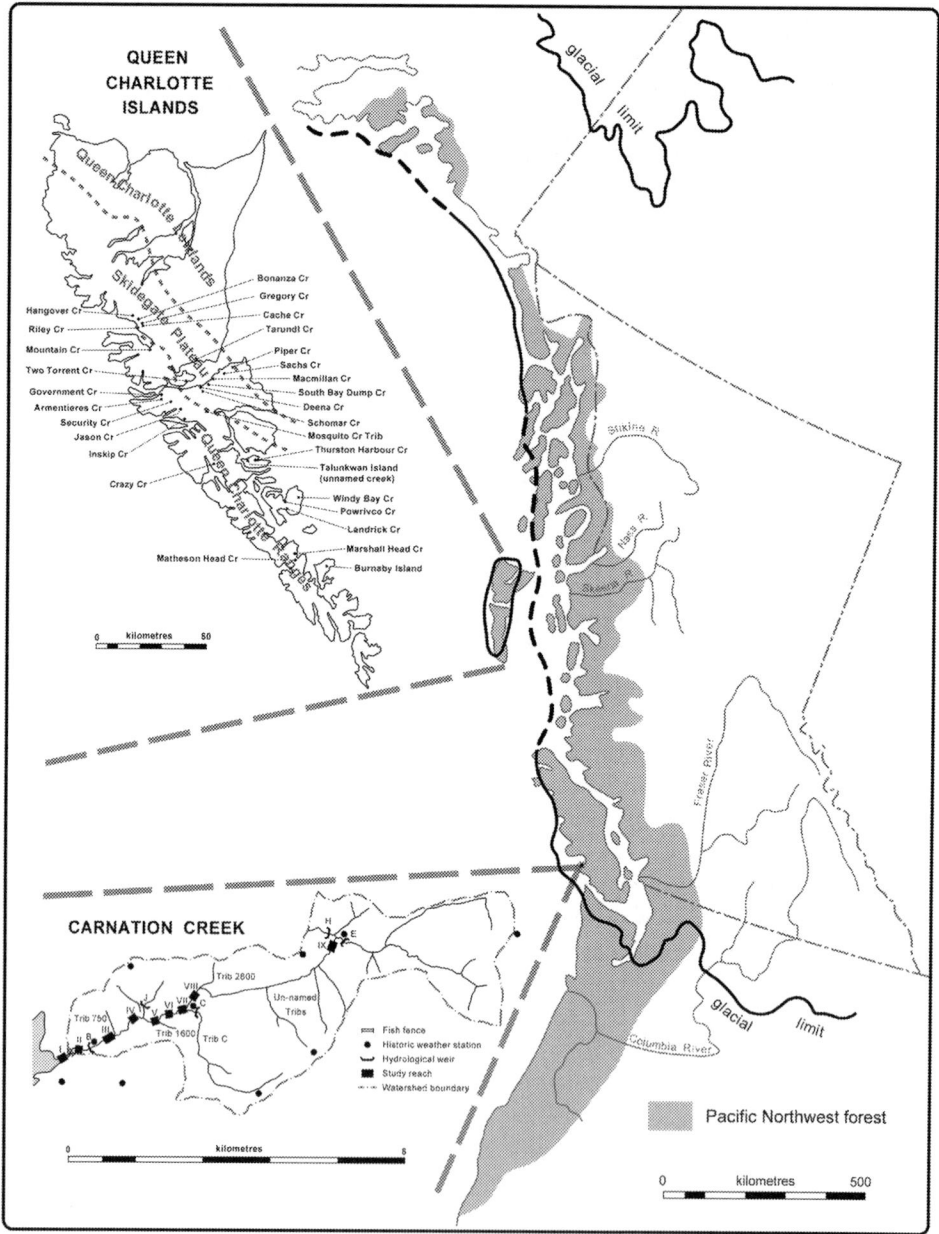

Fig. 18.1 Locations of the Carnation Creek and Queen Charlotte Islands studies within the Pacific Northwest coastal forest.

is dominated by mountainous terrain sculpted by glaciation. Most of the region lies within the Coastal Western Hemlock and Mountain Hemlock biogeoclimatic zones (Krajina 1969).

The contemporary environment of this region began to develop about 10 million years ago with the most recent uplift of the Coast Mountain range, which was as much as 4 km during this time. This uplift resulted from the tectonic collision of several oceanic plates and the America plate. Tectonic history, and cordilleran glaciation within the past three million years, have produced the rugged coastal landforms, the area's dominant geomorphic processes and its regional climate. Together, these forces have conditioned the development of the biota of this region.

Dominant geomorphic processes, identified by prevailing sediment transfer mechanisms, are zoned vertically within this coastal landscape from periglacial processes on mountain tops, episodic mass wasting on hill-slopes, to fluvial processes on valley floors (Ryder 1981; Church 1998a). The two study areas considered here are influenced primarily by colluvial and fluvial processes.

Water is the key link connecting all parts of the Pacific Northwest forest ecosystem. The coastal mountains present a high barrier to westerly atmospheric circulation from the Pacific Ocean. This results in high precipitation, primarily during autumn and winter. Local variation in precipitation occurs as a result of topography and orographic enhancement. Coastal windward slopes may receive 1–3 m of precipitation annually, and some montane slopes receive as much as 6 m (Hogan & Schwab 1990).

Regionally, ≥50% of the annual precipitation falls as snow except near sea level. A high proportion of winter rain and melted snow runs off, particularly in outer coastal basins where runoff ratios may exceed 80%. Extreme precipitation and runoff cause erosion. The most damaging floods occur in mid-autumn to early winter during heavy rains or when rain follows snowfall.

With tectonic uplift, the climate from the late Miocene to the Pliocene Epochs (about 2 million years ago) became progressively wetter, thermally moderate, and cooler in summer (Wolfe & Leopold 1967). These conditions, including increasingly marked seasonal extremes in precipitation, favoured the expansion of coniferous forests. Conifers have the ability to photosynthesize outside of the main period of summer growth, and are efficient at scavenging and retaining nutrients from the heavily leached, nutrient-poor podzol soils typical of the region (Church 1998a).

This forest is exceptional for the large size and longevity of its tree species, and the dominance of conifers throughout the region. Dominant trees, commonly living 400 years, can live >1000 years. Biomass in climax stands is high and may exceed 1000 tonnes/ha (Grier & Logan 1977).

Principal species include western hemlock (*Tsuga heterophylla*), western red cedar (*Thuja plicata*), Sitka spruce (*Picea sitchensis*), Douglas fir (*Pseudotsuga menziesii*) and true firs of the genus *Abies* (e.g. amabilis fir, *A. amabilis*). The diverse landscape, forest age and tree size have created a complex forest structure. This structure includes standing and fallen dead wood, up to 30% of the old-growth biomass, as an important component (Grier & Logan 1977).

These forests have been heavily exploited for their valuable timber since the 1890s. The stream and lake networks within these forests contain valuable fish faunas, especially the Pacific salmon (*Oncorhynchus* spp.). Pacific salmon have long been a social and nutritional foundation of Native societies of the coastal Pacific region, and have been heavily exploited by commercial and recreational fisheries since the middle 1800s.

The number of fish species is small, but the relative isolation among coastal drainages has favoured local genetic differentiation within each species into numerous sub-specific taxa (Riddell 1993). Slaney *et al.* (1996) identified 9663 stocks (races or populations) of trout and anadromous salmon species in BC and Yukon. Anadromous salmon have thrived in this region because sub-adult and adult life stages can exploit the productivity of Pacific marine waters to achieve high growth rates, large individual body sizes and high levels of fecundity. Their eggs and other juvenile life stages can develop and rear in freshwater environments relatively free of aquatic predators. High fecundity buffers the effects of egg and juvenile fish mortality that may be caused by variable adverse conditions in freshwater environments. In spite of this adaptation, many stocks have become extinct or have declined sharply over the past century due primarily to human-related activities (Slaney *et al.* 1996). Attributing specific causes to the decline and extinction of individual stocks has been difficult, but overexploitation in commercial and recreational fisheries, and habitat degradation due to forestry, agriculture, roads, urbanization and hydroelectric power development, are major causes (Slaney *et al.* 1996).

Fish–forestry interactions

High value fishing and forest industries in the coastal region of BC depend on overlapping parts of the land base. Conflict between these industries, and between industry and environmentalists, first became prominent in the late 1960s, and has become an important characteristic of recent BC history. Forestry practices may potentially damage freshwater salmonid habitats. However, measures designed to protect aquatic habitats potentially reduce the profitability of the forest industry.

Until the 1950s, there was little regard for fish habitat protection within forestry planning processes, which were designed for rapid extraction and profit. However, forest and fish managers ultimately recognized that land use planning had to incorporate fish habitat protection. By the 1960s, there was little local research information directly applicable to fish–forestry planning in coastal BC. Therefore, the Carnation Creek study was initiated to provide this information from one type of coastal ecosystem.

The Carnation Creek experimental watershed project

The Carnation Creek project, initiated in 1970 by the federal agency currently called Fisheries & Oceans Canada (DFO) and MacMillan Bloedel Ltd, soon expanded into an inter-agency, multidisciplinary programme on the effects of forest harvesting on a coastal watershed and its salmon and trout populations. The objectives of this study were to:

(1)   provide an understanding of the physical and biological processes operating within a coastal watershed;

(2)   reveal how the forest harvesting practices employed in the 1970s and early 1980s changed these processes; and

(3)   apply the results to make reasonable and useful decisions concerning land use management, fish and aquatic habitat protection.

The study site

Carnation Creek is located on the south shore of Barkley Sound in southwestern Vancouver Island (Fig. 18.1). The main stream channel is 7.8 km long, drains an area of 11 km² with elevations ranging from sea level to over 800 m. Coho and chum salmon, steelhead and cutthroat trout, and two species of sculpins inhabit the lowermost 3.1 km, which extends from the stream mouth to a steep gradient canyon. This part of the watershed contains a valley bottom of about 55 ha that is 50–200 m wide where the valley flats contain both perennial tributaries and seasonally flooded, ponds, and depressions used as winter habitat by juvenile salmonids (Tschaplinski & Hartman 1983; Brown & Hartman 1988). A relict population of non-migratory cutthroat trout inhabits about 800 m of Carnation Creek upstream of the steep canyon.

Annual returns of adult anadromous salmonids have historically numbered from 23 to 4186 chum, 74 to 426 coho, and up to 12 steelhead and 9 cutthroat trout. Three other species of salmon occasionally occur in small numbers as strays.

The watershed has coarse-textured surficial materials that form well-drained soils. Precipitation varies from 210 to >500 cm annually. About 95% of the annual precipitation falls as rain, primarily between October and April. Stream discharge ranges widely from 0.03 m³s⁻¹ in summer to 64 m³s⁻¹ in winter. Flows may increase by a factor of 200 in <48 hours, due to rapid runoff from precipitation up to 26 cm during the same time.

The forests, primarily western hemlock, amabilis fir and western red cedar, contain some Sitka spruce on the valley bottom, Douglas fir on the ridges and western white pine (*Pinus monticola*) at higher elevations. Prior to logging, bigleaf maples (*Acer macrophyllum*) and red alders (*Alnus rubra*) provided a substantial portion of the tree cover along the creek (Oswald 1982).

The study design

The Carnation Creek project is a single watershed, intensive case study incorporating pre-harvest, during-harvest, and post-harvest observations. Spatial controls were available for a variety of study components in Tributary C, a sub-basin unharvested for the duration of the investigation. The study design and the methods employed for monitoring physical variables, fish populations and biological processes before, during and after forest harvesting were described by Hartman & Scrivener (1990) and summarized by Tschaplinski (2000).

This study was designed initially to examine the effects of progressive clear-cutting and three different types of streamside forest harvest treatments on stream channels and fish populations. The three treatments applied within the lowermost 3.1 km of the stream were:

(1)   a *leave-strip* 1–70 m wide from the estuary to 1300 m upstream;

(2)   *'intensive' clear-cutting* along 900 m of stream channel immediately upstream from the leave-strip treatment (no riparian trees were left, some trees were felled and yarded across the stream, and merchantable wind-thrown trees were recovered from the stream channel);

(3)   *'careful' clear-cutting* along 900 m of stream immediately upstream from the intensive treatment area (no activity was permitted in the stream, vegetation on the stream bank such as salmonberry (*Rubus spectabilis*) was untouched but red alders were removed).

The study was carried out in three phases:

(1)   pre-harvest monitoring, 1970–1975;

(2)   during-harvest studies, 1976–1981, when about 41% of the watershed (including almost all of the valley bottom) was harvested;

(3)   post-harvest monitoring, currently from 1982–2002. Within the post-harvest phase (1987–1995), about 21% more of the basin was logged in headwater areas remote from the main stream channel.

Data collected historically included comprehensive information on climate; stream temperatures and discharge; groundwater levels (from piezometers); water chemistry; stream channel morphology; large woody debris (LWD) abundance and distribution; streambed particle size composition (frozen-core methods); suspended sediment generation and transport during high flows; streambed scour and deposition; forestry-related ground disturbance; post-harvest revegetation; biomass of aquatic algae (periphyton); abundance and distribution of benthic macroinvertebrates; and fish habitats. Fish population studies have included those on salmonid spawning and migration patterns; seasonal movements, rearing, growth and survival of juvenile salmonids; chum egg incubation; egg survival; fry emergence; and fecundity determinations for female chum and coho for estimates of annual egg-to-fry survival.

While many of the above studies continue today, several new components have been added in recent years. Studies on suspended sediment, water chemistry and groundwater levels were discontinued in the mid-to-late 1980s. Principal study components currently include stream channel morphology, streambed erosion and sedimentation processes, climate, watershed hydrology, hydrologic recovery, forest regeneration and growth, hill-slope processes, and fish populations, migrations and habitats (Tschaplinski 2000).

Effects of forest harvesting on fish and fish habitat at Carnation Creek

Carnation Creek research identified three broad and inter-related categories of forestry-related effects upon fish and aquatic habitats: (1) physical habitat-structure alterations, (2) temperature-related shifts and (3) trophic responses. These categories, separately and in combinations, have had different effects on fish depending upon species, life stage and distribution. Population and habitat responses to harvest practices are thus complex. Moreover, long-term trends in anadromous fish abundance are often difficult to interpret because they may be confounded by changes in climate, ocean conditions and fisheries management strategies (Tschaplinski 2000).

Chum and coho salmon responded differently to forest harvesting. Chum salmon populations exhibited the most marked changes over the past 30 years, although this species uses freshwater habitats much less extensively than do coho salmon. Mean numbers of chum spawners returning to the creek have fallen 3.5-fold from a pre-harvest mean of 2188 to only 626 since 1982 (Fig. 18.2A). Since 1986, annual returns exceeded 1000 only once, and have been as low as 23 (Fig. 18.2A).

Numbers of adult coho (excluding precocious young males called jacks) returning to spawn have also declined during the post-harvest period (Fig. 18.2B). However, the difference between the pre-harvest mean of 165 and the 20-year post-harvest mean of 127 is not statistically significant (Student's t-test, $p > 0.05$). Mean post-harvest returns were significantly lower than pre-harvest levels until the mid-1990s (Tschaplinski 2000) when strict reductions in commercial and recreational coho fisheries followed by complete closure between 1998 and 2001 resulted in several strong returns including the 30-year peak return in 1998 (Fig. 18.2B). This pattern demonstrates that fishing mortality is an important determinant of the abundance of coho spawners returning to Carnation Creek.

Holtby & Scrivener (1989) used a series of sequential life history-based regression models to determine the relative effects of fishing, forest harvesting and climate change on the returns of coho and chum to Carnation Creek. They concluded that most of the post-harvest decline in both species was due to factors other than forestry practices. About 26% of the reduction in the numbers of chum adults, and <10% of the decline in coho during the 1980s, was attributable to forestry operations. Most of the decline for both species was caused by climatic shifts affecting both marine and freshwater environments. In spite of the prominence of climatic factors, forestry has been an important cause of the observed declines in chum and coho abundance at Carnation Creek (Holtby & Scrivener 1989).

Structural habitat alterations

The main effects of forest harvesting on the stream were:

(1)   bank erosion causing the channel to become wider and shallower;
(2)   loss of large woody material and the accumulation of woody debris into recently formed logjams; and
(3)   movement of sand and pea-sized gravel to the lower reaches of the stream and estuary.

Some of these changes, affecting one-half or more of the lower 3070 m of Carnation Creek, were a consequence of riparian clear-cutting and were initiated shortly after harvesting concluded. However, longer-term, basin-wide processes reflecting critical linkages between steep hill-slopes and the stream channel network have overwhelmed the effects of the riparian treatments. These longer-term changes to the stream channel

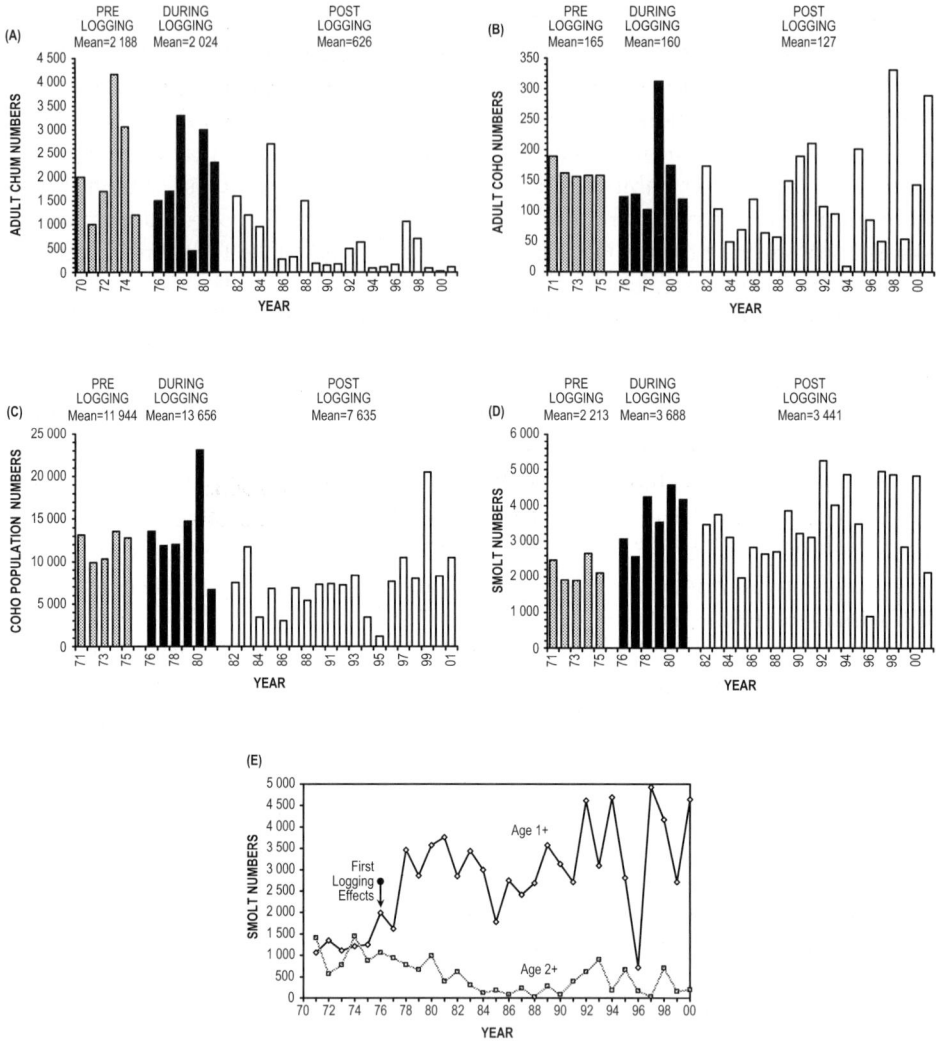

Fig. 18.2 Coho and chum salmon population patterns between 1970 and 2001: (A) numbers of adult chum salmon returning to Carnation Creek, 1970–2001; (B) numbers of adult coho salmon returning, 1971–2001; (C) late summer populations of juvenile coho salmon rearing in fresh water, 1971–2001; (D) numbers of coho salmon smolts leaving Carnation Creek 1971–2001; (E) numbers of 1-year-old versus 2-year-old coho salmon smolts produced in Carnation Creek, 1971–2000.

resulted from increased frequencies of landslides and debris torrents after logging. Over 80 small landslides and three major debris torrents have been observed, all in the logged portions of the watershed. The overall volume of landslide material has increased by 12-fold after logging (Hartman *et al.* 1998). Large logjams and associated sediments deposited by debris torrents in 1984 have moved progressively downstream through the clear-cut riparian treatments to the leave-strip zone, and continue to cause major channel alterations and fish habitat loss 18 years after their initiation.

The post-logging widening of the channel, accelerated scour and deposition, and loss of stable LWD have been largely due to the stream moving around these logjams and sediment deposits, and redistributing materials downstream (Hogan *et al.* 1998a). This complex of short-term and long-term habitat changes has affected all salmonids in the watershed.

Two-thirds of the effect on chum salmon related to forestry practices was explained by impaired embryo development and reductions in egg and alevin survival due to elevated sedimentation of spawning gravel in Carnation Creek (Holtby & Scrivener 1989). Between 68 and 100% of all chum salmon spawn in a channel segment about 100 m long in the upper estuary and immediately upstream where fine sediments are deposited. Egg-to-fry survival declined there from 20.3% before logging to 10.9% after logging (Hartman & Scrivener 1990). The accumulation of fine material reduced intra-gravel water flow and oxygen delivery to the eggs. Sand and pea-gravel deposition also buried alevins and prevented their emergence.

These same processes have affected juvenile coho salmon. The survival of coho eggs has also declined by about one-half after forest harvesting from 28.8 to 15.6% (Hartman & Scrivener 1990). This decline was partly caused by increases in the amount of streambed sand and pea-sized gravel downstream of the clear-cut portions of the creek ($r = 0.81, p < 0.001$; Scrivener & Brownlee 1989). However, most of this reduction was caused by increased streambed scour and deposition in clear-cut areas during freshets after logging (Holtby & Scrivener 1989). Stream bank erosion and loss of LWD caused further impacts on juvenile coho salmon.

Changes in channel morphology and cover also reduced the quality and amount of main channel rearing habitat available for coho, and thus reduced the capacity of the stream to support populations of coho fry and yearlings during summer (Hartman & Scrivener 1990; Tschaplinski 2000). Habitat complexity in both the careful and intensive clear-cut treatments decreased after logging due to reductions in the amount, size and stability of LWD within the stream channel (Toews & Moore 1982; Hartman & Scrivener 1990). Volumes of LWD declined by at least 50% in sections of stream adjacent to the clear-cut treatments (Toews & Moore 1982; Hartman *et al.* 1998). Pools became shallower due to bedload deposition (Hartman & Scrivener 1990). Long stretches of channel were filled with large sediment deposits upstream of new logjams, thus creating ephemeral channels and reducing stream wetted area and salmonid rearing habitat in the main channel (Tschaplinski 2000). The freshwater habitats in Carnation Creek that supported $11,944 \pm 2117$ coho juveniles in late summer before harvesting have sustained only 7635 ± 1924 afterwards (Fig. 18.2C; Student's t-test, $p < 0.05$).

The stream structure and fish habitat characteristics in the lower reaches of the creek have been partially buffered from riparian forestry practices for decades by the leave-strip treatment, and have remained similar to that observed in pre-logging years. However, this zone is presently being degraded as excess sediment and debris move downstream from the intensive and careful treatment sections. This process will likely cause further reductions in the rearing capacity for coho in the future. Therefore, the full extent of the harmful effects of logging on the stream channel, fish habitats and juvenile coho abundance are yet to be observed nearly 20 years after most harvesting ended.

Water temperature alterations

Riparian clear-cutting increased stream temperatures in all seasons. Mean monthly temperatures in small, logged tributaries increased by as much as 4°C in summer (Hartman & Scrivener 1990). Mean temperatures in the main creek increased by as much as 3.2°C in August, 0.7°C in December and 2°C in April (Holtby 1988). Increases during autumn and winter were relatively small, but had the most profound effects on salmonid populations.

Warmer stream temperatures in autumn and winter accelerated egg and alevin development rates that caused sequential changes in emergence timing, growth, survival and seaward migration timing in both chum and coho salmon. Chum fry emerged and migrated seaward earlier in spring, and were also smaller than they were in pre-logging years. Both reduced fry size and earlier seaward migration were strongly correlated with reduced ocean survival (Holtby & Scrivener 1989). Links between freshwater and marine life history processes were thus identified. Increased mortality of chum fry early in their ocean life history was attributed to increased susceptibility to predation due to small size and early-season entry into near-shore waters during winter-like conditions of relatively low salinity and biological productivity (Holtby & Scrivener 1989).

Post-harvest temperature increases had more complex effects on coho salmon life histories. During and after logging, coho fry emerged from the streambed up to 6 weeks earlier in spring (Holtby 1988) permitting 6 more weeks for summer growth than had been available to fry in pre-logging years. The lower numbers of fry rearing in Carnation Creek in most years after logging also resulted in increased growth rates due to density-dependent reductions in competition for food (Scrivener & Andersen 1984; Holtby 1988). Coho fry consequently grew 11 mm longer on average by the end of their first summer after logging compared with sizes during pre-logging years (trout fry also increased in mean length by 18 mm after logging; Hartman & Scrivener 1990). This larger body size was positively associated with improved over-winter survival after logging ($r = 0.91$, $p < 0.001$; Holtby 1988). Larger coho are apparently better able to survive winter conditions that include frequent scouring freshets (Tschaplinski & Hartman 1983). Increased over-winter survival was responsible for the elevated numbers of smolts produced at Carnation Creek after logging.

Although the creek sustained fewer fry after logging, smolt production increased during and after harvesting by 1.6-fold (Fig. 18.2D). Smolt size and biomass also increased. These counter-intuitive relationships are examples of several temperature-related effects upon coho juveniles at least partially attributable to logging. Increased water temperatures also radically changed the age structure of coho smolt populations (Fig. 18.2E). Prior to logging, approximately one-third to one-half of all Carnation Creek coho required 2 years to grow large enough to transform into smolts and migrate seaward (Fig. 18.2E; 1971–1975). Increased seasonal growth due to earlier emergence during and after riparian logging has resulted in most coho reaching smolt size and emigrating seaward after just 1 year in fresh water. Age 2 smolts have become relatively rare after logging (Fig. 18.2E).

These temperature-related effects upon coho salmon that began 26 years ago still persist. Because much of the Carnation Creek channel is more than twice its pre-logging width, these effects will likely continue for several years until a new riparian forest canopy is established over the stream, and both water temperatures and fish growth decline toward pre-harvest levels.

Despite increased smolt abundance and size of 1-year-olds, adult coho returns to Carnation Creek have clearly not increased after logging. Much of this trend appears to be due to reduced marine survival in coho resulting from long-term shifts in marine climate, decreased ocean productivity, and increased predator abundance (Holtby & Scrivener 1989; Tschaplinski 2000). However, as for chum fry, marine survival variations are also linked with the temperature-related effects of forest harvesting. Seasonal increases in water temperatures after logging have also shifted the timing of the spring seaward migration of coho smolts from the stream to about 10 days earlier than during the pre-logging period (Holtby *et al.* 1990). While this shift appears trivial, most mortality in salmonids in marine environments is known to occur soon after they enter the ocean (Mathews & Buckley 1976; Healey 1982; Holtby & Scrivener 1989; Holtby *et al.* 1990). Migration timing and ocean conditions in late winter and spring appear to be critical in determining how many chum fry or coho smolts survive to be adults (Healey 1982; Holtby *et al.* 1990).

Trophic shifts

Forest harvesting including riparian clear-cutting, post-harvest slash burning and ground preparation has had a variety of effects on aquatic communities. However, some relationships were unclear while other interactions were either beneficial or harmful, particularly to periphyton and aquatic macroinvertebrates. Riparian harvesting has increased both water temperatures and the amount of solar radiation reaching the stream. These changes together with short-term increases in nutrient availability due to ground disturbance and slash burning allowed aquatic primary production to increase briefly at Carnation Creek within the during-logging phase of the study (Shortreed & Stockner 1983). However, this increase was limited by stream channel alterations, which increased stream bank erosion, streambed mobility and sediment transport that consequently scoured periphyton from the streambed during freshets (Shortreed & Stockner 1983).

Riparian canopy removal also changed the composition and seasonal pattern of organic litter entering the stream by virtually eliminating the supply of conifer needles between May and December and increasing the supply of deciduous leaf litter from red alders and salmonberry between September and November (Neaves 1978). The consequences of these changes have been unclear. Culp & Davies (1983) concluded that benthic macroinvertebrate populations were reduced in the during-harvest phase of this study in areas where riparian clear-cutting occurred because of reduced leaf litter input and retention, and increased erosion, transport and deposition of sand in the benthos. These results are not consistent with those from some other studies (e.g. Murphy & Hall 1981). Also, the harvest-associated trophic changes at Carnation Creek have not reduced the size or abundance of coho smolts produced from the watershed. The wide variation evident in the literature on the responses of aquatic communities to riparian clear-cutting suggests the strong influence of site-specific conditions including stream size, hydraulic characteristics, gradient, channel and canopy type, and other circumstances.

Fish response to the effects of forestry practices is also influenced strongly by life history strategy. Multiple life history strategies for salmonids during their juvenile stages in fresh water may provide population stability for a species (Brown & Hartman 1988). With one life history strategy each, chum salmon and steelhead trout are influenced exclusively by habitat conditions in the main channel of the creek. Such species have been more strongly affected by forest harvesting than coho salmon, which have employed three life history strategies including main channel habitation, estuary use and seasonal use of refuge habitats in the floodplain (Tschaplinski & Hartman 1983; Tschaplinski 1988; Hartman & Scrivener 1990). Coho have thus persisted more successfully at Carnation Creek than the other species, because they are less sensitive to main channel habitat loss and alteration.

Advantages and limitations of the Carnation Creek study design

Carnation Creek's long-term, continuous datasets on several biological, climatic and watershed physical parameters have allowed the determination of the relative impacts of forest harvesting, climate change, variation in ocean conditions and fisheries management on its salmonid populations (Tschaplinski 2000). However, there are well-known limitations and disadvantages of case studies. The level of impact on Carnation Creek salmonids has been relatively small compared with other factors such as climate change, and may in part be related to physical attributes of the basin. Unlike many other small coastal streams, Carnation Creek has a wide floodplain which contains important winter habitats consisting of seven tributaries and side channels. Despite some deposition of debris and loss of aquatic vegetation, these habitats remained largely intact after logging due to a lack of roads or stream crossings in the floodplain, and contributed to high levels of over-winter survival in juvenile coho and cutthroat trout after logging (Brown & Hartman 1988). Coho smolt abundance after logging may have been substantially lower than observed had these valley bottom features sustained more damage.

The effects of road construction and use were not severe in Carnation Creek. Forestry roads more typical of those in most other coastal watersheds would have had more adverse effects, particularly on the supply of both fine-textured and coarse sediments to the stream. No roads or bridges were built across the creek, and only one short section of road entered the floodplain bordering the lower 3.1 km of stream. All other roads were located on relatively stable hill-slopes on both sides of the basin. Harmful effects on egg-to-fry survival were probably minimized by the absence of construction and use impacts characteristic of other road systems (Cederholm & Reid 1987; Everest *et al.* 1987).

The study was limited by the absence of an external, unlogged, control watershed. Long-term patterns in spawner numbers, smolt production per spawner, and the numbers of spawners produced per smolt from control streams would have strengthened the project. An external control watershed would have helped clarify the complications associated with applying the intensive riparian treatment immediately upstream of the leave-strip treatment. Other watershed morphometric characteristics could have also been resolved, such as the influence of sediment delivery from the canyon immediately upstream of the careful riparian treatment. Comparable studies done on streams with different morphometry, aspect and elevation might have yielded very different results to those from Carnation Creek.

Despite limitations, much has been learned about forestry practices and their effects upon biological and physical processes within a small coastal watershed. More will be learned as this project continues. Main channel habitats at Carnation Creek continue to deteriorate, and soon large portions of the stream will lack stable LWD for many decades. The riparian forest canopy will soon be re-established and begin to reduce water temperatures. Some of the increases in fish growth and survival associated with elevated stream temperatures after logging may disappear. With poorer habitat quality and lower stream temperatures, smolt production from Carnation Creek may soon exhibit different trends.

Carnation Creek results provide guidance for understanding processes elsewhere in spite of variability among streams. However, before such extrapolations can be made with confidence, an appreciation of the effects of landscape variability is required. A valuable measure of regional variability in landscapes, natural disturbance regimes and responses to forestry practices has been provided by multi-watershed synoptic research conducted in the Queen Charlotte Islands. The QCI-FFIP study has demonstrated that conditions prevailing at Carnation Creek, including its forests, fish populations, moderate relief and steep slopes, fall well within the range encountered among the approximately 30 streams included in the synoptic programme. Together, these two studies provide in-depth management guidance applicable to much of the British Columbia coast.

The Queen Charlotte Islands Fish–Forestry Interaction Program

Conflicts between forestry and fishery resource managers on the Queen Charlotte Islands increased in frequency and intensity in the 1960s, and peaked in the late 1970s after much of the forests in low-relief terrain had been harvested, and operations moved into areas containing steeper and less stable hill-slopes. Two large rainstorms in 1978 triggered hundreds of landslides throughout the islands (Hogan & Schwab 1990). These slides, and specifically those in the Riley Creek watershed (Fig. 18.1), led to the implementation of the interdisciplinary Fish–Forestry Interaction Program in 1981 by federal and provincial agencies (Poulin 1984).

The objectives of QCI-FFIP were to:

(1)   understand the effects of steep-slope harvesting on streams, fish habitats and forest sites;
(2)   develop alternative forestry methods to minimize logging-induced slope failures; and
(3)   develop and test rehabilitation techniques for forest sites and streams to mitigate damage caused by landslides (Poulin 1984).

The goal of this study was to provide resource managers with improved knowledge on where and how to conduct forestry operations on steep, marginally stable terrain with minimal environmental damage. This review focuses solely on objective (1).

The study area

The Queen Charlotte Islands are located off the north coast of British Columbia about 250 km north of Vancouver Island (Fig. 18.1). This archipelago consists of two main islands, Graham (north) and Moresby (south), and 148 smaller ones. The islands have typical Pacific coast forest attributes including a mild and extremely wet climate, and valued fish and timber resources (Poulin 1984). About one-third of the area consists of terrain that is inherently unstable due to steep slopes, highly erodible and deeply weathered bedrock, intense rainstorms and frequent seismic activity (Schwab 1998).

Most component studies were located primarily within the Skidegate Plateau physiographic region (Fig. 18.1) which is characterized by steep and gullied hill-slopes with shallow, unconsolidated surficial materials. These areas are subject to severe mass wasting, the regionally dominant geomorphic process (Schwab 1998). Inventories based upon historical aerial photographs have identified over 8000 large landslides, including open-slope failures (debris slides, debris avalanches, debris flows and slump earth flows) and gully/stream failures (debris torrents) on the islands (Gimbarzevsky 1988).

The study plan

This programme used a synoptic study design, an example of an 'extensive, post-treatment' approach (Hall *et al.* 1978). Thirty watersheds were selected to represent a wide range of sizes, stream channel types, forest harvest histories and hill-slope and channel disturbances. Each was classified according to age of logging (old or recent) and extent of mass wasting (Church 1998b). Harvesting in some research watersheds dated back to 1948 (Hartman *et al.* 1998). Other biophysical conditions (bedrock and surficial materials, hydrology, basin morphometry and forest type) were considered to ensure that watersheds were comparable.

The FFIP study was conducted in two phases. In phase 1 (1981–1986), properly matched watersheds were selected and landslide inventories were obtained. Studies emphasized hill-slope and stream channel interactions and processes. Additionally, alternatives were examined for stream channel and forest rehabilitation to mitigate the effects of landslides. Phase 2 (1988–1994) continued the research on hill-slope–channel interactions, and included field demonstrations and information transfer for resource managers and operational foresters.

The synoptic programme comprised a mix of individual study designs including observational studies across a broad spectrum of logged and unlogged watersheds, multiple-basin case studies and paired-basin experiments. This mixture collectively made significant contributions to our understanding of the effects of roads and harvesting on mass wasting, streams and fish habitats (Lewis 1998).

Effects of forestry on watershed processes in the Queen Charlotte Islands

Four major findings emerged from FFIP research: (1) the local watersheds experienced numerous landslides but their occurrence was highly episodic; (2) forestry activities accelerated landslide rates; (3) stream channel morphology was spatially and temporally controlled by the episodic delivery of sediment and debris from hill-slope failures; and (4) the adjustment of stream channels to landslide inputs led ultimately to complex and diverse fish habitats.

Landslide-prone terrain

Schwab (1998) reviewed all landslides that occurred in the study area over the last two centuries. From records including historic aerial photographs and field-verified dates of occurrence, he showed that 85% of all landslide-derived sediment and debris delivered to streams occurred during four large storms in 1891, 1917, 1935 and 1978 (Fig. 18.3a). Only the last storm occurred after logging. The episodic pattern of these natural disturbances is typical of the Pacific Northwest coast (Schwab 1998).

The FFIP study quantified a dramatic increase in hill-slope instability and the rate of landslides due to forestry operations in steep terrain. Mass wasting rates were 15-fold higher in areas with forestry operations compared with areas without forestry-related activity (Schwab 1998). Compared with unharvested sites, the area impacted by debris avalanches alone increased by 43 times due to clear-cutting, and by 17 times due to problems associated with forestry roads. Correspondingly, the volume of mass-wasted materials due to clear-cuts and roads increased by 46 and 41 times (Schwab 1998).

From a study of 1337 landslides in 27 Queen Charlotte Islands watersheds, Rood (1984) established that logging greatly increased the material yield from landslides as well as their frequency. He found that forestry operations dating back on average 7.3 years increased landslide frequency 34-fold over that in terrain without forestry. The volume of mass wasted material from unaltered forested areas was $1.6 \, \text{m}^3 \text{ha}^{-1} \text{y}^{-1}$. Clear-cuts and roads increased this yield between 32- and 90-fold to 50.7 and 144 $\text{m}^3 \text{ha}^{-1} \text{y}^{-1}$, respectively. The connection between hill-slope processes and the stream channel network was also clearly illustrated. About 39% and 47% of the total volume of sediment and woody debris generated by landslides entered streams in unlogged and logged terrain, respectively. In contrast with Carnation Creek, many small streams in the Queen Charlotte Islands have narrow floodplains and are thus closely or directly coupled to hill-slopes and gullies.

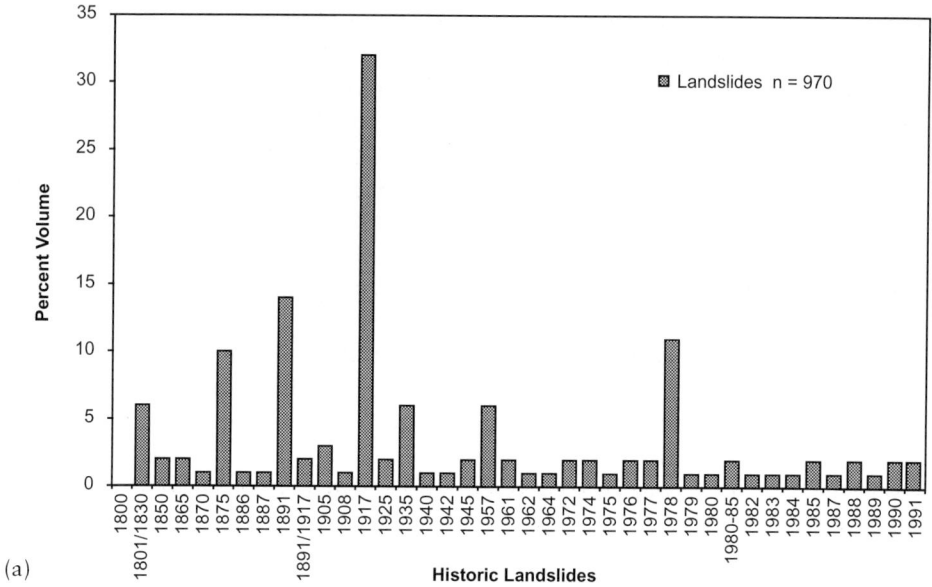

(a)

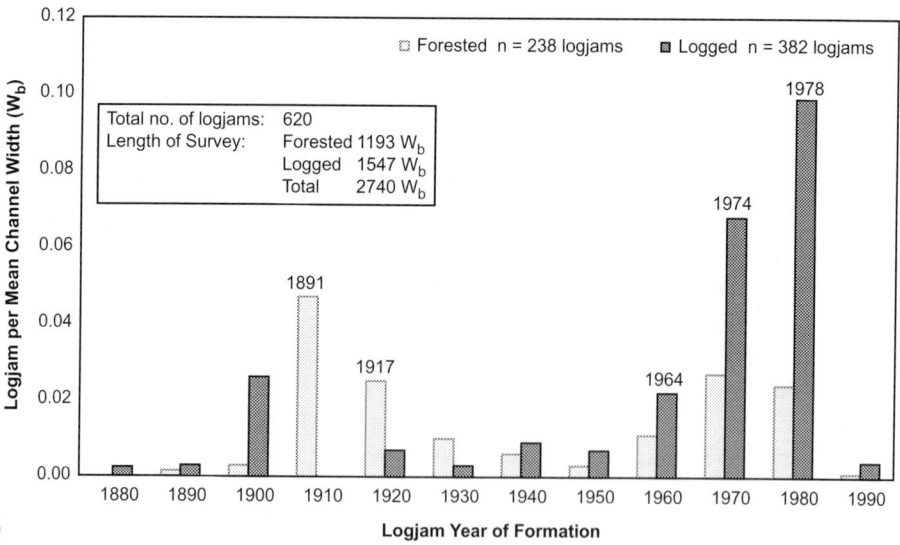

(b)

Fig. 18.3 Two centuries of Queen Charlotte Islands landslides with the percentage of the historic total volume of sediment and woody debris delivered by each event (a). Number of logjams per unit stream length (W_b = mean-channel-width equivalent) in forested and logged drainage basins (b).

Much of the sediment and woody debris produced on hillsides is transferred down-slope through gully systems. Coast-wide observations in the 1980s showed that forest harvesting often caused large accumulations of logging debris in gullies that resulted in increased frequencies and magnitudes of debris flows (Wilford & Schwab 1983; Rood

1984; Roberts & Church 1986). However, Bovis *et al.* (1998) concluded that the need for debris removal immediately following harvest depends on many factors, particularly the steepness and surface area of gully sidewalls and headwalls, the degree of gully disturbance during harvest, and the sensitivity of stream channels below. Bovis *et al.* (1998) studied 26 gullies and determined that sediment output was highest in (1) logged gullies where woody debris was cleared after harvesting, followed by (2) logged gullies that had at least one debris flow (torrent) since harvest, (3) unlogged gullies with no evidence of recent debris flows, and (4) logged gullies with a typical fill of logging debris (uncleared). These investigators concluded that geological differences do not significantly influence gully sediment output; rather, land treatment is the dominant influence, and similar gully responses can be expected in other parts of the coastal forest region.

Fluvial environments

Headwater streams are usually coupled directly to adjacent hillsides. Debris flows that enter these steep channels often travel downstream and scour the entire channel clean of sediment and debris (Hogan *et al.* 1998b). Channels impacted by these torrents could be swept clear of stable LWD or logjams, thus reducing the structural complexity of aquatic rearing habitats needed for juvenile salmonids, and the capacity of the channel to retain spawning and egg-incubation substrates (Tripp & Poulin 1986b, 1992).

Hogan *et al.* (1998a) surveyed nearly 44 km of lower gradient alluvial streams including distances equivalent to 1193 and 1547 mean channel widths (W_b) in forested and logged watersheds, respectively. They documented a direct link between landslides and channel morphology by showing that landslides initiate the formation of logjams in streams. Logging on steep hill-slopes accelerated landslide frequency, and correspondingly increased the number of recently formed logjams (Fig. 18.3b). Specific channel morphology changes occurred both upstream and downstream of the logjams that affected fish spawning, egg incubation and rearing habitats. Streams with these young logjams were characterized by extensive riffles, shallow pools, less stable gravel bars, and increased frequency, extent and temporal persistence of dry beds (Fig. 18.4; Hogan 1988b).

Hogan *et al.* (1998b) reported that the effect of logjams on channel morphology changes with time because of debris deterioration. Morphology was radically altered during the first decade following landslide inputs, but began to resemble undisturbed conditions after approximately 35 years. Complex and diverse channels were typical after 50 years (Fig. 18.4). Because channel morphology in steep coastal terrain is largely controlled by linked hill-slope and stream processes, historic forest management in these watersheds has caused widespread channel disturbances by shifting the age distribution of logjams to favour young ones in logged basins compared with those in forested streams.

Fish and fish habitats

The highest levels of mortality of salmon eggs in the Queen Charlotte Islands were attributed directly to scouring by debris flows. In contrast with the Carnation Creek

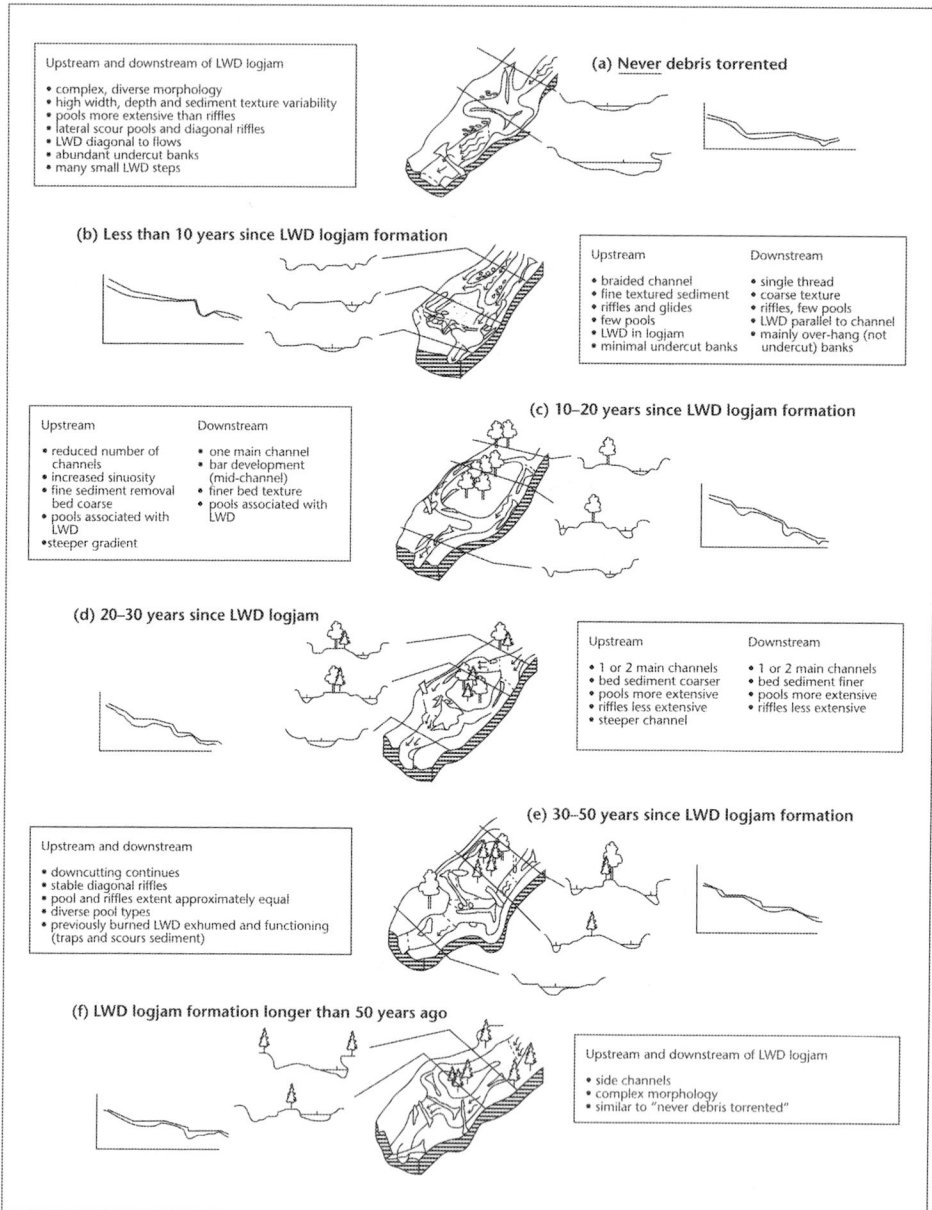

Upstream and downstream of LWD logjam

(a) <u>Never</u> debris torrented

- complex, diverse morphology
- high width, depth and sediment texture variability
- pools more extensive than riffles
- lateral scour pools and diagonal riffles
- LWD diagonal to flows
- abundant undercut banks
- many small LWD steps

(b) Less than 10 years since LWD logjam formation

Upstream	Downstream
• braided channel	• single thread
• fine textured sediment	• coarse texture
• riffles and glides	• riffles, few pools
• few pools	• LWD parallel to channel
• LWD in logjam	• mainly over-hang (not
• minimal undercut banks	undercut) banks

(c) 10–20 years since LWD logjam formation

Upstream	Downstream
• reduced number of channels	• one main channel
• increased sinuosity	• bar development (mid-channel)
• fine sediment removal bed coarse	• finer bed texture
• pools associated with LWD	• pools associated with LWD
•steeper gradient	

(d) 20–30 years since LWD logjam

Upstream	Downstream
• 1 or 2 main channels	• 1 or 2 main channels
• bed sediment coarser	• bed sediment finer
• pools more extensive	• pools more extensive
• riffles less extensive	• riffles less extensive
• steeper channel	

(e) 30–50 years since LWD logjam formation

Upstream and downstream

- downcutting continues
- stable diagonal riffles
- pool and riffles extent approximately equal
- diverse pool types
- previously burned LWD exhumed and functioning (traps and scours sediment)

(f) LWD logjam formation longer than 50 years ago

Upstream and downstream of LWD logjam

- side channels
- complex morphology
- similar to "never debris torrented"

Fig. 18.4 Stream channel morphology and fish habitat changes resulting from logjam formation and deterioration.

study, egg mortality was not assessed directly; rather, it was based upon measurements of the depth of streambed scour during winter compared with the depth distribution of salmon eggs in the streambed. By this method, Tripp & Poulin (1986b) estimated that egg-to-fry mortality varied between 66 and 86% for chum salmon and 45 and

70% for coho salmon in selected QCI streams. These investigators speculated that egg losses of 90–100% in torrent-impacted streams could occur for these species due to scouring in years of severe storms. These losses explained why some of these streams contained juvenile coho in one year but not the next.

Smaller populations of juvenile salmonids were thus sustainable in these simplified channels that featured shallow pools, extensive riffles, little LWD and coarse streambeds consisting of bedrock, boulders and large cobbles (Tripp & Poulin 1986a, 1986b, 1992; Tripp 1998). For example, over-winter survival of juvenile salmonids was lowest in logged systems impacted by debris flows. Over-winter survival of coho in three of these streams ranged from 2.0% to 12.1% compared with a range of 3.6–34.7% in logged systems without debris torrents (Tripp & Poulin 1992). Similarly, over-winter survival of steelhead in one torrent-impacted stream was only 7% compared with 13.7% and 22.4% survivals in logged systems without debris torrents.

Fish habitat impacts resulted from the increased abundance of young logjams associated with forestry. Studies within QCI-FFIP demonstrated that recently formed logjams caused fundamental changes in stream channel morphology that strongly reduced the amount and quality of spawning and rearing habitats available for fish (Hogan 1986; Tripp & Poulin 1986a, 1986b, 1992; Hogan *et al.* 1998a, 1998b). Hogan (1986) documented that young logjams, particularly within their first decade, effectively trap sediments causing channel aggradation, bank erosion, increased channel widths, reduced gradients and finer sediment textures upstream of the logjam. Extensive channel scour occurred downstream of the logjam because much of the sediment supply from sources upstream was blocked. Impacts to fish habitat upstream of the logjam included buried spawning areas (riffles), filled rearing habitats in pools, and smothered egg-incubation substrates. Downstream of the logjam, streambeds were scoured down to bedrock and boulders thus impacting salmonid habitats for all freshwater life stages.

Before the QCI-FFIP study, the importance of logjams for stream channel formation, riparian zones and fish habitat was not well understood. The shift from stream channels with an even distribution of young, medium-age and old logjams in unlogged watersheds to a distribution of predominantly young logjams in logged areas has critical impacts on fish habitat. This was a fundamental finding of the QCI-FFIP study (Hogan *et al.* 1998a; Tripp 1998). Linked to these changes, increases in fine sediments related in roughly equal measure to local (streamside) harvest impacts and mass wasting from locations upstream caused an estimated 15–20% decline in coho salmon egg-to-fry survival (Tripp & Poulin 1986b). Also, a strong negative relationship was observed between depth of scour and the densities of coho fry in spring (Tripp & Poulin 1992). Fry densities were lower in streams impacted by mass wasting than in streams without mass wasting (Tripp & Poulin 1992; Scrivener & Tripp 1998).

However, harmful logging-related effects upon the abundance and survival of stream salmonids were not always observed. Studies in 16 reaches of 8 logged drainages and 11 reaches in 3 unlogged ones revealed that coho fry densities in late summer were much higher in the logged sites (Hartman *et al.* 1998). Differences in density in these low gradient reaches were nearly eliminated by the following spring, suggesting that over-winter survival was lower in the logged areas. Differences in the densities of age 1 coho between seasons and logged versus unlogged drainages were smaller than for fry. This finding further obscured any effect of harvesting but suggested that older and larger fish survived the winter better. Similar age-related trends were apparent for juvenile steelhead trout, but in contrast with coho salmon, juvenile trout of all ages always survived the winter better in logged sites than in unlogged ones (Hartman *et al.* 1998).

These variations in densities between streams and species were difficult to explain on the basis of habitat characteristics or effects of logging. The study design likely confounded the results. Twelve of the 16 logged sites were located on the eastern side of the islands, and all of the unlogged study areas were located on the western side. These areas differed in several attributes that affect fish productivity (Church 1998b; Hartman *et al.* 1998). The effects of geography on salmonid production were illustrated by observations that salmonid densities and over-winter survival were higher in eastern streams than western ones. These differences obscured any harvest-related effects (Hartman *et al.* 1998).

Advantages and limitations of the QCI-FFIP study design

Geographic differences in specific conditions, including bedrock types, rainfall patterns, soil nutrients and other variables, appear to have confounded some of the effects of forestry practices on physical processes, streams and fish habitats in the QCI-FFIP studies. Another confounding factor was the wide variation in logging practices among the study watersheds. Some practices were obsolete or atypical. In 1994, the mean age of harvest blocks was 27 years, with some harvesting dating back a half century. In combination, these variations in geography and forestry operations imposed limitations upon interpretations of the study results, particularly for study components that would have benefited from long-term continuous monitoring such as those focused upon fine sediments. The episodic nature of fine sediment generation and transport, together with their rapid clearing from the beds of the study streams, made conclusions difficult from a synoptic, short-term study approach (Church 1998b).

The minimal availability or complete absence of pre-harvest data was another disadvantage. Data variability due to harvest treatments was difficult to separate from natural variability or that imposed by other factors. For example, the habitat-related effects of forestry operations upon anadromous salmonid abundance were impossible to isolate from the impacts of fisheries management and other short-term factors affecting populations. Also, differences in drainage basin size and hydrological attributes may well have confounded comparisons of sediment generation, transport, and impact among streams.

An appreciation of the variability in the regional coastal landscape in terms of both natural disturbances and forestry effects was provided by QCI-FFIP studies. The range of conditions within this study are arguably representative of the entire BC coast, and include those found at Carnation Creek (Church 1998a). The spatial and temporal perspectives offered by this multi-watershed approach thus allowed researchers and resource managers to extrapolate results to other areas of the coastal forest region with increased confidence.

Significantly, the QCI-FFIP study provided information on the responses of stream channels and fish in an environment containing small streams that were closely coupled to their hill-slopes. It dealt with situations in which channels and fish habitat were more immediately sensitive to hill-slope disturbances, and where fish populations did not have the benefit of wide floodplains with extensive tributary and off-channel networks. The latter, in Carnation Creek, served as seasonal refuges and buffered the effects of forestry practices.

Application and extension of Carnation Creek and Queen Charlotte Islands results

The Carnation Creek and QCI-FFIP studies have made landmark contributions toward our current understanding of the effects of forest harvesting on watersheds and fish in the Pacific coastal region. The results of these two programmes have been widely used to develop forest management practices, regulations and guidelines in BC and elsewhere in the Pacific Northwest. The BC Coastal Fisheries Forestry Guidelines (CFFG) which were first implemented in 1987, and guided forestry practices in coastal forest districts until 1994, were primarily based upon research information from Carnation Creek and the Queen Charlotte Islands. Subsequently, the two studies played a critical role in the development of the hill-slope, stream and riparian management provisions of the Forest Practices Code of BC which replaced the CFFG in 1995, applied to the entire province, and contained legally binding regulations together with guidelines and recommended best management practices.

Together, the influence and value of the Carnation Creek and Queen Charlotte Islands programmes, with complementary and cumulative benefits, have been greater than their individual contributions. The long-term results from Carnation Creek have provided a fundamental understanding of watershed and fish–forestry interaction processes, and have confirmed that the synoptic results provided by QCI-FFIP are realistic (Lewis 1998). Conversely, the multi-watershed FFIP has justified the extrapolation of many Carnation Creek findings to other coastal drainages. The two studies have together demonstrated the importance of natural disturbance regimes, drainage network connectivity, and the need for watershed-level planning and management. Critical connectivities have been identified, including the role of hydro-riparian corridors for aquatic and riparian wildlife communities, and the linkages between hill-slopes and stream channels for LWD and sediment budgets of streams. Important priorities for further studies have also been identified, such as expanded research into the role and importance of small, low-order tributaries within both floodplains and upland headwaters.

References

Bovis, M.J., Millard, T.H. & Oden, M.E. (1998) Gully processes in coastal British Columbia: The role of woody debris. In: *Carnation Creek and Queen Charlotte Islands Fish/Forestry Workshop: Applying 20 Years of Coastal Research to Management Solutions*, (eds D.L. Hogan, P.J. Tschaplinski & S. Chatwin), pp. 49–75. Land Management Handbook No. 41. BC Ministry of Forests.

Brown, T.G. & Hartman, G.F. (1988) Contribution of seasonally flooded lands and minor tributaries to coho (*Oncorhynchus kisutch*) salmon smolt production in Carnation Creek, a small coastal stream in British Columbia. *Transactions of the American Fisheries Society*, **117**, 546–51.

Cederholm, C.J. & Reid, L.M. (1987) Impact of forest management on coho salmon (*Oncorhynchus kisutch*) populations in the Clearwater River, Washington. In: *Streamside Management: Forestry and Fishery Interactions*, (eds E.O. Salo & T.W. Cundy), pp. 373–98. Contribution 57. Institute of Forest Resources, University of Washington, AR-10, Seattle.

Church, M. (1998a) The landscape of the Pacific Northwest. In: *Carnation Creek and Queen Charlotte Islands Fish/Forestry Workshop: Applying 20 Years of Coastal Research to Management Solutions*, (eds D.L. Hogan, P.J. Tschaplinski & S. Chatwin), pp. 13–22. Land Management Handbook No. 41. BC Ministry of Forests.

Church, M. (1998b) Fine sediments in small streams in coastal British Columbia: a review of research progress. In: *Carnation Creek and Queen Charlotte Islands Fish/Forestry Workshop: Applying 20 Years of Coastal Research to Management Solutions*, (eds D.L. Hogan, P.J. Tschaplinski & S. Chatwin), pp. 119–33. Land Management Handbook No. 41. BC Ministry of Forests.

Culp, J.M. & Davies, R.W. (1983) An assessment of the effects of streambank clearcutting on macroinvertebrate communities in a managed watershed. *Canadian Technical Report Fisheries and Aquatic Sciences*, **1208**.

Everest, F.H., Beschta, R.L., Scrivener, J.C., Sedell, J.R. & Cederholm, C.J. (1987) Fine sediment and salmon production: a paradox. In: *Streamside Management: Forestry and Fishery Interactions*, (eds E.O. Salo & T.W. Cundy), pp. 98–143. Contribution 57. Institute of Forest Resources, University of Washington, AR-10, Seattle, WA.

Gimbarzevsky, P. (1988) Mass wasting in the Queen Charlotte Islands; a regional inventory. *BC Ministry of Forests, Land Management Report*, **29**.

Grier, C.C. & Logan, R.S. (1977) Old growth *Pseudotsuga menziesii* communities of a western Oregon watershed: biomass distribution and production budgets. *Ecological Monographs*, **47**, 373–400.

Hall, J.D., Murphy, M.L. & Aho, R.S. (1978) An improved design for assessing impacts of watershed practices on small streams. *Verhandlungen Internationale Vereinigung für Limnologie*, **20**, 1359–65.

Hartman, G.F. & Scrivener, J.C. (1990) Impacts of forestry practices on a coastal stream ecosystem, Carnation Creek, British Columbia. *Canadian Bulletin of Fisheries and Aquatic Sciences*, **223**.

Hartman, G.F., Tripp, D.B. & Brown, T.G. (1998) Overwintering habitats and survival of juvenile salmonids in coastal streams of British Columbia. In: *Carnation Creek and Queen Charlotte Islands Fish/Forestry Workshop: Applying 20 Years of Coastal Research to Management Solutions*, (eds D.L. Hogan, P.J. Tschaplinski & S. Chatwin), pp. 141–54. Land Management Handbook No. 41. BC Ministry of Forests.

Healey, M.C. (1982) Timing and relative intensity of size-selective mortality of juvenile chum salmon (*Oncorhynchus keta*) during early sea life. *Canadian Journal of Fisheries and Aquatic Sciences*, **39**, 952–7.

Hogan, D.L. (1986) Channel morphology of logged, unlogged, and torrented streams in the Queen Charlotte Islands. *BC Ministry of Forests, Land Management Report*, **49**.

Hogan, D.L. & Schwab, J.W. (1990) Precipitation and runoff characteristics, Queen Charlotte Islands. *BC Ministry of Forests, Land Management Report*, **60**.

Hogan, D.L., Bird, S.A. & Hassan, M.A. (1998) Spatial and temporal evolution of small coastal streams: influence of forest management on channel morphology and fish habitats. In: *Gravel-Bed Rivers in the Environment*, (eds P.C. Klingeman, R.L. Beschta, P.D. Komar & J.B. Bradley), pp. 365–92. Water Resources Publications, LLC.

Hogan, D.L., Bird, S.A. & Rice, S. (1998) Stream channel morphology and recovery processes. In: *Carnation Creek and Queen Charlotte Islands Fish/Forestry Workshop: Applying 20 Years of Coastal Research to Management Solutions*, (eds D.L. Hogan, P.J. Tschaplinski & S. Chatwin), pp. 77–96. Land Management Handbook No. 41. BC Ministry of Forests.

Holtby, L.B. (1988) Effects of logging on stream temperatures in Carnation Creek, British Columbia, and associated impacts on the coho salmon (*Oncorhynchus kisutch*). *Canadian Journal of Fisheries and Aquatic Sciences*, 45, 502–15.

Holtby, L.B. & Scrivener, J.C. (1989) Observed and simulated effects of climatic variability, clearcut logging, and fishing on the numbers of chum salmon (*Oncorhynchus keta*) and coho salmon (*O. kisutch*) returning to Carnation Creek, British Columbia. In: *Proceedings of the National Workshop on Effects of Habitat Alteration on Salmonid Stocks*, (eds C.D. Levings, L.B. Holtby & M.A. Henderson), pp. 61–81. *Canadian Special Publication of Fisheries and Aquatic Sciences*, 96.

Holtby, L.B., Andersen, B.C. & Kadowaki, R.K. (1990) Importance of smolt size and early ocean growth to interannual variability in marine survival of coho salmon (*Oncorhynchus kisutch*). *Canadian Journal of Fisheries and Aquatic Sciences*, 47, 2181–94.

Krajina, V.J. (1969) Ecology of forest trees in British Columbia. In: *Ecology of Western North America*, (eds V.J. Krajina & R.C. Brooke), pp. 1–146. University of British Columbia, Vancouver.

Lewis, C.P. (1998) Introduction: workshop outline and experimental design. In: *Carnation Creek and Queen Charlotte Islands Fish/Forestry Workshop: Applying 20 Years of Coastal Research to Management Solutions*, (eds D.L. Hogan, P.J. Tschaplinski & S. Chatwin), pp. 5–11. Land Management Handbook No. 41. BC Ministry of Forests.

Mathews, S.B. & Buckley, R. (1976) Marine mortality of Puget Sound coho salmon (*Oncorhynchus kisutch*). *Journal of the Fisheries Research Board of Canada*, 33, 1677–84.

Murphy, M.L. & Hall, J.D. (1981) Varied effects of clearcut logging on predators and their habitat in small streams of the Cascade Mountains, Oregon. *Canadian Journal of Fisheries and Aquatic Sciences*, 38, 137–45.

Neaves, P.I. (1978) Litter fall, export, decomposition and retention in Carnation Creek, Vancouver Island. *Fisheries and Marine Service Technical Report*, 809.

Oswald, E.T. (1982) Preharvest vegetation and soils of Carnation Creek watershed. In: *Proceedings of the Carnation Creek Workshop, a 10-Year Review*, 24–26 February 1982, (ed. G.F. Hartman), pp.17–35. Pacific Biological Station, Nanaimo, BC.

Poulin, V.A. (1984) A research approach to solving fish/forestry interactions in relation to mass wasting on the Queen Charlotte Islands. *BC Ministry of Forests, Land Management Report*, 27.

Riddell, B.E. (1993) Spatial organisation of Pacific salmon: what to conserve? In: *Genetic Conservation of Salmonid Fishes*, pp. 23–42. Plenum Press, New York.

Roberts, R.G. & Church, M.C. (1986) The sediment budget in severely disturbed watersheds, Queen Charlotte Islands Ranges, British Columbia. *Canadian Journal of Forestry*, 16, 1092–106.

Rood, K.M. (1984) An aerial photograph inventory of the frequency and yield of mass wasting on the Queen Charlotte Islands, British Columbia. *BC Ministry of Forests, Land Management Report*, 34.

Ryder, J.M. (1981) Geomorphology of the southern part of the Coast Mountains of British Columbia. *Zeitschrift für Geomorphologie, Supplement*, 37, 120–47.

Schwab, J.W. (1998) Landslides on the Queen Charlotte Islands: processes, rates, and climatic events. In: *Carnation Creek and Queen Charlotte Islands Fish/Forestry Workshop: Applying 20 Years of Coastal Research to Management Solutions*, (eds D.L. Hogan, P.J. Tschaplinski & S. Chatwin), pp. 41–7. Land Management Handbook No. 41. BC Ministry of Forests.

Slaney, T.L., Hyatt, K.D., Northcote, T.G. & Fielden, R.J. (1996) Status of anadromous salmon and trout in British Columbia and Yukon. *Fisheries*, 21, 20–34.

Scrivener, J.C. & Andersen, B.C. (1984) Logging impacts and some mechanisms that determine the size of spring and summer populations of coho salmon fry (*Oncorhynchus kisutch*) in Carnation Creek, British Columbia. *Canadian Journal of Fisheries and Aquatic Sciences*, 41, 1097–105.

Scrivener, J.C. & Brownlee, M.J. (1989) Effects of forest harvesting on spawning gravel and incubation survival of chum (*Oncorhynchus keta*) and coho salmon (*O. kisutch*) in Carnation Creek, British Columbia. *Canadian Journal of Fisheries and Aquatic Sciences*, **46**, 681–96.

Scrivener, J.C. & Tripp, D.B. (1998) Changes in spawning gravel characteristics after forest harvesting in the Queen Charlotte Islands and Carnation Creek watersheds and the apparent impacts on incubating salmonid eggs. In: *Carnation Creek and Queen Charlotte Islands Fish/Forestry Workshop: Applying 20 Years of Coastal Research to Management Solutions*, (eds D.L. Hogan, P.J. Tschaplinski & S. Chatwin), pp. 135–40. Land Management Handbook No. 41. BC Ministry of Forests.

Shortreed, K.S. & Stockner, J.G. Periphyton biomass and species composition in a coastal rainforest stream in British Columbia: effects of environmental changes caused by logging. *Canadian Journal of Fisheries and Aquatic Sciences*, **40**, 1887–95.

Toews, D.A. & Moore, M.K. (1982) The effects of streamside logging on large organic debris in Carnation Creek. *BC Ministry of Forests, Land Management Report*, **11**.

Tripp, D. (1998) Evolution of fish habitat structure and diversity at log jams in logged and unlogged streams subject to mass wasting. In: *Carnation Creek and Queen Charlotte Islands Fish/Forestry Workshop: Applying 20 Years of Coastal Research to Management Solutions*, (eds D.L. Hogan, P.J. Tschaplinski & S. Chatwin), pp. 97–108. Land Management Handbook No. 41. BC Ministry of Forests.

Tripp, D.B. & Poulin, V.A. (1986a) The effects of mass wasting on juvenile fish habitats in streams on the Queen Charlotte Islands. *BC Ministry of Forests and Lands, Land Management Report*, **45**.

Tripp, D.B. & Poulin, V.A. (1986b) The effects of logging and mass wasting on salmonid spawning habitat in streams on the Queen Charlotte Islands. *BC Ministry of Forests and Lands, Land Management Report*, **50**.

Tripp, D.B. & Poulin, V.A. (1992) The effects of logging and mass wasting on juvenile salmonid populations in streams on the Queen Charlotte Islands. *Fish/Forestry Interaction Program Report, March 1992*.

Tschaplinski, P.J. (1988) The use of estuaries as rearing habitats by juvenile coho salmon. In: *Proceedings of the Workshop: Applying 15 Years of Carnation Creek Results*, (ed. T.W. Chamberlin), pp. 123–42. Pacific Biological Station, Nanaimo, BC.

Tschaplinski, P.J. (2000) The effects of forest harvesting, fishing, climate variation, and ocean conditions on salmonid populations of Carnation Creek, Vancouver Island, British Columbia. In: *Sustainable Fisheries Management: Pacific Salmon*, (eds E.E. Knudsen, C.R. Steward, D.D. MacDonald, J.E. Williams & D.W. Reiser), pp. 297–327. CRC Press, Boca Raton, FL.

Tschaplinski, P.J. & Hartman, G.F. (1983) Winter distribution of juvenile coho salmon (*Oncorhynchus kisutch*) before and after logging in Carnation Creek, British Columbia, and some implications for overwinter survival. *Canadian Journal of Fisheries and Aquatic Sciences*, **40**, 452–61.

Wilford, D.J. & Schwab, J.W. (1983) Soil mass movements in the Rennell Sound area, Queen Charlotte Islands, British Columbia. *Canadian Hydrological Symposium*, **82**, 521–41.

Wolfe, J.A. & Leopold, E.B. (1967) Neogene and Early Quaternary vegetation of Northwestern North America and Northeastern Alaska. In: *The Bering Land Bridge*, (ed. D.M. Hopkins), pp. 192–206. Stanford University Press, Stanford, CA.

Chapter 19
Forestry and fish in the boreal region of Canada

R.J. STEEDMAN, W.M. TONN, C.A. PASZKOWSKI
AND G.J. SCRIMGEOUR

Introduction

This chapter deals with the continental boreal forest of Canada from the Yukon Territory to Quebec, a 500–1500-km wide band, 3 million km², larger than all other forest types in Canada combined. Other boreal forest regions with distinctive ecological and management contexts will be discussed elsewhere in this book, including interior Alaska in Chapter 17, Cape Breton, and Newfoundland and Labrador in Chapter 20, and Fennoscandia in Chapter 24.

Boreal forest ecozones

The Canadian boreal forest has been classified into three ecozones: the Boreal Cordillera, the Boreal Plains, and the Boreal Shield (Fig. 19.1). The forests of the Boreal Cordillera, Plains and Shield share a common suite of trees, and are predominantly coniferous (e.g. spruce, pine and fir), with a significant deciduous component (e.g. aspen and birch). Boreal forest trees tend to be younger (typically <100 years old) than those in areas such as the Pacific Northwest, due to the dominance of relatively short-lived tree species and frequent disturbance by wildfires, particularly before contemporary fire suppression regimes. However, the relative homogeneity of boreal forest vegetation at the bioregional scale belies local and regional diversity in physiography, hydrology and aquatic ecology across this huge area (Table 19.1). The northern reaches of the boreal forest region are partially forested taiga, thinning northward into the arctic tundra. To the south are the subalpine and montane forests of British Columbia, the aspen parkland and grasslands of the Prairie Provinces, and the Great Lakes-St Lawrence forests of Ontario and Quebec. Permafrost may occur in northern or alpine locations of all three ecozones. Canada's boreal forest includes the headwaters of many large rivers, including the Nelson and Churchill in Manitoba, the Severn, Albany, Moose and St Lawrence in Ontario, and the Nottaway, Eastmain and La Grande in Quebec. The Boreal Shield and Boreal Plains generally have lower relief, runoff, hydraulic energy and erosion potential than more humid and mountainous areas, including the Boreal Cordillera.

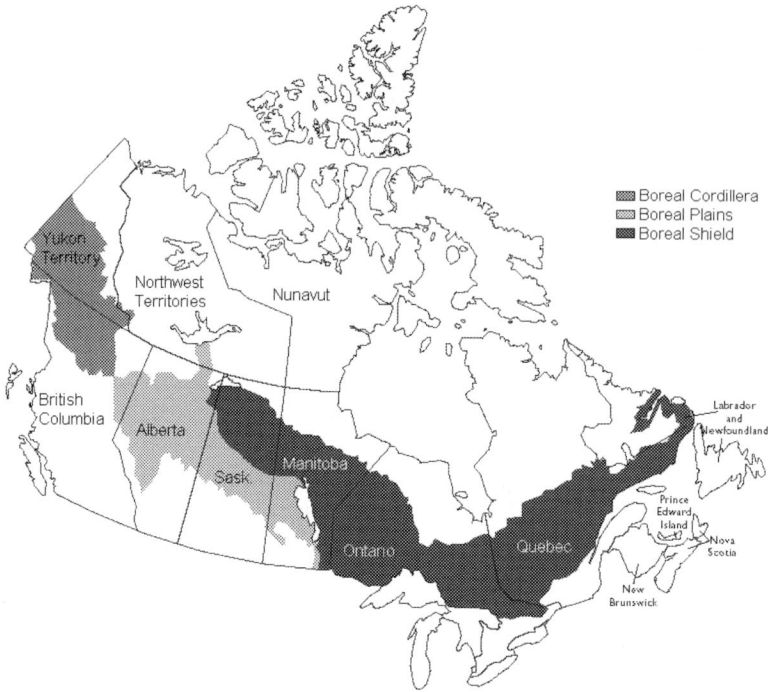

Fig. 19.1 Map of Canada showing the three boreal forest ecozones, the Boreal Cordillera, the Boreal Plains and the Boreal Shield (modified from Natural Resources Canada, Canadian Forest Service, http://www.nrcan.gc.ca/cfs-scf/national/what-quoi/sof/common/maps_e.html).

The **Boreal Cordillera** includes 444,000 km² of mountains, valleys and wide low-lands in the southern Yukon and northern British Columbia. It is sparsely populated, with about 30,000 people centred in Whitehorse and Dawson, Yukon. The ecozone is bordered by the Coast Mountains to the west, and extends east as far as the Peace River country of Eastern British Columbia. The area has an interior subalpine climate, with long cold winters and brief cool summers. Moist Pacific air is associated with frequent severe storms during summer. About 60% of this ecozone is forested, predominantly with coniferous trees such as white spruce (*Picea glauca*), black spruce (*Picea mariana*) and lodgepole pine (*Pinus contorta*), with some white birch *(Betula papyrifera)*.

The **Boreal Plains** are a 650,000 km² forested northern extension of the Great Plains, with low-lying valleys and plains stretching across the mid portions of Manitoba and Saskatchewan, and northern Alberta. About 750,000 people live in the ecozone. Edmonton, Alberta and Prince Albert, Saskatchewan are the major centres. The Boreal Plains are in the rain shadow of the Rocky Mountains, and have a relatively dry continental climate with short, warm summers and long, cold winters. The Boreal Plains have a relatively high amount of deciduous or mixed wood forest (about 50%; mainly trembling aspen *Populus tremuloides* and balsam poplar *Populus balsamifera*), compared with the Boreal Shield (30%) and Boreal Cordillera (<15%). Large river valleys of the Boreal Plains and Cordillera may support extensive stands of white spruce on alluvial floodplain deposits.

The **Boreal Shield** (excluding Newfoundland and Labrador) is the largest ecozone of the Canadian boreal forest, spanning 1.8 million km² in five provinces, including northeastern Alberta, northern Saskatchewan, northern Manitoba, northern Ontario (including the north shore of Lake Superior and the 'claybelt' south of James Bay) and north-central Quebec. More than 2 million people live on the Boreal Shield, primarily in small cities and towns along the southern edge of the region. This ecozone has a continental climate with long cold winters and short warm summers. Cold air masses over Hudson Bay bring relatively high levels of precipitation to most of the Boreal Shield, in a moisture gradient that increases from west to east. Black spruce, jack pine (*Pinus banksiana*), trembling aspen and balsam fir (*Abies balsamea*) are the dominant tree species. White spruce is relatively uncommon on the Shield, relative to the Boreal Plains and Cordillera.

Although extensively managed for forest products, Canada's boreal forest is intact across most of its range. However, in the southwestern portion, about 80,000 km² or 3% has been cleared for farmland in Alberta, Saskatchewan and Manitoba. About 30% of the Canadian boreal forest is within 1 km of an access road (Global Forest Watch 2000). Boreal forest trees tend to be younger (typically <100 years old) than those in areas such as the Pacific Northwest, due to the dominance of relatively short-lived tree species and frequent disturbance by wildfires, particularly before contemporary fire suppression regimes. The Boreal Plains has the most agricultural land (13%) of the boreal forest ecozones (Minister of Public Works and Government 1996). Substantial portions of the Boreal Plain are vulnerable to further deforestation and conversion to agriculture. The human population, although still low, is increasing in the western boreal regions. Between 1971 and 1991, the population of the Boreal Plains increased by 26%, while that of the Boreal Cordillera increased by 50%. The population on the Boreal Shield, although considerably larger, grew by only 11% in that period (Minister of Public Works and Government 1996).

Boreal waters and aquatic biota

Boreal lake and river systems, drainage patterns and fish associations are geologically very young. Most developed within the last 10,000 years, as the Laurentide and Cordilleran ice sheets receded at the end of the Wisconson glaciation. The post-glacial waterscape was strongly influenced by regional topography, bedrock and surficial deposits of tills and sediments. As the continental glaciers melted 5000–9000 years ago they created complex and dynamic drainage systems. Freshwater fishes recolonized the area from a number of unglaciated refugia, primarily the Mississippi and Missouri river systems to the south, the Atlantic seaboard to the east, north-central Alaska and western Yukon to the northwest, and the Pacific seaboard to the southwest (Crossman & McAllister 1986; Lindsey & McPhail 1986). The present fish fauna in the Hudson Bay and Mackenzie River drainages consists of about 110 native fish species. There are 52 fish species in waters draining to the Arctic Ocean from the Mackenzie and associated rivers, including 20 species not found in the Hudson Bay drainage. There are no

Table 19.1 Comparative overview of Canadian boreal forest ecozones

Ecozone	Boreal Cordillera	Boreal Plain	Boreal Shield
Terrain			
Elevation (m)	500–3000	200–1000	0–750
Relief	gentle to extremely rugged	gentle, rolling	moderately rugged
Bedrock type	complex sedimentary, intrusive, volcanic, metamorphic	sedimentary	complex intrusive, sedimentary, volcanic, metamorphic
Surficial deposits	complex glacial, glaciofluvial, lacustrine, alluvial, Aeolian	till, lacustrine, outwash, organic	bedrock, clay till, organic
Climate			
Type	interior subalpine	continental	continental
January, July temperature means (°C)	−15 to −25, 12 to 15	−15 to −25, 15 to 17	−15 to −25, 15 to 17
Annual precipitation, mm (% snow)	400–800 (35–60)	400–500 (30–35)	400–1000 increasing eastward (40–50)
Hydrology			
Annual runoff (mm)	100–750	50–150	200–700
Ionic strength of surface waters (mg/L)	<150	50 to >300	<150

Major river systems	Yukon (Pacific Ocean); Liard (Mackenzie R.–Arctic Ocean)	Peace, Athabasca, Slave (Mackenzie R.–Arctic Ocean); Saskatchewan (Hudson Bay); Beaver (Hudson Bay)	Churchill, Saskatchewan, Winnipeg, Nelson, Severn, Winisk, Attawapiskat, Albany, Missinabi, Abitibi, Moose, (Hudson Bay), Ottawa, St Maurice, Saguenay, Manicouagan (St Lawrence–Atlantic Ocean)
Drainage efficiency	low–high	low	low
Erosion potential, suspended sediment (mg/L)	high, up to 1000	low–moderate, up to 700	low, up to 400
Aquatic habitat and biota			
Predominant game or commercially harvested fish species (lakes)	Dolly Varden, bull trout, lake trout, burbot, inconnu, Arctic grayling, kokanee, pike, walleye, lake and mountain whitefish	northern pike, walleye, yellow perch, lake trout, Arctic grayling, lake whitefish, burbot, walleye, goldeye	walleye, pike, lake trout, brook trout, smallmouth bass (introduced), lake whitefish, yellow perch, burbot
Forestry			
Predominant tree species	white spruce, black spruce, trembling aspen, jack pine, balsam fir, white birch, balsam poplar, lodgepole pine, subalpine fir, tamarack/larch		white spruce, black spruce, trembling aspen, jack pine, balsam fir, white birch, balsam poplar, lodgepole pine,
Predominant forest disturbance	fire, wind, insects		
Harvest systems	even-aged management based on clear-cuts of various size distributions and levels of residual, followed by natural regeneration, seeding or planting; shelterwood harvest systems are also used, but are less common		
Products	kraft and groundwood pulp, dimensional lumber, oriented strand board, veneer		

Data from *Hydrological Atlas of Canada* (1978) and various other sources.

endemic fishes, and the region's fish species tend to be ecological generalists, flexible in their trophic needs and abilities (Beaudoin *et al.* 2001). Fish introductions by humans include cutthroat (*Oncorhynchus clarki*), brown (*Salmo trutta*) and brook trout (*Salvelinus fontinalis*) into rivers of the Arctic drainage. Rainbow trout (*O. mykiss*) continue to be widely stocked for sport-fishing in boreal lakes outside of their native range. Lake Winnipeg has been invaded by exotic species such as carp (*Cyprinus carpio*), smallmouth bass (*Micropterus dolomieu*), black crappie (*Pomoxis nigromaculatus*), white bass (*Morone chrysops*) and rainbow smelt (*Osmerus mordax*). On the Boreal Shield, smallmouth bass have been introduced intentionally, and rainbow smelt accidentally, across the southern part of the region.

Boreal aquatic habitats are strongly influenced by their topographic setting. Channel form, lake morphology, seasonal hydrology and water quality differ greatly in the three boreal ecozones and, in turn, determine the nature and productivity of aquatic biota. Runoff in many parts of the boreal forest is dominated by snowmelt, although summer and fall storms are also associated with significant flow peaks. Summer and multi-year droughts are common and can produce significant fluctuations in river flows and lake levels within and among years. Most small lakes and rivers are regulated to some extent by beaver activity, and humans regulate many larger lakes near developed areas. Most major rivers in the boreal forest have at least one dam for hydroelectric generation. Most notable in the western boreal is the Bennett Dam on Peace River in British Columbia, completed in 1967, which affects water levels and the flooding regime at least as far downstream as the Peace-Athabasca delta. There are several hydroelectric reservoirs on Cordilleran rivers in the Yukon. On the Shield, most suitable dam locations were developed early in the twentieth century to power nearby pulp mills.

In the Boreal Cordillera, aquatic habitats are dominated by large, cold, high gradient rivers draining broad mountain valleys. Peak flows in spring and early summer are associated with snowmelt at higher elevations. Small alpine wetlands and lakes occur, but are frequently isolated from lowland rivers by falls and cataracts. Surface waters may carry high loads of suspended sediment, but have low levels of dissolved organic material due to thin soils and rapid drainage.

On the Boreal Plains, bedrock is generally buried under deep tills and lacustrine soils. Large rivers carry mountain runoff across the Plains, but local river systems are only moderately abundant due to limited precipitation. The Plains ecozone itself has only 50–150 mm of runoff in an average year. This semi-arid climate, combined with low relief, deeper surficial glacial deposits and larger groundwater flow systems than the Shield or Cordillera, creates interactions between surface and subsurface water systems that are more complex than those in the other Boreal ecozones. Dry years may produce almost no runoff on the Boreal Plains, dramatically affecting wetted areas, water renewal time and connectivity of lakes and wetlands (Devito *et al.* 2000). Nevertheless, this low relief, poorly drained landscape has abundant and extensive wetlands and wet forests. Surface waters may be relatively turbid due to the abundance of deep, highly erodible soils, and often have high levels of dissolved organic materials due to prolonged contact with wetlands and organic soils. Lakes are relatively uncommon, and generally shallow and highly productive where they occur. These characteristics render many of the lakes susceptible to oxygen depletion during the long ice-covered period, often resulting in winterkill of fish (Danylchuk & Tonn 2003).

On the Boreal Shield, rugged, poorly drained, granitic bedrock has created extensive lakes and wetlands connected by complex, low gradient river networks. A series of large lakes are associated with the southern boundary of the Canadian Boreal Shield. These include Lake Athabasca in Alberta and Saskatchewan, Lake Winnipeg in Manitoba, and Lake of the Woods, Lake Superior and Lake Huron in Ontario. As is the case on the Plains, shallow surface waters on the Shield become quite warm during summer. Cold, groundwater-fed springs and streams are abundant only where the glaciers left piles of sand or gravel in moraines, eskers or drumlins. Some lake basins, both large and small, were scoured deeply enough to allow thermal stratification during the summer, preserving deep coldwater habitat year-round. While deep organic soils and peat occur in many areas, mineral soils are generally thin or patchy, and plant nutrients are scarce, limiting both terrestrial and aquatic productivity. Surface waters are generally free of turbidity, but stain brown with dissolved organic carbon wherever wetlands occur. Where the Shield is buried under the lacustrine clays of the Hudson Plains (the 'claybelt' of northeastern Ontario and northwestern Quebec), lakes are much less common, river networks more highly developed and waters more turbid.

Boreal forest lakes are dominated by coolwater species, particularly northern pike (*Esox lucius*), white sucker (*Catostomus commersoni*), walleye (*Stizostedion vitreum*) and yellow perch (*Perca flavescens*). Most lakes on the Plain are small, relatively shallow, and isolated, due to the limited number of permanent streams and widespread occurrence of beaver. These small lakes generally have fewer than 6 fish species (out of a pool of about 11 species), and many are fishless (Robinson & Tonn 1989; Paszkowski & Tonn 2000). Small-bodied fishes such as brook stickleback (*Culaea inconstans*), fathead minnow (*Pimephales promelas*) and other cyprinids are absent from pike lakes, but dominate shallow lakes that lack pike. Because of the oxygen depletion that may occur in small boreal lakes during the long winters, winterkill is frequent and populations are unstable (Fox & Keast 1990; Danylchuk & Tonn 2003). Lakes deeper than about 15 m tend to be more common on the Shield. These usually stratify in summer and also provide habitat for coldwater species, including lake trout (*Salvelinus namaycush*), lake whitefish (*Coregonus clupeaformis*), cisco (*C. artedii*) and burbot (*Lota lota*).

Boreal streams typically contain two to five species of fish and, outside of the Cordillera, are usually dominated by sticklebacks, sucker and cyprinids (e.g. Merkowsky 1998; Scrimgeour *et al.* 2002). As in other regions, species richness tends to increase with stream size, e.g. 29 fish species have been recorded from the lower Slave River (Little *et al.* 1998). Headwater streams are generally fishless. In the Cordillera, coolwater and coldwater species, including Arctic grayling (*Thymallus arcticus*), mountain whitefish (*Prosopium williamsoni*), bull trout (*Salvelinus confluentus*), Dolly Varden (*S. malma*), rainbow trout, chinook salmon *(Oncorhynchus tshawytscha)*, chum salmon (*O. keta*) and introduced brook trout may also occur in streams and rivers (Nelson & Paetz 1992; Scrimgeour *et al.* 2002). On the Shield, small groundwater-fed streams may support native brook trout.

A variety of mammals and birds interact directly with fishes and fish habitats in boreal lakes and streams. Loons, pelicans, cormorants, mink and otter prey on small fish, primarily cyprinids and small catostomids. The beaver (*Castor canadensis*) is a key species affecting the distribution and nature of freshwater habitats in the boreal forest. Beaver dams may affect >13% of the landscape, and involve over half of flowing water reaches (Johnston & Naiman 1990). In general, beaver activity increases local biodiversity by creating favourable habitats for a variety of fish species (Schlosser & Kallemeyn 2000), as well as birds, amphibians and invertebrates (e.g. Naiman *et al.* 1986; Clifford *et al.* 1993). Logging practices that replace conifer stands with deciduous species and rejuvenate old aspen stands may benefit beaver populations (Slough & Sadleir 1977; Barnes & Mallik 2001). The abundant lakes and wetlands of the boreal forest provide important breeding habitat for waterfowl and other aquatic birds (Paszkowski & Tonn 2000). The number of ducks nesting in the western boreal forest has been estimated at 13 million (Morrison 2002).

Recreational and commercial fisheries

Recreational fishing is important in all jurisdictions of Canada's boreal forest, and direct expenditures on sport-fishing far exceed the total value of commercial fisheries. In 1995, 40% of all fish caught by recreational anglers in Canada were caught in Ontario, and 20% were caught in Quebec. Trout accounted for 30% of all fish retained, along with perch (17%), walleye (15%), smelt (8%) and northern pike (5%) (Canada Department of Fisheries & Oceans 1998).

Many of Canada's commercial inland fisheries are on large boreal lakes. Compared with marine fisheries, these inland or freshwater commercial fisheries are relatively small in terms of catches and value. In 1997, 39,000 t of fish were landed from Canada's commercial inland fisheries (including Canadian waters of the Great Lakes), worth US$51 million (Food and Agriculture Organization 2001).

Forest management patterns and practices

The high economic value of boreal forests has traditionally been derived from the homogeneous nature of their extensive even-aged, fire-origin coniferous stands, and the high quality of boreal coniferous sawlogs and fibre. Recently, *Populus* spp. has become an important source of pulp. Unmanaged areas of the boreal forest are often dominated by only one to three coniferous species, and the harvestable crop is relatively high in spite of relatively small annual increments in wood volume (Kuusela 1990).

The composition of the boreal forest is changing in the face of fire suppression, logging and climate change. The abundance of deciduous trees, primarily trembling aspen and white birch, is thought to be increasing as a result of fire suppression and increasing harvest rates (Carleton 2000; Hobson *et al.* 2002). In some Yukon stands, reduced fire frequency has expanded permafrost, reduced growth by lodgepole pine, and subsequently resulted in an increased dominance by white spruce (Oswald & Senyk 1977).

Rate and pattern of forest disturbance

Forests of the relatively arid Boreal Plains experience wildfire more frequently than Shield forests (Weir *et al.* 2000). Before twentieth century fire suppression programmes, Boreal Shield tree stands were destroyed by wildfire, wind-throw and insect infestation at intervals of 50–100 years (Bonan & Shugart 1989; Kuusela 1990). Evolving fire suppression regimes in recent decades have generally increased fire recurrence intervals in protected areas of Shield and Plains forests. Estimates vary widely, but are in the range of 40–250 years, depending on the region. In recent years forest disturbance by wildfire is about 0.1% per year on the Boreal Shield, and about 0.4% per year on the Boreal Plain (Minister of Public Works and Government 1996; Perera & Baldwin 2000).

Logging now accounts for most forest disturbance on managed areas of the Boreal Shield. In the 500,000 km² boreal forest management zone of Ontario, the logged area increased from about 0.1% per year in the decade starting in 1951, to about 0.4% per year in the first half of the 1990s. On the Boreal Plains, the average annual area harvested increased from 0.02 to 0.04% per year during the 40-year period between 1925 and 1965. Following the establishment of pulp mills in the 1950s and 1960s, annual harvesting rates increased further to about 0.1% per year in recent years. Insects may kill about the same volume of wood as logging, and disease (mainly fungi) kills about twice that much (Ontario Environmental Assessment Board 1994; Fleming *et al.* 2000).

In the Boreal Cordillera, logging has recently expanded to include upland spruce and pine from the southeastern Yukon Territory. About 10 km² were logged in the Yukon in 1999, and about 80 km² burned (Natural Resources Canada 2001). Published forest statistics for the Boreal Cordillera region of British Columbia cannot be reported as they are pooled with statistics for the rest of the province. There is a substantial logging and sawmill industry in the Boreal Cordillera of British Columbia, centred at Fort Nelson.

Early exploitation of the forest

Early logging in Canada's boreal forest was facilitated by the extensive wetlands, lakes and waterways of the Boreal landscape. Like the railways and settlers, both land clearing and commercial timber harvest moved westward from Quebec and Ontario late in the nineteenth century. The pulp and paper industry on the Boreal Shield of Quebec and Ontario developed early in the twentieth century. Rapid growth of US newspaper circulation after 1870 had depleted US pulpwood supplies, and led to removal of US tariffs on Canadian wood by 1913 (Rajala 1997). By the 1870s sawmills were operating across the southern Boreal Plains, usually located on large rivers and their tributaries, with lumber transported by water during spring flood. The influx of settlers to western Canada around 1900 led to a significant demand for lumber in the Boreal Plains and Cordillera and the establishment of large conifer-based sawmills (Minister of Public Works and Government 1996). Many of those sawmills were closed by the 1930s and replaced by portable sawmills, as easily accessed timber became scarce. In the Boreal Cordillera, initial wood demands were driven by the need for lumber and steamboat fuel during the Klondike Gold Rush. Wood was generally cut along the major rivers, and stockpiled locally (Yukon Department of Renewable Resources 2000). Historically, forestry in the Boreal Cordillera focused primarily on high value white spruce from alluvial flats along the major rivers.

Harvest systems and silvicultural treatments

Boreal forest management is dominated by clear-cut harvest systems. Clear-cut sizes and patterns vary widely depending on jurisdiction and management objectives. Silvicultural objectives and techniques depend on site conditions, but typically include site preparation to expose mineral soil, followed by planting or aerial seeding, and vegetation control by herbicide spraying or manual thinning. Natural forest regeneration after logging generally results in high dominance of deciduous species (Carleton 2000). Retention of seed trees is infrequently practised on the Boreal Plains, but may be used in black spruce partial cutting systems, and in areas where red or white pine is present as a minor component of the forest. Active management of mixed wood stands on the Boreal Plains and Shield is relatively recent, and may involve protection and underplanting of spruce in clear-cut aspen stands, mechanical site preparation, partial cutting systems and understory protection.

Both the Ontario Environmental Assessment Board (Ontario Environmental Assessment Board 1994) and the Federal Parliamentary Standing Committee on Natural Resources (Canadian Forestry Association 1994) have noted that clear-cut silvicultural systems were appropriate for boreal forest management. Most authorities recognize that forest management cannot fully replicate the complex disturbance patterns and mechanisms associated with historical wildfire regimes. However, this does not imply that clear-cut systems should be used universally, nor does it suggest that methods of clear-cutting cannot evolve and improve. In particular, careful landscape-scale planning and design are required to ensure that sustainable forest management objectives are achieved. Clear-cutting systems may be inappropriate in steep, geologically unstable areas, or areas where excessive heat, frost or a potential rise in the water table may lower the probability of successful regeneration. Careful attention to clear-cut size, pattern, and residual vegetation is required to help sustain plant biodiversity and complex wildlife habitat.

Best management practices

A variety of policies and practices are used to protect boreal aquatic habitat from harmful alteration, disruption or destruction by forest management activities. Two approaches are typically used: (1) 'best management' practices to reduce erosion associated with stream crossings and road construction near water bodies; and (2) protection of riparian vegetation and soils during forestry activities.

Construction and maintenance techniques developed in the last 25 years or so have the potential to greatly reduce the frequency and intensity of erosion and sedimentation. Most jurisdictions now employ systems of mandatory design standards, good practice guidelines and mitigation techniques designed to minimize the risk of damage to water quality and aquatic habitats during road construction and maintenance (e.g. Ontario Ministry of Natural Resources 1990; Québec Ministère des Forèts 1992; Alberta Environment 2000). The benefits of these systems are difficult to quantify (Park *et al.* 1994), and road failures still occur due to improper construction techniques and extreme precipitation or runoff events.

Reliable protection of receiving waters from logging road runoff requires adequate spatial separation of roads and shorelines, attention to runoff management during road construction, and proper design, construction and maintenance of water crossings. Sediment and turbidity are transported down-slope in stormwater and snowmelt runoff from forestry roads, but do not appear to originate from unroaded areas of boreal forest clear-cuts (Steedman & France 2000). Since it is the forest floor, not the forest canopy, that disperses road runoff and traps waterborne sediment, unharvested forest is not necessarily better than a clear-cut at trapping road sediment. A number of studies suggest that suspended sediment and bed load in road runoff is attenuated within 10–40 m down-slope, depending on topography, road drainage and slope characteristics. Sediment attenuation occurs in shorter distances where the slope below the road has abundant obstructions and depressions, but is almost as effective on clear-cuts as in forested reserves (Trimble & Sartz 1957; Packer 1967; Plamondon 1982; France 2002).

As in other forest regions of Canada, boreal forest managers routinely prescribe shoreline reserves of undisturbed forest. Reserve width depends on jurisdiction and terrain, and may vary from 10 m to 100 m or more. Preservation of shoreline forest is intended to provide a range of benefits to aquatic habitats. Fish habitat and water quality protection are common design objectives, but shoreline visual aesthetics and terrestrial wildlife habitat requirements are also frequently addressed through this approach. Other objectives include shade, bank stability, inputs of leaves, branches and boles, and retention of sediments and nutrients from upland runoff.

Forested riparian buffer strips of at least 15 m width are generally sufficient to minimize stream temperature increases after logging, but the other expected benefits have proven more difficult to quantify (Clinnick 1985; Ribe 1989; Norris 1993; Castelle *et al.* 1994; Haider & Hetherington 2000). Riparian buffer strips do not appear to be effective in preventing increased water yield and exports of dissolved substances associated with watershed disturbance (Norris 1993; Carignan *et al.* 2000; Steedman 2000). Microclimate edge effects may fully penetrate forested buffer strips <40 m in width, altering vegetation structure and composition (Harper & Macdonald 2002). Narrow 20-m buffers may be adequate to conserve lakeside populations of small mammals and amphibians following harvesting, but forest songbird communities may require buffers of at least 100 m (Hannon *et al.* 2002).

Wind-throw is known to be a chronic problem with riparian buffer strips, because non-wind-firm trees from interior forest become exposed to higher wind velocities on edges of clear-cuts (Steinblums *et al.* 1984). Local topographic control of wind velocity appears to be more important than buffer width in controlling wind-throw in buffer strips (Ruel 2000). Wind-throw is an important source of large wood debris for lakes and streams, and creates complex pit and mound microtopography that may enhance sediment trapping and infiltration.

Harvest and regeneration of some shoreline forests may become a component of landscape-scale emulation of natural disturbance in managed Boreal Shield forests, as it has in the Pacific northwestern and eastern deciduous forests of North America (Gregory 1997; Palik *et al.* 2000). For example, boreal forest managers in Ontario are now required by law to 'emulate natural disturbances and landscape patterns while minimizing adverse effects on plant life, animal life, water, soil, air and social and economic values …' (Ontario Crown Forest Sustainability Act [CFSA] 2[3]2).

Although wildfire, wind-throw and other agents regularly disturb shorelines of boreal forest lakes, rivers and wetlands, there are many uncertainties associated with forest management designed to emulate natural shoreline disturbance. Operational uncertainties may relate to procedures and techniques that prevent undesirable levels of soil and vegetation disturbance. Ecological uncertainties may relate to landscape-scale design criteria that create a desirable mix of forest ages and types adjacent to water bodies. Science to adaptively reduce these uncertainties should be based on monitored, replicated, multi-region assessment of shoreline forest management impacts, under operational conditions.

Fish–forestry interactions

Experimental and comparative studies in the Pacific Northwest and Midwest (primarily stream studies), and boreal forest (primarily lake studies) of North America suggest that forest management has mixed impacts on fish and other biota, depending on location, scale and logging methods.

Forested streams may exhibit temporarily increased production of algae, invertebrates and fish after forest disturbance that removes riparian shade (Murphy & Hall 1981; Behmer & Hawkins 1986). Site-specific details of this response may be strongly influenced by other factors, such as nutrient availability, disturbance of large wood debris, sediment inputs and temperature preferences of resident biota (Moring & Lantz 1975; Murphy *et al.* 1986; Culp 1988; Hartman & Scrivener 1990; Anderson 1992; Young *et al.* 1999). Forest drainage, as practised primarily in Fennoscandia, appears to be particularly disruptive to stream biota (Holopainen *et al.* 1991; Vuori *et al.* 1998). Long-term or large-scale studies of forest management impacts generally confirm the persistence of diverse fish and invertebrate communities after logging treatments that in some cases were highly disruptive by modern standards (Moring & Lantz 1975; Hartman & Scrivener 1990; Rutherford *et al.* 1992; Stone & Wallace 1998; Young *et al.* 1999; Williams *et al.* 2002). Amphibian populations in streams may be less resilient (Corn & Bury 1989).

Lake biota have shown subtle responses to increased nutrient concentrations after forest disturbance (e.g. Planas *et al.* 2000; Scrimgeour *et al.* 2000). However, this effect seems short-lived unless forest regeneration is delayed (Carignan *et al.* 2000; Laird & Cumming 2001; Scrimgeour *et al.* 2001). Minor, equivocal or potential impacts of logging have been reported for zooplankton (Patoine *et al.* 2000, 2002) and fish (Rask *et al.* 1998; St-Onge & Magnan 2000; Marcogliese *et al.* 2001; Steedman 2003; Tonn *et al.* 2003). In shallow lakes of the Boreal Plain, forest harvesting may increase the frequency or severity of winterkill brought about by increased primary production and winter oxygen depletion. However, inter-annual variation in timing of ice cover is also an important factor in over-winter fish survival (Danylchuk & Tonn 2003).

Forest management may influence fish and other aquatic biota through both watershed and riparian mechanisms. The relative importance of these depends on the amount of logging in the watershed, and on the location and design of clear-cuts, roads, water crossings and machinery operations in and around riparian areas. Other key forest management impacts on aquatic ecosystems include road networks and wastewater discharge from processing mills. These four types of impacts are discussed in the following sections.

Access roads and water crossings

Roads almost always accompany human activity in the boreal forest. In the last 50 years, most of the boreal forest has been accessed by roads associated with forestry, mineral and energy development. Although forest management operations may be locally constrained by scarce mineral soil for fill, construction of summer roads is generally economic except in the most remote and rugged regions of the forest (Kuusela 1990). Where water, wetlands and soft organic soils limit logging operations in the summer, as in much of the Boreal Plains, winter operations may often be conducted after these areas freeze over and a snowpack is established.

Roads are a primary source of destructive erosion and sedimentation in managed forest landscapes, particularly in riparian areas or at water crossings (Megahan & Kidd 1972; Swift 1988). The potential for erosion is particularly high during construction of roads, culverts and bridges, because these structures require extensive excavation and disturbance of vegetation and mineral soils. Old, abandoned or poorly maintained road networks may also pose a long-term threat to aquatic habitat. When old culverts fail or roads wash out, large volumes of sediment may be introduced into streams and lakes. Considerable uncertainty remains about long-term, landscape-scale impacts on aquatic biota in managed forests, including habitat fragmentation and sedimentation by roads and culverts, particularly in steep or unstable terrain (Jones *et al.* 2000).

Poor road layout and maintenance, and inappropriate selection of crossing structures, can potentially degrade fish habitat. Habitat degradation may be caused by physical disruption or burial of stream channels and lakeshores, by sedimentation, or by introduction of hanging culverts or other structures that limit fish movement or migration (Warren & Pardew 1998).

Easy forest access facilitated by new roads almost invariably leads to increased exploitation of fish populations, intentional and unintentional introduction of exotic species, and pressure for cottage development in previously inaccessible areas. Angling pressure may be more of a threat to boreal fish populations than habitat degradation associated with road construction (Gunn & Sein 2000). Non-native fishes may be introduced to newly accessed waters accidentally through use of live bait, or intentionally to provide angling opportunities. Such introductions have reduced native fish biodiversity in Boreal Shield lakes (MacRae & Jackson 2001).

Riparian disturbance

In the following discussion, the word 'riparian' refers to the area of wet soils and distinctive vegetation immediately adjacent to boreal streams, rivers and lakes. In low relief boreal landscapes, riparian areas commonly take the form of wide, fringing wetlands dominated by herbaceous or shrubby wetland vegetation, rather than merchantable timber. In areas of locally higher relief, the riparian zone may be narrow and behave more like an upland forest, providing shade, forest litterfall and wood debris inputs to aquatic habitats. We use the word 'shoreline' in a more inclusive way to refer to both upland and riparian areas in the vicinity of a water body, typically within 100 m or so. Shoreline areas are typically designated as 'riparian' buffer strips which often comprise both upland forest and riparian vegetation.

Riparian disturbance is often associated with water crossings or machine activity too close to streams or lakes. Unlike watershed-scale impacts, riparian disturbance may directly influence aquatic habitats through shade loss, physical disturbance or changes to forest litter inputs, particularly along streams. Shorelines with upland vegetation are more likely to be disturbed by wildfire than shorelines with broad wetlands.

Shade and wind

In streams, loss of riparian shade causes changes (usually increases) in water temperature and temperature variability. These changes are usually attributed to decreased shading, although other factors, including wind, stream discharge, channel form and groundwater inputs, have all been used in predictive models (Brown & Krygier 1967; Holtby 1988). Boreal stream temperature response to riparian shade loss is probably similar to that reported elsewhere (Krause 1982), but will be influenced by slope, vegetation and groundwater dominance. Increased exposure to ultraviolet light (UVB) after riparian canopy removal has been shown to influence benthic algae and invertebrates in small streams, but this effect is counteracted to some extent by increased DOC in runoff from the logged watershed (Bothwell *et al.* 1993; Kelly *et al.* 2001).

Lake water temperature is influenced by a complex and interacting suite of factors, including air temperature, solar energy, wind, water clarity, lake depth and morphology. Different factors may dominate at different locations and over different spatial or temporal scales. Shade associated with shoreline forest influences only a small portion of the water surface, on all but the smallest lakes. Shoreline logging did not significantly increase average littoral water temperatures in two small Boreal Shield lakes in northwestern Ontario, but was associated with increases of 1–2°C in maximum littoral water temperature, and increases of 0.3–0.6°C in average diurnal temperature range, compared with undisturbed shorelines or shorelines with 30-m riparian reserves (Steedman *et al.* 1998, 2001). These small, transient impacts did not influence whole-lake thermal regimes (Steedman & Kushneriuk 2000), and are unlikely to affect the distribution or growth of aquatic biota. Shoreline logging may, however, increase wind exposure, mixing and dissolved oxygen of very small lakes and ponds (Scully *et al.* 2000; Steedman & Kushneriuk 2000).

Forest litter

Shoreline forest management may alter the amount and quality of forest litter (leaves, flowers, pollen) entering streams and lakes. In Alaska, regenerating streamside clear-cuts with a high alder component were associated with higher inputs of macroinvertebrates and forest litter into small streams, relative to clear-cut, young conifer, or old conifer stands (Piccolo & Wipfli 2002). In northwestern Ontario, shoreline logging temporarily reduced inputs of leaves and small woody debris to lake littoral zones by over 90% (France & Peters 1995). Based on estimates of airborne litter input from forested and clear-cut shorelines, and laboratory measurement of leaf leachate, France *et al.* (1996) suggested that shoreline logging on the Shield could reduce soluble organic carbon inputs in forest litter from about 18 g/m of shoreline per year, to <1 g/m, and reduce total P inputs from about 3 g/m to <1 g/m. In the absence of increased run-off inputs, such reductions could theoretically decrease primary production of lake plankton communities by as much as 9%, and reduce respiration by as much as 17%. These effects could last more than 10 years in northwestern Ontario, depending on the rate of riparian forest regeneration. Because it decomposes slowly in boreal lakes, littoral forest litter will probably persist until significant inputs resume during forest regeneration (France 1998).

Littoral macroinvertebrates colonize forest litter for both food and habitat, and may respond to temporarily reduced or altered litter inputs after logging or wildfire (France 1997a, 1998). Other parts of lake food webs are also strongly linked to the forest via litter inputs. Lake trout in small Boreal Shield lakes depend on a combination of forest litter and littoral food webs to satisfy their energy requirements (France & Steedman 1996; France 1997b). Stable nitrogen isotope ratios in the flesh of juvenile lake trout indicate omnivory, including predation on the opossum shrimp (*Mysis relicta*) (41% of samples), zooplankton (35% of samples) and littoral organisms (25% of samples). Carbon isotope ratios indicate that these juvenile lake trout obtain on average about half of their carbon from terrestrial sources. In the more productive lakes of the Boreal Plains, however, initial analyses using stable isotopes suggest that significant contributions to aquatic food webs by terrestrial sources were limited primarily to lakes with short water residence times (Beaudoin *et al.* 2001).

Wood debris

Fallen tree trunks and large branches may persist for centuries when submerged in fresh water (Guyette & Cole 1999). The size and durability of large wood debris creates and maintains complex stream channels, including deep pools important to fish production (Swanson *et al.* 1982; Murphy *et al.* 1986; Bisson *et al.* 1987). Large wood debris may have a similar role in boreal lakes, particularly in littoral waters (Mallory *et al.* 2000).

Little is known about abundance or recruitment of wood debris into boreal lakes and streams. Wood may enter boreal streams and lakes through natural forest disturbances, primarily wind-throw of dead or damaged trees, particularly after shoreline wildfire. Such wood inputs are probably episodic and highly variable temporally and spatially. Beavers also fell and drag significant quantities of wood into boreal waters. Natural wood inputs were augmented through much of the twentieth century in some boreal lakes and rivers by losses from log booms and log drives, particularly on the Shield (e.g. Kelso & Demers 1993). Due to relatively low topographical relief and a smaller, younger forest, abundance of wood in boreal waters may be lower than that reported for other regions. In the Pacific Northwest, wood abundance may reach 40 kg/m^2 or more in small streams of the Cascade and Coast Ranges (Sedell *et al.* 1988). In four 30–50-ha northwestern Ontario lakes with 80–100-year-old fire-origin shoreline forests, 10–20 pieces of large wood debris (i.e. >10 cm diameter, >1 m length) were present per 100 m of littoral zone (Ontario Ministry of Natural Resources/CNFER, unpublished data).

Depending on configuration and timing, shoreline forest harvesting has the potential to reduce future supply of large wood debris unless appropriate measures are taken. To contribute to aquatic habitat, retained trees must be close enough to the shoreline to have a reasonably high probability of falling into the water, i.e. within half a tree height of the shoreline, or about 7 m in the boreal forest. Partial shoreline logging may either decrease wood inputs to lakes (due to reduced tree abundance), or increase inputs (due to increased wind-throw).

Watershed disturbance

In the boreal forest, watershed-scale forestry impacts may increase water yield and peak flow by temporarily reducing forest evapotranspiration (Stednick 1996), increasing snowpack and speed of snowmelt (Buttle *et al.* 2000), and by improving drainage efficiency via roads and ditches (Jones & Grant 1996). Additional impacts may also occur due to localized soil disturbance, rutting and compaction, which alter infiltration capacity, microtopography and soil biogeochemical processes (Grigal 2000). Wildfire and logging also temporarily increase cations, dissolved organic carbon, nutrient and mercury loading to boreal aquatic ecosystems, with impacts strongly modulated by watershed characteristics, drainage position and morphometry of the water body (Schindler *et al.* 1980; Bayley *et al.* 1992; Rask *et al.* 1998; Carignan *et al.* 2000). For a given watershed disturbance, streams will show the greatest change in water quality. Shallow lakes with short water renewal times will show an intermediate response, and deep lakes with long water renewal times will have the smallest response.

At any watershed scale, extensive forest disturbance can cause significant but temporary changes in groundwater, stream and lake water quality, including increased concentrations of dissolved material such as organic carbon, cations (particularly potassium) and plant nutrients (nitrogen and phosphorus). These temporary water quality changes are driven by watershed-scale changes in forest hydrology and soil chemistry, and involve normal groundwater, wetland and stream flow pathways. Runoff from clear-cut watersheds often contains more dissolved organic carbon (DOC) and less nitrate and bioavailable phosphorus than runoff from burnt watersheds (Carignan *et al.* 2000). Nutrient export after wildfire is probably influenced by fire intensity and the degree to which organic soil layers are combusted. Water colour is strongly influenced by DOC, and has significant influence on surface heating and stratification of small lakes (Fee *et al.* 1996). Forest disturbance, like forest flooding, is also associated with increased movement of terrestrially deposited mercury into aquatic biota (Garcia & Carignan 1999, 2000).

Present levels of temporary watershed disturbance associated with boreal forest management are generally low, and appear unlikely to cause unnatural or harmful impacts on water yield or peak flows, particularly at the landscape scale (Buttle & Metcalfe 2000; Verry 2000). Extensive synchronous forest disturbance by clear-cutting (i.e. 25–50% or more of a watershed, within a 5-year period) is most likely to occur on relatively small watersheds (i.e. generally much smaller than 1 km²), under modern sustainable forestry scenarios. In contrast, synchronous disturbance by wildfire occasionally causes much larger disturbances, up to several hundred km².

Point-source impacts: mills and other wood-processing facilities

Approximately 30–35 pulp mills of various types currently operate in the Boreal Shield and Plains ecozones (Forest Products Association of Canada 2002). Most discharge effluent into large lakes or rivers, but some have closed process cycles with little or no wastewater effluent. The mills produce bleached kraft pulp, bleached chemithermomechanical pulp, thermomechanical pulp (TMP) and deinked paper. No pulp mills operate in the Boreal Cordillera ecozone.

Historically, pulp mills across the boreal forest discharged large quantities of wood fibre, oxygen-demanding effluent and chlorinated organic compounds into receiving waters. Large areas of river and lake bottom near some mills were blanketed with fibre and organic sludge, obliterating benthic biota, reducing dissolved oxygen levels and generating hydrogen sulphide in sediments and bottom waters. Some of these historical deposits, as well as bark and wood associated with log booms, will persist for many decades. Dramatic improvements in the control of fibre and oxygen demand in mill effluent, however, have occurred at Canadian mills in the last 20 years. Between 1989 and 1999, Canadian mills reduced the oxygen demand of mill effluents by >90%, to about 2 kg/t of pulp produced. Over the same period, suspended solids and fibre were reduced by >70%, to about 3 kg/t (e.g. Québec Ministère des Ressources Naturelles 2000). Adsorbable organic halide (AOX), dioxins and furans have also been greatly reduced in Canadian mill effluents in recent years, due to the elimination of elemental chlorine from Canadian papermaking (see Chapter 16).

Until the 1970s, some pulp mills used mercury chlor-alkali plants to produce chlorine onsite for bleaching kraft pulp. At Dryden, Ontario, 10 t of mercury were discharged from a pulp mill into the Wabigoon River between 1962 and 1970. Mercury levels in biota downstream reached very high levels, and appear likely to remain high for many years (Parks 1988). Other Boreal Shield waters contaminated in this manner included Peninsula Harbor near Marathon and the Kaministiquia River at Thunder Bay (both on Lake Superior).

Impacts on waters presently receiving pulp mill discharges vary with the type of pulp mill, mill capacities, size of the receiving water and distance to other pulp mills that determines effluent dilution (e.g. McCubbin & Folke 1993; Scrimgeour & Chambers 2000; Wrona *et al.* 2000; see also Chapter 16). Toxicological effects of pulp mill

discharges on fish communities within the Boreal Plains and Cordillera have probably been reduced because of pulping and effluent process upgrades since the 1980s.
Such process upgrades include chlorine substitution, oxygen delignification, peroxide
bleaching, secondary effluent treatment, extended water residence times in aerated
stabilization basins and the establishment of new mills that use such technologies
(McCubbin & Folke 1993). Most mills in these regions discharge into large receiving
waters that result in moderate to high dilution ratios, even during periods of low flow
(Culp *et al.* 2000).

The effect of pulp mill effluent on fish health is still debated (Kovacs *et al.* 1997;
Cash *et al.* 2000; see also Chapter 16). A recent review of fish health in the Peace and
Athabasca Rivers in the Boreal Plains reported that elevated concentrations of mixed
function oxidase (MFO), a biological marker for exposure to environmental stress,
in mountain whitefish and longnose suckers (*Catostomus catostomus*) were restricted
to areas immediately downstream of pulp mills (Cash *et al.* 2000). Levels of MFO in
spoonhead sculpin (*Cottus ricei*) exposed to pulp mill effluent were also higher than in
individuals collected from upstream references sites (Gibbons *et al.* 1998). Concentrations of dioxins and furans in burbot tended to be highest near pulp mill discharges,
but declined between 1992 and 1994 after improvements in the pulp bleaching process. Cash *et al.* (2000) also reported significantly lower levels of sex steroids in female
burbot and longnose sucker at nearfield sites (<100 km downstream of a pulp mill discharge) compared with farfield sites (>100 km downstream of a pulp mill discharge).
Assessments of external abnormalities have also indicated increased prevalence of tu-
mours, lesions, scars, skin discoloration and deformities in fish captured immediately
downstream of pulp mills.

Effects of pulp mill effluents on the structure of lotic fish communities are poorly
understood in the boreal forest. New Zealand studies have shown strong differences
related to proximity of the effluent outfall in rivers. Scrimgeour (1989) reported reduced abundance and diversity of fish communities at sites exposed to bleached kraft
pulp mill effluent compared with adjacent reference sites. Sites near outfalls were
dominated by species tolerant of low dissolved oxygen concentrations.

The future of boreal fish and forests

A shift in forest management, from a focus on sustainable timber yield to a sustainable
forestry approach, has occurred in Canada largely during the last decade. A practical definition of sustainable forestry remains elusive, but most formulations specify
that forest harvesting should not degrade social, cultural and biological aspects of the
forest.

Although sustainability principles increasingly guide boreal forest management,
quantitative monitoring of sustainability has been difficult to implement. In 1992, the
Canadian Council of Forest Ministers defined a set of criteria and indicators (C&I)
designed to structure monitoring of Canadian forests (Canadian Council of Forest
Ministers 1995). The C&I included elements of aquatic biota and habitat, in addition
to a broad range of ecological, cultural and economic values.

The Canadian C&I framework specified a classification of ecological concepts associated with forest sustainability, but did not prescribe quantitative monitoring techniques. In practice, mutually exclusive objectives for broad-scale, high-resolution indicators may impair practical implementation of the C&I approach. The best available test of the Canadian C&I for Sustainable Forest Management (Woodley *et al.* 1998) identified significant technical barriers to meaningful monitoring of most proposed indicators. This problem is reflected in a recent report on the process (Canadian Council of Forest Ministers 2000), which includes only two indicators for Criterion 3 (Conservation of Soil and Water Resources), both of the 'policy and protection' type, rather than the 'physical environmental' type. In many cases, practical, implemented biophysical C&I may prove to be based on:

(1)   mappable remote sensing data, such as forest cover and road networks;
(2)   data reported routinely as a regulatory requirement;
(3)   process proxies, such as requirements for 'best practices' described in regulatory guidelines; or
(4)   detailed monitoring at research sites representative of larger areas.

Research reviewed in this chapter suggests that boreal forestry operations need not be associated with deleterious effects on aquatic habitat and biota. If this is a general and reliable result, it implies that both watershed-scale and shoreline forest management should be considered in landscape-scale sustainable forest management designs. Remaining uncertainties will be best resolved through operational adaptive-management trials, coupled with appropriate long-term ecosystem monitoring.

Acknowledgements

We are extremely grateful for the research assistance provided by Shelly Boss, for discussions on specific aspects with Kevin Devito and Chris Davis, and for constructive criticism provided by Erland MacIsaac and Tom Brown.

References

Alberta Environment (2000) *Guide to the Code of Practice for Watercourse Crossings, Including Guidelines for Complying with the Code of Practice.* Edmonton, Alberta.
Anderson, N.H. (1992) Influence of disturbance on insect communities in Pacific Northwest stream. *Hydrobiologia*, **248**, 79–92.
Barnes, D.M. & Mallik, A.U. (2001) Effects of beaver, *Castor canadensis*, herbivory on streamside vegetation in a northern Ontario watershed. *Canadian Field-Naturalist*, **115**, 9–21.
Bayley, S.E., Schindler, D.W., Beaty, K.G., Parker, B.R. & Stainton, M.P. (1992) Effect of multiple fires on nutrient yields from streams draining boreal forest and fen watersheds: nitrogen and phosphorus. *Canadian Journal of Fisheries and Aquatic Sciences*, **49**, 584–96.

Beaudoin, C.P., Prepas, E.E., Tonn, W.M., Wassenaar, L.I. & Kotak, B.G. (2001) A stable carbon and nitrogen isotope study of lake food webs in Canada's Boreal Plain. *Freshwater Biology*, **46**, 465–77.

Behmer, D.J. & Hawkins, C.P. (1986) Effects of overhead canopy on macroinvertebrate production in a Utah stream. *Freshwater Biology*, **16**, 287–300.

Bisson, P.A., Bilby, R.E., Bryant, M.D., *et al.* (1987) Large woody debris in forested streams: past, present and future. In: *Streamside Management: Forestry and Fishery Interactions*, (eds E.O. Salo & T.W. Cundy), pp. 143–90. Contribution Number 57. University of Washington, Institute of Forest Resources, Seattle.

Bonan, G.B. & Shugart, H.H. (1989) Environmental factors and ecological processes in boreal forests. *Annual Review of Ecology and Systematics*, **20**, 1–28.

Bothwell, M.L., Sherbot, D., Roberge, A. & Daley, R.J. (1993) Influence of natural UV-radiation on lotic periphytic diatom community growth, biomass accrual and species composition: short-term versus long-term effects. *Journal of Phycology*, **29**, 24–35.

Brown, G.W. & Krygier, J.T. (1967) Changing water temperatures in small mountain streams. *Journal of Soil and Water Conservation*, **22**, 242–4.

Buttle, J.M. & Metcalfe, R.A. (2000) Boreal forest disturbance and streamflow response, northeastern Ontario. *Canadian Journal of Fisheries and Aquatic Sciences*, **57** (Suppl. 2), 5–18.

Buttle, J.M., Creed, I.F. & Pomeroy, J.W. (2000) Advances in Canadian forest hydrology. *Hydrological Processes*, **14**, 1551–78.

Canada Department of Fisheries & Oceans (1998) *1995 Survey of Recreational Fishing in Canada*. http://www.dfo-mpo.gc.ca/communic/statistics/rec_e.htm

Canadian Council of Forest Ministers (1995) *Defining Sustainable Forest Management*. Ottawa.

Canadian Council of Forest Ministers (2000) *National Status Report on Criteria and Indicators for Sustainable Forest Management in Canada*. Ottawa.

Canadian Forestry Association (1994) *Special Report on Clearcutting: Excerpts of Parliamentary Standing Committee on Natural Resources Hearings*, April 12–20, 1994. Ottawa.

Carignan, R., D'Arcy, P. & Lamontagne, S. (2000) Comparative impacts of fire and forest harvesting on water quality in Boreal Shield lakes. *Canadian Journal of Fisheries and Aquatic Sciences*, **57** (Suppl. 2), 105–17.

Carleton, T.J. (2000) Vegetation responses to the managed forest landscape of central and northern Ontario. In: *Ecology of a Managed Terrestrial Landscape: Patterns and Processes of Forest Landscapes in Ontario*, (eds A.H. Perera, D.J. Euler & I.D. Thompson), pp. 179–97. University of British Columbia Press, Vancouver.

Cash, K.J., Gibbons, W.N., Munkittrick, K.R., Brown, S.B. & Carey, C. (2000) Fish health in the Peace, Athabasca and Slave river systems. *Journal of Aquatic Ecosystems Health and Recovery*, **8**, 77–86.

Castelle, A.J., Johnson, A.W. & Conolly, C. (1994) Wetland and stream buffer size requirements – a review. *Journal of Environmental Quality*, **23**, 878–82.

Clifford, H.F., Wiley, G.M. & Casey, R.J. (1993) Macroinvertebrates of beaver-altered Boreal streams of Alberta, Canada. *Canadian Journal of Zoology*, **71**, 1439–47.

Clinnick, P.F. (1985) Buffer strip management in forest operations: a review. *Australian Forestry*, **48**, 34–45.

Corn, P.S. & Bury, R.B. (1989) Logging in western Oregon: responses of headwater habitats and stream amphibians. *Forest Ecology and Management*, **29**, 39–57.

Crossman, E.J. & McAllister, D.E. (1986) Zoogeography of freshwater fishes of the Hudson Bay drainage, Ungava Bay and the Arctic Archipelago. In: *The Zoogeography of North American Freshwater Fishes*, (eds C.H. Hocutt & E.O. Wiley), pp. 53–104. John Wiley & Sons, New York.

Culp, J.M. (1988) The effect of streambank clearcutting on the benthic invertebrates of Carnation Creek, British Columbia. In: *Applying 15 Years of Carnation Creek Results, Proceedings of a workshop held January 13–15, 1987, Nanaimo, BC*, (ed. T.W. Chamberlin), pp.

87–92. Carnation Creek Steering Committee, c/o Pacific Biological Station, Nanaimo, British Columbia.

Culp, J.M., Cash, K.J. & Wrona, F.J. (2000) Integrated assessment of ecosystem integrity of large northern rivers: the Northern River Basins Study example. *Journal of Aquatic Ecosystems Health and Recovery*, **8**, 1–5.

Danylchuk, A.J. & Tonn, W.M. (2003) Natural disturbances and fish: local and regional influences on winterkill of fathead minnows, *Pimephales promelas*, in boreal lakes. *Transactions of the American Fisheries Society*, **132**, 289–98.

Devito, K.J., Creed, I.F., Rothwell, R.L. & Prepas, E.E. (2000) Landscape controls on phosphorus loading to Boreal lakes: implications for the potential impacts of forest harvesting. *Canadian Journal of Fisheries and Aquatic Sciences*, **57**, 1977–84.

Fee, E.J., Hecky, R.E., Kasian, S.E.M. & Cruikshank, D.R. (1996) Effects of lake size, water clarity, and climatic variability on mixing depths in Canadian Shield lakes. *Limnology and Oceanography*, **41**, 912–20.

Fleming, R.A., Hopkin, A.A. & Candau, J.-N. (2000) Insect and disease disturbance regimes in Ontario's forests. In: *Ecology of a Managed Terrestrial Landscape: Patterns and Processes of Forest Landscapes in Ontario*, (eds A.H. Perera, D.J. Euler & I.D. Thompson), pp. 141–62. University of British Columbia Press, Vancouver.

Food and Agriculture Organization (2001) *Fishery Country Profile (Canada), January, 2001.* http://www.fao.org/fi/FCP/FICP_CAN_E.ASP

Forest Products Association of Canada (2002) http://www.fpac.ca.

Fox, M.G. & Keast, A. (1990) Effects of winterkill on population structure, body size, and prey consumption patterns of pumpkinseed in isolated beaver ponds. *Canadian Journal of Zoology*, **68**, 2489–98.

France, R.L. (1997a) Macroinvertebrate colonization of woody debris in Canadian Shield lakes following riparian clearcutting. *Conservation Biology*, **11**, 513–21.

France, R.L. (1997b) Stable carbon and nitrogen isotopic evidence for ecotonal coupling between boreal forests and fishes. *Ecology of Freshwater Fish*, **6**, 78–83.

France, R.L. (1998) Colonization of leaf litter by littoral macroinvertebrates with reference to successional changes in Boreal tree composition expected after riparian clear-cutting. *American Midland Naturalist*, **140**, 314–24.

France, R.L. (2002) Factors influencing sediment transport from logging roads near boreal trout lakes (Ontario, Canada). In: *Handbook of Water Sensitive Planning and Design*, (ed. R.L. France), pp. 635–44. Lewis Publishers, Boca Raton, FL.

France, R.L. & Peters, R.H. (1995) Predictive model of the effects on lake metabolism of decreased airborne litterfall through riparian deforestation. *Conservation Biology*, **9**, 1578–6.

France, R. & Steedman, R.J. (1996) Energy provenance for juvenile lake trout in small Canadian Shield lakes as shown by stable isotopes. *Transactions of the American Fisheries Society*, **125**, 512–18.

France, R., Culbert, H. & Peters, R. (1996) Decreased carbon and nutrient input to Boreal lakes from particulate organic matter following riparian clear-cutting. *Environmental Management*, **20**, 579–83.

Garcia, E. & Carignan, R. (1999) Impact of wildfire and clear-cutting in the Boreal forest on methyl mercury in zooplankton. *Canadian Journal of Fisheries and Aquatic Sciences*, **56**, 339–45.

Garcia, E. & Carignan, R. (2000) Mercury concentrations in northern pike (*Esox lucius*) from Boreal lakes with logged, burned, or undisturbed catchments. *Canadian Journal of Fisheries and Aquatic Sciences*, **57** (Suppl. 2), 129–35.

Gibbons, W.N., Munkittrick, K.R.E. & Taylor, W.D. (1998) Monitoring aquatic environments receiving industrial effluents using small fish species. 1. Response of spoonhead sculpin (*Cottus ricei*) downstream of a bleached kraft mill. *Environmental Toxicology and Chemistry*, **17**, 2227–37.

Global Forest Watch (2000) *Canada's Forests at a Crossroads: An Assessment in the Year 2000.* World Resources Institute, Washington, DC.

Gregory, S.V. (1997) Riparian management in the 21st century. In: *Creating a Forestry for the 21st Century: The Science of Ecosystem Management,* (eds K.A. Kohm & J.F. Franklin), pp. 69–85. Island Press, Washington, DC.

Grigal, D.F. (2000) Effects of extensive forest management on soil productivity. *Forest Ecology and Management,* **138,** 167–85.

Gunn, J.M. & Sein, R. (2000) Effects of forestry roads on reproductive habitat and exploitation of lake trout (*Salvelinus namaycush*) in three experimental lakes. *Canadian Journal of Fisheries and Aquatic Sciences,* **57** (Suppl. 2), 97–104.

Guyette, R.P. & Cole, W.G. (1999) Age characteristics of coarse woody debris (*Pinus strobus*) in a lake littoral zone. *Canadian Journal of Fisheries and Aquatic Sciences,* **56,** 496–505.

Haider, W. & Hetherington, J. (2000) Effects of forest regeneration practices on resource-based tourism and recreation. In: *Regenerating the Canadian Forest: Principles and Practice for Ontario,* (eds R.G. Wagner & S.J. Colombo), pp. 557–70. Fitzhenry & Whiteside, Markam.

Hannon, S.J., Paszkowski, C.A., Boutin, S., *et al.* (2002) Abundance and species composition of amphibians, small mammals and songbirds in riparian forest buffer strips of varying widths in the boreal mixedwood of Alberta. *Canadian Journal of Forest Research,* **32,** 1784–800.

Harper, K.A. & Macdonald, S.E. (2002) Structure and composition of edges next to regenerating clearcuts in mixedwood boreal forest. *Journal of Vegetation Science,* **13,** 535–46.

Hartman, G.F. & Scrivener, J.C. (1990) Impacts of forestry practices on a coastal stream ecosystem, Carnation Creek, British Columbia. *Canadian Bulletin of Fisheries and Aquatic Sciences,* **223.**

Hobson, K.A., Bayne, E.M. & Van Wilgenburg, S.L. (2002) Large-scale conversion of forest to agriculture in the Boreal Plains of Saskatchewan. *Conservation Biology,* **16,** 1530–41.

Holopainen, A.-L., Huttunen, P. & Ahtianen, M. (1991) Effects of forestry practices on water quality and primary productivity in small forest brooks. *Internationale Vereinigung für Theoretische und Angewandte Limnologie, Verhandlungen,* **24,** 1760–6.

Holtby, L.B. (1988) Effects of logging on stream temperatures in Carnation Creek, British Columbia, and associated impacts on the coho salmon (*Oncorhynchus kisutch*). *Canadian Journal of Fisheries and Aquatic Sciences,* **45,** 502–15.

Hydrological Atlas of Canada (1978) Canadian National Committee for the International Hydrological Decade, Ottawa.

Johnston, C.A. & Naiman, R.J. (1990) Browse selection by beaver: effects on riparian forest composition. *Canadian Journal of Forest Research,* **20,** 1036–43.

Jones, J.A. & Grant, G.E. (1996) Peak flow responses to clear-cutting and roads in small and large basins, western Cascades, Oregon. *Water Resources Research,* **32,** 959–74.

Jones, J.A., Swanson, F.J., Wemple, B.C. & Snyder, K.U. (2000) Effects of roads on hydrology, geomorphology, and disturbance patches in stream networks. *Conservation Biology,* **14,** 76–85.

Kelly, D.J., Clare, J.J. & Bothwell, M.L. (2001) Attenuation of solar ultraviolet radiation by dissolved organic matter alters benthic colonization patterns in streams. *Journal of the North American Benthological Society,* **29,** 96–108.

Kelso, J.R.M. & Demers, J.W. (1993) *Our Living Heritage: The Glory of the Nipigon.* Mill Creek, Echo Bay, Ontario.

Kovacs, T.G., Voss, R.H., Megraw, S.R. & Martel, P.H. (1997) Perspectives on Canadian field studies examining the potential of pulp and paper mill effluent to affect fish reproduction. *Journal of Toxicology and Environmental Health,* **51,** 305–52.

Krause, H.H. (1982) Effect of forest management practices on water quality – a review of Canadian studies. In: *Proceedings of the Canadian Hydrology Symposium '82 – Hydrological Processes of Forested Areas,* pp. 15–29. National Research Council, Fredericton.

Kuusela, K. (1990) *The Dynamics of Boreal Coniferous Forests*. Finnish National Fund for Research and Development, SITRA 112.

Laird, K. & Cumming, B. (2001) A regional paleolimnological assessment of the impact of clear-cutting on lakes from the central interior of British Columbia. *Canadian Journal of Fisheries and Aquatic Sciences*, **58**, 492–505.

Lindsey, C.C. & McPhail, J.D. (1986) Zoogeography of fishes of the Yukon and Mackenzie basins. In: *The Zoogeography of North American Freshwater Fishes*, (eds C.H. Hocutt & E.O. Wiley), pp. 639–74. John Wiley & Sons, New York.

Little, A.S., Tonn, W.M., Tallman, R.F. & Reist, J.D. (1998) Seasonal variation in diet and trophic relationships within the fish communities of the lower Slave River, Northwest Territories, Canada. *Environmental Biology of Fishes*, **53**, 429–45.

McCubbin, N. & Folke, J. (1993) *A Review of Literature on Pulp and Paper Mill Effluent Characteristics in the Peace and Athabasca River Basins*. Report 15 of the Northern River Basins Study, Edmonton.

MacRae, P.S.D. & Jackson, D.A. (2001) The influence of smallmouth bass (*Micropterus dolomieu*) predation and habitat complexity on the structure of littoral zone fish assemblages. *Canadian Journal of Fisheries and Aquatic Sciences*, **58**, 342–51.

Mallory, E.C., Ridgway, M.S., Gordon, A.M. & Kaushik, N.K. (2000) Distribution of woody debris in a small headwater lake, central Ontario, Canada. *Archiv für Hydrobiologie*, **148**, 589–606.

Marcogliese, D.J., Ball, M. & Lankester, M.W. (2001) Potential impacts of clearcutting on parasites of minnows in small boreal lakes. *Folia Parasitologica*, **48**, 269–74.

Megahan, W.F. & Kidd, W.J. (1972) Effects of logging and logging roads on erosion and sediment deposition from steep terrain. *Journal of Forestry*, **80**, 136–41.

Merkowsky, A. (1998) Predicting the importance of Boreal forest streams as fish habitat. In: *Forest-Fish Conference: Land Management Practices Affecting Aquatic Ecosystems*, (tech. co-ords M.K. Brewin & D.M.A. Monita), pp. 169–76. Information report NOR-X-356. Natural Resources Canada, Canadian Forest Service, Northern Forestry Centre, Edmonton.

Minister of Public Works and Government (1996) *The State of Canada's Environment – 1996*. EN21–54/1996E. Ottawa.

Moring, J.R. & Lantz, R.L. (1975) *The Alsea Watershed Study: Effects of Logging on the Aquatic Resources of Three Headwater Streams of the Alsea River, Oregon, Part I. Biological Studies*. Fishery Research Report Number 9. Oregon Department of Fish and Wildlife, Corvallis, OR.

Morrison, D. (2002) The western Boreal forest. *Birdscapes*, **Winter**, 14–15.

Murphy, M.L. & Hall, J.D. (1981) Varied effects of clear-cut logging on predators and their habitat in small streams of the Cascade Mountains, Oregon. *Canadian Journal of Fisheries and Aquatic Sciences*, **38**, 137–45.

Murphy, M.L., Heifetz, J., Johnson, S.W., Koski, K.V. & Thedinga, J.F. (1986) Effects of clear-cut logging with and without buffer strips on juvenile salmonids in Alaskan streams. *Canadian Journal of Fisheries and Aquatic Sciences*, **43**, 1521–33.

Naiman, R.J, Mellillo, M. & Hobbie, J.E. (1986) Ecosystem alteration of Boreal forest streams by beaver (*Castor canadensis*). *Ecology*, **67**, 1254–69.

Natural Resources Canada (2001) *The State of Canada's Forests*. Natural Resources Canada, Canadian Forest Service, Ottawa, Ontario.

Nelson, J.S. & Paetz, M.J. (1992) *The Fishes of Alberta*. University of Alberta Press, Edmonton, Alberta.

Norris, V. (1993) The use of buffer zones to protect water quality: a review. *Water Resources Management*, **7**, 257–72.

Ontario Environmental Assessment Board (1994) *Reason for decision and decision, Class Environmental Assessment by the Ministry of Natural Resources for Timber Management on Crown Lands of Northern Ontario*. Toronto.

Ontario Ministry of Natural Resources (1990) *Environmental Guidelines for Access Roads and Water Crossings.* Toronto.

Oswald, E.T. & Senyk, J.P. (1977) *Ecoregions of Yukon Territory.* Information Report BC-X-164. Canadian Forestry Service.

Packer, P.E. (1967) Criteria for designing and locating logging roads to control sediment. *Forest Science*, 13, 2–18.

Palik, B.J., Zasada, J.C. & Hedman, C.W. (2000) Ecological principles for riparian silviculture. In: *Riparian Management in Forests* (eds E.S. Verry, J.W. Hornbeck, & C.A. Dolloff), pp. 233–54. Lewis Publishers, Boca Raton, FL.

Park, S.W., Mostaghimi, S., Cooke, R.A. & McClennan, P.W. (1994) BMP impacts on watershed runoff, sediment, and nutrient yields. *Water Resources Bulletin*, 30, 1011–23.

Parks, J.W. (1988) Selected ecosystem relationships in the mercury contaminated Wabigoon-English River system, Canada, and their underlying causes. *Water, Air and Soil Pollution*, 42, 267–79.

Paszkowski, C.A. & Tonn, W.M. (2000) Community concordance between the fish and aquatic birds of lakes in northern Alberta, Canada: the relative importance of environmental and biotic factors. *Freshwater Biology*, 43, 421–37.

Patoine, A., Pinel-Alloul, B., Prepas, E.E. & Carignan, R. (2000) Do logging and forest fires influence zooplankton biomass in Canadian boreal shield lakes? *Canadian Journal of Fisheries and Aquatic Sciences*, 57 (Suppl. 2), 155–64.

Patoine, A., Pinel-Alloul, B. & Prepas, E.E. (2002) Effects of catchment perturbations by logging and wildfires on zooplankton species richness and composition in Boreal Shield lakes. *Freshwater Biology*, 47, 1996–2014.

Perera, A.H. & Baldwin, D.J. (2000) Spatial patterns in the managed forest landscape of Ontario. In: *Ecology of a Managed Terrestrial Landscape: Patterns and Processes of Forest Landscapes in Ontario,* (eds A.H. Perera, D.J. Euler & I.D. Thompson), pp. 74–99. University of British Columbia Press, Vancouver.

Piccolo, J.J. & Wipfli, M.S. (2002) Does red alder (*Alnus rubra*) in upland riparian forests elevate macroinvertebrate and detritus export from headwater streams to downstream habitats in southeastern Alaska? *Canadian Journal of Fisheries and Aquatic Sciences*, 59, 503–13.

Plamondon, A.P. (1982) Augmentation de la concentration des sédiments en suspension suite à l'exploitation forestière et durée de l'effet. *Canadian Journal of Forest Research*, 12, 883–92.

Planas, D., Desrosiers, M., Groulx, S-R., Paquet, S. & Carignan, R. (2000) Pelagic and benthic algal responses in eastern Canadian Boreal Shield lakes following harvesting and wildfires. *Canadian Journal of Fisheries and Aquatic Sciences*, 57 (Suppl. 2), 136–45.

Québec Ministère des Forèts (1992) *Modalités d'intervention en Milieu Forestier.* Montreal.

Québec Ministère des Ressources Naturelles (2000) *Québec Forest Resources and Industry: A Statistical Report, 2000 Edition.* Charlesbourg.

Rajala, R.A. (1997) The forest industry in eastern Canada: an overview. In: *Broadaxe to Flying Shear: The Mechanization of Forest Harvesting East of the Rockies,* (ed. C.R. Silversides), pp. 121–56. Transformation Series No. 6. National Museum of Science and Technology, Ottawa.

Rask, M., Nyberg, K., Markkanen, S.-L. & Ojala, A. (1998) Forestry in catchments: effects on water quality, plankton, zoobenthos and fish in small lakes. *Boreal Envionment Research*, 3, 75–86.

Ribe, R.G. (1989) The aesthetics of forestry: what has empirical preference research taught us? *Environmental Management*, 13, 55–74.

Robinson, C.L.K. & Tonn, W.M. (1989) Influence of environmental factors and piscivory in structuring fish assemblages of small Alberta lakes. *Canadian Journal of Fisheries and Aquatic Sciences*, 46, 81–9.

Ruel, J.-C. (2000) Factors influencing windthrow in balsam fir forests: from landscape studies to individual tree studies. *Forest Ecology and Management*, 135, 169–78.

Rutherford, D.A., Echelle, A.A. & Maughan, O.E. (1992) Drainage-wide effects of timber harvesting on the structure of stream fish assemblages in southeastern Oklahoma. *Transactions of the American Fisheries Society*, 121, 716–28.

Schindler, D.W., Newbury, R.W., Beaty, K.G., Prokopowich, J., Ruszczynski, T. & Dalton, J.A. (1980) Effects of a windstorm and forest fire on chemical losses from forested watersheds and on the quality of receiving streams. *Canadian Journal of Fisheries and Aquatic Sciences*, 37, 328–34.

Schlosser, I.J. & Kallemeyn, L.W. (2000) Spatial variation in fish assemblages across a beaver-influenced successional landscape. *Ecology*, 81, 1371–82.

Scrimgeour, G.J. (1989) Effects of bleached kraft pulp mill effluent on macroinvertebrate and fish populations in weedbeds in a New Zealand hydro-electric lake. *New Zealand Journal of Freshwater Research*, 23, 65–71.

Scrimgeour, G.J. & Chambers, P.A. (2000) Cumulative effects of pulp mill and municipal effluents on epilithic biomass and nutrient limitation in a large northern river ecosystem. *Canadian Journal of Fisheries and Aquatic Sciences*, 57, 1342–54.

Scrimgeour, G.J., Tonn, W.M., Paszkowski, C.A. & Aku, P.M.K. (2000) Evaluating the effects on littoral benthic communities within a natural disturbance-based management model. *Forest Ecology and Management*, 126, 77–86.

Scrimgeour, G.J., Tonn, W.M., Paszkowski, C.A. & Goater, C. (2001) Benthic macroinvertebrate biomass and wildfires: evidence for enrichment of Boreal subarctic lakes. *Freshwater Biology*, 46, 367–78.

Scrimgeour, G.J., Hvenegaard, P.H. & Kendall, S. (2002) *Empirical Relationships Between Fish Communities and Watershed Characteristics in the Notikewin Watershed, Alberta*. Report by the Northern Watershed Project to the Northern Watershed Stakeholders Committee. Alberta Conservation Association, Edmonton.

Scully, N.M., Leavitt, P.R. & Carpenter, S.R. (2000) Century-long effects of forest harvest on the physical structure and autotrophic community of a small temperate lake. *Canadian Journal of Fisheries and Aquatic Sciences*, 57 (Suppl. 2), 50–9.

Sedell, J.R., Bisson, P.A., Swanson, F.J. & Gregory, S.V. (1988) What we know about large trees that fall into streams and rivers. In: *From the Forest to the Sea: A Story of Fallen Trees*, (tech. eds C. Maser, R.F. Tarrant, J.M. Trappe & J.F. Franklin), pp. 47–81. Forest Service General Technical Report PNW-GTR-229. Pacific Northwest Research Station, US Department of Agriculture Forest Service, Portland, OR.

Slough, B.G. & Sadleir, R.M.F.S. (1977) A land capability classification system for beaver (*Castor canadensis*). *Canadian Journal of Zoology*, 55, 1324–35.

Stednick, J.D. (1996) Monitoring the effects of timber harvest on annual water yield. *Journal of Hydrology*, 176, 79–95.

Steedman, R.J. (2000) Effects of experimental clearcut logging on water quality in three small Boreal forest lake trout (*Salvelinus namaycush*) lakes. *Canadian Journal of Fisheries and Aquatic Sciences*, 57 (Suppl. 2), 92–6.

Steedman, R.J. (2003) Littoral fish response to experimental logging around small Boreal Shield lakes. *North American Journal of Fisheries Management*, 23, 392–403.

Steedman, R.J. & France, R.L. (2000) Origin and transport of aeolian sediment from new clearcuts into Boreal lakes, northwestern Ontario, Canada. *Water, Air, and Soil Pollution*, 122, 139–52.

Steedman, R.J. & Kushneriuk, R.S. (2000) Effects of experimental clearcut logging on thermal stratification, dissolved oxygen, and lake trout (*Salvelinus namaycush*) habitat volume in three small Boreal forest lakes. *Canadian Journal of Fisheries and Aquatic Sciences*, 57 (Suppl. 2), 82–91.

Steedman, R.J., France, R.L., Kushneriuk, R.S. & Peters, R.H. (1998) Effects of riparian deforestation on littoral water temperatures in small Boreal forest lakes. *Boreal Environment Research*, 3, 161–9.

Steedman, R.J., Kushneriuk, R.S. & France, R.L. (2001) Littoral water temperature response to experimental shoreline logging around small Boreal forest lakes. *Canadian Journal of Fisheries and Aquatic Sciences,* **58,** 1638–47.

Steinblums, I.J., Froelich, H.A. & Lyons, J.K. (1984) Designing stable buffer strips for stream protection. *Journal of Forestry,* **92,** 49–52.

Stone, M.K. & Wallace, J.B. (1998) Long-term recovery of a mountain stream from clear-cut logging: the effects of forest succession on benthic invertebrate community structure. *Freshwater Biology,* **39,** 151–69.

St-Onge, I. & Magnan, P. (2000) Impact of logging and natural fires on fish communities of Laurentian Shield lakes. *Canadian Journal of Fisheries and Aquatic Sciences,* **57** (Suppl. 2), 65–174.

Swanson, F.J., Gregory, S.V. Sedell, J.R. & Campbell, A.G. (1982) Land-water interactions: the riparian zone. In: *Analysis of Coniferous Forest Ecosystems in the Western United States,* (ed. R.L. Edmunds), pp. 267–91. US/IBP Synthesis Series 14. Hutchinson Ross Publishing Co., Stroudsburg, PA.

Swift, L.W., Jr (1988) Forest access roads: design, maintenance, and soil loss. In: *Forest Hydrology and Ecology at Coweeta* (eds W.T. Swank & D.A. Crossley Jr), pp. 313–24. Springer-Verlag, New York.

Tonn, W.M., Paszkowski, C.A., Scrimgeour, G.J., *et al.* (2003) Effects of harvesting and forest fire on fish assemblages in Boreal Plains lakes: a reference condition approach. *Transactions of the American Fisheries Society,* **132,** 514–23.

Trimble, G.R. & Sartz, R.S. (1957) How far from a stream should a logging road be located? *Journal of Forestry,* **53,** 339–41.

Verry, E.S. (2000) Water flow in soils and streams: sustaining hydrologic function. In: *Riparian Management in Forests of the Continental Eastern United States* (eds E.S. Verry, J.W. Hornbeck & C.A. Dolloff), pp. 99–124. Lewis Publishers, Boca Raton, FL.

Vulori, K.-M., Joensuu, I., Latvala, J., Jutila, E. & Ahvonen, A. (1998) Forest drainage: a threat to benthic biodiversity of Boreal headwater streams? *Aquatic Conservation,* **8,** 745–60.

Warren, M.L., Jr & Pardew, M.G. (1998) Road crossings as barriers to small-stream fish movement. *Transactions of the American Fisheries Society,* **127,** 637–44.

Weir, J.M.H., Johnson, E.A. & Miyanishi, K. (2000) Fire frequency and the spatial age mosaic of the mixed-wood Boreal forest in western Canada. *Ecological Applications,* **10,** 1162–77.

Williams, L.R., Taylor, C.M., Warren, M.L., Jr & Clingenpeel, J.A. (2002) Large-scale effects of timber harvesting on stream systems in the Ouachita mountains, Arkansas, USA. *Environmental Management,* **29,** 76–87.

Woodley, S., Alward, G., Gutierrez, L.I., *et al.* (1998) *North American Test of Criteria and Indicators of Sustainable Forestry, Final Report,* Vol. 1. US Agency for International Development and USDA Forest Service.

Wrona, F.J., Carey, J., Brownlee, B. & McCauley, E. (2000) Contaminant sources, distribution and fate in the Athabasca, Peace and Slave River Basins, Canada. *Journal of Aquatic Ecosystems Health and Recovery,* **8,** 39–51.

Young, K.A., Hinch, S.G. & Northcote, T.G. (1999) Status of resident coastal cutthroat trout and their habitat twenty-five years after riparian logging. *North American Journal of Fisheries Management,* **19,** 901–11.

Yukon Department of Renewable Resources (2000) *Yukon State of the Environment Report 1999.* Whitehorse.

Chapter 20
Fish–forestry interactions in freshwaters of Atlantic Canada

R.A. CUNJAK, R.A. CURRY, D.A. SCRUTON AND K.D. CLARKE

Introduction

The commercial harvesting of trees for wood products in Atlantic Canada (see inset Fig. 20.1) dates back to the 1700s when lumber was exported to Britain, largely for use in shipbuilding (MacNutt 1963). During the Napoleonic Wars (1800–1815), the demand for white pine for masts and spars in the British Navy fuelled the forestry industry in New Brunswick (Wright 1944). In 2000, the value of the forestry exports

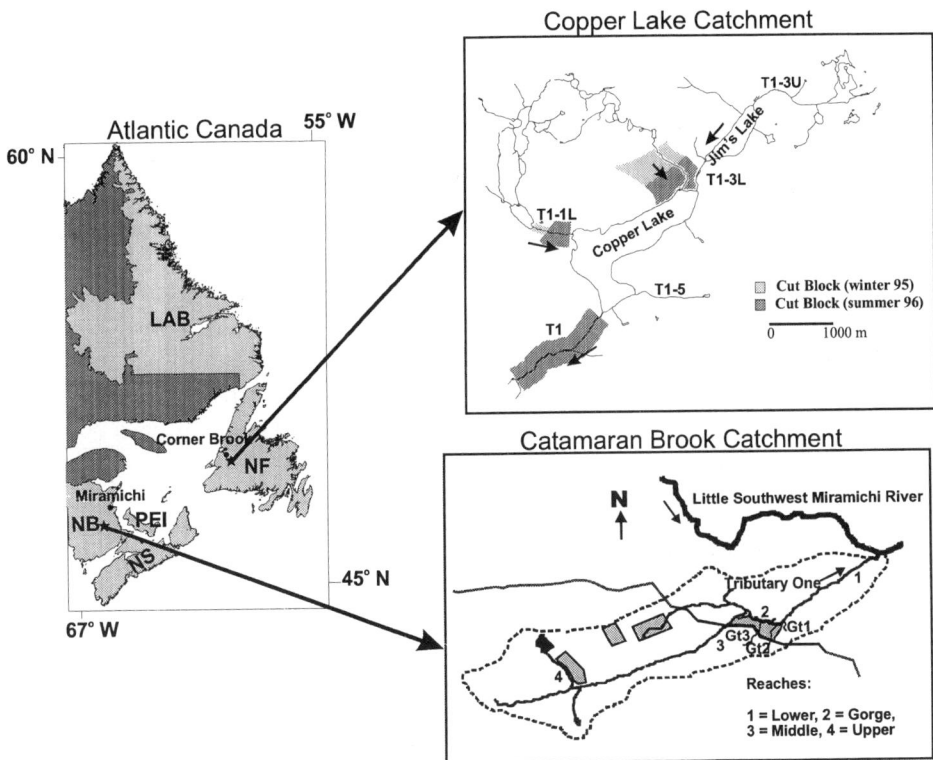

Fig. 20.1 Map of Atlantic Canada showing Catamaran Brook (New Brunswick) and Copper Lake (Newfoundland) basins with timber harvest blocks (shaded areas), study sites and reaches (numbered). Arrows along watercourses represent direction of flow.

from Atlantic Canada was $4.7 billion (Cdn) with approximately 35,000 jobs stemming directly from the industry (Canadian Forest Service 2002).

The rivers and lakes of the region have supported recreational angling since the 1700s. By the nineteenth century, the rivers of New Brunswick were the most prized destination for Atlantic salmon angling among Europeans and Americans (Thomas 2001). The value of recreational fisheries in the region is less impressive than that for forestry but still significant. According to 1995 statistics, direct expenditures made by anglers in the Atlantic provinces amounted to approximately $100 M (Cdn), or 4% of the value of recreational fishery for Canada (EPAD 1997).

Despite the long and concurrent histories of forestry and fisheries in the region (longer than anywhere else in North America), and the often-cited accusation of negative impacts of forestry activities on freshwater fish populations (see for example Elson 1974; Grant *et al.* 1986; and more recently Fairchild *et al.* 1999), few studies have comprehensively addressed the potential linkages between forestry and fishes in eastern Canada. Most of the existing knowledge of fish–forestry interactions is based on research carried out on the west coast of North America (e.g. Hartman & Scrivener 1990; Ralph *et al.* 1993) or the eastern United States (Bilby & Likens 1980; Flebbe & Dolloff 1995) and may be of limited applicability to the maritime and boreal forest ecosystems of Atlantic Canada.

Pertinent research in eastern Canada has focused on hydrological responses to forestry operations and not potential changes in fish populations. In the 1970s, the Nashwaak Experimental Watershed Project (Dickison *et al.* 1981) in central New Brunswick quantified hydrological and chemical changes associated with clear-cutting in a small (3.9 km^2) stream basin (Jewett *et al.* 1995). The St Mary's River Forestry-Wildlife Project in Nova Scotia (Milton & Towers 1989) was carried out in the late 1980s to assess the potential impacts of timber harvest within riparian zones on wildlife and aquatic habitats (principally, water temperature).

Beginning in 1990, the Department of Fisheries and Oceans with several partner agencies initiated the Catamaran Brook Habitat Research Project in the Miramichi region of central New Brunswick. The primary goal was to quantify the impact of clear-cut forestry on a stream with a wild population of Atlantic salmon (Cunjak *et al.* 1993; Cunjak 1995). Using a multi-disciplinary approach to quantifying potential impacts, the study was the first of its kind in Atlantic Canada where fish populations were specifically monitored for their response to forest harvesting.

Intensive forest harvesting in Newfoundland since the early 1900s concentrated harvestable wood in riparian zones and headwater sub-basins. The Copper Lake Buffer Zone Study of western Newfoundland began in 1993 to address the potential for interactions between timber harvesting and fish populations. A primary objective was to test the value of buffer strips in riparian zones for protecting fish and wildlife resources and water quality (Scruton *et al.* 1995). Both the Catamaran Brook and Copper Lake projects will be described in detail in this chapter.

The geographic region

Politically, the Atlantic Region of Canada is comprised of Newfoundland and Labrador and the three 'maritime provinces' of New Brunswick, Nova Scotia and Prince Edward Island (PEI). Population density is low (13.2 people/km²) with less than 2.3 million people living in the region (7.7% of the population of Canada). Geographically, the land mass covers an area of 50.3 million hectares or about 5.5% of Canada's land mass (Canadian Forest Service 2002). Biomes range from boreal taiga in Labrador and the northern peninsula of Newfoundland to boreal forests of insular Newfoundland, Cape Breton and northern New Brunswick to temperate, mixed forests covering much of the three maritime provinces. Mountain ranges are old and weathered. These include the northeastern extension of the Appalachian Range (with Mount Carleton at 820 m, the highest point of the mainland provinces) and the Long Range Mountains (800 m) of western Newfoundland. The Torngat Mountains of northern Labrador have peaks in excess of 1500 m. Bounded on the west by the continental landscape, the eastern boundary is the Atlantic Ocean, Gulf of St Lawrence and Bay of Fundy.

Overview of forestry regulations in Atlantic Canada

In Canada, the management of forest resources is the responsibility of the provincial governments. Each province has recognized the potential for forestry operations to damage fish habitats. Consequently, they have enacted legislation to protect watercourses and have adopted guidelines that establish best management practices for the protection of aquatic ecosystems. In all jurisdictions, legislation is based on the Fisheries Act of Canada (Constitution Act 1982) that clearly defines fish habitat as those parts of the environment on which fish depend, directly or indirectly, in order to carry out their life processes and these habitats cannot be altered, disrupted, or destroyed by chemical, physical or biological means. Each province has adapted the guidelines of the federal Policy for the Management of Fish Habitat (Anonymous 1986) for their own jurisdictions. Similarly, each jurisdiction has its own cultural interpretation and degree of enforcement of the Fisheries Act, provincial acts and best management practices.

New Brunswick has a complex regulatory process for protecting fish habitats. There are two sets of regulations defined by operations taking place on either Crown or privately owned lands. The most stringent regulations exist for operators (licensees) on Crown lands. The Department of Natural Resources and Energy's regulations require annual Operating Plans that describe all proposed activities for forested lands including identifying defined watercourses and appropriate buffer zones (Crown Lands and Forest Act 1982). A watercourse is any natural drainage feature that has a discernible channel. Watercourse buffer zones are defined as areas of relatively undisturbed vegetation maintained between watercourses and adjacent forestry operations for the purpose of managing water quality and aquatic habitats and other resource values (e.g. recreational or aesthetic). The guidelines clearly explain how to establish buffer zones, but also indicate that approval of the operating plan will be based on site-specific criteria. Watercourses draining areas of <600 ha receive a 15-m wide buffer zone on both sides of the channel. Larger drainages receive a 30-m buffer zone unless slopes are extreme (>25%) where the zone is extended to 60 m. For small watercourses (<0.5 m in width), a 3-m buffer of undisturbed ground is required. Operations can still occur within the zone if they satisfy a list of approved activities, e.g. trees can be removed from the buffer zone, but vehicles cannot travel within 15 m or 3 m of the watercourse for large and small streams, respectively.

On privately owned lands, the Clean Water Act (1989) administered by the Department of the Environment and Local Government regulates forestry operations. Activities crossing a watercourse or heavy machinery working within 30 m of a watercourse require a Watercourse Alteration Permit. Although details of the requirements for operations are clear, implementation of the regulation is highly variable and left to the interpretation of individual agents of the government. The result has been mixed application and enforcement, particularly for small streams (<1 m in width) which are not clearly defined as watercourses.

In Nova Scotia, the Wildlife Habitat and Watercourses Protection Regulations of the Forests Act (2001) protect watercourses during forestry operations. For streams >50 cm wide, special management zones (riparian buffer strips) varying from 20 to 60 m in width are required along all boundaries of the watercourse. Smaller watercourses are afforded no protection. Within the special management zone, some trees can be removed, no vehicle can travel within 7 m of the watercourse, and no activity can occur within 20 m if activities would result in sediment being transported to the stream. Similar to New Brunswick, a second regulation, the Environment Act (1995), is used for watercourse crossings and alterations. This regulation is similarly vague in terms of definitions and enforcement.

Forestry operations in Newfoundland and Labrador are regulated by the Environmental Protection Guidelines for Ecologically Based Forest Management (Stand Level Operations) established in 1994. This use of no-cut buffer zones was the first provincial protection of watercourses. The guidelines require a minimum 20 m, no harvest buffer zone with allowance for increased width for increasing slope of the terrain, along all water bodies that appear on a 1:50,000 scale topographic map. In special cases only (e.g. main branch of salmon rivers, fish spawning areas, protected water supply areas, pesticide application areas, areas of significance for wildlife), wider buffer zones are established.

On Prince Edward Island (PEI), watercourses are not defined by size. Rather, forest watercourses are defined in the Environmental Protection Act (1988) as any flowing water including intermittent streams and springs that have a defined sediment bed and flow-defining banks that connect with a larger watercourse, or exhibit continuous flow during any 72-hour period from July 1 to October 31 of any year. The protection is 20- or 30-m riparian buffer strips depending on the slope of the adjacent upland. Construction activity is not allowed within the buffer strip, but harvesting can occur in accordance with established guidelines. Crossing a watercourse requires a permit that identifies construction procedures including protection and mitigation measures. The current concern for watercourse protection in PEI is the expansion of agriculture. The recent acceptance of riparian buffer guidelines associated with this land use (2001) should address this concern.

Clearly, each jurisdiction in Atlantic Canada is making an effort to protect watercourses during forestry operations. Each province has established buffer zones along watercourses adjacent to harvesting, yarding areas and roadways as one best management practice designed to protect the watercourse (e.g. Anonymous 1986). The buffer zone provides direct protection from vehicles such as log skidders whose tracks can create channels that extenuate sediment transport to watercourses. Keeping vehicles away from watercourses also protects them from the direct addition of pollutants derived from the vehicles like fuel oil and coolants. Riparian buffer zones also provide an area to dissipate overland water flow, particularly during rain events, and thereby direct the deposition of transported sediment outside the watercourse.

The riparian buffer zone can provide effective protection for aquatic ecosystems, but appropriate application requires accurate information regarding the location of watercourses. Today's application of GIS (Geographic Information System) technologies has greatly enhanced the forest management process and identification of watercourses can be a routine operation for planners and their computers. However, the data derived from remote sensing information (typically at scales of 1:10,000) are not always reliable in defining watercourses, particularly smaller streams (e.g. Curry *et al.* 1997). Therefore, scale is clearly a significant issue when establishing protection guidelines for fishes, particularly coldwater species that dominate the small, forested streams of eastern Canada.

The Catamaran Brook Habitat Research Project

Description of the study area

Catamaran Brook is a third order tributary of the Little Southwest Miramichi River (Fig. 20.1) located at 46° 52.7′ N, 66° 06.0′ W and 70 m above sealevel (stream mouth). The drainage area covers 52 km^2 and mean annual discharge is 1.2 m^3 s^{-1} (Cunjak *et al.* 1993; Cunjak 1995).

The 14,000 km^2 Miramichi River basin of east-central New Brunswick lies in the Appalachian physiographical region. Three main tributaries – the Northwest, Little Southwest and Southwest – rise in the predominantly granitic and volcanic highlands of western New Brunswick. The lower reaches cross the sedimentary plain of the New Brunswick lowlands and combine to form the main channel near the head of tide. The climate is continental throughout most of the basin with a mean annual temperature of 4.3°C. On average, 1130 mm of precipitation falls in 160 days throughout the year (about 30% as snow).

Water chemistry data from the main stem and tributaries indicate a near-neutral, soft-water river that is typically low to moderate in productivity (Komadina-Douthwright *et al.* 1999; Cunjak & Newbury, unpublished data). Excessive loading of P or N is rare, reflecting the relatively sparse human population in the river's upper and middle reaches, and the lack of agricultural or industrial inputs upstream of the estuary.

The Miramichi basin is predominantly forested with balsam fir (*Abies balsamea*), white spruce (*Picea glauca*) and black spruce (*P. mariana*), especially in the highlands and northern portions of the basin. Deciduous species such as maple (*Acer* spp.), yellow birch (*Betula lutea*) and beech (*Fagus grandifolia*) are relatively abundant in the floodplains and ridges of the lower reaches. The Catamaran Brook basin is mostly comprised of second-growth species as a result of timber harvest since the mid-nineteenth century and forest fires. Clear-cutting has occurred in 15.1% of the basin; 22.6% has been selectively cut. Species composition is about 65% coniferous and 35% deciduous.

Randall *et al.* (1989) listed 21 species of fishes (representing 9 families) that were found entirely in the freshwater reaches of the Miramichi River system. In Catamaran Brook, 17 species have been identified; Atlantic salmon (*Salmo salar*), brook trout (*Salvelinus fontinalis*), slimy sculpin (*Cottus cognatus*), lake chub (*Couesius plombeus*) and blacknose dace (*Rhinichthys atratulus*) are the most common species.

Experimental design and monitoring activities

The study design involved three phases as described by Cunjak *et al.* (1993). The pre-logging (or calibration) phase (1990–1995) was a period when no timber harvest occurred and natural environmental variability was established. The logging phase (1995–2000) was the period when access roads to harvest blocks were built and clear-cutting occurred in pre-designated blocks (Fig. 20.1). The final, post-logging phase (2001–2005) will incorporate experimental manipulations and specific prediction testing from the second phase. This three-phase design is similar to that employed by Hartman and Scrivener (1990) for the Carnation Creek study in British Columbia.

The area harvested in 1996 and 1997 represented 7% of the stream basin, with no cutting in the Lower reach (Fig. 20.1). Four harvest blocks ranging in size from 36 ha to 73 ha were clear-cut, yielding between 4300m^3 and 11,600m^3 of wood (pulp wood, sawlogs, hardwood and firewood). Timber harvesting involved a tracked feller-buncher doing the initial felling and piling of trees on site. Where suitable hardwood was available, a second-pass harvest with chainsaw was performed. All slash material was left on site. For harvest blocks in the headwater reaches, riparian buffers of 30 m were retained, whereas a 60-m buffer was retained along the main stem in the Gorge reach (Fig. 20.1).

Monitoring of water discharge, precipitation and air/water temperature has been continuous since 1989 in the Lower, Middle and Upper reaches of Catamaran Brook (Cunjak *et al.* 1993; Caissie *et al.* 2002). A suite of water chemistry parameters has been measured in water samples taken from the Middle and Lower reaches of the stream (Caissie *et al.* 1995; Komadina-Douthwright *et al.* 1999), as well as from piezometers and shallow groundwater wells inside and outside harvest blocks (MacQuarrie & Maclean 1993; Jones & Bray 1994). Suspended sediment concentrations were measured annually during storm events.

Coarse or large woody debris (LWD) has long been recognized as an important fish habitat feature within streams and an important component in organic matter processing in stream ecosystems (Bilby & Likens 1980). However, the majority of this research has been done on the west coast of North America where tree size and stream power (ability to move in-channel items) are generally greater than in east coast watercourses. To assess possible changes in quality and quantity of LWD entering the bankfull channel from the riparian zone adjacent to harvest blocks, woody debris surveys were conducted annually in the four stream reaches in Catamaran Brook. As well as counting LWD items and estimating volume of each item, LWD was tagged to estimate retention in the various reaches.

Fish monitoring and enumeration involved a fish-counting fence located near the stream mouth, electrofishing surveys conducted semi-annually in all habitat types in the four study reaches, drift-netting to capture emigrant fry of several species in the spring and assessments of egg/alevin survival (Cunjak *et al.* 1993; Cunjak 1995; Johnston 1997; Hardie *et al.* 1998). Models of salmon population dynamics in the brook during the pre-harvest years (1990–1996) demonstrated significant inter-annual fluctuations in juvenile abundance largely attributable to density-independent factors such as low water conditions reducing colonization of upstream reaches (Cunjak & Therrien 1998) and mid-winter ice break-up leading to bed scour and egg/parr mortality in the stream (Cunjak *et al.* 1998).

Results and discussion

Hydrology

Caissie *et al.* (2002) carried out a detailed analysis of hydrological parameters in Catamaran Brook pre- and post-timber harvest (annual water yield and seasonal runoff, stream flow timing, storm flow and peak flow). Their study used a comparative approach by measuring responses of one sub-basin (Tributary One) where >23% of the area was clear-cut relative to the Middle reach where 2% of that sub-basin area was harvested (Fig. 20.1). For both sub-basins, no changes were detected in the annual water yield, seasonal runoff or stream flow timing. However, storm flow peaks were significantly higher in Tributary One in the 2 years immediately following forestry activity; no such relation was measured in the Middle reach (Fig. 20.2).

Similar hydrological findings have been noted in other Atlantic region studies. In the Nashwaak River study, a 59% increase in summer peak flows was measured the year after clear-cutting 90% of the stream basin (Dickison *et al.* 1981). Unlike the situation in Catamaran Brook, annual discharge regimes in the Nashwaak study stream increased; 5–10 years following the harvest, hydrological and chemical parameters returned to pre-harvest levels (Jewett *et al.* 1995). The Catamaran Brook results were consistent with findings that clear-cutting >20% of a (sub)catchment will create detectable, hydrological changes (Hibbert 1967; Bosch & Hewlett 1982).

Pre-impact monitoring of the groundwater in the Catamaran Brook basin indicated a bicarbonate-carbonate chemical composition with pronounced recharge periods in the spring and autumn, coincident with peaks in the stream hydrograph. Post-harvest results to date show higher water levels in the wells situated within the harvest block, suggesting that infiltration is greater than in the wooded control. During precipitation events, the magnitude of the groundwater response was three times greater in wells

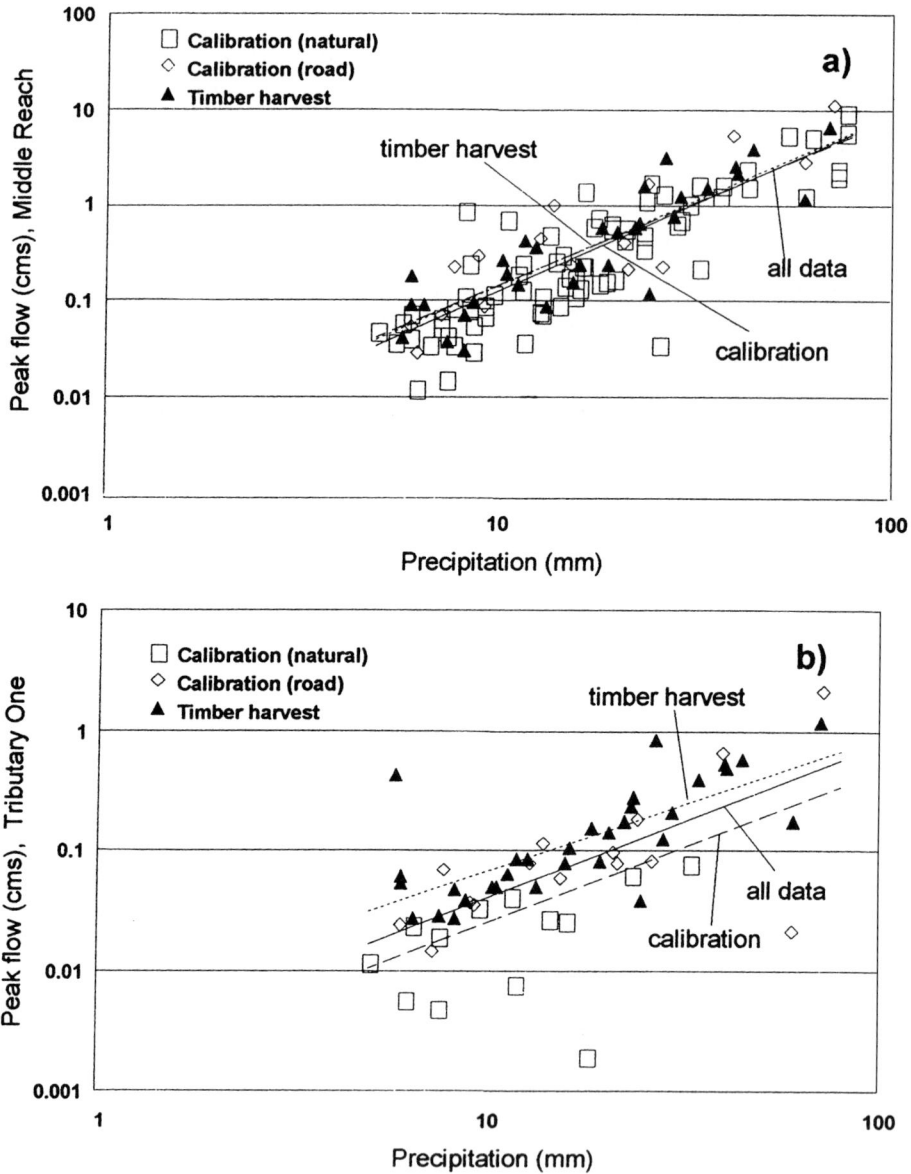

Fig. 20.2 Relationship between peak flow and precipitation during summer storm events for pre-forestry (calibration) and timber harvest periods in the Middle reach (a) and Tributary One (b) sub-basins of Catamaran Brook (redrawn from Caissie *et al.* 2002). Note that both axes use $\log_{10}$ scales.

within the harvest blocks (Alexander 2000). In their review of watershed studies of forestry impacts, Bosch & Hewlett (1982) noted that groundwater levels typically rose and responded faster to precipitation events when the interception of rainfall was eliminated by removal of the forest canopy.

Water quality and nutrient chemistry

In Carnation Creek (British Columbia), conductivity (an indicator of total ion concentration) and nitrates increased following logging, being most evident at high discharges (Hartman & Scrivener 1990). At Hubbards Brook (New Hampshire), Likens *et al.* (1970) measured increases in major ions following clear-cutting. Similarly, Jewett *et al.* (1995) measured post-harvest increases in concentrations of K and NO_3-N in a small New Brunswick stream that persisted for 5 years following the treatment.

No apparent water chemistry changes were noted in Catamaran Brook. Komadina-Douthwright *et al.* (1999) found no evidence of significant flushing or 'leaking' of nutrients and ions from the timber harvest blocks and logging roads during the period immediately following clear-cutting. However, the location of the water sampling at the mouth of Catamaran Brook and at the bridge crossing in the Middle reach may have limited the detection of chemical changes as both sampling sites were at least 5 km downstream from the nearest harvest block.

To measure nutrient loss from the disturbed ground within timber harvest areas, leachate chemistry was monitored at different soil depths within and outside the Gorge reach harvest block from 1993 to 1998 using a system of lysimeters (Titus *et al.* 2000). Preliminary results indicate that soils were very resilient within the sub-basin and that no detectable changes were found in nitrates, manganese or aluminium following timber harvest (T. Mahendrappa, Canadian Forest Service, pers. comm.).

Terrestrial studies

The Upper reach had more LWD items, but smaller unit volumes than those found in the Lower reach, which was 6 km from the nearest harvest block (Table 20.1). Much of this difference can be explained as a longitudinal trend where LWD in headwaters tends to have a higher retention due to the size of stream and the relative power of stream flow to displace items. Indeed, this was evidenced by the 86–95% retention rate of marked LWD items in the Upper reach compared with 58–60% in the Lower reach (P. Hardie, DFO, unpublished data). To date, the number, volume and retention of LWD items have not changed along this stream (Table 20.1).

Table 20.1 Abundance (number of items per 100-m length of stream) and volume (m^3 per 100-m length of stream) of large woody debris in two reaches of Catamaran Brook pre-harvest (1996) and post-harvest (1997)

	Pre-harvest		Post-harvest	
Site	Number of items	Volume	Number of items	Volume
Upper reach (control)	39	5.2	39	4.9
Upper reach (30-m buffer)	36	5.0	36	4.5
Lower reach (no treatment)	30	6.6	26	6.7

From unpublished data of P. Hardie.

A comparative survey of wind-thrown trees (number, volume and orientation) in riparian zones adjacent to the Gorge reach (treatment, 60-m buffer) versus a control zone (opposite side of brook, no adjacent harvest block) was carried out by Lloyd (1999). Trees in the riparian control were randomly oriented. By contrast, 73.5% of wind-thrown trees in the riparian zone adjacent to the harvest block were oriented northward, the predominant wind direction. The number and volume of wind-thrown trees was greater, and the trees were younger, in the riparian treatment zone than in the control (Lloyd 1999). Therefore, there was reasonable evidence that 60-m riparian buffer strips may be insufficient to protect streams and riparian zones from unnatural increases in LWD.

Water temperature and suspended sediment monitoring

Changes in stream temperature associated with the timber harvesting were not apparent in the main stem of Catamaran Brook or in Tributary One (Edwards 2001). Changes in temperature at a finer scale are being monitored in the first order brooks of GT1, GT2 and GT3 (Fig. 20.1). Catchment areas are small (1.5 km², 1.0 km² and 0.7 km² for GT3, GT2 and GT1, respectively) and discharge is <12 L/s at mid-summer flows. Brook trout and slimy sculpin inhabit the streams. Only GT2 and GT3 have road crossings. Monitoring began in the autumn of 1996 coincident with the harvesting of GT2 and GT3 sub-basins. This is a long-term experiment, but preliminary results suggest that during the summers there is a warming of these small streams correlated with removal of forest cover. That is, maximum temperatures are substantially greater in the reaches downstream of the roads and harvest blocks (Fig. 20.3).

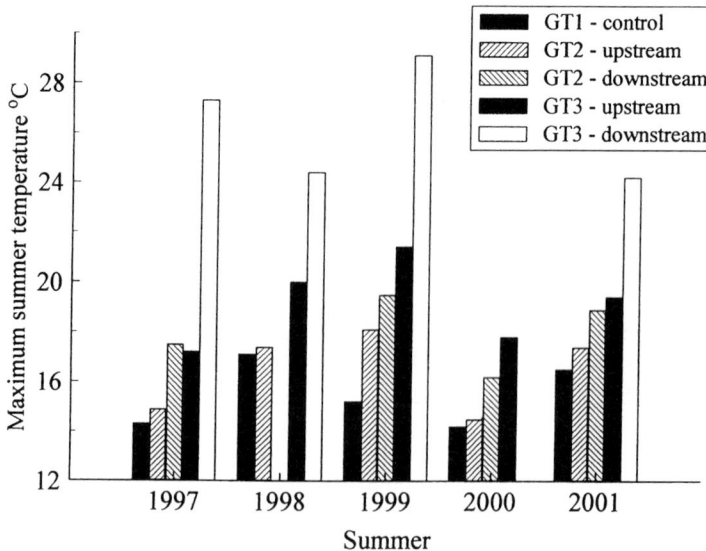

Fig. 20.3 Maximum summer water temperatures (°C) in three first order streams tributary to Catamaran Brook, 1997–2001. The upstream sites are in areas of no roads or forestry activities.

In October and November of 1996, Kull (1997) measured suspended sediment levels in these first order streams during two autumnal rain events. Despite small volumes of rain (<25 mm accumulation), suspended sediment peaks of 113–145 mg L^{-1} were measured in one of the treatment streams, GT2, during both rain events. Relatively little change in suspended sediment was measured in the other treatment stream, GT3, or the control stream, GT1 (both had storm peaks of <45 mg L^{-1}). It was noted that the main source of suspended sediment in GT2 appeared to be runoff from the main logging roads crossing the brooks rather than the recently built (1995) timber access trails within the blocks. Generally, stream-dwelling fishes like brook trout are quite tolerant of suspended sediment concentrations of <100 mg L^{-1}, but some biological impairment (e.g. reduced feeding, avoidance behaviour) has been noted at the peak concentrations observed in GT2 (Meehan 1991). At the GT3 site, there was relatively little sediment disturbance inside the harvest block where the harvester moved about on a mat of branches, and skid trails were aligned parallel to the streams with undisturbed strips between trails (Kull 1997). Such procedures appear to reduce the likelihood that sediments would be transported to the stream.

Less diligence to protective harvesting methods was followed in another sub-catchment, Tributary One (Fig. 20.1), during late October 1996. Poorly designed access roads and a highly disturbed ground surface in the harvest block resulted in the transport of significant amounts of fines to Tributary One and the main stem of Catamaran Brook during a 30-mm rain event. The sediment plume was clearly visible 6 km downstream at the mouth of Catamaran Brook. Peak suspended sediment concentrations of 440 mg L^{-1} were measured in Tributary One while levels were <5 mg L^{-1} upstream during the same event. Such high values (>400 mg L^{-1}) are within the range where negative impacts are measured on stream biota. The event occurred during the spawning period of Atlantic salmon that were concentrated in the main stem, downstream of Tributary One. Therefore, newly cleaned gravel substrates used for incubating eggs would have received a slug of fine sediments that have well-established negative impacts on salmonid incubation survival (e.g. Chapman 1986; Cunjak *et al.* 2002). Despite this extreme event, no effect on production of young-of-the-year was apparent in the year following the event (R. Cunjak, unpublished data).

As was also noted for first order streams, the main source of sediment entry to Tributary One was poorly constructed logging roads. Grant *et al.* (1986) assessed timber harvest impacts for streams in New Brunswick and Nova Scotia. Their study showed that salmonid biomass decreased significantly in two streams downstream of logging roads, presumably because of increased loading of fine sediments. The contribution of fine sediments to streams from forest roads has also been documented in North Carolina (Swift 1988) and British Columbia (Hartman & Scrivener 1990).

Salmon population changes (whole basin-scale)

An assessment of the effect of forestry activity on the Atlantic salmon population in Catamaran Brook was carried out at three life stages: egg survival (winter), juvenile (parr) abundance in the river during summer and smolt production based on enumeration at the counting-fence each spring.

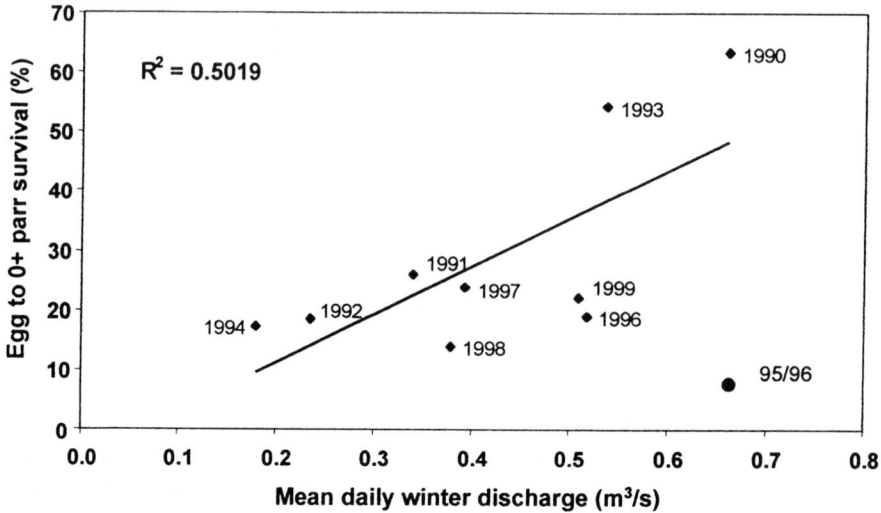

Fig. 20.4 Relationship between mean winter discharge (November–March) and egg survival of Atlantic salmon in Catamaran Brook, New Brunswick. The regression line was plotted after exclusion of the 1995/96 survival estimate as this was a statistical outlier resulting from an atypical mid-winter ice scour event (see Cunjak *et al.* 1998). The value for year represents the start of winter (e.g. 90 = 1990/91).

Egg survival is largely a winter-dependent phenomenon, as Atlantic salmon lay their eggs in the gravel by early November and emerge as free-swimming alevins in early June. Egg survival was based on the number of eggs deposited each autumn by female spawners relative to the abundance of fry enumerated in the brook the following summer (see Cunjak & Therrien 1998 for description of model). In the years following the timber harvest (1997–2000), estimated egg survival has been relatively low (<25%) but similar to that found in 4 of 6 years pre-harvest (Fig. 20.4). The lowest survival (7%) was found in 1995–1996 before timber harvest and was attributed to severe streambed disturbance from a mid-winter ice-scour event (Cunjak *et al.* 1998).

A positive correlation between egg survival and winter stream discharge was found in the pre-harvest period when the ice scour year of 1995/96 was excluded (Cunjak *et al.* 1998). Although post-harvest, winter discharge values (November–March) were within the range of discharges measured before 1997, it is noteworthy that survival in the post-harvest period was less than the average defined by a regression line (Fig. 20.4). This suggests that lower than expected egg survival was found in winters following timber harvest despite apparently acceptable winter stream flows and the absence of density-independent phenomena such as a mid-winter ice break-up.

The annual number of smolts emigrating from Catamaran Brook has declined but this appears to be unrelated to forestry activity. The trend probably reflects the steady decline in recruitment associated with reduced numbers of spawners entering the stream (Cunjak, unpublished data). From a high of >160,000 parr in 1991, the number of juvenile (pre-smolt) salmon in summer has declined to a low of approximately 27,000 parr in 2001. This recruitment trend, combined with density-dependent and -independent factors influencing egg/juvenile mortality and emigration (Cunjak *et al.* 1998) accounts for the relatively low number of salmon parr and smolts enumerated in recent years.

Overall, the Atlantic salmon population in Catamaran Brook may be following a recent trend of declining numbers of adults and juveniles common to many rivers in Atlantic Canada. Although factors in freshwater (e.g. natural barriers to migration, mid-winter ice scour) can account for some of the apparent mortality and emigration, marine-related phenomena (or at least, post-smolt factors) have also been implicated as causal factors to explain the broad geographic trends (Fairchild *et al.* 1999).

Other fishes (slimy sculpin and brook trout)

Populations of slimy sculpin in the Middle and Lower reaches of Catamaran Brook were apparently unaffected by forestry activities (Edwards 2001). However, these sites were well downstream of the harvested portions of the catchment. In Tributary One where peak flows increased after harvesting (Caissie *et al.* 2002) and one high suspended sediment event occurred during harvesting (autumn of 1996), no associated population-level changes in slimy sculpin were noted in the 2 years following timber harvest (Edwards 2001). There was a significant decline in abundance in 1996, presumably as a result of a mid-winter ice break-up and disturbance to the streambed in Catamaran Brook (Cunjak *et al.* 1998).

The headwater tributaries (GT1, GT2 and GT3) are coldwater brooks inhabited by brook trout and slimy sculpin. The brooks may provide spawning and incubation habitats as well as nursery habitats, at least for brook trout spawned in the main stem (e.g. Curry *et al.* 1997). These small brooks are experiencing increases in summer water temperatures to >20°C (approaching lethal levels for both species) and also appear more susceptible to sediment inputs from harvesting operations.

In Catamaran Brook, some physico-chemical changes occurred in response to forest clear-cutting; biological responses were less discernible. In harvest blocks, water level (yield) was higher, and there was a stronger response to precipitation events. There was a heightened peak discharge during summer storm events in the sub-basin where >20% of the catchment was harvested. Suspended sediment concentrations and water temperature increased in low order tributaries draining harvest blocks but, to date, no obvious impact on fish density has been detected.

The Copper Lake Buffer Zone Study

Description of the study area

The Copper Lake sub-basin is a small, 13.5-km² headwater system located approximately 17 km southeast of Corner Brook, Newfoundland, Canada (Fig. 20.1). The sub-basin is located 350–650 m above sealevel and has an average annual rainfall of 1186 mm. Streams in the catchment have moderate to high gradients ranging from 2.5% to 23.8%. Soils are predominantly moderate to coarse glacial tills derived from intensely deformed and highly metamorphosed rocks (Kennedy 1981), which have a relatively large moisture content. Overburden depths are shallow, <2 m. These factors and the steep hill-side slopes create an increased potential for erodibility within the basin (van Kesteren 1992).

The forest cover of the study area is largely composed of mature and over-mature balsam fir (*Abies balsamea*) with interspersed black spruce (*Picea mariana*) and white birch (*Betula papyrifera*). A detailed description of the area is provided in Scruton *et al.* (1995).

Freshwater systems in Newfoundland are typically dilute with low nutrient concentrations, low primary production potential (Scruton 1983; Knoechel & Campbell 1988) and low biological production estimates (Clarke 1995; Clarke & Scruton 1999). The fish communities are simple with only hyposaline species having recolonized the freshwaters of insular Newfoundland after the Wisconsin glaciation (Scott & Crossman 1964).

Brook trout is the sole fish species present in the Copper Lake catchment, and they are distributed from the headwaters through the main stem, lakes, and into the estuary (anadromous form). Generally, the small primary streams are used for reproduction and rearing of 0+ to 2+ trout and the lakes are used by older fish (Clarke *et al.* 1997).

Experimental design and monitoring activities

Within the Copper Lake drainage basin, there are five headwater tributaries (T1-1 to T1-5, Fig. 20.1). The outlet of Copper Lake (T1) is a second order stream that drains into Corner Brook Lake. Road construction began in 1993 (crossing T1) and was completed in June 1994, crossing both T1-1 and T1-2. Harvesting commenced during the winter of 1994/95 using mechanical harvesters, and included a small area of the T1-1 sub-basin and the portion of the T1-2 sub-basin above the road (Fig. 20.1). The remainder of the cutting was conducted by chainsaw during the summer of 1996 and included all other cut blocks within the catchment. Riparian buffer strips were maintained as follows: T1, no buffer; T1-1, no buffer; T1-2, 20 m; T1-3L, 20 m; and T1-3U, at least 100 m (Fig. 20.1). Therefore, 1993–1995 were considered the pre-harvest years although road construction and a minor clear-cut occurred during this time. The major treatment year was 1996 with the subsequent 2 years (1997 and 1998) being monitored for post-harvest responses.

Physical habitat attributes monitored in streams included large woody debris dynamics, sediment accumulation and summer water temperature. Methods employed in these monitoring programmes have been outlined in detail in Scruton *et al.* (1995) and Clarke *et al.* (1997). Biological monitoring focused on fish (brook trout) density and biomass which was estimated via electrofishing (following Scruton & Gibson 1995) in August of each year. These population estimates were used to calculate production and P/B ratios (see Clarke & Scruton 1999). Two directed sub-projects focused on the effect of forest harvesting on brook trout incubation habitats (i.e. redds) and movement/migration patterns. Methods employed in these projects are outlined in Curry *et al.* (2002) for the redd habitats, and in McCarthy (1997) and McCarthy *et al.* (1998) for the movement studies.

Results and discussion

Large woody debris

In the streams located in the lower part of the catchment (T1 and T1-1L), LWD volume was significantly reduced in 1995 ($p < 0.05$) and remained at this level for the duration of the study (Fig. 20.5). This flushing of LWD was most likely due to an intense spring

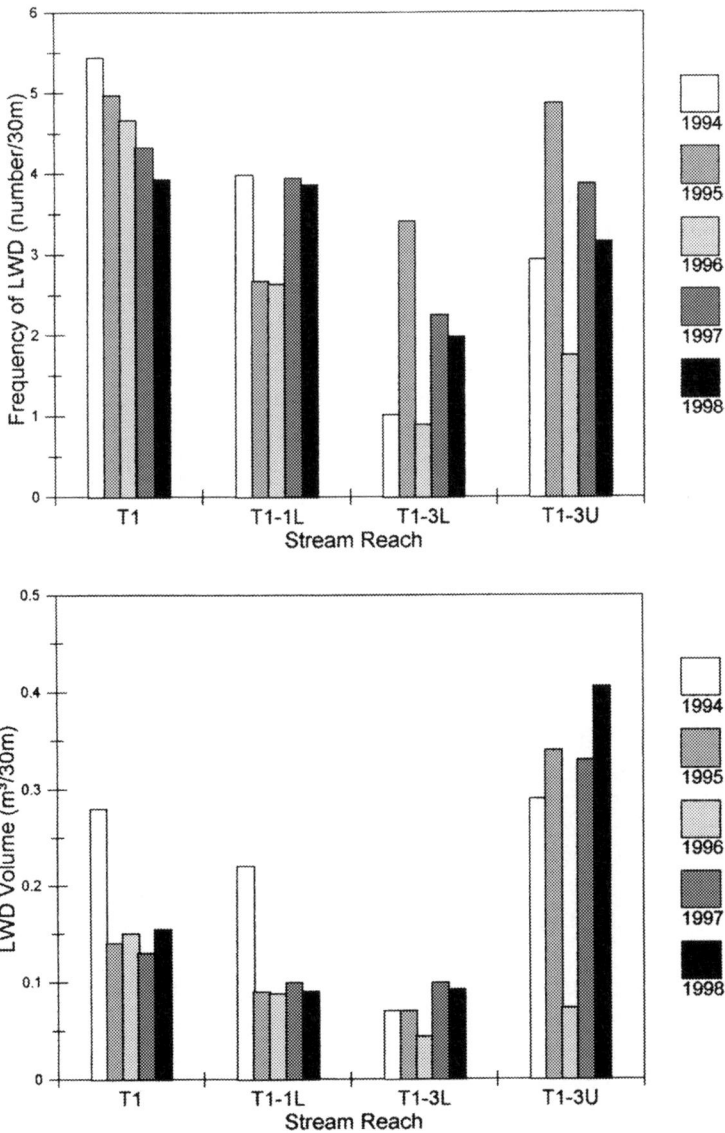

Fig. 20.5 Large woody debris (LWD) frequency and volume in the streams of the Copper Lake Buffer Zone study, western Newfoundland.

runoff that occurred in the catchment in early June 1995 (McCarthy 1997). The subsequent freshet removed larger pieces of LWD, that were replaced by smaller LWD in 1996, derived from the harvesting activities. A similar trend was not observed in the larger river (T1) where these smaller LWD pieces were flushed from the system.

Streams located in the upper parts of the catchment had different trends in LWD volume to those described above. Both T1-3L (20-m buffer) and T1-3U (unharvested) experienced a reduction in LWD volume during forestry activity in 1996 (Fig. 20.5) but returned to pre-harvest levels the following year, indicating a readily available supply of source wood for these streams.

Sedimentation

Sediment accumulation (summer) was significantly higher and more variable in T1 and T1-1L than in the other streams during the study (Table 20.2). Both these streams had a stream crossing and were harvested to the water's edge (no buffer). The stream with the 20-m buffer (T1-3L) had a significant increase in sediment accumulation during 1996 (the disturbance year) but this increase did not persist (Table 20.2). The control stream (T1-3U) has had little inter-annual variation in sediment accumulation (Table 20.2). The high variation observed in T1 and T1-1L was due to the differences in sediment accumulation relative to road crossings. The amount of sediment collected from samplers below the road crossings was two- and seven-fold greater than above crossings in T1 and T1-1L, respectively. This difference in sediment yield did not change post-harvesting (after 1996), indicating that the majority of the sediment influx to these streams during the summer months was derived from the road crossings.

Table 20.2   Mean sediment accumulation (g) in the Copper Lake study streams over the low flow period (October)

	Stream reach			
Year	T1 (no buffer, road)	T1-1L (no buffer; road)	T1-3L (20-m buffer)	T1-3U (control)
1993	21.0 (10.5–32.0)	15.7 (9.0–22.1)	–	–
1994	10.0 (2.8–21.8)	6.4 (4.0–8.6)	1.2 (0.9–1.9)	1.6 (0.9–2.3)
1995	0.7 (0.5–0.9)	26.5 (5.8–52.2)	0.4 (0.3–0.6)	2.3 (0.6–4.7)
1996	8.8 (1.6–32.5)	11.3 (5.3–18.0)	5.1 (0.6–11.0)	4.8 (0.4–11.6)
1997	15.3 (1.1–50.7)	27.1 (1.3–81.9)	2.2 (1.2–3.2)	2.6 (0.6–6.5)

Numbers in parentheses are 95% confidence intervals; size of riparian buffers and presence of road crossings are indicated for study reaches.

Water temperature

An initial analysis of the study streams found that these small headwater streams were high quality thermal habitats for brook trout (Scruton *et al.* 1998). In general, streams harvested to the water's edge (T1-1L and T1) had increased daily maximum temperatures, higher diel fluctuations and exceeded stressful limits (>21°C) more frequently than the control stream in the post-harvesting years. Streams with a 20-m buffer strip (T1-2 and T1-3L) retained a cooling benefit to the stream water from the upper to the lower reaches in post-harvesting years (see also Clarke *et al.* 1999).

Brook trout studies

The density of young-of-the-year brook trout was consistently highest in headwater stations (T1-3U) relative to more downstream sites (Fig. 20.6). There was no discernible effect of harvest activity on trout density; rather, inter-annual differences in trout abundance seemed to account for the fluctuations (Fig. 20.6). The absence of trout in station 6 (immediately downstream of the road crossing on T1-1L) in 1994 and 1995 was coincident with road construction (1994). This reduction continued throughout the study as compared with the lower stations within this stream and with the other small streams sampled within the catchment (T1-3L and T1-3U) (Fig. 20.6). The reduced quality of habitat, and hence fish populations, within this area appears to be related to the high sediment yield below the road crossing (see above). This was also the reason cited for the reduced utilization of this area by larger brook trout in the movement study (McCarthy *et al.* 1998).

Despite the localized effect of forestry practices observed in T1-1L, the overall health of the brook trout populations, as expressed by production:biomass (P:B) estimates showed no discernible trends that could be clearly associated with forestry practices (Table 20.3). P:B did decline to lowest values in T1 and T1-3L during the first post-harvest year before rebounding to the highest values (all four sites) in the second year post-harvest (Table 20.3). In T1-1L, the increase in P:B was related to an increased recruitment, especially in the lower sections of the river, from 1996 to 1998 (Fig. 20.6). This increase may have been due to a cleaning of the gravel beds within this river following the 1995 summer storm event.

From the data collected during 1996/97, a relationship between redd temperatures and surface water temperature was developed (Curry *et al.* 2002). Predicted values were then compared with observed data for the pre-harvesting period (1994/95) and two post-harvesting years (1995/96 and 1996/97). Daily mean temperatures significantly increased in T1-1L (no buffer) post-harvesting while no differences were observed in the buffered streams (Curry *et al.* 2002). These warmer temperatures translated into shorter periods to the eyed, 50% hatched and first-feeding stages of development during the year of harvesting and 1 year post-harvesting (Curry *et al.* 2002). This shorter duration of the incubation periods suggests that brook trout alevins would be emerging from the gravel substantially earlier in clear-cut catchments without buffer strips. This early emergence could impact the survival and/or growth of alevins (see discussion in Curry *et al.* 2002).

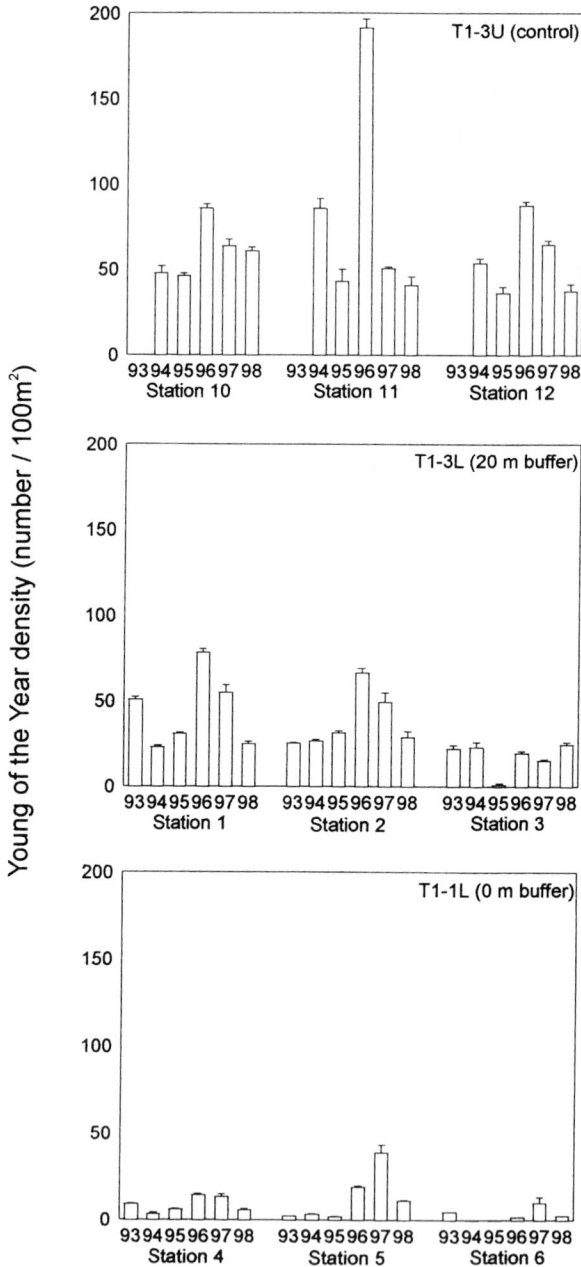

Fig. 20.6 Young-of-the-year brook trout density in three streams (three stations per stream) in the Copper Lake catchment, western Newfoundland, 1993–1998.

For the Copper Lake Buffer Zone Study in general, those streams with no riparian buffers could not replenish LWD, had higher summer and incubation temperatures and had increases in sedimentation. These physical changes resulted in localized reduc-

Table 20.3   Mean production:biomass (P:B) ratios for the Copper Lake study

Stream	Pre-harvest		Disturbance	Post-harvest	
	1993–1994	1994–1995	1995–1996	1996–1997	1997–1998
T1	1.19	0.91	0.73	0.55	1.25
	(1.09–1.30)	(0.83–1.00)	(0.67–0.79)	(0.40–0.70)	(0.81–1.69)
T1-1L	0.97	0.88	0.73	0.97	1.29
	(0.84–1.10)	(0.77–1.00)	(0.52–0.94)	(0.79–1.15)	(1.15–1.43)
T1-3L	1.01	1.42	0.96	0.88	1.35
	(0.94–1.08)	(1.30–1.54)	(0.89–1.03)	(0.79–0.97)	(1.24–1.47)
T1-3U	–	1.06	0.91	0.98	1.11
		(0.99–1.13)	(0.86–0.96)	(0.94–1.01)	(1.05–1.17)

Numbers in parentheses are 95% confidence intervals.

tion in habitat quality, which in turn reduced brook trout recruitment and the use of affected areas. The 20-m buffers, for the most part, provided adequate protection from the physical habitat changes induced by forest harvesting. Sediment accumulation appeared to be the most detrimental of the physical changes observed, and the source of most of the sediment was from road crossings.

Summary and recommendations

The Atlantic Canada research suggests that the effects of timber harvesting on freshwater habitats and fish populations can be minimal when riparian buffer zones ≥20 m are employed, <10% of the forest in a catchment is removed, and roads and watercourse crossings are constructed and maintained appropriately. These findings are consistent with other North American studies (Hibbert 1967; Hartman & Scrivener 1990). In the Nashwaak (NB) study, despite harvesting 90% of the 3.9-km² basin (75-m riparian buffers), the only measurable changes were an increased stream discharge post-harvest (due to increased runoff) and higher concentrations of K and NO_3-N in the stream water (Jewett *et al.* 1995). This physico-chemical response to clear-cutting is generally consistent with the Catamaran Brook studies. However, responses at larger scales (>50 km²) are more difficult to assess.

Localized, small-scale effects were typically resolved as the streams discharged into larger watercourses where dilution of chemicals, temperatures or sediment (or their settling due to reduced stream flow) took place. The challenge is measuring the cumulative effect of small catchment impacts on the entire river ecosystem. The ecological significance or role of headwater streams in river ecology is poorly understood. How many headwater catchments can be impacted before a whole-basin response occurs? Is there a spatial or temporal gradation of importance among headwater catchments? Is it probable that impacts from clear-cutting forests on small streams are resolved in 5–10 years (e.g. Jewett *et al.* 1995)? Such recovery times may appear positive, but are problematic for headwater species like sculpin and brook trout that have generation times <5 years (Scott & Crossman 1973).

Another result emerging from the studies in Atlantic Canada is the efficacy of existing 'best management practices'. No-harvest, riparian buffer zones are an established, successful method for protecting watercourses from direct impacts of machine operations and increased solar radiation. Such protective zones are essential for conservation of aquatic ecosystems. However, a fixed buffer zone width does not protect all components of aquatic ecosystems. In Newfoundland, a change in the thermal environment of brook trout incubating in streambeds during winter was detected even with a 20-m no-harvest buffer zone. Similar changes in summer surface water temperatures were observed in New Brunswick streams with 30-m buffer zones. Even wide buffer strips and careful wood-cutting practices do not preclude damage to aquatic habitats. One New Brunswick buffer zone (60 m) experienced significant blow-down, which compromises its ability to protect the watercourse and increases the potential blockage of fish passage and energy flows.

In general, the size of riparian buffers maintained during forestry operations is directly proportional to stream size; i.e. the larger the stream, the wider the buffer. The problem is that the smallest (first order) streams have the least riparian protection, 10–15 m in many cases, assuming they are even classified as streams. As noted by Vannote *et al.* (1980), streams of order 1–3 are the most numerous streams within catchments, represent the maximum interface with the landscape, and are principal processors of terrestrially derived organic matter. In addition, headwaters are typically the hydrological recharge zones in forested catchments. They regulate groundwater flows and pathways and they respond to forest removal, as observed in Catamaran Brook. These groundwater regimes can be directly linked to fishes in some instances (Curry & Devito 1996). Such relationships are poorly understood, but an appreciation of their ecological significance is growing rapidly (e.g. Curry *et al.* 1997; Power *et al.* 1999).

For terrestrial species, protection from land use activity such as timber harvest is often biologically based. Winter deer-yards, pine marten habitat, spotted owl nesting sites are identified and designated as protected areas during forest management discussions. In the case of aquatic species, little biological information is used in establishing current timber management plans. Rather, physical-logistic criteria are the basis for habitat protection (i.e. arbitrary buffer widths and how dry or frozen is the ground for equipment to work effectively). Although special considerations can be used to invoke more environmentally sound operation protocols, these are rarely applied even though the biological foundations exist that could enhance aquatic habitat protection. As one example, Atlantic salmon and brook trout ascend small tributaries to spawn from mid-October to mid-November in Atlantic Canada. This is often coincident with the autumn rainy season when most forest soils are saturated and subject to high rates of erosion if disturbed. Avoidance of these 'high-risk' areas until soils are frozen and stabilized could protect aquatic habitats and biota from intrusion of fine sediments via erosion.

Finally, in order to protect a watercourse, it has to be identifiable, for digital interpretation. The typical map scale of 1:10,000 (1:50,000 in Newfoundland) is not a fine enough resolution to protect smaller, forested streams. Most headwater streams are not identified at this scale and, therefore, may be omitted from protection measures (Curry *et al.* 1997). Even when regulations and guidelines can be applied, the willingness of regulators, and in some instances industry, to implement existing regulations has been questionable. The forest industry is clearly the economic engine of eastern Canada and the actions required to protect aquatic ecosystems require financial expenditures. In the past, regulatory agencies have made broad interpretations and have been lenient in enforcing regulations, but current global economic pressures for environmental sustainability have strengthened industries' resolve to apply protection regulations. Nonetheless, stronger action by regulators to apply legislation and guidelines to fruition would be a positive step towards the protection of streams of eastern forests.

Acknowledgements

Many individuals have contributed to acquiring the information presented in this review. We are most grateful to the students and researchers who have worked in the Catamaran Brook and Copper Lake projects, especially P. Hardie and D. Caissie. A. Fraser assisted with the research for this paper. M. O'Connell and K. DeVito kindly reviewed the manuscript. Funding was provided by grants from the Meighen-Molson Foundations, and the Canada Research Chairs program to R. Cunjak. This is contribution # 69 of the Catamaran Brook Habitat Research Project.

References

Alexander, M.D. (2000) *Methodologies and preliminary results for monitoring effects of timber harvest on shallow groundwater temperature, seepage flux, and flow paths within a gravel stream bed: Catamaran Brook, New Brunswick.* BSc (Hon) thesis, St Francis Xavier University, Antigonish, Nova Scotia.

Anonymous (1986) *Policy for the Management of Fish Habitat.* Department of Fisheries and Oceans, Ottawa, Canada.

Bilby, R.E. & Likens, G.E. (1980) Importance of organic debris dams in the structure and function of stream ecosystems. *Ecology*, **61**, 1107–13.

Bosch, J.M. & Hewlett, J.D. (1982) A review of catchment experiments to determine the effects of vegetation changes on water yield and evapotranspiration. *Journal of Hydrology*, **55**, 3–23.

Canadian Forest Service (2002) *Report on the State of Canada's Forests.* Natural Resources Canada web site (http://www.nrcan.gc.ca/cfs-scf).

Chapman, D.W. (1986) Critical review of variables used to define effects of fines in redds of large salmonids. *Transactions of the American Fisheries Society*, **117**, 1–21.

Caissie, D., Pollock, T. & Cunjak, R.A. (1995) Variation in stream water chemistry and hydrograph separation in a small drainage basin. *Journal of Hydrology*, **178**, 137–57.

Caissie, D., Jolicoeur, S., Bouchard, M. & Poncet, E. (2002) Comparison of streamflow between pre- and post-timber harvesting in Catamaran Brook (Canada). *Journal of Hydrology*, **258**, 232–48.

Clarke, K.D. (1995) *Numerical, growth and secondary production responses of the benthic macroinvertebrate community to whole lake enrichment in insular Newfoundland.* MSc thesis, Memorial University of Newfoundland, St John's, NF.

Clarke, K.D. & Scruton, D.A. (1999) Brook trout production dynamics in the streams of a low fertility Newfoundland watershed. *Transactions of the American Fisheries Society*, **128**, 1222–9.

Clarke, K.D., Scruton, D.A., Cole, L.J., *et al.* (1997) The Copper Lake Buffer Zone Study: pre-harvest conditions of the aquatic habitat. *Canadian Technical Report of Fisheries and Aquatic Sciences*, **2181**.

Clarke, K.D., Scruton, D.A., McCarthy, J.H. & Curry, R.A. (1999) The Copper Lake Buffer Zone Study: A case study assessing the impacts of forest harvesting on salmonid populations and their habitats within a small headwater system. In: *Assessment and Impacts of Mega-projects. Proceedings of the 38th Anual Meeting of the Canadian Society of Environmental Biologists in collaboration with the Newfoundland and Labrador Environment Network*, (ed. P.M. Ryan), pp. 197–212. St John's, Newfoundland, October 1–3, 1998.

Cunjak, R.A. (1995) Addressing forestry impacts in the Catamaran Brook basin: an overview of the pre-logging phase. *Canadian Special Publication of Fisheries and Aquatic Sciences*, **123**, 191–210.

Cunjak, R.A. & Therrien, J. (1998) Inter-stage survival of wild juvenile Atlantic salmon, *Salmo salar* L. *Fisheries Management and Ecology*, **5**, 209–23.

Cunjak, R.A., Caissie, D., El-Jabi, N., *et al.* (1993) The Catamaran Brook (New Brunswick) habitat research project: biological, physical, and chemical conditions (1990–1992). *Canadian Technical Report of Fisheries and Aquatic Sciences*, **1914**.

Cunjak, R.A., Prowse, T.D. & Parrish, D.L. (1998) Atlantic salmon in winter: 'the season of parr discontent'? *Canadian Journal of Fisheries and Aquatic Sciences*, **55** (Suppl. 1), 161–80.

Cunjak, R.A., Guignion, D., Angus, R.B. & MacFarlane, R. (2002) Using incubation baskets to assess salmonid egg/alevin survival in relation to fine sediment deposition. In: Effects of land use practices on fish, shellfish, and their habitats on Prince Edward Island, (ed. D.K. Cairns). *Canadian Technical Report of Fisheries and Aquatic Sciences*, **2408**, 82–91.

Curry, R.A. & Devito, K.J. (1996) Hydrogeology of brook trout (*Salvelinus fontinalis*) spawning and incubation habitats: implications for forestry and land use development. *Canadian Journal of Forest Research*, **26**, 767–77.

Curry, R.A., Brady, C., Noakes, D.L.G. & Danzmann, R.G. (1997) Use of small streams by young brook trout spawned in a lake. *Transactions of the American Fisheries Society*, **126**, 77–83.

Curry, R.A., Scruton, D.A. & Clarke, K.D. (2002) The thermal regimes of brook trout, *Salvelinus fontinalis*, incubation habitats and evidence of changes during forestry operations. *Canadian Journal of Forestry Research*, **33**, 1200–7.

Dickison, R.B.B., Daugharty, D.A. & Randall, D.R. (1981) Some preliminary results of the hydrological effects of clearcutting a small watershed in central New Brunswick. In: *Proceedings of the 5th Canadian Hydrotechnical Conference*. Canadian Society of Civil Engineering, Fredericton, NB, May 1981.

Economic and Policy Analysis Directorate (EPAD) (1997) *1995 Survey of Recreational Fishing in Canada*. Economic and Commercial Analysis Report, 154.

Edwards, P. (2001) *An investigation of the potential effects of natural and anthropogenic disturbance on the density and distribution of slimy sculpin (Cottus cognatus) in Catamaran Brook, New Brunswick*. MSc thesis, University of New Brunswick, Fredericton, New Brunswick.

Elson, P. (1974) Impact of recent economic growth and industrial development on the ecology of Northwest Miramichi Atlantic salmon (*Salmo salar*). *Journal of the Fisheries Resource Board of Canada*, **31**, 521–44.

Fairchild, W.L., Swansburg, E.O., Arsenault, J.T. & Brown, S.B. (1999) Does an association between pesticide use and subsequent declines in catch of Atlantic salmon (*Salmo salar*) represent a case of endocrine disruption? *Environmental Health Perspectives*, **107**, 349–57.

Flebbe, P.A. & Dolloff, C.A. (1995) Trout use of woody debris and habitat in Appalachian wilderness streams of North Carolina. *North American Journal of Fisheries Management*, 15, 579–90.

Grant, J.W.A., Englert, J. & Bietz, B.F. (1986) Application of a method for assessing the impact of watershed practices: effects of logging on salmonid standing crops. *North American Journal of Fisheries Management*, 6, 24–31.

Hardie, P., Cunjak, R.A. & Komadina-Douthwright, S. (1998) Fish movement in Catamaran Brook, N.B. (1990–1996). *Canadian Data Report of Fisheries and Aquatic Sciences*, 1038.

Hartman, G.F. & Scrivener, J.C. (1990) Impacts of forestry practices on a coastal stream ecosystem, Carnation Creek, British Columbia. *Canadian Bulletin of Fisheries and Aquatic Sciences*, 223.

Hibbert, A.R. (1967) Forest treatment effects on water yield. In: *Symposium on Forest Hydrology*, (eds W.E. Sopper & H.W. Lull), pp. 527–43. Pergamon Press, New York.

Jewett, K., Daugharty, D., Krause, H.H. & Arp, P.A. (1995) Watershed responses to clearcutting: effects on soil solutions and stream water discharge in central New Brunswick. *Canadian Journal of Soil Science*, 75, 475–90.

Johnston, T.A. (1997) Downstream movement of young-of-the-year fishes in Catamaran Brook and the Little Southwest Miramichi River, New Brunswick. *Journal of Fish Biology*, 51, 1047–62.

Jones, A.R.M. & Bray, D.I. (1994) *The Catamaran Brook Groundwater Study: Summary of Field Activities and Results of Field Data Analysis for April 1993 to March 1994*. University of New Brunswick Groundwater Studies Group. Report to Forestry Canada – Maritimes Region.

Kennedy, D.P.S. (1981) *Geology of the Corner Brook Lake Area, Western Newfoundland*. MSc thesis, Memorial University of Newfoundland, St John's, Newfoundland.

Knoechel, R. & Campbell, C. (1988) Physical, chemical, watershed and plankton characteristics of lakes on the Avalon Peninsula, Newfoundland, Canada: a multivariate analysis of interrelationships. *Verhandlungen Internationale Vereinigung Limnologie*, 23, 282–96.

Komadina-Douthwright, S.M., Pollock, T., Caissie, D., Cunjak, R.A. & Hardie, P. (1999) Water quality of Catamaran Brook and the Little Southwest Miramichi River, N.B. (1990–1996). *Canadian Data Report of Fisheries and Aquatic Sciences*, 1051.

Kull, S.J. (1997) *The effects of forest harvesting activities on suspended sediment loading in two streams in the Catamaran Brook watershed, New Brunswick*. BScF thesis, University of New Brunswick, Fredericton, New Brunswick.

Likens, G.E., Bormann, F.H., Johnson, N.M., Fisher, D.W. & Pierce, R.S. (1970) The effects of forest cutting and herbicide treatment on nutrient budgets in the Hubbard Brook watershed ecosystem. *Ecological Monographs*, 40, 23–47.

Lloyd, G.A. (1999) *Examination of windfall in riparian buffers, Catamaran Brook, NB*. BScF thesis, University of New Brunswick, Fredericton, New Brunswick.

McCarthy, J.H. (1997) *Brook charr (Salvelinus fontinalis Mitchell) movement and habitat use in the Copper Lake Watershed, Corner Brook, Newfoundland, and a preliminary assessment of the impact of forest harvesting activity*. MSc thesis, Memorial University of Newfoundland, St John's, Newfoundland.

McCarthy, J.H., Scruton, D.A., Green, J.M. & Clarke, K.D. (1998) The effect of logging and road construction on brook trout movement and habitat use in the Copper Lake Watershed, Newfoundland, Canada. In: *Forest-Fish Conference: Land Management Practices Affecting Aquatic Ecosystems, Proceedings of Forest-Fish Conference, Calgary Alberta, May 1–4, 1996*, (eds M.K. Brewin & D.M. Montia), pp. 345–52. Information Report NOR-X-356. Natural Resources Canada, Canadian Forest Service, Northern Forest Center, Edmonton, Alberta.

MacNutt, W.S. (1963) *New Brunswick, A History: 1784–1867*. Macmillan (Canada), Toronto.

MacQuarrie, KT.B. & Maclean, A.R. (1993) *The Catamaran Brook Groundwater Study: Methodology and Initiation of Data Collection*. University of New Brunswick Groundwater Studies Group. Report to Forestry Canada – Maritimes Region.

Meehan, W.R. (1991) Influences of forestry and rangeland management on salmonid fishes and their habitats. USDA Forest Service. *American Fisheries Society Special Publication*, **19**.

Milton, G.R. & Towers, J. (1989) *An Initial Assessment of the Impacts of Forestry Practices on Riparian Zones*. Report No. 1. St Mary's River Forestry Wildlife Project, Nova Scotia Department of Natural Resources, Antigonish, NS.

Power, G., Brown, R.S. & Imhof, J.G. (1999) Groundwater and fish – insights from North America. *Hydrological Processes*, **13**, 401–22.

Ralph, S.C., Poole, G.C., Conquest, L.L. & Naiman, R.J. (1993) Stream channel morphology and woody debris in logged and unlogged basins of Western Washington. *Canadian Journal of Fisheries and Aquatic Sciences*, **51**, 37–51.

Randall, R.G., O'Connell, M.F. & Chadwick, E.M.P. (1989) Fish production in two large Atlantic coast rivers: Miramichi and Exploits. *Canadian Special Publication of Fisheries and Aquatic Sciences*, **106**, 292–308.

Scott, W. B. & Crossman, E.J. (1964) *Fishes Occurring in the Fresh Water of Insular Newfoundland*. Life Sciences Division, Royal Ontario Museum, Toronto, Ontario.

Scott, W. B. & Crossman, E.J. (1973) *Freshwater Fishes of Canada*. Fisheries Research Board of Canada, Ottawa. Bulletin 184.

Scruton, D.A. (1983) A survey of headwater lakes in insular Newfoundland, with special reference to acidification. *Canadian Technical Report of Fisheries and Aquatic Sciences*, **1195**.

Scruton, D.A. & Gibson, R.J. (1995) Quantitative electrofishing in Newfoundland and Labrador: result of workshops to review current methods and recommend standardization of techniques. *Canadian Technical Report of Fisheries and Aquatic Sciences*, **2308**.

Scruton, D.A., Clarke, K.D., McCarthy, J.H., *et al.* (1995) The Copper Lake buffer zone study: project site description and general study design. *Canadian Technical Report of Fisheries and Aquatic Sciences*, **2043**.

Scruton, D.A., Clarke, K.D. & Cole, L.J. (1998) Water temperature dynamics in small headwater streams in the boreal forest of Newfoundland, Canada: quantification of 'thermal' brook trout habitat to address effects of forest harvesting. In: *Forest-Fish Conference: Land Management Practices Affecting Aquatic Ecosystems, Proceedings of Forest-Fish Conference, May 1–4, 1996, Calgary Alberta*, (tech. coordinators M.K. Brewin & D.M.A. Montia), pp. 325–36. Information Report NOR-X-356. Natural Resources Canada, Canadian Forest Service, Northern Forest Center, Edmonton, Alberta.

Swift, L.W., Jr (1988) Forest access roads: design, maintenance, and soil loss. In: *Forest Hydrology and Ecology at Coweeta*, (eds W.T. Swank & D.A. Crossley, Jr), pp. 313–24. Springer-Verlag, New York.

Thomas, P. (2001) *Lost Land of Moses: The Age of Discovery on New Brunswick's Salmon Rivers*. Goose Lane Publications, Fredericton, New Brunswick.

Titus, B., Kingston, D.G.O., Pitt, C.M. & Mahendrappa, M.K. (2000) A lysimeter system for monitoring soil solution chemistry. *Canadian Journal of Soil Science*, **80**, 219–26.

van Kesteren, A.R. (1992) An application of ecosite mapping to assess land sensitivity to forest harvesting in the Corner Brook Lake Watershed, western Newfoundland. *Forestry Canada Information Report*, **N-X-280**.

Vannote, R.L., Minshall, G.W., Cummins, K.W., Sedell, J.R. & Cushing, C.E. (1980) The river continuum concept. *Canadian Journal of Fisheries and Aquatic Sciences*, **37**, 130–7.

Wright, E.C. (1944) *The Miramichi: A Study of the New Brunswick River, and of the People Who Settled Along It*. Tribune Press.

Chapter 21
Interactions between forests and fish in the Rocky Mountains of the USA

K.D. FAUSCH AND M.K. YOUNG

Introduction

The US Rocky Mountains are an ecosystem of substantial ecological and economic importance. Although the fish fauna is relatively depauperate compared with much of the USA, the region is the centre of evolution for cutthroat trout (Behnke 1992; see Table 21.1 for scientific names) and is recognized internationally for the quality of its recreational fishing. And, although other portions of the USA offer better conditions for tree growth, forestry in the Rocky Mountains has had a significant effect on aquatic ecosystems. We focus here on the linkages among forest ecology, forestry practices and habitat for aquatic biota, emphasizing the role of anthropogenic and natural disturbances on large woody debris (LWD) in forested streams because of its fundamental role in producing fish habitat. We also consider other processes including sediment delivery and allochthonous input of terrestrial invertebrates that affect habitat and fish abundance, and conclude by recommending research and management planning at landscape scales to sustain native fishes.

Ecology of forests in the Rocky Mountain region

The Rocky Mountains of the western USA extend from Montana and Idaho to northern New Mexico. We also consider the high mountains of Arizona and southern New Mexico because of their similarity in forest cover, geology and aquatic fauna, which is related to glaciation and large-scale climatic patterns that led to latitudinal and elevational advances and retreats of the flora and fauna (Betancourt *et al.* 1990).

This region defies simple characterization because of its often abrupt variation in geology, soil, topography and climate (Hauer *et al.* 1997). For example, the seasonality and form of precipitation in the Rocky Mountains varies regionally (Baker 1944), with the northwestern portion receiving most of its precipitation from late autumn to late winter, the central portion from late winter to early spring, and the southern portion from late summer to early autumn. Throughout the region, winter snowfall is derived from storms originating from the North Pacific Ocean, whereas summer monsoons moving north from the Pacific Ocean or Gulf of Mexico influence landscapes south of southeastern Wyoming. Superimposed on these seasonal climatic patterns are longer-term cycles caused by shifts in surface temperature and pressure of the Pacific Ocean, including the Southern Oscillation (Enfield 1989) and Pacific Interdecadal Climate

Table 21.1   Species of fishes mentioned in the text and their scientific names

Common name	Scientific name
Apache trout	*Oncorhynchus gilae apache*[1]
Arctic grayling	*Thymallus arcticus*
Bonneville cutthroat trout	*Oncorhynchus clarki utah*[2]
brook charr	*Salvelinus fontinalis*
brown trout	*Salmo trutta*
bull charr	*Salvelinus confluentus*[1]
Chinook salmon	*Oncorhynchus tschawytscha*
Colorado River cutthroat trout	*Oncorhynchus clarki pleuriticus*[2]
Columbia River redband trout	*Oncorhynchus mykiss gairdneri*
cutthroat trout	*Oncorhynchus clarki*
Gila trout	*Oncorhynchus gilae gilae*[3]
greenback cutthroat trout	*Oncorhynchus clarki stomias*[1,2]
kokanee	*Oncorhynchus nerka*
lake charr	*Salvelinus namaycush*
mountain whitefish	*Prosopium williamsoni*
rainbow trout and steelhead[4]	*Oncorhynchus mykiss*
Rio Grande cutthroat trout	*Oncorhynchus clarki virginalis*[2]
Snake River cutthroat trout	*Oncorhynchus clarki* subsp.[2]
westslope cutthroat trout	*Oncorhynchus clarki lewisi*[2]
yellowfin cutthroat trout	*Oncorhynchus clarki macdonaldi*[2,5]
Yellowstone cutthroat trout	*Oncorhynchus clarki bouvieri*[2]

[1] Listed as a threatened species under the US Endangered Species Act.
[2] One of eight subspecies of cutthroat trout native to watersheds of the US Rocky Mountains.
[3] Listed as an endangered species under the US Endangered Species Act.
[4] Steelhead refers to anadromous life history types of rainbow trout.
[5] Extinct subspecies.

Oscillation (Mantua *et al.* 1997). In the US Southwest (Andrade & Sellers 1988) and in Colorado (Kiladis & Diaz 1989), the large-scale sea-surface warming of the El Niño phase is associated with greater winter-spring moisture, and the reverse, the La Niña phase, with below-normal precipitation in those seasons.

There are many systems for classifying forested ecosystems in the Rocky Mountain region, based on climate, elevation, geology, topography and vegetation (e.g. Kuchler 1964; Bailey 1995). The coniferous forest types of primary ecological and economic interest can be divided into four classes along an increasing elevational and moisture gradient: pinyon-juniper woodlands, ponderosa pine (*Pinus ponderosa*), mixed conifer (typically dominated by lodgepole pine (*Pinus contorta*) north of central Colorado), and spruce-fir (Knight 1994; Dahms & Geils 1997). The cold winter temperatures and short growing season explain the dominance of conifers (Peet 1988), but regional climatic patterns produce additional variation. Although aspen (*Populus tremuloides*) may be found among any of these coniferous forest types, its abundance in portions

Table 21.2 Mean fire frequencies (in years) in major coniferous forest types by state in the US Rocky Mountains

Type	Fire frequency	Location
Pinyon pine-juniper	10–30	AZ, NM
Ponderosa pine	2–70	AZ, CO, ID, MT, NM, UT, WY
Mixed conifer	5–25	AZ, NM
	22–300	WY
	13–350	MT
Spruce-fir	150+	AZ, NM
	100–400	CO
	63–150+	MT
	300	WY

State abbreviations are: AZ, Arizona; CO, Colorado; ID, Idaho; MT, Montana; NM, New Mexico; UT, Utah; WY, Wyoming. Results are taken from Loope & Gruell (1973), Arno (1980, 2000), Peet (1981), Barrett *et al.* (1991), Dahms & Geils (1997), Veblen *et al.* (2000) and Baker & Ehle (2001).

of Wyoming and Utah and southward is attributable to greater summer precipitation there (Knight 1994).

Besides climate, regimes of natural disturbances such as fire, insects and wind shape forested ecosystems in the region, and the rates and outcomes of these disturbances vary by forest type. Fire is probably the most-studied disturbance in Rocky Mountain forests, but its intensity, return interval and spread are controlled by many variables (Pyne 1996). Historically, many fires in ponderosa pine forests (occasionally mixed with Douglas fir, *Pseudotsuga menziesii*) burned as cooler surface fires recurring every 5–25 years (Table 21.2) that frequently resulted in open, park-like stands primarily composed of older trees (Arno 1980; Covington & Moore 1994). In contrast, high elevation forests dominated by Engelmann spruce (*Picea engelmannii*), subalpine fir (*Abies lasiocarpa*) or lodgepole pine tended to have stand-replacing fires at intervals often exceeding 200 years (Peet 1981; Romme & Knight 1981). Surface fires probably caused few detectable changes in aquatic ecosystems, whereas stand-replacing fires may have had major effects. Gresswell (1999) summarized many of these, which include higher water temperatures, increased light and nutrient availability, greater propensity for debris torrents, higher delivery and transport of sediment, and increased mobility of large woody debris (LWD).

Insect- or pathogen-induced mortality of forests is also a frequent disturbance. For example, the spruce beetle (*Dendroctonus rufipennis*) may kill nearly all Engelmann spruce >20 cm diameter at breast height (dbh), but does not attack other species or spruce <10 cm dbh (Baker & Veblen 1990). Beetle kill of Engelmann spruce was widespread throughout the Rocky Mountains from the mid-to-late 1800s to the 1940s, affecting >10^5 ha (Veblen *et al.* 1994). Similar patterns are observed for many other insect–tree complexes at similar scales (Dahms & Geils 1997). In contrast to the effects of fires on streams, little is known about the consequences of pathogen-caused tree mortality, although an accelerated contribution of LWD would be expected (Bragg 2000).

In contrast to snags produced by fire and insects that may take years or decades to enter stream channels (Lyon 1984; Harrington 1996), wind-throw produces immediate and often large-scale changes in LWD loads. In August 1987, winds associated with a microburst in the Teton Wilderness Area in northwestern Wyoming toppled most mature trees on 6000 ha (Knight 1994), and in October 1997 storm-related winds blew down 8000 ha of subalpine forest in north-central Colorado (USDA Forest Service 1998). Smaller-scale wind-throw is also commonplace throughout the Rocky Mountains (M. K. Young, personal observation), but studies of the effects of wind-throw on Rocky Mountain aquatic ecosystems are lacking.

Combinations of these disturbances, which vary in return interval and extent, produce distinctive patterns on the landscape. For example, Veblen *et al.* (1994) identified stands regenerating after disturbances from avalanches, beetle kill and fires in a subalpine study site in central Colorado. They concluded that, since 1663, about 9% of the forested area had been affected by snow avalanches, 39% by spruce beetle outbreaks and 59% by fire. Avalanches were a nearly annual event (which many trees survived), whereas the mean return intervals were 117 years for beetle outbreaks and 202 years for fire. Furthermore, because of a burn in one part of the study area in 1874, spruce there had not yet grown enough to be vulnerable to a 1949 beetle outbreak that killed older spruce in adjacent stands. We hypothesize that similar interactions among disturbances in different forest types will also produce distinctive patterns in aquatic ecosystems, but this has yet to be investigated.

Finally, temporal changes in aquatic ecosystems deserve recognition because the relations between forest types and aquatic ecosystems change through time. Forest succession, the temporal transition of dominant vegetation types following disturbance, is a widely accepted paradigm (Attiwill 1994). This concept has also been applied to aquatic ecosystems (Detenbeck *et al.* 1992), and successional links between riparian forests and aquatic ecosystems have been posited (Trotter 1990; Benda *et al.* 1998). For example, Minshall *et al.* (1989) forecasted trends in stream temperature, macroinvertebrate abundance and community composition, suspended sediment composition and transport, and LWD delivery and abundance in the 300 years following stand-replacing fire in coniferous forests in Yellowstone National Park. The theory of stream succession and its spatial analogue, the river continuum concept (Vannote *et al.* 1980), should lead to predictions of reach-specific patterns in fish habitat (Benda *et al.* 1998) and possibly fish communities or particular fish life history strategies (Schlosser 1995; Rieman & Dunham 2000). Nevertheless, other disturbances, whether natural or anthropogenic, may alter these successional pathways and lead to far greater complexity. In particular, timber harvest in riparian zones is likely to alter LWD dynamics, decoupling this process from other stream–forest successional trends.

Aquatic habitats

Aquatic habitats in the US Rocky Mountains have developed within landscapes shaped primarily by orogeny that built the mountains, but were also strongly influenced by Pleistocene glaciation during the last 2 million years (Wohl 2001). Glaciers that descended from the mountains carved high elevation lake basins in cirques, and left U-shaped valleys with broad floors of thick sediments bounded by lateral and downstream moraines. The channels now traversing these valleys are typically low gradient, meandering streams primarily composed of lateral scour pools. Downstream from the terminal moraines, rivers descend steep bedrock canyons created by shear zones along fault lines, and have a step-pool morphology influenced by boulders or large woody debris. Although glacial meltwaters transported boulders and coarse sediments in these steep mountain channels, now only the largest floods are competent to shape channel morphology. As a result, the general morphology of Rocky Mountain rivers and streams that forms much of the physical template for fish habitat was set by the end of the Pleistocene, approximately 10,000 years ago.

Although there are thousands of high elevation lakes in the Rocky Mountains, few of them supported native populations of fish or large invertebrates (Hauer *et al.* 1997). Most are small, ultraoligotrophic, and upstream from waterfalls or high gradient barriers that prevented colonization. As a result, fish gained access to only a few of the larger lakes at lower elevations, which supported relatively diverse fish assemblages (e.g. Flathead Lake, MT; Yellowstone Lake, WY; Twin Lakes, CO; Bear Lake, UT). Because a number of the native species in these lakes were adfluvial salmonids that spawned and reared in tributaries, we focus here on fluvial habitats which provide most of the living space for fishes in the forested zones of the region.

Within the basic geomorphology of mountain valleys set by tectonics and glaciation, the flow regime is a major feature that has shaped channel morphology at the reach scale, and thus physical habitat for fish in streams. In most of the region, flow regimes are primarily driven by melting snow originally deposited by winter and early spring storms, although the central and southern Rockies also receive thunderstorms during the summer 'monsoon' season (Hauer *et al.* 1997). In rivers of the Front Range along the eastern slope of the Rocky Mountains in Colorado, such summer storms have produced flash floods with peaks six times greater than floods from snowmelt. These floods are estimated to have 100–500-year return intervals, and are capable of reworking the coarse sediments left after Pleistocene glaciation (Wohl 2001). In contrast to these catastrophic flow events, the annual flood from snowmelt runoff is highly predictable, producing peaks in May and June that are 10–100 times greater than base flow and discharging ca. 70% of the annual water yield from mountain basins during the 2–3-month early summer runoff period (Stanford & Hauer 1992). Flows then decline to base flow through autumn and winter until snowmelt the next summer. Regional analysis of flow regimes showed that nearly all streams in the US Rocky Mountains exhibit this pattern (Poff 1996).

Water temperature, another major feature of the habitat template that influences aquatic biota, generally decreases with elevation in streams at a given latitude (Hauer *et al.* 1997). Yet within this framework, water temperature may vary widely among reaches at the same elevation because of differences in basin orientation, basin morphology, riparian vegetation, lakes and groundwater–surface water interactions in alluvial valleys (Fausch 1989; Baxter & Hauer 2000). Nevertheless, variation in temperature at many different scales may be critically important because it influences life history events of all aquatic biota.

Nearly all Rocky Mountain watersheds have undergone a similar chronology of human disturbances that altered water and sediment regimes. For example, Wohl (2001) described the progressive changes in rivers of the Front Range of Colorado, starting with early beaver trapping that resulted in the loss of thousands of small beaver dams. This was followed by railroad tie drives, mining, grazing and water diversions that progressively scoured channels of most woody debris and boulders, reworked most bed sediment, destabilized channel banks, and drastically altered flow regimes. The end results, she argues, are 'virtual rivers' that may look pristine but are actually greatly altered. Although all reaches are susceptible to flow diversions which cause dry channels or fluctuating flows, the low gradient reaches in glaciated valleys are most sensitive to increased sediment supply produced by many human land use disturbances.

The fish fauna

The aquatic vertebrates of the forested zone in the Rocky Mountain region comprise relatively depauperate assemblages of native fishes and amphibians. Native fishes are largely represented by salmonids (salmon, trout, whitefish and grayling), catostomids (suckers), cottids (sculpins) and a few coldwater cyprinids (minnows). Both early summer-spawning trout of the genus *Oncorhynchus* (rainbow, redband, cutthroat, Gila and Apache trout; see Table 21.1 for scientific names) and autumn-spawning bull charr are prevalent, although endemic subspecies of cutthroat trout are most widespread. Arctic grayling occur in a few northern basins, and mountain whitefish extend southward to streams of the central Rockies. Various sculpins (*Cottus*) that radiated from the Mississippi and Columbia river basins extend as far south as Colorado, and several species of suckers (*Catostomus, Pantosteus*) from the Colorado and Mississippi river basins typically occupy lower elevation rivers. A few cyprinids adapted to cold or cool water occupy the transition zones between mountains and plains on both sides of the Continental Divide. On the eastern side of the Front Range of Colorado, many of these are glacial relicts from cooler, wetter periods, with centres of their distribution far to the north and east in North America (Fausch & Bestgen 1997). The richest fish assemblages in the region occupy these transition zones, which are often bordered by cottonwoods (*Populus* spp.). However, these fishes may also have been drastically affected by tie drives from the mountains to cities downstream, most of which are located in this transition zone. A variety of native amphibians were primarily distributed in subalpine lakes upstream from fish barriers, but were drastically reduced by stocking fish in these habitats (Tyler *et al.* 1998).

The widespread degradation of aquatic habitats by early settlers due to mining, tie drives, grazing and water diversions, coupled with over-harvest, caused depletion or loss of most native fish populations within a few decades and led to calls for stocking non-native fishes (Wiltzius 1985). After sources of native cutthroat trout for culture were exhausted or reduced from poor land management or excessive harvest (Young *et al.* 2002), eggs of brook charr, rainbow trout and brown trout were imported during 1860–1880 when the new railroads reached the region and the fish propagated were distributed throughout mountain basins. For example, Bahls (1992) reported that 95% of the 16,000 lakes in the 11 western US states (including those in the Rocky Mountains) were estimated to originally have been fishless, but 60% have fish now and 45% continue to be stocked. Yellowstone cutthroat trout from Yellowstone Lake were also cultured and widely stocked into basins where other cutthroat trout subspecies were endemic, causing introgressive hybridization (Behnke 1992). Some of these salmonid introductions produced self-sustaining populations (i.e. invasions). Invasion success may often be controlled by the match between timing of spawning and fry emergence and flow and temperature regimes. For example, rainbow trout that spawn in spring and emerge in early summer are only moderately successful invaders in Colorado rivers because their fry emerge during snowmelt runoff and are susceptible to displacement by floods (Nehring & Anderson 1993; Fausch *et al.* 2001). In contrast, autumn-spawning brook charr and brown trout are highly successful invaders, probably because their young emerge before runoff and are only depleted in years of high flow (Latterell *et al.* 1998). Cold water temperatures at high elevation may also be insufficient to allow spawning and first year growth of spring-spawning salmonids (Harig & Fausch 2002). Lake charr and kokanee have been widely stocked in Rocky Mountain lakes and reservoirs, creating important fisheries. However, the lake charr are highly piscivorous and long-lived, creating problems for managers (Johnson & Martinez 2000). Those illegally introduced into Yellowstone Lake have the potential to decimate the largest Yellowstone cutthroat trout population in the world (Ruzycki *et al.* 2003).

The upshot of the history of habitat degradation and fish introductions throughout the region since the mid-1800s is that most native taxa are at considerable risk of extinction. For example, the eight subspecies of cutthroat trout native to the region (Table 21.1) have been reduced to <5% of their historical range by habitat loss and biotic interactions with non-native salmonids. One is extinct, one is listed as threatened under the Endangered Species Act (Harig *et al.* 2000; Young & Harig 2001), and conservation plans are completed or pending for most others. Forest practices have contributed to these cumulative effects through short- and long-term changes in channel morphology, loss of habitat complexity, water diversion and sedimentation of streambeds.

Forestry practices in the Rocky Mountain region

The Rocky Mountains have a long history of timber removal. In the Front Range of Colorado, significant logging began with the 1859 gold rush (Wohl 2001), and by the late 1860s the railroad had arrived in many locations throughout the Rocky Mountains. Both industries generated a huge demand for wood to make sluices, flumes, stamp mills, mine timbers, houses, charcoal, firewood and vast numbers of railroad ties. Because few roads existed, wood products were either removed by railroad (Schubert 1974) or transported in stream channels. The driving of railroad ties down streams began in the late 1860s, and the numbers transported indicate the intensity of historical logging. For example, 85,000–350,000 railroad ties and mine timbers were floated down the Laramie River in southeastern Wyoming annually from 1876 to 1938. During spring 1902 alone, 500,000 railroad ties were floated down another stream in southern Wyoming (Rosenburg 1984). Although such harvest was concentrated near Euroamerican settlements, its magnitude should not be underestimated. A traveller through the Front Range of Colorado in 1872 wrote 'The axe has swept through the mountains and left them a wilderness of stumps' (Wohl 2001). Human-caused fires also increased where resource extraction was active (Veblen *et al.* 2000).

The public outrage over unregulated harvest, widespread burning and degraded water quality led to establishing what would become the national forest system (Wilkinson 1992). Although harvest probably proceeded apace on most non-federal lands, legislation in 1891 and 1897 established a system of federal forest reserves and national forests on over 8×10^7 ha. Relatively little harvest occurred on these lands until about 1950, when a booming post-war economy, depletion of private timberlands, and a rapidly expanding road system led to vastly increased harvest. There was little awareness, however, of the ecosystem consequences of such deforestation (Meehan 1991; Wilkinson 1992). Recently, forest harvest on federal lands has declined nationwide, including in many portions of the Rocky Mountains (Fig. 21.1). Forces responsible for this decline are unclear because the issue is highly politicized, but it is partly attributable to increased environmental regulation and more active public involvement in land management decisions, coupled with reduced economic incentives for harvest. The highest value timber has been logged and less expensive timber is available from Canada and the US Pacific Northwest and Southeast.

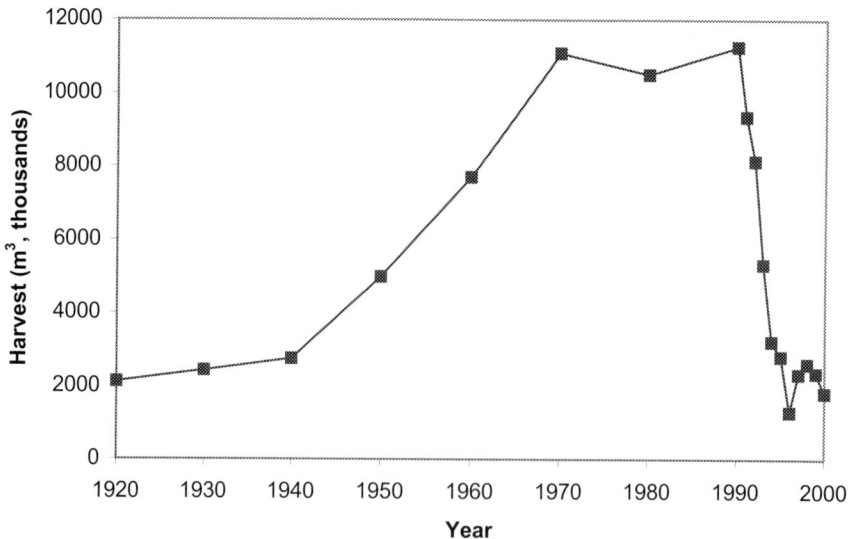

Fig. 21.1 Timber harvest (in thousands of m³) from national forests in Arizona and New Mexico. Data for 1920–1990 are decadal averages; data for 1991–2000 are annual totals (Dahms & Geils 1997; USDA Forest Service, Washington, DC, unpublished data).

Regulation of forestry on national forest lands in the Rocky Mountains differs among regions. For example, the Inland Native Fish Strategy (USDA Forest Service 1995) established forest management guidelines only for those portions of national forests in the Columbia River basin. It was developed in conjunction with national plans addressing effects of forest management on the northern spotted owl (*Strix occidentalis*) and several anadromous fishes, all listed under the US Endangered Species Act. The strategy established quantitative objectives for desired characteristics of stream habitat for fish (e.g. pool frequency, LWD abundance), and riparian habitat conservation areas of different sizes were defined on the basis of permanence of flow and presence of fish. Timber harvest is prohibited in these conservation areas (with a few exceptions), roads contributing excess sediment are to be redesigned or closed, and stream crossings are to accommodate 100-year floods and permit fish passage. In contrast, national forests in Colorado and most of Wyoming are guided by 'watershed conservation practices' (*Forest Service Handbook* 2509.25–99–1), which make more general recommendations for forestry practices, lack quantitative objectives related to stream habitat quality, and designate smaller riparian habitat conservation areas. More importantly, the guidelines do not prohibit riparian logging. These regional discrepancies in forestry practices do not seem to reflect biological, hydrological or geomorphological differences, but instead perhaps different states in the evolution of knowledge and regulations, as well as litigation and the influence of the Endangered Species Act.

Fire suppression is one of the most prevalent forestry practices that influences streams, although it is often overlooked. Suppression of forest fires in the Rocky Mountains began in the early 1900s (Dahms & Geils 1997; Veblen *et al.* 2000), but was probably effective only in low elevation ponderosa pine forests accessible by roads. Increased fire monitoring, a massive road network (currently >370,000 km on national forest lands in this region; Coghlan & Sowa 1998), technological advances, and a large force of professional fire fighters have continued to reduce fires in this forest type. Ironically, the resulting forests are now regarded as overly dense and far more susceptible to stand-replacing fire than they were historically (Covington & Moore 1994), and widespread thinning and prescribed fire have been recommended (Covington *et al.* 1997). Because summer floods following stand-replacing fires in New Mexico ponderosa pine forests produced debris torrents and suspended sediment concentrations that eliminated several populations of federally endangered Gila trout, Brown *et al.* (2001) argued that reductions in the frequency of such intense fires through forest management were the only strategy likely to increase persistence of this species. Similarly, fire suppression may also be altering the structure of aquatic ecosystems by permitting the infiltration of conifers into aspen stands throughout the western US (Bartos 2001), thus affecting a source of food and dam materials for beaver (*Castor canadensis*) which shape aquatic habitats in many Rocky Mountain ecosystems.

Although there is considerable agreement that fire suppression has substantially reduced fire frequency in ponderosa pine ecosystems (Arno *et al.* 1995; Dahms & Geils 1997), fire frequencies in higher elevation forest types apparently have changed little (Romme & Knight 1981; Veblen *et al.* 1994). Despite this, Baker & Ehle (2001) urged caution in interpreting fire frequencies in ponderosa pine stands. They contended that because of sampling biases or inadequacies in many studies, the mean fire interval in these stands may be 22–308 years, instead of the 2–70 years typically reported (Table 21.2), and that densities of historical stands probably were more heterogeneous than commonly presumed. In addition, Veblen *et al.* (2000) concluded that both surface fires and stand-replacing fires were common in mid-elevation ponderosa pine stands in Colorado's Front Range, and that many current stands are a legacy of higher fire frequencies caused by Euroamerican settlers and by a climate in the late nineteenth century that favoured more frequent fire. Consequently, calls for increased harvest and prescribed fire (including in riparian zones) to maintain 'natural' ecosystems must be examined in the context of historical alteration of forest stands through harvest and fire suppression.

Large woody debris recruitment and function in stream channels

Trees killed by fire or other disturbances contribute LWD to terrestrial and aquatic ecosystems when they fall. LWD affects the physical template of habitat for fish by shaping channel morphology (Beechie & Sibley 1997), influencing the abundance and distribution of pools (Fausch & Northcote 1992; Bilby & Bisson 1998), storing organic matter and inorganic sediment (Adenlof & Wohl 1994), providing log steps that dissipate energy (Heede 1985; Wohl *et al.* 1997), and creating habitat complexity at various scales that provides critical summer and winter habitat for fishes and the invertebrates on which they feed (Murphy *et al.* 1986). However, most research on LWD has been conducted in streams of the Pacific Northwest and Alaska (cf. Harmon *et al.* 1986; Gurnell *et al.* 1995), which pre-dated work in the Rocky Mountains by at least 15 years (e.g. Swanson *et al.* 1976; Potts & Anderson 1990; Young *et al.* 1990). To date, much less is known about the characteristics and function of LWD in Rocky Mountain streams than those of the Pacific coast.

The recruitment of LWD from riparian zones, and its interaction with stream chan-nels, can be viewed across at least three different scales: individual pieces, the channel units they occupy or create, and the stream reaches they influence. Functions of LWD over longer segments and whole basins have only recently been studied, and then only in the Pacific Northwest (Nakamura *et al.* 2000; Martin & Benda 2001). Minimum dimensions of LWD pieces in most studies are typically 10 cm diameter and 1 m long. In the Rocky Mountains, density of LWD (pieces/100 m of stream) in relatively undis-turbed streams is similar to the Pacific Northwest, but LWD volume is often 2–10 times less (Bragg *et al.* 2000), largely due to slower tree growth in the Rocky Mountains. For example, average abundance in 11 streams draining old-growth forests in northern Colorado (43 pieces/100 m; Richmond & Fausch 1995) was similar to densities in five undisturbed coastal streams of southeast Alaska (33 pieces/100 m; Robison & Beschta 1990) and a relatively undisturbed stream in British Columbia (42 pieces/100 m; Fausch & Northcote 1992). In contrast, median LWD diameter and length in the Colorado streams (19 cm, 3.3 m, respectively) were much smaller than those in the Alaska streams (53 cm, 7.4 m) and the British Columbia stream (26 cm, 7.4 m). As a result, total LWD volume per 100 m of stream averaged only 13 m^3 in the Colorado streams versus 58 m³ in the Alaska streams and 43 m³ in the British Columbia stream (Richmond & Fausch 1995). However, LWD in the 11 Colorado streams had a similar size distribution to trees in undisturbed Englemann spruce-subalpine fir forests in southeastern Wyoming (Young *et al.* 1990), where few trees exceeded 50 cm diameter. Although the minimum LWD age based on dendrochronology of saplings growing on it was estimated to be 100–300 years in Pacific Northwest streams (Murphy & Koski 1989; Martin & Benda 2001), the drier Rocky Mountain climate rarely produces favourable conditions for seed germination on LWD, so little of it can be dated using this method.

At the channel unit scale, relatively few of the LWD pieces that fall into stream channels form pools at any given time (Berg *et al.* 1998), nevertheless most pools in many channels may be formed by LWD (Murphy *et al.* 1986; Ralph *et al.* 1994). For example, in the 11 Colorado streams draining old-growth forests, only about 10% of LWD pieces (*n* = 1412) functioned to form pools, but 76% of pools present (*n* = 110) were formed by LWD (Richmond & Fausch 1995). The majority of pool-forming LWD pieces spanned the channel (53%) and were oriented perpendicular to stream flow (57%), forming a log step that created a dammed pool upstream and a plunge pool downstream. In contrast, only 23% of all other pieces that did not form pools were perpendicular to flow. Although the diameter of pool-forming pieces was similar to all other LWD (22 vs 19 cm), median length (4.4 vs 2.6 m) and total volume (0.19 vs 0.08 m³) were about twice as large for pieces that formed pools. Young *et al.* (1990) also found that LWD formed about 70% of pools in an undisturbed Wyoming stream.

At the scale of different reaches among streams, many features of LWD change with channel or watershed size because larger streams capture and transport more debris, and scour deeper pools around debris. For example, in old-growth forests not subject to catastrophic disturbances, mortality of individual trees may supply most LWD to small streams (Bragg 2000). In contrast, in larger streams with channels wider than 10 m recruitment of LWD from bank erosion may exceed this mortality, such as Martin & Benda (2001) reported for a southeastern Alaska watershed. Richmond & Fausch (1995) also found that fewer LWD pieces spanned the channel or lay perpendicular to flow in the Colorado streams that were wider than 5.0 m at bankfull (range 3.7–10.2 m for all streams). Instead, more pieces were moved into clumps or jams as channel width exceeded median LWD length, a pattern Hauer *et al.* (1999) also reported for 20 reaches of 8 northwestern Montana streams. As a result, a lower percentage of LWD pieces functioned to form pools in the Colorado streams with larger watershed area. Nevertheless, streams with more LWD pieces/100 m had more pools overall, because influential pieces spanned the channel and formed pools (smaller streams) or captured debris around which pools scoured (larger streams). Pool area and maximum depth at baseflow (i.e. residual depth, Lisle 1987) were not related to diameter, length or volume of LWD pieces forming them, but instead were correlated with stream size (bankfull width and drainage area), indicating that stream depth and gradient that control stream power are more important in determining the depth to which pools will scour (Wohl *et al.* 1993).

Empirical data and models show that watershed disturbances can have drastic effects on LWD recruitment and retention, and thereby exert strong controls on channel morphology and fish habitat in Rocky Mountain streams. Clear-cutting riparian forests for railroad ties and subsequent tie drives during 1868–1940 in streams of the Colorado Front Range and southeastern Wyoming simplified channels by removing most LWD and boulders (Wohl 2001), and has prevented recruitment of new LWD until forests regenerate and are disturbed or senesce. When measured in the 1990s, streams subjected to tie drives had about 2–20 times less LWD than adjacent undisturbed streams, resulting in more riffles and fewer pools (Young *et al*. 1994). Models of LWD recruitment for streams in Alaska (Murphy & Koski 1989) and Wyoming (Bragg 2000; Bragg *et al*. 2000) showed that even without tie drives, clear-cutting of riparian forests resulted in minimum LWD loads in streams 80–100 years later (approaching zero in the Wyoming stream) after residual debris from before harvest eroded and washed away, and LWD remained low for 50 more years before recovery began. In contrast, in the model based on a Wyoming stream, trees killed by moderately intense fire or insect outbreaks created large inputs of LWD that peaked 30 years after these natural disturbances (Fig. 21.2). Moreover, standing dead trees after fire or tree species not killed by beetles provided a low but steady supply of LWD as forests regenerated, so that stream loads never reached levels as low as those caused by clear-cutting. However, both models indicate that at least 200–250 years are required for LWD loads to reach steady state after anthropogenic or natural disturbances.

Fire may also destabilize channels in highly erodible watersheds, causing more debris to be transported out of stream reaches than is recruited into them. Young (1994) reported that a stream traversing a watershed adjacent to Yellowstone National Park

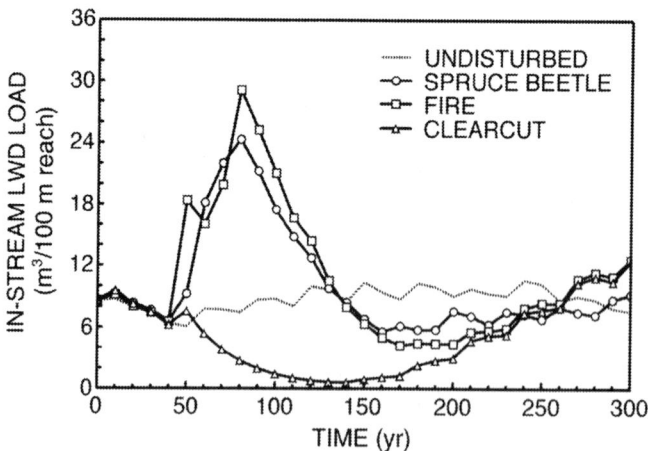

Fig. 21.2 Model predictions of large woody debris loads in a southwestern Wyoming stream based on four scenarios of disturbance (reproduced from Bragg 2000, with permission). Loads were simulated for (1) an undisturbed forest where only individual tree mortality caused input, (2) an outbreak of spruce beetle that killed primarily Engelmann spruce, (3) a fire of moderate intensity, and (4) clear-cut that removed most trees to the stream bank. Disturbances were initiated in year 50, and scenarios progressed for the next 250 years without further catastrophes. Scenarios were simulated using the Forest Vegetation Simulator model coupled to a Coarse Woody Debris model, provided with data from the riparian forest along a 250-m stream reach at 2565 m elevation.

that burned in 1988 had about half the LWD density 2 years afterwards compared with a stream in a similar adjacent watershed that did not burn (15 vs 32 pieces/100 m). Three times as many LWD pieces were transported during a 1-year interval in the burned stream (58% vs 18%), and more than twice as many LWD pieces were buried or displaced and could not be relocated compared with the unburned stream (66% vs 28%). Moreover, sediment mobilized by a summer thunderstorm suffocated fish in the burned stream (Bozek & Young 1994). Minshall *et al.* (1997) also reported substantial loss of LWD pieces (defined as >2 cm diameter, >40 cm long) during a high flow event in 1991 from reaches of 17 streams in Yellowstone National Park that were burned in 1988, compared to 3 unburned reference streams.

Response of fishes to forests and forestry

Habitat for fishes in Rocky Mountain streams may be enhanced by natural or anthropogenic inputs of large woody debris, or degraded by forest practices such as clearcutting and building logging roads. The effects of these elements and processes have been studied at several spatial scales using either observational or experimental approaches, the former providing the most realism and the latter establishing causation. However, we know of no large-scale studies in the Rocky Mountain region comparing fish in streams draining logged versus unlogged watersheds or segments, like the studies carried out in the Pacific Northwest (e.g. Murphy & Hall 1981; Hall *et al.* 1987; Hartman *et al.* 1996).

Salmonids in Rocky Mountain streams have often been observed to hold positions near LWD in streams during winter and summer. Several authors reported that during winter, interior subspecies of cutthroat trout (e.g. westslope cutthroat trout, Colorado River cutthroat trout), Columbia River redband trout, bull charr and brown trout in streams of the northern US and southern Canada Rocky Mountains held concealed positions beneath LWD and boulders in deep pools during the day (Smith & Griffith 1994; Brown & Mackay 1995; Young 1998; Jakober *et al.* 1998, 2000; Muhlfeld *et al.* 2001). In contrast, these fish emerged at night to use positions in shallow water with little cover, a strategy that presumably avoids daytime predation from endothermic vertebrate predators (Jakober *et al.* 2000). Similarly, Young (1996) reported that during summer Colorado River cutthroat trout in a Wyoming stream used deep pools, especially those formed by LWD. These studies, conducted using either underwater observations or radio telemetry, suggest that LWD and the pools it creates are critical habitat features required by stream salmonids.

The positive effects on trout abundance of adding LWD that creates pools has also been established experimentally. An 8-year field experiment in six Colorado mountain streams showed that abundance of adult wild brook charr, brown trout and rainbow trout more than doubled (111% increase, on average) in 250-m 'treatment' sections where 10 perpendicular logs were placed to create pools (Riley & Fausch 1995; Gowan & Fausch 1996a). Abundance in treatment sections and adjacent 250-m control sections without logs was similar before log placement (4% higher in treatment sections), whereas abundance was 44% higher, on average, in treatment sections than controls at the end of the study (abundance also increased in control sections during the study), a statistically significant increase overall ($p = 0.02$). Schmetterling & Pierce (1999) reported that all but 1 of 23 (96%) of such log structures installed in a 9–10-m wide stream of similar gradient and laterally confined channel type (2–3.5%, B channel, after Rosgen 1996) withstood a flood of 50-year recurrence in a western Montana stream, suggesting that such LWD additions can provide long-term benefits to fish populations.

Although the mechanism driving increased trout abundance in the Colorado experiment was expected to be largely higher reproduction and survival within these sections, neither of these hypotheses was supported by the data in most cases (Gowan & Fausch 1996a). Further research showed that immigration of fishes from relatively long distances accounted for most of the increases (Gowan *et al.* 1994; Gowan & Fausch 1996b). High fish movement rates from up to 2 km away or more in these small streams (3–6 m wide) were capable of causing rapid colonization of the new pools within 2–3 months after installing the logs, and some fish marked in the study sections also emigrated long distances. Overall, these data indicate that natural or anthropogenic changes in habitat in forested watersheds can influence fish populations and communities long distances upstream and downstream (Fausch *et al.* 2002).

Sediment and human disturbances brought by roads are among the more important effects of forestry on streams and fish habitat in the Rocky Mountains, as in other regions (see Trombulak & Frissell 2000 for review). For example, Platts *et al.* (1989) reported that slope failures caused primarily by poorly built logging roads in the South Fork Salmon River, Idaho, watershed during 1945–1965 delivered about 2×10^6 m^3 of sediment to the river channel. Along with downstream hydroelectric development this caused about a 90% decrease in abundance of spawning chinook salmon and a 75% decrease in spawning steelhead through 1985, even after a logging moratorium in 1965 and 20 years of watershed rehabilitation. Eaglin & Hubert (1993) reported that the amount of fine substrate in reaches of 28 southern Wyoming streams was positively correlated with the number of road crossings and proportion of the watershed logged, and that trout abundance was also negatively related to the number of road crossings. Baxter *et al.* (1999) found that the number of spawning redds (nests) produced by adult bull charr declined with increased road density in tributaries of the Swan River in northwestern Montana, and that the relationship grew stronger with time, suggesting a lag effect on bull charr populations. In addition to sediment, these authors proposed that other anthropogenic disturbances brought by roads such as angling or poaching, and non-native species introductions, may combine to explain the decline in these large adfluvial spawners that ascend the tributaries.

Past forestry practices have left legacies in streams that continue to have large effects on stream habitat and fish populations today (Wohl 2001). For example, the tie drives in Colorado and Wyoming mountain streams left channels largely devoid of pools for the last 80–130 years, and only now are forests beginning to produce LWD that creates fish habitat (Bragg 2000). The results of studies showing that LWD loads are 2.6 times higher in old-growth streams than disturbed streams (Richmond & Fausch 1995), and that only 10 well-placed pieces of LWD in 250-m reaches have the potential to more than double trout abundance (Gowan & Fausch 1996a), suggest that restoring natural

LWD loads to streams could greatly increase average fish abundance. Rehabilitation of watersheds to reduce sediment supply that decreases fish recruitment and over-winter survival (Platts *et al.* 1989), and closing logging roads that deliver sediment to channels and bring human disturbances to remote areas, could further increase fish abundance toward undisturbed conditions.

New information from other biomes suggests that a patchy mosaic of different vegetation types may provide optimum habitat for fishes in Rocky Mountain forested streams. Wipfli (1997) reported that salmonids in Alaska streams ingested a higher proportion of fallen terrestrial insects relative to drifting aquatic insects in reaches bordered by young-growth deciduous trees (red alder, *Alnus rubra*) than those with old-growth conifers, suggesting that deciduous vegetation may provide more allochthonous prey. Nakano & Murakami (2001) and Kawaguchi & Nakano (2001) also found that about 50% of the annual energy budget of salmonids in a stream in northern Japan bordered by deciduous forest was supplied by terrestrial insects that fell into the stream. Moreover, Kawaguchi & Nakano (2001) reported that input of terrestrial insects, and fish biomass, was about twice as high in forested reaches compared with reaches traversing grassy meadows. Experimentally excluding input of terrestrial insects from forested reaches with a greenhouse reduced fish biomass to half that in control reaches, confirming that this prey subsidy caused the differences in fish biomass (Kawaguchi *et al.* 2003). However, other investigators have reported that wide shallow meadow reaches with little riparian vegetation may supply important substrate for spawning trout (Knapp *et al.* 1998). Therefore, a patchy mosaic of stream reaches, including some bordered by coniferous forests contributing LWD that creates deep stable pools, others by deciduous trees and shrubs that supply terrestrial invertebrates, and others in grassy meadows that provide spawning habitat and autochthonous production, may provide the mixture of habitat required to sustain spatially and temporally dynamic fish populations (Fausch *et al.* 2002). Moreover, Nakano & Murakami (2001) also showed that productive stream habitat supplies emerging aquatic invertebrates that can make up about 25% of the annual energy in the diet of riparian birds, thereby maintaining terrestrial fauna as well (Power 2001). Understanding the mixture of riparian forests and stream habitat needed to sustain animals in both terrestrial and aquatic environments across this forest–stream ecotone is a major research and management challenge in the Rocky Mountain region and elsewhere.

Recommendations for improved forestry practices

Aquatic ecosystems are the spatial and temporal integrators of landscape pattern and process. A heritage of over 150 years of Euroamerican use of forested ecosystems in the Rocky Mountains has rendered many rivers and streams geomorphologically and biologically dysfunctional (Wohl 2001). Unfortunately, additional timber harvest that removes riparian LWD or increases sediment delivery will reinforce this tendency. The prescriptions of the Inland Native Fish Strategy (USDA Forest Service 1995) are intended to retain buffer zones adequate to provide LWD and filter sediment from uphill sources, when the guidelines are applied. Yet salvage logging in the wake of fire may form a substantial component of future harvest and constitutes one of the circumstances under which riparian logging might proceed (such as was done in western Montana in 1996; USDA Forest Service 1996), but virtually no studies have considered the effects of logging on post-fire aquatic ecosystems. Hence, monitoring the application and effectiveness of these guidelines, and evaluating their applicability to Rocky Mountain watersheds, is a critical need.

Basin-wide planning to sustain mosaics of habitat patches in different successional states will be essential to the long-term persistence of many elements of the aquatic biota (Lee *et al.* 1997). Seasonal movements of individuals among these patches are often required to place specific life history stages in the unique but spatially separated habitats that are needed to complete their life history (Schlosser 1995; Rieman & Dunham 2000; Fausch *et al.* 2002). However, this requires a connectivity among habitats that is lacking in most Rocky Mountain watersheds. For example, most populations of native cutthroat trout east of the Continental Divide are isolated from conspecifics due to degradation of downstream reaches (Young & Harig 2001). Habitat loss from fire-related events, toxic spills, stream diversion or drought are likely to lead to further loss of these indigenous populations. Compounding this problem is the fact that invasions by non-native fishes, particularly brook charr, brown trout and rainbow trout, often cause similar fragmentation (Harig *et al.* 2000). Therefore improved forestry practices must be devised that enhance habitats for native fishes over time rather than reducing them, and that avoid conduits like roads that promote introduction of non-native species (Baxter *et al.* 1999), if forest managers are to prevent worsening the tenuous existence of many of these populations.

Acknowledgements

We thank E. Wohl, C. Baxter and R. Hauer for reviews that improved our manuscript. Research reported here by KDF and colleagues has been supported by the Colorado Division of Wildlife, US Forest Service Fish Habitat Relationships Program, Trout Unlimited, and the National Science Foundation.

References

Adenlof, K.A. & Wohl, E.E. (1994) Controls on bedload movement in a subalpine stream of the Colorado Rocky Mountains, U.S.A. *Arctic and Alpine Research*, **26**, 77–85.
Andrade, E.R. & Sellers, W.D. (1988) El Niño and its effects on precipitation in Arizona and western New Mexico. *Journal of Climatology*, **8**, 403–10.
Arno, S.F. (1980) Forest fire history in the northern Rockies. *Journal of Forestry*, **78**, 460–5.
Arno, S.F. (2000) Fire in western forest ecosystems. In: *Wildland Fire in Ecosystems: Effects of Fire on Flora*, (eds J.K. Brown & J.K. Smith), pp. 97–120. *USDA Forest Service General Technical Report*, **RMRS-GTR-42-vol. 2**.
Arno, S.F., Scott J.H. & Hartwell, M.G. (1995) Age-class structure of old growth ponderosa pine/Douglas fir stands and its relationship to fire history. *USDA Forest Service General Technical Report*, **INT-42**.

Attiwill, P.M. (1994) The disturbance of forest ecosystems: the ecological basis for conservative management. *Forest Ecology and Management*, **63**, 247–300.

Bahls, P. (1992) The status of fish populations and management of high mountain lakes in the western United States. *Northwest Science*, **66**, 183–93.

Bailey, R.G. (1995) *Description of the Ecoregions of the United States*, 2nd edn. Miscellaneous Publication 1391. USDA Forest Service, Washington, DC.

Baker, F.S. (1944) Mountain climates of the western United States. *Ecological Monographs*, **14**, 225–54.

Baker, W.L. & Ehle, D. (2001) Uncertainty in surface-fire history: the case of ponderosa pine forests in the western United States. *Canadian Journal of Forest Research*, **31**, 1205–26.

Baker, W.L. & Veblen, T.T. (1990) Spruce beetles and fires in the nineteenth century subalpine forests of western Colorado. *Arctic and Alpine Research*, **22**, 65–80.

Barrett, S.W., Arno, S.F. & Key, C.H. (1991) Fire regimes of western larch-lodgepole pine forests in Glacier National Park, Montana. *Canadian Journal of Forest Research*, **21**, 1711–20.

Bartos, D.L. (2001) Landscape dynamics of aspen and conifer forests. In: *Sustaining Aspen in Western Landscapes: Symposium Proceedings*, (compiled by W.D. Shepperd, D. Binkley, D. Bartos, T.J. Stohlgren & L.G. Eskew), pp. 5–14. *USDA Forest Service Proceedings*, **RMRS-P-18**.

Baxter, C.V. & Hauer, F.R. (2000) Geomorphology, hyporheic exchange, and selection of spawning habitat by bull trout (*Salvelinus confluentus*). *Canadian Journal of Fisheries and Aquatic Sciences*, **57**, 1–12.

Baxter, C.V., Frissell, C.A. & Hauer, F.R. (1999) Geomorphology, logging roads, and the distribution of bull trout spawning in a forested river basin: implications for management and conservation. *Transactions of the American Fisheries Society*, **128**, 854–67.

Beechie, T.J. & Sibley, T.H. (1997) Relationships between channel characteristics, woody debris, and fish habitat in northwestern Washington streams. *Transactions of the American Fisheries Society*, **126**, 217–29.

Behnke, R.J. (1992) *Native Trout of Western North America*. Monograph 6. American Fisheries Society, Bethesda, MD.

Benda, L.E., Miller, D.J., Dunne, T., Reeves, G.H. & Agee, J.K. (1998) Dynamic landscape systems. In: *River Ecology and Management: Lessons from the Pacific Coastal Ecoregion*, (eds R.J. Naiman & R.E. Bilby), pp. 261–88. Springer-Verlag, New York.

Berg, N., Carlson, A. & Azuma, D. (1998) Function and dynamics of woody debris in stream reaches in the central Sierra Nevada, California. *Canadian Journal of Fisheries and Aquatic Sciences*, **55**, 1807–20.

Betancourt, J.L., Devender, T.R.V. & Martin, P.S. (1990) *Packrat Middens: The Last 40,000 Years of Biotic Change*. University of Arizona Press, Tucson, AZ.

Bilby, R.E. & Bisson, P.A. (1998) Function and distribution of large woody debris. In: *River Ecology and Management: Lessons from the Pacific Coastal Ecoregion*, (eds R.J. Naiman & R.E. Bilby), pp. 324–46. Springer-Verlag, New York.

Bozek, M.A. & Young, M.K. (1994) Fish mortality resulting from delayed effects of fire in the Greater Yellowstone Ecosystem. *Great Basin Naturalist*, **54**, 91–5.

Bragg, D.C. (2000) Simulating catastrophic and individualistic large woody debris recruitment for a small riparian system. *Ecology*, **81**, 1383–94.

Bragg, D.C., Kershner, J.L. & Roberts, D.W. (2000) Modeling large woody debris recruitment for small streams of the central Rocky Mountains. *USDA Forest Service General Technical Report*, **RMRS-GTR-55**.

Brown, D.K., Echelle, A.A., Propst, D.L., Brooks, J.E. & Fisher, W.L. (2001) Catastrophic wildfire and number of populations as factors influencing risk of extinction for Gila trout (*Oncorhynchus gilae*). *Western North American Naturalist*, **61**, 139–48.

Brown, R.S. & Mackay, W.C. (1995) Fall and winter movements and habitat use by cutthroat trout in the Ram River, Alberta. *Transactions of the American Fisheries Society*, **124**, 873–85.

Coghlan, G. & Sowa, R. (1998) *National Forest Road System and Use*. Engineering Staff, USDA Forest Service, Washington, DC.

Covington, W.W. & Moore, M.M. (1994) Southwestern ponderosa forest structure: changes since Euro-American settlement. *Journal of Forestry*, 92, 39–47.

Covington, W.W., Fulé, P.Z., Moore, M.M., *et al.* (1997) Restoring ecosystem health in ponderosa pine forests of the Southwest. *Journal of Forestry*, 95, 129–64.

Dahms, C.W. & Geils, B.W. (1997) An assessment of forest ecosystem health in the Southwest. *USDA Forest Service Rocky Mountain Research Station General Technical Report*, RM-GTR-295.

Detenbeck, N.E., DeVore, P.W., Niemi, G.J. & Lima, A. (1992) Recovery of temperate-stream fish communities from disturbance: a review of case studies and synthesis of theory. *Environmental Management*, 16, 33–53.

Eaglin, G.S. & Hubert, W.A. (1993) Effects of logging and roads on substrate and trout in streams of the Medicine Bow National Forest, Wyoming. *North American Journal of Fisheries Management*, 13, 844–6.

Enfield, D.B. (1989) El Niño, past and present. *Reviews of Geophysics*, 27, 159–87.

Fausch, K.D. (1989) Do gradient and temperature affect distributions of, and interactions between, brook charr (*Salvelinus fontinalis*) and other resident salmonids in streams? *Physiology and Ecology Japan*, Special Vol. 1, 303–22.

Fausch, K.D. & Bestgen, K.R. (1997) Ecology of fishes indigenous to the central and southwestern Great Plains. In: *Ecology and Conservation of Great Plains Vertebrates*, (eds F.L. Knopf & F.B. Samson), pp. 131–66. Ecological Studies 125. Springer-Verlag, New York.

Fausch, K.D. & Northcote, T.G. (1992) Large woody debris and salmonid habitat in a small coastal British Columbia stream. *Canadian Journal of Fisheries and Aquatic Sciences*, 49, 682–93.

Fausch, K.D., Taniguchi, Y., Nakano, S., Grossman, G.D. & Townsend, C.R. (2001) Flood disturbance regimes influence rainbow trout invasion success among five Holarctic regions. *Ecological Applications*, 11, 1438–55.

Fausch, K.D., Torgersen, C.E., Baxter, C.V. & Li, H.W. (2002) Landscapes to riverscapes: bridging the gap between research and conservation of stream fishes. *BioScience* 52, 483–98.

Gowan, C. & Fausch, K.D. (1996a) Long-term demographic responses of trout populations to habitat manipulation in six Colorado streams. *Ecological Applications*, 6, 931–46.

Gowan, C. & Fausch, K.D. (1996b) Mobile brook trout in two high-elevation Colorado streams: re-evaluating the concept of restricted movement. *Canadian Journal of Fisheries and Aquatic Sciences*, 53, 1370–81.

Gowan, C., Young, M.K., Fausch, K.D. & Riley, S.C. (1994) Restricted movement in resident stream salmonids: a paradigm lost? *Canadian Journal of Fisheries and Aquatic Sciences*, 51, 2626–37.

Gresswell, R.E. (1999) Fire and aquatic ecosystems in forested biomes of North America. *Transactions of the American Fisheries Society*, 128, 193–221.

Gurnell, A.M., Gregory, K.J. & Petts, G.E. (1995) The role of coarse woody debris in forest aquatic habitats: implications for management. *Aquatic Conservation: Marine and Freshwater Ecosystems*, 5, 143–66.

Hall, J.D., Brown, G.W. & Lantz, R.L. (1987) The Alsea watershed study: a retrospective. In: *Streamside Management: Forestry and Fishery Interactions*, (eds E.O. Salo & T.W. Cundy), pp. 399–416. Contribution 57. College of Forestry, University of Washington, Seattle.

Harig, A.L. & Fausch, K.D. (2002) Minimum habitat requirements for establishing translocated cutthroat trout populations. *Ecological Applications*, 12, 535–51.

Harig, A.L., Fausch, K.D. & Young, M.K. (2000) Factors influencing success of greenback cutthroat trout translocations. *North American Journal of Fisheries Management*, 20, 994–1004.

Harmon, M.E, Franklin, J.F., Swanson, F.J., *et al.* (1986) Ecology of coarse woody debris in temperate ecosystems. *Advances in Ecological Research*, 15, 133–302.

Harrington, M.G. (1996) Fall rates of prescribed fire-killed ponderosa pine. *USDA Forest Service Research Paper*, **INT-RP-489**.

Hartman, G.F., Scrivener, J.C. & Miles, M.J. (1996) Impacts of logging in Carnation Creek, a high-energy coastal stream in British Columbia, and their implication for restoring fish habitat. *Canadian Journal of Fisheries and Aquatic Sciences*, **53** (Suppl. 1), 237–51.

Hauer, F.R., Baron, J.S., Campbell, D.H., *et al.* (1997) Assessment of climate change and freshwater ecosystems of the Rocky Mountains, USA and Canada. *Hydrologic Processes*, **11**, 903–24.

Hauer, F.R., Poole, G.C., Gangemi, J.T. & Baxter, C.V. (1999) Large woody debris in bull trout (*Salvelinus confluentus*) streams of logged and wilderness watersheds in northwest Montana. *Canadian Journal of Fisheries and Aquatic Sciences*, **56**, 915–24.

Heede, B.H. (1985) Channel adjustments to the removal of log steps: an experiment in a mountain stream. *Environmental Management*, **9**, 427–32.

Jakober, M.J., McMahon, T.E., Thurow, R.F. & Clancy, C.G. (1998) Role of stream ice on fall and winter movements and habitat use by bull trout and cutthroat trout in Montana headwater streams. *Transactions of the American Fisheries Society*, **127**, 223–35.

Jakober, M.J., McMahon, T.E. & Thurow, R.F. (2000) Diel habitat partitioning by bull charr and cutthroat trout during fall and winter in Rocky Mountain streams. *Environmental Biology of Fishes*, **59**, 79–89.

Johnson, B.M. & Martinez, P.J. (2000) Trophic economics of lake trout management in reservoirs of differing productivity. *North American Journal of Fisheries Management*, **20**, 127–43.

Kawaguchi, Y. & Nakano, S. (2001) Contribution of terrestrial invertebrates to the annual resource budget for salmonids in forest and grassland reaches of a headwater stream. *Freshwater Biology*, **46**, 303–16.

Kawaguchi, Y., Nakano, S. & Taniguchi, Y. (2003) Terrestrial invertebrate inputs determine the local abundance of stream fishes in a forested stream. *Ecology*, **84**, 701–8.

Kiladis, G.N. & Diaz, H.F. (1989) Global climatic anomalies associated with extremes in the Southern Oscillation. *Journal of Climate*, **2**, 1069–90.

Knapp, R.A., Vredenburg, V.T. & Matthews, K.R. (1998) Effects of stream channel morphology on golden trout spawning habitat and recruitment. *Ecological Applications*, **8**, 1104–17.

Knight, D.H. (1994) *Mountains and Plains: The Ecology of Wyoming Landscapes*. Yale University Press, New Haven, CT.

Kuchler, A.W. (1964) *Potential Natural Vegetation of the Conterminous United States*. Special Publication 36. American Geographical Society, New York.

Latterell, J.J., Fausch, K.D., Gowan, C. & Riley, S.C. (1998) Relationship of trout recruitment to snowmelt runoff flows and adult trout abundance in six Colorado mountain streams. *Rivers*, **6**, 240–50.

Lee, D.C., Sedell, J.R., Rieman, B.E., *et al.* (1997) Broadscale assessment of aquatic species and habitats. In: *An Assessment of Ecosystem Components in the Interior Columbia Basin and Portions of the Klamath and Great Basins*, (eds T.M. Quigley & S.J. Arbelbide), pp. 1057–496. *USDA Forest Service General Technical Report*, **PNW-GTR-405** Vol. 3.

Lisle, T.E. (1987) Using 'residual depths' to monitor pool depths independently of discharge. *USDA Forest Service Research Note*, **PSW-394**.

Loope, L.L. & Gruell, G.E. (1973) The ecological role of fire in Jackson Hole, northwestern Wyoming. *Quaternary Research*, **3**, 425–43.

Lyon, L.J. (1984) The Sleeping Child burn – 21 years of post-burn change. *USDA Forest Service Research Paper*, **INT-330**.

Mantua, N.J., Hare, S.R., Zhang, Y., Wallace, J.M. & Francis, R. (1997) A Pacific Interdecadal Climate Oscillation with impacts on salmon production. *Bulletin of the American Meteorological Society*, **78**, 1069–79.

Martin, D.J. & Benda, L.E. (2001) Patterns of instream wood recruitment and transport at the watershed scale. *Transactions of the American Fisheries Society*, **130**, 940–58.

Meehan, W.R. (1991) *Influences of Forest and Rangeland Management on Salmonid Fishes and their Habitats*. Special Publication 19. American Fisheries Society, Bethesda, MD.

Minshall, G.W., Brock, J.T. & Varley, J.D. (1989) Wildfires and Yellowstone's stream ecosystems. *BioScience*, **39**, 707–15.

Minshall, G.W., Robinson, C.T. & Lawrence, D.E. (1997) Postfire responses of lotic ecosystems in Yellowstone National Park, U.S.A. *Canadian Journal of Fisheries and Aquatic Sciences*, **54**, 2509–25.

Muhlfeld, C.C., Bennett, D.H. & Marotz, B. (2001) Fall and winter habitat use and movement by Columbia River redband trout in a small stream in Montana. *North American Journal of Fisheries Management*, **21**, 170–7.

Murphy, M.L. & Hall, J.D. (1981) Varied effects of clear-cut logging on predators and their habitat in small streams of the Cascade Mountains, Oregon. *Canadian Journal of Fisheries and Aquatic Sciences*, **38**, 137–45.

Murphy, M.L., Heifetz, J., Johnson, S.W., Koski, K.V. & Thedinga, J.F. (1986) Effects of clear-cut logging with and without buffer strips on juvenile salmonids in Alaskan streams. *Canadian Journal of Fisheries and Aquatic Sciences*, **43**, 1521–33.

Murphy, M.L. & Koski, K.V. (1989) Input and depletion of woody debris in Alaska streams and implications for streamside management. *North American Journal of Fisheries Management*, **9**, 427–36.

Nakamura, F., Swanson, F.J. & Wondzell, S.M. (2000) Disturbance regimes of stream and riparian systems – a disturbance-cascade perspective. *Hydrological Processes*, **14**, 2849–60.

Nakano, S. & Murakami, M. (2001) Reciprocal subsidies: dynamic interdependence between terrestrial and aquatic food webs. *Proceedings of the National Academy of Science USA*, **98**, 166–70.

Nehring, R.B. & Anderson, R.M. (1993) Determination of population-limiting critical salmonid habitats in Colorado streams using the Physical Habitat Simulation system. *Rivers*, **4**, 1–19.

Peet, R.K. (1981) Forest vegetation of the Colorado Front Range. *Vegetatio*, **45**, 3–75.

Peet, R.K. (1988) Forests of the Rocky Mountains. In: *North American Terrestrial Vegetation*, (eds M.G. Barbour & W.D. Billings), pp. 63–101. Cambridge University Press, New York.

Platts, W.S., Torquemada, R.J., McHenry, M.L. & Graham, C.K. (1989) Changes in salmon spawning and rearing habitat from increased delivery of fine sediment to the South Fork Salmon River, Idaho. *Transactions of the American Fisheries Society*, **118**, 274–83.

Poff, N.L. (1996) A hydrogeography of unregulated streams in the United States and an examination of scale-dependence in some hydrological descriptors. *Freshwater Biology*, **36**, 71–91.

Potts, D.F. & Anderson, B.K.M. (1990) Organic debris and management of small channels. *Western Journal of Applied Forestry*, **5**, 25–38.

Power, M.E. (2001) Prey exchange between a stream and its forested watershed elevates predator densities in both habitats. *Proceedings of the National Academy of Science USA*, **98**, 14–15.

Pyne, S.J. (1996) *Introduction to Wildland Fire*, 2nd edn. John Wiley & Sons, New York.

Ralph, S.C., Poole, G.C., Conquest, L.L. & Naiman, R.J. (1994) Stream channel morphology and woody debris in logged and unlogged basins of western Washington. *Canadian Journal of Fisheries and Aquatic Sciences*, **51**, 37–51.

Richmond, A.D. & Fausch, K.D. (1995) Characteristics and function of large woody debris in subalpine Rocky Mountain streams in northern Colorado. *Canadian Journal of Fisheries and Aquatic Sciences*, **52**, 1789–802.

Rieman, B.E. & Dunham, J.B. (2000) Metapopulations and salmonids: a synthesis of life history patterns and empirical observations. *Ecology of Freshwater Fish*, **9**, 51–64.

Riley, S.C. & Fausch, K.D. (1995) Trout population response to habitat enhancement in six northern Colorado streams. *Canadian Journal of Fisheries and Aquatic Sciences*, **52**, 34–53.

Robison, E.G. & Beschta, R.L. (1990) Characteristics of coarse woody debris for several coastal streams of southeast Alaska, U.S.A. *Canadian Journal of Fisheries and Aquatic Sciences,* **47**, 1684–93.

Romme, W.H. & Knight, D.H. (1981) Fire frequency and subalpine forest succession along a topographic gradient in Wyoming. *Ecology,* **62**, 319–26.

Rosenberg, R.G. (1984) Handhewn ties of the Medicine Bows. *Annals of Wyoming,* **56**, 39–53.

Rosgen, D.L. (1996) *Applied River Morphology.* Wildland Hydrology, Pagosa Springs, CO.

Ruzycki, J.R., Beauchamp, D.A. & Yule, D.L. (2003) Effects of introduced lake trout on native cutthroat trout in Yellowstone Lake. *Ecological Applications,* **13**, 23–37.

Schlosser, I.J. (1995) Critical landscape attributes that influence fish population dynamics in headwater streams. *Hydrobiologia,* **301**, 71–81.

Schmetterling, D.A. & Pierce, R.W. (1999) Success of instream habitat structures after a 50-year flood in Gold Creek, Montana. *Restoration Ecology,* **7**, 369–75.

Schubert, G.H. (1974) Silviculture of southwestern ponderosa pine: the status of our knowledge. *USDA Forest Service Research Paper,* **RM-123**.

Smith, R.W. & Griffith, J.S. (1994) Survival of rainbow trout during their first winter in the Henry's Fork of the Snake River, Idaho. *Transactions of the American Fisheries Society,* **123**, 747–56.

Stanford, J.A. & Hauer, F.R. (1992) Mitigating the impacts of stream and lake regulation in the Flathead River catchment, Montana, USA: an ecosystem perspective. *Aquatic Conservation: Marine and Freshwater Ecosystems,* **2**, 35–63.

Swanson, F.J., Lienkaemper, G.W. & Sedell, J.R. (1976) History, physical effects, and management implications of large organic debris in western Oregon streams. *USDA Forest Service General Technical Report,* **PNW-56**.

Trombulak, S.C. & Frissell, C.A. (2000) Review of ecological effects of roads on terrestrial and aquatic communities. *Conservation Biology,* **14**, 18–30.

Trotter, E.H. (1990) Woody debris, forest-stream succession, and catchment geomorphology. *Journal of the North American Benthological Society,* **9**, 141–56.

Tyler, T., Liss, W.J., Ganio, L.M., *et al.* (1998) Interaction between introduced trout and larval salamanders (*Ambystoma macrodactylum*) in high-elevation lakes. *Conservation Biology,* **12**, 94–105.

USDA Forest Service (1995) *Inland Native Fish Strategy.* Intermountain, Northern, and Pacific Northwest Regions, Washington, DC.

USDA Forest Service (1996) *Spruce Beetle Control Project Environmental Assessment.* Flathead National Forest, Tally Lake Ranger District, Kalispell, MT.

USDA Forest Service (1998) Routt Divide blowdown analysis. Medicine Bow-Routt National Forest, Hahns Peak/Bears Ears Ranger District, Routt County, CO. *Federal Register,* **63**, 3078–80.

Vannote, R.L., Minshall, G.W., Cummins, K.W., Sedell, J.R. & Cushing, C.E. (1980) The river continuum concept. *Canadian Journal of Fisheries and Aquatic Sciences,* **37**, 130–7.

Veblen, T.T., Hadley, K.S., Nel, E.M., Kitzberger, T., Reid, M. & Villalba, R. (1994) Disturbance regime and disturbance interactions in a Rocky Mountain subalpine forest. *Journal of Ecology,* **82**, 125–35.

Veblen, T.T., Kitzberger, T. & Donnegan, J. (2000) Climatic and human influences on fire regimes in ponderosa pine forest in the Colorado Front Range. *Ecological Applications,* **10**, 1178–95.

Wilkinson, C.E. (1992) *Crossing the Next Meridian: Land, Water, and the Future of the West.* Island Press, Washington, DC.

Wiltzius, W.J. (1985) Fish culture and stocking in Colorado, 1872–1978. *Colorado Division of Wildlife Report,* **12**. Denver, CO.

Wipfli, M.S. (1997) Terrestrial invertebrates as salmonid prey and nitrogen sources in streams: contrasting old-growth and young-growth riparian forests in southeastern Alaska, U.S.A. *Canadian Journal of Fisheries and Aquatic Sciences*, **54**, 1259–69.

Wohl, E.E. (2001) *Virtual Rivers: Lessons from the Mountain Rivers of the Colorado Front Range*. Yale University Press, New Haven, CT.

Wohl, E., Madsen, S. & MacDonald, L. (1997) Characteristics of log and clast bed-steps in step-pool streams of northwestern Montana, USA. *Geomorphology*, **20**, 1–10.

Wohl, E.E., Vincent, K.R. & Merritts, D.J. (1993) Pool and riffle characteristics in relation to channel gradient. *Geomorphology*, **6**, 99–110.

Young, M.K. (1994) Movement and characteristics of stream-borne coarse woody debris in adjacent burned and unburned watersheds in Wyoming. *Canadian Journal of Forest Research*, **24**, 1933–8.

Young, M.K. (1996) Summer movements and habitat use by Colorado River cutthroat trout (*Oncorhynchus clarki pleuriticus*) in small, montane streams. *Canadian Journal of Fisheries and Aquatic Sciences*, **53**, 1403–8.

Young, M.K. (1998) Absence of autumnal changes in habitat use and location of adult Colorado River cutthroat trout in a small stream. *Transactions of the American Fisheries Society*, **127**, 147–51.

Young, M.K. & Harig, A.L. (2001) A critique of the recovery of greenback cutthroat trout. *Conservation Biology*, **15**, 1575–84.

Young, M.K., Haire, D. & Bozek, M.A. (1994) The effect and extent of railroad tie drives in streams of southeastern Wyoming. *Western Journal of Applied Forestry*, **9**, 125–30.

Young, M.K., Schmal, R.N. & Sobczak, C.M. (1990) Railroad tie drives and stream channel complexity: past impacts, current status, and future prospects. In: *Proceedings of the Annual Meeting of the Society of American Foresters*, pp. 126–30. Publication 89–02. Society of American Foresters, Bethesda, MD.

Young, M.K., Harig, A.L., Rosenlund, B. & Kennedy, C. (2002) Recovery history of greenback cutthroat trout: population characteristics, hatchery involvement, and bibliography. Version 1.0. *USDA Forest Service Rocky Mountain Research Station General Technical Report*, **RMRS-GTR-88WWW**. Available at http://www.fs.fed.us/rm/pubs/rmrs_gtr88

Chapter 22
Fish–forestry interface: an overview of Mexico

A. SÁNCHEZ-VÉLEZ AND R.M. GARCÍA-NÚÑEZ

A tree planted by the water, that extends its roots by the stream ... will not be anxious in a year of drought, since its leaves will be green and without fatigue will give fruits (Jeremiah 17:8)

Introduction

In this chapter the relationships between forestry activities and fish will be considered. Such relationships do not, however, stand alone. Therefore we have attempted to consider forestry effects on fish, but the whole matter of forestry, fish habitat and fish must be set in the context of a wider complex of land uses and human population change. The land uses include farming and urbanization with attendant pollution, and river damming and diversion for hydroelectric and agricultural purposes. Many aspects of each of these are dramatically influenced by human population change and they may interact with forestry activities to affect fish habitat and fish.

Forest management practices in communal, private and even federal lands in Mexico have been a point of conflictive dispute. Contrasting value opinions, imprecise data, an impoverished population living in highland forest areas, and above all, political influence, have compromised the decision-making process of balancing traditional timber harvest-oriented management and the conservation of natural ecosystem policies. Wood is the main commodity influencing forest management plans, and the importance of fish habitat has been seen so far as an academic subject that only theoretical ecologists might find interesting. This fact, among others, shapes the dimension of Mexico's environmental problems. Deforestation in Mexico, at 1.2 million hectares per year of diverse types of vegetation, is the second highest in all of Latin America. It has the potential to affect fishery resources.

Fish–forestry protection planning has not advanced as far in Mexico as in other parts of North America. The issue of riparian forest protection through establishment of streamside management zones, well known in northern countries, receives little consideration in Mexico. While scientists from developed countries discuss the scope of riparian areas to protect water and fish habitat quality, in countries such as Mexico, the battle is how to preserve at least the last remaining portions of fluvial forest. Even now, in some regions of the Mexican high plateau, the last remnants of riparian forest are being illegally cut down; 200-year-old ahuahuetes (*Taxodium mucronatum*), the national tree that ancient cultures named the Elder of the River, are being felled and converted to planks to sell in local markets. Vegetation depletion is not a new phenomenon in Mexico, but in the last years the free market-oriented economy and population growth have expanded human consumption. They have also imposed the idea of converting anything possible into cash to satisfy in many cases, not basic needs, but rather, luxury tastes.

The outcome of the scenario described above is of course catastrophic. Water quality is being reduced throughout the country. Springs have dried up and river flow regimes have been shifted to extremes of periodic destructive floods with intervals of desiccation in which fish populations have been extirpated. In some cases species have disappeared forever. A similar scenario prevails in inland and salt marshes, mangrove forests, potholes, swamps and wetlands around the country, where developers drain and compact land for condominiums, hotels and farming, leaving behind impoverished and bleak lives, along with a ruined fishing economy.

There is a set of federal Mexican laws and regulations dealing with forests in general. First of all, the *Ley Forestal* (Forestry Law) prohibits harvesting of riverside forests. After this, the *Ley General del Equilibrio Ecológico y Protección al Ambiente* (General Law of Ecological Equilibrium and Environmental Protection) and the *Ley de Aguas Nacionales* (Law of National Waters) consider the protection of natural ecosystems. However, no enforcement of these laws is possible. In Mexico, many things remain to be done concerning economic, social, environmental and regulatory issues. The new government has shown some prudent and modest signs of political will in working towards an organized research and governmental policy with respect to restoration, protection and management practices on aquatic and forest ecosystems. However, as stated by Aber & Melillo (1991) regarding watershed management 'methods cannot be separated from their social and economic systems of which they are part'. Mexico illustrates this well, due to its environmental diversity and social and economic constraints. This chapter reviews these issues.

Mexico as a geographic region

Mexico is a multicultural and biophysically diverse country. Located between two major life zones, the neo-tropical and neo-boreal, the country is one of the 12 nations containing 60% of all the world's biodiversity (SEMARNAP 2000). Due to its continental position between the tropics of Cancer and Capricorn, it possesses almost all climatic types. Such conditions have created a multiplicity of land forms, soils and biological diversity. A variety of forest ecosystems occur: tropical, temperate, subtropical and arid, possessing a tremendous variety of aquatic and subaquatic vegetation, each with particular features and in many instances containing specific flora and fauna (Fig. 22.1). The aquatic habitats and forest diversity of Mexico are among the most outstanding in the world and the occurrence of many endemic species results in several hot spots of biodiversity (Sánchez & García 2000).

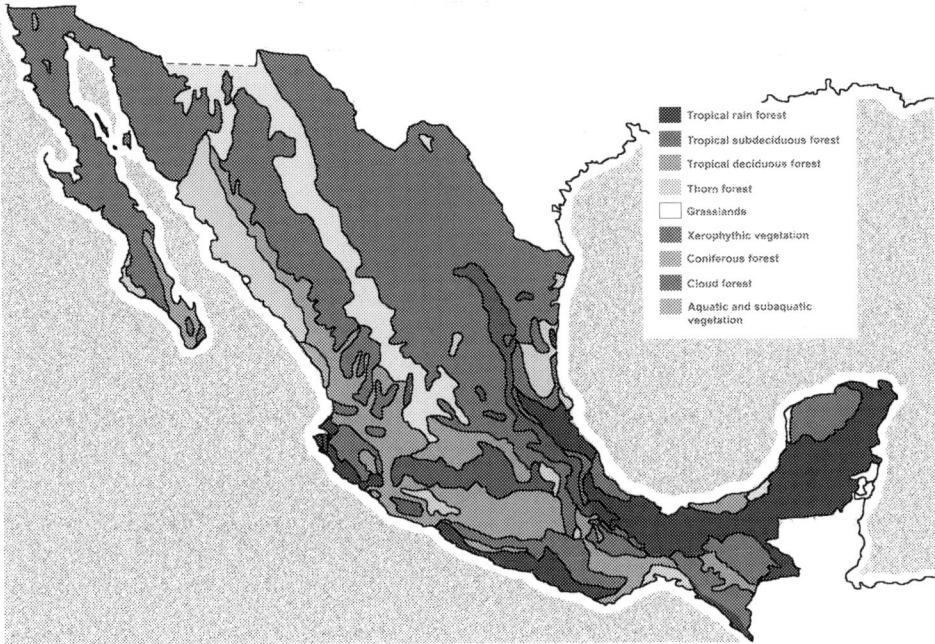

Fig. 22.1 Main forest types in Mexico, according to Rzedowski (1978). So far over 22,000 vascular plants have been classified.

Topographic and geological features

The terrestrial landscape of Mexico is one of the most uneven of the globe. Anywhere from the extensive plains of the Yucatan Peninsula to the northern plains of the state of Tamaulipas, one may see ranges and volcanoes in the landscape. Only a third of the country's surface lies below 500 m in altitude, and more than 50% of the country is more than 1000 m above sea level. From north to south, there are many mountain ranges and volcanoes that are more than 4000 m high.

Climate and weather

High altitudinal variability, complex orographic conditions, narrowness of the country in the south with a marine influence, and its location between North and South America all influence Mexico's climate. Solar radiation, diel temperature fluctuations and atmospheric humidity are vital to the distribution and abundance of organisms. Weather conditions are highly variable year-round, and even during the rainy season, sunny days are common.

Northern Mexico is a transitional zone with humid and semi-humid weather influenced by elysian winds and by cyclones going to the south, making it an ecotonal region between temperate and tropical zones. The rainy regime, June to September, dominates most of the country, is caused by the latitudinal effect. Complex mountain-

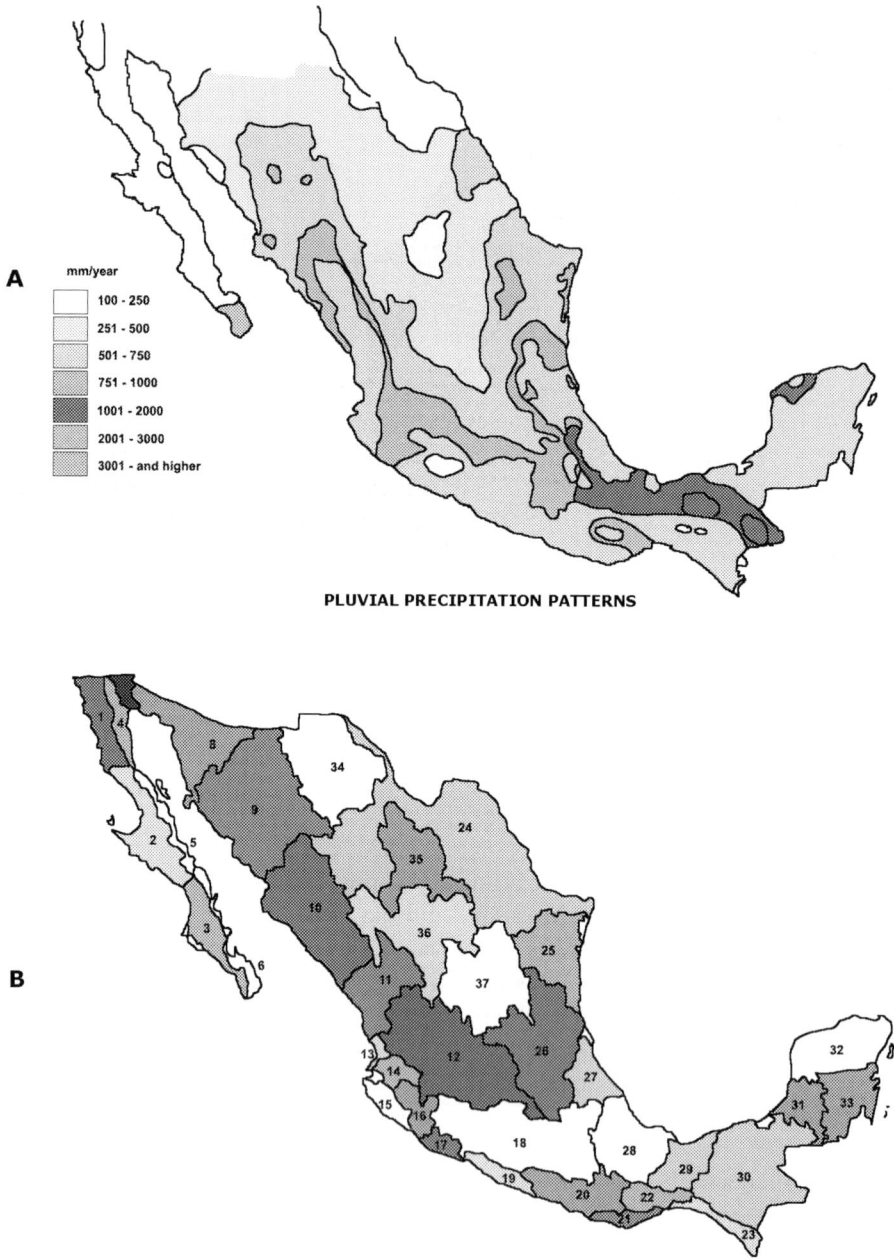

Fig. 22.2 (A) Precipitation patterns in Mexico (CNA 2001). (B) The 37 watersheds of Mexico (CNA 2002).

ous areas, added to differentiated conditions of latitude and seaboard extension, result in a climatic mosaic, with a number of variants in temperature and humidity. These are not yet fully understood. The precipitation patterns are shown in Fig. 22.2A.

The climatic zones of Mexico are not easy to group. The climate, both precipitation and temperature, is variable over the whole country. According to García (1978), three climatic types characterize the country: (1) wet and hot where tropical rain forests and tropical dry forests are present; (2) dry, which embraces most of the northern part of the country, where the matorral and grasslands vegetation coexist; and (3) temperate and cold, with enough rain, where coniferous forest and mixed groves of *Quercus,* in particular, occur close to the high ranges and volcanic slopes.

Hydrography

Mexico was divided by CNA (2002) into 13 hydrologic administrative regions containing 37 watersheds and 314 sub-watersheds (Fig. 22.2B). Most of the drainage networks flow into the sea, along the coasts (Fig. 22.3). In the Peninsula, however, with karstic geological formations, subterranean flows reach the ocean underground through a complex web of caves. Many coastal plains, in the north along the Pacific Ocean and central part of the Gulf, form deltas where varied vegetation of marshes, swamps and mangroves support a plentiful avian fauna. In the central north of the country closed or endorrheic watersheds are present.

The country's average precipitation is about 771.8 mm per year, but it is very unevenly distributed (PNH 2001); 50% of the north central region, the *Altiplano,* receives 9% of the annual average precipitation. However, 75% of the population lives there in industrial settlements, and 40% of the rain-fed agriculture is there. The *Sureste,* the southeast, tropical and subtropical lands, receive 70% of the rainfall, but this region is inhabited by only 24% of the population. This situation challenges planners and

Fig. 22.3 Main rivers of Mexico (CNA 2001).

water administrators, since 85% of the water caught in over 4000 dams is located in altitudes about 500 m above sea level.

One of the most outstanding engineering projects is the Sistema Cutzamala. This network of five dams supplies water to Mexico City, and pumps it over 1000 m up to do so. Of all of the water used in Mexico, 83% goes for agriculture, 12% for domestic use, 3% for industry and 2% for aquaculture (CESPEDES 1998).

Mexico's fish–forestry historical background

A review of Mexican history reveals that the flourishing of pre-Hispanic cultures was strongly associated with the different aquatic environments around them. Aside from food and water values, there were spiritual considerations involved in living near lakes and streams. They were regarded as sacred places. Historical documents about the native aboriginal Mexicans reveal natural resource management techniques and ways of living that now represent a longed-for utopia (Toledo 1980; Del Campo 1984). Early Mexicans had a deep respect for all living things and a cosmic vision that governed man–nature relationships. They knew and applied sustainable, efficient and equitable ways of fish and waterfowl harvesting through observation of reproductive cycles. Chinampas, a multi-cropped system under aquatic conditions, are well known throughout the world.

Disruption of equitable and sustainable systems of resource use was inevitable with the arrival of the Spaniards. The harmonious nature–culture relationship was cut short. The European conquerors started by cutting down forests, and they used the wood in diverse industries. The lacustrine valleys of Mexico were transformed by construction of canals, in-filling of wetlands, introduction of exotic species, rerouting of creeks, rejection of former local water control methods and abandonment of fishing, hunting and forestry regulations. These changes, including the effects of deforestation and the consequent desiccation of lakes, made the area into a bleak plateau with the indigenous people as slaves.

Forest administration and silvicultural techniques

Mexico's forest cover and timber production

In Mexico there are about 141.8 million hectares of forested land. Within them, steep slopes, shallow soils and extreme climates require a dense forest cover to protect soils from erosion and consequent landslides that threaten the lives of people living down-slope. Such conditions cover 72% of Mexico, a mountainous country with an abundance of stream networks. Within the mountainous stream network areas of Mexico, only 56.8 million hectares are covered by some kind of tree stands, and within this, more than 32.6 million hectares (23% of the nation's forests) are severely perturbed and fragmented (SARH 1994).

Deforestation ranges from 615,000 to 1.2 million hectares per annum depending on the author's research methods. From 1985 to 1994, total land degradation and desert formation rose from 17.8 to 22.2 million hectares (PEF 1996). Paradoxically, even though Mexico is 10th in the world ranking for forest cover, and is 26th for production (Tellez 1993), the forest sector contributed only 1% to the national Gross Domestic Product in 1994. Mexico is an importer of pulp cellulose for paper production (PEF 1996). In summary, only 7 million hectares are under forestry management. Most are pine or other coniferous forests, and are strongly oriented to timber harvest. On the other hand, 80% of the forest lands are social properties known as 'ejidos', and are communally owned (CONAFOR 2002). Ejido communities hold almost 70% of all forest (Muñoz-Piña *et al.* 2001). Besides, over 80% of village inhabitants of jungles and temperate forests are considered to be within a high or very high marginality index. Their livelihoods depend basically on wood sales. Poverty is one of the social forces resulting in deforestation.

Forest management

Forestry management science in Mexico is largely concentrated in its temperate forests. Forest harvest regulation implies the application of silviculture treatments known as regulation by area (surface area of forest to timber) or by volume (cubic metres of wood products), or both. The main treatments applied in Mexican forests are 'regeneration cuts'. These consist of the removal of forest for log production. So far, little forest area has been converted to plantations, but those that do exist contain exotic trees, mainly eucalyptus.

Hydrologic–forestry problems

Mexico presents many problems in upland forest use. The rate of deforestation, as noted, is over 600,000 hectares per year (CONAFOR 2002). In the last 50 years, forest lands have lost 50% of their rain-catching capacity and over 50 million people suffer from water shortages. Erosion, from severe to moderate, affects >70% of the hilly lands, previously covered by forests, in such a way that 21% of the territory has been converted into deserts. The rate of desert formation is about 250,000 hectares per year. Soil pollution, by acid rain and other emissions, affects almost 50% of the country in some season of the year. As the forest cover is reduced, annual cyclones cause flooding that razes productive agricultural valleys, destroys communication infrastructure and causes mud slide obstructions. Water saturates hilly urban settlements on slopes that were once covered by forest groves. This situation results in frequent landslides and many fatalities. In 1998 the Niño's wind caused not only a multitude of forest fires but also devastating flooding in the country, with high environmental and economic costs.

Anthropogenic disturbances of forest and aquatic ecosystems have been the most pernicious. Governmental programmes for rural development continue to bring sewage collection systems and services to small villages without corresponding treatment plants. This increases the pollution of surface waters.

Table 22.1   Major sources of pollution in Mexico

Type of industry	Discharge, % of total
Sugar refinery	42
Chemical	23
Petroleum	9
Iron and steel	7
Paper	6
Beverages	6
Textiles	3
Mining	2
Foods	2

Source: SEMARNAP (2000).

In the forest industry, official reports indicate that there are 2076 plants. Of these, 2058 are sawmills, 11 are telegraph pole facilities and 7 are cellulose factories (SEMARNAP 1999). These all contribute to the production of wastewater (Table 22.1), the total of which is about 80 cubic metres per second in Mexico.

Population growth and consumption rates

According to the census for the year 2000, Mexico's population grew to over 100 million people. Population growth since the 1980s, as well the rise in consumption of goods and services in some socioeconomic strata, fuelled by a free market economy, are well documented. On 13 August 2002, the Federal Government recognized the existence of 53.7 million Mexicans who are living at different levels of poverty. The inter-related cycle of increasing population and poverty may be seen as a primary cause of fish–forest ecosystem destruction. In some areas of headwater streams and associated forests, the lack of a clear definition of property rights on the marginal lands is a key factor in the destruction. Planning to regulate the mining of forestry resources does not exist, and there are no real policies or enforcement mechanisms to change land use methods. There is a lack of policy on 'environmental services' payment with respect to mountain forests, and the water for irrigation is given for practically nothing. Forest/water problems and the change of land use may be summed up as follows:

(1)   Deforestation of headwater basins alters the hydrodynamics, composition and structure of the forest ecosystems.
(2)   Overexploitation of water wells for irrigation lowers the water table by about 2 or 3 m per annum and causes riparian forest decline.
(3)   Elimination of riparian vegetation by cultivation along stream banks results in an increase in floodwater sediments in the water used for irrigation.
(4)   Construction of small dams and channels to open new lands for growing cash crops de-waters streams.

(5)   Pollution of streams from uplands occurs as a result of intensive agricultural use of pesticides and artificial fertilizers.

Diversity of aquatic ecosystems in Mexico

Mexico's aquatic ecosystems comprise a tremendous diversity. They can be divided into inland and coastal ecosystems. However, all are influenced in one way or another by upland forest management, as will be discussed later.

There are about 3.8 million hectares of natural lakes, reservoirs and wetlands. Of this total, 2.9 million are coastal and include estuaries, marshes, offshore lagoons, inlets or bays. The remaining 0.9 million hectares are inland lakes with fresh water. At present some coastal lagoons and marshes are used for aquaculture production, with potential for more utilization.

The fluvial forest ecosystems in Mexico

Given the degree of disturbance, some forest streams represent the only corridors of life for many local and migratory species. Controversial interaction between forestry and fish frequently occurs in such zones. Creeks and riparian zones form a network throughout the landscape, serving as both distinctive habitats and transportation systems for water, sediment, and aquatic and terrestrial organisms. The flow of materials is dominantly downstream and unidirectional; however, movement of organisms can be either upstream or downstream. Given all of these considerations, the stream–riparian network in the upland forest should be given full management consideration because of its distinctive attributes and the associated movement of materials, organisms and disturbances through it (Swanson & Franklin 1992).

The fluvial forests of Mexico face the following problems:

(1)   *Point and non-point sources of pollution.* The tropical rivers formed big valleys that are occupied by sugar cane and many other commercial plantations. These developments, as well as their refineries, represent large sources of aerial and aquatic pollution. Forest industries, distilleries, thermoelectric plants and domestic sewage worsen the problem.

(2)   *Introduction of alien species.* About 90 species of exotic fish species (Contreras-Balderas 1999) have been introduced to Mexico's waterways and have caused reductions in native fish populations. Invaders such as tilapia (*Oreochromis mossambicus*), carp (*Cyprinus carpio*), bass (*Micropterus salmoides*) and rainbow trout (*Oncorhynchus mykiss*) wreak havoc among native species. Aquatic plant species such as water hyacinth (*Eichhornia crassipes*), water lettuce (*Pistia estratiotes*) and water lentil (*Hydrilla verticillata*) (Novelo & Martínez 1987) have pervasive effects on indigenous species. As Lugo (1994) indicated, aquatic invaders like the water hyacinth in their native Amazonian riverine habitat are fairly inconspicuous, but exhibit explosive growth in the slow moving, highly eutrophic streams and reservoirs in their newly colonized habitats. In many

tropical regions, the riparian forests are rapidly being replaced by African grasses, e.g. *Cynodon plectostachyus*, known as star or giant grass, and *Pennisetum
purpureum*, elephant grass. These have spread from ranches. During the rainy
season floods transport vegetative parts of these plants, and little by little, many
kilometres of river banks are covered with a uniform 'lawn' incapable of retaining either soil particles or agricultural pollutants.

(3)   *Timber harvest-oriented forest practices.* One of the primary forestry problems
in Mexico is a traditional vision of the forest as an area, the sole purpose of which
is to supply wood products. Only some river sections in remote areas still contain
well-conserved riparian vegetation. But in tropical lands, where wide floodplains
are crossed by big rivers, the fluvial forests have disappeared almost completely
along some river stretches. This loss is due to logging.

Lakes in Mexico experience different changes. There are about 69 natural lakes but
only a few of them have caught the politicians' and researchers' attention, and only 12
of them have been even partially studied (Arredondo & Aguilar 1983). One of the most
nationally significant lakes is Chapala Lake in the central west region of the country.
The lake is located close to Guadalajara, the second largest city of the country. In this
city, industry has grown until it accounts for one-third of the national domestic product. However, this economic development has brought about a loss in environmental
quality concerns, and the area now occupies third place in the depletion of river and
aquatic ecosystems.

Wetlands in Mexico

Wetland areas have unique soils that are distinguished from adjacent uplands and they
support vegetation adapted to moist conditions. There are about 60 major wetlands in
Mexico. Those occurring close to offshore zones (Fig. 22.4) include a diversity of types
including marshes, deltas, estuaries, and potholes, all with mangrove forest. Many
commercial fish, clams, crabs and amphibian species reproduce in the mangroves.
However, the high productivity of this open ecosystem is being seriously impacted by
overexploitation of some trees for firewood.

There are a number of local terms for the different wetlands in Mexico. The names
are based on the type of flooding regime, seasonal waterlogging, and vegetation composition (Mitsch & Gosselink 1986; Mitchell 1992):

- *Swamps*: Dominated by trees (*Taxodium* and *Cupressus* spp.) similar to Florida's
cypress and shrubs as in the state of Tamaulipas.
- *Salt marshes*: Coastal vegetation, inundated almost year-round, characterized by
emergent herbaceous vegetation adapted to saturated soil conditions.
- *Bog or peatlands*: Saturated land with accumulations of partially decayed plant
matter.
- *Riverine bottomlands*: Lowlands along streams, rivers and floodplains that are
seasonally inundated.

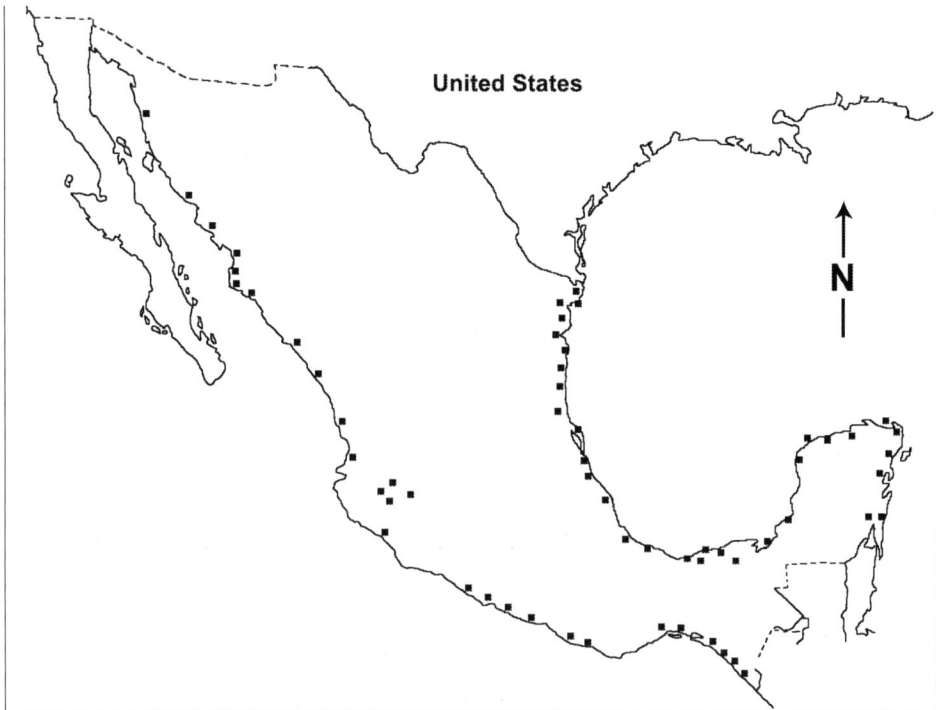

Fig. 22.4 The major wetlands of Mexico (■)distributed along its 11,593 km of coastline (Scott & Monserrat 1986).

- *Reed swamps*: Marshes dominated by *Pragmites communis, Arundo donax*, and other herbaceous species.
- *Sloughs*: Inland wetlands composed of aquatic plants such as *Thalia geniculata, Calathea* sp. and *Heliconia* sp., *Typha* spp., *Scirpus* spp. and *Cyperus* spp., or shallow pond networks in alluvial zones with slowly flowing water.
- *Potholes*: Inland waterlogged vegetation surrounded by tropical rainforest, present in some areas of Quintana Roo and Yucatán states.
- *Mangroves*: Coastal vegetation composed of four typical woody species: *Rhizophora mangle, Avicennia germinans, Laguncularia racemosa* and *Conocarpus erecta*. Mangroves occupy extensive zones along deltas in river mouths and are important nursery habitats for shrimps, fish, molluscs and many other species.

Importance of fisheries and aquaculture in Mexico

According to Ramamoorthy (1998), the inland Mexican ichthyofauna is composed of 384 fish species out of a country-wide total of 2122 (8.5% of the total number of fish species). However, Contreras-MacBeath (2002) says that there are 506 species (about 5% of the total number of freshwater species). Of these, 185 are threatened with extinction and 20 others have disappeared (CONABIO 1998). Many Mexican species remain undescribed. Many are vanishing rapidly due to fish habitat destruction. Inland fisheries were economically significant until 1970. With the explosive growth of the population there was an introduction of alien species and an increase in pollution. Nowadays few rivers and lakes are of economic importance in terms of native fishing.

Freshwater aquaculture is poorly developed in Mexico and only a few alien species are being used commercially. These include rainbow trout (*Oncorhynchus mykiss*), carp (*Cyprinus carpio*), goldfish (*Carassius auratus*), grass carp (*Ctenopharyngodon idella*), silver carp (*Hypophthalmichtys molotrix*), large-headed carp (*Aristichthys nobilis*) and black carp (*Mylopharyngodon piceus*). The African cichlids from Alabama, USA (*Tilapia nilotica, T. mossambica* and *T. melanpleura*), have successfully spread to reservoirs, but their impact on native fish communities has not been analysed.

The catfish species such as *Ictalurus punctatus, I. melas* and *I. dugesi,* although highly appreciated for food, are vanishing due to habitat destruction. Non-native species like black bass (*Micropterus* spp.) have begun to be used in sport-fishing. Some native species are being recovered by intensive efforts of artificial reproduction. In salty estuarine lagoons along the Pacific Ocean and the Gulf of Mexico the reproduction of many species of centropomids (*Centropomus poeyi, C. undecimalis, C. nigrecens, C. medias, C. robalito*), mullet (*Mugil cephalus*) and curvina (*Cynoscion noviles*) are being studied (Secretaría de Pesca 1994).

Mangroves, coastal lagoons, estuaries, salt marshes and related ecosystems face serious problems. Such areas are used indiscriminately as sinks for industrial pollutants in coastal zones. These problems were well documented in the state of Veracrúz, where petroleum refineries drain their effluents without any restriction (Bozada & Paez 1986a, 1986b, 1986c).

In the state of Sinaloa, which contains 11 irrigation dams, 690 km of seashore are used for aquaculture and fisheries. The most important culture groups are the shrimps (*Penaeus*), tuna and cichlids (*Tilapia*). However, there are problems for aquaculture in these waters. Pesticides and metal pollution have made fish unsafe for consumption in the coastal zone of Sinaloa (Galindo 2000). Such pollution has affected aquaculture, altered natural food webs, changed predator–prey relationships and threatened human health (Cruz 2002). In addition to these effects, water quality is being degraded by fertilizer application and release of sugar refinery wastes.

Game fish populations are at risk. They are under extreme pressure for recreational and subsistence fishing, and they are threatened by by-catch damage. In coastal areas 15 tonnes of non-food species may be killed in catching 1 tonne of shrimp (SEMARNAP 2000). Regulations are poorly enforced. Many of these problems occur because of a lack of public education, poor regulation of tourism and public insensitivity.

A forest–agriculture–fish stalemate in Sinaloa, Mexico

The effects of forestry activities on fish and fish habitat are interconnected to other land uses. In this complex mix of land uses, forestry has effects on fish, but its total impact may be hard to separate from the effects of other activities. The following section describes the three land use issues that are all interconnected to fish–forest management problems. The issues are: aquaculture and its water uses, agriculture and its water uses, and upland forest management and effects on water.

The first part of the land use problem complex involves shrimp farming. The state of Sinaloa is representative of many coastal zones of Mexico. Due to the decline of open sea fishing from 30,300 tons in 1987 to only 19,200 in 1999, commercial shrimp farming in estuarine clusters emerged during the last 15 years in the state of Sinaloa. Bays of the reservoirs were fenced to raise shrimp intensively. Then shoreline plains 2 or 3 m above sea level have been converted into primitive hatching reservoirs. This system depends upon both marine and terrestrial systems in natural condition. The ponds receive enriched runoff coming from uplands and irrigation channels, and tidal currents provide them with salt water.

From 1984 to 1999, some 266 sites, covering an area of 24,000 hectares out of a potential 272,440 hectares, were developed for aquaculture. Shrimp production reached 13,511 tons, as the second most important export product from Sinaloa, after vegetables (Lyle *et al.* 2001). Shrimp production in Sinaloa is 46% of the national total (SEMARNAP 1999). Despite this situation, only 194 of the original 266 shrimp farms are now in operation.

Agriculture is the second part of the land use complex in Sinaloa. It is carried on close to the coastal lagoons where one of the most successful irrigation developments is located. Several large rivers coming from the Sierra Madre Occidental Range rise in the states of Durango and Chihuahua (Fig. 22.5). The 11 rivers that flow to Sinaloa valley annually discharge 17,707 million cubic metres. This water is stored in reservoirs and

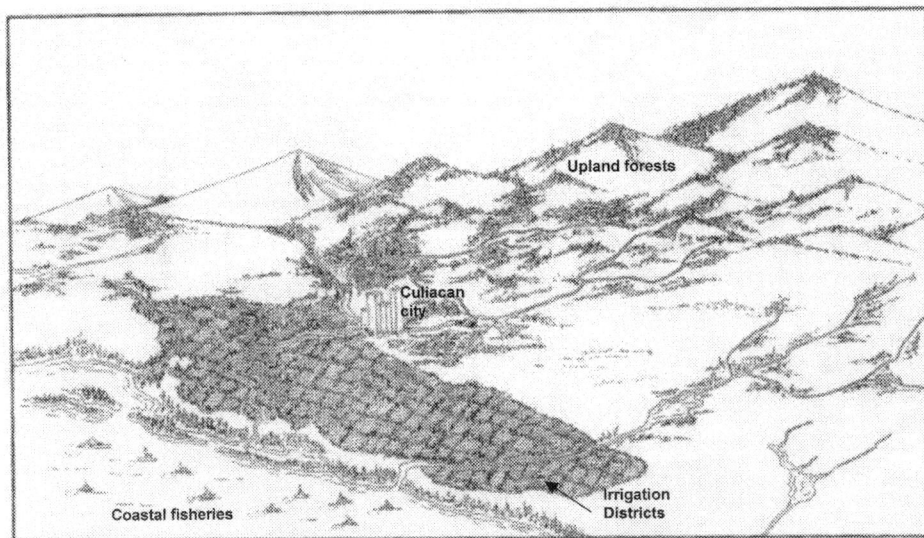

Fig. 22.5 The three elements of the ecosystem stalemate: the forest upland in Durango and Chihuahua states, the horticultural irrigation districts with their network of dams in the Sinaloa plains, and the fisheries in the coastal zone.

used for hydroelectric generation, domestic purposes and the irrigation of 715,721 hectares of export crops (Cruz 2002).

Sinaloa's water resources are administered through Irrigation Districts. However, the systems present many problems in distribution, since in this process over 50% of the water is lost due to infiltration and watering inefficiency. The residual water from irrigation, plus that from sewage systems, reaches the sea by surface or underground runoff. Shrimp farms utilize and recycle such runoff water. Shrimp farming problems have stimulated discussion of water pollution of the coastal lagoons. However, aquaculture itself creates water pollution problems with the application of about 1700 kg of fertilizer per hectare along with antibiotics. The large percentage of this material that is not converted to shrimp biomass is wasted, affects water quality and causes eutrophication. Most operations produce nitrogen and phosphorus above acceptable levels (INE 1992), and only 5% of the operations have treatment facilities that satisfactorily restore water quality.

Shrimp farming has encouraged the discussion of water pollution in coastal lagoons. Pond seeding of shrimp requires about 3 million post-larvae to satisfy pond operations. Most of these come from the USA. This creates a double problem of dependency on an outside country, and risks of pathogenic pollution from local sources. In addition to the heavy demand for water for aquaculture, there is intensive application of limestone, nitrogen and phosphorus fertilizers, plus antibiotics, in the order of 1700 kg/ha. Only 50% of this input is converted into animal biomass. The rest is not utilized, and it impacts water quality and ecosystem function. It is known that discharge of nutrients from shrimp-farm effluents can consist of 8.5 kg of phosphorus and 52.1 kg of nitrogen per hectare. Both these concentrations are indicators of eutrophication processes and most of the operations are above acceptable limits (INE 1992). In addition, only 5% of the installations have sedimentary and oxidation pools to restore residual water quality (Table 22.2).

The third part of the land use complex involves treatment of the watersheds that provide the water to the lowland areas. These forested upland watersheds lie in other state jurisdictions where further human activities occur. As shown in Fig. 22.5, the flourishing prosperity of the valley is based on the use of water from the rangelands in the states of Durango and Chihuahua. Silviculture practices in these areas do not take into account the timber harvesting effects on streams. Furthermore, in these for-

Table 22.2 Average demand for inputs in shrimp production in the Sinaloa farms of about 24,000 hectares

Input	Average amount (tons) applied per hectare	Total (tons) for 24,000 hectares
Food	1.5	36,000
Lime	2.55	61,000
Nitrogen and phosphorus	0.165	3,960
Antibiotics	0.035	126

Data courtesy of A.C. Hernandez.

est lands there are people with no means of making a living other than through forest harvesting. However, erosion caused by forestry activities is causing reservoirs to fill up with sediment. Reduction in storage capacity is diminishing water availability for fish and fisheries.

Forestry impacts on fish are an interconnected part of this complex of adverse effects on fish and their habitat.

It is additionally unfortunate that this complex 'coastal-upland stalemate' has synergistic negative effects on marine populations such as the little sea cow. The little sea cow (*Phocoena sinus*) is an endemic species from the Gulf of California, which is in danger of extinction due to flow reduction and chemical pollution of the Colorado River in the USA. As is the case with dolphins that are incidentally caught in fish nets, this synergistic effect can bring these cetaceans to extinction.

The scenario presented in preceding paragraphs indicates that Mexican natural resources are in peril and this state of environmental emergency demands a holistic approach for solution. It must involve, above all, the people who are creating impacts, and those that determine government policies and allocate funding. The question is: how can we change societal values regarding nature and eliminate destructive practices while promoting a truly equitable, sustainable use of forest uplands and aquatic ecosystems?

Integrated watershed management initiatives

Aber & Melillo (1991) stated that 'multiple and more intensive use of forest land is inevitable as the world population multiplies, and the more thorough and comprehensive our knowledge of woodland ecology, the better will be the prospect for wise use and long-term conservation of the woodland resource'. We may classify needs and recommendations for improvement of forestry practices and standards into:

(1) emergency measures (short term)
(2) urgent measures (middle term)
(3) long-term strategies.

Hammond (1997) proposes as a silvicultural principle a new vision of forest management named ecoforestry. According to this author:

> Ecoforestry is about protecting the forest and related ecosystems that sustain life on Earth. Ecoforestry is not about canny ways to cut more trees or clever ways to combine logging with berry-picking and bird watching. Ecoforestry is about understanding our place as humans in ecosystems and acting on that understanding. Ecoforestry is about future generations of people, about taking the time to provide intergenerational equity for all life.

But operatively, how can that be done in a country such as Mexico? Some viable initiatives are posed.

Hydrographic and environmental services

Scientists around the world have generated enough evidence to show that one key aspect in the area of fish–forestry degradation has been the failure to fully account for the full value of 'ecosystem services'. In any restoration technique or social participatory strategy, policy enforcement measures must include a serious, quantifiable reckoning of all the resources and resource values associated with forests. There is some hope for the protection of watersheds if such environmental costs can be internalized in production process costs. One of the most promising schemes is to analyse the total value of the highlands still covered by forest, and to restore degraded areas.

The importance of placing an option value and an environmental services value on the forests relies on a social understanding with respect to natural resource economics, considering what environmental quality of life means, versus material quality of life. Theoretically speaking, an option value is the price that beneficiaries (individuals, producers, organizations, industry owners and societies) would be willing to pay in order to preserve forest cover and associated resources. These might include such values as gene pools as natural capital, and environmental services such as water supply, flood protection, erosion control, etc. Putting a value on environmental services, such as option value, may safeguard the possible future use of these lands. At present, timber is produced only as a secondary raw material. It may therefore be more valuable (economically) to maintain cover forests for environmental services. It is due to current circumstances in which only the direct value of forest products, established by conventional market rules, is counted. This undervalues the other important assets of the ecosystems. Applying the methodological approaches proposed by Pearce & Moran (1994), the total value of the forest could be formulated as presented in Table 22.3.

The recognition of the value of a forest, as a supplier of water, is gaining support among many social and political sectors in Mexico. In the last 3 years there have been significant contributions in academic and legislative arenas to establish methods to collect monetary returns from those beneficiaries of hydrographic and environmental services offered by forests. Scientists have identified a potential opportunity of getting a cash flow from water coming from natural and artificial reservoirs supplied by forest cover. Its utility can be permanently justified economically, and it is a way to protect and maintain upland forests to guarantee water supply. Five elements are set and should be added to give an economic value to the watershed services and internalization of environmental costs:

(1) The potential value of catching water by the upland forest. It is given by the annual hydrological productivity. It includes the volume of water produced by certain conditions of tree density, rates of evapotranspiration, canopy features (needle or wide leaves), understory vegetation, soil conditions (slope, depth and texture), precipitation patterns and basin form.

(2) The protection value established by the capacity of the canopy and soil in stopping floods and erosion.

(3) The value of the water harvested from a watershed, which is used as input in the agriculture and industry production.

Table 22.3 Categories of economic value recognized in mountain forests in Mexican watersheds

Use values			Non-use values	
Direct consumption (market-oriented); primary net production	Indirect (non-market-oriented)	Optional or future assets	Bequest values	Existence values
Firewood	Water supplies	Gene pools to support new crops, drugs	Trans-generation enjoyment (present generations have the moral responsibility to preserve and maintain the best possible conditions of the forest for the next generations)	Evolution continuity
Timber	Flood control	Generic improvement of current crops		Universal appreciation (benefit of knowing that species will persist over time)
Traditional medicines	Erosion control			
Forages	Carbon fixation	Ecotourism		
Aromatic plants	Nutrient cycling			
Hunting and fishing	Waste absorption			
Fibres	Oxygen generation			
Ritual plants	Beneficial fauna			
Construction materials	Migratory species habitat	Environmental prospecting		
Mulching for gardens	Visual quality and aesthetic contemplation			
Resins	Food chain fluxes			

(4)   The value of streamside forest as both receptacle and withholder of point and non-point source pollutants.
(5)   The value as a medium for freshwater fish populations.

The idea is to promote a new era of forestry practice, a new vision of the forest. A keystone principle is to recognize that many upland forests yield more economic benefits through environmental services (indirect use of value), such as water supply and erosion control, than timber harvesting or extensive livestock raising. A second requirement is to recognize that it is possible to manage the density, composition and structure of a forest so that an equilibrium is kept between a high quality water flow and erosion control by cutting only a residual mass of forest. It requires maintenance of only the needed density of tree stands, since a high rate of evaporation will yield low runoff discharges. Therefore, the issue is to develop a model on each watershed base to catch enough water, but not so much that it fails to satisfy downstream irrigation, industrial and human water needs. A third consideration is to take into account the other economic and non-economic values such as carbon fixation, biodiversity continuity and so on, whose opportunity cost ought to be considered.

This approach to forestry requires legislative, technical and social changes. One technical challenge is to determine the yield of water per hectare of forest under a given density of trees, environmental conditions and forestry management strategies by switching the conventional timber-oriented management into environmental service plans. This approach may require compensation of forest owners for the amount of timber not harvested, so that they can change into a diversified production compatible with water production, fish population maintenance, ecotourism, carbon sequestration and so on.

The approach that we have discussed, which puts emphasis on the value of ecosystem services and internalizes the cost of loss of such services in forestry production, may provide a different way of considering forestry–fish interactions. In this approach fishery values may be protected by special land use planning methods, but they are also protected by having resource exploitation set within the context of a fuller system of valuing and managing all ecosystem elements.

Agriculture reconversion and productive diversification

Parallel to the above-mentioned measures, the hydrological rehabilitation of marginal lands in priority watersheds is an economic imperative for the conversion of conventional mountain agriculture practices, based on chemical fertilization and pesticides, into cleaner organic techniques. Community-based organizations should be developed to participate in the farm programmes that are intended to diminish flooding and soil erosion and concurrently protect fish and increase food production and mountain farm incomes.

To address this situation all beneficiaries must be persuaded to participate in the rehabilitation and conservation of the forest as a whole. All the stakeholders must work to accomplish three specific ends:

(1) Design appropriate methods to provide economic assessment of the present and future values of ecosystems services from functional forest watersheds.
(2) Establish policies, regulations and enforcement based on people's commitments through education, training and participation.
(3) Authorize only sustainable ways to exploit forest resources that still provide economic incentives for peasants. In these instances, with agrosilvopastoralism introduced as a long-term strategy, competitive land uses will need to be coordinated through a participatory process aimed at restoring degraded watersheds (Vandermeer & Perfecto 1995).

Educational institutions, government agencies and society, above all, have a stake in such an approach. Any effort to restore harmony between the people and the forests will have to rely, to a large extent, on accurate knowledge of local species and aboriginal people, adaptable to adverse conditions.

Regional planning of water resources

Watersheds are divided into an array of hierarchical categories, from micro-watersheds composed of 5000 hectares to a few hectares of drainage areas. The Comisión Nacional del Agua (CNA, National Water Commission) is in charge of water administration and management. The CNA, along with other governmental and non-governmental institutions, has been working on a Watershed Council in each micro-watershed. However, so far social involvement is minimal considering the size and urgency of the problem.

With the challenge of managing watersheds and associated resources in a modern, sustainable and equitable way, the CNA has designed strategic, watershed-level, planning mechanisms to prevent environmental damage and operate in a fashion that is economically and socially sustainable. Watershed Councils were established based on such concepts. This requires participation of all water users and interested parties. To date, 26 watershed councils have been established and they have been involved in such issues as over-use of aquifers, sewage treatment plant construction and sale of underground water rights.

Green and blue revolutions reframed

An integrated fish and forestry vision is needed in Mexico. Eco-farming of fresh and salt water is a promising industry. Aquaculture around the world currently accounts for 22% of global production of fishery products for human nourishment (Mathias 1995). To feed the population, Mathias states that fish must grow at only 4% per year in order to keep within the capacity of marine and inland waters. Fish may be divided, by their food habits, into carnivorous and non-carnivorous. The latter forms, that feed low on the food chain, and consume algae, aquatic weeds and insects, offer advantages for Mexico's conditions.

A successful experience of sustainable use of forest and water

For many decades traditional communities in Putla de Guerrero, Oaxaca, used an artisan-manufactured substitute for refined sugar, called *panela*. In 1991, a group of sugar cane producers in the southern state of Oaxaca initiated a sustainable project to replace polluting, diesel-based mills with those that use hydropower and to operate old machinery again to grind sugar cane stalks for *panela* production. This initiative let the producers see that, to maintain water flow, it was essential to protect forest cover in upland areas of the agricultural valley. In this way the watermill owners started a campaign to reduce destructive practices in the forests and promote reforestation and protection of riparian forests. Later on new projects integrated with the *panela* process emerged. One of them involved catching water, flowing through the waterwheels, and running it into two natural pools to grow native fish under organic certified procedures. The water, free from pollutants, was returned to the river system. This scenario is an example of what can be done if technology, producers' talents and cooperative participation are put together.

Such new projects are emerging around the country. One of the most promising in the fish–forestry realm is the Valle de Bravo Watershed management plan. Valle de Bravo is a basin composed of about 60,000 hectares of uplands. This watershed contributes 6 cubic metres of water per second to the Cutzamala systems which supply thirsty Mexico City. Since most of the land is still covered by coniferous forests under different stages of management and perturbation, protection of the watershed is considered one of the priorities of water production. The forest management plan for the Valle de Bravo watershed seeks to impose an ecosystem services approach through the internalization of environmental benefits in the costs of water production for Mexico City's conurbations.

Conclusions: a race against time

Mexico, as well as many other Latin American nations, is torn between economic, social and political pressures, and environmental crisis. Time is not on the side of a growing population. The decline of forest and aquatic ecosystems is already under way. To address these threats to our immediate future rationally, a massive mobilization of monetary resources and political action is urgently needed. There is no choice but to launch a national effort to restore and care for the interconnected resources of forests, fishes, aquifers, soil and environmental life-support systems. Otherwise, we will leave our descendants a country without any hope. Professionals, politicians, leaders and constituencies together have to respond to the greatest issue ever faced by our country.

Over 50% of the Mexican biodiversity is associated with streams and aquatic ecosystems. An aggressive and socially oriented programme is needed to protect the last fluvial forests and the biodiversity within them. Protecting and managing Mexico's forest ecosystems requires a holistic and international effort, including help through NAFTA partnerships and cooperative programmes. The need for a comprehensive approach is clear if the Lake Chapala situation is considered. Burton (1997) noted that 'It would be socially irresponsible and naïve to suggest protecting Lake Chapala without first trying to ensure improved living standards for all the basin's 8 million inhabitants.'

Mexican forestry practices, for at least a century, have only considered timber production. They are now in a transitional process. A new policy and social perceptions are emerging from economic pressures and ecological problems. We recognize that changes take time, and much work lies ahead for scientists, technicians and policy makers. The challenge is there for forestry schools to establish academic and research programmes concerning fish–forestry relationships. Fish–forestry issues are new realities in this country where traditionally rivers have been used for depositing sewage.

Despite the chaotic and deplorable scenario for Mexican forests, there are enough reasons to be optimistic. At the end of March 2001, the Federal Government announced that it was putting into gear a 'National Crusade for Forest and Water'. This proclamation was put into action by the Comisión Nacional Forestal (National Commission of Forest). One of the main goals of this federal institution was to establish a national policy to value environmental services. This goal has to involve participation by all the other governmental agencies, particularly the Secretaría de Hacienda (the equivalent of the Treasury Department of the USA), in order to change forest regulations linked to water and forest soil conditions. It is encouraging that government subsidies, applied to promote certain crops, are being redesigned to allocate federal money towards sustainable projects associated with reforestation of hilly lands. This is being done to replace slash and burn practices with a diversified local agrobiodiverse germ plasm, and to protect and grow new forest stands so important to fish and fisheries.

Acknowledgements

Our grateful thanks go to Dr Thomas G. Northcote and Dr Gordon F. Hartman for their patience and help in correcting our drafts. They made many significant contributions to improve this chapter. Finally, we want to thank Dr Denis Hoffman from the Universidad Autónoma Chapingo for his editing work on the first draft.

References

Aber, J.D. & Melillo, J.M. (1991) *Terrestrial Ecosystems*. Sounders College Publishing, Philadelphia, PA.

Arredondo, F.J.I. & Aguilar, D.C. (1983) *Bosquejo Histórico de las Investigaciones Limnológicas Realizadas en los Lagos Mexicanos*. Contribuciones en Hidrobiología. Inst. Biol. UNAM, pp. 91–133.

Bozada, L. & Paez, M. (1986a) La fauna acuática del Río Tonalá. Centro de Ecodesarrollo, Universidad Veracruzana. *Serie Medio Ambiente en Coatzacoalcos*, 7, 82–7.

Bozada, L. & Paez, M. (1986b) La fauna acuática del Río Coatzacoalcos. Centro de Ecodesarrollo, Universidad Veracruzana. *Serie Medio Ambiente en Coatzacoalcos*, 8, 70–6.

Bozada, L. & Paez, M. (1986c) La fauna acuática del litoral. Centro de Ecodesarrollo, Universidad Veracruzana. *Serie Medio Ambiente en Coatzacoalcos*, **14**, 101–7.

Burton, T. (1997) Can Mexico's largest lake be saved? An international rehabilitation effort gets under way. *Ecodecision*, **23**, 68–71.

CESPEDES (1998) *Eficiencia y Uso Sustentable del Agua en México: Participation del Sector Privado*. México, Cámara Mexicana de la Industria de la Construcción (CMIC)-Centro de Estudios del Sector Privado para el Desarrollo Sustentable, pp. 10, 26, 34, 37.

CNA (2002) Programa Nacional Hidráulico 2001–2006. *Compendio Básico del Agua en México*. CNA PND 2001–2006. SEMARNAT. p. 28.

CONABIO (1998) *La Diversidad Biológica de México: Estudio de País, 1998*. Comisión Nacional para el Conocimiento y Uso de la Biodiversidad de México.

CONAFOR (2002) *La Comisión Nacional Forestal: Su programa Nacional de Educación, Capacitación, Investigación y Cultura Forestal*. Work paper.

Contreras-Balderas, S. (1999) Annotated checklist of introduced invasive fish in Mexico, with example of some recent introductions. In: *Nonindigenous Freshwater Organisms: Vectors, Biology, and Impacts*, (eds R. Claudi & J.H. Leach), pp. 33–54. Lewis Publishers, Boca Raton, FL.

Contreras-MacBeath, T. (2002) *La carencia de criterios de sustentabilidad en el manejo de los ecosistemas acuáticos dulceacuícolas*. Unpublished paper submitted at the Johannesburg, South Africa, Summit Conference on Sustainable Development.

Cruz, H.A. (2002) *Contribucion al Estudio de las Condiciones de Contaminacion de Aguas Superficiales y Lagunas Costeras en el Estado de Sinaloa*. Unpublished paper.

Del Campo, R.M. (1984) Productos biológicos del Valle de México. *Revista Mexicane de Estudios Antropológicos*, **14**, 53–77.

Galindo, R.J.G. (2000) *Condiciones Ambientales y de Contaminación en los Ecosistemas Costeros*. UAS-SEMARNAP.

Garcia, E. (1978) *Modificaciones al Sistema de Clasificación Climática Kopen*. Instituto de Geografía UNAM.

Hammond, H. (1997) What is ecoforestry? *Global Biodiversity*, **7**, 3–7.

INE (1992) *Programa de Ordenamiento Ecologico para el Desarrollo Acuícola de la Regio Costera de Sinaloa y Nayarit*. OEA-SEDESOL.

Lugo, A.E. (1994) Maintaining an open mind on exotic species. In: *Principles of Conservation Biology*, (G.K. Meffe & C.R. Carroll). Sinauer Associates Inc., Sunderland, MA.

Lyle, F.L.J., Romero, B.E. & Bect, V.J. (2001) *Características y Desarrollo de la Camaronicultura en el Estado de Sinaloa*. Centro Regional de Investigaciones Pesqueras, Mazatlán, Mexico.

Mathias, J.A. (1995) Aquaculture: the blue revolution. *Ecodecision*, **18**, 66–70.

Mitchell, J.G. (1992) Our disappearing wetlands. *National Geographic Society*, **182**, 3–45.

Mitsch, W.J. & Gosselink, J.G. (1986) *Wetlands*. Van Nostrand Reinhold, New York.

Muñoz-Piña, C., de Janvry, A. & Sadoulet, E. (2001) *Recrafting Rights over Common Property Resources in México: Divide, Incorporate, and Equalize*. Department of Agricultural and Resource Economics, University of California at Berkeley, CA.

Novelo, A. & Martínez, M. (1989) *Hydrilla verticillata* (Hydrocharitaceae), problemática de una maleza acuática de reciente introducción en México. *Anales del Instituto de Biologia de la UNAM*, **53**, 97–102.

Pearce, D. & Moran, D. (1994) *The Economic Value of Biodiversity*. IUNC-Earthscan, London.

PEF (Poder Ejecutivo Federal) (1996) *Programa Forestal y de Suelo 1995–2000*. Secretaria de Medio Ambiente, Recursos Naturales y Pesca, México, DF.

PNH (2001) *Programa Nacional Hidráulico 2001–2006*. CNA, SEMARNAT.

Ramamoorthy, J. (1998) *Diversidad Biológica de México*. Instituto de Biología de la UNAM.

Rzedowski, J. (1978) *Vegetación de México*. LIMUSA, México.

Sánchez, V.A. & García, R.M. (2000) The dying Mexican tropical dry forest: finding treasures among its ruins. *Tropical Conservancy*, **1**, 16–26.

SARH (1994) *Inventario Nacional Forestal Periódico*. Memoria Nacional. SARH-SFF, México, DF.

Scott, D.A. & Monserrat, C.A. (1986) *Dictionary of Neotropical Wetlands*. International Union for Conservation of Nature and Natural Resources.

Secretaria de Pesca (1994) *Desarrollo Científico y Tecnológico del Cultivo del Robalo*. Dirección General de Acuacultura, México.

SEMARNAP (1999) *Estadísticas de Producción Forestal*. Dirección General Forestal. Dirección de Desarrollo Forestal.

SEMARNAP (2000) *Areas Naturales Protegidas de Mexico*. Instituto Nacional de Ecología, Secretaria de Medio Ambiente, Recursos Naturales y Pesca.

Swanson, F.J. & Franklin, J.F. (1992) New forestry principles from ecosystem analysis of Pacific Northwest forests. *Ecological Applications*, **2**, 262–74.

Tellez, K.L. (1993) El sector forestal y el tratado de libre comercio. In: *Reunión Nacional de Economía Forestal 1993 y II Seminario Nacional: Tratado de Libre Comercio y Sector Forestal*. Colegio de Postgraduados, Montecillo, México.

Toledo, V. *et al.* (1980) Los purépechas de patzcuaro: una aproximación ecológica. *América Indígena*, **15**.

Vandermeer, J. & Perfecto, I. (1995) *Breakfast of Biodiversity*. Food First, Oakland, CA.

Part VII
Non-North American Fish–Forestry Interactions

Chapter 23
Fishes–forestry interactions in tropical South America

C.A.R.M. ARAUJO-LIMA, N. HIGUCHI AND W. BARRELLA

Introduction

South America is 17.8 million km² in area. It is composed of 12 nations and has a population of 310 million people (1990 census). The continent has very large river basins (Welcomme 1985) and extensive forests (FAO 1999), most of them located in the tropics. Tropical fish and plant biodiversity are regarded as being among the richest of the planet (Hueck 1978; Schaefer 1998). Many of these countries have been cattle and crop producers and these activities have modified the original vegetation pattern, particularly during the last 400 years.

Most of the environmental changes have concentrated in the southern and eastern side of the continent. The Amazon rainforest, which covers regions of seven nations, is still relatively pristine. However, the deforestation rate of the Amazonian forest, due to development programmes, including logging, is high (0.5% per year) (INPE 2002) and invariably will lead to dramatic changes in the next decades.

In this chapter we gathered the available information about the fishes and forestry resources and how these two interact. Because interactions between trees and fish tend to be more important in large forested areas we have concentrated our efforts on tropical regions. Savannas and grassland are mentioned mostly for comparison purposes.

Description of the South America region

Geology, soils, topography and geographic features

Two Precambrian shields and a recent mountain range represent the geological structure of the continent. They are the Andes mountain range and the Guyana and Brazilian plateaus (Petri & Fulfaro 1983; Lundberg *et al.* 1998). Others that are relatively unimportant in area, when compared with the two main plateaus, are the Macarena massif in Colombia, Sierras Pampeana in Northwest Argentina and the Somuncura and the Deseado, the two Patagonian shields.

The main sedimentary basins of the continent are located between the cratons and between cratons and the Andes and often are segmented by structural arches (Salo & Rasanen 1989; Jordan 1995; Souza Filho & Stevaux 1997). East of the Andes there is a foreland sedimentary deposit that received sediments from the shields and the Andes, and it is nowadays segmented to form the modern drainage of the western Paraná basin and headwaters of some mid-Amazon tributaries. The Amazonas/Solimões sedimentary basin received deposits from both shields.

The continent is sustained by only one tectonic plate, which has been in an East–West compression state of push against the Nazca Plate. This has resulted in the lift of the Andes. The complex history of the Andean mountain building, the consequent foreland sedimentation and the sea level changes are specially important in understanding the evolution of the river drainages in South America and their fish fauna (see Lundberg *et al.* 1998 for a summary on recent geological history of the continent).

Excluding the Andes, with an average altitude of 1000 m and a maximum of 6960 m (Aconcagua peak), most of the continent consists of plains and highly eroded low altitude highlands (<500 m). The central parts of the Guyana Shield and the eastern part of the Brazilian Shield have altitudes between 1000 and 2000 m, with a few peaks reaching 3000 m.

Nowadays, the river basins draining the relatively young Andean formation or its foothills erode lithosols and acrisols leading to alluvium rich in nutrients and with high cation exchange capacity (Stallard *et al.* 1990; Furch 1997). Examples of the latter are found in the Amazon basin, Orinoco basin and Paraná basin. Rivers draining the old Precambrian shields or their sedimentary basins erode ultisols or red-yellow podzolics, oxisols or yellow latosols and ferrasols and carry less sediment (Junk & Furch 1980; Thomaz *et al.* 1997). The alluvium eroded from them has low nutrient reserves and cation exchange capacity and is high in aluminium. There are a few patchy limestone and volcanic extrusions in the Precambrian shields, which lead to an increase in the nutrient load of highland rivers (Petri & Fulfaro 1983; Souza Filho & Stevaux 1997). Highland rivers make up the most common river type in the continent, many of which have been dammed to generate hydropower.

Climate and weather

The Equatorial, Tropical and Atlantic-Polar Air Masses influence South American climate. In the Amazon, Orinoco and Guyana basins (Fig. 23.1), the mean yearly air temperature ranges from 24° to 27°C and seasonal variation in the lowlands and in the foothills of mountains is small (Lowe-McConnell 1964; Roche 1982; Weibezahn 1990; Gonçalves *et al.* 1997). In the interior of the forest the microclimate is even more stable, especially in the low strata that receive no direct sunlight. Cool weather is found in the Andes and in the mountain range of the central Guyana Shield. The mean precipitation ranges from 1500 to 3600 mm y^{-1} (Roche 1982; Marengo & Nobre 2001). The highest values of annual precipitation are found near the ocean and along the Andes foothills, while the lowest rainfall values are in the centre-south Amazon (2000 mm y^{-1}), south Amazon (1700 mm y^{-1}) and the north (1500 mm y^{-1}). Water vapour coming from the ocean moves inland and is continually recycled. Up to 50% of the water vapour generating rainfall is recycled from plant evapotranspiration. The rainy season runs from October to June, with a south–north displacement. Dry climate areas (~500 mm y^{-1}) lie in the north of Venezuela and Colombia and the northeast of Brazil. These regions are subject to prolonged droughts.

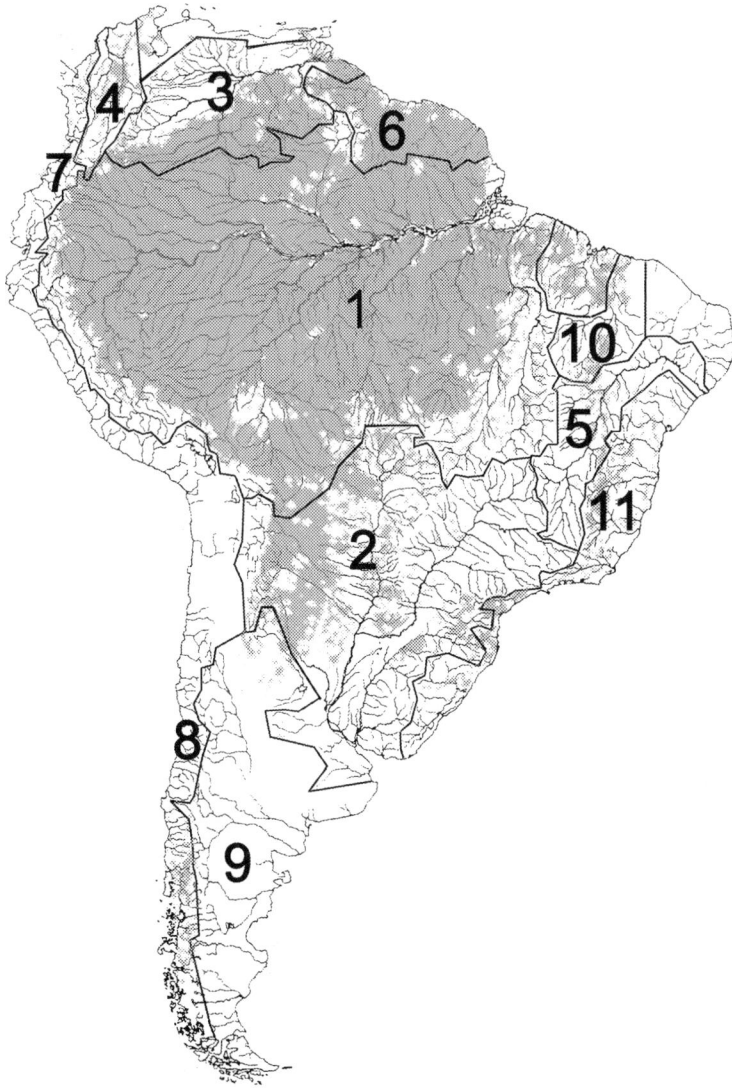

Fig 23.1 Main river basins of South America and forest resources (shaded). Source: FAO, 2000. 1, Amazon; 2, Paraná; 3, Orinoco; 4, Magdalena; 5, São Francisco; 6, Guianas; 7, San Juan; 8, Pacific; 9, Patagonia; 10, Parnaiba; 11, Atlantic or Coastal.

The climate is tropical to subtropical with marked seasons in a band of approximately 20° around the Tropic of Capricorn, where most of the Atlantic rainforest is located. This band includes the headwaters of the Paraná, São Francisco watersheds and the Atlantic basin. Winter months are from May to August and summer extends from November to February. Mean annual temperature in the lowlands is 12°C in the south and 24°C in northern edges of the zone. Maximum and minimum temperatures of 42°C and –4°C occur in the summer and winter, respectively (Gonçalves *et al.* 1997; Julio *et al.* 1997). Summer storms caused by polar fronts are common and deliver

30–50% of the yearly precipitation (Gonçalves *et al.* 1997). The mean precipitation ranges from 1250 to 2000 mm y^{-1} but higher values (3600 mm y^{-1}) occur in a few sites (Drago 1989; Julio *et al.* 1997). In central South America it rains less than it does along the coast.

General forest cover features

Broadly, the vegetation distribution corresponds closely to the climatic zones (Hueck 1978). The areas of tropical climate with precipitation >1500 mm y^{-1} and less than 6 months of dry season have a dense cover of rain forest (Amazon and Guyana basins and parts of Orinoco and Atlantic basins). Very dry areas in the extreme north and northeast are covered with scrubland. In between these are zones of tall grass (savannahs, tropical grassland). Mixed (containing both deciduous and evergreen trees) and semi-deciduous forests occur in the Paraná basin, south of the Amazon basin and along the slopes of the Andes. From the flat Pampas of east central Argentina to Patagonia the vegetation is mostly grassland type.

The Brazilian part of the Paraná basin has been largely deforested and only small fragments of the original cover remain near the main river and on its floodplain (Campos & Souza 1997). The situation is slightly better in Argentina and Paraguay (Neiff 1990; Mereles *et al.* 1992). Deforestation of the Atlantic basin (Mata Atlantica or Atlantic Forest) has also been intense, and its conservation status is defined as critical (Saatchi *et al.* 2001).

The Amazon, Orinoco and Guyana basins still have large fractions of their original vegetation. Pires & Prance (1985) grouped the Amazon forests as upland and flooded forests, and their classification can be extended to the other two basins. Upland (Terra Firme) forests include dense forest (or lowland forest) and open forest (dry or semi-deciduous forest, liana forest and montane forest). Dense forest, the most common type, occurs in low altitude relief, e.g. below 250 m above sea level. Its floristic composition is highly variable, and a high diversity of Leguminosae is common (Hammond *et al.* 1996). The trees of this forest can be very tall (~ up to 50 m). Liana forests, which have tall trees as well, are abundant in the eastern Amazon, and are relatively open compared with the dense forest types. Dry forests occur between montane forest and the dense forest and the savannahs, at altitudes between 800 and 2500 m asl. Montane forests of Peru and Ecuador have been severely deforested (Kvist & Nebel 2001).

The upland forest formation is very diverse with more than 3000 species (TCA 1995), but many have no current market value. *Swietenia macrophylla,* Meliaceae (mahogany) is by far the most valuable species of the region. Other important hardwood trees in Brazil, Peru, Bolivia, Venezuela and Guyana markets are: *Cedrela odorata, Carapa guiensis, Torresea acreana, Dinizia excelsa, Hymenaea courbaril, Cordia apurensis* (Hammond *et al.* 1996; Howard *et al.* 1996; Kammesheidt 1998).

The flooded forests (gallery or riparian forests) include floodplain, seasonally flooded, mangrove, tidal and permanent swamp forests (Pires & Prance 1985). Floodplain forests are on low-lying ground, which is flooded by irregular rainfall. Seasonally flooded forests are flooded by regular annual flood cycles of rivers (Fig. 23.2). Mangrove and tidal forests are flooded twice a day by salt water and fresh water.

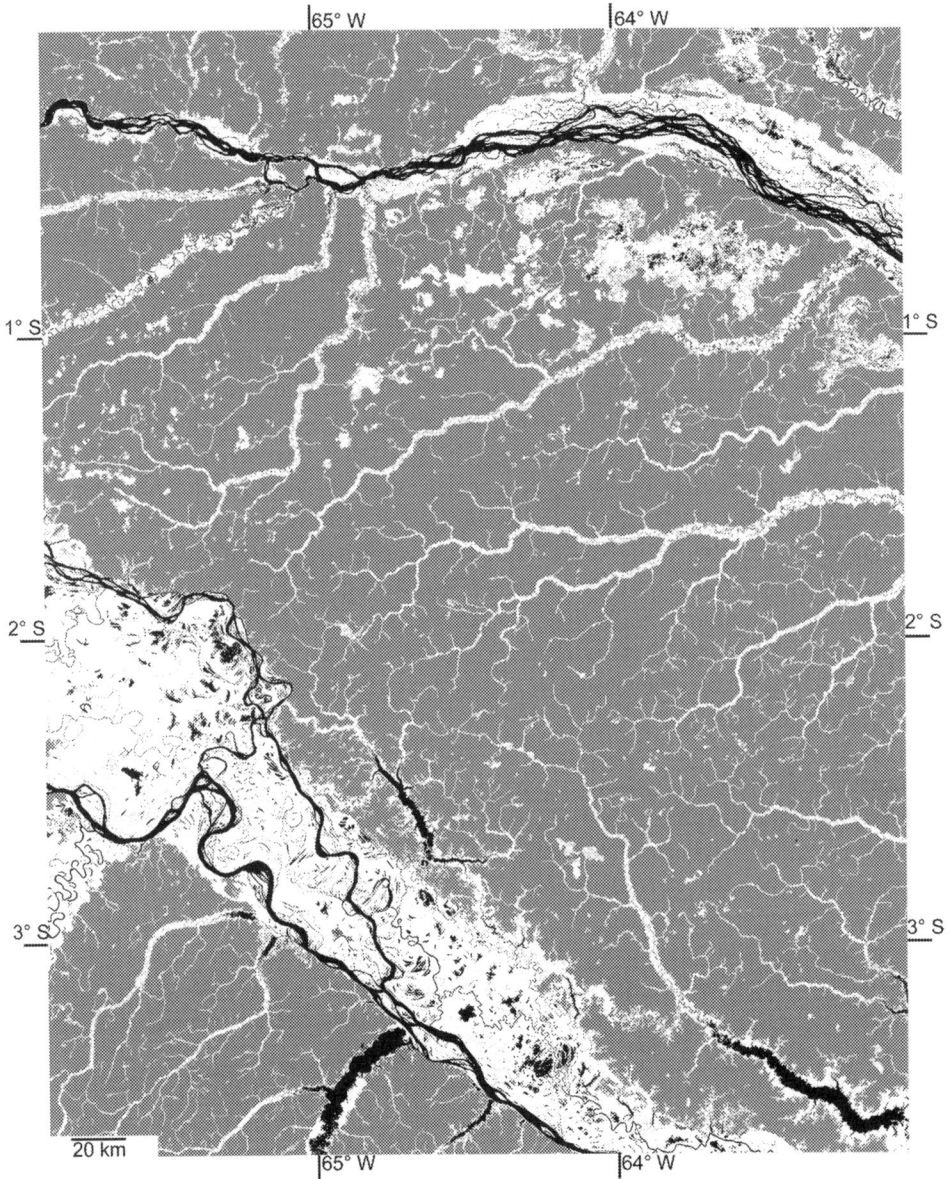

Fig 23.2 Different types of vegetation (see text for definitions) of a central section of the Amazon basin during the wet season. The vegetation classification was based on a radar image (JERS-1 SAR). Grey = upland forest and aquatic herbaceous vegetation; white = flooded forest; black = open water.

Seasonally flooded forests further differ in composition and production rates, in response to the cation exchange capacity of the alluvial soils, e.g. 'varzea' and 'igapó' forests in the Amazon. They also vary in relation to tree distribution in the floodplain (Ayres 1993; Furch 1997; Ferreira 1997; Worbes 1997). Some Amazonian trees can withstand as much as 7 months of flood (Worbes 1997) and are distributed in the floodplain up to 6 m depth. However, the trees are thinner at the deeper edge. A slightly similar pattern of tree distribution has been identified in the Paraná River basin (Campos & Souza 1997). Trees usually found in the seasonally flooded forest are listed in Ayres (1993), Tito de Morais *et al.* (1995), Campos & Souza (1997), Godoy *et al.* (1999) and Kvist & Nebel (2001).

Major river basins and aquatic habitats

There are 11 main river basins in South America (Fig. 23.1). Five of them (Amazon, Paraná, S. Francisco, Magdalena, Orinoco) drain to main channels. The others (Essequibo or Guyana, San Juan, Pacific, Patagonian, Parnaiba and Coastal or Atlantic basins) are composed of independent and relatively small watersheds that are close to each other and usually flow in the same direction. They have been grouped for classification purposes and may share similar fish fauna (Menezes 1972; Ringuelet 1975; IBGE 1992; Lundberg *et al.* 1998; Lowe-McConnell 1999).

The basins owe their flow and chemical characteristics to their stream networks. The streams are subject to short-term local flooding because of their small water storage/retention capacity. Low order streams and small rivers usually harbour floodplain forest, but seasonally flooded forest occurs only in mid and high order rivers. Streams and small rivers are important fish habitat used seasonally by some fish species of commercial value.

The Amazon basin is by far the largest of the continent, covering an area of up to 7.05 million km² (including Tocantins River) and approximately 39% of South America (Table 23.1). The Amazon River, the main artery of the basin, has 19 major tributaries, 14 of which are in Brazil. Their headwaters are on the Andes, sedimentary basin or on the plateaus, which give each of them distinctive water quality. The streams of the lower and central Amazon drain intensely weathered soils and are generally poor in sediments and nutrients (black and clear waters), contrasting with the sediment-rich streams draining the Andes and their foothills. Stream density can reach 2 km km⁻². The riparian vegetation covers about 1 million km² or 14% of the Amazon basin (Junk 1993).

The main fish habitats are the river channels, the lakes and the floodplain. The latter can be up 80 km wide. Lakes occupy on average 15% of the maximum area flooded by the Amazon River; the remaining area is covered by herbaceous vegetation (43%) and seasonally flooded forest (41%) (Melack & Forsberg 2001).

Other types of water bodies in the Amazon basin are the man-made lakes and reservoirs. There are presently four large and three small reservoirs in the Amazon. The total area of these reservoirs is 5700 km² (IBGE 1992).

The Paraná basin (including Prata, Uruguai and Salado rivers) is the second largest in South America (Fig. 23.1). It flows from north to south and therefore changes from a tropical to a subtropical regime. The main river is the Paraná, which has 10 large tributaries. The Paraguai River, the largest tributary of Paraná, contains two enormous floodplains: the Pantanal and the Chaco. Both floodplains are covered mostly by grass/savannah vegetation.

Table 23.1 Main hydrological features of the six most important river basins of tropical South America and the approximated number of fish species in each basin

River basin	Area (10^3 km^2)	Discharge (10^3 m^{3}s^{-1})	High water (months)	Low water (months)	Headwaters	Gauge change (m)	Number of species
Amazon	7050	210	May–June	Oct–Nov	1+2	8–12	4000
Paraná	3700	22	Feb–Mar	Sep–Oct	1+2	2–5	600
Orinoco	990	36	July–Aug	March–April	1+2	9–12	1000
São Francisco	630	3.1	Nov–Dec	June–July	2		150
Guyana	440	–	May–July	Nov–Dec	2	3–6	430
Magdalena	324	7	Nov–Dec	Feb–Mar	1	4–6	150

Headwaters: 1 = Andes Cordillera; 2 = Precambian shields. Sources: Lewis & Weibezahn 1981; Bonetto 1986; Di Persia & Neiff 1986; Drago 1990; Hastenrath 1990; Neiff 1990; Stallard *et al.* 1990; Tito de Morais *et al.* 1995; Planquette *et al.* 1996; Furch 1997; Mozeto & Albuquerque 1997; Richey *et al.* 1997; Thomaz *et al.* 1997; Britski *et al.* 1999; Lasso *et al.*1999; Lowe-McConnell 1999; Mojica 1999; Ponte *et al.* 1999.

Many rivers within this basin are regulated. There are more than 130 reservoirs in the Upper Paraná covering an area of 14,000 km^2 (Agostinho *et al.* 1995). The basin has the highest demographic density of the continent. The seasonally flooded forest has been largely removed, and only patches still persist (Neiff 1990; Campos & Souza 1997).

The floodplain of the upper and middle Paraná and lower Paraguai rivers, as well as some western tributaries, is 25 km wide in some reaches (Drago 1989). The seasonally flooded forest covers up to ~50% of the floodplain (rivers excluded) (Neiff 1990; Campos & Souza 1997).

The Orinoco watershed has four main tributaries (Fig. 23.1). The floodplain below the mouth of the Meta, much of which is grassland (Los Llanos), covers an area of 11,860 km^2 (Lewis 1988). Seasonally flooded forest is found along the river, near the delta (Colonnello *et al.* 1986) and in some tributaries and floodplain forest is common along streams (Lewis 1988; Winemiller *et al.* 1996; Lasso *et al.* 1999). At least two important reservoirs have been built in the Orinoco basin.

The Guyana basin is composed of five mid-sized and several small watersheds (Lundberg *et al.* 1998). Seasonally flooded forest and floodplain forests occur in the middle reaches. They are relatively pristine, and usually are <100 m wide. Floodplains covered by savannah occur in the headwaters and middle reaches of the Essequibo River (Lowe-McConnell 1964) and others.

The rivers of the Atlantic basin have their headwaters on the Brazilian plateau. They flow through formerly dense Atlantic rainforest. The region has been heavily deforested during centuries of crop farming. Streams are, however, often still fringed by floodplain forest. The Magdalena and the São Francisco basins have large areas (Table 23.1), but little flooded forests, flowing mostly through highlands or crossing savannahs and scrubland areas.

Major fish community components

The number of freshwater fishes of South America is still disputed. Schaefer (1998) registered 6187 species for the continent up to 1996, but projected that the number of species would exceed 8000. This species richness means that at least 25% of the world's 24,600 fish species are present in South America. Approximately 83% of the South American fish are ostariophysans (Characiformes, Siluriformes and Gymnotiformes), 14% are Perciformes and 1–3% belong to the orders Clupeiformes, Osteoglossiformes, Rajiformes and a few others (Nelson 1994). The number of species per basin is only approximated, but the Amazon basin is by far the most diverse (Table 23.1). Forest streams should be the main contributor to the increase in the number of fish species of South America. Most research efforts in the Amazon, Orinoco, Guyana basins have been concentrated in the main rivers. Stream inventories have been neglected because of their remoteness and lack of economic importance.

The seasonally flooded forests have a high predictability and periodicity, optimizing fish–forest interactions. In some regions of the Amazon basin, for example, large extensions of forests are flooded yearly for 6–7 months. In these areas, fish swim among the trees, feeding on seeds and dispersing them as well. In low order streams the predictability of the flood is low, water unexpectedly spills over the banks and fish–forest contacts occur mostly along the edges of the riparian zone. In the latter case interactions between fish and trees decrease as stream order increases.

Seasonally flooded forests form important habitat for fish, by providing food and shelter. A large fraction (60% of the species) of fish of the Amazon River, 29% of the species of the Apure River (Orinoco basin) and 39% of the species of the Sinnamay River (Guyana basin) use this habitat (Goulding *et al.* 1988; Tito de Morais *et al.* 1995; Crampton 1999; Lasso *et al.* 1999).

Trees also are substrate for algae and support invertebrates eaten by the fish (Goulding 1980). Periphytic algae do not grow well in the flooded forest (Alves 1993) due to water level change and low transparency that causes light limitation. However, phytoplankton from the floodplain lakes accumulates on the leaves and provides protein-rich food for substrate feeders, such as the highly abundant prochilodontid and curimatid fish.

Few species forage on tree leaves, despite their high availability. Many flooded forest trees keep their leaves under water and are available to aquatic herbivores. The presence of polyphenols (tannins) and other plant defence chemicals and the poor nutritional value has been used to explain the low electivity of the leaves (Goulding *et al.* 1988). Flowers are eaten by some species during the low water (Goulding *et al.* 1988), however, the most important food resources provided by the forest are seeds and fruits.

Fruit and seed consumers are concentrated in a few phylogenetic groups. Most species belong to the families Characidae and Anostomidae (characiforms) and Pimelodidae and Doradidae (siluriforms). Frugivory is rare among perciforms. Frugivorous fish are found in many basins and good examples are *Colossoma macropomum*, *Mylossoma duriventre*, *Piaractus brachypomus*, *Myleus torquatus*, *M. schomburgkii*, *Metynnis argenteus*, *Leporinus fasciatus*, *Phractocephalus hemioliopterus*, *Lithodoras dorsalis* in the Amazon and Orinoco basins, and *Piaractus mesopotamicus*, *Leporinus friderici*, *Pterodoras granulosus*, *Pimelodus maculatus* in the Paraná basin (Table 23.2).

Frugivory in fish has been better studied in the Amazon basin, where most fruits and seeds produced by the forest are eaten by fish (Kubitzki & Ziburski 1994; Waldhoff *et al.* 1996; Araujo-Lima & Goulding 1997). Fish appear to select seeds with high energy and low water content such as *Pseudobombax munguba* and *Nectandra amazonum* (Table 23.3). In fact some of these seeds have the highest energetic content ever described among plants and also have high protein and vitamin content. Although fish may prefer seeds, they also eat fleshy fruits.

Fish are dispersers as well as consumers of tree seeds. Catfish do not break the shells or pierce the protective layers of the seeds, while characins normally crush and destroy them. Fish are the most important seed-dispersing animal in the flooded forest, followed by turtles and monkeys (Gottsberger 1978; Kubitzki & Ziburski 1994). Many plant seeds are carried to new locations by water but fish can also disperse them. Some plants, however, are strictly ichthyochorous species. Seeds of *Cecropia latifolia*, *Astrocaryum jauari*, *Crataeva benthamii*, *Crescentia amazonica* and *Simaba orinocensis* taken from the terminal part of the intestinal tract of fish had a high germination rate.

Table 23.2   Main herbivorous and omnivorous fish of four river basins and their rank in the fisheries yield (tonnes per year) in the Amazon

Species	Basin	Main food	Yield and rank
Colossoma macropomum	Amazon + Orinoco	Fruits and zooplankton	2000*
Mylossoma duriventre	Amazon + Orinoco	Fruits and insects	3600*
Piaractus brachypomus	Amazon + Orinoco	Fruits and leaves	2554*
Phractocephalus hemioliopterus	Amazon	Fruits and fish	237
Lithodoras dorsalis	Amazon	Fruits and leaves	<100
Triportheus elongatus	Amazon	Fruits and insects	1045*
Myleus torquatus	Amazon	Fruits and leaves	<100
Myleus schomburgkii	Amazon	Fruits and leaves	<100
Metynnis argenteus	Amazon	Fruits and invertebrates	<100
Leporinus fasciatus	Amazon	Fruits and invertebrates	1000*
Piaractus mesopotamicus	Paraná	Fruits and invertebrates	1172*
Leporinus friderici	Paraná	Fruits and invertebrates	<100
Pterodoras granulosus	Paraná	Fruits and molluscs	100*
Pimelodus maculatus	Paraná	Fruits and fish	82
Leporinus elongatus	São Francisco	Fruits and leaves	<100*
Brycon orbignyanus	Paraná	Fruits and invertebrates	<100

*Species that rank among the top ten fish in their respective markets. Sources: Novoa 1986; Agostinho *et al*. 1995; IBAMA 1999; Lowe-McConnell 1999.

Characids masticate the seeds, and it is generally agreed that small seeds have better survival chances than do large seeds. Exceptions to this rule do occur. The seeds of Sapotacae, for example, are large and fish usually do not crush them. The seeds are swallowed whole and are passed out in the faeces in a viable germination condition.

The flooded forest is also the spawning site for some species of fish. Tree trunks and roots form nest substrate for large cichlids such as *Cichla* spp., *Astronotus ocellatus* and *Crenicichla* spp. These species deposit their eggs and guard them for a few days, until the larvae are able to swim with the parents to open areas where they feed.

Forest cover affects the species diversity in streams. Fish assemblages in streams with dense forest have higher diversity and feeding specialization than those in deforested streams where the assemblage contains more generalist species (Barrella *et al*. 1994, 2000; Montag *et al*. 1997).

The species composition in low order streams was related to canopy cover in some studies. Mérigoux *et al*. (1998) have shown that canopy cover is an important variable affecting community structure in streams in French Guyana. However, other factors such as altitude/slope, flow and sediment load were more important in streams of the Paraná and Amazon river basins (Barrella & Petrere 1994; Mendonça 2002). Therefore, the effect of canopy cover on fish assemblages cannot yet be generalized.

Riparian trees can be an important source of food in tropical streams (Araujo-Lima *et al*. 1995; Tito de Morais *et al*. 1995). Knöppel (1970) reported that fish of

Table 23.3   Approximate composition of fruits and seeds eaten by fish in the Amazon

Tree	Water	Protein	Carbohydrate	Lipids	Energy (kJ/g)
Alchornea schomburgkiana (all)	75%	17%	66%	10%	23
Annona montana (pericarp)	82%	9%	55%	2%	18
Astrocaryum jauary (pericarp)	60%	8%	78%	8%	7
Astrocaryum jauary (seed)	42%	6%	71%	26%	16
Carapa guianensis (seed)	60%	8%	37%	20%	–
Catreva benthami (pericarp)	64%	7%	43%	2%	19
Cecropia latiloba (all)	52%	4%	68%	1%	13
Escheweilera tenuifolia (seed)	40%	10%	78%	3%	19
Hevea spruceana (seed)*	39%	16%	20%	45%	26
Ilex inundata (all)	64%	7%	67%	12%	21
Macrolobium acaciifolium (seeed)	51%	20%	65%	1%	–
Nectandra amazonum (seed)*	78%	12%	72%	9%	20
Piranhea trifoliata (seed)*	50%	16%	–	–	–
Pouteria glomerata (seed)	78%	9%	60%	5%	19
Pseudobombax munguba (seed)	14%	21%	26%	32%	22
Symeria paniculata (all)	64%	20%	51%	1%	–
Vitex cymosa (seed)*	71%	17%	43%	8%	–

Water content is relative to total fresh weight, other components are in relation to dry weight.
Sources: Waldhoff *et al.* 1996; Araujo-Lima & Goulding 1997.
*Species are commercially logged.

an Amazonian stream depended heavily on terrestrial insects falling from the canopy. Other vertebrates feeding in the riparian trees discard fruits which are consumed by fish. Association between fish and monkeys has been demonstrated in a stream of the Paraná basin, where shoals of *Brycon micropterus* follow monkeys, eating discarded fruits (Sabino & Sazima 1999).

Trees in artificial reservoirs are less important as a direct food source. Sharp reduction in the abundance of frugivorous fish has been reported in rivers of the Amazon and Paraná basins after impoundment (Northcote *et al.* 1985; Araujo-Lima *et al.* 1995) suggesting that the reduction of seeds directly affects the fish community. In most Amazonian hydroelectric reservoirs, the trees were not removed before the dam closure, because it was not technically and/or economically feasible. Trees left standing in the reservoirs die and their trunks remain submerged, becoming substrata for the growth of algae and insects and for fish nests. Not surprisingly, the fish populations benefiting from this modified habitat are those that build nests in the flooded forest (e.g. cichlids) and species which forage on algae (hemiodontids and prochilodontids). This pattern has been observed in the Balbina and Tucuruí reservoirs in the Amazon basin (Ribeiro *et al.* 1995; Santos & Oliveira 1999).

Frugivorous and detritivorous fishes are abundant species. Many attain relatively large size, migrate seasonally and are commercially exploited by fisheries (Table 23.2) in the Orinoco, Amazon, Magdalena and Paraná basins. Frugivorous fish probably were abundant in other South American rivers before deforestation took place. *Colossoma macropomum*, a well-known Amazonian frugivorous species, reached the highest yields in the central Amazon until 1980, before over-fishing reduced its abundance (Araujo-Lima & Goulding 1997). Nowadays, other frugivores/omnivores (*Brycon cephalus* and *Mylossoma duriventre*) rank fifth and seventh in Amazon fish landings (IBAMA 1999). *Piaractus brachypomus* and *P. mesopotamicus* rank among the top ten species landed in the fish markets of the Orinoco and Paraná river basins, respectively (Novoa 1986), and *Brycon moorei* is an important fish in the Magdalena fishery (Barreto & Turriago 1995).

Cichla spp. are harvested from reservoir fisheries and, despite not feeding directly on trees, benefit from the increase of nesting substrate offered in the submerged trees (Petrere 1996; Santos & Oliveira 1999). The density of these species has increased in most reservoirs and has given rise to a very specialized commercial fishery and an active sport fishery.

Fisheries

Fisheries play an important social role in South America, especially in the floodplains of large rivers. In the early 1990s logging was the second and fishing was the third most important source of income of Amazonian households (Ruffino *et al.* 1997). After the mid-1990s the fisheries started to lead household income. Fish is also the main source of protein in the rural areas of Brazilian, Colombian and Peruvian Amazon (Batista *et al.* 1998; Fabré & Alonso 1998; Cerdeira et *al.* 2000). This activity is a means of support for 780,000 people, which is equivalent to 15% of the total population or 34% of the rural population of Pará, one of the largest states of Brazilian Amazon (Ruffino *et al.* 1997). In the whole Amazon, the number of people depending on fisheries may be at least twice this figure. The commercial fisheries in the Amazon basin landed 110,000 tonnes of fish in 1995 (Barreto & Turriago 1995; IBAMA 1999; Tello 1999). However, if subsistence fisheries are considered the yield will triple. The catch is dependent on fish fauna of the Amazon River and its Andean tributaries.

Reservoirs dominate the upper Paraná and many important fisheries are concentrated on them. The yield of Itaipu reservoir, for example, is near 1500 tonnes per year (Okada *et al.* 1996). Fisheries in the river are less well documented. However, the fisheries in the Paraguai River and lower Paraná River are based mainly on the river and floodplain. Their yield summed up to 12,000 tonnes per year in the mid-1990s (CIH 1997; IBAMA 1999). Sport fishing is important in this densely human populated basin, and provides many jobs for the rural population. Other basins also have important fisheries with annual landings of a few thousands tonnes per year (Novoa 1986; Barreto & Turriago 1995).

General features of forestry practices

Historical background

The focus in this section is on tropical South American forest cover based on the FAO (1999) classification. It is assessed as percentage crown cover, i.e. productive forests from the tropical hardwood market point of view (Fig. 23.1). The primary tropical forests of South America are subsumed under the Amazon Cooperation Treaty (TCA 1995), and this region (including the Orinoco and Guyana basins) is referred to as the 'greater' Amazon. According to Nepstad *et al.* (1999), 10,000–15,000 km² of undisturbed forests are selectively logged each year in the Brazilian Amazon. Brazil alone contains 66% of the Amazonian territory, and is also responsible for nearly 100% of deforestation and hardwood production in the region.

The forests of the Guyanan countries were used so little before mid-1980s that they were almost absent in the deforestation statistics (Hammond *et al.* 1996). Only 60% of the Brazilian Amazon (roughly 3 million km²) could be classified as productive tropical moist forestland. During the last 20 years the annual deforestation of primary forests has averaged 17,000 km² (INPE 2002). Combining non-productive forests, selectively logged and deforested areas, South American forest cover is probably nearly 4 million km². For this reason, it is reasonable to extrapolate Brazilian statistics for forest extraction to the whole of South America.

Although the continent is 50% forested, the forestry industry in most South American nations is small and oriented toward domestic markets (FAO 1999). Some subtropical hardwoods and softwoods are exported; however, much of the wood comes from the Amazon Basin. Pine lumber is also exported from southern Brazil and south central Chile, together with some pulpwood. In these countries commercial pine forests have been planted. The widespread planting of eucalyptus trees for firewood, timber and use in rough construction is historically important in Brazil.

Since 1988, the International Tropical Timber Organization (ITTO 2002) has been the primary source for world tropical timber statistics. These include data on domestic consumption and import and export of tropical hardwood (non-coniferous). Tropical world annual production averaged 182 million m³ (round wood + sawn wood + veneer + plywood) during the period 1988–2000.

Bolivia, Brazil, Colombia, Ecuador, Guyana, Peru, Suriname and Venezuela are the main producers in tropical America (Table 23.4). The contribution of tropical America to the total tropical world production during the period 1988–2000 has averaged 24%, 31%, 16% and 11% of round wood, sawn wood, veneer and plywood production, respectively. Brazil is the major South American tropical hardwood producer. Its annual contribution has averaged 79%.

The domestic consumption in tropical America is nearly 100%. During the last 12 years the contribution of South American countries to the international timber market has been only 3.2%. However, this situation may change in the future. Malaysia and Indonesia have reduced their production to the international timber market by 40% and 20%, respectively. In 1989 these two countries were responsible for 82% of all exported tropical hardwood. However, today their joint contribution has dropped to

Table 23.4 Contribution (%) of each country to the annual hardwood production of tropical America, from 1988 to 2000

Country	1988	1990	1992	1994	1996	1998	2000
Bolivia	1.0	1.2	1.5	1.5	1.5	2.4	1.9
Brazil	81.6	81.6	78.1	78.9	81.6	76.5	78.1
Colombia	5.0	4.3	4.1	2.8	3.8	6.7	6.7
Ecuador	8.7	9.4	8.0	7.6	4.7	5.1	4.7
Guyana	0.0	0.4	0.6	1.3	1.4	1.2	1.1
Peru	3.7	3.1	3.5	4.4	4.4	5.6	5.0
Suriname	0.0	0.0	0.0	0.0	0.7	0.4	0.5
Venezuela	0.0	0.0	4.2	3.4	1.9	2.0	2.0

Source: ITTO (2002).

68% (Schmidt 1991). Based on the last 12 years' market dynamics, on the stability of global wood product consumption and economic simulations, it is possible that tropical America may become the major tropical hardwood supplier (Putz *et al.* 2000).

Recent and current practices

In Malaysia there is a discrepancy between the sustainable forest management (SFM) as conceived, which is sound, and the SFM as practised, which is not sound (Tang 1987). However, this is not exclusively a Malaysian phenomenon because the same statement also applies to Indonesia (Daryadi 1994), India (Shah 1994), Ghana and Nigeria (Asabere 1987) and Brazil (Silva 1996). So far, nearly 100% of tropical hardwood production has relied on dense forests, which are selectively logged under some kind of SFM. SFM was conceived almost 150 years ago (Higuchi *et al.* 2000). With 20–100-year cutting cycles, it is expected to have two to four harvests, but this is not happening. In this sense, the harvests are not repeated because loggers move to exploit new areas (Poore 1989; Laurance 2000). SFM is also applied in Venezuela and Guyana (Barreto *et al.* 1998; Ochoa 1998).

Tropical forests usually are transformed to other land uses such as cattle ranching and agriculture after selective logging. This invariably results in a total forest removal. Even though the tropical American population is not as dense as it is in other tropical regions, many are involved in extensive cattle production, which is a real threat to the regrowth of forests. In the Brazilian Amazon, 569,269 km^2 of primary forests were deforested up to 1999 in an attempt to increase national food production (INPE 2002). Deforestation for agriculture has been especially intense near roads and rivers with a high cation exchange capacity in the Amazon, Orinoco and Paraná basins because of the access and soil quality (Winemiller *et al.* 1996; Araujo-Lima & Goulding 1997; Laurance *et al.* 2001).

Reforestation practices in tropical America are insignificant. In the Brazilian Amazon, the Jari Project is the only reforestation project carried out on a commercial scale. It represents <2% of the deforested area in the region, and is only for pulpwood production, based mainly on *Eucalyptus* spp. In the south and southeast regions of Brazil the reforestation area totals almost 40,000 km^2, and is dominated by *Eucalyptus* spp. and *Pinus* spp. Plantations of indigenous species have been carried out in many Latin American countries, but only on a research scale.

Forestry practices in the upland forests of the Brazilian Amazon are based on SFM projects and on authorized/licensed deforestation. According to Hummel (2001), these practices account for 30% of the annual round wood production, of approximately 9 million m^3. The remaining 70% comes from unknown (probably illegal) sources.

SFM projects in Amazon contribute 2 million m^3 per year. This practice is characterized as a 'log and leave' form of logging in which, after selective logging has taken place, access to the forest is closed and it is protected against encroachment and/or further logging. Presently, only three SFM projects are certified in the Brazilian Amazon.

Traditional timber harvesting is the most common procedure in Amazonian dense forests. Workers have no formal education or training and the extraction of timber is unplanned. Logging begins with a chainsaw operator walking into the area to be exploited and selecting and felling trees with a deemed market value (>45 cm DCH). There is little concern for collateral damage. The logging roads and landings are then opened by bulldozers, and logs are skidded out. This type of logging is destructive and during the past decade it has been demonstrated that planned logging, which causes less collateral damage, is also more economical (Barreto *et al.* 1998; Putz *et al.* 2000).

So far, there are no SFM projects in the Peruvian and Brazilian Amazon's seasonally flooded forest. In this forest logging is a typical example of exploitation without any concern for succession, replacement or sustainability (Anderson 1991). Even so, such logging produced 90% of all wood until 1989. Production by such means is now down to 15%.

Logging in the seasonally flooded forests is based on a 'wood procurement', or 'dealer system', which is similar along all the Amazon Basin's rivers. This system involves timber-based industries, buyers, middlemen and loggers. Middlemen may act not only as an intermediary between buyer and logger, but also as a logger. They deal with loggers, in general, through a patronage system, prepare and tow logs to a lake or deeper river to be bunched into rafts, and supervise the raft transportation to the sawmills/industry. The loggers (tree cutters) are the basis of the round wood production chain. Most loggers have no formal training and are involved in other extractive activities. They work only during the felling and extracting phases.

The logging is low cost, depends on water levels and relies heavily on floating the wood to the market, and cannot exploit large quantities of non-floating wood. The extraction methods consist of open trails, selecting and felling the trees during low water season, then yarding and transporting the logs during high water season. Trees are selected based on their DCH (>50 cm), buoyancy, position in the floodplain, stem quality and the demands of the timber industry. Three species account for almost 100% of the annual production taken from flooded forests in Brazil (46% *C. petandra*; 33% *N. caloneura* and 20% *Copaifera* sp.). The damage caused by felling is impressive and the gap size after logging in seasonally flooded forest averages 845 m². It is twice the average gap obtained in upland forests (Vieira & Higuchi 1990).

Yarding and transporting the trees is the most labour-intensive phase of the extraction process, and causes substantial damage to the forest. When the log is semi-buoyant a float/raft is constructed of lightwood trees (*Hevea spruceana*, *Styrax* sp). Occasionally, a log is lost and recovery requires diving. As the log lengths average 15 m, the raft, which passes through the watery trails, can reach up to 200 m.

Some farmers in the Peruvian and Brazilian flooded forests practise small-scale forest management principles nowadays. They combine agricultural crops and fast growing trees, planting both on the floodplain (Bahri 1993; De Jong 2001).

General effects of forestry practices on fish and fisheries

Recent and current effects

We are not aware of any publications that have examined the effect of forestry practices on fish and fisheries in tropical South America, and they probably are rare. The evidence of the effect is indirect (Waldhoff *et al.* 1996; Araujo-Lima *et al.* 1998). It seems remarkable, given the size of river basins, the total area covered by tropical forest and the diversity of its fish fauna, that so little attention has been given to this issue. Evaluation studies are now being conducted in at least two basins (Amazon and Paraná) and preliminary results will be available soon. Therefore, at this time, the impact of forestry on fish populations can only be discussed hypothetically.

Several tree species regularly felled and/or damaged by loggers in the Amazonian floodplain produce seeds that feed the fish. The seasonally flooded forest is the second most important primary producer of the Amazonian floodplain and contributes at least 30% of the carbon fuelling the Amazonian aquatic food chain (Melack & Forsberg 2001). It is reasonable to assume that reduction in the number of trees will have an impact on the availability of food for fish and consequently on fishery yields. Tree carbon has been traced in fish flesh using stable isotopes. Araujo-Lima *et al.* (1998) estimated that 60% of the *Colossoma macropomum* (a prized frugivorous fish) flesh that ends up in the Amazonian fish market of Manaus is based on seed carbon from the forest. They further estimated the value of the forest based on the fish yield and found that just one species of tree alone (*Hevea spruceana*) was able to generate as much as one million dollars per year in fish meat. This tree is now being selectively logged to produce plywood and floats to yard semi-buoyant logs. Other authors suggested even higher contributions of the forest to the fisheries (Waldhoff *et al.* 1996).

Despite the evidence of the contribution of tree seeds to fish production and to fisheries yields, one can argue that the impact of selective logging on fisheries yields may not be severe because: (1) most Amazonian fish are not specialized frugivores and may turn to other food sources if seeds/fruits become scarce; (2) selective logging of the seasonally flooded forest allows the growth of highly productive softwood/fast growing trees; and (3) the abundance of fish populations is controlled by predation (top-down) rather than by resources (bottom-up). However, such selective logging should have an effect on specialized frugivorous fishes. In controlled rivers of the Amazonian and Paraná basins where flooded forests have been reduced, the abundance of herbivores was greatly reduced (Araujo-Lima *et al.* 1995). Selective logging has been shown to affect the species composition of understory birds in Venezuela (Mason 1996), and the same may happen to fish assemblages.

Clear-cutting is often employed when the land is prepared for agricultural and livestock use. Cattle ranching is the most common use for cleared land in the Brazilian and Colombian Amazon by large landlords. The practice of slash and burn, for logging associated with plantations, is especially common among small farmers (Nepstad *et al.* 1991). In the south of Brazil and Peru, clearing the forest to make charcoal is also widespread. Clearing the forests affects the flux of water, sediments and biological material to river basins (Alegre *et al.* 1986; Walling & Webb 1996). The consequent change in the water quality of Amazonian floodplains is not yet clear, but in streams the impact on transparency, sediment load and flow will be severe (Richey *et al.* 1997). Because the composition of fish assemblages of streams seems to be affected by these environmental variables, strong impacts should be expected.

Reproduction among stream fish may also be influenced by changes in water quality. Many fish of South American streams are pelagophils (Balon 1975), but nest builders and hiders also occur in the assemblages. Siltation of streams can affect egg incubation, especially of cichlids, by reducing gas exchange between egg and water, by reducing the efficiency of nest protection, or even by burying them.

Most countries of South America have laws that regulate logging (Putz *et al.* 2000) and gallery forests are also covered in the legislation. The Brazilian Forestry Law, for example, prohibits logging activity in gallery forest, which has the status of a Protected Area (Article 2). The gallery forest is defined not by its floristic composition, but by its dimension. Along rivers and lakes, wider than 600 m, the gallery forest is located in a 500-m wide band from the maximum water level at each margin. In streams <10 m wide it occupies a 30-m band. Also logging is prohibited above 1800 m altitude, on the tops of mountains, on slopes with >45° incline, in mangrove forests, in national parks and in Indian reserves, or in any area specified by the government as one of public interest. Additionally, at least 20% of the area in private property (80% in the Amazon) should be in native vegetation, although there is currently political pressure to reduce this percentage.

In spite of the restrictions on deforestation of the riparian forest there is little law enforcement. The best example of lack of political will to enforce the law is the fact that most timber exported from the Amazon came, until recently, from the edges of seasonally flooded forest, an area protected by law. In other countries institutional failures are also common (Putz *et al.* 2000).

Another forestry-related activity that affects fish assemblages is the pollution of rivers by the paper mill industry, especially in the Paraná basin. Effluents are normally discharged into the rivers without any treatment. The industry is notorious for water demand. It uses excessive quantities, which also contribute to reduced river levels and riparian vegetation. Water recycling is the exception, not the rule.

Needs and recommendations for improvement of forestry practices and standards

The logging activity in South America will develop mainly in the upland forests. A better knowledge of the fish fauna of Amazonian forest streams is urgently needed. There are less than 15 publications dealing with composition of the fish fauna of streams and most of them are in the region of Manaus. It would not be an overstatement to mention that the stream fish fauna of the Amazon basin is unknown.

Scientific research aimed at measuring the impact of forestry practices on fish and fisheries is also needed in South America. These works should be highly focused and, preferably, experimental. It would be beneficial to conduct the experimental work within the framework of an ongoing logging operation. This would alleviate the need to cause any more damage than necessary.

One issue to be addressed is the current forest legislation. Laws regulating forestry activities in most South American countries have not yet contributed effectively to forest conservation. Most of the laws consider only the physical environment and they neglect the biota. It is important to determine whether the forestry regulations, as currently presented, are effective in protecting river fauna, including the fish assemblages.

Riparian forests are key habitats for the function and protection of aquatic ecosystems. Logging prohibitions in these forests must be enforced by environmental agencies. Regular law enforcement may work in the Paraná basin where human population density is high. However, it is impossible to implement in the Amazon region, for obvious reasons. Remote sensing, using radar imagery, which is not sensitive to cloud cover, may be an effective way to monitor logging operations and deforestation. The technology is generally available and can identify deforested patches as small as 5 ha. Enforcement of the rules can work with large ranchers and industries, but it is ineffective for the activities of small farmers and floodplain loggers. In the latter case, only development policies for land use and environmental education may help.

Protecting the riparian forest is important. However, if large forest areas are cleared, it probably will not prevent the impacts from the erosion. Also, roads seem to be the main cause of hydrological damage. The timber industry should apply logging practices that reduce erosion, particularly in sloping landscapes, which are so common in the Amazon uplands. Poor logging practices are notorious in the tropics, and reduced impact logging (RIL) or planned logging have been barely implemented (Barreto *et al.* 1998; Putz *et al.* 2000). Planned logging as presented by Barreto *et al.* (1998) aims to reduce collateral damage to trees. It does not address hydrological problems directly, but because RIL reduces the number of logging roads, it can ultimately reduce erosion. In addition, forestry planning that zones extraction areas should consider stream density. This planning element seems to have been neglected in the past.

These procedures should be recommended to concession holders. It should not be very difficult to achieve a compromise since the concessions are constantly evaluated by environmental agencies, which advise local environmental agencies about licence renewal.

Acknowledgements

We are grateful to Dr. Juan Neiff and Dr. Bernard d'Merona for providing useful information on the Paraná and Guyana basins, respectively. We also thank Dr. A. Agostinho, Dr. P. Hazelton and both editors for their helpful comments on the draft.

References

Agostinho, A.A., Vazzoler, A.E.A.M. & Thomaz, S.M. (1995) The high river Paraná basin: limnological and ichthyological aspects. In: *Limnology in Brazil*, (eds J.G. Tundisi, C.E.M. Bicudo & T. Matsumura-Tundisi), pp. 59–104. ABC/SBL, Rio de Janeiro.

Alegre, J.C., Cassel, D.K. & Bandy, D.E. (1986) Effects of land clearing and subsequent management on solid physical properties. *Soil Science Society of America Journal*, 50, 1379–84.

Alves, L.F. (1993) *The fate of stream water nitrate entering littoral areas of an Amazonian floodplain lake: the role of plankton, periphyton, inundated soils and sediment*. PhD thesis, University of Maryland.

Anderson, A. (1991) Forest management strategies by rural inhabitants in the Amazon Estuary. In: *Rain Forest Regeneration and Management*, (eds A. Gómez-Pompa, T.C. Whitmore & M. Hadley), pp. 351–60. Parthenon Publishing Group, New Jersey.

Araujo-Lima, C.A.R.M. & Goulding, M. (1997) *So Fruitful a Fish. Ecology, Conservation and Aquaculture of the Amazon's Tambaqui*. Columbia University Press, New York.

Araujo-Lima, C.A.R.M., Agostinho, A.A. & Fabré, N.N. (1995) Trophic aspects of fish communities in Brazilian rivers and reservoirs. In: *Limnology in Brazil*, (eds J.G. Tundisi, C.E.M. Bicudo & T. Matsumura-Tundisi), pp. 105–36. ABC/SBL, Rio de Janeiro.

Araujo-Lima, C.A.R.M., Goulding, M., Forsberg, B.R., Victoria, R. & Martinelli, L. (1998) The economic value of the Amazon flooded forest under a fisheries perspective. *Verhandlungen Internationale Vereinigung für Theoretische und Angewandte Limnologie*, 26, 2177–9.

Asabere, P.K. (1987) Attempts at sustained yield management in the tropical high forests of Ghana. In: *Tropical Management of Tropical Moist Forests: Silvicultural and Management Prospects of Sustained Utilization*, (eds F. Mergen & J.R. Vincent), pp. 47–69. Yale University Press, New Haven, CT.

Ayres, J.M. (1993) *As Matas de Várzea do Mamirauá*. Sociedade Civil Mamirauá, Belém.

Bahri, S. (1993) Les systèmes agroforestiers de l'ile do Careiro. *Amazoniana*, XII, 551–63.

Balon, E.K. (1975) Reproductive guilds of fishes. A proposal and definition. *Journal of Fisheries Research Board of Canada*, 32, 821–64.

Barrella, W. & Petrere, M., Jr (1994) The influence of environmental factors on a fish community structure in the Jacaré Pepira river, Brazil. In: *Rehabilitation of Freshwater Fisheries*, (eds I.G. Cowx), pp. 161–70. Fishing News Books, Oxford.

Barrella, W., Beaumord, A.C. & Petrere, M., Jr. (1994) Comparison between the fish communities of Manso River (MT) and Jacaré Pepira River (SP), Brazil. *Acta Biologica Venezuelica*, 15, 11–20.

Barrella, W., Petrere, M., Jr, Smith, W.S. & Montag, L.F.A. (2000) As relações entre as matas ciliares, os se os peixes. In: *Matas Ciliares: Conservação e Recuperação*, (eds R.R. Rodríguez & H.F. Leitão Filho), pp. 187–207. EDUSP, São Paulo.

Barreto, C.G. & Turriago, R. (1995) *Boletin Estatistico Pesquero*. INPA, Bogotá.

Barreto, P., Amaral, P., Vidal, E. & Uhl, C. (1998) Cost and benefits of forest management for timber production in eastern Amazonia. *Forest Ecology and Management*, 108, 9–26.

Batista, V.S., Inhamuns, A.J., Freitas, C.E.C. & Freire-Brasil, D. (1998) Characterization of the fishery in river communities in the low-Solimões/high-Amazon region. *Fisheries Management and Ecology*, 5, 419–35.

Bonetto, A.A. (1986) The Paraná river system. In: *The Ecology of River Systems*, (eds B.R. Davies & K.F. Walker), pp. 541–55. Dr. Junk Publishers, Dordrecht, the Netherlands.

Britski, H.A., Silimon, K.Z.S. & Lopes, B.S. (1999) *Peixes do Pantanal. Manual de Identificação*. EMBRAPA, Brasília.

Campos, J.B. & Souza, M.C. (1997) Vegetação. In: *A Planície de Inundação do Alto Rio Paraná*, (eds A.E.A.M. Vazzoler, A.A. Agostinho & N.S. Hahn), pp. 331–42. EDUEM, Maringá.

Cerdeira, R.G P., Isaac, V.J. & Ruffino, M L. (2000) Fish catches among riverside communities around Lago Grande de Monte Alegre, Lower Amazon, Brazil. *Fisheries Management and Ecology*, 7, 355–74.

CIH (1997) *Evaluacion del Impacto Ambiental del Mejoramiento de la Hidrovia Paraguay-Parana*. UNOPS, Proyeto UNOPS, Buenos Aires (unpublished report).

Colonnello, G., Castroviejo, S. & Lopez, G. (1986) Comunidades vegetales asociadas al rio Orinoco en el sur de Monagas y Anzoategui. *Memoria Sociedad de Ciencias Naturales la Salle*, **XLVI**, 127–66.

Crampton, W.G.R. (1999) Os peixes da reserva Mamirauá: diversidade e história natural na planície alagável da Amazônia. In: *Estratégias para o Manejo de Recursos Pesqueiros em Mamirauá*, (eds H.L. Queiroz & W.G. Crampton), pp. 10–36. Sociedade Civil Mamirauá, Belém.

Daryadi, L. (1994) Indonesia's experience in sustainable forest management. *FAO Forestry Paper*, **122**, 201–13.

De Jong, W. (2001) Tree and forest management in the floodplains of the Peruvian Amazon. *Forest Ecology and Management*, **150**, 125–34.

Di Persia, D.H. & Neiff, J.J. (1986) The Uruguay river system. In: *The Ecology of River Systems*, (eds B.R. Davies & W.F. Walker), pp. 599–621. Dr. Junk Publishers, Dordrecht, the Netherlands.

Drago, E.C. (1989) Morphological and hydrological characteristics of the floodplain ponds of the Middle Paraná River (Argentina). *Revue d'Hydrobiologie Tropicale*, **22**, 183–90.

Drago, E.C. (1990) Hydrological and geomorphological characteristics of the hydrosystem of Middle Paraná river. *Acta Limnologica Brasiliensia*, **3**, 907–30.

Fabré, N.N. & Alonso, J.C. (1998) Recursos ícticos no alto Amazonas: sua importância para as populações ribeirinhas. *Boletim do Museu Paraense Emílio Goeldi, Ser Zoologia*, **14**, 19–55.

FAO (1999) *State of the World's Forests*. Food and Agriculture Organization of the United Nations, Rome.

FAO (2000) *Forest Resource Assessment*. http://www.fao.org/forestry/FO/fra/Index.jsp.

Ferreira, L.V. (1997) Effects of the duration of flooding on species richness and floristic composition in three hectares in the Jaú National Park in floodplain forest in central Amazonia. *Biodiversity and Conservation*, **6**, 1353–63.

Furch, K. (1997) Chemistry of varzea and igapó soils and nutrient inventory of their floodplain forest. In: *The Central Amazon Floodplain. Ecology of a Pulsing System*, (ed. W.J. Junk), pp. 47–68. Springer, Berlin.

Godoy, J.R., Petts, G. & Salo, J. (1999) Riparian flooded forest of the Orinoco and Amazon basins: a comparative review. *Biodiversity and Conservation*, **8**, 551–86.

Gonçalves, C.S., Monte, I.G. & Câmara, N.L. (1997) Clima. In: *Recursos Naturais e meio Ambiente. Uma Visão do Brasil*, (ed. IBGE), pp. 191–6. IBGE, Rio de Janeiro.

Gottsberger, G. (1978) Seed dispersal by fish in the inundated regions of Humaitá, Amazônia. *Biotropica*, **10**, 170–83.

Goulding, M. (1980) *The Fishes and the Forest: Explorations in Amazonian Natural History*. University of California Press, Los Angeles.

Goulding, M., Carvalho, M.L. & Ferreira, E.J.G. (1988) *Rio Negro. Rich Life in Poor Water*. SPB Academic Publishing, The Hague.

Hammond, D.S., Gourlet-Fleury, S., van der Hout, P., ter Steege, H. & Brown, V.K. (1996) A compilation of known Guianan timber trees and the significance of their dispersal mode, seed size and taxonomic affinity to tropical rain forest management. *Forest Ecology and Management*, 83, 99–116.

Hastenrath, J. (1990) Diagnostic and prediction of anomalous river discharge in Northern South-America. *Journal of Climate*, 3, 1080–96.

Higuchi, N., Ribeiro, J.S.R.J., Silva, R.P. & Rocha, R.M. (2000) Sustentabilidade na produção de madeira dura tropical. *Silvicultura*, 83, 32–7.

Howard, A.F., Rice, R.E. & Gullinson, R.E. (1996) Simulated financial returns and selected environmental impacts from four alternative silvicultural prescriptions applied in the neotropics: a case study of the Chimanes Forest, Bolivia. *Forest Ecology and Management*, 89, 43–57.

Hueck, K. (1978) *Los Bosques de Sudamérica*. GTZ, Eschborn.

Hummel, A.C. (2001) *Normas de acesso ao recurso florestal na Amazônia brasileira: o caso do manejo florestal madereiro*. Master thesis, INPA/FUA.

IBAMA (1999) *Estatísticas da Pesca de 1998. Brasil-grandes Regiões e Unidades da Federeção*. CEPENE/IBAMA, Tamandaré.

IBGE (1992) *Anuário Estatístico do Brasil*. IBGE, Rio de Janeiro.

INPE (2002) Instituto Nacional de Pesquisas Espaciais. *www.inpe.br*.

ITTO (2002) International Tropical Timber Organization. *www.itto.org.jp*.

Jordan, T.E. (1995) Retroarc foreland and related basins. In: *Tectonics of Sedimentary Basins*, (eds C.J. Busby & R.V. Ingersoll), pp. 331–62. Blackwell Science, Cambridge, MA; Oxford, UK.

Julio, H.F., Jr, Bonecker, C.C. & Agostinho, A.A. (1997) Reservatório de Segredo e sua inserção na bacia do rio Iguaçú. In: *Reservatório de Segredo: Bases Ecológicas para o Manejo*, (eds A.A. Agostinho & L.C. Gomes), pp. 1–17. EDUEM, Maringá.

Junk, W.J. (1993) Wetlands of tropical South America. In: *Wetlands of the World*, (ed. D.F. Whigham), pp. 679–739. Kluwer, The Hague.

Junk, W.J. & Furch, K. (1980) Quimica da água e macrófitas aquáticas de rios e igarapés na Bacia Amazônica e nas áreas adjacentes. *Acta Amazonica*, 10, 611–33.

Kammesheidt, L. (1998) Stand structure and spatial pattern of commercial species in logged and unlogged Venezuelan forest. *Forest Ecology and Management*, 109, 163–74.

Knöppel, H.A. (1970) Food of Central Amazonian fishes. Contribution to the nutrient ecology of Amazonian rain forest streams. *Amazoniana*, II, 257–351.

Kubitzki, K. & Ziburski, A. (1994) Seed dispersal in floodplain forest of Amazonia. *Biotropica*, 26, 30–43.

Kvist, L.P. & Nebel, G. (2001) A review of Peruvian flood plain forest: ecosystems inhabitants and resource use. *Forest Ecology and Management*, 150, 3–26.

Lasso, C.A., Rial, A. & Lasso-Alcala, O.M. (1999) Composicion y variabilidad espacio–temporal de las comunidades de pesces en ambientes inundables de los llanos de Venezuela. *Acta Biologica Venezuelica*, 19, 1–28.

Laurance, W.F. (2000) Cut and run: the dramatic rise of transnational logging in the tropics. *TREE*, 15, 433–4.

Laurance, W.F., Cochrane, M.A., Bergen, S., *et al.* (2001) The future of the Brazilian Amazon. *Science*, 291, 438–9.

Lewis, W.M., Jr (1988) Primary production in the Orinoco River. *Ecology*, 69, 679–92.

Lewis, W.M., Jr & Weibezahn, F. (1981) The chemistry and phytoplankton of the Orinoco and Caroni rivers, Venezuela. *Archives für Hydrobiologie*, 91, 521–8.

Lowe-McConnell, R.H. (1964) The fishes of the Rupununi Savanna district of British Guiana, South America. 1. Ecological groupings of fish species and effects of the seasonal cycles on the fish. *Journal of the Linnean Society (Zoology)*, 45, 103–44.

Lowe-McConnell, R.H. (1999) *Estudos Ecológicos em Comunidades de Peixes Tropicais*. EDUSP, São Paulo.

Lundberg, J.G., Marshall, L.G., Guerrero, J., Horton, B., Malabarba, M.C.S.L. & Wesselingh, F. (1998) The stage for neotropical fish diversification: a history of tropical South American rivers. In: *Phylogeny and Classification of Neotropical Fishes*, (eds L.R. Malabarba, R. Reis, R.P.Vari, Z.M.S. Lucena & C.A.S. Lucena), pp. 13–48. EDIPUCRS, Porto Alegre.

Marengo, J.A. & Nobre, C. (2001) General characteristics and variability of climate in the Amazon basin and its links to the global climate system. In: *The Biogeochemistry of the Amazon Basin*, (eds M.E. McClain, R.L. Victoria & J.E. Richey), pp. 17–41. Oxford University Press, Oxford.

Mason, D.J. (1996) Responses of Venezuelan understory birds to selective logging, enrichment strips, and vine cutting. *Biotropica*, **28**, 296–309.

Melack, J.M. & Forsberg, B.R. (2001) Biogeochemistry of Amazon floodplain lakes and associated wetlands. In: *The Biogeochemistry of the Amazon Basin*, (eds M.E. McClain, R.L. Victoria & J.E. Richey), pp. 235–74. Oxford University Press, Oxford.

Mendonça, F.P. (2002) *Ictiofauna de igarapés de terra firme: estrutura das comunidades de duas bacias hidrográficas, Reserva Florestal Adopho Ducke, Amazônia Central*. Master thesis, INPA/FUA.

Menezes, N. (1972) Distribuição e origem da fauna de água doce das grandes bacias fluviais do Brasil. In: *Poluição e Piscicultura*, (eds C.I. D. B. Paraná-Uruguai), pp. 216–21. Instituto de Pesca/USP, São Paulo.

Mereles, F., Degen, R. & Kochalca, N.L. (1992) Humedales en el Paraguay: Breve reseña de su vegetacion. *Amazoniana*, **XII**, 305–16.

Mérigoux, S., Ponton, D. & Mérona, B. (1998) Fish richness and species-habitat relationships in two coastal streams of French Guiana, South America. *Environmental Biology of Fishes*, **51**, 25–39.

Mojica, J.I. (1999) Lista preliminar de las especies de peces dulceaquicolas de Colombia. *Revista de la Academia Colombiana de Ciencias Exatas, Fisicas e Naturales*, **XXIII** (Suppl.), 547–66.

Montag, L.F.A., Smith, W.S., Barrella, W. & Petrere, M. Jr (1997) As influências das matas ciliares nas comunidades de peixes do Estado de São Paulo. *Revista Brasileira de Ecologia*, **1**, 76–80.

Mozeto, A.A. & Albuquerque, A.L.S. (1997) Biogeochemical properties at the Jataí Ecological Station wetlands (Mogi-Graçú river, São Paulo, SP). *Ciência e Cultura*, **49**, 25–33.

Neiff, J.J. (1990) Aspects of primary productivity in the lower Paraná and Paraguai riverine system. *Acta Limnologica Brasiliensia*, **III**, 77–114.

Nelson, J.S. (1994) *Fishes of the World*. John Wiley & Sons, New York.

Nepstad, D.C., Uhl, C. & Serrão, E.A.S. (1991) Recuperation of a degraded Amazonian landscape: forest recovery and agricultural restoration. *Ambio*, **20**, 248–55.

Nepstad, D.C., Veríssimo, A., Alencar, A., *et al.* (1999) Large-scale impoverishment of Amazonian forests by logging and fire. *Nature*, **398**, 505–8.

Northcote, T.G., Arcifa, M.S. & Froehlich, O. (1985) Effects of impoundment and drawdown on the fish community of a South America river. *Verhandlungen Internationale Vereinigung für Theoretische und Angewandte Limnologie*, **22**: 2704–11.

Novoa, D. (1986) Una revision de la situacion actual de las pesquerias multiespecificas del rio Orinoco y una propuesta del ordenamiento pesquero. *Memoria Sociedad de Ciencias Naturales La Salle*, **XLVI**, 167–91.

Ochoa, J. (1998) Preliminary assessment of the effects of logging on the composition and structure of forest in Venezuelan Guayana region. *Interciencia*, **23**, 197–201.

Okada, E.K., Agostinho, A.A. & Petrere, M., Jr (1996) Catch and effort data and the management of the commercial fisheries of Itaipu reservoir in the upper Parana river, Brazil. In: *Stock Assessment in Inland Fisheries*, (ed. I.G. Cowx), pp. 154–61. Fishing News Books, Oxford.

Petrere, M., Jr (1996) *Sintesis Sobre las Pesquerias de los Grandes Embalses Tropicales de America del Sur*. COPESCAL, Habana.

Petri, S. & Fulfaro, V.J. (1983) *Geologia do Brasil*. EDUSP, São Paulo.

Pires, J.M. & Prance, G.T. (1985) The vegetation types of the Brazilian Amazon. In: *Amazonia*, (eds G.T. Prance & T.E. Lovejoy), pp. 109–45. Pergamon Press, Oxford.

Planquette, P., Keith, P. & Le Bail, P.-Y. (1996) *Atlas des Poissons d'Eau Douce de Guyane*. MNHN/INRA, Paris.

Poore, D. (1989) *No Timber without Trees: Sustainability in the Tropical Forest*. Earthscan Publications, London.

Ponte, V., Machado-Allison, A. & Lasso, C.A. (1999) La ictiofauna del delta del Rio Orinoco, Venezuela: una aproximacion a su diversidad. *Acta Biologica Venezuelica*, 1, 25–46.

Putz, F.E., Dykstra, D.P. & Heinrich, R. (2000) Why poor logging practices persist in the tropics. *Conservation Biology*, 14, 951–6.

Ribeiro, M.C.L.B., Petrere, M., Jr, & Juras, A.A. (1995) Ecological integrity and fisheries ecology of the Araguaia-Tocatins river basin, Brazil. *Regulated Rivers: Research & Management*, 11, 325–50.

Richey, J.E., Wilhelm, S.R., McClain, M.E., Victoria, R.L., Melack, J.M. & Araujo-Lima, C.A.R.M. (1997) Organic matter and nutrient dynamics in river corridors of the Amazon basin and their response to anthropogenic change. *Ciência e Cultura*, 49, 98–110.

Ringuelet, R.A. (1975) Zoogeografia y ecologia de los peces de aguas continentales de la Argentina y consideraciones sobre las áreas ecologicas de America del Sur. *Ecosur*, 2, 1–122.

Ruffino, M.L., Lima, D.M., Petrere, M., Jr, McGrath, D. & Vieira, R.S. (1997) *Floodplain Natural Management Project*. Washington, World Bank (unpublished report).

Roche, M.A. (1982) Compared hydrological behaviours and erosion of the amazonian forest ecosystem at Ecerex, French Guyana. *Cahiers ORSTOM, Sér Hydrologie*, XIX, 81–114.

Saatchi, S., Agosti, D., Alger, K., Delabie, J. & Musinsky, J. (2001) Examining fragmentation and loss of primary forest in the Southern Bahian Atlantic Forest of Brazil with radar imagery. *Conservation Biology*, 15, 867–75.

Sabino, J. & Sazima, I. (1999) Association between fruit eating fish and foraging monkeys in western Brazil. *Ichthyological Explorations in Freshwaters*, 10, 309–12.

Salo, J. & Rasanen, M. (1989) Hierarchy of landscape patterns in western Amazon. In: *Tropical Forests*, (eds L.B. Holm-Nielsen, I.C. Nielsen & H. Balslev), pp. 35–45. Academic Press, London.

Santos, G.M. & Oliveira, A.B., Jr (1999) A pesca no reservatório da hidrelétrica de Balbina (Amazonas, Brasil). *Acta Amazonica*, 29, 145–63.

Schaefer, S. (1998) Conflict and resolution: impact of new taxa on phylogenetic studies of the Neotropical cascudinhos (Siluroidei: Loricariidae). In: *Phylogeny and Classification of Neotropical Fishes*, (eds L.R. Malabarba, R.E. Reis, R.P. Vari, Z.M.S. Lucena & C.A.S. Lucena), pp. 375–400. EDIPUCRS, Porto Alegre.

Schmidt, R.C. (1991) Tropical rainforest management: a status report. In: *Rainforest Regeneration and Management*, (eds A. Gomez-Pompa, T.C. Whitmore & M. Hadley), pp. 181–203. Parthenon Publishing, New Jersey.

Shah, S.A. (1994) Sustainable forestry philosophy, science and economics. *The Indian Forester*, June, 471–6.

Silva, J.N.M. (1996) *Diagnóstico dos Projetos de Manejo Florestal no Estado do Pará. Fase Paragominas*. EMBRAPA, Belém.

Souza Filho, E.E. & Stevaux, J.C. (1997) Geologia e geomorfologia do complexo rio Baía, Curutuba. In: *A Planície de Inundação do Alto Rio Paraná*, (eds A.E.A.M. Vazzoler, A.A. Agostinho & N.S. Hahn), pp. 3–46. EDUEM, Maringá.

Stallard, R.F., Koehnken, L. & Johnsson, M.J. (1990) Weathering processes and the composition of inorganic material transported through the Orinoco river system, Venezuela and Colombia. In: *The Orinoco River as an Ecosystem*, (eds F.H. Weibzahn, H. Alvarez & W.M. Lewis Jr), pp. 81–120. FEACV, Caracas.

Tang, H.T. (1987) Problems and strategies for regenerating Dipterocarp forests in Malaysia. In: *Tropical Management of Tropical Moist Forests: Silvicultural and Management Prospects of Sustained Utilization,* (eds F. Mergen & J.R. Vincent), pp. 23–41. Yale University Press, New Haven, CT.

TCA (Amazon Cooperation Treaty) (1995) *Amazonia without Myths.* Commission on Development and Environment for Amazonia, Quito.

Tello, S. (1999) *Analysis of a multispecies fishery: The commercial fishery fleet of iquitos, Amazon Basin, Peru.* Master thesis, Oregon State University, Newport.

Thomaz, S.M., Roberto, M.C. & Bini, L.M. (1997) Caracterização limnológica dos ambientes aquáticos e influência dos níveis fluviométricos. In: *A Planície de Inundação do Alto Rio Paraná,* (eds A.E.A. Vazzoler, A.A. Agostinho & N.S. Hahn), pp. 179–208. EDUEM, Maringá.

Tito de Morais, L., Lointier, M. & Hoff, M. (1995) Extent and role for fish populations of riverine ecotones along the Sinnamary river (French Guiana). In: *The Importance of Aquatic-Terrestrial Ecotones for Freshwater Fish,* (eds F. Schiemer, M. Zalewski & J.E. Thorpe), pp. 163–79. Kluwer Academic Publishers, London.

Vieira, G. & Higuchi, N. (1990) Efeito do Tamanho de Clareira na Regeneração Natural em Floresta Mecanicamente Explorada na Amazônia Brasileira. In: *Proceedings of the 6th Brazilian Forest Congress,* pp. 666–72. SBS, Belo Horizonte, Brazil.

Waldhoff, D., Saint-Paul, U. & Furch, B. (1996) Value of fruits and seeds from the floodplain forest of central Amazon as resource for fish. *Ecotropica,* 2, 143–56.

Walling, D.E. & Webb, B.W. (1996) Water quality 1. Physical characteristics. In: *River Flows and Channel Forms,* (eds G. Petts & P. Calow), pp. 77–101. Blackwell Science, Oxford.

Weibezahn, F.H. (1990) Hidroquimica e solidos suspendidos en el alto y Medio Orinoco. In: *The Orinoco River as an Ecosystem,* (eds F.H. Weibzahn, H. Alvarez & W.M. Lewis Jr), pp. 151–210. FEACV, Caracas.

Welcomme, R.L. (1985) River fisheries. *FAO Fisheries Technical Paper,* 262, 1–330.

Winemiller, K.O., Marrero, C. & Taphorn, D. (1996) Perturbaciones causadas por hombre a las poblaciones de peces de los llanos y del piedemonte andino de Venezuela. *Biollania,* 12, 13–48.

Worbes, M. (1997) The forest ecosystem of the floodplain. In: *The Central Amazon Floodplain. Ecology of a Pulsing System,* (ed. W. J. Junk), pp. 223–66. Springer, Berlin.

Chapter 24

Europe – with special reference to Scandinavia and the British Isles

D.T. CRISP, T. ERIKSSON AND A. PETER

General introduction

The coverage of this chapter is uneven. Analysis of Blackie *et al.* (1980) and an extensive electronic literature search for later years showed that most of the critical quantitative European studies on forestry and fisheries inter-relationships were made in Scandinavia and the British Isles. Therefore, the bulk of the chapter refers to Scandinavia and the British Isles with passing reference to Europe as a whole.

Relevant features of the geographical region

General

Europe has a land area (2,260,128,000 ha) c. 67% of that of Asia and larger than that of North and Central America, South America, Africa or Oceania. South America has the highest proportion of forest cover (49.7%) closely followed by Europe (41.3%) (Forestry Commission 1999).

Figure 24.1 shows the major political boundaries of the countries of Scandinavia and the British Isles.

Relief and rivers

Denmark and southern Sweden consist mainly of glacial marine deposits. The remainder of Scandinavia consists of Achaean crystalline rock sloping gently down southeastwards. A mountain chain runs in a north-south direction along the Swedish/Norwegian border and has altitudes up to 5500 m (Lundegårdh *et al.* 1964). Nordic rivers have small catchment areas and are relatively short. The longest is the eighth order Göta älv (Sweden) at 720 km. Although short, they are numerous, especially in the northern boreal and alpine regions of Norway-Sweden. There are four main types: (1) southern deciduous forest rivers, (2) southern mixed coniferous forest rivers, (3) boreal rivers and (4) alpine and arctic rivers (Petersen *et al.* 1995).

The relief of the British Isles is summarized in Fig. 24.2A. Ireland has a large central plain surrounded by smaller upland areas. In Great Britain the upland areas lie mainly to the north and west of a line drawn from the mouth of the River Tees to the mouth of the River Exe (the 'Tees-Exe line'). Few of these mountains reach more than 1000 m.

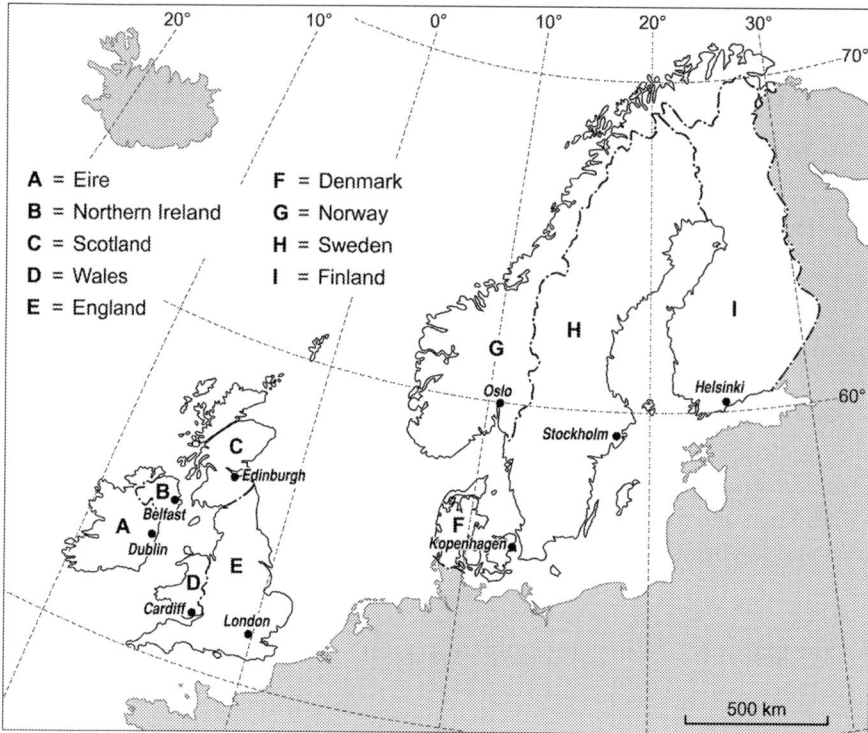

Fig. 24.1 Outline map of Scandinavia and the British Isles to show political boundaries.

The main river systems of the British Isles are outlined in Fig. 24.2B. Even the larger rivers of the British Isles are small compared with major continental rivers. The longest is the River Shannon (Eire) at 385 km.

Climate

Annual precipitation in Europe varies between 100 and 2000 mm year⁻¹. Western Britain, Norway, the Alps and Adriatic areas are the wettest. Long-term mean winter air temperatures vary from –15°C (Russia) to +15°C (Spain and southern Italy).

The closeness of Scandinavia to the Atlantic in the west and to the large land masses in the east gives wide climatic variation. Norway and Denmark have an oceanic climate while Sweden and Finland are influenced by a continental climate (Bogren *et al.* 1999). Annual precipitation varies from 400 to 2000 mm y⁻¹ (Fig. 24.3A). The highest values are along the western side of the Scandinavian mountains, and the lowest in the inland parts of Sweden and the northernmost parts of Sweden and Finland. Winter precipitation is mainly as snow. The inland areas of northern Sweden have a snow cover of 70–100 cm depth and of 150–200 days duration. Mean winter temperature is close to zero in southern Scandinavia and –10° to –15°C in the north (Fig. 24.3B). Mean summer temperature is 17–18°C in the south and 10–12°C in the north (Fig.

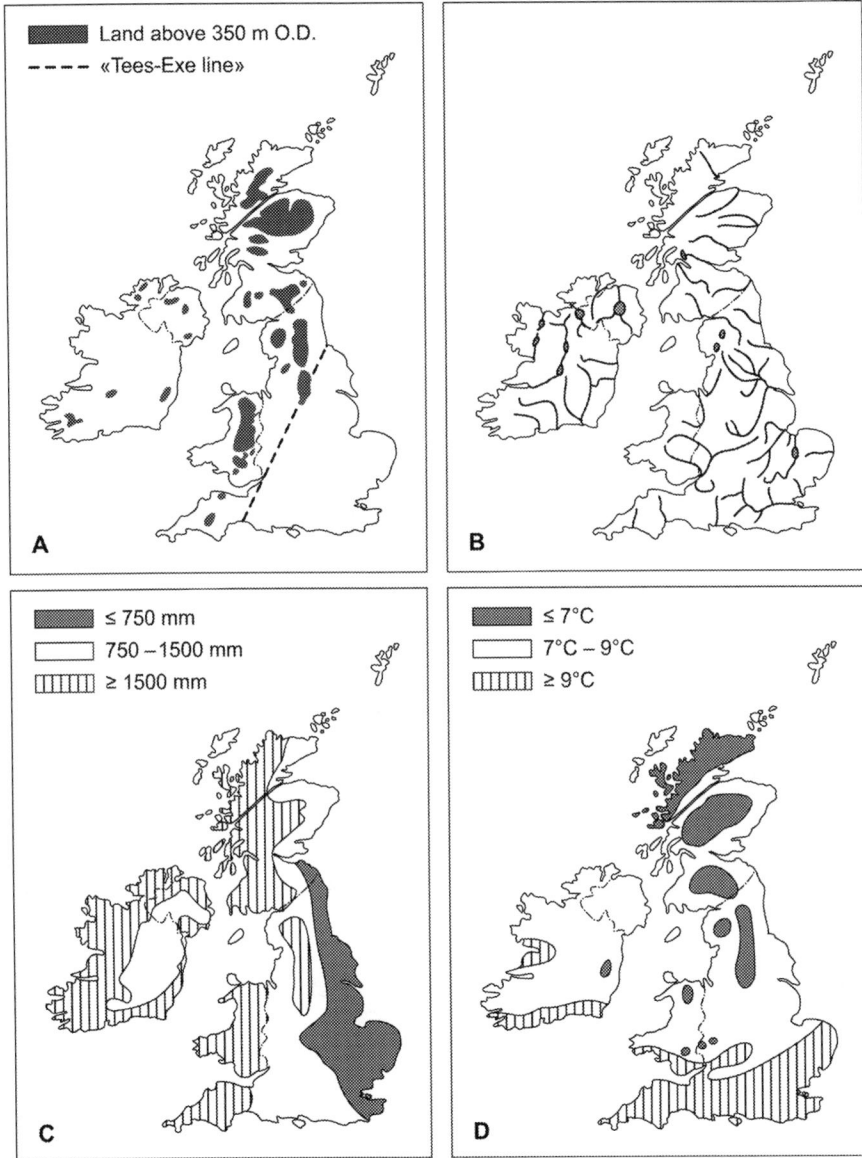

Fig. 24.2 Outline maps of the British Isles to show: (A) relief; (B) main river systems; (C) annual precipitation (mm y⁻¹); (D) mean annual temperature (°C).

24.3C). Annual vegetation growth period varies in different parts of Scandinavia from over 240 days in Denmark to below 120 days in the northernmost parts of Sweden and Finland. The oceanic influence is clearly seen along the Norwegian coast with a comparably longer growth period (Fig. 24.3D).

Annual precipitation over the British Isles varies from <750 mm y⁻¹ to >1500 mm y⁻¹ (Fig. 24.2C). The highest values are in the west and over high ground, largely because

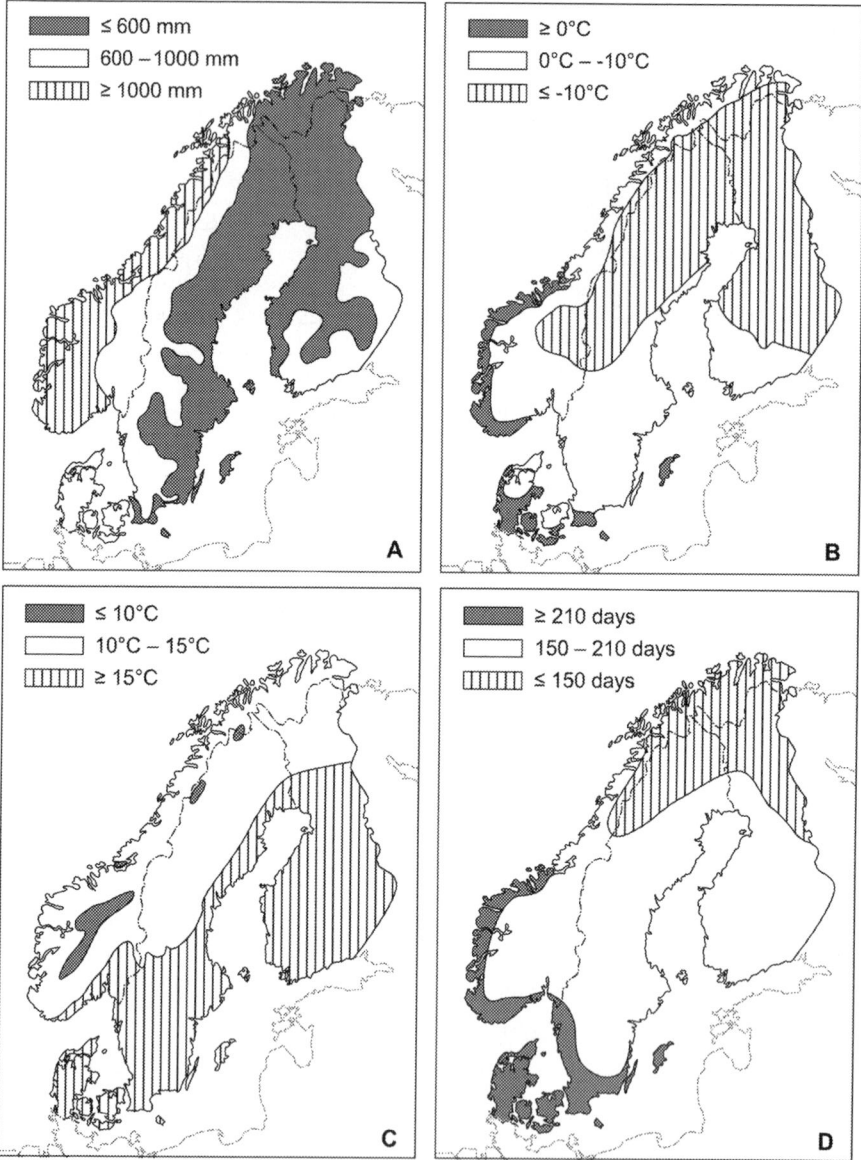

Fig. 24.3 Outline maps of Scandinavia to show: (A) mean annual precipitation (mm y⁻¹); (B) mean winter (January) air (0°C); (B) mean summer (July) air temperature (°C); (D) growing period for plants as (days y⁻¹).

the prevailing winds are from the southwest. The climate is oceanic and temperatures are less extreme in terms of maxima and minima than those observed on mainland Europe. There is a general tendency within the British Isles for the temperatures to be lower as one proceeds north and as altitude increases (Fig. 24.2D). Winter temperatures are generally higher in the west than in the east because of the influences of warm North Atlantic drift and cold continental Europe.

Vegetation

Over half of Europe's historic forest cover has been lost. On average, forest covers 41% of the total area with values ranging from 8% in Ireland to 72% in Finland. European forest cover is expanding by c. 1 million ha y^{-1} (United Nations 1999). The most extensive pristine forests are in the northern taiga. In the more densely populated southern parts pristine forest is largely confined to mountains such as the Alps, Dinaric Alps, Iberian mountains and sierras (World Wide Fund for Nature 2001).

Lundegårdh *et al.* (1964) described the vegetation zones of Scandinavia. The northern temperate zone includes Denmark and the coastal area of western Sweden. North of this is the hemi-boreal zone, characterized as a middle zone between the true temperate deciduous forests and the northern coniferous forests. The boreal zone covering most of Finland, most of Norway and central and northern Sweden, described as the 'taiga', is dominated by spruce (*Picea* spp.) and pine (*Pinus* spp.). The northern boreal zone at higher altitudes on the Scandinavian mountain chain is characterized by small irregular scattered stands of forests or small trees, largely birch (*Betula* spp.) (Ahti *et al.* 1968). The highest parts of the Scandinavian mountain chain are treeless.

The natural climax vegetation of most of the British Isles is broad-leaved woodland, chiefly oak (*Quercus* spp.), elm (*Ulmus* spp.) and ash (*Fraxinus excelsior* L.). In parts of northern Scotland the climax is pine forest dominated by Scots pine (*Pinus sylvestris* L.) and birch. Much of this natural forest was long ago cleared and more recent replanting has been mainly with conifers, especially exotics such as Sitka spruce (*Picea sitchensis* Bong. Carr). Forest, mainly coniferous, now covers 8.3% of Eire and 10.2% of the UK (Forestry Commission 1999).

Forest and fisheries

The Scandinavian freshwater fish fauna has c. 50 species. Only c. 10 of these are important for fisheries (Borgström 1987; Nyberg 1998). Typical fish species of smaller forest streams are resident trout (*Salmo trutta* L.) and bullhead (*Cottus gobio* L.). Species of special importance for fisheries, relative to forestry, are *S. trutta* in resident and anadromous forms, Atlantic salmon (*S. salar* L.) and grayling (*Thymallus thymallus* (L.)). Angling is a popular recreation in the Nordic countries (Toivonen *et al.* 2000). In a Swedish national questionnaire 55% of the population expressed some interest in angling and recreational fishing was an important leisure activity for 14% (Anonymous 2000a). Nordic waters generally are oligotrophic, with mesotrophic or eutrophic waters along coastal borders, and highly eutrophic waters around the heavily populated areas in the south (Bengtsson *et al.* 1991). In northern and mountainous areas salmonids such as Arctic charr (*Salvelinus alpinus* (L.)), whitefish (*Coregonus* spp.), grayling and brown trout are the main target species. In the forested inland areas of the north the Eurasian perch (*Perca fluviatilis* L.) and pike (*Esox lucius* L.) are also important. The pikeperch (*Sander lucioperca* (L.)) is important in eutrophicated waters. Along the coasts migratory trout and Atlantic salmon are the most important species. See Chapter 15 for more on still waters.

The British Isles have a more limited freshwater fish fauna than continental Europe. A checklist (Maitland 1972) includes c. 40 native species. The main species impinged upon by forestry are the Atlantic salmon and the trout in both resident and anadromous forms. These are particularly important for sport and in economic terms in the northern and western parts of the British Isles.

During the period 1987–1996 the declared world catch of Atlantic salmon was 9.54 thousand metric tons. Of this catch 53.8% (Finland 19.7%, Norway 16.2%, Sweden 10.5%, Denmark 7.4%) was taken in Scandinavia and 16.9% (Eire 9.4%, UK 7.5%) in the British Isles. These two regions together took 70.7% of the declared world catch (FAO/FIDI Statistical Data Base). The commercial catches of *S. trutta* cannot be so quantified but are probably significant.

Radford *et al.* (1991) assessed the nett economic value of recreational and commercial fisheries for migratory salmonids in Great Britain, based on 1988 returns. They estimated 75.0–77.5 million pounds sterling for England and Wales and 261.0–266.0 million pounds sterling for Scotland. Approximately 96% of this was attributed to sport fishing.

General features of forestry practice

Brief history

As early as the thirteenth century, the mining industry became an important timber consumer in Scandinavia. Forest materials were used in the production of iron and steel and in making houses and ships. By the mid-1800s a growing forest products industry generated an increasing demand for sawn logs and, some 50 years later, for raw materials for manufacture of pulp and paper.

In the British Isles forest clearance began as early as 1500 BC. Most of the native hardwoods that were the climax vegetation of much of the British Isles were cleared for agriculture and for use as fuel and in house and shipbuilding from c. 1300 AD onwards. In Ireland, for example, by 1900 AD only 1.5% of the land remained forested (Giller *et al.* 1993). In Great Britain commercial forest activity and major planting, mainly with exotic conifer species, began after the First World War in response to a perceived lack of self-sufficiency in forest production. These trees approached harvestable size during the latter part of the twentieth century.

Forest distribution and production

European forests can be divided into four main regions:

(1)   a boreal region, mainly in Sweden, Finland, Norway and the former USSR
(2)   a central European region
(3)   a mountain region in the Alps
(4)   a Mediterranean region.

Table 24.1   European forest area and forest as a percentage of total land area, by countries

Country	Forest area (thousand hectares)	Forest as percentage of total land area
Austria	3877	46.9
Belgium/Luxembourg	709	21.6
Denmark	417	9.8
Finland	20,029	65.8
France	15,034	27.3
Germany	10,740	30.7
Greece	6513	50.5
Ireland	570	8.3
Italy	6496	22.1
Netherlands	334	9.8
Norway	8073	26.3
Portugal	2875	31.4
Spain	8388	16.8
Sweden	24,425	59.3
Switzerland	1130	28.6
UK	2469	10.2
Others (north and west Europe)	17	0.2
Estonia	2011	47.6
Latvia	2822	46.4
Lithuania	1976	30.5
Russian Federation	763,500	45.2
Others (E. Europe)	50,940	25.9
Totals	933,405	41.3

Derived from Forestry Commission (1999).

The area of forest in Europe is rather less than 100 million hectares (Table 24.1). The forests have a growing volume of c. 13,500 million m^3 of which 69% is contributed by conifers. European wood production in 1999 was c. 500 million m^3 solid volume excluding bark. Details of production in Sweden and Finland are given in Anonymous (2000b, 2001b).

Main methods

Forestry practice in Britain was described by Hibbert (1991). The mechanization of forestry began around 1950. In Sweden this led, over a period of c. 25 years, to an eight-fold increase in the quantity of round timber produced per man-hour.

The basic forestry cycle can include felling, ground scarification and (re)planting.

A major effect of forestry is to change the age distribution of the trees. In Swedish forests, at present, c. 5% is clear-felled, 25% is young forest, 30% is medium-aged forest and 40% older forest (over 40 years in S. Sweden and over 60 years in N. Sweden).

Timber transport and processing

In northern Europe, timber transportation was originally largely by log driving in rivers (Östlund 2000). This ended in the 1960s and road transport now dominates.

The Scandinavian countries produce four times more paper pulp than they consume. The surplus is mainly exported to other parts of Europe.

Detritus deposition by pulp mills and the use of chlorine in bleaching have been major concerns and measures have been taken to reduce these problems.

Effects of forestry practices on fish

Introduction

This section mainly covers coniferous forests and salmonid fishes, although deciduous forests and other fish groups are mentioned where appropriate.

Table 24.2 lists possible effects upon the aquatic environment arising from the planting, management, harvesting and processing phases of the forestry cycle and summarizes possible consequential impacts upon salmonid populations.

Long-term studies on the hydrological, chemical and physical effects of forestry are summarized by Robinson *et al.* (1998), Kirby *et al.* (1991) and Neal (1997). Forestry influences acidification and summary works, although not related entirely to forestry effects, include Edwards *et al.* (1990), Curtis *et al.* (1999) and Monteith & Evans (2000).

Ideally detailed studies would have linked each forestry phase, quantitatively, through its environmental effects, to the consequences for fish populations. In practice there has been a much more fragmented approach. Many studies have been based on spatial or temporal correlations. These often tell us little about the details of the underlying mechanisms and interactions and they do not, necessarily, prove causation. There have been few fully integrated quantitative studies and few critical field experiments, despite the fact that the relevant mechanisms are complex.

One pragmatic line of approach is to:

(1)   take the results of quantitative studies of the effects of various forestry activities upon water quality and quantity and on habitat structure;
(2)   abstract from the literature quantitative information on the requirements of the fish as, for example, in Grandmottet (1983), Crisp (1996) and Mann (1996);
(3)   compare these two information sets.

This procedure has been broadly followed below. Most major aspects of the environment (Table 24.2) have been considered, although there are important interactions between different variables.

Table 24.2　List of forestry activities, their possible environmental effects and possible consequential effects on salmonid fishes

Forestry activities	Possible environmental effects	Possible consequential effects, mainly on salmonid fishes
1. Road construction damage and ploughing and planting	a. Modified patterns of runoff, bed movement, silt transport and deposition, increased culvert density	Changes in rates of egg/alevin washout, egg/alevin asphyxiation by silt, concentrations of >80 mg L^{-1} are considered harmful to salmonids. Impediment of fish movements
2. Application of fertilizer and pesticides	a. Enrichment of runoff	Changes in species and/or growth of algae and/or macrophytes
	b. Pollution of runoff	Lethal and/or sublethal effects on fish and/or their invertebrate food
3. Growth of trees	a. Reduced light input	Changes in species and/or growth of algae and/or macrophytes
	b. Evapotranspiration and canopy interception	Various flow-related effects
	c. Acidification by differential absorption and by canopy interception of acid deposition	Where little or no buffering this can lead to lethal concentrations of monomeric aluminium
	d. Modified allochthonous inputs	Possible changes in available food quantity and/or quality

Table 24.2 (*Continued.*)

Forestry activities	Possible environmental effects	Possible consequential effects, mainly on salmonid fishes
3. Growth of trees	e. Changed water temperature regime. Usually summer reductions and sometimes winter elevations	Modified growth rate, modified incubation rate, reduction of frequency of high summer temperatures that may be stressful to fish
	f. Accumulation of large woody debris	Cover for fish
	g. Accumulation of fine woody debris	Cover for 0-group fish
4. Felling	a. Modified patterns of runoff, bed movement, silt transport and deposition	Changes in rate of egg/alevin washout and of egg/alevin asphyxiation by silt
5. Transport	a. Increased suspended solids load in forest streams	Egg/alevin asphyxiation
6. Processing	a. Increased suspended solids load in streams	Egg/alevin asphyxiation
	b. Inputs of toxic materials	Materials such as chlorine compounds can be lethal to fish, especially at low pH

Based, in part, on Egglishaw (1985).

Environmental effects of forestry activities

Water yield, discharge pattern and movements of silt and bed material

The water yield of a catchment is modified by the presence of tree cover via the effects of evapotranspiration and canopy interception. Water yield and the temporal pattern of runoff may be modified by the construction and maintenance of forest roads and drains and by the ploughing associated with drainage and tree planting. The magnitude of these effects varies between sites.

At a site in northern England the interception loss via trees averaged 28% of rainfall (Robinson *et al.* 1998), although higher values (c. 35%) have been observed elsewhere (Calder 1990). In mid-Wales evaporative losses amounted to 29–32% of rainfall in an afforested catchment and 15–17% on moorland (Kirby *et al.* 1991).

After general ploughing, evaporative losses may diminish and stream flow has been observed to increase by c. 8%. The effects of ploughing generally diminish with time (Robinson *et al.* 1998). Ploughing, drainage and road construction lead to higher spate peaks (Kirby *et al.* 1991) with peak flows increased by c. 15% (Robinson *et al.* 1998). Müller (2000) observed peak discharges of six times the annual mean in a predominantly (97%) afforested catchment in western Ireland.

Generally, afforestation decreases annual water yield (hence lowers mean stream discharge) and increases the height of spate peaks. The probable consequences for fish populations vary between sites. Salmonid fishes are sensitive to flow regime via wetted area and water velocity and depth. These variables influence the area available for spawning and juvenile habitat and may modify incubation success via changes in intra-gravel flow. Flow-related effects via siltation and bed movement are likely to be most significant.

An evaluation of buffer zones (Nyberg & Eriksson 2001) showed that increased sediment load was a major negative effect from forestry. Suspended solids fluxes may increase 60-fold as a result of ploughing (Robinson *et al.* 1998). These loss rates generally decrease with time, although sometimes the cutting of forestry drains through peat into underlying mineral soils can cause major continuing erosion problems (e.g. Fig. 5.11 in Crisp 2000). In the Coalburn catchment (N. England) suspended solids concentration averaged 4 mg L^{-1} before drainage. During ploughing values varied with weather conditions from 30 to 150 mg L^{-1} but, exceptionally, up to 7000 mg L^{-1} (Robinson *et al.* 1994). Mean values of 3 mg L^{-1}, 13 mg L^{-1} and 30.5 mg L^{-1} were observed from pasture, mature forest and mature forest being felled in mid-Wales (Kirby *et al.* 1991). The importance of ditching and consequent soil erosion was noted in western Finland (Jutila *et al.* 1998) and the effects of clear-cutting and scarification were studied in small Finnish brooks (Ahtianen 1992). There was a small increase in suspended solids load after clear-cutting, while scarification led to a more than 200-fold increase in the average concentration over a period of 3 years.

Construction and use of forest roads can give large increases in suspended solids loads. Concentrations may increase by an order of magnitude during road construction (Duck 1985). Construction of a loading bay increased suspended sediment yield by 20% (Ferguson *et al.* 1987) and drain clearance gave suspended solids concentrations of almost 2000 mg L⁻¹ (Leeks 1992). Fording of streams by heavy machines can raise concentrations from about 1 mg L⁻¹ to 380 mg L⁻¹ (Leeks & Roberts 1987). Suspended solids loading may increase by an order of magnitude during and after tree harvesting (Leeks & Marks 1997). It is clear from these examples that forestry activities can raise suspended solids concentrations from low background levels (<10 mg L⁻¹) to values above the levels of 80 mg L⁻¹ that are claimed to be harmful to fisheries (Alabaster & Lloyd 1982). Although not toxic to fish and invertebrates, these inert suspended solids can damage or clog delicate membranes such as fish gills and cause irritation and abrasion leading to secondary infections. They reduce light input to the aquatic system. During spate recession they are deposited in the interstices of streambed gravel and may reduce living space for small fish and invertebrates. Vuori & Joensuu (1996) studied the impact of forest ditching on macroinvertebrates of small boreal streams and concluded that drainage gave significantly lower species richness in *fontinalis* habitats below ditched sections compared with control sections. They suggested that this reflected deposition of particles on benthic habitats. Deposition of 'fines' (particles of <1 mm diameter) in spawning gravels can reduce intra-gravel flows and cause asphyxiation of eggs/alevins and entrapment of emerging alevins. High concentrations of peat in the gravel (c. 40%) gave low survival (c. 65%) and sand concentrations of between 10 and 20% reduced survival from c. 100% to c. 30% (Olsson & Perssen 1986, 1988). Premature emergence of alevins also increased in finer gravel and at higher peat or sand concentrations. Deposition within the gravel of fine materials, either organic or inorganic, has been shown to reduce the survival of trout embryos in the field (Jutila *et al.* 1998; Nyberg & Eriksson 2001), although sand deposited on top of the gravel may have little effect on emergence (Crisp 1993). Although there are disagreements on points of detail between different studies, it is likely that successful incubation and emergence require gravels with fines contents of <10–20%, an intra-gravel dissolved oxygen concentration of 7.0 mg L⁻¹ or more and a seepage velocity of at least 250–650 cm h⁻¹ (Crisp 2000).

A combination of modified flow regime and miscellaneous destabilizing effects from road making and use, from drainage and from felling and removal of timber may modify the pattern of bed movement in forested streams. Kirby *et al.* (1991) quote bedload movements of 6.1 m³ km² y⁻¹ for a pasture catchment, 38.4 m³ km² y⁻¹ for mature forest and 24.4 m³ km² y⁻¹ for mature forest during the first year of felling operations. The eggs and alevins of salmonid fishes develop within the streambed for variable times, from several weeks to several months, dependent upon temperature and they are vulnerable to washout as a result of bed movements during spates. This can cause physical damage to eggs/alevins, redeposition in habitats unsuitable for further development and death from mechanical shock. We lack adequate knowledge to predict the sites or depth of gravel disturbance likely from any given spate in any given stream. Field experiments have shown that spates of reasonably frequent occurrence can wash out artificial eggs from 0 to 10 cm burial depth and larger spates (10–20 years return period) may disturb eggs at 15 cm (Crisp 1989). Harris (unpublished thesis, Liverpool University, 1970) noted that c. 27% of sea trout redds were washed out by spates in some tributaries of a Welsh river. When trout eggs at an early stage of development drifted along 10 m of experimental stream they suffered 50% mortality (Crisp 1990). It is, therefore, clear that any increase in the amplitude of spate peaks and in disturbance of streambed gravels arising from forestry activity could increase the mortality of the intra-gravel stages of salmonids.

Water quality

Suspended solids have already been covered. Acidification is a major issue and has, therefore, been given its own section. This section covers other aspects of water quality.

Gray & Edington (1969) noted that felling of deciduous woodland near one small stream increased summer water temperature by up to 6.5°C. Smith (1980) studied afforested and unafforested portions of a small stream in Scotland. The former was cooler in summer and warmer in winter and had smaller diel fluctuations than the latter. Weatherley & Ormerod (1990) compared several afforested and unafforested streams in Wales. Daily means were lower in afforested streams in spring and summer and higher in winter. Clearance of bankside vegetation brought forest stream temperatures closer to those of moorland. Crisp (1997) studied an afforested and an unafforested stream with similar catchment areas. In the afforested stream annual mean water temperature was c. 0.4°C lower, mainly in summer through depression of daily maxima and minima (chiefly the former). There was no clear evidence of differences in winter temperatures. Each of these last three studies had some inherent weaknesses. Smith (1980) compared different reaches of the same stream, Weatherley & Ormerod (1990) refer to a set of streams with disparate catchment areas and aspects and there was doubt about the possible influence of groundwater inputs in Crisp's (1997) data.

A more recent study (Stott & Marks 2000) compared temperatures at the same stream station before and after clear felling of 20% of the southern side of the catchment. Felling led to a 0.58°C rise in mean stream water temperature and the monthly means of daily maxima showed an increase of 7.0°C in July and 5.3°C in August. There was a statistically significant change in the linear relationship between observed air and water temperatures such that, in the 10–15°C range, an increase of c. 1°C in stream temperature would be expected after felling. At lower temperatures the effect would be negligible. This experimental design avoided problems arising from differences in catchment size and aspect and possible effects of groundwater inputs. The results confirmed the general conclusions drawn from the earlier studies.

Fishes are poikilotherms; therefore, temperature has a large influence upon such activities as embryonic development, growth, behaviour and survival. There are mathematical models to predict, from temperature, embryonic development (Crisp 1981, 1988; Jungwirth & Winkler 1984; Humpesch 1985) and growth on maximum rations (Elliott 1975; Elliott & Hurley 1997) for *Salmo trutta* and *S. salar*. Thermal tolerance polygons have been developed for both species (Elliott 1981, 1991).

Salmonid spawning time at any given site is largely a function of day length (Bye 1984) and is, therefore, relatively independent of temperature, whereas rate of embryonic development is highly temperature sensitive. Weatherley & Ormerod (1990) used observed temperatures to predict embryonic development for trout in afforested and unafforested streams in two different years. Relative to the unafforested stream, predicted hatching in the forested stream was delayed by 7–9 days and predicted swim-up was advanced by 7 days in 1 year and delayed by 5 days in another year. Crisp & Beaumont (1997) predicted that trout embryos in their afforested stream would have median eyeing delayed by 1 day, median hatch by 2 days and median swim-up by 3 days. The effects of temperature changes caused by afforestation upon embryonic development of trout will, therefore, be small. The magnitude and direction of the change at any given site will depend upon the timing of spawning relative to the annual temperature cycle. Weatherley & Ormerod (1990) used the equation of Elliott (1975) to show that, at the end of the first growing season, given maximum rations, brown trout would be 22–39% heavier in an unafforested than in an afforested stream and 31% heavier at the end of the second season. Crisp & Beaumont (1997) showed that the depression of summer temperature in their afforested stream would reduce predicted weight by 9–14% and mean instantaneous growth rate by 5–10% in the second growing season.

The upper critical temperature range for *Salmo trutta* is 19–30°C, dependent upon acclimation temperature (Elliott 1981). In this range fish cease feeding, become distressed and may die. Over 4 years there were no days on which a temperature of 19°C was attained in an afforested stream, whereas this value was exceeded on 1 or 2 days in each year and on 10 days in 1 year in an unafforested stream. Thus, high temperature stress is less likely in forested than in unafforested streams (Crisp & Beaumont 1997).

There are two major chemical aspects of water quality whereby forestry activity may impinge upon fish. First, various pesticides and herbicides used in forest management can harm fish and/or their invertebrate prey. Misuse or accidental spillages may cause fish mortalities in recipient streams. Second, nutrients and other chemicals may be released after fertilizer application or other management. Inputs of phosphate or nitrate have the potential to modify the flora of recipient streams and, hence, trophic relationships. Application of 375 kg ha^{-1} of rock phosphate and 200 kg ha^{-1} of potash to a forested catchment in Wales did not give any detectable change in phosphate or potassium concentrations in the efferent stream. In contrast, application of 375 kg ha^{-1} of phosphate on a young forest in northern England resulted in 2% of the phosphate passing downstream and concentrations of 15–20 times the natural background over a period of 5 months (Kirby *et al.* 1991).

Long-term effects of forestry management on water quality were studied in a number of small Finnish brooks (Ahtianen 1992; Ahtianen & Huttenen 1999). Clear-cutting and scarification gave a four-fold increase in phosphorus concentration. After clear-cutting and scarification of small lake catchments, concentrations of total phosphorus and iron increased (Rask *et al.* 1998). Increases in water colour and biochemical oxygen demand suggested an increased organic load. The limnological responses to catchment forestry were modest and this may reflect the presence of protective zones round the lake shores (Rask *et al.* 1998). The effects of forest fertilization on water chemistry, benthic invertebrates and brown trout were followed in a small creek in central Sweden (Göthe *et al.* 1993). The toxicities of iron, aluminium, dissolved humic material and acidity were tested in laboratory exposures with grayling and brown trout (Vuorinen *et al.* 1998). Grayling were more sensitive than brown trout to both iron and aluminium. Waters draining forests or peat bogs may contain concentrations of iron and aluminium harmful to fish.

Acidification

Consumption of fossil fuels in vehicles, homes and power stations pollutes the air with oxides of sulphur and nitrogen. These are deposited on the needles of conifers and can also fall directly in rainfall. Conifers acidify the soil in which they grow by absorbing positive ions (e.g. calcium) through the roots and releasing hydrogen ions from their leaves. The strong acids that accumulate on the needles are washed down by rainfall. Many soils contain ample carbonate and other bases to neutralize these acids. In poorly buffered soils heavy rainfall may rapidly carry them through or over the soil to cause 'acid pulses' in watercourses. Aluminium is a widespread and abundant element. Under most circumstances it is harmless to fish but in water of low pH it is converted to the labile form Al^{3+} and at low concentrations this can kill salmonids. Forests do not cause this problem, but they do aggravate it (Jenkins *et al.* 1990).

Acid inputs have caused extensive acidification over large areas of Scandinavia in the surface waters of areas with base-poor bedrock. Lakes and flowing waters within one-quarter of Sweden have critically low pH values. Similar problems, with consequent effects upon aquatic biota, also occur in Norway and Finland (e.g. Hesthagen *et al.* 2000). Work in Scandinavia has focused mainly upon the chemical and biological effects of acidification *per se* (Overrein *et al.* 1981; Anonymous 1991; Wahlström *et al.* 1992; Appelberg *et al.* 1993; Staaf & Tyler 1995). The role of forests has received less attention, although the restoration of acid forest soils is included in recent research programmes. The importance of natural acidity in northern Scandinavia is also being debated (Laudon 2000). In contrast, work in the British Isles has given major attention to forestry effects.

Numerous papers describe temporal and/or spatial correlations between acidification/low pH/high labile aluminium concentration and the decline or extinction of fish and invertebrate populations. Examples include: Harriman & Morrison (1981, 1982), Stoner *et al.* (1984), Egglishaw *et al.* (1986), Harriman *et al.* (1987), Waters & Jenkins (1992*)*, Giller *et al.* (1993), Wright *et al.* (1994), Clenaghan *et al.* (1998a, 1998b), Johnson *et al.* (2000), Lehane *et al.* (2000). A study of 113 streams in Wales showed that with increasing forest cover there was a decrease in pH value and an increase in aluminium concentration (Ormerod *et al.* 1989). In a subset of 13 sites the aluminium was present chiefly in the labile form. Comparisons of pH and dissolved aluminium values for two afforested and two unafforested streams in mid-Wales show pH values between 6.0 and 7.0 during base flow in all four streams. During storm flow, pH was above 4.5 in the unafforested streams but below 4.5 in the forested ones. Similarly, during storm flow, aluminium concentration was <0.1 mg L^{-1} in the unafforested streams but >0.4 mg L^{-1} in the afforested streams (quoted by Crisp & Beaumont 1996). At pH >5.0 only a small percentage of the aluminium is in the labile form but as pH falls below 5.0 the percentage in this form rises sharply to reach values close to 75% at pH 4.2 (Neal *et al.* 1992).

Models have been developed to evaluate and predict acidification in streams and lakes (Cosby *et al.* 1990; Waters & Jenkins 1992).

Water at pH 3.5 is lethal to trout and salmon eggs but no effect of pH alone can be demonstrated at pH 4.5 and above (Carrick 1979). Similarly, free-swimming trout may have salt balance problems at pH 4.5 and below (McWilliam 1982) and, where concentrations of calcium, sodium and chloride are low, pH values between 4.5 and 5.0 can be harmful (Alabaster & Lloyd 1982). In combination with aluminium, pH values of >4.5 can be harmful. At these low pH values aluminium may occur at concentrations in excess of 0.4 mg L^{-1} and, below pH 5.0, a large percentage is present as the labile, monomeric form (Al^{3+}) which is lethal to fish at concentrations of only a few micrograms per litre (Alabaster & Lloyd 1982).

Possible methods of ameliorating acidity problems have been considered (Hornung *et al.* 1990; Jenkins *et al.* 1991; Rundle *et al.* 1995). Chemical conditions improved after liming but Rundle *et al.* noted that these changes were not matched by sustained responses among the biota. It is probable that, in practice, 'reduction in the emissions of acid pollutants is the only way of solving the general problem of surface acidification' (Forestry Commission 1993).

Habitat structure and incident light

The structure of the habitat is important to fish. The composition and form of the streambed is significant in the incubation of salmonid eggs and alevins. The pool and riffle sequence and the presence of cover in the form of deep pools, undercut banks and other hiding places influences the distribution and territory size of free-swimming fishes. The amount of light reaching the stream may influence the type and amount of aquatic vegetation and hence the invertebrate fauna and, thus, the fish (Mann 1971). Some fish species may use adventitious tree roots as spawning sites (Mann 1996).

These aspects of the requirements of fish have been little studied relative to the effects of afforestation, but it is clear that the growth of trees will have implications for trophic relations within the stream and changes in habitat structure arising from forestry may influence fish population density.

Large woody debris (LWD) is added to streams by natural processes (dead branches and trees) and by forestry activities (trimmings, bark and brash). It traps sediment, creates a stepped channel profile and gives bed stability and cover for invertebrates and fishes from predators and spate flows. Despite emphasis on the significance of LWD in North American studies, European studies are few and recent. In a Swedish study, only 5.5% of a set of 286 stream sections of 100–200 m length and 4.5 m mean width had 5% or more bottom cover of LWD, and this reflects the fact that there is routine clearance of dead wood (on land or in water) (Markusson 1998). The percentage of stream sections with LWD is generally low in Swedish forests (2–3% of standing stock) but values of c. 40% can occur in old-growth spruce forest in Northern Sweden (Svensson *et al.* 1989; Linder 1984). In the presence of predators LWD increased the numbers of brown trout (Markusson 1998). It had a significant positive effect on the distribution of trout in summer but not in autumn (Sundbaum 2001). It modified behaviour patterns but not growth. Probably the addition of LWD increased the number of summer feeding territories but had less influence on trout distribution in the autumn, when spawning areas were likely to be more important.

Past use of Scandinavian rivers for log driving caused considerable modification to river channels, giving channelized courses with high water velocities and little variation of bed topography. In addition, dams sometimes created barriers to fish migration (Näslund 2000). Attempts are now being made to restore some of this damage.

Reintroduction of the beaver (*Castor fiber* L.) to Swedish forests increased the proportion of wetlands and open water and decreased mean water velocity leading to an increase in numbers of minnows (*Phoxinus phoxinus* (L.)) and a decrease in numbers of trout, although beaver ponds formed refuges for larger trout during droughts (Hägglund & Sjöberg 1999).

Stream margins

The stream environment is heavily influenced by the management of the catchment and especially the stream margins.

The importance of leaving or creating buffer zones beside forest streams has been stressed. In Sweden such zones are a legal requirement but without formal specification of width or structure. Buffer zones have general value in allowing streams to develop natural sinuosities and pool-riffle sequences. It is claimed that in forests streamside zones not planted with conifers, but possibly with a sparse growth of broad-leaved trees, can reduce the runoff of suspended solids, protect banks from erosion, give intermittent shade, give protective cover to aquatic fauna and improve allochthonous inputs from leaf fall.

Studies in Sweden (Nyberg & Eriksson 2001) on small trout streams at logged and unlogged sites showed that logging led to increased aquatic vegetation cover but no change in number of fish species. Numbers of brown trout and minnow decreased (Degerman & Sers 1995) and numbers of pike and burbot (*Lota lota* (L.)) increased. More detailed studies on a smaller number of sites showed clear correlations between fish fauna, LWD, forest type and the presence or absence of a buffer zone (Markusson 1998). The results of detailed studies on the effects of buffer zones of two different widths during the logging phase (Table 24.3) demonstrate the value of buffer zones and the importance of them being of adequate width.

Needs and recommendations for improvement of forestry practices and standards

Introduction and legal frameworks

Parts of northen Europe are dominated by forest land. The major economic importance of this is reflected in national forestry policies. Sweden formulated a forestry policy as early as 1903. Recent updating in 1994 suggested greater liberalization of business aspects and gave more attention to conservation issues (Wallin *et al.* 1996; Anonymous 2001b). The Swedish strategy features a combination of multiple-use

Table 24.3 Summary of a comparison between the effects of buffer zones of 20 m and 5 m width during the logging phase of the forestry cycle

Variable	Effects 20-m zone	5-m zone
Water temperature	Variable/small change	Variable/small change
Sediment deposition	Unchanged – increased if steep slopes	Increased
Detritus deposition	Unchanged	Increased
LWD quantity	Unchanged, but felling remnants for first season after clear-cutting	Unchanged, but felling remnants for first season after clear-cutting
Aquatic vegetation	Unchanged	Increased amount of filamentous green algae and higher plants
Benthos	Unchanged	Unchanged
Freshwater pearl mussel (*Margaritifera margeritifera* (L.))	Unchanged	Unchanged
Fish	Unchanged	Trout – more 0+, fewer older fish Pike – short-term increase Burbot – decrease
Hatching success of brown trout eggs	Unchanged	Reduced

After Nyberg & Eriksson (2001).

forestry and protection of areas with very high conservation values (Wallin *et al.* 1996). Scandinavian forestry laws include paragraphs to prevent damage by nutrients, fertilizers and pesticides to groundwaters, lakes and streams. Other regulations relate to drainage and road construction and the need for buffer zones.

Similar trends, on a less formal level, are apparent in the British Isles. There is no legal framework specific to forestry but more general regulations cover the conservation and quality of soil and water. This framework is summarized, together with guidelines, by Ministry of Agriculture, Fisheries and Food & Welsh Office Agriculture Department (1998a, 1998b).

Forest certification

In recent years the international forest products market has sought to give assurances about the quality and environmental impacts of forest management, with particular reference to sustainability, via various certification schemes. Two international schemes are in operation, 'Forestry Stewardship Certification' (FSC) and 'Pan-European Forestry Certification' (PEFC). Approximately 16 million hectares of European forest are certified by the former and 41 million hectares by the latter (c. 6% of total forest). The goal of FSC is to promote environmentally responsible, socially beneficial and economically viable management of the world's forests, by establishing a worldwide standard of recognized and respected Principles of Forest Stewardship. The PEFC scheme seeks to establish an internationally credible framework for forest certification schemes and initiatives in European countries that will facilitate mutual recognition of such schemes. The PEFC Technical Document and Statutes define basic requirements of forest certification standards and schemes and the set-up of institutional arrangements at Pan-European and national and sub-national levels. The certification criteria cover the whole of forest functions including all economic, ecological and social functions.

Guidelines and education

Good standards of forestry management for environmental purposes (including the environment of fish) depend upon both the forestry managers and the forestry workers clearly understanding the issues and the underlying principles. Guidelines for good forestry practice have been produced in several European countries (e.g. Hänninen *et al.* 1996; Forestry Commission 1993; Henrikson 2001). In Britain there are a number of more general guidelines that are relevant to forestry, even though mainly aimed towards agriculture. These include such topics as buffer zones (Environment Agency 1997) and river bank erosion (Environment Agency 1998). Recent studies in the UK were reviewed to examine the likely effectiveness of forest guidelines (Nisbett 2001). This showed that the best modern management practices can be efficient in limiting soil erosion and consequent turbidity and sedimentation.

Tourism

In some forests the growth in tourism has exceeded their ecological capacity. In Spain there are 3,000,000 visitors per year to National Parks (World Wide Fund for Nature 2001). Such levels of pressure may require some form of access rationing to protect the terrestrial and aquatic environments.

Science base

Forestry activities have complex effects upon aquatic environments (Hartman & Scrivener 1990) and some of these were described in detail by Mills (1989). Effective management/restoration needs to be based on sound science. Hartman *et al.* (1996) stressed the weakness of the science base in North America and a similar problem has been noted in Europe (Crisp 2000). Points raised include: (1) lack of detailed, quantitative understanding of the complex mechanisms and interactions in the forestry–environment–fisheries chain; (2) some effects of forestry may not be reversible; (3) some remedial measures may not be long-lasting and may even have harmful side effects; (4) ideally, project teams should be interdisciplinary.

There has, generally, been a lack of scientific rigour in the design of studies and remedial programmes and in assessing the long-term effectiveness of the latter. In general, the science base is weak and is too dependent upon short-term *ad hoc* studies.

Acknowledgements

The authors are grateful for information provided by Paul Giller (University College, Cork), Ron Harriman (Scottish Office Agriculture and Food Department), Jim Hudson, Alan Jenkins and Brian Reynolds (Centre for Ecology and Hydrology) and Tom Nisbett (Forestry Commission). Diane Crisp did the word processing and the figures were made by Lydia Zweifel. David Le Cren, Richard Mann, Tom Northcote and Gordon Hartman reviewed the draft and made helpful comments.

References

Ahti, T., Håmet-Ahti, L. & Jalas, J. (1968) Vegetation zones and their sections in northwestern Europe. *Annales Botanici Fennici*, 5, 168–211.

Ahtianen, M. (1992) The effects of forest clear-cutting and scarification on the water quality of small brooks. *Hydrobiologia*, **243/244**, 465–73.

Ahtianen, M. & Huttenen, P. (1999) Long-term effects of forestry management on water quality and loading in brooks. *Boreal Environmental Research*, 4, 101–14.

Alabaster, J.S. & Lloyd, R. (eds) (1982) *Water Quality Criteria for Freshwater Fish.* Butterworths, London.

Anonymous (1991) Försurning och kalkning av svenska vatten. *Monitor 12 Naturvårdsverket informerar.*

Anon (2000a) *Fiske 2000 – En undersökning om Svenskarnas sport – och husbehovsfische.* [*An investigation about the recreational fishery in Sweden.*] Finfo 2001:1. [In Swedish with English summary.]

Anon (2000b) *Finnish Statistical Yearbook of Forestry 2000,* (editor-in-chief Y. Sevola). Finnish Forest Research Institute.

Anon (2001a) *The Swedish Forestry Model.* Swedish Academy of Agriculture and Forestry.

Anon (2001b) *Statistical Yearbook of Forestry – 2001.* Swedish National Board of Forestry.

Appelberg, M., Henrikson, B-I., Henrikson, L. & Svedäng, M. (1993) Biotic interactions within the littoral community of Swedish forest lakes during acidification. *Ambio*, 22, 290–7.

Bengtsson, B., Norling, I. & Pierrou, U. (1991) National survey on sport fishing, economics and environmental changes in Sweden. *American Fisheries Symposium*, 12, 74–83.

Blackie, J.R., Ford, E.D., Horne, J.E.M., Kinsman, D.J.J., Last, F.T. & Moorhouse, P. (1980) Environmental effects of deforestation: an annotated bibliography. *Freshwater Biological Association Occasional Publication*, 10, 1–73.

Bogren, J., Gustafsson, T. & Loman, G. (1999) *Climatology – Meteorology.* Student litteratur. [In Swedish.]

Borgström, R. (1987) Fish resources in freshwater. In: *Fish in Freshwater – Ecology and Management*, (eds R. Borgstrøm & L-P. Hansen), pp. 11–19. Landbruksforlaget, Oslo. [In Norwegian.]

Bye, V.J. (1984) The role of environmental factors in the timing of reproductive cycles. In: *Fish Reproductive Strategies and Tactics,* (eds R.W. Potts & R.J. Wootton), pp. 187–205. Academic Press, London.

Calder, I.R. (1990) *Evaporation in the Uplands.* John Wiley & Sons, Chichester.

Carrick, T.R. (1979) The effect of acid water on the hatching of salmon eggs. *Journal of Fish Biology*, 14, 165–72.

Clenaghan, C., Giller, P.S., O'Halloran, J. & Hernan, R. (1998a) Stream macroinvertebrate communities in a conifer afforested catchment in Ireland: relationships to physico-chemical and biotic factors. *Freshwater Biology*, 40, 175–93.

Clenaghan, C., O'Halloran J., Giller, P.S. & Roche, N. (1998b) Longitudinal and temporal variation in the hydrochemistry of streams in an Irish conifer afforested catchment. *Hydrobiologia*, **389**, 63–71.

Cosby, B.J., Jenkins, A., Ferrier, R.C., Miller, J.D. & Walker, T.A.B. (1990) Modelling stream acidification in afforested catchments: long-term reconstructions at two sites in central Scotland. *Journal of Hydrology*, **120**, 143–62.

Crisp, D.T. (1981) A desk study of the relationship between temperature and hatching time for the eggs of five species of salmonid fishes. *Freshwater Biology*, **11**, 361–8.

Crisp, D.T. (1988) Prediction from water temperature, of eyeing, hatching and swim-up times for salmonid embryos. *Freshwater Biology*, **19**, 41–8.

Crisp, D.T. (1989) Use of artificial eggs in studies of washout depth and drift distance for salmonid eggs. *Hydrobiologia*, **178**, 155–63.

Crisp, D.T. (1990) Some effects of the application of mechanical shock at varying stages of development of British salmonid eggs. *Hydrobiologia*, **194**, 57–65.

Crisp, D.T. (1993) The ability of UK salmonid alevins to emerge through a sand layer. *Journal of Fish Biology*, **43**, 656–8.

Crisp, D.T. (1996) Environmental requirements of common riverine European salmonid fishes in fresh water with special reference to physical and chemical aspects. *Hydrobiologia*, **323**, 201–22.

Crisp, D.T. (1997) Water temperature of Plynlimon streams. *Hydrology and Earth System Sciences*, **1**, 535–40.

Crisp, D.T. (2000) *Trout and Salmon: Ecology, Conservation and Rehabiliatation*. Fishing News Books, Oxford.

Crisp, D.T. & Beaumont, W.R.C. (1996) The trout (*Salmo trutta* L.) populations of the headwaters of the rivers Severn and Wye, mid-Wales, UK. *The Science of the Total Environment*, **177**, 113–23.

Crisp, D.T. & Beaumont, W.R.C. (1997) Fish populations in Plynlimon streams. *Hydrology and Earth Sciences*, **1**, 541–8.

Curtis, C., Murlis, J, Battarbee, R., *et al.* (1999) *Acid Deposition in the UK: A Review of Environmental Damage and Recovery Prospects*. National Society for Clean Air and Environmental Protection.

Degerman, E. & Sers, B. (1995) Små vattendrags funktion och värde för fisk. *Skog och Forskning*, **4**, 32–43. [In Swedish.]

Duck, R.W. (1985) The effect of road construction on sediment deposition in Loch Earn, Scotland. *Earth Sediment Processes and Landforms*, **10**, 401–6.

Edwards, R.W., Gee, A.S. & Stoner, J.H. (eds) (1990) *Acid Waters in Wales*. Kluwer, Dordrecht, the Netherlands.

Egglishaw, H.J. (1985) Afforestation and fisheries. In: *Habitat Modification and Freshwater Fisheries*, (ed J.S. Alabaster). Butterworths, London.

Egglishaw, H.J., Gardiner, R. & Foster, J. (1986) Salmon catch decline and forestry in Scotland. *Scottish Geographical Magazine*, **102**, 57–61.

Elliott, J.M. (1975) The growth of brown trout, *Salmo trutta*, fed on maximum rations. *Journal of Animal Ecology*, **44**, 805–21.

Elliott, J.M. (1981) Some aspects of thermal stress in freshwater teleosts. In: *Stress and Fish*, (ed A.D. Pickering), pp. 209–45. Academic Press, London.

Elliott, J.M. (1991) Tolerance and resistance to stress in juvenile Atlantic salmon, *Salmo salar*. *Freshwater Biology*, **52**, 61–70.

Elliott, J.M. & Hurley, M.A. (1997) A functional model for maximum growth of Atlantic salmon parr, *Salmo salar*, from two populations in northwest England. *Functional Ecology*, **11**, 592–603.

Environment Agency (1997) *Understanding Buffer Zones*. Environment Agency.

Environment Agency (1998) *Understanding Riverbank Erosion from a Conservation Perspective*. Environment Agency.

Ferguson, R.I., Stott, T.A. & Johnson, R. (1987) Forestry and sediment yields in upland Scotland. *International Symposium on Erosion and Deposition in Forested Steplands.* Poster Presentation. Oregon State University.

Forestry Commission (1993) *The Forest and Water Guidelines,* 3rd edn. HMSO, London.

Forestry Commission (1999) *Forestry Facts and Figures 1998–99.* Statistics Unit, Forestry Commission, Edinburgh.

Giller, P.S., O'Halloran, J., Hernan, R., *et al.* (1993) An integrated study of forested catchments in Ireland. *Irish Forestry,* 50, 70–83.

Göthe, L., Sjölander, E. & Nohrstedt, Ö. (1993) Effects on water chemistry, benthic invertebrates and brown trout following forest fertilization in central Sweden. *Scandinavian Journal of Forest Research,* 8, 81–93.

Grandmottet, J.P. (1983) Principals exigences des téléostéens dulcicoles vis-à-vis de l'habit aquatique. *Annals Scientific Université Besançon,* 4, 3–32.

Gray, J.R.A. & Edington, J.M. (1969) Effect of woodland clearance on stream temperature. *Journal of the Fisheries Research Board of Canada,* 26, 399–403.

Hägglund, A. & Sjöberg, G. (1999) Effects of beaver dams on the fish fauna of forest streams. *Forest Ecology and Management,* 115, 259–66.

Hänninen, E., Kärhä, S. & Salpakivi-Salornaa, P. (1996) *Forestry and the Protection of Watercourses. Conservation of Water Courses and Water Ecosystems in Relation to Forestry.* Finnish Forest Industries Federation.

Harriman, R. & Morrison, B.R.S. (1981) Forestry, fisheries and acid rain in Scotland. *Scottish Forestry,* 36, 89–95.

Harriman, R. & Morrison, B.R.S. (1982) Ecology of streams draining forested and non-forested catchments in an area of central Scotland subject to acid rain precipitation. *Hydrobiologia,* 88, 251–63.

Harriman, R., Morrison, B.R.S., Caines, L.A., Collen, P. & Watts, A.W. (1987) Long-term change in fish populations of acid streams and lochs in Galloway, South West Scotland. *Water, Air and Soil Pollution,* 32, 89–112.

Hartman, G.F. & Scrivener, J.C. (1990) Impacts of forestry practices on a coastal stream ecosystem, Carnation Creek, British Columbia. *Canadian Bulletin of Fisheries and Aquatic Science,* 223.

Hartman, G.F., Scrivener, J.C. & Miles, M.J. (1996) Impacts of logging in Carnation Creek, a high energy coastal stream in British Columbia, and their implication for restoring fish habitat. *Canadian Journal of Aquatic Sciences,* 53 (Suppl), 237–51.

Henrikson, L. (2001) *Skogsbruk vid vatten* [*Forestry along Watercourses*]. National Board of Forestry, Sweden.

Hesthagen, T., Aastorp, G., Langåker, R.M., Fastad, M. & Berger, H.M. (2000) Responses of brown trout (*Salmo trutta* L.) to acidification and excess critical loads in lakes of western Norway with low ionic content. *Verhandlungen Internationale Vereinigung für Theoretische und Angewandte Limnologie,* 27, 2079–89.

Hibbert, B.G. (ed.) (1991) *Forestry Practice.* Forestry Commission Handbook 6. HMSO, London.

Hornung, M., Brown, S.J. & Ranson, A. (1990) The role of geology and soils in controlling surface water acidity in Wales. In: *Acid Waters in Wales,* (eds R.W. Edwards, A.S. Gee & J.H. Stoner), pp. 311–28. Kluwer, Dordrecht, the Netherlands.

Humpesch, U.H. (1985) Inter- and intra-specific variation in hatching success and embryonic development of five species of salmonids and *Thymallus thymallus. Archiv für Hydrobiologie,* 104, 129–44.

Jenkins, A., Cosby, B.J., Ferrier, R.C., Walker, T.A.B. & Miller, J.D. (1990) Modelling stream acidification in afforested catchments: an assessment of the relative effects of acid deposition and afforestation. *Journal of Hydrology,* 120, 163–81.

Jenkins, A., Waters, D. & Donald, A. (1991) An assessment of terrestrial liming strategies in upland Wales. *Journal of Hydrology,* 124, 243–61.

Johnson, M, O'Gorman, K., Gallagher, M., Clenaghan, C., Giller, P.S. & O'Halloren, J. (2000) The identification and amelioration of the impacts of clearfelling operations on lotic macroinvertebrates and salmonids in Ireland. *Verhandlungen Internationale Vereinigung für Theoretische und Angewandte Limnologie*, 27, 1565–70.

Jungwirth, M. & Winkler, H. (1984) The temperature dependence of embryonic development of grayling (*Thymallus thymallus*), Danube salmon (*Hucho hucho*), arctic char (*Salvelinus alpinus*) and brown trout (*Salmo trutta fario*). *Aquaculture*, 38, 315–27.

Jutila, E., Ahvonen, A., Laamanen, M. & Koskiniemi, J. (1998) Adverse impact of forestry on fish and fisheries in stream environments of the Isojoki basin, western Finland. *Boreal Environmental Research*, 3, 395–404.

Kirby, C., Newson, M.D. & Gilman, K. (1991) Plynlimon research: the first two decades. *Institute of Hydrology Report*, 109, 1–188.

Laudon, H. (2000) *Separating natural acidity from anthropogenic acidification in the spring flood of northern Sweden.* Doctoral thesis, Swedish University of Agricultural Sciences, Umeå.

Leeks, G.J.L. (1992) Impact of planting forestry on sediment transport processes. In: *Dynamics of Gravel Bed Rivers*, (eds P. Bill, R.D. Hey, C.R. Thorne & P. Tacconi). John Wiley & Sons, Chichester, UK.

Leeks, G.J.L. & Marks, S.D. (1997) Dynamics of river sediments in forested headwater streams: Plynlimon. *Hydrology and Earth System Sciences*, 1, 483–97.

Leeks, G.J.L. & Roberts, G. (1987) The effects of forestry on upland streams with special reference to water quality and sediment transport. In: *Environmental Aspects of Plantation Forestry in Wales*, (ed. J.E.G. Good). Institute of Terrestrial Ecology, Grange-over-Sands, UK.

Lehane, B.M., Giller, P.S., O'Halloran, J. & Walsh, P.M. (2000) Conifer forest location and fish populations in southwest Ireland. *Verhandlungen Internationale Vereinigung für Theoretische und Angewandte Limnologie*, 27, 116–21.

Linder, P. (1984) Kirjesålandet. *En skogsbiologisk inventering av ett fjällnära urskogsområde i Västerbottens län.* Inst. F. Skoglig ståndortslära, SLU, Umeå.

Lundegårdh P.H., Lundqvist, J. & Lindström, M. (1964) *Berg och Jord i Sverige. Almqvist och Wicksell.*

McWilliam, P.G. (1982) A comparison of physical characteristics in normal and acid exposed populations of brown trout *Salmo trutta. Comparative Biochemistry and Physiology*, 73, 515–22.

Maitland, P.S. (1972) Key to British freshwater fishes. *Freshwater Biological Association Scientific Publications*, 27, 1–139.

Mann, R.H.K. (1971) The populations, growth and production of fish in four small streams in southern England. *Journal of Animal Ecology*, 40, 155–90.

Mann, R.H.K. (1996) Environmental requirements of European non-salmonid fish in rivers. *Hydrobiologia*, 32, 223–35.

Markusson, K. (1998) Omgivande skog och skogsbrukets betydelse för fiskfaunan i små skogsbäckar. *Skogsstyrelsens, Rapport* (*National Board of Forestry, Report*) 8, 1–35. [In Swedish.]

Mills, D.H. (1989) *Ecology and Management of Atlantic Salmon.* Chapman & Hall, London.

Ministry of Agriculture, Fisheries and Food & Welsh Office Agriculture Department (1998a) *Code of Good Agricultural Practice for the Protection of the Soil.* Ministry of Agriculture, Fisheries and Food & Welsh Office Agriculture Department.

Ministry of Agriculture, Fisheries and Food & Welsh Office Agriculture Department (1998b) *Code of Good Agricultural Practice for the Protection of Water.* Ministry of Agriculture, Fisheries and Food & Welsh Office Agriculture Department.

Monteith, D.T. & Evans, C.D. (eds) (2000) *Acid Waters Monitoring Network: 10 Year Report.* Ensis Publishing, London.

Müller, M. (2000) Hydrological studies in the Burrishole catchment, Newport, Co. Mayo, Ireland: effects of afforestation on the run-off regime of small mountain spate river catchments.

Verhandlungen Internationale Vereinigung für Theoretische und Angewandte Limnologie, **27**, 1146–8.

Näslund, I. (2000) Log driving and fishing. In: *Log Driving, Watercourses, Work and Stories*, (eds E. Törnlund & L. Östlund). *Skrifter om skogs-och lantsbrukshistoria nr* **14**.

Neal, C. (1997) Water quality of the Plynlimon catchments (UK). *Hydrology and Earth System Sciences*, **1**, 381–764.

Neal, C., Smith, C.J. & Hill, S. (1992) Forestry impact on upland waters. *Institute of Hydrology Report*, **119**, 1–50.

Nisbett, T.R. (2001) The role of forest management in controlling diffuse pollution in UK forestry. *Forest Ecology and Management*, **143**, 215–26.

Nyberg, P. (1998) Fish and Fisheries in Swedish inland waters. Fisk och Fiske i Sveriges inlandsvatten. In: *The Fishery Board 50 Years Anniversary Publication*, pp. 67–75. [In Swedish.]

Nyberg, P. & Eriksson, T. (2001) The SILVA project – buffer zones and aquatic biodiversity. *Finfo 2001*, **6**.

Olsson, T. & Persson, B-G. (1986) Effects of gravel size and peat material concentrations on embryo survival and alevin emergence of brown trout, *Salmo trutta* L. *Hydrobiologia*, **135**, 9–14.

Olsson, T. & Persson, B-G. (1988) Effects of deposited sand on survival and alevin emergence in brown trout (*Salmo trutta* L.). *Archiv für Hydrobiologie*, **113**, 621–7.

Ormerod, S.J., Donald, A.P. & Brown, S.J. (1989) The influence of plantation forestry on pH and aluminium concentration of upland Welsh streams: a reexamination. *Environmental Pollution*, **62**, 47–62.

Östlund, L. (2000) Early log driving in a watercourse in Central Norland in northern Sweden. In: *Log Driving, Watercourses, Work and Stories*, (eds E. Törnlund & L. Östlund). *Skrifter om skogs-och lantbrukshistoria*, **14**.

Overrein, L.N., Scip, H.M. & Tollan, A. (1981) *Acid Precipitation – Effects on Forest and Fish*. Final report on the SNSF-project 1972–1980.

Petersen, R.C., Jr, Gislason, G.M. & Vought, L.M. (1995) Rivers of the Nordic countries. In: *Ecosystems of the World: River and Stream Ecosystems*, pp. 295–342. Elsevier, Amsterdam.

Radford, A.F., Hatcher, A. & Whitmarsh, D. (1991) An economic evaluation of salmon fisheries in Great Britain. *Report prepared by the University of Portsmouth Centre for Economics and Management of Aquatic Resources for the Ministry of Agriculture, Fisheries and Food*, Vol. I, 1–292; Vol. II, 1–15; and Summary, 1–32.

Rask, M., Nyberg, K., Markkanen, S.L. & Ojala, A. (1998) Forestry in catchments: effects on water quality, plankton, zoobenthos and fish in small lakes. *Boreal Environmental Research*, **3**, 75–86.

Robinson, M., Jones, T.K. & Blackie, J.R. (1994) Coalburn catchment experiment – a 25 year review. *National Rivers Authority R.D. Note*, **270**, 1–54.

Robinson, M., Moore, R.E., Nisbett, T.R. & Blackie, J.R. (1998) From moorland to forest: the Coalburn catchment experiment. *Institute of Hydrology Report*, **133**, 1–64.

Rundle, S.D., Weatherley, N.S. & Ormerod, S.J. (1995) The effects of catchment liming on the chemistry and biology of upland Welsh streams: testing model predictions. *Freshwater Biology*, **34**, 165–75.

Smith, B.D. (1980) The effects of afforestation on the trout of a small stream in Southern Scotland. *Fisheries Management*, **11**, 39–58.

Staaf, H. & Tyler, G. (1995) Effects of acid deposition and tropospheric ozone on forest ecosystems in Sweden. *Ecological Bulletins*, **44**.

Stoner, J.H., Gee, A.S. & Wade, K. (1984) The effects of acidification on the ecology of streams in the upper Tyvie catchment in West Wales. *Environmental Pollution*, **35**, 125–57.

Stott, T. & Marks, S. (2000) Effects of plantation forest clearfelling on stream temperatures in the Plynlimon experimental catchments. *Hydrology and Earth System Sciences*, **4**, 95–104.

Sundbaum, K. (2001) *Importance of woody debris for stream dwelling brown trout* (*Salmo trutta L.*). Ph Licentiate thesis, Swedish University of Agricultural Sciences, Umeå, Report 32.

Svensson, S.A., Toet, G. & Kempe, G. (1989) *Riksskogstaxeringen 1978–82. Skogstillstånd, tillväxt och avverkning.* Inst. F. Skogstaxering, SLU, Umeå, Rapport 47.

Toivonen, A-L., Appelblad, H., Bengtsson, B., *et al.* (2000) Economic value of recreational fisheries in Nordic countries. *Tema Nord, 2000–604.* Nordic Council of Ministers, Copenhagen.

United Nations (1999) *State of European Forests and Forestry. United Nations Report.* ECE/TIM/SP/16. United Nations, New York, Geneva.

Vuori, K.M. & Joensuu, I. (1996) Impact of forest drainage on the macroinvertebrates of a small boreal headwater stream: Do buffer zones protect lotic biodiversity? *Biological Conservation*, 77, 87–95.

Vuorinen, P.J., Keinänen, M., Peuranen, S.A. & Tigerstedt, C. (1998) Effects of iron, aluminium, dissolved humic material and acidity on grayling (*Thymallus thymallus*) in laboratory exposures and a comparison of sensitivity with brown trout (*Salmo trutta*). *Boreal Environmental Research*, 3, 405–19.

Wahlström, E., Reinikainen, T. & Hallanaro, E-I. (1992) *Miljöns tillstånd i Finland, Vatten och Miljöstyrelsen i Finland.* Gaudeamus Kirja.

Wallin, B., Wester, J. & Johansson, O. (1996) *Action Plan for Biological Diversity and Sustainable Forestry – A Summary with Examples of Landscape Analysis.* National Board of Forestry, Sweden.

Waters, D. & Jenkins, A. (1992) Impacts of afforestation on water quality trends in two catchments in mid-Wales. *Environmental Pollution*, 77, 167–72.

Weatherley, N.S. & Ormerod, S.J. (1990) Forests and the temperature of upland streams in Wales: a modelling exploration of the biological effects. *Freshwater Biology*, 24, 109–22.

World Wide Fund for Nature (2001) *Insight into Europe's Forest Protection.* WWF Report, WWF Gland, Switzerland.

Wright, R.F., Cosby, B.J., Ferrier, R.C., Jenkins, A., Bulger, A.J. & Harriman, R. (1994) Changes in acidification of lochs in Galloway, southwestern Scotland, 1979–1988: the MAGIC model used to evaluate the role of afforestation, calculate critical loads and predict fish status. *Journal of Hydrology*, 161, 257–85.

Chapter 25
Freshwater fishes and forests in Japan

M. INOUE AND F. NAKAMURA

The Japanese archipelago

Japan is an island country, which exhibits undulated, mountainous landscapes. Forested mountains and narrow spaces of flatlands occupied by rice paddies are major components of a typical landscape of the country. Wood/forest and rice/paddy have fostered Japanese culture, and the country is sometimes referred to as 'the green archipelago'(see Totman 1998). On the other hand, the country has also been blessed with abundant aquatic resources, owing to surrounding ocean and thousands of rivers and streams on the land. Fishes, both marine and freshwater, have long been an important food resource for Japanese people. In this chapter, we provide an overview on forests, fishes, human activities and their relationships in Japan.

Geography, climate and landscape

The Japanese archipelago lies at the far-east margin of the Eurasian Continent, extending approximately 3000 km from northeast (46° N, 146° E) to southwest (24° N, 123° E) (Fig. 25.1). The archipelago, which is 378,000 km² in total area, consists of four major islands, Hokkaido, Honshu, Shikoku and Kyushu, and numerous other small islands such as Ryukyu (Okinawa) and Ogasawara (Bonin) Islands. The climate is generally wet (annual precipitation 1000–3000 mm), and exhibits a great variation among regions (Fig. 25.2). Hokkaido, the northernmost main island, has a cool, subarctic climate, and its northeastern coast (the Sea of Okhotsk) is often blocked by drifting sea ice during late winter. On the other hand, Ryukyu and Ogasawara Islands are surrounded by beautiful coral reefs under a subtropical climate. Although the other islands (Honshu, Shikoku, Kyushu and their associated islands) are categorized as temperate regions, the precipitation regime differs distinctly between northwestern (the Sea of Japan coast) and southeastern sides (the Pacific coast). In winter, regions facing the Sea of Japan receive a large amount of precipitation as snowfall, which is brought by the northwest monsoon, while the Pacific Ocean sides receive less precipitation. In contrast, the Pacific Ocean sides, especially southwestern parts, receive heavy rainfall in summer, owing to a seasonal stationary front during June to July (the rainy season called 'tsuyu' in Japan) and typhoons during August to September.

The archipelago is formed on a region where continental plates and oceanic ones meet, and various geological phenomena associated with tectonism and volcanism are active. Three-quarters of the total land area is categorized as mountain ranges, and

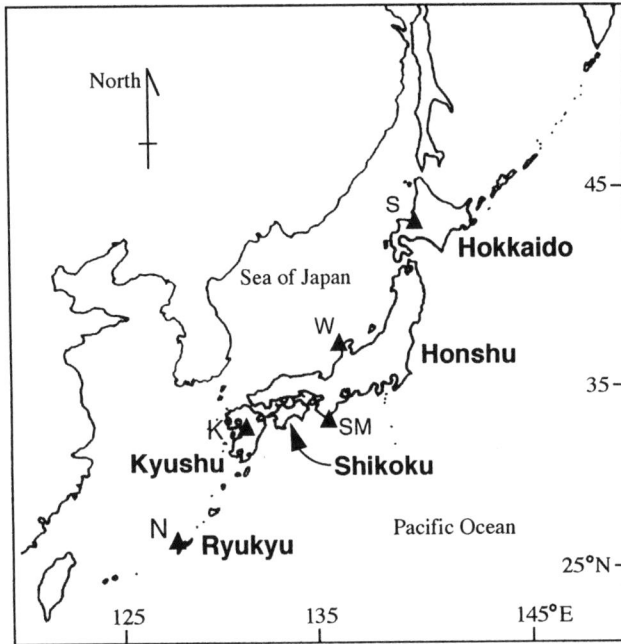

Fig. 25.1 Map of the Japanese archipelago, which consists of four major islands, Hokkaido, Honshu, Shikoku and Kyushu. Solid triangles indicate the location of the sites for which seasonal patterns of temperature and precipitation are shown in Fig. 25.2. S, Sapporo; W, Wajima; SM, Shionomisaki; K, Kumamoto; N, Naha.

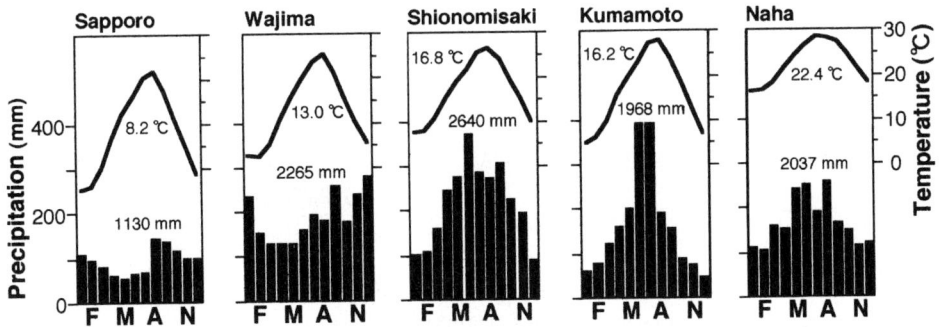

Fig. 25.2 Seasonal patterns of air temperature and precipitation at Sapporo, Wajima, Shiono-misaki, Kumamoto and Naha. Bars and lines indicate monthly total precipitation and mean monthly temperature, respectively. F, February; M, May; A, August; N, November. Figures in each graph indicate annual mean air temperature and annual total precipitation. Data are mean of 30 years (1961–1990).

flatlands are confined to alluvial fans and plains formed around rivers. Such mountainous islands support a 126-million human population (332 people/km²), resulting in one of the most heavily populated societies in the world. Furthermore, such a large population is concentrated on flatlands, which are spatially limited and also used for agriculture and industries. Therefore, population density in terms of the area appropriate for residence (e.g. flatland) is much higher than 332/km², which is based on the total land area. Climatic, geological and social conditions in Japan make people prone to natural disasters. Destructive earthquakes and volcanic eruptions are not rare in Japan. Heavy precipitation in combination with steep topography causes landslides, debris flows and floods, which have harmed the dense population on alluvial fans and plains seriously. Therefore, people have made much effort to prevent such natural disasters. As a result, most streams and rivers now have been highly regulated.

Although intensive land alterations and industrial development proceeded rapidly via civil engineering after World War II, the present landscapes in Japan also reflect a long history of human–nature interactions prior to the recent industrialized age. For example, irrigation systems for paddy fields, including complex networks of ditches and artificial ponds, have been developed over a millennium, resulting in a unique aquatic ecosystem (see Yuma *et al.* 1998). The present forested landscapes have also been shaped by histories of exploitation and afforestation, and intact forests are now very rare in Japan. The historical background of Hokkaido, however, is quite different from those of the other three main islands. Early Japanese rulers had established a capital in Kinki district (Nara, Kyoto), central Honshu, and had expanded their territories over Honshu, Shikoku and Kyushu (except northern Honshu and southern Kyushu) by the early eighth century (Amino 1997). However, it was after the late nineteenth century that Hokkaido actually became integrated into 'Japanese society'. Thereafter the colonization from the southwestern islands (Honshu, Shikoku and Kyushu) was promoted, and land alterations, such as logging and agricultural development, began. Until then, the land of Hokkaido had not been exploited intensively, because the lifestyle of the 'Ainu' people, the original inhabitants of Hokkaido, was hunter-gatherer. Therefore, the history of extensive and intensive alterations in Hokkaido is rather short, the land is not so highly developed, and human population density is relatively low (68 people/km²: less than a quarter of the mean density). Owing to such social conditions as well as climatic factors, landscapes in Hokkaido are quite different from those in Honshu, Shikoku and Kyushu.

Forest vegetation

Throughout the Japanese archipelago, potential vegetation of most areas is basically forest, owing to wet climates. Various vegetation types are distributed along latitudinal and altitudinal gradients, and the potential forest vegetation can be broadly categorized into four types (Fig. 25.3): evergreen broad-leaved forest, deciduous broad-leaved forest, mixed conifer and deciduous broad-leaved forest and subarctic (subalpine) coniferous forest.

Evergreen deciduous forest, of which representative trees are evergreen oak (*Quercus* spp.), is distributed over warm temperate regions, such as Kyushu, Shikoku and southwestern Honshu. In more southern, Ryukyu Islands, subtropical species (e.g. *Ficus retusa*) are included in the evergreen forests, and also mangrove forests are developed in some estuaries. On the other hand, deciduous broad-leaved forest occupies cool temperate regions, such as northeastern Honshu, southwestern Hokkaido and high

Fig. 25.3 Distribution of the potential vegetation types in Japan (modified from Yoshioka 1973).

altitude parts (>500–1000 m) of southwestern Honshu, Shikoku and Kyushu. These regions are often referred to as beech (*Fagus crenata*) zones. Beech is distributed up to southwestern Hokkaido, and replaced by other deciduous species (e.g. oak, *Quercus crispula*; maple, *Acer mono*) mixed with conifers (spruce, *Picea jezoensis, P. glehnii*; and fir, *Abies sachalinensis*) in the other parts of Hokkaido. Such a type of mixed conifer and deciduous broad-leaved forest is considered as a transition from temperate broad-leaved forest to subarctic coniferous forest. The latter type of forest, which is dominated by spruce (*Picea* spp.) or fir (*Abies* spp.), is limited to northern and eastern Hokkaido and subalpine zones of central Honshu and Hokkaido (subalpine coniferous forest).

Even now, approximately 67% (25 million hectares) of the total land area of Japan is covered by forests. In many regions, however, the actual state of the forests is quite different from their potential vegetation described above, owing to various effects of human activities. Representative forested landscapes resulting from human activities include pine woodlands and conifer plantations, both being very familiar to Japanese people. Woodlands dominated by pine (*Pinus thunbergii* or *P. densiflora*) have long prevailed throughout Japan except Hokkaido. Both pines dominate at an early successional stage, because these species prefer sunny sites and have tolerance for infertile soils. Therefore, the pines would be gradually replaced by climax species, such as evergreen oak, under natural conditions. However, natural succession processes of woodlands around villages had been interfered with by human activities. Until the recent industrialized age, people had often exploited woods, brush, weed and leaf litter from woodlands for fuel and agricultural use (fertilizer). As a result, there were many places where pine woodlands, an early seral stage of succession, had been maintained constantly. At present, although pine woodlands can be seen in many places, pines are now being replaced by evergreen broad-leaved trees because the exploitation for fuel and fertilizer has ceased owing to the prevalence of fossil fuel and chemical fertilizer.

Plantation forests that consist of monotonic, even-aged stands of conifers (usually *Cryptomeria japonica* or *Chamaecyparis obtusa*) are a major component of a typical landscape in Honshu, Shikoku and Kyushu. Because these species have been favoured as building lumber, silvicultural techniques for these were developed centuries ago. The area of the plantation forests has now reached 10 million hectares, which contribute 40% of the total forested area. In other words, a quarter of the total land area of Japan is occupied by the conifer plantation forests.

Freshwater fishes and their habitats

Fifteen orders, 35 families and 96 genera with 211 species and subspecies of freshwater fishes (including lampreys) have been recorded in Japan (see Yuma *et al.* 1998). Their main habitats are rivers and lakes, and now in rice paddies and their associated irrigation systems including ditches and artificial ponds. Japanese rivers are short and steep; the length of the longest river (the Shinano River, located in central Honshu) is 367 km, with 11,900 km² drainage area. Large lakes are also rare, although there are more than 450 natural lakes. Lake Biwa, located in central Honshu (near Kyoto), is the largest with 671 km² in surface area, and also is the oldest lake in Japan. This 'ancient lake' supports a highly diverse fish fauna including many endemic species (see Kawanabe 1996; Yuma *et al.* 1998; Nakai 1999).

A steep, boulder-bed stream shaded by riparian canopy is a typical feature of freshwater habitat at upper reaches of rivers. Such coldwater streams are generally inhabited by char (*Salvelinus leucomaenis*), masu salmon (*Oncorhynchus masou*) and/or minnow (*Phoxinus* spp.), which are gradually replaced by dace (*Tribolodon* spp.) and chub (*Zacco* spp.) as stream size and water temperature increase downstream. In middle to lower reaches, the dominant fish fauna includes many diadromous species, such as ayu (*Plecoglossus altivelis*), eel (*Anguilla japonica*), gobies (*Rhinogobius* spp., *Tridentiger* spp., *Chaenogobius* spp., *Sicyopterus* spp.) and sculpin (*Cottus* spp.). Ayu, which is an amphidromous grazer fish, is one of the most popular fish and is stocked for recreational and commercial fisheries throughout Japan except Hokkaido. Also eel is important for commercial freshwater fisheries. Such diadromous fish number 77 species and subspecies (Yuma *et al.* 1998), which contribute 37% of the fauna, this large contribution being characteristic of the freshwater fish fauna of Japan.

Rice paddies are irrigated from rivers and artificial ponds through web-like ditches. These paddy-associated aquatic habitats harbour various fish, of which representative species are medaka (*Oryzias latipes*), loach (*Misgurnus anguillicaudatus*), catfish (*Silurus asotus*) and crucian carp (*Carassius* spp.). In particular, medaka and the loach are highly dependent on paddies. Catfish and crucian carp are also common inhabitants of ponds, lakes and lower reaches of rivers. Many other species inhabiting middle to lower reaches of rivers also use paddy-associated habitats seasonally or opportunistically (Saitoh *et al.* 1988; Katano *et al.* 2001). The rice paddy in combination with its associated irrigation system is considered as an important agent forming the freshwater fauna of Japan (Yuma *et al.* 1998).

The freshwater fauna of Hokkaido and Ryukyu Islands are quite different from that described above, which is typical of Honshu, Shikoku and Kyushu. Many species that are very common in the latter regions (e.g. ayu, chub, eel, catfish and *Rhinogobius* gobies) are minorities or absent from Hokkaido, where salmonids, *Tribolodon* species and sculpins dominate. In addition, rice paddies are absent from northern and eastern Hokkaido owing to its cool climate. Instead, farmland for other crops and pasture are prevalent, affecting freshwater habitats (Nagasaka & Nakamura 1999; Inoue & Nakano 2001). In Ryukyu Islands, the freshwater fauna is dominated by amphidromous (e.g. *Rhinogobius* and *Sicyopterus* gobies) and saltwater species that penetrate upstream, owing to the subtropical and oceanic nature of the islands.

Several exotic species have been introduced. Among them, largemouth bass (*Micropterus salmoides*) and bluegill (*Lepomis macrochirus*) have become widespread throughout Japan, especially southwestern regions. In most cases, they have been illegally introduced for recreational angling, and often dominate in irrigation ponds and reservoirs (e.g. Azuma & Motomura 1998). Their ecological effects on aquatic communities are of great concern (Nakai 1999). In Hokkaido, distributions of rainbow trout (*Oncorhynchus mykiss*) and brown trout (*Salmo trutta*), which inhabit streams and lakes, are expanding (Takami & Aoyama 1999), and their potential for negative effects on native salmonids has also been pointed out (Takami & Aoyama 1999; Taniguchi *et al.* 2000).

Forestry in Japan

Historical background: from forest exploitation to regenerative forestry

It is believed that ancient people in the Japanese archipelago had lived on forest products, such as acorns (Fagaceous species) and chestnuts (*Castanea crenata*), until rice culture was introduced and prevailed 2000–2500 years ago (Anonymous 1997). The introduction of paddy culture, which involves forest clearance, is considered as the beginning of intensive and extensive land alterations in the archipelago. Since then, forest exploitation for fuel, fertilizer and building lumber has been intensified and the land has experienced several critical phases of deforestation during the past two millennia. In particular, deforestation in the seventeenth century was widespread throughout the archipelago (except Hokkaido) and caused serious problems, which consequently led to the emergence of regenerative forestry. A good review of the history of the shift from forest exploitation to regenerative forestry in Japan is provided by Totman (1998) in English. Here we summarize the history on the basis of Totman (1998) and other Japanese literature (e.g. Yorimitsu 1984; Anonymous 1997).

During the seventh to ninth centuries, when 'Japan' was being established as a nation, a building boom occurred in Kinki district (Nara, Kyoto) where a capital was established. Tremendous quantities of timber were consumed for construction of palaces, mansions, temples and shrines. As a result, forests around the city became devastated, timber sources were lost, and finally the building boom tapered off. This period is considered as the first era of intensive deforestation in Japan (Totman 1998). Although the severe deforestation during the era was limited to the Kinki district, it became widespread throughout Japan during the subsequent deforestation era of the sixteenth to seventeenth centuries. This period included a civil war and the beginning of the 'Tokugawa regime', a stable regime which continued for more than 250 years (1603–1867: called 'Edo period'). During the civil war (late in the sixteenth century), large castles and towns, which needed great quantities of timber, were constructed throughout Honshu, Shikoku and Kyushu by each regional lord, and battles also required fuel wood and sometimes burned woodlands. After the strife was pacified by Toyotomi Hideyoshi and his successor Tokugawa Ieyasu, both rulers triggered a building boom, which further accelerated deforestation. Since the rulers' supremacy prevailed all over the archipelago (except Hokkaido) and trafficability was far more advanced than the previous deforestation era (fifth to seventh centuries), they could gather timber from throughout the country. Furthermore, the human population rapidly increased under the stable Tokugawa regime, resulting in an inevitable expansion of forest exploitation (residential and agricultural development, fuel wood, fertilizer). Consequently, by the end of the seventeenth century, severe deforestation was widespread throughout Japan except Hokkaido, where the supremacy of the rulers had not yet reached.

In the face of serious deforestation problems, rulers established various regulations to protect forests, and promoted afforestation. Initially, the afforestation was focused on the 'protection forest' that could prevent earth surface erosion. Concern for a 'production forest' that yields timber and fuel subsequently grew. During the eighteenth to nineteenth centuries, afforestation for timber production became widespread throughout the country, and silvicultural techniques (e.g. seedling and slip cultures) were well advanced and prevailed through silvicultural literature. By the beginning of the nineteenth century, Japan had entered the era of plantation forestry. The basis of present forestry practices in Japan was established during this period.

Current state of production forestry and timber resources

In general, the objectives of forestry are broadly categorized into two aspects, environmental protection and production of raw materials (e.g. fuel wood, timber). Conifer plantation for timber production is a representative of the current forestry system in Japan. The plantation species is usually *Cryptomeria japonica* or *Chamaecyparis obtusa*, both having been favoured as building lumber. In places where the climate is too cool for these species (i.e. Hokkaido and high altitude parts of Honshu), larch (*Larix leptolepis*), spruce (*Picea* spp.) and fir (*Abies* spp.) are used as plantation species instead. The conifer plantation forestry generally aims to establish even-aged, monoculture stands, which are to be harvested by clear-cutting. Typical operations of the conifer plantation (*Cryptomeria japonica* or *Chamaecyparis obtusa*) include ground preparation, transplanting of seedlings, weeding, limbing, thinning and harvest, which are summarized as follows (Anonymous 1980).

Seedlings that have grown for 2–3 years in seedbeds are transplanted to a plantation site after ground preparation (e.g. removal of weed and brushes), with the seedling density being generally 2500–4500 per hectare. The seedlings are then protected from competitive effects of other plants (grasses, brush) by weeding, which is generally conducted once or twice per year. Weeding is continued until the seedlings outgrow other plants (5–7 years after the transplantation), although creeper plants (e.g. ivy) need to be removed even after the seedlings outgrow. After a young forest stand is established by weeding, limbing and thinning become necessary to develop a mature stand that will produce high quality timber. Thinning is conducted to prevent the stand from spindly growth, while limbing is to enhance timber quality (e.g. lumber without knots). A typical example of a thinning plan for building lumber production is to adjust the stand density to 1400 trees per hectare at 20 years old, 1000 trees per hectare at 30 years old, and 700 trees per hectare at 40 years old. Timber is generally harvested at 35–50 years old by clear-cutting. In the cleared site, ground preparation is conducted for the subsequent plantation.

Forestry practices for fuel wood (including charcoal) production were also prevalent until the 1960s. Broad-leaved trees, typically oaks, were exploited for fuel wood, the stands being maintained by regeneration of sprout from stumps. Such coppice forestry often had been conducted in combination with agriculture over centuries, and forest products were exploited for agricultural use (e.g. leaf litter for fertilizer) as well as for fuel (over-exploitation of broad-leaved coppices made them pine woodlands, and further exploitation resulted in bald mountains). In the early 1950s, approximately 30% (in money costs) of domestic fuel consumption was provided by wood, and the annual demand for fuel wood was nearly 20 million m^3, which contributed approximately 30% of the total timber demand (Anonymous 2000). However, the coppice forestry rapidly declined during the 1960s, being taken over by conifer plantations or abandonment. The demand for fuel wood sharply declined during the 1960s and has been levelling off since 1972 at a volume of less than 1 million m^3 (<1% of the total timber demand).

The present prevalence of the conifer plantations, which occupy a quarter of the total land area of Japan, is largely due to an afforestation boom that occurred during the 1950s in the context of environmental, social and economic conditions post World War II. The period during World War II and the subsequent decade was the most recent era of severe deforestation. Throughout the archipelago including Hokkaido, usable timbers were cleared for war-related demand and subsequent rebuilding of destroyed cities after the war. By 1950, forests in Japan were exhausted. Being faced with this critical phase, the government vigorously promoted afforestation from 1950. The annual afforested area rapidly increased from <50,000 hectares in 1946 to >300,000 hectares in 1950 and peaked in 1954 at 430,000 hectares (Anonymous 2000). Reforestation of the wastelands generated by logging for the war and rebuilding was completed in 1956. The government forestry policy, however, promoted further conversion of forests (coppices, natural forests) to conifer plantations, because the demand for building lumber increased while that for fuel wood declined. This movement was also facilitated by the growing demand of the paper industry for broad-leaved trees as wood pulp. Large areas of coppices and natural broad-leaved forests were harvested by clear-cutting and the cleared sites were afforested to establish conifer plantations from the 1960s. Owing to the afforestation boom, the establishment of new conifer plantations was kept on at >300,000 hectares per annum during the 1950s and 1960s.

During the last three decades, the total forested area in Japan has been maintained at about 25 million hectares and the stock of stand biomass has increased by 70% to 3500 million m³, which is now increasing by 70 million m³ per annum (Anonymous 2000). These figures sound as if the forest stands were advancing toward favourable conditions. However, although the quantity has increased by the growth of conifer plantations, the quality of the latter has generally deteriorated for economic reasons, such as raised labour costs and the low market price of timber (cheap timber imports). In the conifer plantation system, aftercare, such as weeding, limbing and thinning, is indispensable to maintenance of healthy stands and production of commercial timber. In the market, however, such high-cost timber was not able to compete with imported timber, the amount of which rapidly increased from the 1960s. As a result, many conifer plantations have grown to be spindly stands under insufficient management, because aftercare of the stands became fruitless labour. Under the present situation, it is difficult for the plantation forestry to proceed as an economic activity. The self-sustaining level of timber supply in Japan has decreased to 19% and the rest of the timber needed for domestic use is imported from abroad. Moreover, the degradation of the monoculture stands of conifer plantations is not only a problem of timber quality but also of other aspects, such as vulnerability to forest destruction (e.g. blow-down), environmental protection (e.g. decline of the erosion prevention function), wildlife habitats and visual amenity. Overall, although the land has now been covered by deep green through afforestation efforts, Japanese forestry faces serious problems.

Linkages of forest with fishes and their habitats

Historical background: protection forestry and river management

In present forest management, the institution of 'Protection Forest' is a representative of the legislative institutions that are potentially related to fish and fisheries. On the basis of the Forest Law, which was established in 1897, forested areas of approximately 8.8 million hectares in total are now assigned as 'Protection Forest' (Anonymous 2000), where logging and land use are regulated to exert protective functions of forests. Although most of them are assigned for headwaters protection, erosion prevention or windbreak, some areas are assigned for fish habitat conservation (generally wooded coasts for marine fish). Such a management practice for environmental protection is also underlaid by the past deforestation history.

Historically, the problems of the severe deforestation included not only the deficiency of timber resources but also the devastation of headwaters, surface erosion and river sedimentation, all of which could alter the flow regime of rivers and frequently cause natural disasters (e.g debris flows, floods). The severe deforestation made the rulers recognize the protective functions of forests. In the eighth to ninth centuries, the rulers issued several orders against logging and for afforestation, with headwaters protection and erosion prevention being clearly stated as their objectives (Anonymous 1997; Totman 1998). Similarly, during the seventeenth to eighteenth centuries, the government of the Tokugawa regime continued to issue regulations on forest and river management (e.g. prohibitions against logging and agricultural development, promotion of afforestation) to prevent surface erosion and river sedimentation for river navigation as well as disaster prevention, and established the Office of Erosion Control for administration (in 1667). Such a concept of protection forestry was succeeded by the present Forest Law.

Despite the prevalence of forest management for environmental protection and regenerative production forestry in the eighteenth century, afforestation did not appear to be able to catch up with the forest devastation. It is inferred that the total area of the devastated lands (e.g. bald mountains) peaked at the beginning of the twentieth century at about 3 million hectares (Tsukamoto 2001). Although a great deal of afforestation efforts had gone into such wastelands, it was not until the recent afforestation boom of the 1950s that reforestation of the wastelands was completed. Although both fish and fisheries have been poorly considered in forest management, the past efforts of afforestation for timber production and disaster prevention may have played an important role in maintaining and restoring proper habitats for fish through prevention of erosion and sedimentation. Other aspects of disaster prevention, however, have apparently harmed fish habitats.

As engineering technology advanced, the use of concrete structures prevailed among disaster prevention works. During the past century, erosion control dams have been installed in most of the high gradient streams throughout the archipelago. In middle to lower reaches of most rivers, channel alteration including short-cutting, straightening, revetment work and levee construction have been conducted for flood control and residential, agricultural and industrial development. Moreover, after World War II, a drastic increase in water and electricity demands required construction of numerous large dams. Such intensive alterations resulted in serious habitat degradation such as habitat fragmentation and simplification, while management efforts concentrated on disaster prevention and economic efficiency. Diadromous fishes disappeared upstream from erosion control dams (Shimoda *et al.* 1993; Nakano *et al.* 1995), channel straightening decreased fish abundance (Takahashi & Higashi 1984; Inoue & Nakano 1994), and reservoirs altered community compositions of lotic fish (Mizuno & Nagoshi 1964).

Such great losses or changes of wildlife habitats were also apparent in forest ecosystems. Although the afforestation boom from the 1950s can be considered as a restoration activity for degraded ecosystems, it was succeeded by conversion from natural forests to conifer plantations, which actually involved extensive destruction of natural forests by clear-cutting. As such environmental degradations became apparent, the general public came to advocate environmental conservation from the 1970s, and subsequently, both forestry and river management policies gradually followed the movement. Concern for fish–forest relations was at last generated in the late 1980s in the context of the above movement.

Research on linkages between forests and fishes

As described above, it has long been recognized that conditions of streams and rivers can be affected by surrounding forests through hydrological and earth surface processes, and such topics of forest functions have been a traditional research theme in the field of forestry or forest sciences (e.g. Nakano 1976; Tsukamoto 1998). Despite the sufficient recognition of the substantial effects of forests on streams and rivers, effects on fishes (or other aquatic animals), the inhabitants of streams and rivers, had received little attention until recently, perhaps because fish had such little priority in forest management. Also, in the field of fisheries or fish ecology, fish–forest relations had rarely been concerned, while various aspects of freshwater fishes have been studied since the 1950s (see Fausch & Nakano 1998; Taniguchi *et al.* 2001, for review). However, as severe alterations of streams and rivers (e.g. channelization, construction of dams) prevailed, their effects were assessed by researchers (Onodera 1957; Mizuno & Nagoshi 1964; Takahashi & Higashi 1984). Subsequently, ecological interactions between forests and streams came to attract attention in the context of reconsideration of riparian management and also in the field of community ecology.

Research activities on forest–stream interactions and ecological effects of anthropogenic disturbances rapidly grew in Hokkaido during the 1990s. In Hokkaido, considerable areas of hills and plains that were originally covered by forests have been converted to farmland, with streams and rivers being channelized to develop and conserve the latter (Nakamura *et al.* 1997; Nagasaka & Nakamura 1999). Effects of such a typical land use regime on fishes and macroinvertebrates have been assessed by several studies (Takahashi & Higashi 1984; Inoue & Nakano 1994, 2001; Toyoshima *et al.* 1996; Yanai *et al.* 1996; Nunokawa & Inoue 1999; Nagasaka *et al.* 2000; Watanabe *et al.* 2001; Negishi *et al.* 2002). These studies generally focused on streams (first to fourth order) with fish assemblages typically dominated by masu salmon (*Oncorhynchus masou*), dace (*Tribolodon* spp.), Siberian stone loach (*Noemacheilus barbatulus*) and sculpin (*Cottus nozawae*). The results of the studies outlined a scenario of how fish assemblages are changed by the land development processes: forest clearance would cause replacement of masu salmon by dace by raising water temperature (Inoue & Nakano 2001), channelization following farm development would further eliminate dace and sculpin through habitat simplification (Inoue & Nakano 1994; Toyoshima *et al.* 1996; Watanabe *et al.* 2001), and consequently, only stone loach would survive through the land development processes (Takahashi & Higashi 1984; Inoue & Nakano 1994; Toyoshima *et al.* 1996; Watanabe *et al.* 2001). Through such habitat assessment work, critical roles of riparian forest in habitat of juvenile masu salmon were also emphasized: maintenance of cold water by canopy and supply of cover habitats (e.g. woody debris).

Like other salmonids, masu salmon is a typical coldwater species. A laboratory experiment in combination with field data indicated that their feeding activity and growth declined when water temperatures reached 24°C (Sato *et al.* 2001). A field study in northern Hokkaido streams suggested that their abundance is primarily limited by water temperature (Inoue *et al.* 1997). Masu salmon density in grassland reaches, of which summer maximum water temperature was generally higher than 20°C, was well below that in forest reaches. Effects of riparian canopy on stream water temperature have been examined in streams flowing through deciduous broad-leaved forests (Nakamura & Dokai 1989; Sugimoto *et al.* 1997). Nakamura & Dokai (1989) estimated the effect by a heat budget analysis and showed that 86% of the daily total input of solar radiation is intercepted by riparian canopy and its removal would result in a 4°C increase of water temperature. A relationship between summer maximum water temperature and the extent (in channel length) of canopy removal has been presented by Sugimoto *et al.* (1997), indicating a 2–4°C increase by a 1-km removal and a 5–8°C increase by a 6-km removal. On the basis of the data, they further estimated a historical change of summer maximum water temperature in a small stream, and showed a 6°C increase from 1947 (22°C) to 1990 (28°C) with a sharp increase during the 1950s, owing to rapid agricultural development. The results suggest historical shrinkage of suitable habitat for masu salmon.

Woody debris had been considered as no more than a disaster source in forest and river management, and most studies on it also had focused on disaster prevention (e.g. Mizuhara *et al.* 1979). At present, however, ecological aspects of woody debris have been examined in some Hokkaido streams, with subjects including its effects on organic matter retention (Kishi *et al.* 1999), habitat structure (Abe & Nakamura 1996) and fish abundance (Inoue & Nakano 1998; Urabe & Nakano 1998; Abe & Nakamura 1999). The importance of woody debris to fish habitat formation (channel morphology) was first emphasized by data from primeval, old-growth forests in North America (Keller & Swanson 1979; Bisson *et al.* 1987). In Japan, however, both primeval and old-growth forests are very rare, owing to the past deforestation. Thus, in most streams, the size and amount of woody debris are both quite small (see Abe & Nakamura 1996; Inoue & Nakano 1998; Urabe & Nakano 1998). Despite such a situation, positive effects of woody debris on salmonids (masu salmon or rainbow trout) through pool and/or cover have shown to be similar to those in North America (Inoue & Nakano 1998; Urabe & Nakano 1998; Abe & Nakamura 1999), although effects on channel morphology (e.g. pool formation) were not always detected, probably because of the small debris size (Inoue & Nakano 1998). A few studies included information on non-salmonid fishes and their results suggested that effects on benthic fishes (i.e. stone loach, *N. barbatulus* and sculpin *C. nozawae*) were not so great as those on salmonids. The density of neither stone loach nor sculpin was related to woody debris abundance (Inoue & Nakano 2001), and loach abundance was not affected by experimental removal of woody debris (Abe & Nakamura 1999).

Another related topic of anthropogenic disturbances is fine sediment pollution. Several field studies have suggested detrimental effects of fine sediment deposition on fish (Watanabe *et al.* 2001) as well as on periphyton (Yamada & Nakamura 2002) and benthic macroinvertebrates (Nagasaka *et al.* 2000). Fine sediment fills interstitial spaces between boulders and cobbles, resulting in a loss of essential habitat for benthic fish (Watanabe *et al.* 2001). Moreover, an experimental study suggested that fine sediment decreases survival rate of masu salmon embryo in redds through reduced permeability and oxygen supply (Yamada & Nakamura 2001). These findings also emphasize the importance of riparian forest to fish habitat in filtering of fine sediment. However, such filtering effects have not yet been evaluated sufficiently (but see Nanba *et al.* 1976; Tsukamoto & Arai 1991; Okura *et al.* 1997).

Numerous interesting findings on community ecology including forest–stream interactions have derived from Horonai Stream in the Tomakomai Experimental Forest of Hokkaido University. In 1995, Shigeru Nakano and his colleagues launched an ambitious research project on biodiversity and ecological function in forest–stream ecotone. The scope of the project included physical factors associated with fluvial processes (Urabe & Nakano 1999; Miyake & Nakano 2002) and competitive and predator–prey interactions (Kuhara *et al.* 1999; Miyasaka & Nakano 1999; Taniguchi & Nakano 2000), especially in relation to allochthonous inputs from riparian forest (woody debris: Urabe & Nakano 1998; leaf litter: Konishi *et al.* 2001; Motomori *et al.* 2001; terrestrial arthropods: Nakano *et al.* 1999a, 1999b; Kawaguchi & Nakano 2001). In particular, they focused on terrestrial arthropod inputs from riparian canopy as a resource subsidy for stream fishes, and conducted a large-scale manipulation of the input by using greenhouse-type cover (Nakano *et al.* 1999a). The results demonstrated a strong effect of the terrestrial arthropod input on trophic interactions among fish, benthic invertebrates and periphyton. When terrestrial arthropod inputs were reduced by the greenhouse-type cover, predation pressure by fish (Dolly Varden, *Salvelinus malma*) shifted dramatically from terrestrial arthropods (drift) to benthic invertebrates. The ensuing depletion of benthic invertebrates resulted in a subsequent increase in periphyton biomass. After this work, they further focused on the emergence of aquatic insects from streams as a resource subsidy for forest birds, and showed reciprocal subsidies between forest and stream by monthly surveys of terrestrial and aquatic prey fluxes for stream fishes and forest birds, respectively, over 14 months (Nakano & Murakami 2001). The aquatic insect emergence peaked around spring, when terrestrial prey abundance for forest birds was low. In contrast, terrestrial arthropod input to stream occurred primarily during summer, when aquatic resource for fishes was scarce. The reciprocal prey fluxes across the forest–stream interface represented 25.6% and 44.0% of the total energy budget of the forest birds and stream fishes, respectively. Their work clearly demonstrated interdependence between forest and stream systems through seasonal dynamics of material flows.

The growing attention to forest–stream interactions promoted collaboration among researchers of different fields. The Hokkaido Forestry Research Institute began collaborative studies with the Hokkaido Fish Hatchery, the Hokkaido Fisheries Experimental Station and the Hokkaido Agricultural Experiment Station to seek for better management of riparian forests. Their project first focused on masu salmon habitat, and examined roles of riparian vegetation in food provision (Nagasaka *et al.* 1996) and several aspects of habitat, such as spawning (Yanai *et al.* 1996), young-of-the-year in summer (Sato *et al.* 2001) and winter (Yanai *et al.* 2001). Information from these studies was further applied to habitat improvement practices (e.g. installation of large woody debris, provision of spawning gravel; Yanai *et al.* 2000). Their scope has also included macroinvertebrate habitats (Ito 1996; Nagasaka *et al.* 2000), dynamics of particulate organic matter in streams (Yanai & Terazawa 1995a) and material transport from forest to the sea (Yanai & Terazawa 1995b). Their recent collaboration has focused on association of riparian forests with fisheries resources in the estuary and marine coastal zones (Hokkaido Forestry Research Institute *et al.* 2001).

Research activities for fish habitat conservation have also increased in regions other than Hokkaido. The subjects include habitat requirements of stream fishes (e.g. Mori 1994), abundance–habitat relationships (e.g. Yagami & Goto 2000), assessment of anthropogenic disturbances (e.g. Nakamura *et al.* 1994) and improvement of engineering work (e.g. Nakamura *et al.* 1991; Nakamura 1997). In particular, concern for conservation of paddy-associated habitats has grown (Fujisaku *et al.* 1999; Tanaka 1999;

Katano *et al.* 2001; Suzuki *et al.* 2001). However, less research has been conducted on fish–forest relations, although some aspects of forest–stream interactions, such as dynamics of riparian forests under fluvial processes (Ishikawa 1991; Sakio 1997; Shin *et al.* 1999) and leaf litter processing by benthic macroinvertebrates in forest streams (Kagaya 1990), have been examined.

A potential subject for fish–forest relations in Honshu, Shikoku and Kyushu would be to examine effects of conifer plantations on stream fishes. As described above, mono-culture stands of conifer are now widespread, with their quality both as timber resources and wildlife habitats declining. Such conifer plantations are expected to be quite different from natural broad-leaved forests in ecological functions for stream communities. For example, an investigation in north-central Honshu indicated that input of terrestrial invertebrates, which are a major food resource for stream salmonids (Kawaguchi & Nakano 2001), in a stream flowing through a conifer plantation was less than half that in a comparable stream in a natural, deciduous broad-leaved forest during summer (Kawaguchi 1996). A difference in seasonal patterns of litter input is evident between conifer plantations (evergreen conifer) and deciduous broad-leaved forests, input from the latter having a peak in autumn. This difference, as well as differences in litter quality, may affect stream communities. In addition, degraded conifer plantations under insuf-ficient management tend to induce surface erosion (Tsukamoto 1998). Consequently, it is generally suspected that conifer plantations have detrimental effects on stream fish (e.g. Nagoshi 1998). However, these effects and associated forestry activities (e.g. clear-cutting, road construction) on fishes have rarely been assessed rigorously, possibly owing to the lack of suitable study sites. In Honshu, Shikoku and Kyushu, it is rather difficult to find a reference site for a comparative study design, because forests other than conifer plantations generally consist of complicated stands with different histories of various anthropogenic effects. In addition, there are so many complex interacting factors that may mask effects of forestry, such as effects of agriculture (e.g. irrigation), engineering work (e.g. dams), fisheries management (stocking) and over-fishing.

Recent movements and needs for better management

During the last two decades, river management policy gradually shifted to consider ecological aspects of streams and rivers. The River Law was revised in 1997, and then 'conservation and improvement of riparian environment' was included in the purpose statement of the law. Engineering work has come to take account of fish habitat. For ex-ample, 'natural river engineering', which originated in Europe but includes techniques similar to old, Japanese traditional ones, is now widely applied. This engineering work aims to mimic natural channel and flow to conserve (or restore) fish and wildlife habitats, using natural materials such as wood and stone (Göldi & Fukutome 1994). Habitat restoration work, generally removing concrete from reveted banks and install-ing large woody debris, on previously simplified channels has increased (Toyoshima *et al.* 1996; Yanai *et al.* 2000). Installation of fishways on dams for fish migration is also becoming prevalent, and research efforts have been made on more effective fishways and passable weirs (Nakamura 1997; Takahashi 1998). Overall, efforts for better fish habitat have been directed to conservation or restoration of habitat diversity within channels and maintenance of river continuity for migration.

Although most cases of habitat restoration and fishway installations are no more than seeking a better way within a limited space (i.e. within a channel at a particular location), some large-scale projects have recently been generated in places where ample space is available. For example, two large conservation plans have been launched as pilot projects in eastern Hokkaido. One is the conservation of Kushiro Marsh (started in 1999), the largest marsh in Japan, that had a problem of sedimentation and subsequent forest expansion (see Nakamura *et al.* 1997, 2002). The other is the re-meandering project of the Shibetsu River (started in 2001), on a channel previously straightened. The project aims to restore the original meandering channel and the floodplain wetlands. However, our present knowledge on riparian ecosystems is limited and ecosystem responses to the restoration project are unpredictable. The conservation committees, therefore, introduced an idea of adaptive management, which implements restoration measures while examining ecosystem responses by field experiments.

Forestry policy has also changed. Through the 1990s, non-production roles of forest (e.g. for wildlife habitats, visual amenity, field education, as well as environmental protection) came to acquire higher priorities than in the past. In 2001, the Forest and Forestry Basic Law was revised and the National Forestry Agency shifted its management policy from timber production to environmental functions of forests. Currently, 80% of national forests are assigned for headwaters protection, erosion prevention and recreation, and only 20% are used solely for timber production. Moreover, the agency has recently promoted establishment of 'green corridors' that link forest reserves fragmentarily located throughout the archipelago, for wildlife conservation (Anonymous 2000). The concept of corridors also has been applied to management of riparian forest (see below for an example in the Bekanbeushi River). On the other hand, action of afforestation for fish was first triggered by fisheries-related groups (marine fisheries rather than inland) in the late 1980s. Fishermen began to plant tree seedlings in mountain, riparian or coastal zones for future fisheries environment of their own. Such afforestation for fish and/or fisheries has become a kind of boom. As far as we know, more than 30 fisheries, forestry or civic groups throughout Japan performed afforestation activities during 2 recent years (1999–2000). Forest and river managers of local governments also came to take an interest in afforestation and conservation of riparian environment for fish and/or fisheries.

Although several decades will be needed to see consequences of the present afforestation, there are several examples of successful efforts of past afforestation. One is in the Bekanbeushi River watershed, eastern Hokkaido. In 1956, the National Forestry Agency, Obihiro Branch, launched an afforestation project on a 10,000-ha area of the watershed, which had been a vast wasteland as a result of logging and burns by early colonists. Stands of larch forest covering 7000 ha have been established through more than 40 years of effort. This reforested area includes an upstream portion of the Bekanbeushi Marsh, which is designated by the Ramsar Convention (Convention on Wetlands of International Importance Especially as Waterfowl Habitat) in 1993, and the Bekanbeushi River flows into Akkeshi Bay, which is famous for oyster fisheries. Therefore, although the project was initially aimed at timber production and farm protection, the present management policy has expanded its scope for conservation of riparian environment, to a downstream marsh, and to fisheries resources in marine coastal zones and biodiversity of the whole watershed. In particular, riparian forests all along the streams are planned to be conserved or restored to maintain various forest–stream interactions and functions of riparian corridors. Such a management policy for the afforested stands is regarded as innovative, and hoped to be an advanced model of 'landscape management' that concerns the ecosystem health of the whole watershed, from headwaters to the sea (Nakamura 1999).

As described above, the needs for better management of forests and rivers have become recognized widely among managers, policy makers and the general public, and new concepts for better riparian management have been proposed and promoted by literature (Nakamura 1999; Japan Society of Erosion Control Engineering 2000; Riparian Forest Research Network 2001). Not only for fish but also for various aspects of ecosystem health, conservation or restoration of proper riparian zones should be a general principle in forestry and river management. However, such ideal riparian management plans usually conflict with disaster prevention and economic interests (e.g. residential, agricultural and industrial development). Although space is needed to resolve the conflicts, available land is desperately restricted, which is the reality in Japan (the above cases of the Kushiro Marsh and the Shibetsu and Bekanbeushi rivers are exceptional in respect to availability of space). Before the importance of environmental conservation was seriously recognized, most flatlands had been highly developed for residential, agricultural and industrial use. Various functions and facilities, such as disaster prevention, water demand, public amenities and environmental conservation, have been imposed on confined riparian spaces. Although a high priority has recently been put on environmental conservation, space for it has been largely lost.

Ultimately, it is most essential to reserve and expand space as riparian management zones for conservation and restoration. In this context, there are problems related to social and legislative institutions. Although within-channel zones have been under the management of the River Law, government policy until now has never had the concept that riparian zones (expanded beyond the channel) are to be managed for diverse functions of public needs and ecosystem health. Therefore, riparian zones, which include private lands, are managed under various regulations related to agriculture, forestry, urban planning and land use. This makes it difficult to establish riparian management zones that should be treated under consistent management objectives (Kakizawa 2000). Although still insufficient, scientific and technical backgrounds of forest–fish relations and riparian reforestation have begun to be developed (Japan Society of Erosion Control Engineering 2000; Riparian Forest Research Network 2001). To make such knowledge reflected in actual management, coordination and collaboration among official managers, specialists, land owners and other related local residents are indispensable.

Acknowledgements

We thank S. Yanai, A. Nagasaka and K.D. Fausch for valuable comments on the manuscript and T.G. Northcote for linguistic improvements during editing. We also thank

Y. Kawaguchi, T. Abe and M. Nunokawa for providing information on the topics and Shin Sone for drawing the figures.

References

Abe, T. & Nakamura, F. (1996) Pool and cover formation by coarse woody debris in a small low-gradient stream in northern Hokkaido. *Journal of Japanese Forestry Society*, 78, 36–42 [in Japanese with English summary].

Abe, T. & Nakamura, F. (1999) Effects of experimental removal of woody debris on channel morphology and fish habitats. *Ecology and Civil Engineering*, 2, 179–90 [in Japanese with English summary].

Amino, Y. (1997) *Nihon-shakai no rekishi* (*History of Japanese Society*). Iwanami Shoten, Tokyo [in Japanese].

Anonymous (1980) *Nihon no ringyo, rinsan-gyo* (*Forestry and Forest Products in Japan*). Nihon Mokuzai Bichiku Kikou, Tokyo [in Japanese].

Anonymous (1997) *Nihon no mori to ki to hito no rekishi* (*History of Forest, Trees and Human in Japan: A Chronological Table*). Nihon Ringyo Chosakai (Japan Forestry Investigation Committee), Tokyo [in Japanese].

Anonymous (2000) *Ringyo hakusho, Heisei 11* (*White Paper on Forestry in 1999*). Nou-rin Toukei Kyokai, Tokyo [in Japanese].

Azuma, M. & Motomura, Y. (1998) Feeding habits of largemouth bass in a non-native environment: the case of a small lake with bluegill in Japan. *Environmental Biology of Fishes*, 52, 379–89.

Bisson, P.A., Bilby, R.E., Bryant, M.D., *et al.* (1987) Large woody debris in forested streams in the Pacific Northwest: past, present, and future. In: *Streamside Management: Forestry and Fishery Interactions*, (eds E.O. Salo & T.W. Cundy), pp. 143–90. Contribution no. 57. College of Forest Resources, University of Washington, Seattle, WA.

Fausch, K.D. & Nakano, S. (1998) Research on fish ecology in Japan: a brief history and selected review. *Environmental of Biology of Fishes*, 52, 75–95.

Fujisaku, M., Jinguji, H., Mizutani, M., Goto, A. & Watanabe, S. (1999) Relations of fish fauna to environmental structures of a small stream and agricultural ditches. *Ecology and Civil Engineering*, 2, 53–61 [in Japanese with English summary].

Göldi, C. & Fukutome, S. (1994) *Kin-shizen kasen kou-hou no kenkyu* (*Researches on Natural River Engineering*). Shinzan-sha, Tokyo [in Japanese].

Hokkaido Forestry Research Institute, Hokkaido Fisheries Experimental Station & Hokkaido Fish Hatchery (2001) *Shinrin ga kakou-iki no suisan-shigen ni oyobosu eikyou no hyouka* (*Effects of forest on fisheries resources in the estuary*). Collaborative Research Report 2000 (Heisei 12), Hokkaido Forestry Research Institute, Bibai [in Japanese].

Inoue, M. & Nakano, S. (1994) Physical environment structure of a small stream with special reference to fish microhabitat. *Japanese Journal of Ecology*, 44, 151–60 [in Japanese with English summary].

Inoue, M. & Nakano, S. (1998) Effects of woody debris on the habitat of juvenile masu salmon (*Oncorhynchus masou*) in northern Japanese streams. *Freshwater Biology*, 40, 1–16.

Inoue, M. & Nakano, S. (2001) Fish abundance and habitat relationships in forest and grassland streams, northern Hokkaido, Japan. *Ecological Research*, 16, 233–47.

Inoue, M., Nakano, S. & Nakamura, F. (1997) Juvenile masu salmon (*Oncorhynchus masou*) abundance and stream habitat relationships in northern Japan. *Canadian Journal of Fisheries and Aquatic Sciences*, 54, 1331–41.

Ishikawa, S. (1991) Floodplain vegetation of the Ibi River in central Japan II. Vegetation dynamics on the bars in the river course of the alluvial fan. *Japanese Journal of Ecology*, 41, 31–43 [in Japanese with English summary].

Ito, T. (1996) A preliminary survey of macroinvertebrates associated with submerged leaves and stems of terrestrial plants in a headwater stream of northern Japan. *Biology of Inland Waters*, **11**, 12–19.

Japan Society of Erosion Control Engineering (2000) *Mizube-iki kanri (Management of Riparian Zone)*. Kokon-shoin, Tokyo [in Japanese].

Kagaya, T. (1990) Processing and macroinvertebrate colonization of leaf detritus in a mountain forest stream in Japan. *Bulletin of the Tokyo University Forest*, **82**, 157–76 [in Japanese with English summary].

Kakizawa, H. (2000) Mizube-kanri-iki no settei ni kan-suru shakai-teki seido-teki mondai (Social and legislative problems with establishment of riparian management zones). In: *Mizube-iki kanri (Management of Riparian Zone)*, (ed. Japan Society of Erosion Control Engineering), pp. 258–308. Kokon-shoin, Tokyo [in Japanese].

Katano, O., Hosoya, K., Iguchi, K. & Aonuma, Y. (2001) Comparison of fish fauna among three types of rice fields in the Chikuma River basin. *Japanese Journal of Ichthyology*, **48**, 19–25 [in Japanese with English summary].

Kawaguchi, Y. (1996) *Effects of riparian forest type on terrestrial invertebrate inputs to streams: contrasting deciduous forest and planted cedar forest in central Japan*. Master's thesis, Niigata University, Niigata [in Japanese].

Kawaguchi, Y. & Nakano, S. (2001) Contribution of terrestrial invertebrates to the annual resource budget for salmonids in forest and grassland reaches of a headwater stream. *Freshwater Biology*, **46**, 303–16.

Kawanabe, H. (1996) Asian great lakes, especially Lake Biwa. *Environmental Biology of Fishes*, **47**, 219–34.

Keller, E.A. & Swanson, F.J. (1979) Effects of large organic material on channel form and fluvial processes. *Earth Surface Processes*, **4**, 361–80.

Kishi, C., Nakamura, F. & Inoue, M. (1999) Budgets and retention of leaf litter in Horonai Stream, southwestern Hokkaido, Japan. *Japanese Journal of Ecology*, **49**, 11–20 [in Japanese with English summary].

Konishi, M., Nakano, S. & Iwata, T. (2001) Trophic cascading effects of predatory fish on leaf litter processing in a Japanese stream. *Ecological Research*, **16**, 415–22.

Kuhara, N., Nakano, S. & Miyasaka, H. (1999) Interspecific competition between two stream insect grazers mediated by non-feeding predatory fish. *Oikos*, **87**, 27–35.

Miyake, Y. & Nakano, S. (2002) Effects of substratum stability on diversity of stream invertebrates during baseflow at two spatial scales. *Freshwater Biology*, **47**, 219–30.

Miyasaka, H. & Nakano, S. (1999) Effects of drift- and benthic-foraging fish on the drift dispersal of three species of mayfly nymphs in a Japanese stream. *Oecologia*, **118**, 99–106.

Mizuhara, K., Minami, T. & Takei, A. (1979) Fundamental study on the check of drifting woods (I): on the moving form of the group of drifting woods. *Shin-sabo (Journal of the Japan Society of Erosion Control Engineering)*, **113**, 10–16 [in Japanese with English summary].

Mizuno, N. & Nagoshi, M. (1964) Fishes of the Sarutani Reservoir in Nra Prefecture, Japan – II: an outline of their abundance in the reservoir. *Japanese Journal of Ecology*, **14**, 61–5 [in Japanese with English summary].

Mori, S. (1994) Nest site choice by three-spined stickleback, *Gasterosteus aculeatus* (form *leiurus*), in spring-fed waters. *Journal of Fish Biology*, **45**, 279–89.

Motomori, K., Mitsuhashi, H. & Nakano, S. (2001) Influence of leaf litter quality on the colonization and consumption of stream invertebrate shredders. *Ecological Research*, **16**, 173–82.

Nagasaka, A. & Nakamura, F. (1999) The influences of land-use changes on hydrology and riparian environment in a northern Japanese landscape. *Landscape Ecology*, **14**, 543–56.

Nagasaka, A., Nakajima, M., Yanai, S. & Nagasaka, Y. (2000) Influences of substrate composition on stream habitat and macroinvertebrate communities: a comparative experiment in a forested and an agricultural catchment. *Ecology and Civil Engineering*, **3**, 234–54 [in Japanese with English summary].

Nagasaka, Y., Yanai, S. & Sato, H. (1996) Relation between fallen insects from riparian forest and stomach content of masu salmon (*Oncorhynchus masou*). *Bulletin of the Hokkaido Forestry Research Institute*, 33, 70–7 [in Japanese with English summary].

Nagoshi, M. (1998) Kirikuchi no seitai to hozen-jyou no mondai (Ecology of kirikuchi, *Salvelinus leucomaenis japonicus*, and problems with its conservation). In: *Sakana kara mita mizu-kankyou (Aquatic Environment Viewed by Fishes)*, (ed. S. Mori), pp. 107–19. Shinzan-sha, Tokyo [in Japanese].

Nakai, K. (1999) Recent trends in fish diversity of Lake Biwa, central Japan: drastic decline in many endemic species and explosive increase in some exotic species. *Japanese Journal of Limnology*, 60, 407–12.

Nakamura, F. (1999) *Ryu-iki ikkan (Watershed Management: Consistency throughout a River System)*. Tsukiji-shokan, Tokyo [in Japanese].

Nakamura, F. & Dokai, T. (1989) Estimation of the effect of riparian forest on stream temperature based on heat budget. *Journal of Japanese Forestry Society*, 71, 387–94 [in Japanese with English summary].

Nakamura, F., Sudo, T., Kameyama, S. & Jitsu, M. (1997) Influences of channelization on discharge of suspended sediment and wetland vegetation in Kushiro Marsh, northern Japan. *Geomorphology*, 18, 279–89.

Nakamura, F., Jitsu, M., Kameyama, S. & Mizugaki, S. (2002) Changes in riparian forests in the Kushiro Mire, associated with stream channelization. *River Research and Application*, 18, 65–79.

Nakamura, S. (1997) Fishway for check (sabo) dams. *Journal of the Japan Society of Erosion Control Engineering*, 50, 52–57 [in Japanese].

Nakamura, S., Mizuno, N., Tamai, N. & Ishida, R. (1991) An investigation of environmental improvements for fish production in developed Japanese Rivers. *American Fisheries Society Symposium*, 10, 32–41.

Nakamura, T., Maruyama, T. & Nozaki, E. (1994) Seasonal abundance and the re-establishment of iwana charr *Salvelinus leucomaenis* f. *pluvius* after excessive sediment loading by road construction in the Hakusan National Park, central Japan. *Environmental Biology of Fishes*, 39, 97–107.

Nakano, H. (1976) *Shinrin suimon-gaku (Forest Hydrology)*. Kyo-ritsu Shuppan, Tokyo [in Japanese].

Nakano, S., Inoue, M., Kuwahara, T., *et al.* (1995) Freshwater fish fauna in the Teshio and Nakagawa Experimental Forests and adjacent areas with reference to damming effects on their distribution. *Research Bulletin of the Hokkaido University Forests*, 52, 95–109 [in Japanese with English summary].

Nakano, S. & Murakami, M. (2001) Reciprocal subsidies: dynamic interdependence between terrestrial and aquatic food webs. *Proceedings of the National Academy of Science USA*, 98, 166–70.

Nakano, S., Miyasaka, H. & Kuhara, N. (1999a) Terrestrial-aquatic linkages: riparian arthropod inputs alter trophic cascades in a stream food web. *Ecology*, 80, 2435–41.

Nakano, S., Kawaguchi, Y., Taniguchi, Y., *et al.* (1999b) Selective foraging on terrestrial invertebrates by rainbow trout in a forested headwater stream in northern Japan. *Ecological Research*, 14, 351–60.

Nanba, S., Kitamura, Y. & Yanase, H. (1976) Prevention of soil erosion by buffer forests. *The 1975 Report by Forestry and Forest Products Research Institute (Gijyutsu Kaihatsu Shiken Seiseki Houkoku-sho)*, 783–9 [in Japanese].

Negishi, J., Inoue, M. & Nunokawa, M. (2002) Effects of channelisation on stream habitat in relation to a spate and flow refugia for macroinvertebrates in northern Japan. *Freshwater Biology*, 47, 1515–29.

Nunokawa, M. & Inoue, M. (1999) Patterns of benthic insect assemblages in relation to riparian vegetation in a small stream, northern Hokkaido, Japan. *Japanese Journal of Limnology*, 60, 385–97 [in Japanese with English summary].

Okura, Y., Kitahara, H. & Sammori, T. (1997) Forest soil and litter as filtering media for suspended sediment. *Journal of Forest Research*, 2, 9–14.

Onodera, K. (1957) On the change of fishing effort corresponding to that of types of river. *Bulletin of the Japanese Society of Scientific Fisheries*, 23, 410–19.

Riparian Forest Research Network (2001) *Mizube-rin kanri no tebiki* (*Manual of Riparian Forest Management*). Nihon Ringyo Chosakai (Japan Forestry Investigation Committee), Tokyo [in Japanese].

Saitoh, K., Katano, O. & Koizumi, A. (1988) Movement and spawning of several freshwater fishes in temporary waters around paddy fields. *Japanese Journal of Ecology*, 38, 35–47 [in Japanese with English summary].

Sakio, H. (1997) Effects of natural disturbance on the regeneration of riparian forests in Chichibu Mountains, central Japan. *Plant Ecology*, 132, 181–95.

Sato, H., Nagata, M., Takami, T. & Yanai, S. (2001) Shade effect of riparian forest in controlling summer stream temperature: impact on growth of masu salmon juveniles (*Oncorhynchus masou* Brevoort). *Journal of Japanese Forestry Society*, 83, 22–9 [in Japanese with English summary].

Shimoda, K., Nakano, S., Kitano, S., Inoue, M. & Ono, Y. (1993) Present condition of stream fish assemblage in Shiretoko Peninsula with special reference to human impacts. *Bulletin of the Graduate School of Environmental Science*, 6, 17–27 [in Japanese with English summary].

Shin, N., Ishikawa, S. & Iwata, S. (1999) The mosaic structure of riparian forest and its formation pattern along the Azusa River, Kamikochi, central Japan. *Japanese Journal of Ecology*, 49, 71–81 [in Japanese with English summary].

Sugimoto, S., Nakamura, F. & Ito, A. (1997) Heat budget and statistical analysis of the relationship between stream temperature and riparian forest in the Toikanbetsu River basin, northern Japan. *Journal of Forest Research*, 2, 103–7.

Suzuki, M., Mizutani, M. & Goto, A. (2001) Trial manufactures and experiments of small-scale fishways to ensure both upward and downward migration of freshwater fishes in the aquatic area with paddy fields. *Ecology and Civil Engineering*, 4, 163–77 [in Japanese with English summary].

Takahashi, G. (1998) Study on fishway function of the method of low dams series. *Journal of the Japan Society of Erosion Control Engineering*, 50, 43–50 [in Japanese with English summary].

Takahashi, G. & Higashi, S. (1984) Effects of channel alteration on fish habitat. *Japanese Journal of Limnology*, 45, 178–86.

Takami, T. & Aoyama, T. (1999) Distributions of rainbow and brown trouts in Hokkaido, northern Japan. *Wildlife Conservation Japan*, 4, 41–8 [in Japanese with English summary].

Tanaka, M. (1999) Influence of different aquatic habitats on distribution and population density of *Misgurnus anguillicaudatus* in paddy fields. *Japanese Journal of Ichthyology*, 46, 75–81 [in Japanese with English summary].

Taniguchi, Y. & Nakano, S. (2000) Condition-specific competition: implications for the altitudinal distribution of stream fishes. *Ecology*, 81, 2027–39.

Taniguchi, Y., Miyake, Y., Saito, T., Urabe, H. & Nakano, S. (2000) Redd superimposition by introduced rainbow trout, *Oncorhynchus mykiss*, on native charrs in a Japanese stream. *Ichthyological Research*, 47, 149–56.

Taniguchi, Y., Inoue, M. & Kawaguchi, Y. (2001) Stream fish habitat science and management in Japan: a review. *Aquatic Ecosystem Health and Management*, 4, 357–65.

Totman, C. (1998) *The Green Archipelago: Forestry in Pre-Industrial Japan*. Ohio University Press, Athens, OH.

Toyoshima, T., Nakano, S., Inoue, M., Ono, Y. & Kurashige, Y. (1996) Fish population responses to stream habitat improvement in a concrete-lined channel. *Japanese Journal of Ecology*, 46, 9–20 [in Japanese with English summary].

Tsukamoto, Y. (1998) *Shinrin, mizu, tsuchi no hozen* (*Conservation of Forest, Water and Soil*). Asakura Shoten, Tokyo [in Japanese].

Tsukamoto, Y. (2001) Devastation process of forests and surface soils by human impacts: devastation and denudation of low relief mountains in Japan. *Journal of the Japan Society of Erosion Control Engineering*, **54**, 82–92 [in Japanese with English summary].

Tsukamoto, Y. & Arai, M. (1991) A guideline for forest management to prevent sediment production from steep forested slopes. *Bulletin of Experimental Forests, Tokyo University of Agriculture and Technology*, **29**, 43–54 [in Japanese with English summary].

Urabe, H. & Nakano, S. (1998) Contribution of woody debris to trout habitat modification in small streams in secondary deciduous forest, northern Japan. *Ecological Research*, **13**, 335–45.

Urabe, H. & Nakano, S. (1999) Linking microhabitat availability and local density of rainbow trout in low-gradient Japanese streams. *Ecological Research*, **14**, 341–9.

Watanabe, K., Nakamura, F., Kamura, K., Yamada, H., Watanabe, Y. & Tsuchiya, S. (2001) Influence of stream alteration on the abundance and distribution of benthic fish. *Ecology and Civil Engineering*, **4**, 133–46 [in Japanese with English summary].

Yagami, T. & Goto, A. (2000) Patchy distribution of a fluvial sculpin, *Cottus nozawae*, in the Gakko River system at the southern margin of its native range. *Ichthyological Research*, **47**, 277–86.

Yamada, H. & Nakamura, F. (2001) Effect of fine sediment deposition on masu salmon (*Oncorhynchus masou*) associated with a decrease in permeability. *Journal of Japanese Forestry Society, Hokkaido Branch*, **49**, 112–14 [in Japanese].

Yamada, H. & Nakamura, F. (2002) Effect of fine sediment deposition and influence of channel works on periphyton biomass in a freshwater catchment in northern Japan. *River Research and Applications*, **18**, 481–93.

Yanai, S. & Terazawa, K. (1995a) Varied effects of the forest on aquatic resources in a coastal mountain stream, in southern Hokkaido, northern Japan (II): autumn leaf processing of nine deciduous trees along the stream. *Journal of Japanese Forestry Society*, **77**, 563–72 [in Japanese with English summary].

Yanai, S. & Terazawa, K. (1995b) Varied effects of the forest on aquatic resources in a coastal mountain stream, in southern Hokkaido, northern Japan (I): suspended sediment and organic suspended sediment transported from a coastal mountain stream into the Tsugaru Strait. *Journal of Japanese Forestry Society*, **77**, 408–15 [in Japanese with English summary].

Yanai, S., Nagata, M. & Shakotan River Joint Research Group (1996) Influence of channel alteration on spawning habitat of masu salmon *Oncorhynchus masou* (Brevoort) in central Hokkaido, northern Japan. *Journal of the Japan Society of Erosion Control Engineering*, **49**, 15–21 [in Japanese with English summary].

Yanai, S., Sakamoto, T. & Baba, H. (2000) Mizube-iki no gutaiteki na tori-atsukai (Management practices of riparian zones). In: *Mizube-iki kanri* (*Management of Riparian Zone*), (ed. Japan Society of Erosion Control Engineering), pp. 103–74. Kokon-shoin, Tokyo [in Japanese].

Yanai, S., Nagata, M., Nagasaka, Y., Sato, H., Miyamoto, M. & Okubo, S. (2001) The role of riparian trees in providing wintering habitat for juvenile masu salmon (*Oncorhynchus masou*) in southwestern Hokkaido, northern Japan. *Journal of Japanese Forestry Society*, **83**, 340–6 [in Japanese with English summary].

Yorimitsu, R. (1984) *Nihon no shinrin, midori shigen* (*Forest and Forest Resources in Japan*). Toyo Keizai Shinpou Sha, Tokyo [in Japanese].

Yoshioka, K. (1973) *Shokubutsu chiri-gaku* (*Vegetation Geography*). Seitai-gaku kouza (Ecology lecture) 12. Kyo-ritsu Shuppan, Tokyo [in Japanese].

Yuma, M., Hosoya, K. & Nagata, Y. (1998) Distribution of the freshwater fishes of Japan: a historical overview. *Environmental Biology of Fishes*, **52**, 97–124.

Chapter 26
Fish–forest harvesting interactions in perhumid and monsoonal southeast Asia (Sundaland)

J.H. DICK AND K. MARTIN-SMITH

Introduction

Sundaland is the southeastern extremity of tropical Asia and comprises the countries of Malaysia, Singapore, Brunei, Cambodia, Laos, Vietnam and Thailand, plus western Indonesia and southern Burma (Fig. 26.1). The approximately 3,461,060 km^2 land area is home to a diverse population of about 335 million people, ranging from urban residents of large modern cities to indigenous, forest-dwelling ethnic groups whose livelihood and food security is inextricably linked to the health and productivity of natural terrestrial and aquatic systems.

The region contains the second largest block of humid tropical forest in the world after the Amazon Basin. These can be separated at a macro-scale into two major types: evergreen rainforest in perhumid (ever-wet) climates between ± 6° of the equator, and monsoon deciduous forests in distinctly seasonal climates from 6° to 22° latitudes in the tropics. These forests are amongst the most biologically diverse on earth and comprise over a dozen major forest formations.

Historically, traditional cultures had little impact on the forest, except in seasonal climates where fire could be used as a means of vegetation management. Such areas have often been degraded by long-term use to simpler secondary forest, open savanna woodlands and treeless grasslands. Commercial forest harvesting began under various colonial regimes towards the end of the nineteenth century and some areas, such as the teak forests of Burma, have been under continuous management for over 100 years. The pace of commercial exploitation began to accelerate after World War II, but the most profound effects occurred during the 1980s and 1990s as the demand for fine hardwoods, veneer woods and pulp wood, coupled with huge increases in harvesting and processing capacity, rapidly exceeded sustainable wood supply. Unfortunately, this period of unprecedented pressure on the forest resource coincided with a period of political unrest and economic dislocation that has severely tested the capacity even for basic governance in many countries of the region – let alone their ability to ensure adequate control over land use and natural resource exploitation. The end result is that Southeast Asia is now experiencing rates of forest depletion and deforestation that are among the highest in the world. These changes to forest ecosystems will have obvious economic implications, as yields of timber and other fibre commodities decline and large areas are degraded and made unproductive. However, the most devastating and

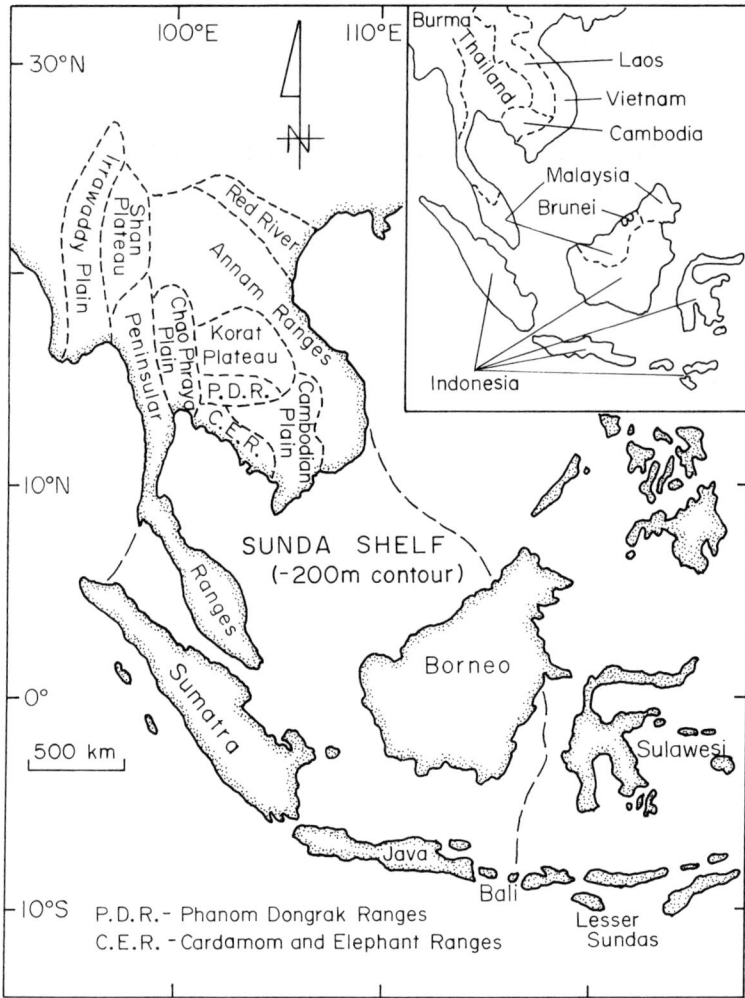

Fig. 26.1 Political and geomorphological units of Sundaland.

long-term impacts will be felt by millions of traditional rural forest-dwelling peoples, who have been marginalized by political processes that are often dominated by ethnic and/or urban elites. Traditional forest-dwellers rely on the forest directly for a great array of services: to replenish the nutrients on land farmed under rotational (swidden) agriculture; for plant foods and medicines; for livestock fodder; for fuel; for building, tool and artisanal materials; for traditional handicrafts (fibres and dyes); and for income generation from activities such as rattan collection and resin tapping. In addition to these direct 'forest' products, many traditional cultures rely on the natural forest ecosystems for a large part of their annual protein supply. Many Southeast Asian rural peoples derive a higher proportion of their annual protein intake from freshwater sources (fish, crayfish, frogs and turtles) than in any other part of the world. Healthy natural forests play a vital role in sustaining the freshwater systems that produce those food resources.

The goal of this chapter is to:

- set the biophysical context for considerations of terrestrial and freshwater aquatic natural resource management in Southeast Asia;
- describe the current state of forestry and fisheries management in the region and current trends in the health and extent of forest and freshwater aquatic ecosystems; and
- suggest future directions and needs in research, inventory and monitoring, and forest management and land use practices.

This discussion concentrates on upland and freshwater forest ecosystems. The very important tidal mangrove communities are dealt with in detail in Chapter 28.

Regional description of Sundaland

Biophysical description

Topography and geology

Sundaland forms a single geomorphological unit, despite considerable internal diversity, because it is underlain by a common, pre-Cambrian crystalline, basement complex (Bridges 1990). This complex is, for the most part, covered by younger materials and is partly submerged as the Sunda Shelf. There are 11 distinct 'geomorphological provinces' (Fig. 26.1).

The landscapes of the Asian mainland are dominated by three major mountain ranges (Arakan Yoma, Peninsular Ranges and Annam Ranges) trending south from the eastern Himalayas. These mountain ranges are separated by four major riparian lowlands (the Irrawaddy/Chindwin, Chao Phraya, Mekong/Cambodian and Red), and two uplifted plateaus (Shan and Korat) and associated minor mountain ranges (the Phanom Dongrak and Cardamom). The geology of the mountain ranges is very complex, consisting of sandstone, karstic limestone, intrusive granite and a variety of acidic to neutral metamorphic formations (schists and gneisses). A large area of the northern Annams is overlain with loess deposits and is the potential source of considerable erosional sediments. The river lowlands contain substantial depths of alluvial sediments, deposited on young sedimentary rock formations. The plateaus and associated mountains consist of uplifted gneiss, slate, limestone, granite, quartzite and sandstone.

The Sunda Shelf comprises a shallow sea, nowhere more than 200 m in depth, and the islands of Borneo and Sulawesi. The backbone of Borneo consists of mountains trending northeast to southwest with peaks ranging from <2000 m to >4000 m. These mountain ranges are a complex mix of sandstones, shales, limestones and intruded granites. Coastal plains extend for considerable distances upstream along major rivers of Kalimantan and for shorter distances along the coasts of East Malaysia. Sulawesi is composed of young sedimentary rocks that have been strongly folded into mountainous terrain rising to >3000 m.

The Indonesian Arc comprises the islands of Sumatra, Java, Bali and the Lesser Sundas and occurs where oceanic material of the Indo-Australian plate is passing below continental materials of the Sunda Shelf. The main arc is formed from volcanic peaks, including seven over 3300 m in height, along the lengths of Sumatra and Java and beyond into the smaller islands. Volcanic lavas and ash tend to be of the calc-alkaline type, which accounts for the very high productivity of most Javanese and Balinese soils.

Regional soils

Tropical soils are, generally, of much lower productivity than temperate soils derived from similar materials because of their age and the effects of high temperature and high precipitation on organic matter decomposition and nutrient leaching. Some discussions of the current status of tropical forests in the region have tended to down-play the differences between tropical and temperate soils (Aiken & Leigh 1992), and thus give the impression that tropical forest ecosystems are more resilient than may be generally believed. This view ignores the fact that, because of the long human history and population pressures in the region, most of the better soils have long since been converted to human use wherever topography allowed; leaving a disproportionate amount of forested ecosystems on unstable, acidic, infertile, problematic and even toxic soils. The common generalization that 'in temperate ecosystems the main reservoir of nutrients is the soil, while in tropical ecosystems the main reservoir of nutrients is above- and below- ground vegetation' probably holds true for most lowland forest sites – at least in perhumid tropical Asia. Thus when forest vegetation is cleared, and particularly when it is burned, the result most often is permanent, or at least very long-term, site degradation.

Soil taxonomy, physical and chemical properties and management potential have been summarized by FAO/UNESCO 1978 soil maps for Southeast Asia, RePPProT (1990), Soil Survey Staff (1990) and Dick (1991b). Sundaland has a wide variety of soil types, with vastly different inherent agricultural capability and capacity to respond to cultural inputs. Upland soils are dominated by deep, highly erodible, acidic and infertile podzolic soils. However, they also include smaller areas of rich and fertile volcanic soils; montane colluvial soils of moderate fertility in drier climates; thin, fertile but droughty and erodible limestone soils, acidic soils with near-toxic metal concentrations on ultra-mafic rocks; and subalpine peat soils. Wetland soils include silty, brackish marine (mangrove) soils; peat swamp soils of varying depth; fertile alluvial freshwater swamp soils; fertile riparian soils and heavy clays. Mangrove, freshwater swamp, riparian and, to a lesser extent, peat swamp soils all support ecosystems with important fishery values.

A rough, subjective estimate of the percentage land area of these soils in the Sundaland region (based on 1978 FAO/UNESCO 1:5 million soil maps) and their potential for long-term, sustainable, intensified use (defined here as including commercial food crops, cash crops, tree-crops and fibre plantations) is as follows:

- 64% of the area contains red and yellow podzols, thin limestone soils, saturated mangrove soils, deep lowland and montane peats or steep montane colluvial soils with little or no potential for intensive sustainable use due to infertility, soil acidity, soil depth and texture, subsidence and flooding, steep topography.
- 8% of the area contains drier mangrove soils, heavy clays, limestone soils or shallow lowland peats with moderate to low potential for intensive sustainable use due to salinity and inundation, soil texture, soil depth, soil chemistry and subsidence.
- 28% of the area contains volcanic soils, shallow freshwater swamp soils, riparian soils or gently sloping colluvial soils with moderate to good potential for intensive sustainable use and no serious limitations.

It is instructive to compare these rough figures with the equally rough figures in Table 26.3 (see later), which show a forest cover across the region of about 44%. If this information is even close to being reliable, it would indicate that (allowing for a relatively small area that perhaps was not naturally forested) somewhere over 20% of the land area has been converted from natural forest to uses that have little or no long-term sustainability. Many of these areas are now degraded and unproductive wastelands.

Climate

The climate of Sundaland, lying in the Humid Tropical Domain (Bailey 1996), is controlled by the movement of continental and oceanic air masses as determined by the passage of the sun back and forth between the tropics of Capricorn and Cancer (the zone of inter-tropical convergence). Throughout this region the climate can be characterized as follows:

- mean temperature above 18°C in all months;
- temperature variation from day to night is greater than that from season to season; and
- average annual rainfall is heavy and exceeds annual evapotranspiration, but varies in amount, season and distribution.

Two types of climate can be differentiated in the region, based on the seasonal distribution of precipitation.

Tropical ever-wet (perhumid) climates with no regular dry season and no monthly precipitation <60 mm. Perhumid climates generally occur between ± 6° of the equator and dominate Sumatra, Peninsular Malaysia, Borneo and the central portion of Sulawesi. Smaller patches of ever-wet climate occur on the seaward slopes of mountains in Java, Bali, Cambodia, Thailand, Laos, Vietnam and Burma. Year-round rainfall is heavy throughout the year, averaging from 2500 to over 5000 mm depending on location and elevation. Perhumid areas receive rainfall from both the southerly and northerly monsoons. Average annual temperatures at low elevations are close to 27°C (with a range from a mean maximum of 35°C to a mean minimum of 26°C), with little or no perceptible variation in season.

Seasonal or monsoon climates with a distinct dry season and 3 or more months with <60 mm of rainfall. Tropical monsoon climates generally occur between 6° and 22° of the equator and dominate parts of Java, Bali, the Lesser Sundas, the tips of the Sulawesi peninsulas, the northern tip of Sumatra, and the central, interior portion of the Asian mainland excluding Malaysia. Seasonality is caused by moist, warm, tropical maritime air masses flowing onshore at times of the year when the sun is overhead, and dry continental tropical or subtemperate air masses flowing offshore at times of the year when the sun is lower in the sky. In the northern hemisphere the maritime, southwest monsoon brings rain from mid-April to October and the continental, northeast monsoon brings dry conditions from November to March. In the southern hemisphere the pattern is reversed, with the southeast monsoon blowing out of continental Australia bringing dry conditions from May to October and the moist, northern, maritime monsoon bringing rain from November to April. Total annual rainfall varies from 2500 to <1000 mm depending on location, elevation and mountain 'rain-shadow' effects.

All areas of Sundaland have been increasingly influenced over the last three decades by the El Niño or Southern Oscillation phenomenon, which has triggered extended and accentuated dry seasons and droughts throughout the region. Historically such events are estimated to have occurred about 14 times a century (Enfield 1989), however, there have been eight since 1970. Each of these has brought water shortages, failed agricultural harvests, widespread and destructive fires exacerbated by poor logging and land clearing, air pollution, and subsequent serious soil erosion.

Major physical features of freshwater aquatic systems

A summary of freshwater systems in Sundaland based on Kottelat *et al.* (1993), MacKinnon *et al.* (1996) is provided below.

Torrents or hill streams

Torrents are characterized by high gradient, fast flow, high oxygenation, clear and cool water, well-developed 'pool and riffle' morphology and permanent to highly ephemeral stream flow. Higher elevation torrents may have little fringe vegetation, boulder-stabilized channels and poor fauna composed mainly of aquatic insects. Lower elevation torrents have more diverse riparian vegetation, channels stabilized by coarse woody debris and a richer fauna including aquatic insects and fish.

Large rivers and floodplains

Large, lowland rivers are low gradient, highly turbid and deoxygenated with variable depth, velocity and extent of inundated habitat. Rivers in perhumid and monsoon climates have significant hydrological differences.

Stream flow in perhumid climates is generally continuous because the large annual water surplus provides ample runoff. Major flow variations involve peak flows and may occur over quite short time frames (i.e. 10–12 m rise over a few days) as a result of intense storm events of short duration. Dense gallery forests line river channels where they have not been cleared for agriculture. Sand bars and mudflats are much less common than in monsoon climates. Floodplains have cut-off meanders (oxbows) and many swamps where meandering river channels have shifted their courses. Except where they arise in relatively recent volcanic formations, perhumid rivers carry little dissolved solids because of the highly leached nature of most of the older soils, which are generally derived from sedimentary materials.

Stream flow in monsoon climates is subject to strong seasonal fluctuations. In the rainy season extensive low-lying areas are inundated and the direction of river flow may reverse if water volume exceeds discharge capacity. In the dry season, substrates of silt, sand and gravel are exposed as water levels recede and stream channels and swamplands dry out. Rivers may retain a residual flow during the dry season or may shrink to a series of pools fed by subsurface river flows. The two great rivers of the monsoonal Asian mainland, the Mekong and the Red, arise in the Tibetan and Chinese Himalayas and thus a substantial dry season flow is maintained by glacier and snowmelt.

Large lakes

Lakes in the region are more or less permanent water bodies with low current velocity, formed either from a widening of a river channel or from cut-off oxbows. They are often shallow (<2 m in depth) and expand and contract in area according to the flows of associated rivers. Lake water typically exhibits strong gradients in oxygen and temperature from surface to lake bed.

Freshwater swamps

Freshwater swamps are most often associated with large river floodplains and lakes, and vary considerably in the area, frequency and period of inundation. Flood waters carry nutrient-rich sediments and organic matter, and generally have a near-neutral pH (about 6). The nutrient turnover resulting from the alternating flooding and drying of these swamps makes them among the most biologically productive of all tropical freshwater ecosystems.

Peat swamps

Peat swamps are found to the landward side of mangrove forests in perhumid or very slightly seasonal climates. They are characterized by very slow rates of decomposition and significant organic matter accumulation. The majority of the water entering the system is from rain and is therefore extremely deficient in mineral nutrients. The rainwater leaches organic acids from the peats and thus the water draining from them is of very low pH (≤4), extremely nutrient-poor, low in oxygen and dark-coloured (which has given rise to the term 'blackwater' streams).

Major fish communities, habitat requirements and migration patterns

Sundaland has a highly diverse freshwater fish fauna, estimated to be greater than 3500 species (Kottelat & Whitten 1996). Three countries in the region (Indonesia,

Thailand and Malaysia) are in the top ten in the world for freshwater fish diversity with over 500 species recognized from each (Table 26.1). Diversity may be similarly great in Cambodia, Lao PDR and Vietnam but there have been large gaps in field surveys and no rigorous taxonomic identification (but see Rainboth 1996 and Kottelat 2001). Within the region, 'hot spots' of biodiversity have been identified in the Mekong Basin and throughout most of Borneo, eastern Sumatra and Peninsular Malaysia (Kottelat & Whitten 1996).

The freshwater fish fauna of the region is dominated by the family Cyprinidae (carps, minnows and barbs) comprising 20–40% of all species. Other important families are the Balitoridae (hillstream loaches), Cobitidae (true loaches), Gobiidae (gobies) and the siluroid families (catfishes) (Table 26.1). Endemism in the region is high and there are extensive radiations within genera, e.g. 50 species in the cyprinid genus *Rasbora* in Indonesia and 55 species in the balitorid genus *Schistura* in Lao PDR (FishBase 2002).

The major freshwater fish communities in the region are generally recognized by their respective preferences for the habitat types whose physical characteristics were described in the previous section.

- *Hill streams*: High algal production but no macrophytes. Dominated by cyprinids (up to 90% of species) and balitorids with morphological adaptations for high current velocity. High levels of endemism and generally poorly sampled compared with larger water bodies. Distinct fish communities in pools and riffles (Inger & Chin 1962; Martin-Smith 1998).
- *Large rivers and floodplains*: Dominated by siluroid catfish and cyprinids. Lower levels of endemism and better taxonomic knowledge than other habitats. High fisheries production.
- *Lakes*: Considerable macrophyte growth in littoral areas. Dominated by cyprinids in upper water layers, siluroids and others on substrate.
- *Freshwater swamps*: Large amounts of organic debris but little autochthonous production. Fish diversity from a variety of families – cyprinids, siluroids, anabantoids and others.
- *Peat swamps*: Some cyprinids, but dominated by siluroid and anabantoid families with adaptations for low oxygen (e.g. labyrinthine organ, air breathing) (Ng *et al.* 1994). Very poorly sampled in most places and diversity often considerably underestimated.

Detailed habitat preferences and/or requirements are unknown or unrecorded for the vast majority of Southeast Asian freshwater fishes. Many species are known from few specimens, and habitat preferences that have been recorded may be indicative of the sampling regime, not the distribution of the fish in question. Local fishers often have detailed knowledge of the habitats of important food fishes but this information is not commonly recorded.

The habitat preferences of species with extreme morphological adaptations are easier to determine. For example, species in the balitorid genus *Gastromyzon* have a ventral sucker formed by the fusion of the pelvic fins and greatly expanded pectoral

Table 26.1 Some elements of freshwater fish diversity in Southeast Asia

Country or area	Kottelat & Whitten (1996)		FishBase (2002): total number of species and breakdown by family				
	Number of species	Estimated percentage of total species	Number of species	Number of Cyprinidae	Number of Balitoridae and Cobitidae	Number of siluroids	Number of Gobiidae
Borneo	440						
Burma (Myanmar)	300	50	325	118 (36%)	41 (13%)	66 (20%)	6 (2%)
Cambodia	215	60	478	153 (32%)	31 (6%)	96 (20%)	43 (9%)
Indonesia	1300	70	1008	203 (20%)	89 (9%)	167 (17%)	112 (11%)
Kapuas River	320						
Lao PDR	262	50	486	190 (39%)	106 (22%)	80 (18%)	9 (2%)
Malaysia	600	85	509	153 (30%)	73 (14%)	92 (18%)	25 (5%)
Mekong River	400+						
Thailand	690	90	632	237 (38%)	93 (15%)	109 (17%)	25 (4%)
Vietnam	450	80	349	122 (35%)	36 (10%)	60 (17%)	12 (3%)

fins. These species are invariably found in high velocity streams where their morphology allows them to resist the current flow (Roberts 1982). Similarly, *Channa bankanensis* has only been recorded from acid blackwaters (Ng *et al.* 1994). However, for the vast majority of species, habitat information is absent or rudimentary.

Further complicating factors are ontogenetic or temporal changes in habitat use. Fish fry and juveniles may occupy different habitats to adults (e.g. Martin-Smith 1998). Sampling techniques often do not collect juveniles or their habitats may not be sampled at all. Extreme seasonal changes in water flow and inundated area are a characteristic of many Southeast Asian aquatic systems and fish migrations are common (Welcomme 1985; Rainboth 1991; Christensen 1992). These migrations may be upstream, downstream or laterally from streams or rivers into flooded riverine forest. Often such movements are related to reproduction and spawning. In the monsoonal areas spawning may be extremely predictable and major fisheries are based on the movement of fishes past particular locations. In perhumid areas, reproduction may or may not be related to seasonal changes in water levels (Martin-Smith & Laird 1999).

Fish migration may also be closely linked with feeding ecology. During times of inundation fish may gain access to large areas of forest and additional food items. Feeding ecology is again poorly known for the majority of fishes in Southeast Asia. Inger & Chin (1962) gave a detailed account of the trophic relationships of fishes in small streams in northern Borneo, identifying two types of herbivores, five types of predators and a group of omnivores. Differences in classification were based on the source of the food – in-stream (autochthonous) or from the surrounding forest (allochthonous) – and the trophic level of prey. Rainboth (1991) summarized information from cyprinids from throughout Southeast Asia and noted that there were substantial differences in reported diet from the same species in different area. This highlights the lack of long-term studies and/or problems in taxonomy within the region. It would appear that the majority of fish species are opportunistic feeders that can consume a variety of material, and that diet is probably more a function of competition with other species than preference for particular items. There are some specialist taxa, usually with highly modified morphology for particular food items, such as *Gastromyzon* which scrapes diatoms from rocks (Roberts 1982) or the obligate planktivore *Thynnichthys* (Rainboth 1991).

Major forest zonation

Several excellent studies provide detailed descriptions of the forest ecosystems of Southeast Asia; however, there has been a notable lack of agreement and consistency in ecosystem classification. Several country-specific classification systems exist, largely based on ecosystem composition and structure, and it is common across the region for several names to apply to the same formation. It is beyond the scope of this section to provide a review of ecosystem nomenclature; rather it presents a very brief and simple description of the major forest formations. This is based on the classifications proposed by Whitmore (1984 and 1990) for the perhumid forests of Peninsular Malaysia, Sumatra, Borneo and Sulawesi, and the monsoon forests of Indonesia, and by Rundel (1999) for the monsoon forests of the Asian mainland.

The region possesses considerable diversity in its forest cover, with over a dozen major forest formations. These forest formations vary considerably in ecological composition, in current condition, in commercial value, in sensitivity to external influences and in the potential for conversion to non-forest uses. It thus makes little sense to speak of forest conservation, national forest cover and deforestation rates without reference to specific forest formations, because such generalizations can effectively mask significant changes in their respective distribution and extent. Unfortunately few comparative or time-series studies on national forest cover in the region distinguish between different forest formations.

The natural diversity and distribution of these forest ecosystems are a function of three primary factors: rainfall regime (particularly seasonality); elevation (as it affects temperature and evapotranspiration) and soil (both soil moisture and inherent fertility). Distinct forest formations and their relationship to these environmental factors are given in Table 26.2. The forestry and fisheries values of the six most important upland and freshwater forest formations are described below. Tidal mangrove forests, which also have significant forestry and fisheries values, are excluded from this discussion because they are dealt with in detail in Chapter 28.

Freshwater wetland forests
Peat swamp forest:

- low elevation, wetland forests of short stature comprised primarily of trees of the family Dipterocarpaceae;
- low species diversity but high endemism;
- domed landforms, with species diversity and tree size decreasing towards the centre of the dome;
- very important hydrological functions in regulating stream flow in adjacent rivers and streams;
- important wood resource and the first forest type exploited in Malaysian Borneo.

Freshwater swamp and gallery forests:

- irregularly inundated areas with mineral-rich water of pH around 6.0;
- periodic soil drying;
- occurs in coastal plains and large river valleys;
- heterogeneous environments with diverse plant communities and high biological productivity supporting an array of subsistence and commercial human uses;
- adjacent to lakes and large rivers thus providing critical feeding and rearing habitat for important fish communities.

Table 26.2 Forest formations and determining ecological influences in Sundaland

Forest formation	Rainfall regime	Elevation	Type soils	Soil water	Dominant influences
Coastal forest	Maritime	Sea level	Psamments	Water table high (daily tidal variation)	Soils and sea spray
Tidal forest	Maritime	+/– Sea level	Aquents	Water table high (daily tidal variation)	Salinity and daily inundation
Peat swamp forest	Perhumid	Lowlands	Histosols	Water table high (rainfall-dependent)	Soil and climate
Freshwater swamp forest	Perhumid	Lowlands	Aquepts	Water table high (seasonal to short-term variations)	Short-cycle and seasonal inundation
Heath forest	Perhumid and seasonal	Lowlands	Spodosols	Great seasonal to short-term variation in soil moisture	Soil texture and impeded subsoil drainage
Lowland evergreen rainforest	Perhumid	Lowlands to 1200 m	Ultisols	Moist soils	Perhumid climate
Forest on limestone	Perhumid	Lowlands to 1200 m	Mollisols	Great seasonal to short-term variation in soil moisture	Soil depth
Forest on ultra-basic rocks	Perhumid	Lowlands to 1200 m	Oxisols	Soil moisture highly variable	Soil depth and chemistry
Lower montane forest	Perhumid	1200–2000 m	Tropepts, humults	Soil moisture variable depending on aspect and soil depth	Elevation, aspect and precipitation
Upper montane forest	Perhumid	2000–3000 m	Orthents, histosols	Moist soils	Elevation and soil depth
Semi-evergreen forest	Slightly seasonally dry	Lowlands to 1200 m	Ultisols	Slight seasonal soil water deficit	Transitional climate
Moist deciduous forest	Seasonally dry	Lowlands to 1200 m	Tropepts	Moderate seasonal soil water deficit	Monsoonal climate
Savanna woodlands	Strongly seasonally dry	Lowlands to 1200 m	Tropepts	Marked seasonal soil water deficit	Monsoonal climate, fire

Adapted from Dick (1991a).

Lowland mixed rainforest
Evergreen rainforest:

- in perhumid areas within ± 6° of equator, and on wet mountain slopes in monsoon climates;
- multi-layered upper canopy up to 45 m high, dominated by trees of the family Dipterocarpaceae;
- ecologically complex, containing up to 1500 genera and 10,000 species of plants;
- principal lowland forest formations of Sumatra, Borneo, Peninsular Malaysia and central Sulawesi.

Semi-evergreen rainforests:

- transitional belt between evergreen rainforests and monsoon forests;
- partially open canopy with lower species diversity;
- comprised of evergreen *Dipterocarp* and deciduous tree species (i.e. *Intsia* and *Koompasia*);
- originally occurred in many parts of the region, and form the richest lowland forests in Cambodia, Laos and southern Thailand.

The last three decades have witnessed a very significant decline in lowland mixed rainforests throughout the region, the result of poor logging practices and conversions to oil palm, rubber and subsistence agriculture. Intact lowland rainforest has now largely disappeared from Thailand, Philippines, Vietnam, Peninsular Malaysia and Sulawesi, is being rapidly degraded in Malaysian Borneo, and is projected to be gone from Sumatra and Kalimantan by the end of this decade.

Lower montane forest
- zone of transition between tropical and subtemperate vegetation;
- lowland rainforest *Dipterocarp* species gradually give way to subtropical species of the families Fagaceae, Lauraceae, Myrtaceae and southern hemisphere conifers;
- increasing evidence of very destructive logging practices in the foothills and mountains of Sumatra, Borneo and Sulawesi resulting in serious site destabilization and erosion as timber production from low elevation rainforests has declined;
- Indonesian and Malaysian forest industries do not yet have the necessary road-building skills or harvesting equipment to log safely in mountainous terrain (Dick & Purwono 1991; Schweithelm & Zuwendra 1991).

Seasonal (monsoon) forests
- more or less open-canopied formations growing in areas between 6° and 22° of the equator with a distinct dry season;

- complex mosaic of plant communities determined by the interaction of rainfall, soil moisture and soil texture with human influences such as cultivation, livestock grazing and the use of fire;
- two broad types, e.g. moist deciduous forest and savanna woodlands, with only the former presently having commercial forestry significance;
- moist deciduous forest is semi-closed forest 30–40 m high composed largely of deciduous species with a bamboo/evergreen shrub understory;
- characteristic trees are teak (*Tectona grandis*) and deciduous *Dipterocarp* species in the northern and southern parts of the region, respectively. Other pan-tropical woodland hardwood genera are common, particularly from the family Fabaceae, and two species of the genus *Pinus*;
- moist deciduous forests are common throughout the Asian mainland and in small pockets on Sumatra, Sulawesi, Java and Bali.

The only comprehensive, consistent, region-wide assessment of forest cover has been carried out by Collins *et al.* (1991) from a wide variety of sources, mapped to a base year of 1987. Despite acknowledged shortcomings of these data, they provide a useful 'snap-shot' against which to measure the impacts of the increasingly destructive land use and forest practices that took place in many parts of the region during the 1990s. Table 26.3 summarizes the results of this assessment as they apply to Sundaland.

Resource use and impacts/interactions

Fisheries resources (ecological importance, commercial and subsistence use)

Freshwater fisheries resources vary considerably in importance through the Southeast Asian region (Table 26.4). Indonesia and Thailand are both in the top ten countries in the world for inland capture fisheries (FAO 1999). Official capture statistics are also high for Vietnam and Cambodia but relatively low for Laos and Malaysia (FAO 1999). The volume of freshwater fishes caught is probably severely underestimated in a number of countries in the region as statistics apply to the commercial catch and subsistence harvest is unrecorded. For example, in Laos annual captures of <30,000 metric tonnes were recorded during the period 1984–1997. However, a recent detailed survey of total freshwater capture fisheries in one province indicates that the total annual production in the whole country is 205,000 metric tonnes, i.e. more than six times the official statistics (Sjorslev 2000). In most countries, catches of freshwater fishes increased over the period 1984–1997. The exception to this was Vietnam where annual catches halved from a peak of over 120,000 tonnes in 1988–1989 to 66,000 tonnes in 1997.

Catch characteristics vary widely within and between countries. Sources of this variation include the use of a large range of fishing gears each with different catch selectivity, seasonal changes in fishing effort, fish migrations, geographical variation in fish communities and cultural values associated with different fish species (Roberts & Baird 1995; Lim *et al.* 1999; Dudley 2000). In most fisheries, catches are dominated

Table 26.3 Forest cover (km²) in Sundaland countries in 1987

Country or political unit	Total land area (km²)	Tidal forest	Freshwater wetland forest	Lowland mixed rainforest	Montane forest	Seasonal (monsoon) forest	Total forest area under forest cover	Percentage forest cover
Burma (Myanmar)	460,420	2,900	170	72,070	26,750	75,770	177,660	47
Thailand	511,700	5,700	–	54,900	16,800	29,500	106,900	21
Laos	230,800	–	–	87,950	14,430	22,220	124,600	54
Cambodia	176,520	250	7,500	55,500	2,250	47,750	113,250	64
Vietnam	352,360	1,610	–	28,040	9,020	18,010	56,680	17
Peninsular Malaysia	131,598	1,200	4,060	57,610	6,880	–	69,750	53
Malaysian Borneo	198,209	4,190	12,315	75,460	16,775	–	108,740	55
Brunei	5,765	200	1,760	2,670	70	–	4,700	81
Kalimantan	534,890	11,500	62,210	289,070	25,540	–	388,320	74
Sumatra	472,610	10,010	65,310	123,150	32,190	–	230,660	49
Java, Lesser Sundas and Bali	228,350	1,340	140	7,500	5,660	–	14,640	6
Sulawesi	184,840	2,170	2,150	77,680	21,920	8,120	112,040	61
Regional total	3,461,062	41,070	155,615	931,600	178,285	201,370	1,507,940	44

Summarized from Collins *et al.* (1991).
*The areas given in this table for Burma (Myanmar) refer only to that portion of the country that lies within Sundaland (i.e. the Irawaddy Lowland, Shan Plateau and Peninsular Mountains).

Table 26.4 Estimates of freshwater capture fisheries and aquaculture from Sundaland countries

| Country | FAO statistics in metric tonnes (1997 unless otherwise indicated) | | |
	Inland capture fisheries	Freshwater aquaculture	Main gear types
Burma (Myanmar)	150,000	108,000 (2000)	Fixed traps, set nets, seines
Cambodia	73,000	>10,000 (1995)	Set nets, fish traps, seines, barrages
Indonesia	339,310	300,000 (1995)	Gillnets, fish traps, barriers, handlines
Lao PDR	26,000	>4,400	Beach seines, gillnets, traps, longlines
Malaysia	3,949	<35,000 (1998)	Gillnets, handlines, traps, seines
Thailand	228,898	228,700 (1996)	Gillnets, longlines, scoop and lift nets, handlines
Vietnam	66,000	270,000 (1996)	Set nets, fish traps, seines, barrages

by cyprinids and/or siluroids, although other species may be locally important such as snakeheads (Channidae), featherbacks (Notopteridae) or herrings (Clupeidae) (e.g. Dudley 2000; Sjorslev 2000).

Forest management practices and forest harvesting

The history of forest management and the forest industry in the Sundaland Region has been described by Whitmore (1984), Hurst (1991), Collins *et al.* (1991), Aiken and Leigh (1992) and Marchak (1995). As a starting point, it is useful to look briefly at four countries that experienced such high rates of forest degradation from the 1950s to the 1970s that by the early to mid-1980s they had ceased to be significant producers of wood from natural sources. Thailand, Vietnam and Peninsular Malaysia are within the Sundaland region and the Philippines lie on the periphery. These countries serve as a stark example of unsustainable forestry to other countries of the region. They now have a much more direct impact on their more forest-rich neighbours – as sources of capital for forest investments, sources of (often bad) forestry and logging technology, and markets for logs for their well-developed processing capacity. A small capsule history of each of these jurisdictions is given below.

Thailand began exploiting its teak forests from the middle of the nineteenth century and continued into the 1970s, when it began to be acknowledged that harvesting was seriously depleting the natural forest resource. Forest area was reduced from about 55% in 1961, to 29% in 1985, and to 21% by 1989. During this period Thailand moved from being a net exporter of wood to a heavy importer. By 1989, uncontrolled illegal exploitation of nearly all forest areas resulted in a total ban over any further natural forest harvesting.

Vietnam is estimated to have had 45% forest cover in 1945. The next 30 years witnessed almost uninterrupted warfare, and by 1975 forest cover had fallen to about 20%. Much of the deforestation resulted from military bombing, use of defoliants, and tactical clearing along roads and canals. In addition, during and immediately after the war large areas of forest were felled for temporary agriculture and for wood products to supply a rapidly growing population. As a result, forest cover is now about 17%, over half of the land is now classed as degraded, there is increased flooding and sedimentation of irrigation/hydroelectric reservoirs, and Vietnam is a significant importer of wood to feed its wood-processing industry. Deforestation may also have played a part in the significant declines in freshwater fisheries referenced in the previous section.

Peninsular Malaysia is estimated to have had over 80% forest cover in 1950. Government statistics (Collins *et al.* 1991) indicate that over the next 40 years there was a steady decline in forest cover: 68% in 1966, 55% in 1977 and 43% in 1990. These figures differ from the forest cover breakdown shown in Table 26.3 and the discrepancies are probably due to inconsistent government record-keeping. Much of this reduction in forest cover was the result of deliberate forest conversion to tree crops (oil palm, rubber and coconut) and cash-crop agriculture by the Malaysian government. These conversion activities kept timber supply artificially high through the 1960s, 1970s and 1980s, but timber from these sources was largely exhausted by the mid-1990s. Much of the remaining natural forest is either already cut over or on steep-land protection forests. Peninsular Malaysia is now a significant net importer of logs.

The Philippines were originally almost fully forested. Commercial logging began in 1904; however, rates of harvest began to accelerate significantly after World War II. By 1980 forest cover had been reduced to 32%, and declined over the next 10 years to 23% in 1990. In 1989 the government banned logging in all provinces with <40% forest cover, which allows harvest in only 9 of 73 provinces. Deforestation has destroyed sources of traditional non-wood forest products for many indigenous peoples, and soil erosion and siltation of lakes and streams have eliminated many freshwater fish stocks (Collins *et al.* 1991).

In the light of declining internal timber harvests, these jurisdictions began to look towards their neighbours to sustain their wood-processing industries and to meet local markets. These demands, on top of those from Japan, China and Taiwan, are putting unsustainable pressures on the remaining forests of the region.

All of the present timber-producing countries, with the exceptions of Laos and Burma, have opted for a system of private industrial forest concessions as the main instrument of commercial forest management. The reasons for adopting such a system are usually given as:

- government funding is limited, and long-term tenures could attract private sector investment;

- technical capacity in government is limited and delegation of responsibility to a private sector company could both reduce the administrative burden on government and enhance the overall level of forest stewardship; and
- a private tenure system offers the possibility of a clear separation between regulation and management, with government setting standards and enforcing performance, and the concessionaire responsible for sustainable management.

Unfortunately, few of these potential benefits of an industrial concession system have been realized anywhere in the region (see Chapter 27). Ill-conceived investment policies by many governments have created processing capacity greatly in excess of sustainable yield. Increasing demands for wood have led to reduced harvesting cycles, premature and multiple re-entries of felling coupes, and a steady erosion of cut-constraints and retention standards – resulting in significant forest depletion. Where even these measures are not sufficient to feed the mills, many concessionaires resort to buying (or even colluding in the cutting of) illegal wood. In fact, the experience of the past few decades suggests that the industrial concession model may be inappropriate for tropical forests (Aiken & Leigh 1992; Marchak 1995). Because of high mechanization and high capitalization, industrial concessions may be incapable of operating economically and still fulfilling their obligations to sustainable forestry, environmental protection and the welfare of local communities. Some countries are now looking at more appropriate, community-based alternatives to industrial concessions and these will be further explored in the final section of this chapter.

Throughout the region, there is adequate information to support prudent, sustainable forest resource management; in other words the many failures have been institutional rather than technical. Several multi-lateral international organizations and at least six countries have contributed to the development of forest management standards and codes of practice in the region, most of which are similar in content. They are usually developed within the context of umbrella legislation that enables the establishment of rules, standards, and a planning, approval and permitting system. Codes are generally formulated at three levels:

- a set of rules or regulations, setting broad management direction and laying out those forest management practices that apply nation-wide;
- a set of legal forest standards for each major forest zone, establishing long-term forest and ecosystem management objectives for the zone and the nature and rates of acceptable use; and
- a set of planning and operational guidelines providing direction to the forest management unit or concession planning process.

Guidelines are usually not mandatory, but become enforceable when stipulated in an *approved* licence, plan, permit or contract. Common guidelines relate to planning, inventory, silviculture, community consultation and participation, biodiversity conservation, traditional NTFP (non-timber forest products) conservation, watershed and riparian zone delineation and protection, reduced impact logging, road design, construction and maintenance, and forest engineering.

Forest planning processes generally function at three to four levels, reflecting a range of spatial scales and time horizons. The most commonly applied forest management planning in technical assistance programmes consists of a three-tiered system (Armitage 1998):

(1) Medium-term (15-year) strategic forest management unit plans, which include:
   - delineation of forest zones to define both protection areas (biodiversity reserves, protected area buffer zones and corridors, watershed protection areas, riparian reserves, fragile areas, critical degraded areas requiring rehabilitation, NTFP production areas, and community forest areas) and the net operable working forest by major forest type;
   - a scientifically calculated medium and long-term sustainable yield;
   - appropriate silvicultural systems and harvesting techniques for different forest and terrain types;
   - environmental and social impact management and monitoring programmes;
   - ongoing community disclosure, consultation and participation programmes; and
   - inventory, monitoring and evaluation programmes.

   The strategic forest management unit plan is often prepared to fulfill the requirements of the national environmental and social impact assessment process as well as the forest concession permitting and approval process.

(2) 'Rolling' 3-year to 5-year compartment plans, which include medium-term access planning, management and decommissioning; more refined environmental and cultural resource zoning; and medium-term silviculture and site rehabilitation operations.

(3) Annual coupe plans, guided by 'codes of best practice', which prescribe harvesting methods and equipment; operational inventory results; timber volumes to be removed; location and design of forest access; tree marking and log tracking methods; onsite environmental and cultural resource protection and biodiversity conservation; regeneration and stand-tending activities; and annual community consultation.

With the possible exception of Laos, which has only recently begun to request and receive international technical assistance, no country in the region lacks a forest management system that, if conscientiously implemented and rigorously enforced, would protect and sustain the forest resource. And yet the region is now experiencing rates of deforestation and forest depletion that are among the highest in the world. The fundamental problems frustrating forest protection and improved forest management in the region are:

- a lack of political and bureaucratic will;
- a serious breakdown in basic governance, particularly at the local level;
- a lack of technical monitoring and enforcement capacity at the field level; and
- pervasive, systemic corruption and illegal acts on a grand scale.

General fish–forest harvesting interactions

Forest degradation is presently by far the most serious threat to freshwater fishes and the habitats they occupy in Southeast Asia (Kottelat *et al.* 1993; Kottelat & Whitten 1996). In general terms, the impacts of forest harvesting are similar in nature to those in temperate systems, but they are exacerbated by high precipitation (both total and intensity) and by the deeply weathered, highly erodible nature of the dominant soil types. The most serious effects of vegetation loss and site disturbance are direct or indirect loss of food resources and habitat, higher temperatures and lower oxygen levels due to decreased shading, altered stream flow and suspended sediment levels, and stream channel destabilization (Kottelat & Whitten 1996).

Many tropical fish species are directly dependent for food on animal and plant material falling into the water from overhanging vegetation. Still others are dependent on the regular contribution of organic materials that form the detritus on which aquatic food chains are built. The clearing or degradation of riparian vegetation, gallery forests and freshwater swamp communities can significantly reduce food inputs to streams, rivers, swamps and lakes, and feeding and rearing habitat.

The rise in water temperature resulting from the removal of vegetative shade usually results in significant drops in dissolved oxygen, particularly if accompanied by increased rates of organic matter decomposition. At the same time higher water temperatures increase the metabolism of fish while depressing haemoglobin oxygen uptake. These combined factors are suspected to have resulted in many mass fish-kills, particularly in monsoonal climates where large numbers of fish concentrate in ponds and pools during the dry season.

Perhaps the most serious and widespread impact of forest harvesting on fish is the alteration of stream flows and the increased generation of suspended sediment due both to vegetation removal and to soil disturbance from access construction and log extraction. Moist tropical forest vegetation intercepts a surprising quantity of rainfall. Various studies quoted in Aiken and Leigh (1992) estimate that lowland evergreen rainforest intercepts on average between 20% and 35% of incoming rainfall, with extremes varying from 10% to 100% depending on storm intensity. Litter is also estimated to intercept up to 10% of gross rainfall. Reduced canopy cover, litter-layer disturbance and soil compaction in logged watersheds causes greater overland flows and much higher peak stream flows, both of which usually result in increased sediment generation. As in many forestry operations, roads are the primary source of sediment, and three 'operational' characteristics of even well-constructed tropical roads result in sediment generation that is disproportionately higher than that from comparable temperate logging roads. First, the deeply weathered nature of tropical landscapes means that there is almost never sufficient coarse aggregate material available to stabilize road surfaces. Second, common practice is to build road surfaces and cleared rights-of-way up to three times wider (i.e. 10–12 m running surface and 20 m cleared width) than is necessary. The reason given for this is to allow sun to reach and dry the running surface, but it is more often an excuse for poor maintenance practices and for obtaining timber in excess of the allowable cut, since timber on road rights-of-way is excluded from annual quotas. Finally, in the perhumid tropics, log extraction continues year-round, stopping only during short periods of very high intensity rainfall. Thus road surfaces are almost continuous sources of sediment.

Sediment impacts on fish and other aquatic fauna in a number of ways. Fine colloidal material and flocculated iron and aluminium salts can accumulate on delicate gill tissue, resulting in death by suffocation (Kottelat *et al.* 1993). Sediment deposition can smother eggs, food resources and spawning habitat, reduce the depth and extent of lakes and swamps, and in-fill deep pools and channels necessary for dry season survival in monsoon climates.

Because of the age and depth of weathered landscapes in Southeast Asia, hill streams at lower elevations are generally stabilized by in-channel vegetation and coarse woody debris. Excessive forest harvesting, particularly when followed by land clearing and fire, can act to destabilize stream channels in two ways. Soon after logging, large amounts of debris may dam stream channels temporarily, resulting in debris flows and channel destruction when the dams give way. Over time, however, the reduced tree cover may actually begin to starve the streams of stabilizing coarse woody debris leading to down-cutting and a loss of channel stability. Both the long- and short-term effects can result in direct habitat destruction and to increased sediment generation.

Recommendations and needs for improved management performance

In the late 1980s and early 1990s, numerous reports and publications – by government agencies, individual authors, international NGOs and bi-lateral and multi-lateral aid agencies – characterized the decade of the 1990s as a watershed in forest and land use management in Southeast Asia. These warnings went essentially unheeded, and the decade witnessed unprecedented rates of deforestation, and levels of forest exploitation that often exceeded calculated sustainable yields by four to five times. As will be discussed in more detail in Chapter 27, it is predicted that the rich and diverse lowland evergreen rainforests of Indonesia have now been lost from Sulawesi, and will disappear from Sumatra and Kalimantan by 2005 and 2010, respectively, if present trends continue. Both Cambodia and Laos, where forests have been depleted rather than lost, will experience significant reductions in annual allowable cut for the next 15–20 years. In all three countries, a large portion of the harvest was illegal and thus, through the worst of the Asian economic crisis of the mid-1990s, national governments received little or no royalty revenue from the liquidation of their forest resources. However, the most critical implications have been for local forest-dwelling communities, often disenfranchised ethnic minorities, whose food security and economic livelihoods have been jeopardized by the impacts of logging on botanical NTFPs (non-timber forest products) and terrestrial and aquatic wildlife. Despite the problems of the last 10 years, Southeast Asia still possesses a remarkable set of forest and aquatic ecosystems that offer many opportunities for sustainable management, but unless wise decisions are taken this decade these remaining resources may also be squandered. The following sections outline the authors' views on priorities for improved management, research, and inventory and monitoring if the remaining natural forest and aquatic ecosystems of Southeast Asia are to be conserved and sustained for future generations.

Forest and fisheries management and land use practices

Delineation of a natural forest estate

All countries of the region should:

- legally establish, delineate and protect a Natural Forest Estate (NFE), which would include all remaining primary forests, secondary forests and all advanced regenerating seral communities;
- prohibit conversion of NFE lands to other permanent uses (i.e. commercial agriculture, tree crops and industrial fibre plantations); but allow traditional rotational cultivation as an integral part of the forest mosaic; and
- prepare natural forest rehabilitation and restoration strategies to provide a context for national programmes and international donor support.

Transparent and honest enforcement of existing laws, regulations and standards

- All countries in the region have forest management systems sufficient to ensure forest conservation and sustainable forest harvesting if implemented prudently and conscientiously.
- Deforestation has been due to a lack of political and institutional will and capacity to enforce laws, political and institutional corruption, large-scale timber theft and illegal land clearing.
- International donors have now made ongoing financial assistance conditional on demonstrated improvements in forest management, environmental protection and regulatory enforcement in Indonesia, Cambodia and Laos, and have formed 'consultative groups on forestry' to coordinate their technical assistance.
- Reducing timber theft will be possible only with the cooperation of West Malaysia, Vietnam, Thailand, Philippines, China, Japan and Taiwan – the main recipients of illegal wood. The IMF and other donors must use their influence to ensure regional cooperation in curtailing the international transport of illegal timber.
- Donor countries must advertise and promote to ensure greater market insistence on certified tropical wood by processors and consumers, which could reduce both the illegal harvest and the administrative/enforcement burden on hard-pressed regional governments.

Establishment of sustainable yields

- There is an urgent need to develop science-based sustainable yields for all forest resources – timber, botanical NTFPs, wildlife and fisheries.
- Few countries of the region have any reliable information on resource productivity, and thus set or manage harvest levels either intuitively, as in the case of timber, or not at all, in the case of botanical NTFPs and terrestrial and aquatic animals.

Reduction of excess wood-processing capacity

- Inappropriate investment policies, linking forest tenures to construction of processing facilities, have resulted in processing capacities that exceed estimated long-term sustained yield by 2.5–3 times across the region.
- There is a need to implement national policies that encourage the down-sizing of the wood-processing industry, both to balance capacity with long-term sustainable yields and to promote efficient value-added.

Improve forest access and harvesting systems for steeper terrain

- As lowland forests are lost or depleted, harvesting pressures are shifting to hill and montane forests.
- Most industrial concessions lack the equipment and experience to log safely on steeper ground, and the common result is significant site disturbance, stream channel destabilization, sediment generation and fish habitat destruction.
- Technical assistance should be provided, particularly to Indonesia, Cambodia and Laos, to develop more environmentally appropriate silvicultural and harvesting systems.

Greater emphasis on village-based forestry and fisheries management

- The government of Laos, with donor assistance, has instituted participatory sustainable forest management programmes involving contracted partnerships between local villagers and government in the management of production forests.
- Harvesting controls involve very light cutting intensities (1–2 trees/ha, <12 m³/ha per felling cycle) and very high retention standards, which are economically feasible because of the low capitalization compared with most commercial operations.
- Log extraction utilizes existing village roads wherever possible and lightly slashed, lightly used temporary 'skid' trails.
- The light 'ecological footprint' of village participatory forestry provides the potential for:
  - significant *in situ* biodiversity conservation;
  - the creation of buffers around, and connectivity between, existing protected areas;
  - the incremental introduction of sustainable management and use of plant and animal NTFPs;
  - zoning to protect wetlands, gallery forests, mineral licks, perennial river pools and channels and other critical habitats;
  - the recovery of depleted forests ecosystems; and
  - the protection of important social and cultural values such as spirit trees, sacred forests and other cultural assets through villager participation in tree selection.

- Existing donor-supported programmes will bring over 65% of Laos's production forests under such schemes over the next 6 years.
- Laos and Cambodia are also involved in promoting village-based fisheries management plans utilizing traditional management knowledge and harvesting controls; i.e. over 530,000 ha of commercial fishing areas have been reallocated to village management in Cambodia's Tonle Sap (I. Baird and P. Evans pers. comm.).
- These plans are guided by regional fisheries management strategies being coordinated by the Mekong River Commission.

Research: forestry, fisheries and watershed geo-hydrology

Priorities for research centre on two broad areas: first, the information necessary to establish sustainable yields of important natural resources; and second, the information necessary to predict the impacts of forest harvesting and land use change on forest and aquatic ecosystem health. Specific priority research topics include the following.

Forestry

Research is required for all major forest formations to establish defensible sustainable yields and annual allowable cuts, the calculation of which has two main components:

- maximum sustainable yield (MSY), based on growth and yield information; and
- rate-of-cut constraints modifying the MSY and designed to maintain an optimum environment for forest regeneration and growth, and to conserve forest composition, structure and ecological function.

Fisheries

- Life history information is required for almost all species; i.e. growth, age, mortality (fishing and natural), reproductive parameters (seasonal/aseasonal, age-specific reproductive output, spawning location, hatching or gestation period).
- Estimates of production and population dynamics modelling, utilizing life history data, are necessary for calculations of potential sustainable yields and to assess the sensitivity of those yield estimates to environmental change.
- Better information on fish movements and migration, including short-term movements and home ranges, and seasonal migrations for various requirements (food, spawning and dry season survival) is required to assess the spatial sensitivity of fish populations to land use change.
- Identification of vulnerable species – due to restricted geographic range, overexploitation or sensitivity to effects of logging and land use change – is required if aquatic biodiversity is to be conserved.
- Information leading to a basic understanding of trophic relations and energy transfer is required for long-term projections of fisheries yields.

Geo-hydrology

- Information on the sources, rates of production, lag-times, pathways and deposition patterns of sediments, particularly those arising from extreme storm events, in both perhumid and monsoonal aquatic systems, is required to assess the impacts of forest harvesting and modify management practice.
- Wise land use planning will depend on an understanding of the effects of land use change – forest harvesting, access development, conversion to tree crop plantations and agriculture – on river and floodplain hydrology (i.e. peak flows, water storage and release, extent and duration of inundation in wetlands), water quality and critical aquatic habitats.

Inventory and monitoring

Forestry

Throughout the 1990s many regional governments were simply in denial of anecdotal information about the extent of deforestation because, although they had baseline information dating from the late 1980s, none had a reliable forest cover monitoring programme in place to document change. Recent GIS-based assessments in Indonesia and Cambodia have concluded that anecdotal information may even have significantly underestimated rates of forest loss. It is now imperative that all governments in the region establish:

- GIS/satellite-based forest cover monitoring systems, broken down by major forest formations, to provide regular (3–5-year) 'snap-shots' of change in forest cover and condition; and
- quarterly, broad-scale satellite imagery analysis to detect major, unauthorized land use change, as a tool of regulatory compliance.

Fisheries

Inventory and monitoring programmes in the region have been haphazard and largely inadequate. Minimum requirements to provide basic information for fisheries management include:

- comprehensive species inventory programmes, with Kalimantan, Laos and Cambodia being the priorities;
- long-term trend monitoring of fish diversity, abundance, seasonality and size-frequency; and
- monitoring of levels and trends of exploitation of freshwater fishes throughout the region, in particular subsistence harvest – information collected to include catches by month and species, by types of gear used, catch-per-unit-effort (CPUE), the size of fishes caught, and the purpose of harvest.

Geo-hydrology

In order to interpret the impacts of land use activities, including forest harvesting, on watershed hydrology and aquatic ecosystems the following are required:

- long-term rainfall, water flow and sediment monitoring networks throughout the region; and
- a small, statistically representative number of stations capable of monitoring (through electronic data-loggers) extreme runoff events, including those triggered by the interaction of intense rainstorms on 'dormant' features such as old logging coupes, roads and abandoned agricultural areas.

References

Aiken, S.R. & Leigh C.H. (1992) *Vanishing Rainforests: The Ecological Transition in Malaysia.* Oxford University Press, Oxford.

Armitage, I. (1998) *Guidelines for the Management of Tropical Forests.* Forest Resources Division, Food and Agriculture Organization of the United Nations, Rome, Italy.

Bailey, RG. (1996) *Ecosystem Geography.* Springer-Verlag, New York.

Bridges, E.M. (1990) *World Geomorphology.* Cambridge University Press, Cambridge, UK.

Christensen, M.S. (1992) Investigations on the ecology and fish fauna of the Mahakam River in East Kalimantan (Borneo), Indonesia. *Internationale Revue Gesampte Hydrobiologie,* 77, 593–608.

Collins, N.M., Sayer, J.A. & Whitmore, T.C. (1991) *The Conservation Atlas of Tropical Forests: Asia and the Pacific.* Produced by the World Conservation Monitoring Centre, World Conservation Union and British Petroleum. Macmillan Press, London.

Dick, J.H. (1991a) *Forest Land Use, Forest Use Zonation and Deforestation in Indonesia: A Summary and Interpretation of Existing Information.* A background paper to the UNCED Conference prepared for the State Ministry for Population and Environment and the Environmental Impact Management Agency, Jakarta, Indonesia.

Dick, J.H. (1991b) *A Simple Primer on Tropical Soil Science for Land Use Planners, Biologists and Engineers.* Unpublished report, Environmental Management and Development Project (EMDI), Jakarta, Indonesia.

Dick, J.H. & Purwono, B. (1991) *A Report on an Environmental Audit of the P.T. Oceaneas Timber Concession, East Kalimantan.* Environmental Impact Management Agency (BAPEDAL), Jakarta, Indonesia.

Dudley, R.G. (2000) The fishery of Danau Sentarum. *Borneo Research Bulletin,* 30, 261–306.

Enfield, D.B. (1989) *El Nino,* past and present. *Review of Geophysics,* 27, 159–87.

FAO (1999) *Review of the State of World Fishery Resources: Inland fisheries.* FAO Fisheries Circular No. 942. FAO, Rome.

FishBase (2002) World Wide Web electronic publication, (eds R. Froese & D. Pauly) www.fishbase.org [accessed 8 March 2002].

Hurst, P. (1991) *Rainforest Politics: Ecological Destruction in South-East Asia.* S. Abdul Masjid & Co. Publishing, Kuala Lumpur, Malaysia.

Inger, R.F. & Chin, P.K. (1962) The fresh-water fishes of North Borneo. *Fieldiana: Zoology,* 45, 1–268 (+47 pp. in supplementary chapter by P.K. Chin 1990).

Kottelat, M. (2001) *Freshwater Fishes of Northern Vietnam.* A preliminary checklist of the fishes known or expected to occur in Northern Vietnam with comments on systematics and nomenclature. World Bank Biodiversity Report.

Kottelat, M. & Whitten, A.J. (1996) *Freshwater Biodiversity in Asia with Special Reference to Fish*. World Bank Technical Report 343.

Kottelat, M., Whitten, A.J., Kartikasari, S.N. & Wirjoatmodjo, S. (1993) *Freshwater Fishes of Western Indonesia and Sulawesi*. Dalhousie University, Canada and Periplus Editions, Hong Kong.

Lim, P., Lek, S., Touch, S.T., Mao, S.O. & Chhouk, B. (1999) Diversity and spatial distribution of freshwater fish in Great Lake and Tonle Sap River (Cambodia, Southeast Asia). *Aquatic Living Resources*, **12**, 379–86.

McKinnon, K., Hatta, G., Halim, H. & Mangalik, A. (1996) *The Ecology of Kalimantan*. Dalhousie University, Canada and Periplus Editions, Hong Kong.

Marchak, M.P. (1995) *Logging the Globe*. McGill-Queen's University Press, Montreal and Kingston, Canada.

Martin-Smith, K.M. (1998) Fish-habitat relationships in rainforest streams in Sabah, Malaysia. *Journal of Fish Biology*, **52**, 458–82.

Martin-Smith, K.M. & Laird, L.M. (1999) Reproductive patterns in some Cypriniformes from Borneo. *Proceedings of the 5th Indo-Pacific Fish Conference, Nouméa*, 1997, (eds B. Séret & J.-Y. Sire), pp. 493–504. Société Française d'Ichtyologie, Paris.

Ng, P.K.L., Tay, J.B. & Lim, K.K.P. (1994) Diversity and conservation of blackwater fishes in Peninsular Malaysia, particularly in the North Selangor peat swamp forest. *Hydrobiologia*, **285**, 203–18.

Rainboth, W.J. (1991) Cyprinids of South East Asia. In: *Cyprinid Fishes. Systematics, Biology and Exploitation,* (eds I.J. Winfield & J.S. Nelson), pp. 156–210. Chapman & Hall, London.

Rainboth, W.J. (1996) *Fish of the Cambodian Mekong*. FAO species identification field-guide for fisheries purposes.

RePPProT (Regional Physical Planning Program for Transmigration) (1990) *The Land Resources of Indonesia*. ODA/Ministry of Transmigration, Jakarta, Indonesia.

Roberts, T.R. (1982) The Bornean gastromyzontine fish genera *Gastromyzon* and *Glaniopsis* (Cypriniformes, Homalopteridae), with descriptions of new species. *Proceedings of the Californian Academy of Science,* **42**, 497–524.

Roberts, T.R. & Baird, I.G. (1995) Traditional fisheries and fish ecology on the Mekong River at Khone Waterfalls in southern Laos. *Natural History Bulletin of the Siam Society*, **43**, 219–62.

Rundel, P.W. (1999) *Forest Habitats and Flora in Laos PDR, Cambodia and Vietnam*. Conservation Priorities in Indochina, WWF Desk Study, December 1999.

Schweithelm, J. & Zuwendra. (1991) *Report on an Environmental Audit of the P.T. Wenang Sakti Timber Concession, North Sulawesi*. EIA (AMDAL) Sub-Directorate, Directorate of Nature Conservation, Ministry of Forests, GOI, Jakarta, Indonesia.

Sjorslev, J.G. (2000) *Luangprabang Fisheries Survey*. LARReC Technical Report No. 0008. AMFC/MRC and LARReC/NAFRI, Vientiane, Lao PDR.

Soil Survey Staff (1990) *Keys to Soil Taxonomy*, 4th edn. SMSS Technical Monograph No. 6. Blacksburg, Virginia.

Welcomme, R.L. (1985) *River Fisheries*. FAO Fisheries Technical Paper 262.

Whitmore, T.C. (1984) *Tropical Rain Forests of the Far East*. Clarendon Press, Oxford.

Whitmore, T.C. (1990) *An Introduction to Tropical Rainforests*. Clarendon Press, Oxford.

Chapter 27
Regional case studies in fish–forest harvesting interactions: Malaysian and Indonesian Borneo and Cambodia

J.H. DICK AND K. MARTIN-SMITH

Introduction

The intent of this chapter is to illustrate some of the general observations and issues on forestry/fisheries interactions raised in Chapter 26 by describing in more detail the current situation in two specific areas of Sundaland that span the range of biogeoclimatic conditions in the region. The two areas chosen are the island of Borneo, focusing on hill streams and smaller rivers, and Cambodia, focusing on the main stem of the Mekong River and the Tonle Sap or Great Lake. This chapter focuses primarily on forest practice and forest condition rather than on impacts on fish and fish habitat. The reason for this is that, with few exceptions, there is little reliable information on the impacts of forest harvesting on fish populations in Sundaland. By default, we fall back on the judgement quoted in Chapter 26: 'Forest degradation is presently by far the most serious threat to freshwater fishes and the habitats they occupy in Southeast Asia' (Kottelat *et al.* 1993; Kottelat & Whitten 1996). In short: as the forests go, so do the fish.

Malaysian and Indonesian Borneo: hill streams and medium-sized rivers in a perhumid climate

Biophysical description

Borneo straddles the equator between latitudes $7°$ N and $4°$ S. With a total area of $738,864$ km², it is the third largest island in the world. It falls under three political jurisdictions: the East Malaysian states of Sabah and Sarawak ($198,209$ km² or 27%), the Sultanate of Brunei (5765 km² or <1%) and Indonesian Kalimantan ($534,890$ km² or 72%) (Fig. 27.1).

Over half the island consists of low coastal and river plains lying below 150 m in elevation, and rivers can be tidal up to 100 km inland. The main mountain ranges, with peaks from 1500 to over 4000 m, are igneous in origin and run from northeast to southwest, with the central portion of the ranges forming the boundary between East Malaysia and Indonesia.

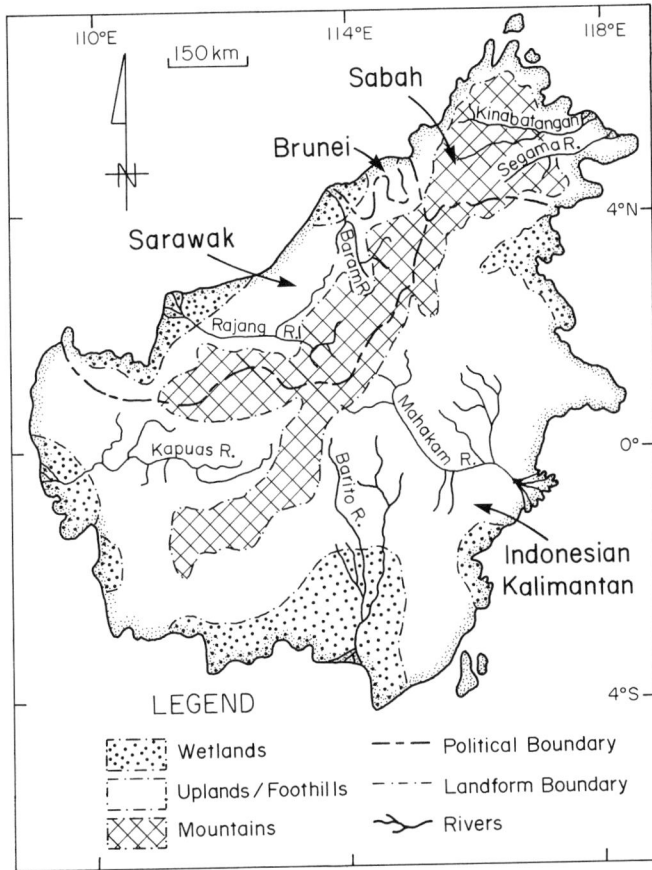

Fig. 27.1 Political and biophysical units of Borneo.

Borneo, within the tropical perhumid zone, has relatively constant temperatures that vary between 25°C and 35°C in low elevation areas. Annual rainfall varies between 2000 and 4000+ mm, and comes from both the northwestern monsoon (November to April) and the southeast monsoon (May to October); however, the former is wetter. Borneo has been influenced periodically by the Southern Oscillation, or El Niño, phenomenon, with prolonged droughts in 1972–1973, 1982–1983 and 1997–1998. During the latter two events a combination of drought and fires affected over 14.6 million ha of forest land.

Because of high temperature, high rainfall and topographic variation, Borneo contains the largest expanse of humid lowland and montane forest formations in Sundaland, and some of the most species-rich ecosystems on earth. Botanically, Borneo is one of the richest areas in the world, both in terms of total species richness and ecosystem diversity, containing between 10,000 and 15,000 species of flowering plants.

The island is dissected by over a dozen major rivers that run from the interior mountains to the coast and still form the principal routes of transportation and communication. Borneo has Indonesia's three longest rivers: the Kapuas (1143 km), the

Barito (900 km) and the Mahakam (775 km). The principal rivers of East Malaysia are much shorter. Many of the main rivers have extensive lake systems along their lowland courses, which support very important fisheries. All of these aquatic systems have the typical hydrology of perhumid climates, with a sustained and constant base flow, and peak flows that are highly variable depending on short-term storm events in the upper watersheds.

Major fish communities and their importance

Borneo is characterized by high species diversity of freshwater fishes – at least 440 species (Kottelat & Whitten 1996). The aquatic communities in the northern part of the island are less diverse than those in the river basins of the south and east. There is considerable change in community composition from source to river estuaries (Roberts 1989; Martin-Smith & Tan 1998). Hill streams and small rivers flowing from the central highlands remain poorly studied and their diversity is probably underestimated (Inger & Chin 1962; Watson & Balon 1984; Martin-Smith 1998a; Nyanti *et al.* 1999).

The fish fauna of hill streams is dominated by cyprinids with up to 70% of species and 90% of biomass (Roberts 1989; Kottelat *et al.* 1993; Martin-Smith 1998a). The majority of hill stream cyprinids are small- to medium-sized species (50–200 mm maximum length), although there a few species that may attain lengths of >400 mm such as *Hampala macrolepidota*, *Leptobarbus melanotaenia* or *Tor* spp. Balitorids (hill stream loaches) also contribute significantly to species diversity, particularly in fast water habitats (riffles). Balitorids are invariably small species (<100 mm maximum length), although they may attain densities as high as $15\,m^{-2}$ (Martin-Smith 1998a). There are a number of other families with one or few species in each, particular the Anguillidae (eels), Bagridae (bagrid catfishes), Clariidae (walking catfishes), Pangasidae (pangasid catfishes), Mastacembelidae (spiny eels) and Osphronemidae (giant gouramis). Although present in low numbers these species may contribute disproportionately to overall biomass due to their larger maximum body size.

Within-stream variation has been shown to be high with distinct faunal assemblages found at the meso-habitat scale of tens to hundreds of metres (Martin-Smith 1998a). These faunal assemblages are associated with distinct habitat types and there is little species overlap between fast water habitats (riffles) and slower, deeper water habitats (pools). Those species found in both habitat types generally show an ontogenetic shift in preference, being found in fast shallow water as juveniles and slow deep water as adults (Martin-Smith 1998a). Riffles have more specialized species with higher abundance and diversity of balitorids, whereas pools have more generalist species and higher numbers of cyprinids. Specialization in riffle species involves morphological adaptations for fast water flow, i.e. an adhesive disc formed by fused pelvic fins, expanded paired fins or a thoracic adhesive organ.

Within a watershed, between-stream variation in species diversity appears to be largely related to barriers to movement and, to a lesser extent, habitat heterogeneity (Martin-Smith & Laird 1998). Streams with significant barriers, such as waterfalls, had a subset of species found in streams with open access, as were small, structurally simple streams (Martin-Smith 1998a; Martin-Smith & Laird 1998). Standing stock for these streams is relatively low with mean estimates of 20,000–54,000 ind ha^{-1} (equivalent to 130–300 kg ha^{-1}) and production has been calculated to be smaller than comparable temperate streams (Watson & Balon 1984; Martin-Smith 1998a).

Feeding habits and trophic relationships of some Borneo hill stream fish were determined by Inger & Chin (1962) from collections over 6 weeks. Eight feeding guilds were proposed – two types of herbivory, five predatory at different trophic levels (nematodes through to decapods and fishes) and one omnivorous. Macrophytes are rare in these streams, thus herbivorous guilds are limited to autochthonous algae or diatoms or terrestrial input (leaves, flowers and fruit). Many species fed on insects or crustaceans (both autochthonous or allochthonous) and there were few piscivorous fish (Inger & Chin 1962). Low numbers per species examined (generally <10) and the limited temporal and spatial scale of collection mean that feeding and trophic relationships remain tentative. Other feeding studies (e.g. Nyanti *et al.* 1999) have been similarly limited in temporal and spatial scope. For example, there is no information on seasonality or ontogenetic shifts in diet.

There is also very little information on reproductive seasonality or behaviour for hill stream fishes. Martin-Smith & Laird (1999) studied ten species of cyprinid and balitorid and found that five species spawned continuously throughout the year while the other five species showed peaks in reproductive output roughly coinciding with the onset of the rainy season. McAdam *et al.* (1999) found no evidence of seasonality in reproduction for a congeneric cyprinid in peninsular Malaysia while Abidin (1986) documented protracted spawning for *Hampala macrolepidota*.

Seasonal migration appears to be limited for hill stream species. Seasonal catches were not observed during a year of semi-quantitative sampling in the Segama River, Sabah (Martin-Smith 1998c). However, *Pangasius macronema* and *Leptobarbus melanotaenia* were reported by local fishers to be caught only at certain times of year (K.M. Martin-Smith, unpublished data). Christensen (1992) noted the movement of two cyprinid species into smaller rivers and reaches from the Mahakam mainstem. Balitorids have been shown to be particularly limited in their movement, while small cyprinids showed less site fidelity (Martin-Smith *et al.* 1999).

Forest practices, forest harvesting and land use

Borneo's forests contain six major commercial types: mangrove, peat swamp forest, freshwater swamp forest, lowland evergreen and semi-evergreen rainforest, and lower montane forest. Although these ecological systems are relatively consistent across the island they have been subject to very different governance and forest management practices between Malaysia and Indonesia. The following discussion will focus on East Malaysia and Kalimantan separately. Brunei is not dealt with here, because its forests have remained virtually unexploited.

East Malaysia: Sabah and Sarawak

The history of forest management in Malaysia has been described in detail in a number of publications, from which the following discussion is summarized (Collins *et al.*

1991; Hurst 1991; Aiken & Leigh 1992; Marchak 1995). Under the Malaysian constitution, each state government has jurisdiction over land and forest management, although implementation must be consistent with a 1997 national forest policy. The policy requires that each state should reserve at least 47% of its land area for sustainable forest management. Both states have created a 'permanent forest estate (PFE)':

- Sarawak's PFE is approximately 46,065 km² (37% of the land area), of which 30% is steep-land watershed protection forest;
- Sabah's PFE is approximately 33,170 km² (45% of the land area), of which 9% is steep-land watershed protection forest.

Sabah has employed a very intrusive harvesting system (the Malaysian Uniform System or MUS) since forest harvesting began in the late 1950s. Originally it was based on a 70–80-year felling cycle, but pressure for greater timber removal in the mid-1970s resulted in a reduction to 40–50 years. As a consequence, the area of intact forest was halved in the 10 years between 1973 and 1983. Log production peaked during the mid-1980s at over 12 million m³ per year, but yields had dropped to 9 million m³ per year by 1990, and continue to decline to the present. In 1989 the legislature imposed a freeze on all concession applications because of concern that timber resources had been exploited too quickly.

Logging in Sarawak focused first on peat swamp forests, all of which were logged over by the early 1980s. By 1978 most timber production came from lowland evergreen rainforest harvested under MUS, over 30,000 km² of which had been logged to 1990. Log production peaked at >18 million m³ in the late 1980s but within 5 years had declined to about 9 million m³. This decline in yield coincided with the logging of the final low elevation forests and a move into the steeper 'hill Dipterocarp' forests. The MUS has now been abandoned in favour of a single-tree selection system.

In both states of East Malaysia, forests have been 'managed' almost exclusively for industrial wood production, most of that for round wood exports. Logging in East Malaysia is proceeding at a pace that is unsustainable, and a significant decline in timber production and wood exports is inevitable in the near future. Inappropriate harvesting systems and methods exacerbate the effects of over-cutting. Recently, heavier, more powerful and more destructive equipment has been employed for timber extraction (Wyatt-Smith 1987). This change has coincided with the move from lowland to hill forests, for which harvesting and access development experience of both industry and government was ill prepared. The result has been extensive residual stand damage, site degradation and soil disturbance, and downstream impacts on aquatic systems.

Statistics on forest cover in Malaysia are highly variable. Those produced by Malaysian government agencies are generally unreliable. The most objective time-series information for Eastern Malaysia has been produced by FAO (1987) and reveals the following trends in forest cover and forest as a percentage of total land area:

- Sabah: 1982 – 46,645 km² (63%); 1985 – 33,130 km² (45%); 1990 – 29110 km² (39%).
- Sarawak: 1982 – 84,000 km (67%); 1985 – 81,910 km (65%); 1990 – 79,630 (64%).

Loss of forest cover is not entirely attributable to forest harvesting operations – smallholder subsistence agriculture, cash and tree crop agriculture, and El Niño-related fires also play a part. However, in many areas logging-induced reduction of both stand density (ease of land clearing) and canopy closure (increased fire-proneness) is the first step towards eventual deforestation. More recently, close observers of the forestry scene in Sarawak report increasing political interventions to grant permission for the clear-felling of managed selection-cut forests for conversion to commercial fibre plantations, because companies are reluctant to wait the 30-year felling cycle for another cut.

Indonesia: Kalimantan

The history of forest management and forest land use in Indonesia has been described in detail in a number of publications, from which the following discussion is summarized (Sutter 1989; Collins *et al.* 1991; Dick 1991; Hurst 1991; Marchak 1995; Sunderlin & Resosudarmo 1996; FWI/GFW 2002; Holmes 2002).

Before the creation of the first forestry law in 1967, Indonesian forestry was a small-scale business conducted mainly by local entrepreneurs to provide wood products for the domestic market. In 1966, total commercial harvest was only 4 million m³, and most of this went to local sale. The 1967 law provided the basis for establishing 20-year industrial logging concessions and over the next 10 years about 600 concessions were established, many of these obtained by the Indonesian military. Joint ventures for concession management during this period led to increased harvesting and by 1977 annual timber harvest had reached 28 million m³, mostly exported as logs.

Throughout the 1980s the government took a series of measures, including log export bans and tax incentives, to encourage greater foreign investment in wood-processing with little attention to sustainable wood supply. By 1998, across the country as a whole, total wood-processing capacity exceeded legal domestic wood production by 2.85 times, resulting in large-scale deforestation, pressure for reduced retention standards on remaining forests, and widespread forest crime.

Ironically, Indonesia has very credible forest policies, regulations and standards – the result of significant technical assistance from multi-lateral and bi-lateral donors. In 1970 the Indonesian government initiated a consensus-based forest function classification to deal with chronic land use conflicts. Forest land was classified into five broad categories: nature reserve, watershed protection forest, limited production forest, regular production forest and potential conversion forest. The distinctions between protection forests and the two categories of production forest were based on a simple erosion index derived from dominant soil type, slope and rainfall intensity. The five functional categories (Table 27.1) were delineated on 1:500,000 maps and have formed the basis for many subsequent land allocation decisions.

While the theory behind forest function zoning may have been sound, in most regions the designations were deeply flawed because adequate information and technical expertise did not exist. Zoning became subjective, inconsistent, inaccurate and vulnerable to political manipulation. In particular, the conversion forest designation was meant only to identify land *potentially* available for conversion, subject to a land use planning process and an economic analysis. In the end, however, little further planning

Table 27.1   Forest function zones in Indonesia

Function	Primary purpose	Permitted timber extraction
Nature reserve	Biodiversity conservation	None
Watershed protection	Watershed protection and biodiversity conservation	None
Limited production	Natural forest management and erosion prevention	Selection felling to a minimum 60 cm diameter
Regular production	Natural forest management	Selection felling to a minimum 50 cm diameter
Potential conversion	Potentially convertible to non-forest uses	Clear-felling

or analysis was undertaken, and conversion forests became a 'free-for-all' with alloca-tion decisions made on an *ad hoc* basis, often as a result of corruption and/or political intervention.

One of the single most valuable contributions to land use planning in Indonesia was the World Bank-funded Regional Physical Planning Program for Transmigration (RePPProT 1990), carried out between 1984 and 1990. Transmigration was an am-bitious (and some would say ill considered) scheme of the government of Indonesia to relocate poor peasant farmers from overpopulated Java and Bali to the sparsely populated 'outer islands'. RePPProT was based on a 1:250,000 scale biogeoclimatic landscape classification, and although its ultimate objective was to identify areas with the greatest suitability for transmigration resettlement, it provides a national, regional-scale natural resource planning database that is the best in Asia. RePPProT had two components with special implications for forest management:

- land use/land cover were mapped for the entire country, providing a 1985 'snap-shot in time' for 19 forest cover types; and
- the forest function zonation, which was now recognized as worthless as a planning tool, was redefined using the best available biogeoclimatic information, and the results plotted on new, accurate map bases.

The most significant result of the second exercise was a great increase in the amount of land under protection forest (78% increase) at the expense of land under limited and regular production forest (together reduced by 45%). The ultimate conclusion was that a considerable proportion of land under forest concessions had too high an ero-sion hazard to be suitable for timber harvesting and the area remaining was insufficient to support processing capacity under any environmentally prudent forest management strategy. The Ministry of Forests largely ignored the results of this exercise.

On paper Indonesia has a number of credible regulations at the operational level. Natural forests in Indonesia (except mangroves) are harvested under the selection cutting system. Trees may be harvested to a minimum diameter of 50 cm for regular production forests and 60 cm for limited production forest and core stand of 25 stems per hectare of commercial species over 20 cm in diameter must be retained *undamaged* after logging. Originally the stipulated felling cycle was 20 years, but this was quickly recognized to over-estimate growth rates and was changed to 35 years. However, a report to the Indonesian government (FAO 1990) severely criticized the Forest Ministry's implementation of the selection cutting system. The FAO report observed that, 'In general, standards of adherence to proper logging plans, planning and maintenance of logging roads, control of operations, attention to silvicultural prescriptions, selection and management of equipment, environmental protection, and the adequacy of manpower and skills have all been compromised. These flaws in the implementation of the (selection system) will seriously jeopardize the effectiveness of the forest to produce another stand in 35 years.' Unfortunately this warning, and many others, went unheeded by the Indonesian government.

Through the 1980s, the annual rate of deforestation in Indonesia is estimated to have ranged from 0.6 to 1.2 million ha. Although the Indonesian government attempted to blame indigenous swidden agriculturists, these people are now believed to have contributed in only in a minor way to forest loss (Dick 1991; FWI/GFW 2002; Holmes 2002). It is now recognized that the overwhelming causes of deforestation during this period were government-supported activities – poor commercial logging practices, ill-conceived large-scale swampland conversion, commercial rubber and oil palm developments, and the transmigration programme (Dick 1991; Sunderlin & Resosudarmo 1996).

In the late 1980s and early 1990s numerous reports predicted that forest and land use practices in Southeast Asia would have to change in the 1990s, or the region would experience serious resource degradation and depletion. A number of events combined to militate against the necessary changes and nowhere was this combination of events more devastating than in Indonesia. The following inter-related factors led to a significant deterioration in natural resource management from the mid-1990s to the present.

- The transmigration programme, implemented from 1969 to 1993, was fundamentally flawed, and it is estimated that transmigrants cleared about 3.5 million ha of forest land for largely unsustainable agriculture, displaced indigenous peoples, and formed a large, mobile labour pool for the legal and illegal forest conversions that were to occur through the remainder of the decade.
- Indonesia suffers from a serious imbalance between sustainable wood supply and processing capacity, and the shortfall has been made up principally from illegal logging. Illegal logging is estimated to have accounted for over 50% of domestic wood supply in 1997 and about 65% in 2000 (Brown 1999; Scotland 2000).
- From the 1980s onward forest concessionaires were required to contribute to a reforestation fund, that by 1990 exceeded US$ 3 billion. Large areas of degraded forest land prompted the government to initiate a programme of industrial plantation establishment in 1987, with generous financial subsidies. Strict forest cover criteria were to determine which areas could be converted, but these were widely ignored (World Bank 1998). Only about 23% of plantation areas have been successfully established to date, which seems to indicate that access to financial

subsidies and residual native wood resources were the participating companies' primary interest.

- The Indonesian economy has contracted by 4–6% each year since the onset of the Asian economic crisis in 1998 (Resosudarmo 1998). The national currency (rupiah) has declined in value by about 80%, crippling the manufacturing sector but making Indonesian commodity prices very competitive on the world market. Increased exports of natural resource commodities are thus seen as a major strategy for economic recovery. This has stimulated pressure for further forest clearing, both to produce wood products for export and for land conversion to agricultural cash crops.

- One of the most serious consequences of forest degradation and deforestation has been the increasing frequency and intensity of El Niño-related fires. Intact and well-managed moist tropical forests are relatively fire-resistant, but excess timber extraction can result in significant ground-level fuel accumulation, making them much more fire-prone. Recent fires, largely to clear land, have damaged approximately 500,000 ha in 1991, almost 5 million ha in 1994, and 9.75 million ha in 1997–1998. The Indonesian government appears to lack the political will and the technical capacity for fire prevention and suppression.

- The 4 years succeeding the fall of the Suharto regime have seen three presidents and much political upheaval, resulting in governmental paralysis that has prevented a serious response to land use problems. Widespread mistrust of the national government has spawned a rapid programme of decentralized decision-making, which is taking place in a policy and regulatory vacuum and for which local governments have inadequate technical capacity. The result has been a breakdown in governance and the rule of law that has further exacerbated illegal logging and land clearance.

Two independent studies have assessed deforestation rates over the 1990s, using the 1985 RePPProT land use/land cover estimates as a baseline and either the author's considerable experience (Holmes 2002) or GIS-based satellite analysis (FWI/GFW 2002). The results are synthesized in Table 27.2. Annual deforestation rates from 1995 onward appear to have risen from about 1.5 million ha to about 2 million ha. The total loss of forest cover from 1985 to 1997 is estimated by these studies to have been 24% in Kalimantan, and 29% in both Sumatra and Sulawesi. However, these gross figures are in a sense misleading, because they relate to total forest cover. Both studies conclude that lowland evergreen and semi-evergreen rainforest, among the richest and most diverse forests on earth, are now gone from Sulawesi and, if current trends continue, will disappear from Sumatra by 2005 and from Kalimantan by 2010.

In late 1999 the Indonesian government and a consortium of international donors, called the Consultative Group on Indonesia (CGI), reached agreement on a forestry action plan. The plan includes: strong enforcement action against illegal logging; a moratorium on all natural forest conversion; a substantial reduction in wood-processing capacity; tenure reform; prevention and control of forest fires; and significant improvements in the forest management system. A CGI report over a year later, however, concludes:

Table 27.2 Deforestation in Indonesian Sundaland, 1985–1997

Island	Land area (ha)	1985		1997		CHANGE	
		Forest cover (ha)	Forest cover (%)	Forest cover (ha)	Forest cover (%)	Forest cover (ha)	Forest cover (%)
Kalimantan	53,489,000	38,832,000	74	29,637,000	55	–9,195,000	–24
Sumatra	47,530,900	23,066,000	49	16,430,500	34	–6,636,000	–29
Sulawesi	18,614,500	11,204,000	61	7,951,000	43	–3,253,000	–29

Source: RePPProT (1990) and FWI/GFW (2002).

In terms of results in the forests, which is the ultimate measure of achievements, there have been no tangible improvements. The rate of forest loss has not abated. The situation in the forests remains grave by any measure, and the donors remain gravely concerned. There are major problems of overall governance that affect the forestry sector particularly severely, including official corruption, weak law enforcement, and a judicial system needing significant reform. (European Commission 2001, quoted in FWI/GFW 2002)

Forestry impacts on aquatic systems, fish populations and fisheries

The only detailed investigations of the effects of logging on fishes in Borneo have been conducted around Danum Valley Field Centre (DVFC), Sabah (457′ 40″ N 11748′ 00″ E). These consisted of a series of quantitative spatial and temporal surveys within a selective logging area (Martin-Smith 1998b, 1998d). Fish assemblages were compared between streams running through undisturbed forest and those running through forest logged 3–18 years previously (Martin-Smith 1998b) and for a period of 24 months during and following logging (Martin-Smith 1998d).

On a community level, there were few differences between streams in logged and unlogged forest that could be unambiguously assigned to timber extraction. Species inventories were similar and within-stream variation was often much greater than between-stream variation. However, there was some evidence of reduced species abundance at one site during road construction and pre-logging operations with five species absent compared with surveys 18 months later (Martin-Smith 1998d). Although present in streams in logged forest, these species showed reduced abundance and biomass detectable up to 18 years post-logging (Martin-Smith 1998b). The most affected species were a guild of fairly specialized herbivores feeding on autochthonous diatom and algal production (particularly *Garra borneensis*, *Lobocheilos bo* and *Osteochilus chini*). It was hypothesized that reduced light levels from increased sediment loads (Greer *et al.* 1995) had resulted in lower food levels and consequently lower production of these fishes (Martin-Smith *et al.* 1999). Calculations of von Bertalanffy growth coefficients for *G. borneensis* supported this hypothesis with reduced growth observed in streams in logged forest (Martin-Smith 1998e). Biodiversity plots of rank-abundance showed small shifts in community structure to less even distributions in both recently logged and old-logged areas, suggesting long-term persistence of the new state. Similar patterns of higher dominance of some species were found during road construction and logging, although recovery of the system was evident after 12 months (Martin-Smith 1998d).

Balitorids have been suggested as a particularly vulnerable group to the effects of logging. The family is highly diverse in Borneo and species often appear to be confined to single watersheds (Roberts 1982; Tan & Martin-Smith 1998). As they are almost exclusively confined to hill streams, which are the most poorly studied freshwater habitats in Borneo, many more species probably await discovery. Although they are often locally abundant, they have limited dispersal as adults (Martin-Smith *et al.* 1999). Six of 13 species from the studies described above were assessed as potentially vulnerable (Martin-Smith 1998e).

In interpreting these results, it should be kept in mind that logging in the area surrounding DVFC was well regulated and controlled. Access to the concession was strictly controlled, there was little or no illegal logging and compliance with forestry standards and prescriptions was generally high. In contrast to many other areas of Borneo, there are few indigenous people, and there has been no in-migration following access development and no forest conversion. These factors make it difficult to generalize to other areas, where illegal logging, poor forest practices, forest conversion and increased exploitation of fisheries resources are common.

No studies have been carried out on the effects of large-scale deforestation and land conversion, such as that occurring in Indonesia, on aquatic systems and fisheries resources. It is likely that these activities will have far greater negative effects on fish communities than controlled forest harvesting. Allochthonous inputs are greatly reduced following forest conversion and these inputs are of greater importance than autochthonous production (Dudgeon 2000). Additional consequences include increased human populations, leading to greater access and higher catch rates, and large inputs of pesticides and fertilizers which run off into the watercourses and will have considerable impacts on fish diversity and abundance.

Cambodia: a large river and lake system in a monsoonal climate

Biophysical description

Cambodia lies between 10° 30′ and 14° 30′ north latitude and abuts the countries of Thailand and Laos to the west and north, and Vietnam to the east and south (Fig. 27.2). It has a relatively short (440 km) coastline on the Gulf of Thailand. Cambodia contains several major biogeoclimatic regions. The landscape of the country is centred on the floodplains of the Tonle Sap (Great Lake) and the lower Mekong River. To the southwest of these floodplains are the Elephant and Cardamom Mountains, which rise to an elevation of approximately 1800 m, and a coastal zone consisting of a plain and offshore islands. To the northwest and northeast of Tonle Sap and the lower Mekong are rolling lowland plains, hills and plateaus rising gently to the Dangrek and Annam Mountains on the borders with Thailand and Laos/Vietnam respectively. The Dangrek and Annam Mountains are much lower than the coastal ranges, with elevations of 400–755 m and 700–800 m, respectively.

The climate is dominated by the southwest monsoon lasting from May to October, which brings the majority of the annual rainfall, and the northeast monsoon from November to March, which brings a pronounced dry season. Because of the direction of the southwest monsoon, the wettest areas of the country, with only a 2-month partial dry season, are the coastal plain, with annual rainfall between 3500 and 4000 mm, and the windward slopes of the Cardamom Mountains with between 4500 and 5000 mm of precipitation. Precipitation is lowest in the Tonle Sap and Mekong floodplains because of their position in the 'rain-shadow' of the southwestern mountains. These lowlands receive annual precipitation of <1500 mm and experience a 5–6-month

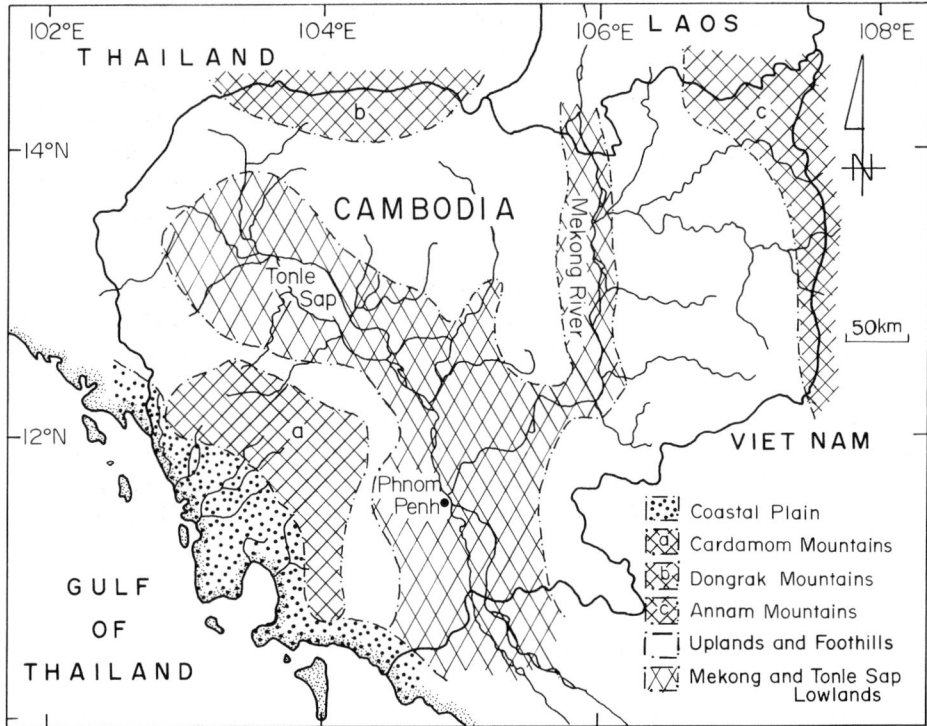

Fig. 27.2 Biophysical units of Cambodia.

drought. Northern and eastern foothills and mountains receive annual rainfall of 1500–2600 mm and experience a 4–6-month drought.

The interaction of climate and topography results in a rich floral mosaic. Wetland communities include mangroves, fresh and brackish water swamps and swamp forests. Dryland forest types include wet and moist evergreen forests, semi-evergreen forests, deciduous (dry Dipterocarp) forests, small areas of savanna forests, and subtemperate montane forests. The swamps and swamp forests of the Tonle Sap and the Mekong floodplain are the most extensive freshwater wetlands in Southeast Asia, providing critical habitat for large numbers of water birds and the basis for a regional freshwater fishery that is the largest in Asia. Of the dryland ecosystems, the highest biodiversity occurs in the evergreen forest, in particular the wet evergreen communities of the southwest mountains.

The 'heart' of Cambodia is the Mekong River and Tonle Sap. The Mekong is one of the great rivers of the world and its hydrology has been thoroughly described by Pantulu (1986). From its origins in the Tibetan Himalayas at over 5000 m elevation, the Mekong flows 4200 km through six countries to its delta on the South China Sea. It is the twelfth longest river in the world and has the sixth highest rate of annual discharge. About 20% of the Mekong's flow comes from the Chinese and Tibetan Himalayas, 50% from Burma, Thailand, Laos and Vietnam, and the remaining 30%

from Cambodia. The river enters northeastern Cambodia from Laos and flows generally southwesterly to its exit into Vietnam. Of the 30% of the flow entering the Mekong from Cambodia, about 10% comes from nine rivers that drain much of the northwestern part of the country and flow to Tonle Sap. The remaining 20% comes from a series of rivers draining east and northeast Cambodia and small adjacent parts of Laos and Vietnam.

The watershed area of Tonle Sap is about 67,600 km^2; however, of the total annual water input to the lake only 38% comes from its own watershed and the remaining 62% from Mekong River flows (ORSTEN & BCEOM 1993). The outlet of the lake is the Tonle Sap River, which joins the Mekong at Phnom Penh. During the wet season the lower Mekong is unable to fully discharge and Mekong waters reverse flow up the Tonle Sap River and into the lake. During this period the lake expands over sixfold in area and about 55-fold in volume (ORSTEN & BCEOM 1993). This flooding transforms the lake and its swamp forests into intensive feeding habitats for fish. It is estimated that fish productivity may reach 138–175 kg/ha; far higher than any other permanent water body in the region (van Zalinge *et al.* 2000). Total catch from the Tonle Sap system is between 170,000 and 210,000 tonnes per year, or about 15–20% of the total Lower Mekong fishery (van Zalinge *et al.* 2000; Jensen 2001). This important fishery is dependent on two unique hydrological characteristics of the Tonle Sap/Mekong System, which sustain important annual migrations (Jensen 2001):

- the alternate flooding and drying of the swamps and swamp forests of the lake, which results in feeding habitat of exceptional productivity; and
- the stable base flow in the river from snowmelt in the Himalayas, which sustains important dry season holding habitat in deep pools and channels below the Lao border.

Although data on suspended sediments are generally poor, what information exists indicates that the major portion of sediment entering the lake comes from the Mekong and that much of this is flushed back out in the dry season. The major source of sediment in the Mekong is thought to be past land-clearing activities in China, but data collected by the Mekong River Commission indicate that levels have been steadily dropping in the Lao section of the river since 1986 (Evans pers. comm., Campbell pers. comm.).

Major fish communities in the Mekong and Tonle Sap

Fish diversity in the entire Mekong River basin has been the subject of considerable debate among taxonomists with estimates ranging from 400 to 1300 species (Kottelat & Whitten 1996; Rainboth 1996; Jensen 2001). Kottelat (1989) suggested that the species list was approximately 300 and this figure was subsequently revised upwards to 400–500 (Kottelat & Whitten 1996). A recent estimate of 221 species from the Mekong and Great Lake in Cambodia was given in Lim *et al.* (1999), compiled from a number of earlier works. Other recent identification guides to fishes from northern Vietnam (Kottelat 2001a) and Lao PDR (Kottelat 2001b) list 268 and 481 species, respectively. FishBase (2002) lists 478 fishes from Cambodian freshwaters. It is clear that fish diversity is very high, although the exact magnitude remains undetermined.

The fish fauna of the Cambodian Mekong is dominated by cyprinids (carps and barbs) and related families – these comprise almost half of the recorded species (Lim *et al.* 1999). Other important groups are siluroids (catfish) and percids (perch-like fish) comprising 21% and 23% of species diversity (FishBase 2002). Clupeids (herrings), pleuronectiformes (flatfish), synbranchiformes (swamp eels), notopterids (featherbacks) and tetraodontids (pufferfish) are also important fishery items in addition to freshwater sharks, rays and sawfish. For the whole Mekong, Taki (1978) estimated that cyprinids and related families represented 49% of species and siluroids a further 20%. Cyprinids become more dominant moving upstream, comprising 60% of the fauna in Luangprabang province in Lao PDR (Taki 1978).

The detailed habitat requirements of many species are not well known or the information is not widely accessible. Traditional fishers' knowledge is generally restricted to certain periods during the year when fish are available to them (Roberts & Baird 1995; Chhuon 2000; Srun & Ngor 2000). However, recent basin-wide investigations of traditional knowledge have produced detailed assessments for important food species (Bao *et al.* 2001). Habitats of cryptic or non-food fish are particularly lacking. The importance of deep pools in the main river as dry season habitat for migratory fishes has been documented recently, with at least 50 species using these habitats (Poulsen & Valbo-Jørgensen 2000, 2001).

There is a huge range in size for Mekong fishes from the dwarf medaka (*Oryzias minutillus*) and dwarf noodlefish (*Sundasalanx praecox*), which reach a maximum length of 20 and 25 mm respectively (Rainboth 1996; Roberts 1998) to the giant barb (*Catlocarpio siamensis*) and giant catfish (*Pangasius gigas*), both of which reach lengths of 3000 mm and weights of 300 kg (Roberts & Warren 1994; Rainboth 1996). The majority of cyprinids are small- to medium-sized fish (50–400 mm length) while the siluroids are generally larger (400–1000 mm length).

A wide variety of feeding habitats has been documented for fishes in the Mekong and Great Lake (Lim *et al.* 1999; references in FishBase 2002). Planktivory, which is a feeding strategy restricted to larger water bodies, is found in the herrings and some cyprinids and gouramis. Considerable numbers of species caught in Cambodian fisheries are piscivorous (34% of 120 species in Lim *et al.* 1999). Feeding on insects, crustaceans and molluscs is also important, as is omnivory (22% and 44% respectively, in Lim *et al.* 1999). Levels of herbivory and detritivory have probably been underestimated in Cambodia as fishes showing these feeding habits are often small and not caught for food.

Levels of knowledge about reproductive seasonality, particularly for important fishery species, are much higher than for comparable species elsewhere in Southeast Asia (Lim *et al.* 1999; Bao *et al.* 2001). Spawning periodicity is known to some degree for 83 of 120 species listed in Lim *et al.* (1999) with almost all species spawning during May–July. This corresponds to the period of rising water. Eggs and larvae are swept downstream into the expanding Great Lake where there is abundant food.

Migration is a dominant aspect of the ecology of fishes in the Mekong and Tonle Sap (Welcomme 1985; Jensen 2001). In the Mekong main stem, cyprinid species dominate the upstream movements during the dry season, but it is mainly catfish species of the Pangasiidae and Siluridae families that migrate upstream during the wet season. Migration during the dry season is hypothesized to be in response to food availability whereas wet season migrations are for spawning (Chhuon 2000; Chomchanta *et al.* 2000). Migration from the Mekong to the Tonle Sap occurs during the expansion of the latter, which inundates huge areas of forest, while the converse occurs when Tonle Sap empties at the end of the wet season. These migrations are the basis of the very important fisheries, both commercial and subsistence (Jensen 2001). For example, at least 4000–6000 tonnes of the small cyprinid *Henicorhynchus siamensis* (Trey riel) are caught in bag nets (Dais) during the filling of Tonle Sap (Jensen 2001). Migrations for some other valuable food species such as the large catfishes are reasonably well studied (Chhuon 2000; Srun & Ngor 2000).

Forest practices, forest harvesting and land use

Forest is currently estimated to occupy approximately 105,000 km² or 58% of Cambodia's total land area, which would indicate a loss of about 6% over the 1990s. Forests comprise six major ecological types: wet evergreen forest, moist semi-evergreen forest, dry (deciduous) Dipterocarp forest, wet montane forest, swamp forest and mangrove. Presently, *permitted* commercial harvesting is directed almost entirely at the moist evergreen and semi-evergreen forest types.

At least on paper, Cambodia has a small number of very credible harvesting regulations. A very conservative annual increment of 0.33 m³/ha over the productive, operable forest area is assumed to determine a *maximum sustained yield* for commercial harvesting, and the designated silvicultural system is single-tree selection over a 30-year felling cycle. In addition to these determinations, there are two further controls that function as *rate of cut constraints*: a minimum diameter limit of 60 cm for Dipterocarps and 50 cm for other species; and a 30% limit to the number of commercial stems of each species that can be removed in any annual felling coupe. If strictly applied, this system should result in a very conservative level of harvest of 3–5 trees and 10–15 m³ per hectare which, given appropriate harvesting methods, should have little long-term impact on stand age structure, canopy structure, aquatic systems or biodiversity. All forest harvesting operations, including road construction, are prohibited during the wet season from May to October. The main forest policy problem is that Cambodia has no overall forest law to provide a context for such regulations, to establish a 'forest estate' and to mandate forest protection. This leaves large areas of Cambodia's forests vulnerable to agricultural encroachment and other forms of unmanaged and unplanned destruction.

Prior to the early 1990s, legal forest harvesting took place under a system of collection permits and quotas that originated in French colonial times. The unstable political and internal security conditions of the 1970s and 1980s effectively precluded organized forestry operations which, ironically, meant that Cambodia's forests entered the 1990s in relatively good condition. With the slowly improving security situation since 1993, normal government operations have gradually been resumed.

Early in the 1990s the Cambodian government introduced private industrial concessions as the primary instrument of commercial forest management in the country (ADB & RGC 2000). However, there was no clear legal context for the concession system and the government simply superimposed it on the existing collection permit system and relied on a set of antiquated and often-conflicting forest regulations. This greatly constrained the government's ability to enforce adequate standards of forest practice (ADB & RGC 2000). Moreover, the older system of collection permits and quotas was not abolished. Various government officials had the authority to dispose of timber without reference to forestry authorities or to any concept of sustainable yield.

The first concession was granted in 1994, and by 1997, 33 concessions (totalling almost 7 million ha) had been created without even a rudimentary resource assessment. Considerable proportions of concession were later found to contain lands that were commercially inoperable or ecologically fragile. By mid-1999, the number of concessions had been reduced to 22. One of the major objectives of the concession system was to curb illegal and uncontrolled harvest. However, the still precarious security situation, and the fact that much illegal harvest was carried out by various armed forces, meant that few companies were able to exercise much control. Estimated annual sustainable yield is 650,000 m^3; however, in each year from 1996 to 1999 actual harvest is thought to have exceeded this by three to four times. In 1997 alone, about 93% of the industrial wood harvest was illegal and government revenues were collected on only about 12% of that total (White & Case 1998). Illegal logging rapidly emerged as the single most serious danger to Cambodia's forests, and threatened, along with the poor concession establishment process and poor harvesting procedures, to undermine the economic viability of the forest sector.

Under these conditions, few concessionaires approached their management responsibilities with any concept of stewardship. Few concessionaires attempted to present required concession management plans, and most did the minimum operational planning necessary for cutting approvals. Most plans were very poor documents, often prepared for a fee by the same forestry officers responsible for reviewing, approving and enforcing the plans – a clear conflict of interest.

In 1997, an International Consultative Group (ICG) on Cambodian forestry was formed by international aid donors to coordinate technical assistance, and the IMF made further economic aid conditional on improved forest management. In consultation with the ICG, the Cambodian government initiated activities in five main areas:

- significant and immediate reduction in illegal logging and other forms of forest crime;
- an effective legal and regulatory basis for forest management;
- improved standards of forest practice;
- rationalization of the number and the commercial viability of concessions; and
- technical assistance to the Cambodian forestry agency to improve its capacity to implement sustainable forest management and to monitor and enforce 'best management practice' by concessionaires.

The following paragraphs describe the chronological order in which these tasks were and are being carried out.

From 1997 to 1999 the World Bank supported a Forest Concession Management Project focused primarily on providing the forest department with a code of best forest practice consisting of an umbrella regulation governing concessions linked to the national environmental assessment process, and a series of standards and guidelines. The guidelines include forest planning, inventory, biodiversity conservation, social forestry, forest harvesting and forest engineering. From a water quality and fisheries perspective, these codes stipulate procedures for: the zoning of watershed protection areas and riparian and stream protection corridors; road construction and maintenance; reduced impact logging operations; stream crossings; and construction timing windows.

In April 1999, FAO/UNDP agreed to assist the Cambodian government in establishing an independent 'Forest Crime Monitoring Unit'. The role of the unit is to undertake formal field inspections of reported violations and to provide a case-tracking system to monitor enforcement actions.

In 1999, the Asian Development Bank agreed to draft a comprehensive Forest Law and an enforceable Model Concession Contract, and to conduct a performance audit on all operating concessions. The draft law and contract were completed by the end of 1999 and the performance audit began in November of that year. The objective of the audit was to evaluate and report on:

(1)   the degree of compliance with existing contracts;
(2)   the degree of compliance with existing forest regulations;
(3)   the extent to which current operations were implementing the new forest practice code; and
(4)   the viability of the concession as determined by the number of intact annual harvesting coupes for the remaining years of the 30-year felling cycle.

The results of the audit were appalling (ADB & RGC 2000). No company had yet fulfilled commitments under existing contracts. Fourteen (of 22) companies had been charged or convicted of serious offences (the most common being unauthorized cutting and transport of timber). In terms of operational performance (for the 11 concessions that operated in 1999 and 2000), one concession was rated satisfactory, four were rated poor (unacceptable in several important areas) and six were rated very poor (unacceptable in all aspects, urgent action required). Only three companies received a passing grade in engineering works (i.e. roads, bridges, culverts, etc.) and only one in harvesting operations. The audit concluded:

> It cannot be over-stressed that no one entity is to blame for the current crisis. It is the result of a total system failure; resulting from greed, corruption, incompetence and illegal acts that were so widespread and pervasive as to defy the assignment of primary blame. Responsibility for the debacle must be shared by national and provincial politicians, government staff, the police and military, concessionaires, private businesses and individuals, and by individuals and or-

ganizations in the neighbouring countries of Thailand, Laos and Vietnam. (ADB and RGC 2000)

The Cambodian government's reaction to the audit was to cancel three concessions outright, and to put the remainder on notice. Initially it decreed that no harvesting permit would be issued for the 2001 season until a company had submitted an acceptable concession management plan by September 2000 and entered into a new concession agreement. Subsequently the government recognized that this was probably an unrealistic expectation and deferred the deadline to September 2001. No company met the September 2001 deadline for management plan submission, and it is the authors' understanding that no cutting permits have been issued for the 2002 season. The government has indicated that, at some point, it will cancel any concessions that do not comply with the planning requirement. The international donor community has put the government firmly on notice that any further financial and technical assistance depends on decisive action to rationalize the concession system.

Despite the problems of the past decade, Cambodian forests are still in fairly good condition, particularly relative to those in other parts of the region. The principal reason for this is that because of depressed market conditions associated with the recent Asian financial crisis, timber harvesting, whether legal or illegal, focused primarily on large veneer peeler logs of only three or four *Dipterocarp* species. Thus, while the forests may be somewhat economically depleted, they still retain much of their ecological function and a significant residual economic value. The challenge to the government will be to improve its capacity and performance to ensure sustainable forestry in the future.

Forestry impacts on aquatic systems, fish populations and fisheries

There have been reports of declining fisheries catches from the Mekong Basin over the past two decades (Roberts & Baird 1995; Sjorslev 2000). There is, however, no direct evidence, particularly in Tonle Sap and the Cambodian sections of the Mekong, linking these effects with poor forestry activities (Campbell pers. comm. and Evans pers. comm.). The data on sedimentation are poor and the Mekong River Commission is currently supporting more detailed studies. Existing data indicate that there is no threat, at least for the immediate future, to the lake or the river from sedimentation. Earlier fears that the lake was filling up at a rate of centimetres per year have now been shown to be baseless. They appear to have been derived from measurements of lake bed elevation taken 10 years apart that inadvertently used two different base elevations. Preliminary results from a modelling study supported by the government of Finland indicate maximum rates of deposition in the order of 0.5 mm per year.

There are several reasons to expect that upland sediment generation is not a threat to Tonle Sap (Campbell pers. comm.). Most of the topography within the lake valley is very flat and streams are of very low energy. Where streams enter the valley from upland areas, they quickly lose velocity and drop non-colloidal material. That zone of deposition is well away from the lake. Significant colloidal material enters the lake during the rainy season, most of it from the Mekong. Some of this will be deposited in the inundation area, but much will pass back out of the lake when normal flows resume, because about 90% of the lake water flushes each year. During the dry season strong winds whip up the lake surface, resuspending much of the deposited sediment, which then passes out down the Tonle Sap River to the main stem Mekong. This flushing action protects the lake from sedimentation, excessive eutrophication and the accumulation of other materials such as pesticides.

As noted earlier, sediment data collected by the Mekong River Commission since 1986 suggest that sediment levels in the main stem river have been in fairly steady decline over the past dozen years and should pose no threat to either fish migrations or to dry season holding (Campbell *et al.* 2002). Most close observers of the lake and the main stem river feel that the most immediate threats to the fishery are the clearing of swamp and riparian gallery forests for fuel wood and subsistence agriculture, excessive fishing pressures, upstream irrigation and hydroelectric dams and, over the longer term, increased water pollution (Jensen 2000; Evans pers. comm.). Any disruption of migration routes and/or spawning habitats has the potential to severely impact fisheries, but the links between forest harvesting practices and these ecological processes are not well understood. It appears that deforestation can substantially alter the water retention capacity of the basin and lead to widespread flooding (Fraser & Jewell 2002). The effect of these events on juvenile fishes is unknown and should be a priority for research.

The most immediate threat to fisheries is probably the removal or conversion of seasonally inundated forest that forms the spawning and 'nursery' habitat in Tonle Sap. Decreases in food availability and increases in predation pressure from such activity will inevitably reduce fish production.

All of this is not to say that poor forest harvesting activities have had an insignificant impact on fisheries – merely that the impact in the lake and main stem river, which is the principal focus of this section, appears to date to have been slight. It is likely in smaller river and stream tributaries that the primary impact has been felt, probably particularly in dry season holding habitat. The concession audit team observed 'extremely poor standards of access development, especially in location, design and construction …' that is '… not only causing significant environmental damage, but is also increasing concessionaires' transport costs' (ADB & RGC 2000). Roads developed for illegal logging are even worse. A common practice in stream crossings is simply to bulldoze a soil causeway, which presumably is then 'blown-out' during the next rainy season. These practices probably have a significant impact on fish migration and on dry season holding in tributary streams, through the silting up of deeper pools and channels, although such impacts remain to be documented.

References

Abidin, A.Z. (1986) The reproductive biology of a tropical cyprinid, *Hampala macrolepidota* (Van Hasselt) from Zoo Negara Lake, Kuala Lumpur, Malaysia. *Malaysian Journal of Fish Biology*, **29**, 381–91.

ADB & RGC (Asian Development Bank & Royal Government of Cambodia) (2000) *Cambodian Forest Concession Review Report.* Asian Development Bank (TA-3152-CAM), Phnom Penh, Cambodia, April 2000.

Aiken, S.R. & Leigh, C.H. (1992) *Vanishing Rainforests: The Ecological Transition in Malaysia.* Oxford University Press, Oxford.

Bao, T.Q., Bouakhamvongsa, K., Chan, S., *et al.* (2001) Local Knowledge in the Study of River Fish Biology: Experiences from the Mekong. Mekong Development Series No. 1. Mekong River Commission, Phnom Penh.

Brown, D.W. (1999) *Addicted to Rent: Corporate and Spatial Distribution of Forest Resources in Indonesia – Implications for Forest Sustainability and Government Policy.* Indonesia-UK Tropical Forest Management Programme, Jakarta.

Campbell, I., Tin Nguyen Thanh & Sien Mya *The Mekong River Commission and Water Quality Monitoring in the Lower Mekong Basin.* Manuscript in preparation.

Chhuon, K.C. (2000) *Fisher's Knowledge about Migration Patterns of Three Important Pangasius Catfish Species in the Mekong Mainstream.* Department of Fisheries, Phnom Penh, Cambodia.

Chomchanta, P., Vongphasouk, P., Soukhaseum, V., Soulignavong, C., Saadsy, B. & Warren, T. (2000) *Migration Studies and CPUE Data Collection in Southern Lao PDR. 1994 to 2000.* LARReC, Vientiane, Lao PDR.

Christensen, M.S. (1992) Investigations on the ecology and fish fauna of the Mahakam River in East Kalimantan (Borneo), Indonesia. *Internationale Revue gesampte Hydrobiologie,* 77, 593–608.

Collins, N.M., Sayer, J.A. & Whitmore, T.C. (1991) *The Conservation Atlas of Tropical Forests: Asia and the Pacific.* Produced by the World Conservation Monitoring Centre, World Conservation Union and British Petroleum. Macmillan Press, London.

Dick, J.H. (1991) *Forest Land Use, Forest Use Zonation and Deforestation in Indonesia: A Summary and Interpretation of Existing Information.* A background paper to the UNCED Conference prepared for the State Ministry for Population and Environment and the Environmental Impact Management Agency, Jakarta, Indonesia.

Dudgeon, D. (2000) The ecology of tropical Asian rivers and streams in relation to biodiversity conservation. *Annual Review of Ecology and Systematics,* 31, 239–63.

European Commission (2001) *Policy Dialogue for Creation of a Conducive Environment for the Sustainable Management of All Types of Forests in Indonesia.* A Position Paper presented on behalf of donors, 11th Consultative Group Meeting on Indonesia, April 2001.

FAO (1987) *Special Study on Forest Management and Utilization of Forest Resources in the Developing Regions. Assessment of Forest Resources in Six Countries, Asia-Pacific Region.* FAO Bangkok Field Document No. 17.

FAO (1990) *Situation and Outlook of the Forestry Sector in Indonesia.* Food and Agriculture Organization of the United Nations and the Ministry of Forestry, Jakarta, Indonesia.

FishBase (2002) World Wide Web electronic publication, (eds R. Froese & D. Pauly) www.fishbase.org [accessed 8 March 2002].

Fraser, A. & Jewell, N. (2002) *The Impact of the Loss of Forest Cover on River System Hydrology and Human Settlements.* SETIS RETA report to Mekong River Commission.

FWI/GFW (Forest Watch Indonesia & Global Forest Watch) (2002) *The State of the Forest: Indonesia.* FWI, Bogor, Indonesia & GFW, Washington, DC.

Greer, T., Douglas, I., Bidin, K., Sinun, W. & Suhaimi, J. (1995) Monitoring geomorphological disturbance and recovery in commercially logged tropical forest, Sabah, East Malysia, and implications for management. *Singapore Journal of Tropical Geography,* 16, 1–21.

Holmes, D. (2002) *Indonesia: Where Have All the Forests Gone?* Discussion Paper, Environment and Social Development, East Asia and Pacific Region, The World Bank, Washington, DC.

Hurst, P. (1991) *Rainforest Politics: Ecological Destruction in South-East Asia.* S. Abdul Masjid & Co. Publishing, Kuala Lumpur, Malaysia.

Inger, R.F. & Chin, P.K. (1962) The fresh-water fishes of North Borneo. *Fieldiana: Zoology*, 45, 1–268 (+47 pp. in supplementary chapter by P.K. Chin 1990).

Jensen, J.G. (2001) Managing fish, flood plains and food security in the Lower Mekong Basin. *Water Science and Technology*, 43, 157–64.

Kottelat, M. (1989) Zoogeography of the fishes from Indochinese inland waters with an annotated check-list. *Bulletin Zoölogisch Museum, Universiteit van Amsterdam*, 12, 1–54.

Kottelat, M. (2001a) *Freshwater Fishes of Northern Vietnam. A Preliminary Check-List of the Fishes Known or Expected to Occur in Northern Vietnam with Comments on Systematics and Nomenclature*. World Bank Biodiversity Report.

Kottelat, M. (2001b) *Fishes of Laos*. World Bank/IUCN.

Kottelat, M. & Whitten, A.J. (1996) *Freshwater Biodiversity in Asia, with Special Reference to Fish*. World Bank Technical Report 343.

Kottelat, M., Whitten, A.J., Kartikasari, S.N. & Wirjoatmodjo, S. (1993) *Freshwater Fishes of Western Indonesia and Sulawesi*. Dalhousie University, Canada and Periplus Editions, Hong Kong.

Lim, P., Lek, S., Touch, S.T., Mao, S-O. & Chhouk, B. (1999) Diversity and spatial distribution of freshwater fish in Great Lake and Tonle Sap River (Cambodia, Southeast Asia). *Aquatic Living Resources*, 12, 379–86.

McAdam, D.S.O., Liley, N.R. & Tan, E.S.P. (1999) Comparison of reproductive indicators and analysis of the reproductive seasonality of the tinfoil barb, *Puntius schwanenfeldii*, in the Perak River, Malaysia. *Environmental Biology of Fishes*, 55, 369–80.

Marchak, M.P. (1995) *Logging the Globe*. McGill-Queen's University Press, Montreal and Kingston, Canada.

Martin-Smith, K.M. (1998a) Fish-habitat relationships in rainforest streams in Sabah, Malaysia. *Malaysian Journal of Fish Biology*, 52, 458–82.

Martin-Smith, K.M. (1998b) Effects of disturbance caused by selective timber extraction on fish communities in Sabah, Malaysia. *Environmental Biology of Fishes*, 53, 155–67.

Martin-Smith, K. M. (1998c) Temporal variation in fish communities from the upper Segama River, Malaysian Borneo. *Polish Archives of Hydrobiology*, 45, 185–200.

Martin-Smith, K.M. (1998d) Biodiversity patterns of tropical freshwater fish following selective timber extraction: a case study from Sabah, Malaysia. *Italian Journal of Zoology*, 65, 363–8.

Martin-Smith, K.M. (1998e) *Impacts of Selective Timber Extraction on Freshwater Fish Populations in Sabah, Malaysia*. Final report to ODA/NERC.

Martin-Smith, K.M. & Laird, L.M. (1998) Depauperate fish communities in Sabah: the role of barriers to movement and habitat quality. *Journal of Fish Biology*, 53 (Suppl. A), 331–44.

Martin-Smith, K.M. & Laird, L.M. (1999) Reproductive patterns in some Cypriniformes from Borneo. *Proceedings of the 5th Indo-Pacific Fish Conference, Nouméa, 1997*, (eds B. Séret & J.-Y. Sire) pp. 493–504. Paris, Société Française d'Ichtyologie.

Martin-Smith, K.M. & Tan, H.H. (1998) Diversity of freshwater fishes from eastern Sabah: annotated checklist for Danum Valley and a consideration of inter- and intra-catchment variability. *Raffles Bulletin of Zoology*, 46, 573–604.

Martin-Smith, K.M., Laird, L.M., Bullough, L. & Lewis, M.G. (1999) Mechanisms of maintenance of tropical freshwater fish communities in the face of disturbance. *Philosophical Transactions of the Royal Society Series B*, 354, 1803–10.

Nyanti, L., Yee, L.T. & Adha, K. (1999) Freshwater fishes from Bario, Kelabit Highlands, Sarawak. *ASEAN Review of Biodiversity and Environmental Conservation*, 4, 1–6.

ORSTEN & BCEOM (1993) *Development Plan for Tonle Sap and Chakdomuk, Phase 1*. Mekong River Secretariat Project No. CAN/167/TSAP. Phnom Penh, Cambodia.

Pantalu, V.R. (1986) The Mekong River System. In: *The Ecology of River Systems*, (eds B.R. Davis & K.F. Walker), pp. 695–719. *Monographiae Biologicae*, 60. Dr. W. Junk, Dordrecht, the Netherlands.

Poulsen, A.F. & Valbo-Jørgensen, J. (2000) *Fish Migrations and Spawning Habits in the Mekong Mainstream – A Survey using Local Knowledge.* AMFC Technical Report, Mekong River Commission.

Poulsen, A.F. & Valbo-Jørgensen, J. (2001) Deep pools in the Mekong. *Catches & Culture,* 7, 1–5.

Rainboth, W.J. (1996) *Fishes of the Cambodian Mekong.* FAO Species Identification Field Guide for Fishery Purposes. FAO, Rome.

RePPProT (Regional Physical Planning Program for Transmigration) (1990) *The Land Resources of Indonesia.* ODA/Ministry of Transmigration, Jakarta, Indonesia.

Resosudarmo, D.P (1998) *The Economic Crisis and Indonesia's Forest Sector.* Unpublished monitoring report, Centre for International Forestry Research, Bogor, Indonesia, May 1998.

Roberts, T.R. (1982) The Bornean gastromyzontine fish genera *Gastromyzon* and *Glaniopsis* (Cypriniformes, Homalopteridae), with descriptions of new species. *Proceedings of the Californian Academy of Science,* 42, 497–524.

Roberts, T.R. (1989) The freshwater fishes of Western Borneo (Kalimantan Barat). *Memoirs of the Californian Academy of Science,* 14, 1–210.

Roberts, T.R. (1998) Systematic observations on tropical Asian medakas or ricefishes of the genus *Oryzias,* with descriptions of four new species. *Ichthyological Research,* 45, 213–24.

Roberts, T.R. & Baird, I.G. (1995) Traditional fisheries and fish ecology on the Mekong River at Khone Waterfalls in southern Laos. *Natural History Bulletin of the Siam Society,* 43, 219–62.

Roberts, T.R. & Warren, T.J. (1994) Observations of fishes and fisheries in southern Laos and northeastern Cambodia, October 1993–Febuary 1994. *Natural History Bulletin of the Siam Society,* 42, 87–115.

Scotland, N. (2000) *Indonesia Country Paper on Illegal Logging.* A Paper Prepared for the World Bank/WWF Workshop on Control of Illegal Logging in East Asia, 28 August 2000, Jakarta.

Sjorslev, J.G. (2000) *Luangprabang Fisheries Survey.* LARReC Technical Report No. 0008. AMFC/MRC & LARReC/NAFRI, Vientiane, Lao PDR.

Srun, P. & Ngor, B.P. (2000) *The dry season migration pattern of five Mekong fish species: Riel (Henicrorhychus spp.), Chhkok (Cyclocheilichthys enoplos), Pruol (Cirrhinus microlepis), Pra (Pangasianodon hypophthalmus) and Trasork (Probarbus jullieni).* Web summary of BSc thesis, Royal University of Agriculture, Phnom Penh [http://www.mekonginfo.org/mrc_en/doclib.nsf].

Sunderlin, W.D. & Resosudarmo, I.A.P. (1996) *Rates and Causes of Deforestation in Indonesia: Towards a Resolution of the Ambiguities.* Centre for International Forestry Research, Bogor, Indonesia, December 1996.

Sutter, H. (1989) *Forest Resources and Land Use in Indonesia.* Food and Agricultural Organization of the United Nations and the Ministry of Forests, GOI, Jakarta, Indonesia.

Taki, Y. (1978) An analytical study of the fish fauna of the Mekong Basin as a biological production system in nature. *Research Institute for Evolutionary Biology, Tokyo Special Publication* 1.

Tan, H.H. & Martin-Smith, K.M. (1998) Two new species of *Gastromyzon* (Teleostei: Balitoridae) from the Kuamut headwaters, Kinabatangan basin, Sabah, Malaysia. *Raffles Bulletin of Zoology,* 46, 361–71.

van Zalinge, N.P., Thuok, N., Tana, T.S. & Loeung, D. (2000) Where there is water, there is fish? Cambodian fisheries issues in a Mekong regional perspective. In: *Common Property in the Mekong: Issues of Sustainability and Subsistence,* (eds M. Ahmed & P. Hirsch), pp. 109–39. ICLARM Studies and Reviews.

Watson, D.J. & Balon, E.K. (1984) Structure and production of fish communities in tropical rain forest streams of northern Borneo. *Canadian Journal of Zoology,* 62, 927–40.

Welcomme, R.L. (1985) *River Fisheries.* FAO Fisheries Technical Paper 262.

White and Case Ltd (1998) *Forestry Legal Counsel Assignment.* World Bank/RCG Technical Assistance Project (IDA CR 2664-KH), Phnom Penh, Cambodia.

World Bank (1998) *Involvement in Sector Adjustments for Forests in Indonesia: The Issues.* A World Bank memorandum, Jakarta.

Wyatt-Smith, J. (1987) *The Management of Tropical Moist Forest for Sustained Production of Timber.* IUCN/IIED Tropical Forest Policy Paper 4. International Institute for Environment and Development, London.

Chapter 28
Interactions: mangroves, fisheries and forestry management in Indonesia

D.G. BENGEN AND I.M. DUTTON

Introduction

Indonesia, the largest archipelagic state in the world, consists of some 17,508 islands with over 81,000 km of coastline. It stretches over 5000 km from Sumatra in the west, to Papua in the east. With coastal and marine waters making up about 75% of the total land/water area, Indonesia has some of the largest and most diverse coastal and marine resource areas in the world (Bengen 2001).

The mangrove forest ecosystem is one of the most important ecological systems in the Indonesian coastal zone. Of the 180,000 km² of total worldwide mangrove area, around 27% is found in Indonesia. Indonesia's mangrove forests are the most extensive in the Southeast Asia region, covering some 2.43 million hectares. About 75% of this area is concentrated in Papua, East Kalimantan and East Sumatra (Anonymous 1996; Hinrichsen 1998; Bengen 2002). While estimates of area and condition vary greatly due to lack of consistent methods for survey/research and the existence of asynchronous and asymmetric data series, the World Resources Institute (WRI 1998) estimates that around one third of these forests are protected. The largest of these is the Bintuni Bay 'Cagar Alam' (Strict Nature Reserve) in Papua Province.

As one of the unique coastal inter-tidal wetlands, mangroves provide breeding grounds for a large number of commercially valuable fish species, crustaceans (crabs and shrimps), bivalves (cockles, mussels and oysters) and gastropods. Fish and other biota also use mangroves as nursery and feeding areas, and as refuge habitat from predators. The mangrove vegetation, the rich detritus and shallow water provide unique habitat conditions that benefit many species of fish and other organisms.

Functionally, mangrove and fish are inextricably linked. This ecological relationship between the fish and mangroves is, however, not generally reflected in mangrove or fisheries management policy. Both mangrove and fish populations are being depleted and face increasing pressure from human exploitation, in part through forestry activities. Both resources would benefit from improved management and conservation. By linking mangrove and fisheries management, sustainable use and better protection for both would be provided.

Mangrove typology and values

Mangrove environmental setting

Mangrove trees and shrubs form conspicuous wetland ecosystems fringing extensive areas of coastline in tropical and subtropical latitudes. In addition to the mangrove forest itself, waterways such as estuaries, creeks, canals, lagoons and backwaters, as well as mudflats, salt pans and islands contribute to the physical dimension of these ecosystems (Kjerfe 1990). True mangroves are mainly restricted to inter-tidal areas between the high water levels of neap and spring tides. Under optimal conditions, such as those found in tropical river deltas, estuaries and lagoons, mangrove trees can reach a height of 45 m, and as such they create a valuable timber resource (Tomlinson 1986; UNEP 1994).

Abiotic characteristics

Mangrove soils in Indonesia are of recent marine alluvial origin, and have been transported as sediment and deposited by rivers and the sea. They have been classified by geographers as the 'Kranji series'. These soils are made up of sand, silt and clay in different combinations. The 'mud' in mangrove areas actually refers to a mixture of silt and clay that is rich in organic detrital matter.

Dissolved calcium from shells and offshore coral makes the brackish mangrove water alkaline. Mangrove soils, however, are neutral to slightly acidic due to the sulphur-reducing bacteria, and the presence of acidic clays.

The amount of dissolved oxygen in mangrove water is generally lower than that of the open sea. In areas of organic pollution, this low content may be depressed further, to the point of creating an anoxic zone in the water column. Decay and respiration by bacteria use up the interstitial oxygen in the soil.

The circulation of tidal water and exchange with the atmosphere replenishes the oxygen content of only the first few millimetres of soil. Below that, the organic content and fine particle size of mud result in anoxic conditions tolerated only by anaerobic bacteria which break down organic materials and produce hydrogen sulphide.

Materials from primary producers are passed on to the community and eventually, to the detrital pool via the breakdown of leaf litter and wood. This is, in part, accomplished by the action of grazing herbivores, which accelerate the transfer of energy to detrital feeders.

However, nutrients are not solely produced from within the ecosystem, autochthonous materials, but are also derived externally from allochthonous materials, imported from rivers and the sea. Rain regularly flushes detritus down from rivers to the mangrove areas. The sea brings in dissolved material, suspended organic matter and microscopic organisms, the latter two of which are consumed by filter feeders during high tide. The receding sea drains through soil, which acts as a sieve, leaving a layer of microscopic organisms deposited on the surface. The emerging terrestrial fauna graze on this layer during the low tide. This is a two-way process, with nutrients also exposed to the sea and lost with the receding tide.

The conditions of light, temperature and humidity within the mangrove forest and on the mudflat are very different. Mudflats are exposed to sunlight during diurnal low tides and become very hot and highly reflective, whereas the forest canopy shades the mangrove floor and keeps it cool. The relative humidity of the mangrove forest does not, however, approach that of an inland forest like 'Bukit Timah'. The mangrove forest is not as dense as an inland forest and is more permeable to wind. It may be dried out. In 1997, the haze over Singapore caused by forest fires in Indonesia reduced light intensity considerably, thus lowering ambient temperatures in the mangroves.

In Indonesia, high water (HW), as rising or flood tides, alternates twice a day with low waters (LW) as receding or ebb tides. Caused by the gravitational pull of the moon and centrifugal forces of the earth's rotation, tides are further modified by local geography.

Indonesia experiences predominantly mixed semi-diurnal tides. The two high tides and two low tides, within the day, are not of equal height. Tide times shift by an additional 50 minutes a day since they are based on a lunar day, which is 24 hours 50 minutes long.

When the moon and sun are aligned every 2 weeks during a full or a new moon, the resultant spring tides (usually experienced 2 days later) are high water spring tides (HWST) and low water spring tides (LWST). The average between the two may be considered the tidal range. In Singapore it reaches a maximum of about 3.5 m. There are shorelines in other countries that are subject to a tidal range of 10 m.

The area of seashore lying between the highest high water spring tide (HHWST) and the lowest low water spring tide (LLWST) is the inter-tidal zone. The level exactly in between the two extreme tides can be taken to be the mid-tide level (MTL), and mangrove forests grow between MTL and HHWST.

Spring tides occur every 2 weeks, and between these, the sun and moon approach at right angles to each other. Their gravitational effects are partially cancelled out and produce alternating neap tides, resulting in high water neap tides (HWNT) and low water neap tides (LWNT). When the moon is in its first and third quarter, the lowest tidal range is reached, and may shrink to as little as 0.6 m.

The degree of salinity may be categorized into oligohaline waters of low salinity (0.5–5 parts per thousand, ppt), mesohaline waters of intermediate salinity (5–30 ppt) and polyhaline water of high salinity (18–30 ppt). The term brackish water actually refers to conditions that range from oligohaline to weakly mesohaline. Specific readings of salinity within a mangrove may range from 0.5 to 35 ppt. This variation is caused by the inflow and uneven mixing of polyhaline and low salinity water in the mangrove forest. Salinity also varies with estuarine depth, because the heavier, more saline water tends to sink. When the sea recedes, tide pools can become hypersaline (>30 ppt), due to evaporation. This may occur especially during the long exposure caused by low water spring tides. Inside the mangrove, however, the influence of freshwater runoff from the land becomes significant, particularly during monsoons. Small streams in the mangrove are oligohaline, and further inland, some are freshwater. In our narrow mangroves, the effect of freshwater inflow is considerable.

Biotic characteristics

It is not easy to define 'mangrove' using objective criteria, as classification of vegetation is subjective. The somewhat arbitrary classification of Tomlinson (1986) has been adopted with species categorized as major and minor elements of the mangrove community or as associates.

According to Soerianegara and Indrawan (1984) key characteristics of mangrove forests in Indonesia are:

(1) uninfluenced by climatic factors
(2) influenced by tidal conditions
(3) located on soils, mainly clay plus mud and sand, flooded by seawater
(4) located in low coastal areas
(5) unstructured in layer/stratification of the forest stand
(6) made up of trees that are up to 30 m high
(7) composed of tree associations that, going from the sea inland, are: *Avicennia, Sonneratia, Rhizophora, Rhizophora/Bruguiera, Bruguiera, Xylocarpus, Lumnitzera* and *Nypa fruticans*
(8) composed of ground cover of the following species: *Acrostichum aureum, Acanthus ilicifolius, A. ebracteatus.*

Mangrove forest vegetation is very diverse in Indonesia with different factors affecting zonation and diversity (Table 28.1). It includes 202 species, consisting of 89 trees,

Table 28.1 Environmental factors that affect the zoning of some mangrove vegetation growth

No.	Botanical name	Salinity (ppt)	Tolerance against wave and wind	Tolerance against mud	Flooding frequency
1	*Rhizophora mucronata*	10–30	High	High	29 days/month
2	*Rhizospora apiculata*	10–30	Moderate	High	A couple of days/month
3	*Bruguiera gymnorrhiza*	10–30	Low	High	A couple of days/month
4	*Lumnitzera littoralis*	10–30	Very low	Moderate	A couple of days/month
5	*Bruguiera parviflora*	10–30	Low	High	<9 days/month
6	*Rhizophora stylosa*	10–30	Moderate	High	<9 days/month
7	*Sonneratia alba*	10–30	Moderate	High	10–19 days/month
8	*Sonneratia caseolaris*	10–30	Moderate	High	10–19 days/month
9	*Avicennia* spp.	10–30	Moderate	High	10–19 days/month
10	*Xylocarpus granatum*	10–30	Low	Moderate	9 days/month

Adapted from Bengen (2002).

5 species of palm and 19 species of liana, 44 species of epiphytes and 1 species of *Cycas*. However, there are only about 47 specific plant species commonly found in the true mangrove forest. There is at least one dominant tree species in the mangrove forest from each of the following families: Rhizophoraceae (*Rhizophora, Bruguiera* and *Ceriops*), Sonneratiaceae (*Sonneratia*), Avicenniaceae (*Avicennia*) and Meliaceae (*Xylocarpus*) (Bengen 2002).

Following Tomlinson (1986), the major mangrove species possess the following attributes:

(1) They are obligate inhabitants of the mangrove ecosystem and cannot be found elsewhere.
(2) They have a major role in the structure of the mangrove community and can form pure stands.
(3) They are morphologically adapted to their environment (e.g. having aerial roots and vivipary of the embryo).
(4) They can withstand saline conditions, with a physiological mechanism for salt exclusion (e.g. by excretion) so that they can grow in sea water.
(5) They are taxonomically distinct from terrestrial relatives, and separated at least at the generic level.

Mangrove distribution and condition

Mangroves are found throughout the Indonesian archipelago in 22 provinces (Fig. 28.1), but are concentrated mainly in Papua, East Kalimantan, South Kalimantan, Riau and

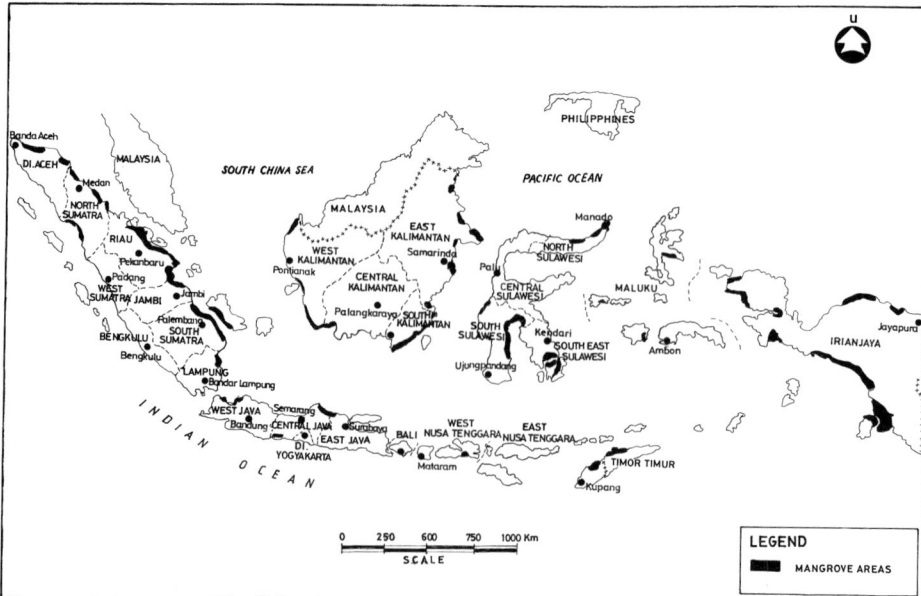

Fig. 28.1 Distribution of mangrove forest in Indonesia (modified from Anonymous 1996).

South Sumatra (Table 28.2). Indonesia's mangroves remain the most extensive in the Southeast Asia region, covering some 2.43 million hectares, with about 75% of the total amount concentrated in Papua, East Kalimantan and East Sumatra (Table 28.2).

Table 28.2 Mangrove forest areas in Indonesia by province (in hectares)

Province	PIPRAN (1)	PHPA-AWB (1987) (1)	RePPPRrot (1985–1989) (1)	GIESEN (1993) (1)	RLPS (1999) (2)
Aceh	54,335	555,000	59,400	20,000	31,503
North Sumatra	60,000	60,000	86,800	30,750	386
Jambi	65,000	50,000	18,000	4,050	3,294
Riau	276,000	470,000	239,900	184,400	63,953
South Sumatra	195,000	110,000	240,700	231,025	262,832
Bengkulu	0	20,000	2,100	<2,000	2,610
West Sumatra	0	0	3,000	1,800	4,850
Lampung	17,000	3,000	31,800	11,000	0
West Kalimantan	40,000	60,000	205,400	40,000	0
Central Kalimantan	10,000	20,000	28,700	20,000	256,109
East Kalimantan	266,800	750,000	667,800	266,800	11,156
South Kalimantan	66,650	90,000	112,300	66,650	0
South Sulawesi	66,000	55,000	67,200	34,000	104,030
Southeast Sulawesi	29,000	25,000	100,900	39,000	70,840
North Sulawesi	4,833	10,000	27,300	4,833	38,150
Central Sulawesi	0	0	42,000	7,100	78,840
Maluku	100,000	46,500	212,100	100,000	148,710
West Java & Jakarta	28,608	5,700	8,200	<5,000	77
Central Java	13,576	1,000	18,700	13,577	2,906
East Java	7,750	500	6,900	500	0
Bali	1,950	500	500	<500	501
West Nusa Tenggara	3,678	0	6,700	4,500	9,295
East Nusa Tenggara	1,830	21,500	20,700	20,700	10,780
Papua	2,943,000	1,382,000	1,583,300	1,382,000	1,326,990
Total	4,251,010		3,790,400	2,490,185	2,427,812

Source: (1) Dahuri *et al.* (1996); (2) Santoso (2001).

Highly accelerated economic development and population growth, within the last few decades, have affected mangrove forests directly and indirectly. The degradation of mangroves has adversely affected habitat quality as well as biodiversity. In the Indonesian coastal zone, where the population growth is twice the national average, rapid degradation of coastal and marine ecosystems has occurred. This is historically evident in the western part of Indonesia in the inter-tidal coastal wetlands where mangrove ecosystems once dominated. In provinces of Java and Sumatra, more than 90% of original mangrove cover has been lost since the 1940s, with most of that loss caused by the proliferation of 'tambak' (shrimp and fish ponds) since the 1970s.

Even in more remote areas of Eastern Indonesia there is still increasing and unprecedented loss of mangrove forest. For example, from 1982 to 1999, the mangrove forest in Papua was drastically reduced from 2,943,000 hectares to a mere 1,326,990 hectares (Table 28.2). This 1.6 million hectares of mangrove degradation or loss has occurred at a rate of about 94,000 hectares per annum. The degradation of Indonesian mangrove has reached an alarming level, particularly in Papua. Worldwide trends indicate that international interest in the conservation of mangrove ecosystems has increased. Among other things, this concern has led to the establishment of the International Society for Mangrove Ecosystem (ISME) with its headquarters in Okinawa, Japan. ISME has proposed a 'Deeds for Mangrove' programme as part of the world commitment for nature, declared in 1982, by the United Nations Organization. Indonesia, having one of the largest mangrove ecosystems in the world, has acknowledged the importance of mangrove management based on the conservation and sustainability aspects of ecological and economic functions of the ecosystem. This principle was based on three important commitments:

(1) to 'Agenda 21' (following the World Summit on Sustainable Development in 1992) and the 'Formulation of Action Plan for Biodiversity in Indonesia';
(2) to establish a long-term development plan for the conservation and management of coastal and marine resources; and
(3) to establish a National Control Board for Mangrove Forest ecosystems with a mandate to produce a National Strategy for Mangrove Forest Management in Indonesia.

Mangrove functions

Some of the most important functions of mangroves in Indonesia's coastal areas are:

(1) protection area from erosion by waves or wind (Bengen 2002);
(2) production of organic matter so that it can enter the food chain for fish, crabs, shrimp (Bengen 2002);
(3) protection area for young fauna such as birds, bats (FAO 1982; Aksornkoe 1995; Bengen 2002), and feeding and spawning habitat for certain fish and shrimp (Martosubroto and Naamin 1978);
(4) source of industrial raw materials;
(5) supply area for larvae of fish, shrimp and other marine biota;

(6)   tourism and recreation areas (Bengen 2002).

Anwar and Subiandono (1997) divided mangrove function into three categories: physical, biological and economic. The physical functions include protecting the coastline and coastal river banks from wind and erosion. Mangroves also protect freshwater areas from intrusion of seawater, decompose organic matter, provide oxygen and absorb carbon dioxide.

The biological functions include provision of a source of nutrients for flora and fauna, and reproductive habitats for fish, shrimp, prawns, birds and other fauna. They also serve as reservoirs for biomass and genetic material. Annual litter-fall normally ranges from 10,000 to 14,000 kg dry weight per hectare. In Indonesia, an input of 6–9 tonnes of dry matter/hectare/year is suggested. Of this, 50% is exported from the mangrove forest to the offshore shallow areas (Anonymous 1996). During decomposition, mangrove litter becomes progressively enriched in protein and serves as a food source for a wide variety of filter, particulate and deposit feeders such as molluscs, crabs and polychaete worms. These primary consumers form the food of the secondary consumer level. This secondary level is usually dominated by small forage fish species, and by juveniles of the larger predatory species that form the third consumer level. In addition, there are important crustacean species, such as shrimps, which feed directly on particulate organic detritus and also feed, to some extent, upon primary consumers.

Economic functions are provision of wood, either for fuel wood or furniture, and supply of industrial materials for pulp, paper, textiles, medicine, alcohol and cosmetics. The biological roles of producing seedling fish, shrimp and prawn, etc., and providing natural habitat, also have economic value.

The organic material, exported from the mangrove habitat, is utilized in one form or another by the inhabitants of the estuaries, the near-coast water, the seagrass meadows and the coral reefs. Many commercial shrimps and fish species are supported by this food source. Because of this, the ecological functions of the mangrove are closely associated with economic values. Many sea animals living in this area (Fig. 28.2) are very dependent on the existence of the mangrove forest.

Mangrove uses

Mangrove forests have been used over wide geographic areas by people who live in or close to them, and have traditionally made a living from them for thousands of years.

In Indonesia mangrove areas have shrunk so much that many original productive uses have ceased. In the rest of the world, however, such values are very significant. The products that can be extracted for sale in local and international markets are numerous.

Sawn timber from *Heritiera* and *Xylocarpus* species is a high quality timber product, but it is becoming scarce. Unsawn poles of *Rhizophora* species (bakau piles) are the most common extraction product in the region; they are easily harvested by manual methods and have a short crop-rotation period in managed forests. Trees may be used directly for fuel or they may be converted into charcoal and then sold as fuel.

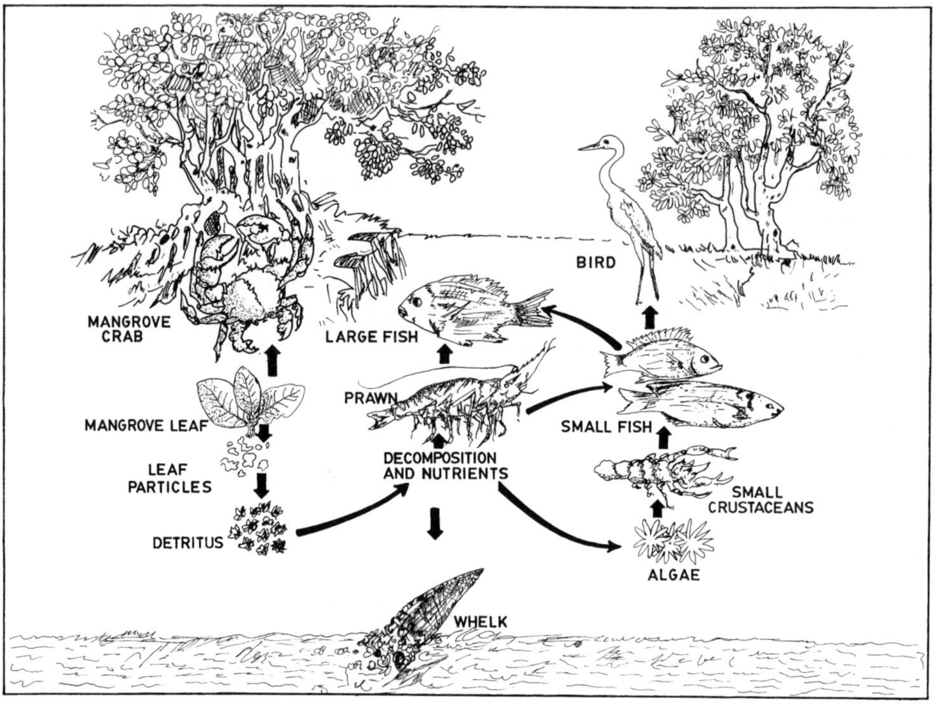

Fig. 28.2 Interdependency relationship between mangrove and fish.

Rhizophora species of wood produce relatively more heat per unit of weight than other species; they are therefore the major species exploited. Charcoal exports from Indonesia in 1980 were reported at 42,920 metric tonnes while the total production of charcoal for domestic purposes and export between 1978 and 1980 was estimated at about 52,000 metric tonnes per year (FAO 1985). Bark of mangrove trees is harvested as a source of tannin for the tanning industry. The high tannin content, found especially in species of the Rhizophoraceae, increases their resistance to herbivores. Traditionally, in Indonesia and Southeast Asia, tannin was only used by fishermen for dyeing their fishing nets. The use of tannin for this purpose has almost ceased since the introduction of nylon nets, and nowadays tannin extraction remains a small-scale operation in Indonesia. Mangrove trees are exploited for the lignocellulose for the manufacture of chipboard, pulp wood (newspaper and cardboard) or synthetic materials (e.g. rayon). In 1978, some 382,737 m³ of mangrove logs (mainly *Rhizophora* spp.) were exported from Indonesia for wood chips. Most of this came from Aceh and Riau provinces in Sumatra. The wood-chipping operation, that was clear-felling areas of mangrove in Bintuni Bay, Papua Province, was stopped by the government of Indonesia in 1990.

As pointed out already, mangrove forests provide food and habitat for prawns, crabs and fish at critical phases of their life cycle. In fact, studies have shown that when mangroves are lost, fishermen suffer substantially decreased catches of prawns and many fish species. However, a recent study of climate change influences by Mathews *et al.* (2000) has shown that, besides mangrove cover, there are marked differences in the factors that affect fisheries production. Their analysis showed two quite contrasting examples of the relative importance of shrimp culture in degraded mangrove systems, and postulated that these are due to much broader influences on fisheries production than near-shore ecosystem status.

The importance of mangrove areas as habitats for commercially important fishery species has become widely accepted. However, it should be recognized that the primary habitats of these organisms are the shallow bays, inlets and channels that are an integral part of the mangrove system. The inter-tidal mangrove forest contains few habitats that are used directly by species important to fisheries (a notable exception being the mangrove oyster). Rather, mangrove forests provide the nutritional inputs to adjacent shallow channel and bay systems that constitute the primary habitat of a large number of aquatic species of commercial, subsistence or recreational importance.

Fishery activities within or dependent upon the mangrove system rarely require any major modification of mangrove forests or associated creeks and lagoons. The mangrove forests contribute nutrients to the ecosystem, provide shelters and nurseries for fish and help support extensive aquatic species. Fish such as mullet and milkfish, which constitute an important source of high quality protein in Southeast Asia, utilize mangrove estuaries as living habitat and feeding areas. Dutton and Bengen (2001) have shown how important such fisheries are to national food security. However, the success or failure of any fishery, dependent upon the mangrove system as a basic food source, is influenced by the impact of numerous activities generally unrelated to fisheries. Many activities that occur in the mangrove catchment areas may, singly or collectively, alter the nature and productivity of the mangrove system that provides the detrital food source for mangrove organisms. This has been well described in the case of the Segara Anakan estuary in Java (Purba 1991).

Fisheries in mangrove areas are usually conducted on a small scale. The major fishery resources in these waters are detritivorous fish, crabs, crustaceans and molluscs. In Indonesia, fishermen exploiting the area use traditional gear such as fixed traps, 'sero', 'kelong', hook and line, cast nets and some gillnets; hence the low level of production per person. In 1973 the mangrove forest of Segara Anakan and its adjacent waters contributed only 7% of the total landing of Cilacap offshore fishery (Martosubroto & Sudradjat 1973). Many of the shrimp obtained from the capture fisheries are mangrove-dependent species (e.g. *Penaeus merguiensis*, *P. monodon* and *Metapenaeus* spp.). In 1977 it was estimated that the total marine capture in Indonesia was about 1,489,000 metric tonnes, of which mangrove-dependent species constituted 3% or 46,638 tonnes. Of this, 7,717 tonnes were *Penaeus monodon*, 24,346 tonnes were *P. merguiensis*, 13,848 tonnes were *Metapenaeus* spp. and 728 tonnes were mangrove crabs, *Scylla serrata*.

Mangroves also support artisanal fisheries. People who live in or near mangrove forests catch fish, shrimps, crabs and molluscs daily from around the estuarine areas. There are no reliable data on the quantities caught, although the predominant species of the fish catch are mullet (*Mugil dussumieri*), sea bass (*Lates calcarifer*), tilapia (*Tilapia mossambica*), snake eel (*Ophichthus microcephalus*), catfish eel (*Plotosus canius*) and milkfish (*Chanos chanos*). The shrimp caught are as described above and the important species of molluscs are cockles (*Anadara* spp.) and oysters (*Crassostrea commercialis*).

Aquaculture is widely practised in the mangrove areas of Asian countries. The most common form of coastal aquaculture in Indonesia is pond culture or 'tambak', which is widely practised in Java, Sumatra, South Sulawesi and Kalimantan. The production of shrimp species from brackish water pond cultivation, in particular, has spread rapidly in Southeast Asia. It rose from an estimated 60,000 tonnes in 1980 to 500,000 tonnes in 1990 (FAO 1994). Brackish water shrimp production in Indonesia (*Penaeus monodon, P. merguiensis* and *Metapenaeus* spp.) increased from 27,595 tonnes in 1983 to 105,906 tonnes in 1990. The average annual growth rate over the period was 24%, peaking in 1987 at 47%. In 1990 growth in production declined to 9%, indicating a slow-down of development in the industry, but has accelerated again since then, particularly since the Asian economic crisis in 1997.

Threats to mangroves

With growing population and economic development, there is increasing pressure to develop mangrove areas for fisheries, residential, commercial and recreational purposes. Mangrove ecosystems may be destroyed directly by removal, or indirectly as the result of activities elsewhere.

Saenger *et al.* (1983) attribute the general cause of mangrove destruction and degradation to the preference for the short-term exploitation for immediate economic benefit, rather than longer-term, sustainable exploitation. Expanding on their basic classification, three specific types of human interference causing mangrove destruction are:

(1) over-exploitation by traditional users, e.g. excessive removal of trees for fuel wood, especially charcoal;
(2) activities requiring maintenance of the mangrove ecosystem, e.g. rotational felling and replanting of mangrove stands for wood production; and
(3) natural resource activities, e.g. coastal agriculture, salt production and intensive shrimp culture.

The conversion of mangrove areas to shrimp ponds or 'tambak' (Fig. 28.3) represents the single largest threat to the mangrove ecosystem in Indonesia. In 1977 it was estimated that tambak covered about 174,605 ha in Indonesia. By 1993, this estimate had risen to 268,743 ha, an increase of 54%. The high price of prawns on the international market, and the need to encourage export commodities in Indonesia in the wake of 'krismon' (the Asian economic crisis), are putting further pressure on the mangrove ecosystem.

The importance of prawn exports to the national economy is recognized. The decline in productivity of many tambak, however, and the possible linkages between this decline and such factors as pond location, pond 'mismanagement' and increased incidence of disease, suggest that any further tambak development in mangrove areas requires detailed land suitability studies. Such studies should help to avoid location in

A

CHANNEL FOR
FISH CULTURE

SLUICE

MANGROVE

DYKE : -TOP SIZE : 1 – 3 M
        - BOTTOM SIZE : 2 – 3 M

B   CHANNEL FOR FISH CULTURE
    MAXIMUM WIDTH 5 METERS

SLUICE

MANGROVE

SLUICE

C

SLUICE   CHANNEL

DYKE

MANGROVE

FISH POND

SLUICE

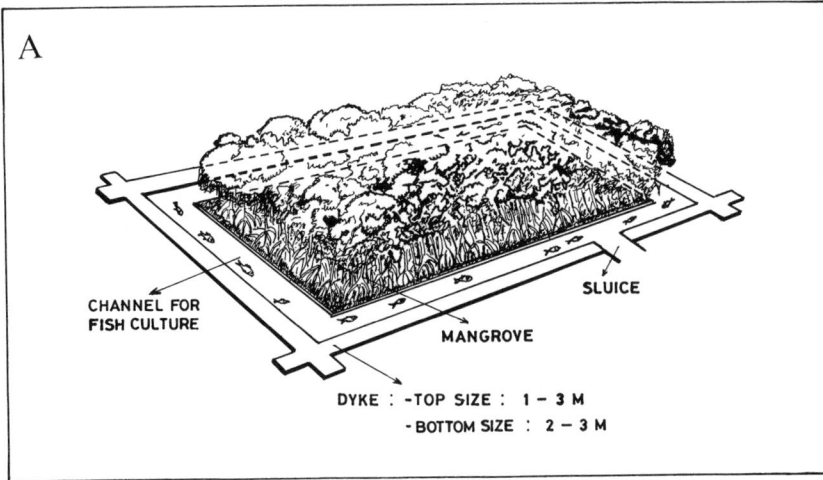

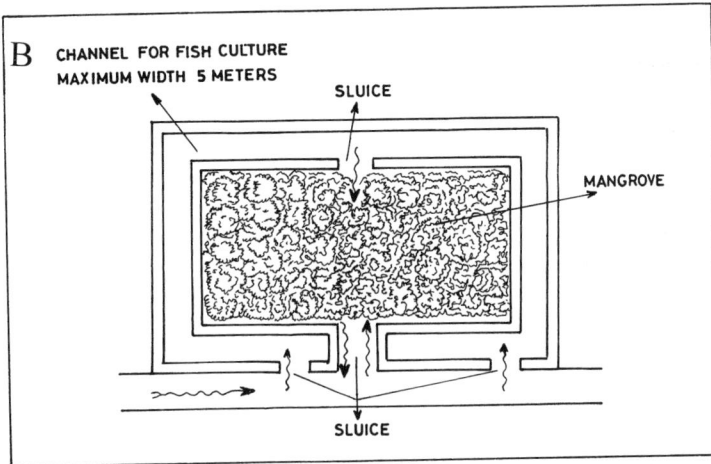

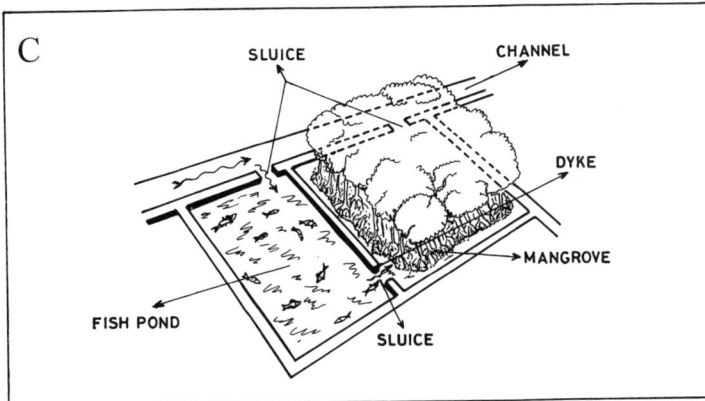

Fig. 28.3 (A) Silvo-fishery of 'Empang Parit' model; (B) silvo-fishery of 'Empang Parit disempurnakan' model; (C) silvo-fishery of 'Komplangan' model.

acid sulphate, or potential acid sulphate, areas. Such conditions are very prevalent in the mangrove areas. A further condition of development must be through cost–benefit analyses of potential tambak areas.

Looking at the degradation of mangrove forest ecosystem, we can conclude that there are two major kinds of pressure causing degradation. The first, an 'external' pressure, is the temptation to convert mangrove forest for other functions. The pressure originates from parties developing projects that do not depend on the mangrove ecosystem, such as government and private companies intending to develop coastal zones for housing, airports, shrimp ponds, industrial zones and recreation resorts. The second, an 'internal' pressure, comes from heavy demand to utilize the natural resources found in mangrove ecosystem. This originates from communities within and surrounding the mangroves. It usually reflects the poor economic conditions of that particular community and its high dependence on the mangrove resources for its livelihood. Furthermore, inadequate education results in poor understanding of the importance of the mangrove forest ecosystem.

Mangrove management strategy

Mangrove protection and rehabilitation

Looking at the framework for the management and conservation of mangrove forest ecosystems, there are two major concepts that can be applied: mangrove forest protection and the mangrove forest rehabilitation. Both concepts basically provide validation and understanding that mangrove ecosystems urgently require protection and proper management in order to make them sustainable.

Mangrove forest protection

One practical method that can be applied to ensure the existence of mangrove forest is to designate certain mangrove forest areas as forest conservation zones, thus making mangrove forest a green belt along the coastline, in estuaries and on river banks. This kind of mangrove protection proved to be successful and quite effective, as demonstrated at Rambut Island and Dua Island (in West Java), which have been assigned wildlife sanctuary status (i.e. as bird habitat). Another method is to designate areas such as Tanjung Puting National Park in Kalimantan and the Sali Zone in West Bali, which include mangrove forest, as protected forest zones (Kusmana & Suhendang 1997).

The designation of mangrove forest areas as protected zones was reinforced by the joint decree between the Minister of Agriculture and the Minister of Forestry dated 30 April 1984 (number KB. 550/264/4/1984), which states that the mangrove forest should be retained in a 200-m coastal green belt. The major goal of this joint decree was to legalize the protection of mangrove forest. However, it was also meant to harmonize regulations on mangrove protection between related government agencies or institutions. Technical guidelines for the implementation of the joint decree were 'issued' by the Ministry of Forestry in its Internal Circular (Surat Edaran) (number 507/IV – BPHH/1990). This would, among other things, regulate the width of green belt in mangrove forest, requiring at least 200 m along the coastline and 50 m along the banks of rivers.

Soerianegara *et al.* (1986) suggested a formula to determine the suitable width of green belt in mangrove forest along the coastline. The formula was derived from ecological studies in conjunction with knowledge of production capacity of mangrove forest in terms of organic substances, as well as productivity in terms of fish and shrimp. The suggested formula was a function between amplitude of the tidal range, multiplied by a constant of 130.

Mangrove forest rehabilitation

Reforestation activity has been carried out on denuded forest areas not only to restore the aesthetic value of the areas, but most importantly to successfully reactivate the ecological function of the area. Perum Perhutani (a state-owned forest company) has used this to preserve or restore the ecological function of several mangrove forests that had been cut down and converted to other uses. Rehabilitation of mangrove forest was pioneered in 1960 in the northern coast of Java (Kusmana & Suhendang 1997).

About 20,000 ha of damaged mangrove in the northern coast of Java has been reported as successfully rehabilitated, with *Rhizophora* spp. and *Avicennia* spp. used as the major species. About 60–70% of the planted vegetation was reported as surviving or alive (Soemodihardjo & Soerianegara 1989). Similar activity had also been carried out on approximately 105 ha of mangrove forest in Cilacap, on the south coast of Central Java, which was successfully rehabilitated with *Rhizophora* spp. and *Bruguiera* spp. as the major species.

In addition, it was reported that 'PT Bina Lestari' achieved good results in the reforestation of its forest concession in Riau, using *Rhizophora apiculata* and *Bruguiera sexangula* as the major species at 2 m × 2 m planting distance. Since 1988/1989, approximately 256 ha have reportedly been reforested.

Community considerations in mangrove management

The conservation of mangrove forest has proven to be highly complex and it requires accommodation of and tolerance towards all parties concerned (i.e. stakeholders), both those living in the surrounding area and those from outside the area. Ultimately mangrove forest conservation is conducted to fulfil the various needs of stakeholders. However, accommodation of stakeholders will be more beneficial if strong consideration is given to parties that are sensitive about the disturbance of mangrove resources; i.e. paying greater attention toward the community needs.

Therefore, it is important to develop, within the community, a good understanding of the benefits and the importance of mangrove forest resources. In this context, the understanding of the importance of mangrove forest ecosystems to coastal communities is an important and current issue. According to research carried out by the Institute for the Study and Development of Mangrove (LPPM) on a community in Segara

Anakan, near Cilacap, Central Java, it was found that the community's perception of the mangrove forest was still limited. This was reflected by the fact that about 90% of the community did not differentiate between the mangrove forest as an ecosystem and the individual mangrove tree species.

The level and type of mangrove forest utilization by a community depends on the needs and the livelihood of the community. An example is the utilization of mangroves for firewood. When it was done for local daily consumption, the use rate was between 0.5 m³ and 1.5 m³ per day. However, when it was cut for trading in the market, the community harvest was between 5 and 12 m³ per day. In regard to conservation activities, LPPM (1998) found that the community did not carry out mangrove reforestation because:

(1)   they did not know how to plant mangrove;
(2)   the location was too distant;
(3)   there were no mangrove seedlings at hand; and
(4)   the community preferred cultivating paddy and other food crops to planting mangrove vegetation.

Based on the above, we can derive a preliminary conclusion that a basic problem of sustainable management of mangrove forest is how to combine the ecological needs (mangrove forest conservation) with the sociological and economic demands of the community. For conservation to be applied, it should have the capacity to address the socio-economic problems of the related community in addition to mangrove conservation.

One of the important strategies widely applied in Indonesia now for natural resources management, including mangrove ecosystems, is *community-based management*. Rahardjo (1996) states that 'community-based management' means direct involvement of the community in managing natural resources in a certain area or zone. This can take many forms as Dutton (2002) notes in an overview of community-based partnerships. The term 'management' here means that the community is actively involved in conceptualization, planning and implementing actions, as well as evaluating and monitoring the progress and results. 'Community-based management' implies that a 'bottom-up approach' is required instead of the usual ' top-down approach'.

By involving the community in the management process, it is more likely that their needs, worries, problems and aspirations can be adequately addressed. Hence, the proposed activities must be in accordance with what is really needed by the concerned community. It is recognized that the nature of communities varies widely, hence the demands of, and approach to, each kind of community will be specific and unique in each case. Therefore, we can accept that there is no single methodology that will be applicable to every kind of community all the time. Thus, we should pay proper attention to the characteristics of each community living in a mangrove forest area. It is often said that one of the causes of destruction of natural resources is the damaging method used by communities in fulfilling their livelihood. Therefore, in this situation some sort of feasible livelihood alternative has to be offered in order to reduce negative disturbance on the natural resources in that area.

A community-based development can be established if people collaborate closely as a group. There is need for an awareness that they cannot do a task or reach their goals as individuals, due to the nature of the job or the goal(s) and the limitation of natural resources. Having unified goal(s), problems and resource scarcity usually makes the community function together. Once a unified feeling has been established, that leads to solidarity, trust, establishment of regulation and social rules, a foundation for a community-based action(s) has been laid for further development. Rahardjo (1996) provided some characteristics of a successful community-based group as follows:

(1) The benefit derived is greater than the costs or 'sacrifice' involved. If the result of the effort is 'negative' or in opposition to the above, then the community will be reluctant to participate, will avoid meetings and will not be intensively involved in conducting the programme. Profit or benefit may take the form of economic returns or sociological values such as knowledge, skills in problem-solving, improved health conditions, etc.

(2) The activity or programme is conducted as a common need or common demand. Whenever the community does not consider an issue to be a common problem, the members will not take part in addressing the issue. For example, the issue may affect only a certain part of the community such as women, non-wealthy or unskilled persons. However, if that particular issue is accepted as a communal problem, then the probability of a successful 'community-based' approach will be much greater.

(3) The community-based group can be attached to the existing institution and social organization. In such a case, there may be less risk of rejection of newly introduced matters.

(4) The particular community-based group acquires the needed capacity, skills leadership, knowledge and capability for carrying out their duties.

(5) The regulations and proper operational procedures are established by the community-based groups. Procedures will work if community members accept, acknowledge and abide by the regulations and operational procedures and if there are forces to have people obey the rules. If, for some reason, its members do not understand or are unwilling to abide by the rules/regulations and operational procedures, this would indicate that the group, as a community group, is fragile.

Other than the above-mentioned characteristics, there is another factor that plays an important role in the management and constitutes the core of community-based development, i.e. human behaviour. We understand that, through human action, people interact with each other as well as with their natural surroundings. Where aspects of human behaviour could endanger the sustainability of natural resource ecosystem, it is necessary to understand how to change human attitudes towards the environment. It is especially important to know how to inculcate a positive and 'eco-friendly' attitude toward the environment, and to encourage people to participate in the conservation of the surrounding ecosystem.

Basically speaking, in order to make the community committed to, and involved in, the conservation of mangrove forests, it is necessary to implement an incentive system that will initiate or accelerate managerial action or efforts for responsible management. The incentive system may take many forms, including the improvement of human resources quality. There are management skills, legal tools, public education and community group structures needed in order to improve the strength and capacity of the community for management of mangrove forest resources:

(1) *Training in mangrove forest management.* Materials needed to improve both the knowledge and the skill of the community in mangrove forest management are: information on benefits and functions of mangrove forests, establishment and uses of various mangrove vegetation species, techniques for the selection of mangrove vegetation fruits and seeds, and maintenance and harvesting of the trees.

(2) *Laws and regulations.* Materials that should be disseminated to the community are: Law no. 1967 on the basic description of forestry; Law no. 24, 1997 on utilization zoning; Law no. 23, 1992 on the Management of Environment; Government Decree no. 28, 1985, on Forestry Protection; Government Decree no. 18, 1994 on Nature Tourism Enterprises in Utilization Zone of all Marine Parks, National Parks, Great Forest Parks, and Nature Tourism Parks.

(3) *Training for fisheries intensification.* Activities carried out should cover services such as training and provision of demonstration sites for fishery aquaculture and mariculture. They should include training on introduction of culture species, rearing methods, construction techniques for containment areas and ponds, silvo-fishery models, feedstock ration delivery, pest and disease control, and harvesting and post-harvest treatment and processing of products.

(4) *Training in agricultural intensification.* In order to improve agriculture productivity, an information distribution system must be established that will help people to choose the most economically beneficial and technically viable production systems in mangrove areas, and in areas developed as an alternative to mangrove exploitation.

(5) *Establishment of community self-help groups.* A community self-help group is needed to ensure local community participation in conducting mangrove forest conservation programmes. Various programmes and activities can be conducted through this kind of group. They include dissemination of information on the following: rehabilitation of mangrove forest ecosystems, laws and regulations, fishery mariculture, silviculture, aquaculture and agricultural cultivation techniques. This group could also play a role in facilitating and initiating the community to commit and actively participate in mangrove forest ecosystem conservation activity.

(6) *Dissemination of data and information on rehabilitation and management plan for mangrove forest.* Precise information on rehabilitation and management of mangrove forests (e.g. exact location, areas, objective, goal, involved group or component, time schedule, etc.) should be delivered to the concerned community, through village authority officers or via community self-help group. Hence,

the community obtains clear and accurate information. In this way a feeling of security develops within the community, and reinforces the active participation of the community in mangrove forest management.

Improving community participation

In order to improve the effectiveness of mangrove forest management, the related community needs to be closely involved in the planning and management of sustainable mangrove forests. By applying community-based and market-based management approaches, it is hoped that every plan formulated benefits from the aspirations of the community. Environmentally friendly social-cultural methods found in the local community can be developed further. This may take the form of extension services, information dissemination and initiating the awareness of the community to participate actively in mangrove forest management. This kind of approach can be carried out in two ways.

Participation Planning Programme for Rural Community Development (abbreviated as P3MD in Indonesia) is one of the methodologies for planning which was developed with the involvement of village community and institutions.

The Participatory Rural Appraisal approach (PRA approach) is intended to increase the participatory role of farmers and fishers in development planning, especially that related to mangrove forest ecosystems. In this case, studying local regulation of customs or cultural behaviour should be undertaken as a high priority activity. In addition, there are facilities and infrastructures that are part of the incentive system, which should be adequately provided. Some of the most important items are listed below:

(1)   Facilities for public services, such as health, transportation, potable water system, environmental sanitation, education, lighting, housing layout and renovation services.
(2)   Preparation and establishment of rules, regulations and laws that will assist the management of mangrove forests.
(3)   The provision of soft capital loans.
(4)   Building or construction of agricultural demonstration plots, involving the community (community-based activity).
(5)   Rights to utilize state-owned lands that are presently not productive and can be cultivated in a more optimal fashion.
(6)   Provision of information systems that clearly explain the mangrove forest utilization with regards to conservation, preservation and utilization principles.

Silvo-fishery model in community-based mangrove management

Community participation in mangrove conservation can be improved if the community involved can reap benefits from the conservation efforts carried out. Therefore, the need to empower the community in mangrove management through a social forestry model is crucial. The effort should be designed to utilize the mangrove forest resources in a sustainable fashion.

One of the models, 'silvo-fishery', has already been developed for quite some time. This model was first implemented in a mangrove rehabilitation site in Cikeong, Kara-wang, West Java. Nowadays, it is also well developed in Sinjai, South Sulawesi. The model presently operated is based on the utilization of mangrove forest, especially the area still under rehabilitation programme, where 20% of the land will be under ponds and the rest (80%) under mangroves.

The silvo-fishery model applied in Indonesia is favoured for mangrove rehabilita-tion programmes. It involves a simple level of silvo-fishery where a channel carrying tidal water with a bund surrounds an area of replanted *Rhizophora* spp. mangroves. This model uses the element of small-scale but usefully productive fish ponds (shrimps, fish and crabs) to encourage the preservation of mangroves. The aim is to have a fish harvest while the mangroves are maintained just as a protective vegetative cover.

In general, there are three major patterns in the silvo-fishery model (Fig. 28.3):

(1)   Silvo-fishery using a simple pond and water channel system, called 'Empang Parit'. This effort is patterned such that mangrove forest and the pond(s) are in the same flatland area and water flow is controlled by a single common sluice or gate (Fig. 28.3a).
(2)   Silvo-fishery using an improved pond and water channel system, called 'Empang Parit disempurnakan'. In this pattern, water flow for the mangrove site and the ponds is conducted through separate water channel systems (Fig. 28.3b).
(3)   Silvo-fishery based on the 'Komplangan' model. The basic feature of the 'Komp-langan' model is that the land for mangroves and the fishponds are constructed on two different sites and the water management is conducted through two dif-ferent, independent water channel systems for each of them (Fig. 28.3c)

So far, the simple pond and water channel system is the most popular one used by the mangrove community due to its simplicity and production of reasonably good returns. Fish production in simple ponds of silvo-fishery (80% mangrove forest and 20% ponds) was carried out in mangrove area around Ciasem, in Pamanukan, West Java. It produced 3.3 ton (3,263.53 kg) of fish (milkfish and shrimp) in 1998. In other words, having 150 ha of pond area in Tegal Tangkil will produce 489.5 ton (489,529.5 kg) of fish (i.e. shrimp, milkfish, tilapia, etc.) a year (Table 28.3).

In general it can be said that mangrove utilization through the silvo-fishery pat-tern is more beneficial than mangrove utilization by traditional or intensive ponds. The Segara Anakan area, in 1998, produced only 0.4 ton of shrimp per hectare. By comparison, a silvo-fishery pattern produced 1.5 ton of shrimp per hectare (almost four times more than the traditional ponds). Also, that production does not include 'incidental' production such as tilapia and milkfish.

Table 28.3   Fish production of silvo-fishery project in mangrove forest area (BKPH) of Ciasem-Pamanukan, Subang, West Java (1998)

Mangrove forest site	Area (ha)	Pond channel area (ha)	Fishery resources	Production kg/ha/year	ton/year
Tegal Tangkil	750	150	Shrimp	1,517.40	277,610.30
			Milkfish	226.26	33,939.00
			Tilapia	539.75	80,963.00
			Others	980.12	147,017.90
			Total	3,263.53	489,530.20
Muara Ciasem	800	160	Shrimp	1,466.56	234,649.80
			Milkfish	216.45	34,632.00
			Tilapia	511.13	81,781.50
			Others	1,482.03	237,125.70
			Total	3,676.17	588,189.00
Poponcol	2,123	424.6	Shrimp	409.16	173,729.10
			Milkfish	24.41	10,366.20
			Tilapia	139.76	59,341.70
			Others	81.71	34,692.70
			Total	655.04	278,129.70
Bobos	1,655.6	331.12	Shrimp	323.91	107,252.00
			Milkfish	1,112.94	368,516.00
			Tilapia	921.86	305,247.00
			Others	1,727.51	572,014.00
			Total	4,086.22	1,353,029.00

Source: Primary data.

Concluding remarks

Indonesia's mangroves remain the most extensive in the Asian region. They are found throughout the archipelago in the 22 provinces, but are concentrated in Papua, East Kalimantan and East Sumatra.

Mangrove forests are one of the most productive and biologically diverse coastal ecosystems. They supply habitats for a large number of commercially valuable fish, shellfish, invertebrates and epiphytic plants. They also provide sanctuary, breeding and nursery grounds for a variety of fish, crustaceans and molluscs.

Unfortunately the remaining large areas of mangrove in Papua, Sumatra and Kalimantan are under pressures from both direct exploitation and competing resource users. There are increasing pressures to develop mangroves for residential purposes, fisheries and agriculture. Over the past 10 years, over one million hectares of Indonesia's mangroves have been destroyed or replaced, particularly with brackish water fish and shrimp ponds (tambaks), and residential or industrial developments.

It is time to change the way we think about mangrove-dependent fisheries, and recognize the critical part of the energy transfer system that occurs within mangroves. We often miss this point when thinking about fisheries management. Well-managed mangrove forests do more than support fish.

Fisheries management that includes proactive mangrove management coupled with adaptive mitigation where necessary will strengthen coastal systems for fish that share these same ecosystems, and for the people who depend upon both.

Management actions for fisheries fall into ecosystem and socio-system management. As mangroves losses accumulate, a disregard for ecosystem management hampers and diminishes socio-system management. Fisheries management must include mangrove management as the primary component. Indonesian approaches to this are still in their infancy and lack adequate inter-sectoral integration. It will be vital to accelerate conservation and education efforts at all levels if we are to be ultimately effective in protecting the many values of Indonesia's vast mangrove estate.

References

Aksornkoe, S. (1995) Ecology and biodiversity of mangrove. *Proceedings of Ecology and Management of Mangrove Reforestation and Regeneration in East and Southeast Asia.* Thailand, 18–22 January, 1995.

Anonymous (1996) *National Strategy for Mangrove Management in Indonesia.* Office of the Minister of Environment, Department of Forestry, Indonesian Institute of Science, Department of Home Affairs and The Mangrove Foundation, Jakarta.

Anwar, M. & Subiandono (1997) *Technical Guidelines of Mangrove Planting.* Cooperation between Research and Development Center of Forest and Natural Conservation in Bogor and Surakarta, Indonesia.

Bengen, D.G. (2001) *Sinopsis Ekosistem dan Sumberdaya Alam Pesisir dan Laut.* Pusat Kajian Sumberdaya Pesisir dan Lautan, Institut Pertanian Bogor.

Bengen, D.G. (2002) *Pedoman Teknis Pengenalan dan Pengelolaan Ekosistem Mangrove.* Pusat Kajian Sumberdaya Pesisir dan Lautan, Institut Pertanian Bogor.

Dahuri, R., Rais, J., Ginting, S.P. & Sitepu, M.J. (1996) *Integrated Coastal and Marine Resource Management.* Pradnya Paramita, Jakarta.

Directorate General of Fishery (1979) *Fisheries Statistics of Indonesia,* Jakarta.

Dutton, I.M. (2002) Engaging communities as partners in conservation and development. *Van Zorge Report on Indonesia,* III, 24–32.

Dutton, I.M. & Bengen, D.G. (2001) The significance of coastal resources to food security in Indonesia. *InterCoast,* 38, 4–5.

FAO (1982) *Management and Utilization of Mangrove in Asia and the Pacific.* FAO Environment Paper 3, Rome, Italy.

FAO (1985) *Mangrove Management in Thailand, Malaysia and Indonesia.* FAO Environment Paper, Rome, Italy.

FAO (1994) *Mangrove Forest Management Guidelines.* FAO Forestry Paper 117, Rome, Italy.

Hinrichsen, D. (1998) *Coastal Waters of the World: Trends, Threats, and Strategies.* Island Press, Washington DC.

Kjerfe, B. (1990) UNESCO/UNDP Region Project. *Manual for Investigation of Hydrological Processes in Mangrove Ecosystem.*

Kusmana, C. & Suhendang, E. (1997) *Kelestarian Dalam Pengelolaan Hutan Mangrove Lestari*. Paper for Training on Sustainable Mangrove Forest Management. Cooperation between Directorate General for Regional Development, Ministry of Home Affairs and CCMRS-IPB, Bogor, 18 August–18 October, 1997.

Lembaga Pengkajian dan Pengembangan Mangrove (LPPM) (1998) *Pengembangan Peran Serta Masyarakat Dalam Pengelolaan Hutan Mangrove di Kawasan Segara Anakan*. LPPM, Jakarta.

Martosubroto, P. & Naamin, N. (1978) Relationship between tidal forests (mangrove) and commercial shrimp production in Indonesia. *Marine Fisheries Research Institute*, **18**, 81–6.

Martosubroto, P. & Sudradjat (1973) *A Study on some Ecological Aspects and Fisheries of Segara Anakan in Indonesia*. Marine Fisheries Research Institute.

Mathews, C., Cholik, F., Badrudin, M. & Willoghby, N.G. (2000) The effects of El Niño on shrimp fisheries in Indonesia. *InterCoast*, **35**, 2–3.

Purba, M. (1991) Impact of high sedimentation rates on the coastal resources of Segara Anakan. In: *Towards an Integrated Management of Tropical Coastal Resources*, (ed. L.M. Chou). National University of Singapore and ICLARM, Manila.

Rahardjo, Y. (1996) *Community Based Management di Wilayah Pesisir*. Pelatihan Perencanaan Wilayah Pesisir Secara Terpadu. Pusat Kajian Sumberdaya Pesisir dan Lautan, Institut Pertanian Bogor.

Saenger, P., Hegerl, E.J. & Daviel, J.D.S. (1983) *Global Status of Mangrove Ecosystems*. IUCN Gland, Switzerland.

Santoso, N. (2001) Neraca Sumberdaya Hutan Mangrove. In: *Pedoman Umum Penyusunan Neraca Sumberdaya Alam Kelautan Spasial*. Pusat Survei Sumberdaya Alam, Badan Koordinasi Survei dan Pemetaan Nasional, Bogor.

Soemodihardjo, S. & Soerianegara, I. (1989) The status of mangrove forests in Indonesia. In: *Symposium on Mangrove Management: Its Ecological and Economic Considerations*, (eds I. Soerianegara, D.M. Sitompul & U. Rosalina). *Biotrop Special Publication*, **37**, 73–114.

Soerianegara, I. & Indrawan, I. (1984) *Ekologi Hutan Indonesia*. Jurusan manajemen Hutan, Fakultas Kehutanan, IPB. Bogor.

Soerianegara, I., Naamin, S. Hardjowigeno, A.A. & Soedomo, M. (1986) *Proceedings of Panel Discussion on the Purpose and the Borderline of the Mangrove Forest Width as Green Belt*.

Tomlinson, P.B. (1986) *The Botany of Mangroves*. Cambridge University Press, Cambridge.

UNEP (United Nations Environment Program) (1994) *Integrated Management Study for The Area of Izmir*. MAP Technical Report Series No. 84, Regional Activity Center for Priority Actions Program.

WRI (World Resources Institute) (1998) *World Resources 1998–1999: A Guide to the Global Environment*. Oxford University Press, New York.

Chapter 29
Forestry interactions – New Zealand

B.J. HICKS, G.J. GLOVA AND M.J. DUNCAN

Description of geographic region

New Zealand is a small country (about 270,000 km²) comprising an elongated archipelago with many offshore islands oriented northeasterly to southwesterly in the southwestern Pacific Ocean, with its main islands spanning a latitudinal range of about 34–47°S (Fig. 29.1). Approximately half of the country lies >300 m above sea level, with slopes often greater than 28° (Statistics New Zealand 1998). Mountain ranges dominate the landscape of the South Island, where there are large areas with permanent snowfields and glaciers. These mountains are quite recent (Pliocene age), having formed from the Indo-Australian continental plate pushing up over the Pacific plate. The North Island is hilly, and much of it is steep terrain. A small chain of volcanic mountains that are snow-covered in winter lies near the centre of the North Island; some of these volcanoes are still active. A succession of major volcanic explosions from Lake Taupo has spread pumiceous tephra over much of the centre of the island.

New Zealand's rainfall patterns are largely a result of its long and narrow land mass, its steep topography and its isolated oceanic position. The country's mountainous backbone lies directly across the path of eastward-moving anticyclones and low pressure troughs (Duncan 1992). Although the passage of these weather systems results in high and regular rainfall over much of the country, winter build-up of snow on the mountains holds back runoff that is released later as snowmelt in spring and summer (Fitzharris et al. 1992). The north of the country can be subject to tropical cyclonic weather in summer and autumn. Rainfall decreases from west (2000–10,000 mm/year) to east (600 mm/year), with evaporation exceeding rainfall in summer, particularly in the eastern areas and the inland basins of the South Island where droughts are common.

The central North Island area, where most of the major forest plantations are situated, has a rainfall of about 1300 mm/year (New Zealand Soil Bureau 1968a; Tomlinson 1992). In the North Island, snowfalls occur only at higher altitudes, and snow accumulates only on the highest peaks. New Zealand's temperate climate results in relatively cool water. The mean temperature of 256 rivers was 12.6°C, ranging from 5.1° to 21.8°C (Mosley 1982). The soils of the central North Island, formed from rhyolitic pumice that erupted about 1700 years ago, are deep, free-draining pumice, with sandy silt and loamy sand textures that developed under the previous natural podocarp forests.

Key

········· regional boundaries
 1 Dons 1986
 2 Dons 1987
 3 Evans 1993a
 4 Evans 1993b
 5 Graynoth 1979, 1992
 6 Hanchet 1990
 7 Harding & Winterbourn 1995
 8 Jowett *et al.* 1996
 9 Jowett *et al.* 1998
10 Main *et al.* 1985
11 McDowall *et al.* 1977
12 Mosley 1981
13 Purukohukohu experimental basins
14 Rowe *et al.* 1999
15 Rowe *et al.* 2002
16 Taylor & Main 1987
17 Whatawhata

Northland

North Island

Auckland

Central North Island

Lake Taupo

East coast

Hawkes Bay

Southern North Island

Nelson-Marlborough

West coast

Canterbury

South Island

N

Otago-Southland

0   200 km

Stewart Island

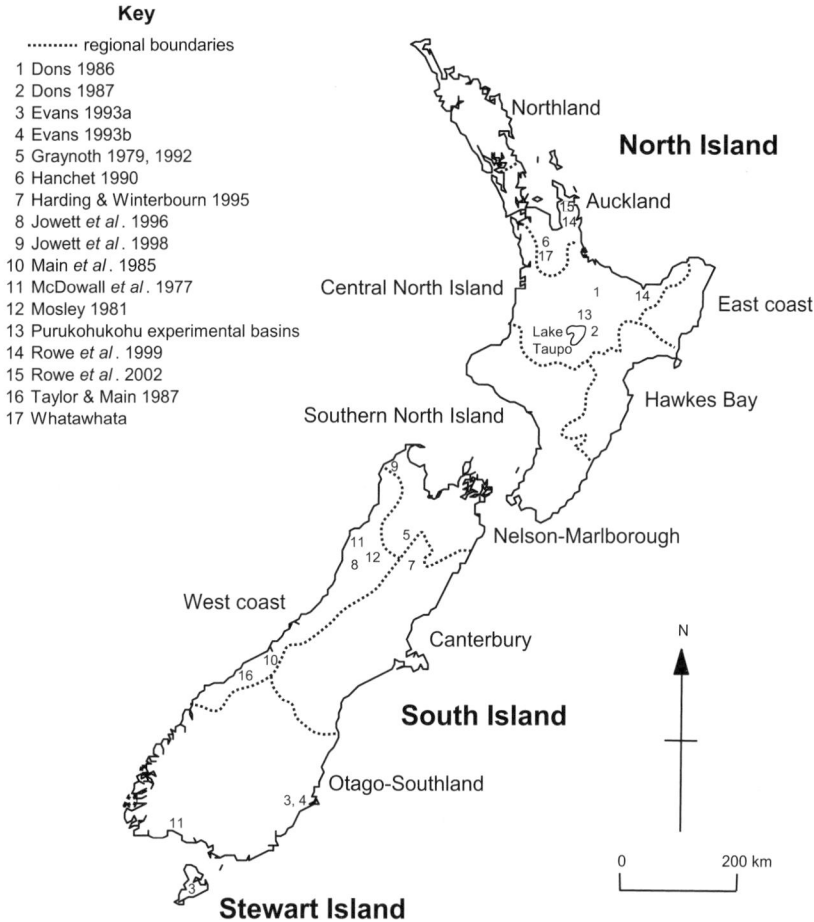

Fig. 29.1 The wood supply regions of New Zealand, showing the sites of selected watershed and aquatic ecological studies mentioned in the text.

The Nelson District of the South Island, where major forest plantations also exist, is predominantly rolling lands and hills, formed mainly from loess-covered, impervious Pleistocene gravels. Rainfall in the area ranges from 1100 to 1300 mm/year (New Zealand Soil Bureau 1968b). The soils are shallow and infertile with stony loam textures and drain slowly.

Major features of aquatic environments

Rivers

In general, the rivers in New Zealand are short, shallow and swift with gravel and boulder beds. The east coasts of both islands have braided rivers. Most rivers flow east or west from the northeast–southwest-lying mountain chains (Duncan 1987). The water is generally poorly buffered, and floods with high sediment loads are frequent. On average, New Zealand's rivers carry 400 million tonnes of sediment from the land to the ocean annually. The intermontane regions of Otago and Canterbury and the low rainfall regions of the North Island have sediment yields of 30–100 tonnes/km^2/year, whereas the other regions have yields of 200–30,000 tonnes/km^2/year (Hicks & Griffiths 1992).

The mountains in the South Island create a wet environment with high river flows on the west coast and a rain shadow with seasonal low flows in parts of the east coast (Tomlinson 1992). The drainages on the western slopes of the mountains of the North Island also periodically experience high flows.

Many of the streams and rivers of the central North Island region are partially spring-fed, with suppressed flood peaks and high, well-sustained base flows. Free-draining tephra soaks up the rainfall, releasing it slowly from springs to provide cool, clear, stable flows. In contrast, the streams draining the Nelson Pleistocene gravels are flashy and have long periods of low to no flow during summer and autumn. Replacement of remnant beech forest, scrub and pasture with plantation forests has reduced the flows in these streams by up to 70% (Duncan 1995).

Lakes

New Zealand has nearly 800 lakes, most of which are small (<5 km^2). They have diverse origins, including tectonism, vulcanism and damming by landslides, dunes and coastal bars (Lowe & Green 1987). Many of the lakes in the South Island have glacial origins, and in the North Island some of the largest lakes have volcanic origins. Lake Taupo (623 km^2), the largest lake in Australasia, is 163 m deep and is oligotrophic. The North Island also contains a diverse array of small lakes associated with sand dunes or floodplains. New Zealand lakes are generally polymictic, and very few ever freeze over in winter.

Wetlands and marshes

About 85% of New Zealand's original 670,000 ha of freshwater wetlands have been drained (Taylor & Smith 1997), mainly for conversion to pasture. The largest remaining wetlands in New Zealand are in the Waikato region where the Kopuatai Peat Dome and the Whangamarino wetland complex are of national significance. In the lower Waikato, wetlands were reduced from their original extent of nearly 200,000 ha to less than 34,000 ha by 1978 (Wardle 1991). Although the rate of loss throughout the country has slowed since this time, some wetland drainage continues with farm improvement.

Important for forestry are the wet heaths, which have ultra-infertile soils underlain by an impervious layer. In the west of the South Island, these wet heaths are known as 'pakihi', and in the north of the North Island as 'kauri gumlands', which formed following the removal of kauri (*Agathis australis*) forest (Wardle 1991).

Estuaries

There are about 300 estuaries around New Zealand (McLay 1976), covering a total area of about 100,000 ha (Taylor & Smith 1997). Most estuaries are short, and many are permanently protected from ocean waves by bars of sand or shingle. Some rivers and streams empty almost directly into the sea, and have no appreciable estuarine area. However, forests of the mangrove *Avicennia marina* var. *australasica* grow in harbours and estuaries of the North Island from about latitude 38° S northwards. Estuaries are sensitive to the accumulation of sediment from tributary streams, and this has been an issue for forestry.

General features of forests

Distribution of major vegetation types

Before human habitation, New Zealand's vegetation was principally evergreen temperate rainforest of conifers and broad-leaved trees. Indigenous forests once covered 23 million ha (85%) of New Zealand's land area; only 2.5 million ha (9%) of the land was above the tree line (Taylor & Smith 1997).

Of New Zealand's forest trees, the most important and widespread are the four endemic beech species (genus *Nothofagus*, family Fagaceae), and the 20 or so endemic species of conifers in the two families Podocarpaceae and Araucariaceae. Forests in the north were dominated by the giant kauri, of which little remains. The kauri is the sole New Zealand representative of the family Araucariaceae. It is the giant of the New Zealand forest, reaching a height of 60 m with a trunk diameter up to 7 m (Wardle 1991). Broad-leaved dominants, puketea (*Laurelia novae-zelandiae*), puriri (*Vitex lucens*), taraire (*Beilschmedia tarairi*), tawa (*B. tawa*) and towai (*Weinmannia silvicola*) also occur in the north. Throughout the country, depending on site characteristics, a variety of coniferous podocarps have become canopy trees (totara, *Podocarpus totara*; rimu, *Dacrydium cupressinum*; miro, *Prumnopitys ferruginea*; matai, *P. taxifolia*; kahikatea, *Dacrycarpus dacrydioides*), usually mixed with broad-leaved and beech species. Forests in the south are dominated by the southern beeches (*Nothofagus* spp.), the broad-leaved kamahi (*Weinmannia racemosa*) and southern rata (*Metrosideros umbellata*; Wardle 1991).

Most of the remaining native forest is in the South Island, and 40% of the total occurs in Southland and West Coast regions. After human habitation, beginning about 900 AD, fire destroyed most of the forest in the eastern half of the South Island and much of the central and eastern North Island (Wardle 1991). More forest was cleared for agriculture from about 1840, with the result that native vegetation was greatly modified in some places.

Ecological and forestry distinctiveness

The islands of New Zealand have been isolated from other land masses for about 70 million years, and successive glaciations created waves of extinctions that resulted in a depauperate flora and fauna with a high degree of endemism. Before the arrival of humans the only native land mammals were bats. With human habitation came rats, mustelids, browsing and grazing mammals such as deer, rabbits, Australian possums, goats, chamois and thar, as well as fish such as the salmonids and cyprinids (Wardle 1991). New Zealand also has numerous exotic land plants and aquatic weeds.

New Zealand's indigenous flora has some 2300 vascular species, about 85% of which are endemic. This flora has biogeographical affinities with southeastern Australia, Chile and South Africa. Genera of the Podocarpaceae with commercially harvestable trees are *Podocarpus*, *Dacrydium*, *Prumnopitys* and *Dacrycarpus*. However, the regeneration and growth rates of the native trees that attain commercial sizes are considered too slow to support commercially viable forestry; many individual trees are from several hundreds to a thousand or more years old. Kauri, for instance, has an annual increment of 4.5–7.5 m³/ha, and rimu only 1.2–1.8 m³/ha. The annual increment for the beech species is greater (5–17 m³/ha), but does not match that of introduced radiata pine (*Pinus radiata*) in New Zealand, which has a normal annual increment of 23–36 m³/ha, with a maximum of 50 m³/ha (Wardle 1991). Radiata pine originated in Monterey, California. Although many other coniferous species have been tried as plantation forest species, none has matched the realized growth and wide range of site suitability of radiata pine. Some forestry species, such as lodgepole pine (*Pinus contorta*) now create environmental problems with their ability to spread by wind dispersal and invade sensitive habitats (Taylor & Smith 1997).

Distribution of logging

For administrative purposes, New Zealand is divided into 10 wood supply regions (Fig. 29.1). The land area in plantation forest is 1.8 million ha, 71% of which is in the North Island, with 33% in the Central North Island region alone (Table 29.1).

Major fish communities

Life history characteristics and ecological features

New Zealand's long isolation from other land masses, the severe cyclical effects of the geologically recent ice ages and vulcanism have limited the indigenous freshwater fish fauna. Up to 1995, New Zealand had 27 recognized species of truly freshwater fishes and 7 marine wanderers that occasionally frequent fresh waters (McDowall 1990). However, since 1995, several new species have been recognized following genetic examination, discoveries of new fish and reinstatement of species. There are 36 indigenous species (McDowall 2000), and there may still be more undescribed taxa (McDowall 2001).

Typically, the fish faunas of New Zealand streams and rivers are characterized by few species and variable abundance, with densities ranging from 4.5 to 362 fish/100 m² (Jowett & Richardson 1996; Rowe *et al.* 1999). In some small pastoral streams, fish

Table 29.1   Areas of plantation forest in New Zealand in 2000 by wood supply region (see Fig. 29.1)

Wood supply region	Land area in plantation forest (ha)
North Island	
Central North Island	575,607
Northland	203,458
Southern North Island	155,777
East Coast	149,722
Hawkes Bay	120,934
Auckland	54,720
South Island	
Otago-Southland	186,638
Nelson-Marlborough	173,606
Canterbury	114,244
West Coast	33,932

Source: Ministry of Agriculture and Forestry (2001), Table A8.

biomass comprising mainly eels can be very high (80–90 g/m^2; Hicks & McCaughan 1997; Rowe *et al.* 1999).

With the exception of eels and salmonids, the fish in New Zealand are relatively small. Eels (*Anguilla* spp.) dominate most fish communities by number and weight. Fish density and species richness in rivers tend to decrease with distance from the coast, reflecting the importance of passage for fish colonizing from the sea. Many riverine fish communities are structured by diadromy (i.e. the migration of fish between the sea and fresh water; McDowall 1993). About 50% of New Zealand's native fish fauna is diadromous, and three patterns of diadromy are common. Southern lampreys (*Geotria australis*) and many populations of the common smelt species (*Retropinna retropinna*) are anadromous, spawning in fresh water, but returning to the sea to rear to adulthood. Eels are long-lived and catadromous (i.e. spawn at sea and rear in fresh water), and occur in virtually all waterways draining to the sea. The bullies and galaxiids are amphidromous, migrating downstream to the sea as larvae and then back upstream some months later as larger juveniles. In lakes, the common smelt, and some bully and galaxiid species, can form populations that never go to sea.

There are 21 introduced freshwater fish species (McDowall 2000). With the exception of sea-run chinook salmon (*Oncorhynchus tshawytscha*) and some brown trout (*Salmo trutta*) populations, the salmonids in New Zealand are wholly freshwater resident; although rainbow trout (*O. mykiss*) can move from river to river by means of the sea, no anadromous stocks are known to exist.

Fish distributions in localized areas are quite well known because of fisheries surveys that have been conducted for a variety of reasons. For example, a scheme to log indigenous beech forests of the West Coast and Southland regions prompted extensive sampling of the fish populations in these regions (McDowall *et al.* 1977). The fish diversity recorded (16 native and 3 introduced species) was not high by world standards, but is typical for New Zealand streams. Similar fish species compositions have been reported by others, with a high proportion of large galaxiid species in forest streams of the West Coast of the South Island (Main *et al.* 1985; Taylor & Main 1987). Streams and rivers of the unmodified forest of the Kahurangi National Park had 12 species of native fish and brown trout, and a lower diversity and abundance of fish than at equivalent elevations in other areas of New Zealand (Jowett *et al.* 1998).

Eels, bullies, galaxiids and salmonids feed mainly on aquatic invertebrates, but also on terrestrial invertebrates, especially the large galaxiids in forest streams (e.g. Main & Lyon 1988). In small North Island streams, 44% of the diet of longfinned eels in native forest comprised terrestrial taxa, including cicadas, harvestmen, spiders and green beetles; 38% of the diet in plantation forest streams was terrestrial (Hicks 1997). Large eels and salmonids can be piscivorous, especially in lakes. There is generally a lack of herbivorous fish in New Zealand among the native fauna, although herbivores such as rudd (*Scardinius erythophthalmus*) have been introduced.

Fisheries values

Eels are highly valued by the native Maori people as a customary fishery and some localized fisheries for lampreys and whitebait (juvenile galaxiids) still remain. In addition, eels and whitebait support major commercial and recreational fisheries for all New Zealanders. The salmonids, principally rainbow trout, brown trout and chinook salmon, provide major recreational fisheries (McDowall 1990). The extensive lakes and their tributaries of the central North Island have world-renowned fisheries for rainbow trout (McDowall 1990), whereas brown trout form the majority of fisheries in the South Island.

Forestry practices in New Zealand

Historical background

The first commercial harvest from New Zealand's forests occurred in the late 1700s, following Captain James Cook's observations of the potential of the trees for mast building (Roche 1990). Young kauri trees (rickers) were cut for ships' spars for the British Royal Navy before 1820, and kauri logging reached a peak of 1 million m³/year⁻¹ in 1907 (Roche 1990). Driving dams played a crucial part in transporting kauri logs down to tidal waters and harbours, where logs were loaded onto ships or made into rafts for transport to sawmills or for export (Halkett 1991). Drives of kauri logs were very destructive, sweeping everything before them, and tearing undergrowth and small trees from the stream banks (McDowall 1990; Halkett 1991). Kauri harvest declined between 1908 and 1922, and was partly replaced by harvest of rimu and kahikatea (or white pine).

Plantation forests were first established in New Zealand in the early 1900s with prison labour (Coker 1992). Following the imminent demise of the kauri harvest and reduction in the amount of merchantable timber from indigenous forests, there was

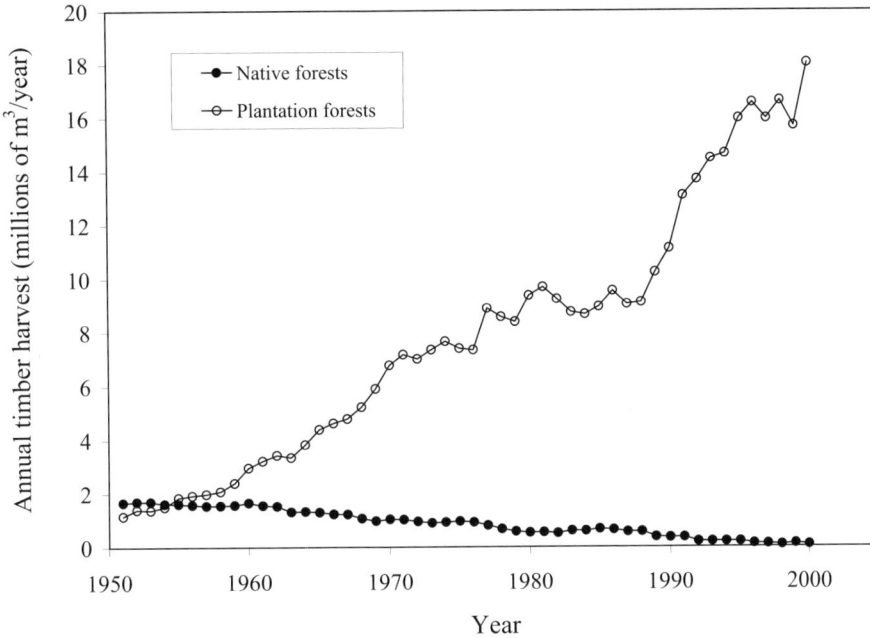

Fig. 29.2 Timber harvest as estimated by annual round wood removals from native and plantation forests in New Zealand between 1951 and 2000. Source: Ministry of Agriculture and Forestry (2001), Table A11.

major growth in afforestation with introduced tree species in the late 1920s to mid-1930s; by 1936, the total area of plantation forest had reached 317,000 ha (Coker 1992). Up until the mid-1970s, native forest was generally cleared to make way for plantation forest. This practice was phased out by the mid-1980s and any expansion of the plantation forest estate is now on marginal farmland. During the past four decades, the harvest from plantation forests has increased dramatically, while the harvest from native forests has fallen (Fig. 29.2). By 2000, the harvest from native forests had dropped to 76,000 m^3 while that from plantation forest had increased to 18 million m^3.

Extensive plantation forests have been developed in both islands of New Zealand, but development has been particularly extensive on the Central Plateau of the North Island. Free-draining volcanic soils and year-round rainfall provide ideal site conditions for the growth of radiata pine. Plantings in the North Island's central volcanic region occurred where 'bush sickness' (cobalt deficiency) in sheep and cattle prevented pastoral farming.

New Zealand's 10.7 million ha of forests and shrub land now cover about 40% of the land area. This area includes 6.3 million ha of indigenous forest in addition to the exotic plantation forests. Radiata pine comprises 90% of the total plantation forest estate, and Douglas fir (*Pseudotsuga menziesii*) about 5%. Hardwoods comprise about 3%, the most important being the Australian *Eucalyptus* species (Ministry of Agriculture and Forestry 2001).

Recent and current forestry practices

Once harvested, almost all areas of forest that are harvested in New Zealand are replanted. In addition, large areas of erosion-prone pasture have been recently afforested with radiata pine to control soil erosion and mass wasting. After intense cyclonic rainfall in 1988, land areas in established plantation forest in the East Coast wood supply region of the North Island had similar rates of landslides to undisturbed native forest (0.048–0.066 landslides/ha). Equivalent areas in pasture had an eight-fold greater rate of landsliding (0.564 landslides/ha; Maclaren 1996).

Preparation for planting

Very little preparation for planting is required on pastoral land, but herbicide treatment, burning, or windrowing has been used on sites that have gorse (*Ulex europeaus*) or other weeds. Sites with hard pan, such as pakihi, are sometimes ripped (deeply scored) to improve drainage and root penetration (Maclaren 1996).

Fire was considered a cheap form of site preparation as late as 1986, but is currently rare because of the costs of fire control, risks to surrounding forest, and environmental effects (Roberts 1994; Maclaren 1996). An alternative method of site preparation is windrowing, where slash left over from forest harvest is mechanically piled into rows or heaps (Maclaren 1996). Weed suppression by chemicals or oversowing with grasses and legumes is also used.

Optimal tree stocking densities for radiata pine depend on site quality, and general forestry practice is to plant more trees than are required for the final crop. Mature stands of radiata pine of ≥200 trees/ha will intercept most of the sunlight falling on a site, and the original stocking density may be about 400 trees/ha. In erosion-prone sites where the aim is to achieve closed canopy as fast as possible, densities of 1000 trees/ha may be planted (Maclaren 1993).

Forest maintenance

Radiata pine in New Zealand generally matures in 25–35 years (Maclaren 1996). During the growing cycle, thinning is usually carried out at least once a rotation to achieve canopy closure without stunting tree growth, and to ensure maximum production of knot-free wood. Early thinning to waste, in which the cut trees are left on the forest floor, usually occurs 3–6 years after planting. Production thinning at 10–16 years yields trees large enough to be merchantable. Pruning to maximize the amount of knot-free wood is often carried out twice a rotation, and is based on tree diameter. On sites with good tree growth, pruning can begin at 3 years of age. Weed control is essential to guarantee high seedling survival and uniform subsequent growth (Maclaren 1993).

Animal browsing can damage foliage and bark, especially during the establishment phase. Pest control to reduce browsing is important where rabbits, hares, possums, or deer are present. Trapping and 1080 poison (sodium monofluoroacetate) are often used to control possums (Maclaren 1993). Persistence of 1080 in the soil or waterways is short (Eason *et al.* 1992). For established trees, grazing can be used to control grass and weed growth, and at stocking rates of 200 trees/ha enough light penetrates the canopy to permit 'agroforestry' (i.e. a combination of grazing and tree growth; Maclaren 1993).

New Zealand is free from most diseases of forest trees, but fungal diseases such as needle blight (*Dothistroma pinii*) affect radiata pine. *Pinus ponderosa* and *P. nigra* are even more susceptible, but these species have now been almost completely replaced by radiata pine and Douglas fir. Regular spraying with low rates of fungicides is required to control *Dothistroma*, depending on season and locality (Coker 1992). The copper-based fungicides that are generally used have negligible effect on waterways (Fish 1968). Insect pests such as pinhole borer (*Platypus* spp.) and the wood wasp (*Sirex noctilio*) can be locally important, and biological control has been partially effective in the case of *Sirex* (Taylor & Smith 1997).

Road building

Road building was a prominent feature of logging in the 1970s, especially in steeper areas such as the northwest of the South Island. Unpaved skid tracks followed hill contours at intervals of 30–40 m during logging in the Golden Downs State Forest (Graynoth 1979).

Roads, tracks and landings (areas used for aggregation of logs during harvesting) that are permanently out of production can account for 3–8% of the total forest area (Coker 1992), and many are unpaved. Unpaved roads are an important source of sediment from forests, and sediment yield increases with use from 2 to 500 tonnes/km for lightly to heavily used roads (Maclaren 1996). Most sediment from roads is caused by traffic during storm events. Reducing tyre pressure from 90 to 30 psi decreases sediment production by as much as 84%.

Harvest methods

The most common harvest practice for plantation forest is clear felling in coupes (harvest blocks) of about 25 ha (Maclaren 1996). This form of harvest is considered the most economic and allows re-establishment of shade-intolerant species such as radiata pine. Ground-based logging operations (e.g. caterpillar tractors and rubber-tyred log skidders) generally work on slopes <25% (Coker 1992). These skidder logging and ground-based extraction systems were once used almost exclusively, and can result in considerable soil disturbance. As steeper slopes were planted and the forests reached maturity, there has been a move towards cable logging systems (O'Loughlin *et al.* 1980). Cable logging systems generally haul logs uphill along the ground to a prominent knob of land that commands a wide area within the economic hauling distance of the machine (generally about 400 m; Coker 1992). However, downhill cable systems where logs are lifted clear of the ground result in lower sediment yields than uphill systems. Sediment yields from downhill yardage systems can be similar to those of unlogged controls (O'Loughlin *et al.* 1980).

Transportation, storage and processing

Transportation of logs is mainly by truck, with some rail transport. During the harvest phase, truck traffic increases considerably, which can lead to increased sediment yields from unpaved roads (Maclaren 1996). There are several large mills in New Zealand, mostly in the North Island because of their proximity to the extensive forest estate in the centre of the island. These mills produce a variety of wood products, from pulp and paper to fibreboard and timber for construction. Effluent from some mills has been a long-standing source of environmental pollution, despite considerable improvements in effluent quality, e.g. the Tarawera River (Kanber *et al.* 2000).

Effects of forestry practices on fish and fisheries

The effects of forestry on fish may be divided into (1) the afforestation phase (including forest growth) and (2) the timber harvest phase. Our knowledge of the impacts of the afforestation phase is quite good, at least from the viewpoint of the physical effects on fish distribution. However, direct evidence of the effects of timber harvest on fish is limited, although inference can be made by considering effects on sediment yield, wood supply to streams, invertebrates and water quality.

Historical background

Growing concern over the effects of forestry practices on fish and fisheries in the mid-1970s prompted a review of the subject (Morgan & Graynoth 1978), and the effects of forestry practices and buffer strips on streams and their faunas were studied in the Golden Downs State Forest, Nelson District, South Island. A 30-m wide buffer strip of unlogged native forest may have protected the native fish fauna from the negative effects of timber harvest, especially the high concentrations of suspended sediment resulting from rainfall on exposed surfaces. Sediment concentrations were 13–860 g m^{-3} in logged reaches without buffer strips, compared with 7–124 g m^{-3} in the reach with buffer strips, and 2–22 g m^{-3} in unlogged stream reaches (Graynoth 1979). The buffer strips may also have protected the invertebrate fauna from the effects of logging, but did not prevent increased nitrate concentrations that occurred after timber harvest. A combination of greater fine and coarse sediment and warmer stream temperatures are thought to have substantially reduced the abundance of dwarf galaxias (*Galaxias divergens*) and eels in the logged reaches without buffer strips.

Several studies in the 1980s focused on basin hydrology and nutrient yield from plantation forests at the Purukohukohu experimental basins of the central North Island (e.g. Cooper *et al.* 1987; Cooper & Thomsen 1988; Fig. 29.1). The concentrations of total phosphorus and dissolved reactive phosphorus from plantation forest were intermediate between pasture and native forest, with more in the pasture stream than in the native forest stream. The median monthly concentration of nitrate over 14 years was, however, much greater in the native forest stream (805 mg m^{-3}) than in the plantation forest stream (176 mg m^{-3}), and was lower still in the pasture stream (13 mg m⁻³; Cooper *et al.* 1987). The researchers ascribed the differences between land uses to different processing rates, concluding that the stream in plantation forest had a high capacity to remove nitrate from the stream waters (Cooper & Thomsen 1988). However, contrary results were observed in streams at Whatawhata (Fig. 29.1), where mean dissolved inorganic nitrogen concentrations were 180, 390 and 800 mg m⁻³ in native forest, plantation forest and pasture, respectively (Quinn *et al.* 1997). The cause of the difference between the Purukohukohu and Whatawhata studies has not been adequately explained, but streams draining native forest in the central North Island appear to have higher nitrate concentrations than native forest streams in other parts of the country.

In 1990, fishery biologists and resource managers met to discuss the impacts and remedies of forestry on physical and biological processes in rivers, mainly from studies in western North America (Hayes & Davis 1992). Since then, this subject has attracted considerable research in New Zealand, mainly on the effects of forestry practices on water quality and benthic invertebrate communities in streams.

Recent and current effects

Watershed hydrology and fluvial geomorphology

Forest harvest affects water yield and stream hydrology. The immediate response to land clearing in preparation for forest planting is an increase in water yield. In the first year after timber harvest, increases in annual water yield have ranged from 168 to 650 mm, depending on precipitation (Fahey & Rowe 1992). In these studies, annual precipitation ranged from 1069 to 2827 mm, and the original vegetation was manuka (*Leptospermum scoparium*), gorse or native forest.

Once a planted forest of radiata pine has become established, water yield is usually reduced significantly. A number of studies have shown that conversion of scrub and indigenous forest to plantation forest reduces stream flow (Fahey & Rowe 1992). Between 1964 and 1981, 28% of the Tarawera River drainage basin was afforested, principally with radiata pine. Tarawera River flows diminished following afforestation, and 13% of the decrease was attributed to the afforestation (Dons 1986); the decrease in flow continued into the early 1990s (Pang 1993).

Annual water yields after afforestation with radiata pine have tended to stabilize at about 150–200 mm below the previous yields from native vegetation (Fahey & Rowe 1992). The effect is greatest in small basins, where water yield from plantation forest is 50% less than the flow from pasture, and 25% less than the flow from native forest (Dons 1987). As the primary mechanism for reduced low flows appears to be interception of rainfall, flow reductions following afforestation with radiata pine are most severe in areas with low rainfall (Taylor & Smith 1997).

Water yields from established plantations of radiata pine 8–21 years of age were 230–290 mm lower than those from pasture, but flood flow was also reduced (Fahey & Rowe 1992). Thus the effect of reduced low flows may be offset somewhat by reduced flood severity.

Afforestation of pakihi wetlands is only possible after ripping or V-blading to break through the hard pan (Fahey & Rowe 1992; Maclaren 1996). This procedure disturbs about two-thirds of the land surface area in the plantation and can lead to higher and more frequent peak flows (Fahey & Rowe 1992). Also, runoff from roading works can increase a stream's drainage network and its peak discharge (Harding *et al.* 2000). During harvest, greater flow variability can be expected until the replanted forest grows.

The influence of forestry on sediment yield is generally much less than that caused by variations in rainfall around the country and, except during harvest phases, plantation forests generate only 20–60% of the sediment yield of pasture. However, in the high rainfall area of the West Coast, sediment yield increased up to 100-fold following clear-felling of native forest (Hicks & Griffiths 1992).

The role of woody debris in controlling channel morphology has not been widely studied in New Zealand, but can be important in small streams. Large woody debris (LWD) controlled channel morphology and bed-load transport in the west coast of the South Island (Mosley 1981), and LWD was retained even after a large flood in a second order stream on the east coast of the South Island (Evans *et al.* 1993b).

Quantitative studies of wood loadings in plantation forests are few, and radiata pine forests tend to have relatively small amounts of LWD compared with some of the higher loadings for streams in coniferous forests of North America. In New Zealand, however, streams in plantation forests >15 years old have similar wood loadings (mean 112–245 m³/ha) to streams in ancient native forests (mean 101 m³/ha; Evans *et al.* 1993a). Streams in younger radiata pine plantations (10 years old) have much less wood (mean 2.4 m³/ha; Evans *et al.* 1993a). Radiata pine logs can remain intact in streams for over 20 years (Collier & Baillie 1999), so are likely to be important for stream structure, especially in the pumice-bed streams of the central North Island (Collier & Halliday 2000).

Streams and rivers

Studies by Graynoth (1979) in the Nelson District indicated that pine harvesting without buffer strips can reduce the abundance of fish in streams. In a re-evaluation of the effects of logging in the Nelson area 16 years later, stream flows appeared to be lower than in 1973–1974. While the most severe effects of sediment on the fish populations appear to have been short-lived, the lowered stream flows that occurred as the plantation forest grew to maturity lasted longer, and reduced some stream reaches to a series of disconnected pools. Brown trout, eels and upland bullies were less abundant in 1990 than in 1973–1974 (Graynoth 1992).

On the other hand, Harding *et al.* (2000) maintain that while headwater streams may become ephemeral or completely dry under plantation forest, the loss of habitat may be compensated to some extent by the greater width of stream channels in forest catchments downstream (Davies-Colley 1997) and improved habitat conditions provided by forest cover compared with pasture.

Recent studies indicate that the native fish faunas of streams in mature pine forests planted on pasture are similar to those in native forests, but not to those of pasture land (Hicks & McCaughan 1997; Rowe *et al.* 1999, 2002). Similarly, benthic invertebrate community composition in plantation forest streams more closely resembles that found in native forest streams than pasture streams (Quinn *et al.* 1997), although there are significant differences in relative abundance of some key species such as mayflies, stoneflies and caddisflies (Harding & Winterbourn 1995). Adult caddisflies caught by light trapping within 2 m of stream margins were lower in species diversity and overall abundance in pine forest than in native forest and pasture streams (Collier *et al.* 1997b).

The effect on fish and invertebrates of converting pasture land to plantation forests was investigated in a suite of studies conducted in hill country near Whatawhata, Waikato region, North Island. Physical habitat characteristics, algal productivity, and fish and invertebrate abundance were compared in Waikato River tributaries in native forest, 18-year-old plantation forest of radiata pine, and pasture. Fish were more abundant in pastoral streams than in forest streams, and therefore afforestation of pasture land is likely to reduce fish abundance. Native fish biomass (mainly eels) was four times greater (Table 29.2), and density was seven times greater, at pasture sites than in plantation sites. The likely cause of greater fish abundance in pasture streams than forested streams at Whatawhata is increased production due to greater light availability, higher mean and maximum water temperatures, and greater dissolved inorganic nitrogen concentrations. There were greater gross photosynthesis and invertebrate densities and biomass in the pasture streams (Quinn *et al.* 1997).

Fish abundance in plantation forest streams showed many similarities to native forest streams. The density and biomass of fish in plantation streams were similar to

Table 29.2 Fish biomass in North Island streams at 11 west coast sites at Whatawhata (Hicks & McCaughan 1997) and 19 east coast sites (Rowe *et al.* 1999)

Species	Fish biomass (g/m²)		
	Native forest	Plantation forest	Pasture
North Island – Whatawhata			
Longfinned eel	11.10	18.50	60.00
Banded kokopu	1.32	0.07	0.00
Shortfinned eel	0.40	0.50	28.80
Cran's bully	0.00	0.23	0.29
Redfinned bully	0.00	0.02	0.00
Common smelt	0.00	0.00	0.70
Total	12.82	19.32	89.79
North Island – east coast sites			
Longfinned eel	9.24	16.50	47.60
Redfinned bully	1.60	1.42	1.22
Banded kokopu	1.22	0.56	0.01
Shortfinned eel	0.58	0.42	30.50
Common smelt	0.12	0.16	1.17
Inanga	0.00	0.60	0.86
Total	12.76	19.66	81.36

those of native forest, and east coast streams had very similar fish biomasses to west coast streams in the North Island (Table 29.2). Stable isotope analyses showed that leaf litter rather than epilithic diatoms was the primary source of carbon and nitrogen for the food webs in both native and plantation forests (Hicks 1997). However, the greater allochthonous inputs to forested streams (Quinn *et al.* 1997) failed to compensate for the reduced light availability. Dissolved inorganic nitrogen concentrations and gross photosynthesis in plantation forest streams were intermediate between pastoral and native forest streams. Light availability was, however, similar in plantation and native forests.

The number of fish species was low in Whatawhata streams (mean number of species 2.3–3.0), possibly because of barriers to the upstream migration of fish. In streams at 19 coastal sites in native forest, plantation forest, and pasture on the east coast of the North Island, there were more fish species (mean number of species 3.8–4.7) and fish densities were 2–10 times greater than at Whatawhata. Fish biomass in equivalent land uses was, however, almost identical (Table 29.2; Rowe *et al.* 1999). This suggests that light and nutrient availability caused by land use was similar in the different regions. In another study of Coromandel streams, the abundance of eels and redfinned bullies increased in the post-harvest period. Streams with undisturbed riparian forest had greater densities of native fish than those lacking such strips (Rowe *et al.* 2002).

Some differences between fish abundance in streams in native and plantation forest are contradictory. Banded kokopu, redfinned bullies and shortjawed kokopu were found at fewer sites in plantation forest compared with native forest sites in the Waikato region, whereas giant kokopu, shortfinned eels and Cran's bullies occurred more frequently in plantation forest (Hanchet 1990). Hanchet postulated that increased fine sediment loading in plantation forest streams was responsible for the differences in fish distribution. These plantation forests were planted on pasture land 18 years before the study, and stream channels had abundant fine sediment, possibly as a result of channel widening (Davies-Colley 1997). However, at the North Island east coast sites there was no significant difference in banded kokopu biomass between forest types (Rowe *et al.* 1999).

In tributaries of the Grey River on the west coast of the South Island, lamprey ammocoetes were more abundant in plantation forest than in adjacent native forest. As lamprey ammocoetes are generally found in silty backwaters, these plantation forest streams might have had more fine sediment than native forest streams, although this was not measured. Dwarf galaxias and bluegilled bullies were more abundant in native forest than in plantation forest sites (Jowett *et al.* 1996).

Forestry activities can have marked impacts on benthic invertebrate community composition, mainly through alteration of light availability, stream hydrology, morphology and water chemistry, especially where riparian buffers are not provided. Benthic invertebrate communities are resilient, however, and communities can resemble their pre-impact state within 10–15 years (Harding *et al.* 2000), or earlier, depending on the state of the riparian vegetation and proximity of sources of recolonists.

Wood in streams can be an important component of fish habitat, and slash overlying streams can reduce water temperature fluctuations (Collier *et al.* 1997a). However, where large amounts of slash (branches and tops) remain submersed following logging, dissolved oxygen can be lower than in reaches with slash removed (Collier *et al.* 1997a). Pine wood in streams can act as a substrate and food for stream invertebrates (Collier & Halliday 2000).

Wetlands

Information on forestry impacts on wetlands is sparse, although some exists for West Coast pakihi swamps. In preparation for planting with pine, these swampy areas are extensively channelized for drainage and the soils are enriched with phosphate fertilizers. Studies comparing modified with unmodified wetlands have shown that conversion to plantation forest has resulted in increased water temperature (Collier *et al.* 1989). The lower pHs and reduced number of invertebrate taxa found at sites that had been recently V-bladed and planted with radiata pines were attributed to the location of these sites in acidic wetlands rather than the effects of afforestation. However, these changes did not appear to be long-lived, and in one case, after 8 years, the species composition of the impacted invertebrate communities was similar to that of native forest streams (Valentine 1995).

Lakes

Nothing is known of the effects of forestry on fish populations in New Zealand lakes despite the extensive plantation forest in the drainage basins of some sand dune lakes and lakes in the central North Island. Forest planting may have reduced water levels in some northern dune lakes in the North Island with dwarf inanga (*Galaxias gracilis*), but there was no relationship between water level changes and dwarf inanga abundance (Rowe & Chisnall 1997).

Estuarine and coastal waters

Virtually no studies have been done on the effects of forestry on estuarine and coastal environments in New Zealand. For example, in the Marlborough Sounds situated at the north end of the South Island, steep hillsides with pine plantations have been logged close to sea level. Such activity almost certainly has considerable negative effect on water quality near-shore. However, coastal environments in New Zealand are reasonably well flushed and such impacts may not persist for extended periods.

Environmental effects from processing mills

Large pulp and paper mills in the central North Island have, in the past, discharged poorly treated effluent into waterways, especially from kraft processing and chemical pulping, which affected fish and invertebrate abundance and community composition (Scrimgeour 1989). More recently there has been a move towards thermomechanical pulping, which improves effluent quality.

Large amounts of pentachlorophenol (PCP) were used by the New Zealand timber industry as an anti-sapstain fungicide that was applied to the timber surface immediately after milling. Sawmills have also produced effluent discharges, and PCPs have built up in soils around treatment sites and in the sediment of Lake Rotorua in the central North Island. PCPs are toxic to a range of zooplankton (Hickey 1989; Willis *et al.* 1995). The technical grade of PCP used contained about 20% tetrachlorophenol and about 5% of other contaminants such as dibenzodioxins and dibenzofurans, and was more toxic than 'pure' PCP, which contained about 5% of these impurities (Willis *et al.* 1995). The larval stages of common smelt, inanga, koaro and common bullies are very sensitive to PCP, showing similar sensitivity to rainbow trout fry (Hannus 1998). Although the use of PCP as a timber treatment is no longer prevalent, many contaminated sites and waterways exist in localized areas (Taylor & Smith 1997).

Needs and recommendations for improvement of forestry practices and standards

The influence of forestry on New Zealand's environment, principally through the trend of conversion of steep pasture to plantation forest, is largely positive, but there are several areas in which improvements can still be made. A code of practice for forestry operations was developed to help forest managers to plan, manage and carry out forestry operations in a sustainable manner (Vaughan *et al.* 1993). Forest Research Ltd, the crown research institute responsible for research into forest management, has continued commitment to sustainable forest management, and current research initiatives include improvements in the treatment of mill effluent and the search for new tree species. Harvest of native forests on crown-owned land was recently halted (Griffiths 2002), but logging of native forests on privately owned land continues.

The timber processing industry in New Zealand has substantially reduced use of elemental chlorine for bleaching wood pulp, and now uses chlorine dioxide, which produces far fewer chlorinated hydrocarbons in the mill effluent. A series of anti-sap-staining chemicals has replaced PCP, and there is current research into biocontrol of sapstaining fungi.

Increased efficiency of wastewater recovery and water use within mills now occurs, so that less water is required. The performance of secondary treatment systems is being optimized by additional treatment systems within mills. In general, New Zealand mills are in the top 25% of mills worldwide for environmental performance. The quality of receiving waters has improved with modernization of pulp and paper mills, especially with improvements in the bleaching process, but a gradient in faunal composition associated with effluent discharge at some sites was still apparent in the early 1990s. There were fewer crayfish (*Paranephrops planifrons*) but more goldfish (*Carassius auratus*) at the site receiving mill effluent compared with more distant sites (Sharples & Evans 1998).

Roading

Barriers to upstream fish migration can be caused by dams, water intake weirs and road culverts. The timber harvest phase requires a well developed roading network that should minimize the number of stream crossings to reduce the potential for creating barriers to fish passage. Where stream crossings are unavoidable, guidelines for improved culvert design and a summary of the swimming and climbing ability of native and introduced fishes are available (Boubée *et al.* 1999). Fish passage is an area of ongoing research.

Sediment control during logging will always be a problem with ground-based harvest techniques. Paradoxically, improved protection of riparian buffers around headwater streams might require more roading and landings than harvest regimes without buffers, which is likely to increase sediment yield.

Planting

Preparation for planting

Burning has been largely phased out, partly to retain more nutrients in the soil between rotations. Chemical weed control is in many cases a better alternative for site preparation than burning. In addition, there has been a move away from systemic hormone-based chemicals to species-specific and spot treatments. Stream edges are an area of special concern during planting, and many first rotation stands were planted to the stream edge following land clearance. Streamside trees can lean out over the water and be difficult to fall back onto the land (Coker 1992). As early plantings are harvested, stream edges should not be replanted with the commercial crop. Most current practices avoid planting commercial trees in the riparian zone.

Tree species

Cypresses, acacias and some native species are being investigated as alternatives to radiata pine. Realistically, because of its fast growth and short rotation time, radiata pine will probably always be favoured by the forestry industry. Improvements in selecting genetic stocks for specific sites and cloning have the potential to maximize production and shorten rotation times. As rotation times are being reduced from 35 to 25 years on average by these methods, the fish fauna may spend a large proportion of the rotation recovering from the effects of timber harvest.

Riparian margins

Dense ground cover in the riparian strips has been demonstrated to reduce sediment ingress to streams during harvesting, and detailed guidelines have been prepared for management of riparian zones in agricultural settings in New Zealand (Collier *et al.* 1995a, 1995b). These guidelines suggest appropriate species for planting, and consider the role of riparian vegetation in providing bank stability, managing water temperatures, terrestrial carbon inputs and attenuating flood flows. In addition, a decision support system is under development by the National Institute of Water and Atmospheric Research and Forest Research to promote objective and effective riparian management to sustain in-stream environments within plantation forests (Murphy *et al.* 1998; Collier 2001). Stream shade and nutrient inputs have also been considered (Rutherford *et al.* 1999). Sediment entry into streams during timber harvest is still a problem where suitable buffer strips are not provided, especially on erosion-prone sites. Sediment can bypass riparian zones by way of ephemeral channels, but slash left on slopes can help prevent sediment mobilization.

The shade provided by riparian forest lowers water temperature, minimizing temperature extremes. Where riparian forest is removed during timber harvest, water temperatures on summer afternoons can rise by 3–8°C in unshaded reaches, emphasizing the importance of maintaining riparian shade through the harvest phase of the forestry cycle (Maclaren 1996). The decrease in light caused by riparian shade under plantation forest is likely to result in lowered aquatic biomass compared with pasture streams, as demonstrated by the lower fish and invertebrate biomass in both native and plantation forest.

Timber harvesting

Harvest techniques

On steep ground, cable logging is an acceptable alternative to the ground-based skidder and tractor operations. Downhill hauling that keeps earthworks away from streams and avoids dragging logs along the ground is preferable to hauling by cable to an uphill landing. Where this is unavoidable, trimming of limbs should occur away from the landing so that an unstable 'bird's nest' of limbs and tops does not accumulate around the landing. Harvest operations should avoid suspending logs above riparian zones, and should instead haul away from streams by making them boundaries. Harvest method influences the volumes of small and large woody debris, with methods that haul across the stream contributing the most wood (Baillie *et al.* 1999). Helicopter logging has been used in some selective logging operations.

The amount of wood that remains in streams after harvest depends to a large extent on harvest method and riparian protection (Baillie *et al.* 1999), and the influence of wood in different settings has yet to be fully evaluated (e.g. Collier & Halliday 2000). Research continues into optimal post-harvest treatment of wood in streams to balance the needs of fish and invertebrates.

Conclusion

Plantation forestry is generally a benign land use in New Zealand. During the growing cycle, fragile hill slopes are protected from erosion, and sediment yields are lowered, especially compared with pasture. During the harvest cycle, however, some impacts on the aquatic environments are inevitable. Good management can reduce these effects, and speed recovery following logging and establishment of the new tree crop. One impact of the growing cycle cannot be mitigated: lowered water yield remains an unavoidable consequence of plantation forests of radiata pine in many areas of New Zealand.

Acknowledgements

Dave Rowe and Kevin Collier of the National Institute for Water and Atmospheric Research (NIWA), Hamilton; Don Jellyman of NIWA, Christchurch; Bruce Clarkson of the Centre for Biodiversity and Ecology Research, Department of Biological Sciences, University of Waikato; and Tom Northcote and Gordon Hartman made constructive criticisms of the manuscript.

References

Baillie, B.R., Cummins, T.L. & Kimberley, M.O. (1999) Harvesting effects on woody debris and bank disturbance in stream channels. *New Zealand Journal of Forestry Science* **29**, 85–101.

Boubée, J., Jowett, I., Nichols, S. & Williams, E. (1999) *Fish Passage at Culverts: A Review, with Possible Solutions for New Zealand Indigenous Species.* Department of Conservation, Wellington. [63 pp. plus appendices and computer program.]

Coker, R. (1992) Forest management considerations in New Zealand. In: *Proceedings of the Fisheries/Forestry Conference, 28 February 1990, Christchurch,* (eds J.W. Hayes & S.F. Davis), pp. 35–44. New Zealand Freshwater Fisheries Report No. 136. MAF Fisheries, Christchurch.

Collier, K. (2001) A riparian DSS for plantation forests: update. *Water and Atmosphere,* **9**, 8.

Collier, K.J. & Baillie, B.R. (1999) Decay state and orientation of *Pinus radiata* wood in streams and riparian areas of the central North Island. *New Zealand Journal of Forestry Science,* **29**, 225–35.

Collier, K.J. & Halliday, J.N. (2000) Macroinvertebrate-wood associations during decay of plantation pine in New Zealand pumice-bed streams: stable habitat or trophic subsidy? *Journal of the North American Benthological Society,* **19**, 94–111.

Collier, K., Winterbourn, M.J. & Jackson, R.J. (1989) Impacts of wetland afforestation on the distribution of benthic invertebrates in acid streams of Westland, New Zealand. *New Zealand Journal of Marine and Freshwater Research,* **23**, 479–90.

Collier, K.J., Cooper, A.B., Davies-Colley, R.J., Rutherford, J.C., Smith, C.M. & Williamson, R.B. (1995a) *Managing Riparian Zones: A Contribution to Protecting New Zealand's Rivers and Streams. Vol. 1: Concepts.* Department of Conservation, Wellington.

Collier, K.J., Cooper, A.B., Davies-Colley, R.J., Rutherford, J.C., Smith, C.M. & Williamson, R.B. (1995b) *Managing Riparian Zones: A Contribution to Protecting New Zealand's Rivers and Streams. Vol. 2: Guidelines.* Department of Conservation, Wellington.

Collier, K., Baillie, B., Bowman, E., Halliday, J., Quinn, J. & Smith, B. (1997a) Is wood in streams a dammed nuisance? *Water and Atmosphere,* **5**, 17–21.

Collier, K.J., Smith B.J. & Baillie, B.R. (1997b) Summer light-trap catches of adult Trichoptera in hill-country catchments of contrasting land use, Waikato, New Zealand. *New Zealand Journal of Marine and Freshwater Research,* **31**, 623–34.

Cooper, A.B. & Thomsen, C.E. (1988) Nitrogen and phosphorus in streamwaters from adjacent pasture, pine, and native forest catchments. *New Zealand Journal of Marine and Freshwater Research,* **22**, 279–91.

Cooper, A.B., Hewitt, J.E., Cooper, A.B. & Thomsen, C.E. (1987) Land use impacts on streamwater nitrogen and phosphorus. *New Zealand Journal of Forestry Science,* **17**, 179–92.

Davies-Colley, R.J. (1997) Stream channels are narrower in pasture than in forest. *New Zealand Journal of Marine and Freshwater Research,* **31**, 599–608.

Dons, A. (1986) The effect of large-scale afforestation on Tarawera River flows. *Journal of Hydrology (NZ),* **25**, 61–73.

Dons, A. (1987) Hydrology and sediment regime of a pasture, native forest, and pine forest catchment in the central North Island, New Zealand. *New Zealand Journal of Forestry Science*, **17**, 161–78.

Duncan, M.J. (1987) River hydrology and sediment transport. In: *Inland Waters of New Zealand*, (ed. A.B. Viner), pp. 113–137. DSIR Bulletin 241. Science Information Publishing Center, Department of Scientific and Industrial Research, Wellington.

Duncan, M.J. (1992) Flow regimes of New Zealand rivers. In: *Waters of New Zealand*, (ed. M.P. Mosley), pp.13–27. New Zealand Hydrological Society, Wellington.

Duncan, M.J. (1995) Hydrological impacts of converting pasture and gorse to pine plantation, and forest harvesting, Nelson, New Zealand. *Journal of Hydrology (NZ)*, **34**, 15–41.

Eason, C.T., Wright, A. & Fitzgerald, H. (1992) Sodium monofluoroacetate (1080) water residue analysis after large scale possum control. *New Zealand Journal of Ecology*, **16**, 47–69.

Evans, B.F., Townsend, C.R. & Crowl, T.A. (1993a) Distribution and abundance of coarse woody debris in some southern New Zealand streams from contrasting forest catchments. *New Zealand Journal of Marine and Freshwater Research*, **27**, 227–39.

Evans, B.F., Townsend, C.R. & Crowl, T.A. (1993b) The retention of woody debris structures in a small stream following a large flood. *New Zealand Natural Sciences*, **20**, 35–9.

Fahey, B.D. & Rowe, L.K. (1992) Land use impacts. In: *Waters of New Zealand*, (ed. M.P. Mosley), pp. 265–84. New Zealand Hydrological Society, Wellington.

Fish, G.R. (1968) The hazard presented to freshwater life by aerial copper spraying. *New Zealand Journal of Forestry*, **13**, 239–43.

Fitzharris, B., Owens, I. & Chinn, T. (1992) Snow, ice, and glacier hydrology. In: *Waters of New Zealand*, (ed. M.P. Mosley), pp. 75–93. New Zealand Hydrological Society, Wellington.

Graynoth, E. (1979) Effects of logging on stream environments and faunas in Nelson. *New Zealand Journal of Marine and Freshwater Research*, **13**, 79–109.

Graynoth, E. (1992) Long-term effects of logging practices in streams in Golden Downs State Forest, Nelson. In: *Proceedings of the Fisheries/Forestry Conference, 28 February 1990, Christchurch*, (eds J.W. Hayes & S.F. Davis), pp. 52–69. New Zealand Freshwater Fisheries Report No. 136. MAF Fisheries, Christchurch.

Griffiths, A.D. (2002) *Indigenous Forestry on Private Land: Present Trends and Future Potential*. MAF Technical Paper No. 01/6. Indigenous Forestry Unit, New Zealand Ministry of Agriculture and Forestry, Wellington.

Halkett, J. (1991) *The Native Forests of New Zealand*. GP Publications Ltd, Wellington.

Hanchet, S. (1990) Effect of land use on the distribution and abundance of native fish in tributaries of the Waikato River in the Hakarimata range, North Island, New Zealand. *New Zealand Journal of Marine and Freshwater Research*, **24**, 159–71.

Hannus, I.M. (1998) *Native New Zealand fish: lethal sublethal effects of pentachlorophenol on early and adult life stages*. Unpublished doctoral thesis, University of Waikato, Hamilton, New Zealand.

Harding, J.S. & Winterbourn, M.J. (1995) Effects of contrasting land use on physico-chemical conditions and benthic assemblages of streams in a Canterbury (South Island, New Zealand) river system. *New Zealand Journal of Marine and Freshwater Research*, **29**, 479–92.

Harding, J.S., Quinn, J.M. & Hickey, C.W. (2000) Effects of mining and production forestry. In: *New Zealand Stream Invertebrates: Ecology and Implications for Management*, (eds K.J. Collier & M.J. Winterbourn), pp. 230–59. Caxton Press, Christchurch, New Zealand.

Hayes, J.W. & Davis, S.F. (1992) *Proceedings of the Fisheries/Forestry Conference, 28 February 1990, Christchurch*. New Zealand Freshwater Fisheries Report No. 136. MAF Fisheries, Christchurch.

Hickey, C.W. (1989) Sensitivity of four New Zealand cladoceran species and *Daphnia magna* to aquatic toxicants. *New Zealand Journal of Marine and Freshwater Research*, **23**, 131–7.

Hicks, B.J. (1997) Food webs in forest and pasture streams in the Waikato region: a study based on analyses of stable isotopes of carbon and nitrogen and fish gut contents. *New Zealand Journal of Marine and Freshwater Research*, 31, 651–64.

Hicks, B.J. & McCaughan, H.M.C. (1997) Land use, associated eel production, and abundance of fish and crayfish in streams in Waikato, New Zealand. *New Zealand Journal of Marine and Freshwater Research*, 31, 635–50.

Hicks, D.M. & Griffiths, G.A. (1992) Sediment load. In: *Waters of New Zealand*, (ed. M.P. Mosley), pp. 229–48. New Zealand Hydrological Society, Wellington.

Jowett, I.G. & Richardson, J. (1996) Distribution and abundance of freshwater fish in New Zealand rivers. *New Zealand Journal of Marine and Freshwater Research*, 30, 239–55.

Jowett, I.G., Richardson, J. & McDowall, R.M. (1996) Relative effects of in-stream habitat and land use on fish distribution and abundance in tributaries of the Grey River, New Zealand. *New Zealand Journal of Marine and Freshwater Research*, 30, 463–75.

Jowett, I.G., Hayes, J.W., Deans, N. & Eldon, G.A. (1998) Comparison of fish communities and abundance in unmodified streams of Kahurangi National Park with other areas of New Zealand. *New Zealand Journal of Marine and Freshwater Research*, 32, 307–22.

Kanber, S.A., Wilkins, A.L. & Langdon, A.G. (2000) Speciation of pulp mill derived resin acids in the Tarawera River, New Zealand. *Bulletin of Environmental Contamination and Toxicology*, 64, 622–9.

Lowe, D. & Green, J.D. (1987) Origins and development of the lakes. In: *Inland Waters of New Zealand*, (ed. A.B. Viner), pp. 1–64. DSIR Bulletin 241. Science Information Publishing Center, Department of Scientific and Industrial Research, Wellington.

McDowall, R.M. (1990) *New Zealand Freshwater Fish: A Natural History and Guide*. Heinemann-Reed, Auckland.

McDowall, R.M. (1993) Implications of diadromy for the structuring and modelling of riverine fish communities in New Zealand. *New Zealand Journal of Marine and Freshwater Research*, 27, 453–62.

McDowall, R.M. (2000) *The Reed Field Guide to New Zealand Freshwater Fishes*. Reed Publishing, Auckland.

McDowall, R.M. (2001) Getting the measure of freshwater fish habitat in New Zealand. *Aquatic Ecosystem Health and Management*, 4, 343–55.

McDowall, R.M., Graynoth, E. & Eldon, G.A. (1977) The occurrence and distribution of fishes in streams draining the beech forest of the West Coast and Southland, South Island, New Zealand. *Journal of the Royal Society of New Zealand*, 7, 405–24.

Maclaren, J.P. (1993) *Radiata Pine Grower's Manual*. FRI Bulletin No. 184. New Zealand Forest Research Institute Ltd, Rotorua.

Maclaren, J.P. (1996) Environmental effects of planted forests in New Zealand. *FRI Bulletin No. 198*. New Zealand Forest Research Institute Ltd, Rotorua.

McLay, C.L. (1976) An inventory of the status and origin of New Zealand estuarine systems. *Proceedings of the New Zealand Ecological Society*, 23, 8–26.

Main, M.L. & Lyon, G.L. (1988) Contribution of terrestrial prey to the diet of banded kokopu (*Galaxias fasciatus* Gray) (Pisces: Galaxiidae) in South Westland, New Zealand. *Internationale Vereinigung für theoretische und angewandte Limnologie, Verhandlungen*, 23, 1785–9.

Main, M.R., Nicoll, G.J. & Eldon, G.A. (1985) *Distribution and Biology of Freshwater Fishes in the Cook River to Paringa River Area, South Westland*. Fisheries Environmental Report 60. Fisheries Research Division, New Zealand Ministry of Agriculture and Fisheries, Christchurch.

Ministry of Agriculture and Forestry (2001) *New Zealand Forestry Statistics 2000*. Ministry of Agriculture and Forests, Wellington.

Morgan, D.R. & Graynoth, E. (1978) The influence of forestry practices on the ecology of freshwater fishes in New Zealand: an introduction to the literature. *Fisheries Research Division Occasional Publication No. 14*, New Zealand Ministry of Agriculture and Fisheries, Wellington.

Mosley, M.P. (1981) The influence of organic debris on channel morphology and bedload transport in a New Zealand forest stream. *Earth Surface Processes and Landforms*, 6, 571–579.

Mosley, M.P. (1982) New Zealand river temperature regimes. *Water and Soil Miscellaneous Publication No. 36*. Ministry of Works and Development, Christchurch.

Murphy, G., Collier, K., Baillie, B., Boothroyd, I., Langer, L. & Quinn, J. (1998) Development of a riparian zone decision support system for production forest environments. *Water and Atmosphere* 6, 26.

New Zealand Soil Bureau (1968a) *Soils of New Zealand. Part 3*. New Zealand Soil Bureau Bulletin, 26. New Zealand Department of Scientific and Industrial Research, Wellington.

New Zealand Soil Bureau (1968b) *Soils of South Island New Zealand*. New Zealand Soil Bureau Bulletin 27. New Zealand Department of Scientific and Industrial Research, Wellington.

O'Loughlin, C.L., Rowe, L.K. & Pearce, A.J. (1980) Sediment yield and water quality responses to clearfelling of evergreen mixed forests in western New Zealand. In: *The influence of man on the hydrological regime with special reference to representative and experimental basins. Proceedings of the Helsinki Symposium, June 1980*, pp. 285–92. IAHS Publication number 130.

Pang, L. (1993) *Tarawera River flow analysis*. Environmental Report 93–2. Unpublished report, Environment Bay of Plenty, Whakatane.

Quinn, J.M., Cooper, A.B., Davies-Colley, R.J., Rutherford, J.C. & R.B. Williamson (1997) Land-use effects on habitat, water quality, periphyton, and benthic invertebrates in Waikato, New Zealand, hill-country streams. *New Zealand Journal of Marine and Freshwater Research*, 31, 579–7.

Roberts, N.J.V. (1994) Past, present, and future forest land management practices in New Zealand. *New Zealand Forestry*, **August**, 22–6.

Roche, M. (1990) *History of Forestry*. New Zealand Forestry Corporation Ltd in association with GP Books, Wellington.

Rowe, D.K. & Chisnall, B.L. (1997) Environmental factors associated with the decline of dwarf inanga *Galaxias gracilis* McDowall in New Zealand dune lakes. *Aquatic Conservation: Marine and Freshwater Ecosystems*, 7, 277–86.

Rowe, D.K., Chisnall, B.L., Dean, T.L. & Richardson, J. (1999) Effects of land use on native fish communities in east coast streams of the North Island of New Zealand. *New Zealand Journal of Marine and Freshwater Research*, 33, 141–51.

Rowe, D.K., Smith, J., Quinn, J. & Boothroyd, I. (2002) Effects of logging with and without riparian strips on fish species abundance, mean size, and the structure of native fish assemblages in Coromandel, New Zealand, streams. *New Zealand Journal of Marine and Freshwater Research*, 36, 67–79.

Rutherford, J.C., Davies-Colley, R.J., Quinn, J.M., Stroud, M.J. & Cooper, A.B. (1999) *Stream Shade: Towards a Restoration Strategy*. Department of Conservation, Wellington.

Scrimgeour, G.J. (1989) Effects of bleached kraft mill effluent on macroinvertebrate and fish populations in weedbeds in a New Zealand hydro-electric lake. *New Zealand Journal of Marine and Freshwater Research*, 23, 373–9.

Sharples, A.D. & Evans, C.W. (1998) Impact of pulp and paper mill effluent on water quality and fauna in a New Zealand hydro-electric lake. *New Zealand Journal of Marine and Freshwater Research*, 32, 31–53.

Statistics New Zealand (1998) *New Zealand Official Yearbook. 101*. GP Publications, Wellington.

Taylor, M.J. & Main M.R. (1987) *Distribution of Freshwater Fishes in the Whakapohai River to Waita River Area, South Westland*. Fisheries Environmental Report 77. Fisheries Research Division, New Zealand Ministry of Agriculture and Fisheries, Christchurch.

Taylor, R. & Smith, I. (1997) *The State of New Zealand's Environment*. Ministry for the Environment, Wellington.

Tomlinson, A.I. (1992) Precipitation and atmosphere. In: *Waters of New Zealand*, (ed. M.P. Mosley), pp. 61–74. New Zealand Hydrological Society, Wellington.

Valentine, D.A. (1995) *The effects of forestry operations and catchment development on lotic ecosystems in North Westland*. Unpublished Master of Forestry Science thesis, University of Canterbury, Christchurch.

Vaughan, L., Visser, R. & Smith, M. (1993) *New Zealand Forest Code of Practice*, 2nd edn. New Zealand Logging Industry Research Organisation, Rotorua.

Wardle, P. (1991) *Vegetation of New Zealand*. Cambridge University Press, Cambridge.

Willis, K.J., Ling, N. & Chapman, M.A. (1995) Effects of temperature and chemical formulation on the acute toxicity of pentachlorophenol to *Simocephalus vetulus* (Schoedler, 1858) (Crustacea: Cladocera). *New Zealand Journal of Marine and Freshwater Research*, **29**, 289–94.

Chapter 30
Australia

W.D. ERSKINE AND J.H. HARRIS

Introduction

Australia is an ancient continent characterized by low relief and rainfall, highly weathered soils, saline surface waters, a small forested area and a low diversity, generalized fish fauna. Furthermore, the effects of regulation of the low, highly variable runoff in southeastern Australia for irrigation and urban water supplies have significantly impacted on the resilience of native fish communities to invasion by alien species (Gehrke & Harris 2001). Accelerated soil erosion following the clearing of native forest by European settlers for agriculture caused detrimental changes to fish habitat due to the influx of large sand volumes to many rivers which infilled pools, forming mobile sand slugs (Erskine 1994, 1999; O'Connor & Lake 1994).

Forestry impacts on fish are still poorly documented. Greater attention has been devoted to determining the hydrologic and soil erosion effects of forest harvesting and road construction. These physical effects are important because they can impact on fish by altering stream flow regimes, reducing aquatic habitat and diversity, obstructing physical connectivity of rivers and degrading riparian zones (Harris & Silveira 1999).

The biophysical characteristics of Australia are described before discussing the fish fauna and forestry activities in Australia. Issues relating to fish–forestry interactions are then briefly discussed before outlining the impacts of integrated logging on fish in the Genoa River basin in southeastern Australia. Lastly, recommendations are made for improvements in forestry practices and standards for fish conservation.

Biophysical characteristics

Geomorphology

Jennings & Mabbutt (1986) identified three major landform divisions which they called the Eastern Highlands, Interior Lowlands and Western Plateau. The Eastern Highlands are a continuous linear upland that extends the complete length of the east coast, including Tasmania. The broad Interior Lowlands parallel the Eastern Highlands on the western side. The Western Plateau is a large area covering the western half of the continent and is highly variable in character, including sand dunes, salt lakes, mountains, plateaux and plains. Some of the oldest landforms in the world are present here (Ollier 1986).

These large-scale landforms have formed over tens to hundreds of millions of years and were not modified, to any significant degree, by Pleistocene glaciation although Cretaceous marine incursions inundated a relatively large area (Jennings & Mabbutt 1986; Ollier 1986). Landform evolution in Australia has occurred on the same timescale as continental drift and biological evolution, and many of the gross landforms have changed little for tens to hundreds of millions of years (Ollier 1986). As a result, regolith (surficial earth materials down to fresh bedrock including transported sediments and *in situ* weathered bedrock, both saprolite and residual material) has experienced very long periods of weathering and is characterized by large areas of infertile ferricrete, silcrete, bauxite and sand dunes (Ollier 1986). Australia has its origins in Gondwanaland and started to drift apart from Antarctica about 125 Mya. Actual separation was progressive between 125 and 10 Mya (Ollier 1986). Australia is a stable continent with little contemporary seismic activity and no currently active volcanoes (Jennings & Mabbutt 1986; Ollier 1986). The dominant landforms are low relief plains, tablelands and plateaux (Jennings & Mabbutt 1986).

Climate

Australia is located between latitudes 10°S and 42°S and is, therefore, subjected to west-to-east passing subtropical high pressure cells which are normally associated with stable climatic conditions of clear skies and low rainfall (Gentilli 1986; Warner 1986). True monsoonal conditions do not always reach north Australia and, as a result, high summer rainfall does not extend far inland (Gentilli 1986). The low relief results in restricted orographic effects on rainfall.

Australia is the driest inhabited continent. Precipitation is dominated by rainfall, with snow being restricted to the highest parts of the Eastern Highlands of southeastern Australia. Median annual rainfall varies greatly from about 100 mm east of Lake Eyre to 3685 mm at Lake Margaret in Tasmania and to 4745 mm near Tully on the north Queensland coast (Gentilli 1986; Warner 1986). About half of the continent receives $<350\,\mathrm{mm\,y^{-1}}$ and over a third receives $<250\,\mathrm{mm\,y^{-1}}$ (Gentilli 1986). Only a part of the coastal margin receives over $800\,\mathrm{mm\,y^{-1}}$. Summer rainfall occurs in the north, northeast and east, particularly on the coastal margins. Winter rainfall dominates in the southwest and southeast coastal mainland areas and in Tasmania due to the repeated passage of low pressure cells and associated cold fronts between at least April and September. A small area of the southeast mainland experiences both of the above conditions and hence receives a fairly uniform rainfall distribution. The inland areas cover two-thirds of the mainland and experience semi-arid and arid conditions.

Average annual pan evaporation varies from <1000 mm for western Tasmania and the highest parts of the Eastern Highlands to >4500 mm for the desert areas in north central Western Australia (Warner 1986). Such potential evaporation greatly exceeds rainfall for all of Australia, except for very small areas (Warner 1986). Actual evaporation is limited by available water. Over most of Australia, evaporation is approximately equal to rainfall and, therefore, runoff is negligible and only occurs during intense storms (Warner 1986). Furthermore, many Australian inland waters are saline due to a range of mechanisms but evaporative concentration of the low runoff is important (Bayly & Williams 1973).

Hydrology

Twelve major drainage divisions have been defined for water resources purposes (Fig. 30.1). There is a narrow peripheral discontinuous zone of coordinated exterior drainage (eight drainage divisions), one integrated internal drainage system with an external outlet to the sea (Murray-Darling), one coordinated internal drainage system (Lake Eyre) and two disconnected drainage divisions (Bulloo-Bancannia and Western Plateau) with no discharge to the ocean (Warner 1986). While the three completely internal drainage systems cover 48.5% of the land surface of Australia, they generate only 0.9% of total runoff (Warner 1986). The eight coordinated external drainage systems cover 37.7% of the land surface but produce 93.9% of total runoff (Warner 1986).

McMahon *et al.* (1992) found that the variability of annual runoff in Australia for the same climate is greater by a factor of two than for the rest of the world, except

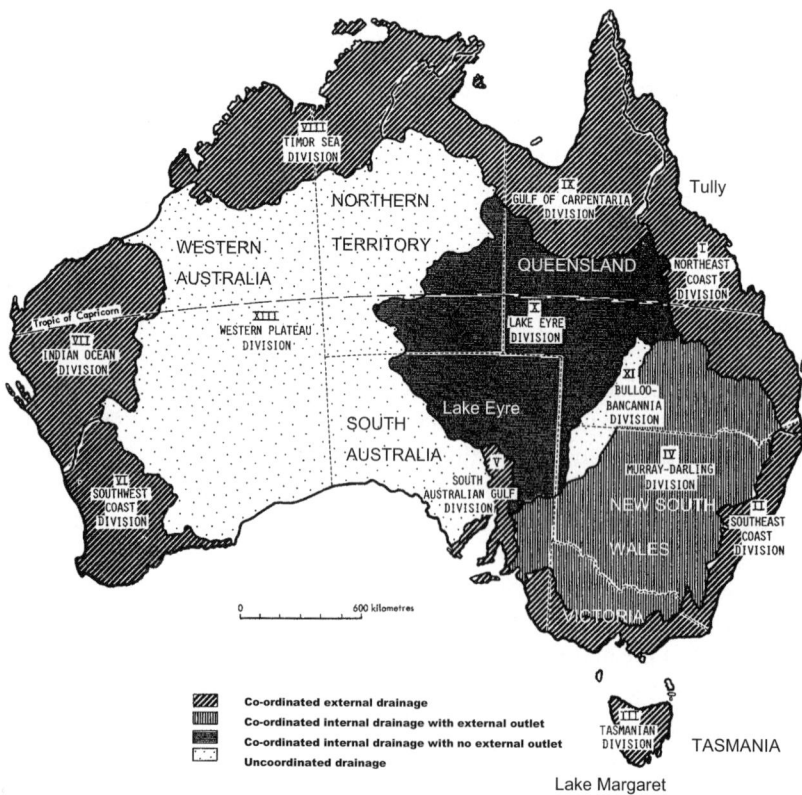

Fig. 30.1 The major drainage divisions of, and drainage types in, Australia (after Warner 1986). There is little surface water in the Western Plateau Division. Reproduced with permission of the author.

southern Africa. Furthermore, variability increased with basin area, which is an unusual result. Flood variability is much greater in Australia for all basin sizes than in other continents and also increases with basin area (McMahon *et al.* 1992; Erskine & Livingstone 1999). Erskine (1996) and Erskine & Livingstone (1999) reported some of the highest values of flood variability ever recorded. The mean flood ratio (flood peak discharge for a specified recurrence interval/mean annual flood) for a 1:100 year flood for all Australian rivers is about 5, the mean for rivers worldwide is <4 and that for European rivers is only 2 (McMahon *et al.* 1992). Many Australian rivers have a flood ratio >10 (McMahon *et al.* 1992). This indicates a high potential for very large floods which cause damage to fish habitat and riparian zones (Erskine 1993, 1996, 1999).

Forest resources

Australia has about 156 Mha (20.3%) of native forest and woodland and 1 Mha (0.13%) of plantations on a landmass of 769 Mha (Montreal Process Implementation Group for Australia – MPIG 1997). Closed forests have a canopy cover >80% and occupy 4.6 Mha (3% of forest and woodland cover); open forests have a canopy cover of 50–80% and occupy 39.1 Mha (25%); and woodlands have a canopy cover of approximately 20–50% and occupy 112.1 Mha (72%) (MPIG 1997). Forestry activities are conducted in all three forest types. Up to 25% of some forest regions consists of woodland that is managed for timber production.

Table 30.1 lists the main forest types of Australia, which are restricted to the coastal margins where median annual rainfall is usually >500 mm. Eucalypts dominate (Table

Table 30.1   Extent of each forest and woodland type in Australia (MPIG 1997)

Forest type	Area (Mha)	%
Rainforest	3.583	2.3
Tall open eucalypt forest	5.475	3.5
Medium open eucalypt forest	22.656	14.4
Low open eucalypt forest	0.385	0.2
Tall eucalypt woodland	1.068	0.7
Medium eucalypt woodland	68.8	43.9
Low eucalypt woodland	14.315	9.1
Eucalypt mallee[1] forest and woodland	11.764	7.5
Callitris forest and woodland	0.867	0.6
Acacia forest and woodland	12.298	7.8
Melaleuca forest and woodland	4.093	2.6
Mangrove forest and woodland	1.045	0.7
Casuarina forest and woodland	1.052	0.7
Unclassified forest and woodland	8.435	5.4
Plantation forest and woodland	1.043	0.7
Total	156.878	100

[1]Mallee is a growth form where several small stems grow from a lignotuberous rootstock.

30.1) being found in about 86% of open forest and woodland and are very diverse with over 700 species, most endemic to Australia. Eucalypts are a light-demanding species and strongly compete for water and nutrients. The difficulty for forest management is to determine environmentally acceptable silvicultural treatments that produce sufficient disturbance to ensure good regeneration and that selectively remove slow growing trees (Nicholson 1999). In 1994 Australia had 883,980 ha of softwood plantations and 158,570 ha of hardwood plantations (MPIG 1997). Only 8.6% of the total forested area in Australia is State forest managed for timber production (MPIG 1997).

Freshwater fish

Only 214 of the approximately 10,000 described freshwater fish species are found in Australia (Allen *et al.* 2002). The freshwater ichthyofauna is depauperate in terms of land mass area (but not fish habitat), mostly derived from marine ancestors, highly endemic and dominated by generalist species (Allen 1989; McDowall 1996; Allen *et al.* 2002). Two main factors have contributed to this situation (Allen 1989). Firstly, Australia separated from the rest of Gondwanaland before the dominant freshwater fish families could reach Australia. Secondly, Australia is so dry that much of the continent lacks permanent rivers and surface water, and this prevented the development of a large diversity of freshwater fish fauna.

Table 30.2 lists the fish families found in the freshwaters of Australia. However, the only families that have evolved entirely in freshwater are Ceratodontidae (the only extant species is *Neocerotodus forsteri*) and Osteoglossidae (two species, *Scleropages leichardti* and *S. jardinii*). Approximately 22 alien species of fish have also become established in Australia (Table 30.3). In addition to these freshwater fish, there are at least another 20 families of estuarine fish with an estimated 150 species which enter the lower part of freshwater streams (Allen 1989).

The conservation status of freshwater fish has declined severely since European settlement (Allen *et al.* 2002). A total of 76 species were listed in various classifications of threatened fishes in 1997 by the Australian Society for Fish Biology (Jackson 1997). Seventeen of these were classed as Endangered or Vulnerable. Following a fish survey in New South Wales, Harris & Gehrke (1997) concluded that riverine ecosystems were rapidly losing their biodiversity and that, as indicators of river ecosystem condition, the fish were in severe decline. Numerous causes were identified, especially habitat degradation through sedimentation following watershed damage, river regulation, coldwater pollution downstream of large dams and barriers obstructing migration pathways.

Forestry in Australia

Timber has been an important resource since Australia was first settled by Europeans in 1788 (Grant 1989), although Aborigines also used timber for firewood for at least 40,000 years. The rainforest softwoods of eastern Australia, particularly *Toona cili-*

Table 30.2   Native fish families found in the freshwaters of Australia (Allen *et al.* 2002)

Family or subfamily	Common name	Genera Australian	Number of species
Mordaciidae	Lampreys	*Mordacia*	2
Geotriidae	Lampreys	*Geotria*	1
Ceratodontidae	Australian lungfish	*Neocerotodus*	1
Anguillidae	Freshwater eels	*Anguilla*	4
Clupeidae	Herrings	*Nematolosa*	1
		Potamolosa	1
Osteoglossidae	Bony tongues	*Scleropages*	2
Engraulidae	Anchovies	*Thryssa*	1
Retropinnidae	Southern smelts	*Retropinna*	2
Aplochitoninae	Tasmanian whitebait	*Lovettia*	1
Lepidogalaxiidae	Salamanderfish	*Lepidogalaxias*	1
Galaxiidae	Galaxiids	*Galaxias*	14
		Paragalaxias	4
		Galaxiella	3
Prototroctidae	S. hemisphere Grayling	*Prototroctes*	1
Ariidae	Forktailed catfish	*Arius*	4
		Cinetodus	1
Plotosidae	Eel-tailed catfish	*Anodontiglanis*	1
		Neosiluroides	1
		Neosilurus	6
		Porochilus	3
		Tandanus	2
Hemiramphidae	Garfishes	*Arrhamphus*	1
		Zenarchopterus	2
Belonidae	Longtoms	*Strongylura*	1
Atherinidae	Hardyheads	*Atherinosoma*	1
		Craterocephalus	13
		Leptatherina	1
Melanotaeniidae	Rainbowfish	*Cairnsichthys*	1
		Iriatherina	1
		Melanotaenia	12
		Rhadinocentrus	1
Pseudomugilidae	Blue-eyes	*Pseudomugil*	4
		Scaturiginichthys	1
Synbranchidae	Swamp-eels	*Monopterus*	1
		Ophisternon	3
Scorpaenidae	Scorpionfish	*Notesthes*	1
Centropomidae	Giant perches	*Lates*	1
Ambassidae	Glassfishes	*Ambassis*	8
		Denariusa	1
		Parambassis	1

Table 30.2　(*Continued.*)

Family or subfamily	Common name	Genera Australian	Number of species
Percichthyidae	Perichthyids	*Bostockia*	1
		Guyu	1
		Maccullochella	3
		Macquaria	4
Terapontidae	Grunters	*Amniataba*	1
		Bidyanus	2
		Hannia	1
		Hephaestus	5
		Leiopotherapon	3
		Pingalla	3
		Scortum	5
		Syncomistes	4
		Variichthys	1
Nannopercidae	Pygmy perches	*Edelia*	1
		Nannatherina	1
		Nannoperca	4
Kuhliidae	Flagtails	*Kuhlia*	1
Apogonidae	Cardinalfish	*Glossamia*	1
Toxotidae	Archerfish	*Toxotes*	3
Gadopsidae	Freshwater blackfishes	*Gadopsis*	2
Mugilidae	Mullets	*Myxus*	1
Bovichtidae	Tupong	*Pseudaphritis*	1
Eleotridae	Gudgeons	*Bostrichthys*	1
		Bunaka	1
		Giurus	1
		Gobiomorphus	2
		Hypseleotris	10
		Kimberleyeleotris	2
		Milyeringa	1
		Mogurnda	6
		Oxyeleotris	5
		Philypnodon	2
Gobiidae	Gobies	*Awaous*	1
		Chlamydogobius	6
		Glossogobius	7
		Pseudogobius	1
		Schismatogobius	1
		Sicyopterus	1
		Stiphodon	1
Cynoglossidae	Tongue soles	*Cynoglossus*	1
Soleidae	Soles	*Aseraggodes*	1
		Brachirus	2
Total: 36		83	214

Table 30.3   Alien fish families found in the freshwaters of Australia (Allen *et al.* 2002)

Family	Common name	Genera	Number of species
Salmonidae	Salmon and trout	*Oncorhynchus*	2
		Salmo	2
		Salvelinus	1
Cyprinidae	Carps and barbs	*Carassius*	1
		Cyprinus	1
		Puntius	1
		Rutilus	1
		Tinca	1
Poeciliidae	Livebearers	*Gambusia*	2
		Phalloceros	1
		Poecilia	2
		Xiphophorus	2
Percidae	Perches	*Perca*	1
Cichlidae	Cichlids	*Cichlasoma*	1
		Oreochromis	1
		Tilapia	1
Cobitidae	Loaches	*Misgurnus*	1
Total: 6		17	22

ata (cedar), were selectively harvested in preference to the eucalypt hardwoods. These 'cedar getters' were the first settlers of most coastal valleys north and immediately south of Sydney (Swain 1912; Grant 1989). Eucalypt forests were progressively used as the rainforest was cleared for agriculture (Swain 1912). Gold mining after 1850 and ringbarking (girdling) after 1870 resulted in the loss of large areas of eucalypt forests. Nevertheless, timber production was an important activity throughout the nineteenth and twentieth centuries (Grant 1989).

Integrated logging, the joint harvesting of saw logs and pulp wood, has developed into a major but controversial industry since the early 1970s (Ovington & Thistlethwaite 1976; Lunney & Moon 1987). Large forest areas in New South Wales, Tasmania and Western Australia are involved. In New South Wales, integrated logging is restricted to southern coastal forests and replaced single-tree selection. Single-tree and group selection (canopy openings averaging no more than 40 m) are the main harvesting methods elsewhere in New South Wales but they usually do not cause sufficient disturbance for forest regeneration (Nicholson 1999). Integrated logging was initially carried out as clear-felling in 800-ha compartments with some corridors of undisturbed forest. In 1972, compartments were reduced to 200 ha and only alternate coupes were logged. In 1976, 15-ha alternate coupes were adopted but the size of alternate coupes was increased to about 50 ha. The basis of alternate coupe logging is that adjacent coupes are harvested at about 20-year intervals. Current licence conditions (Anonymous 1999) constrain harvesting and stipulate the retention of many seed and habitat trees.

Regional forest agreements have been completed for a large part of the forest estate in Australia and are based on the National Forest Policy statement which is the national strategy for achieving ecologically sustainable forest management. The key outcomes of this strategy were improved environmental management, the creation of a comprehensive, adequate and representative (CAR) reserve system and reliable wood supply for the next 20 years at reduced levels for the timber industry. To identify land for the CAR reserve system, a programme of Comprehensive Regional Assessments was implemented across New South Wales, Victoria, Queensland, Tasmania and Western Australia.

Forestry activities have become increasingly regulated over the last 30 years. A single integrated approval for each regional forest agreement area was implemented in New South Wales at the end of 1999. Each approval contains about 1600 conditions (see below) and three individual Licences. This Integrated Forestry Operations Approval (e.g. Anonymous 1999) ratifies the outcomes of the regional forest agreements and includes conditions covering forest planning, soil and water protection, biodiversity, threatened species protection, fish passage, aquatic habitat protection and other issues. As a result of this increasing regulation, larger areas of forest have been excluded from harvesting by special protection prescriptions. Licences are either based on best management practices (for water pollution and fish) or species-specific habitat prescriptions (for other fauna and flora). Compliance with licence conditions is being internally and externally audited.

The fish licence covers a wide range of issues (e.g. Anonymous 1999). Riparian exclusion or buffer zones, in which all forestry activities are prohibited, are defined on the basis of stream order. First, second, third and higher order streams have exclusion zones at least 10, 20 and 30 m wide, respectively. An additional 20-m wide machinery exclusion zone is stipulated on the landward side of buffers on third and higher order streams. Zero order drainage forms (drainage features not depicted on topographic maps) have either 5- or 10-m wide buffers, depending on the degree of channel incision. Wetlands >4 m^2 in area are also protected by exclusion zones between 10 and 40 m wide. The habitat of threatened species is afforded special protection. Road crossings must not act as barriers to fish passage. Targeted fish species surveys are required for threatened species.

Issues on fish–forestry interactions in Australia

The most significant issues in Australia are:

- increased sediment yields from harvested compartments and forest roads;
- water yield reductions due to harvesting-induced eucalypt forest regeneration;
- maintenance of integrity and biogeomorphic functioning of the riparian zone;
- fish habitat protection; and
- maintenance of upstream and downstream fish passage.

Each issue is now discussed briefly and integrated logging is covered in the next section.

Harvesting and road effects on sediment yields

Increased sediment yields from harvested compartments and/or forest roads can impact on fish by in-filling pools, blanketing the benthos with mud, adding fine sediment to gravel substrates, reducing water quality, increasing invertebrate drift, reducing macroinvertebrate diversity and abundance, impacting on the reproductive potential of fish (Davies & Nelson 1993, 1994) and nutrient loading. Roads are usually the most significant sediment source in Australian forests (Davies & Nelson 1993; Croke *et al.* 1999a) and are further discussed under integrated logging.

Riley (1988) found for the tall closed eucalypt forests in New South Wales that cut road batters (excavated into soil and bedrock) generated one order of magnitude more sediment than fill batters (placed soil and bedrock), which, in turn, generated one order of magnitude more than native forest. However, the rate of erosion declined exponentially over time following road construction although sediment production from the road surface was not measured (Riley 1988). In Victoria, Grayson *et al.* (1993) reported that the harvesting of tall open *Eucalyptus regnans* forest according to a strict code of practice with no roads in runoff-producing zones resulted in no effect on storm flow turbidity and suspended solids concentrations. For base flows, small increases were recorded which were of a similar magnitude to the measurement error. Annual sediment yields were 50–90 t ha y^{-1} from the road surface. Gravel reduced sediment generation rates from roads, as did more frequent road maintenance because of increased traffic.

In the Dazzler Range, Tasmania, Davies & Nelson (1993) determined the infiltration of sand and organic matter in the bed of ephemeral and perennial streams following skyline cable logging, roading and post-harvest burning in tall open eucalypt forests on steep slopes. Total sediment infiltration for logged ephemeral first order streams was significantly greater than for unlogged streams. There was also a significant increase in infiltration of sediment with diameters of <125, 125–250 and 250–500 µm by factors of 2.98, 2.52 and 1.86, respectively. The mean organic sediment yield for the <125 and 0.5–1 µm fractions was also significantly greater by factors of 2.4 and 1.7, respectively, for logged than for unlogged streams but both declined with time since logging. The only significant difference between logged and unlogged streams for inorganic sediment yield was for the <125 µm fraction. For perennial streams, Davies & Nelson (1993) found no significant change in fine sediment infiltration on riffles below junctions with logged tributaries. However, road crossings caused large increases in sediment infiltration in downstream riffles, even 30–50 years after construction, when roads were still used.

Davies & Nelson (1994) investigated the effects of cable logging and clear-felling with ground-skidder log haulage of tall closed and open eucalypt forests in Tasmania on stream habitat, riffle macroinvertebrates and fish. Silt cover on the stream bed increased significantly where buffer widths were <30 m. However, such narrow buffer widths did not conform to the relevant Forest Practices Code.

In Tasmania, Wallbrink & Murray (1996) used the fallout radionuclides ^{137}Cs and [210]Pb$_{ex}$ to measure soil losses from different logging treatments in tall open eucalypt forests. Fourteen months elapsed between harvesting and soil sampling. Because of concerns about large spatial variability in atmospheric radionuclide fallout, Wallbrink & Murray (1996) proposed a new method of determining soil loss rates based on the average inventory ratios of fallout ^{210}Pb to [137]Cs. The inventory ratios for the unreplicated harvested plots were significantly different and the average soil loss from the heavily logged plot was 40 ± 5 mm (440 ± 55 t ha^{-1}) while soil loss from the conservatively harvested plot was 17 ± 5 mm (190 ± 55 t ha^{-1}). The standard errors refer to the variation between soil cores for individual treatments which were unreplicated.

Wilson (1999) used a rainfall simulator 6 months after logging and fire on four 300-m^2 plots on 15° slopes in the same forest as Wallbrink & Murray (1996). The two logged plots of Wallbrink & Murray (1996) were also included in the rainfall simulator experiment; another plot had been burnt by an intense wildfire; and the last was an unlogged and unburnt control. Lack of treatment replication limits the applicability of both studies. The standard harvested plot produced the greatest runoff for all rainfall intensities and the conservatively logged plot exhibited greater runoff for all rainfall intensities than the other two plots. The standard harvested plot also had the highest total erosion for all rainfall intensities and the control had the the lowest. A significant result of Wilson's (1999) research was that fire and/or intensive harvesting destroyed a biotic soil crust, increasing sediment production. Such crust destruction confounds the results.

Water yield response to forest harvesting

Fish habitat is discharge-dependent and eucalypt forest regeneration following harvesting has caused significant reductions in water yields in southeastern Australia (Vertessy *et al.* 1998; Cornish & Vertessy 2001; Erskine 2001). Erskine (2001) synthesized the experimental results and produced the three generalized water yield response curves in Fig. 30.2. The time-span covered by the abscissa is variable depending on the growth characteristics and longevity of the tree species but is shown as 50 years for convenience.

Curve a shows the most commonly reported situation where water yield increases immediately after harvesting and subsequently declines logarithmically to the original pre-harvest yield as the forest gradually recovers and ages over time. However, the nature of this yield decline is highly variable, depending on the age of the original forest, the rate of forest recovery/regeneration, its silvicultural treatment, the incidence of insect attacks and the water demand of the vegetation. Water yields often recover to pre-treatment levels in 5–25 years after timber harvesting. The magnitude of the initial increase is dependent on the proportion of tree cover removed. *Curve b* shows a common situation in many eucalypt forests in southeastern Australia (Kuczera 1987) where water yields rapidly decline to significantly below pre-harvest values because of the higher leaf and sapwood areas and hence higher water demand of the regrowth

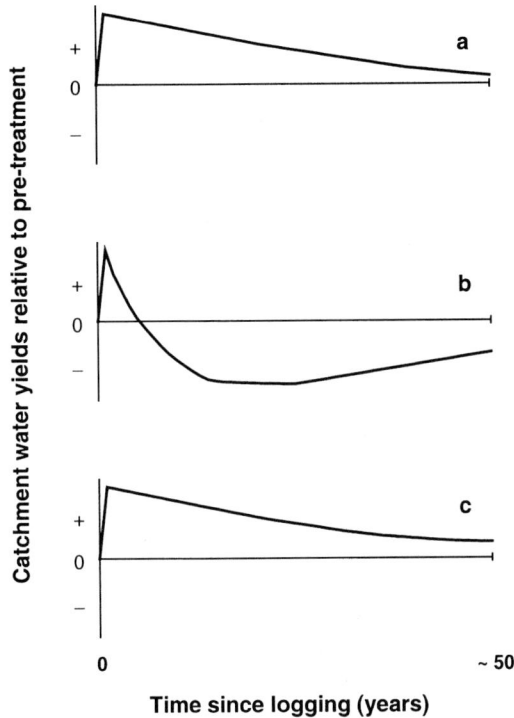

Fig. 30.2 Generalized water yield response curves for the 50 years after forest harvesting (after Erskine 2001). (a) Initial yield increase with gradual recovery to pre-harvest levels. (b) Initial yield increase followed by rapid yield reductions in regenerating eucalypt forests. (c) Initial water yield increase followed by thinning of regrowth eucalypt forest which increases water yields over pre-harvest values. Reproduced with permission of IAHS Press from: Erskine, W.D. (2001) Water yield response to integrated native forest management in Southwestern Australia. In: *Integrated Water Resources Management* (ed. M.A. Marino & S.P. Simonovic) (Proc. Davis Symposium, April 2000), 249–255. IAHS Publ. no. 272. IAHS Press, Wallingford, UK. Copyright IAHS Press 2001.

than the original forest. Such trends were initially documented for tall open *Eucalyptus regnans* forests near Melbourne (Kuczera 1987; Vertessy *et al.* 1998; Watson *et al.* 1999) but have been more recently documented for most eucalypt forest types in New South Wales also (Cornish & Vertessy 2001; Lane & Mackay 2001; Roberts *et al.* 2001). It must be stressed that this trend occurs whenever eucalypt forest is successfully regenerated and was first documented following wildfire (Kuczera 1987). *Curve c* shows a situation produced by active silviculture which increases yield over that shown in curve a. The regenerating vegetation has a lower water demand in comparison to the original forest because of thinning and hence post-treatment water yields are higher than from the original forest. Thinning reduces tree basal area and leaf area. Forest harvesting activities are deliberately dispersed spatially so that usually only 0.5–1% of the total basin area is treated in any 1 year (Erskine 1992, 2001). Water yield increases and reductions have implications for the maintenance of fish habitat and the movement of diadromous species.

Harvesting effects on riparian zones

The important issues for maintenance of the integrity and biogeomorphic functioning of the riparian zone in timber production forests are that adequate buffer widths are maintained to protect streams from sediment inputs and to protect riparian vegetation communities and their associated fauna from damage by harvesting. Dignan (2001) surveyed 10 km of stream buffers in the West Tarago River basin, Victoria, for visual evidence of sediment deposition within buffers. He found only eight unmanaged incursions due to either failure of in-coupe sediment management techniques or road drainage. Only one reached a stream.

Riparian vegetation usually exhibits marked vertical and lateral zonation. Webb *et al.* (2001) and Webb & Erskine (2001) investigated the *Tristaniopsis laurina* riparian forest in New South Wales and Victoria. In Nadgee State Forest (New South Wales) *T. laurina* grows in the bed on riffles, on the face of river banks and in a narrow strip next to the channel (Webb *et al.* 2001). This species rarely extends >30 m from the channel before being replaced by tall open eucalypt forest. A 30-m wide buffer strip is sufficient to protect this riparian community. Furthermore, wood identification of in-stream large woody debris has shown that *T. laurina* is the dominant source tree and, therefore, potential recruitment trees are usually restricted to one tree height from the channel bank (Webb *et al.* 2001; Webb & Erskine 2001). However, similar information for other riparian vegetation communities is lacking.

Fish habitat protection

Davies & Nelson (1994) found that snag volumes and length of open stream increased at sites downstream of logged areas where buffer widths were <10 m. The former effect was caused by the input of logging debris and the latter by the decrease in buffer canopy cover due to harvesting next to the buffer (Dignan 2001). Stream temperatures also increased significantly (by 10%) where stream buffers were <10 m wide. Semi-aquatic macroinvertebrates decreased in abundance with decreasing buffer strip width, with leptophlebiid mayflies and stoneflies being most affected at widths <30 m. They did not find enough native fish to test the effect of harvesting on fish abundance and biomass. We have also found low numbers of native fish in many New South Wales rivers. However, for the alien *Salmo trutta* Davies & Nelson (1994) found both a significant positive correlation between change in abundance between logged and unlogged streams and buffer width; and a statistically significant decrease in abundance between logged and unlogged sites for streams with buffers <30 m wide.

Riparian forest supplies large woody debris to streams in Australia mainly by tree senescence, bank erosion, wind-throw and lightning strikes. This debris is essential for inducing local scour that forms pools that are utilized by a range of fish (Webb & Erskine 2001). As many riparian tree species are hardwoods, the wood survives for long periods of time and accumulates at loadings that are often large by world standards (Webb *et al.* 2001; Webb & Erskine 2001). The pools are relatively small and spaced much closer than free-formed alluvial pools (Webb & Erskine 2001). The maintenance of riparian vegetation is essential for recruitment of large woody debris which then dissipates stream energy, stabilizes channels and forms pools (Webb & Erskine 2001; Webb *et al.* 2001). Therefore, harvesting-induced large woody debris accessions should not be removed from streams.

Riparian buffer zones 30 m wide certainly protect fish habitat on streams with basin areas between 2.5 and 40.7 km^2 in Tasmania (Davies & Nelson 1994). Buffers at least 30 m wide, and occasionally wider, on high order streams are implemented as part of most codes of forestry practice or licences in Australia.

Fish passage

Northcote (1978) defined migration as 'movements resulting in an alternation between two or more separate habitats occurring with regular periodicity and involving a large fraction of the population'. This definition shows a need for understanding of the spatial and temporal scales of migration (Northcote 1998). It is uncommon for fish to complete their life cycle in one location and the simplest general model of alternating habitat use would include movements between separate habitats for spawning and growth (Northcote 1998). Commonly a third class of habitat, nursery areas, is incorporated into the movement pattern during fish life histories. Survival of fish populations depends on maintaining movement between these fundamental habitats (Northcote 1998). Whether the movements are at a river-basin scale over a period of years, or merely at a local scale over days, is irrelevant.

Freshwater fish must be able to move between discrete habitats during their lives. Habitat maintenance, plus free passage between habitats, are required for sustainable fish populations. Furthermore, within adult growth habitats, each fish moves to feed within its home range.

Fish passage processes operate at widely differing scales. Basin-scale migrations, their significance for fish production and the often dramatic consequences of blocking them have been well documented (Northcote 1978, 1998). Some Australian fish, especially the smaller species, have short-range life cycle movements extending only a few kilometres, and are not so readily recognized as 'migratory'. They need to travel either through relatively small lengths of the river, or laterally to floodplains, and their populations may not be greatly affected by individual barriers. Larger species may travel long distances through river systems (Reynolds 1983) and are greatly affected by migration barriers (Harris & Mallen-Cooper 1994).

Fish migrations at the scale of reaches – up to hundreds of kilometres, rather than thousands – have been documented in Australia (Harris 1988; Mallen-Cooper *et al.* 1997). The ecological significance of connectivity at various scales and directions – spatial and temporal, lateral and longitudinal, upstream and down – in Australian rivers has also been documented (Mussared 1997; Schiller & Harris 2001).

The impacts of various types of forest road crossings on fish passage have not received the attention that they deserve in Australia. Of 93 species found in fresh water in southeastern Australia, the migrations of 48 species can be classified. Among these, there were six anadromous, nine catadromous, two amphidromous and 31 potamodromous species. Freshwater fish migrations are classified as:

- *Potamodromous* – fish that migrate wholly within fresh water, and
- *Diadromous* – fish that migrate between fresh water and the sea.

Within the diadromous group there are a further three subdivisions:

- *Anadromous* – diadromous fish that spend most of their life in the sea and migrate to fresh water to breed;
- *Catadromous* – diadromous fish that spend most of their life in fresh water and migrate to the sea to breed; and
- *Amphidromous* – diadromous fish that migrate between the sea and fresh water, but not for the purpose of breeding.

For the catadromous species, the upstream movement of the juveniles is essential for recruitment to maintain populations and they have poorer swimming capabilities than adults because of their smaller size. Barriers as low as 0.5 m in height can block the migration of juvenile catadromous fish like *Macquaria novemaculeata* (Harris 1984). Mallen-Cooper (1994) found for many Australian fish that the maximum velocity they could swim against increased with body size to a maximum velocity of 1.8 m s^{-1} which is less than that for similar-sized salmonids.

Road crossings should not increase natural flow velocities, so that the full range of fish sizes and species can pass these structures. Hydraulic designs for road crossings are often based on the minimum specific head principle (critical flow to produce minimum depth of flow for a given discharge) so as to build the smallest possible structure to pass the design discharge. While this reduces construction costs, such structures significantly restrict fish passage. Road crossing designs need to conform to fish passage principles (Boubee *et al.* 1999; see also Chapter 7).

Surveys of road crossing-related fish barriers have only been conducted for relatively small areas of Australia (e.g. Pethebridge *et al.* 1998). This survey did identify some forest roads as barriers to fish passage but most of the forest road network was not formally surveyed. Road crossings may be much more important barriers to fish passage than has been recognized to date. Furthermore, the ability of different species to pass through different types of road crossings (see Warren & Pardew 1998; Boubee *et al.* 1999) is unknown in Australia. While some Australian species are capable climbers (for example, juveniles of *Galaxias brevipinnis, Anguilla australis, Anguilla reinhardtii, Gobiomorphus coxii, Mogurnda mogurnda*), few jump over barriers like the North American salmonids and most have modest swimming abilities (Mallen-Cooper 1994).

Impacts of integrated logging on native freshwater fish in the Genoa River basin, New South Wales and Victoria

Background

The Genoa River basin covers 1950 km^2 of largely forested land in both New South Wales and Victoria (Fig. 30.3). Community concerns about the impacts of integrated

Fig. 30.3 The Genoa River basin in southeastern Australia.

logging on fish stocks highlight many issues with fish–forestry interactions in Australia. Sand deposition in the upper estuary (Genoa and Wallagaraugh rivers) has been the main issue (Erskine 1992; Brooks *et al.* 2001). Reduced catches of the estuarine *Acanthopagrus butcheri* and the catadromous *Macquaria novemaculeata*, plus an apparent lack of population recruitment, seemed to fuel the issue.

Roughley (1966) described a long-term decline in *A. butcheri* stocks in the Gippsland Lakes, Victoria and a similar decline has probably occurred in the Genoa River. *A. butcheri* is a highly fecund, long-lived (29+ years), slow growing fish that spawns in water with salinities of between 11 and 18 ppt (Roughley 1966). There is also little information on temporal trends in *M. novemaculeata* populations. It is a highly fecund, long-lived (22+ years), slow growing, catadromous fish that spawns in water with salinities of between 8 and 14 ppt (Harris 1985, 1986, 1987). Males are smaller and mostly remain in the upper estuary and lower freshwater reaches of streams, whereas females seasonally disperse throughout the freshwater reaches to utilize the summer peak in insect production that supports the growth needed for egg production (Harris 1986, 1987). However, Raadik (1992, 1994) only recorded *M. novemaculeata* from the Genoa River below the lower gorge (reach 6 in Fig. 30.3) which contains steep bedrock cascades and falls (Grant *et al.* 1990) that form a natural barrier to movement by species that do not climb or jump, such as *M. novemaculeata* (Harris 1984). Similarly, the authors have found a 2–3-m vertical bedrock fall on the Wallagaraugh River upstream of Timbillica which is a natural barrier. Therefore, the distribution of *M. novemaculeata* is naturally restricted to the lower sections of the Genoa (reaches 6, 7 and 8 in Fig. 30.3) and Wallagaraugh rivers.

Estuarine sedimentation and reduction in *A. butcheri* and *M. novemaculeata* stocks apparently occurred at the same time as large-scale integrated logging commenced in New South Wales. However, temporal association does not necessarily imply causation and integrated logging had not occurred in the Genoa River basin when sand aggradation was recorded in the estuary (Erskine 1992). Furthermore, there was never a clear enunciation of how estuarine sedimentation had altered estuarine morpho- and hydrodynamics to the extent that recruitment and population dynamics of *A. butcheri* and *M. novemaculeata* were detrimentally affected. Episodic closing of the estuary mouth (due to prolonged periods of low flows) presumably altered salt dispersion that, in turn, suppressed fish reproduction and recruitment. Initially, integrated logging was blamed for causing closure of the estuary mouth. However, the estuary mouth sand had lower mica and feldspar, and a higher quartz content than the river sand supplied by the Genoa and Wallagaraugh rivers and, in fact, originated offshore (Reinson 1977).

The Genoa and Wallagaraugh estuary is sensitive to sand deposition for three reasons. Firstly, the drainage basin is composed largely of various granitic rocks of the Bega batholith and sandstone which weather to produce large quantities of quartz sand. Secondly, both the Genoa and Wallagaraugh rivers are closely bedrock-confined for most of their length and hence efficiently transport sand to the head of the estuary (Erskine 1992, 1993). However, the bedrock gorges above the estuary also represent significant natural barriers to the passage of many diadromous fish. Thirdly, rainfall intensities over the basin are the highest in Victoria and flood variability is very high by world standards (Erskine 1992, 1993; Erskine & Saynor 1996; Brooks *et al.* 2001). Therefore, the channel and floodplain, where present, are episodically subjected to high energy, catastrophic floods which erode large volumes of alluvium that is then temporarily stored in the channel bed as a sand slug (Erskine 1993, 1999; Erskine & Saynor 1996).

Estuarine sedimentation

Erskine (1992, 1993) found that there had been significant sand storage in the Genoa River for about 7 km downstream of the lower granite gorge to the upper estuary (reaches 7 and 8 in Fig. 30.3). Up to 4 m of bed aggradation, which had displaced the tidal limit 2.3 km downstream, and substantial but localized floodplain deposition had occurred. However, sand storage was restricted to during, and immediately after, the flood of February 1971 (Erskine 1992, 1993; Erskine & Saynor 1996) and did not occur on the neighbouring Wallagaraugh River (Erskine 1992; Brooks *et al.* 2001). Sedimentation actually preceded the start of any integrated logging in the upstream basin (Erskine 1992). The February 1971 flood also caused large-scale channel erosion on the nearby Cann River in East Gippsland, Victoria (Erskine 1999).

Channel erosion by February 1971 flood

Erskine & Saynor (1996) found that the flood of 4–6 February 1971 on the Genoa River was generated by a storm with a maximum 24-hour rainfall of 290 mm and produced a peak instantaneous discharge between 3.15 and 5.00 $m^3 s^{-1} km^{-2}$. The flood peak discharge was 12.4 times greater than the mean annual flood and had a return period greater than 100 years on the annual maximum series (Erskine 1993). The specific sediment yield for this flood on the Genoa River was estimated at 292 $t km^{-2} yr^{-1}$ and the minimum amount of sediment generated by channel erosion (massive widening and floodplain stripping) alone during this flood was at least 3500 $t km^{-2}$ (Erskine & Saynor 1996).

Effects of integrated logging on fish

The upper reaches of Nungatta Creek (Fig. 30.3) are very steep and characterized by bedrock, log and boulder steps, step pools, cascades and rapids (Grant *et al.* 1990). The bedrock, boulder and log steps are important energy dissipators and the step pools eroded on the downstream side are utilized by a large population of the climbing fish, *Galaxias brevipinnis*. An average of 15 specimens per step pool was sampled. The presence of *G. brevipinnis* is important because it demonstrates that the channel and riparian zone are protected from desiccation by the reserved filter strips. O'Connor & Koehn (1997) found that *G. brevipinnis* in the Otway Ranges, Victoria, spawned during high flows over gravelly substrate in mid-autumn. Fertilized eggs were deposited up to 7 m from the low flow water level and remained out of water for days or weeks. Hatching only occurred when the eggs were inundated by a subsequent flood after initial spawning. Riparian cover is important for providing shade and maintaining moist conditions for egg survival. The riparian zone is important for regulating water temperatures, restricting sedimentation and providing cover and terrestrial food resources for fish.

Effects of integrated logging on sand supply to the estuary

Mackay *et al.* (1985) used 350 erosion pins installed on a grid to determine soil erosion rates in Yambulla State Forest following integrated logging (Fig. 30.3). Logged sites exhibited net erosion and, in contrast, undisturbed sites exhibited net deposition. There was substantial redistribution of soil but the net result was 3 mm of deposition. This result was explained by lower bulk densities of deposited material in comparison to *in situ* soil. Similar net deposition following harvesting has also been reported for northern NSW by Saynor *et al.* (1994).

CSIRO Land and Water measured soil erosion and sediment redistribution using radionuclides after integrated logging within a single forest compartment (Bondi State Forest) in the upper Genoa basin (Fig. 30.3). Soil and sediment sampling took place 5 years after harvesting. Wallbrink *et al.* (1997, 2002) and Croke *et al.* (1999a) used the tracers ^{137}Cs and ^{210}Pb$_{ex}$ to construct sediment budgets for the harvested basin (Figs 30.4 and 30.5). There was no net soil loss from the compartment within the uncertainties of the ^{137}Cs technique (Fig. 30.4). Wallbrink *et al.* (2002) accounted for 59.3 ± 4.4 MBq after harvesting of the 61.1 ± 4.3 MBq total ^{137}Cs activity before harvesting ($97 \pm 10\%$), the difference being within the uncertainty limits of the pre-harvest budget.

The ^{210}Pb$_{ex}$ budget (Fig. 30.5) showed total retention of only $78 \pm 12\%$ but no evidence of surface soil was found in sediments in the channel bed within and downstream of the compartment (Wallbrink *et al.* 1997). A greater amount of the total ^{210}Pb$_{ex}$ is bound to organic material and to surface soil than ^{137}Cs (Wallbrink *et al.* 1997). Therefore, the post-harvest burn may have preferentially removed more ^{210}Pb$_{ex}$ than ^{137}Cs. Significant redistribution, storage and transport of sediment occurred within the harvested area (see also Mackay *et al.* 1985; Saynor *et al.* 1994). Net soil loss was recorded from snig or extraction tracks (11% of total initial ^{137}Cs budget) and log landings (2%). This eroded material was trapped in down-slope cross banks, general

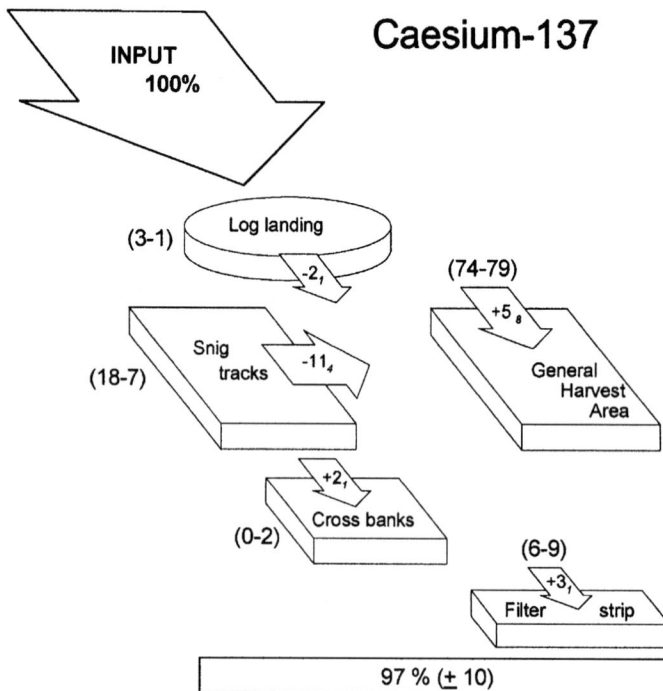

Fig. 30.4 Tracer budget for compartment 1708 in Bondi State Forest, New South Wales based on fallout ^{137}Cs. Values within arrows represent tracer amounts as a fraction of the total initial input. Values in parentheses represent amount of activity before and after harvesting as a percentage of the total inputs. Uncertainties are shown as subscripts. Snig tracks are skidder or extraction tracks. (From Wallbrink *et al.* 2002, with permission)

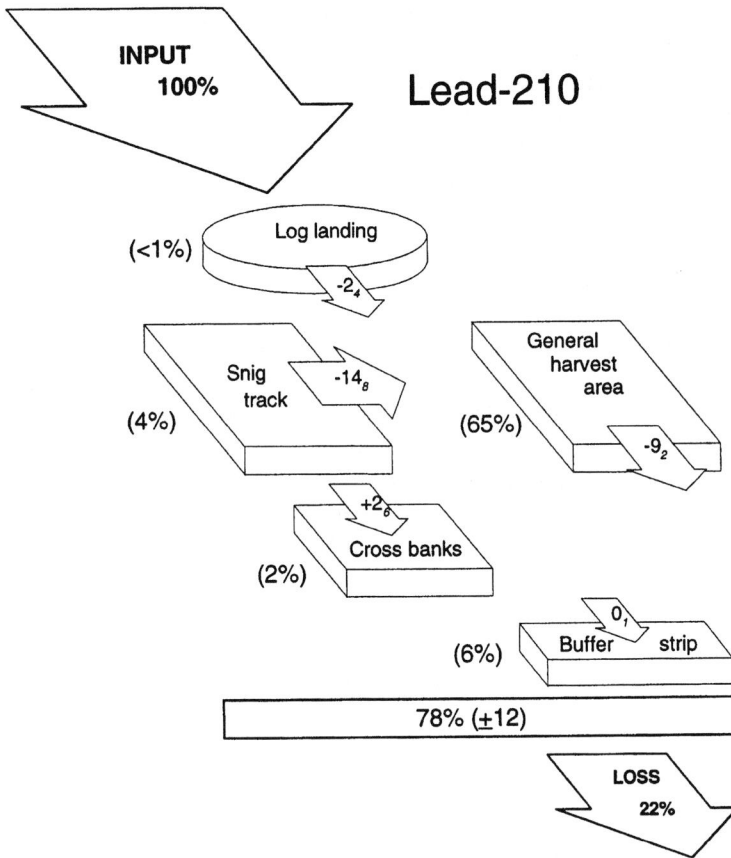

Fig. 30.5 Tracer budget for compartment 1708 in Bondi State Forest, New South Wales based on fallout $^{210}Pb_{ex}$. Values within arrows represent tracer amounts as a fraction of the total initial input. Values in parentheses represent amount of activity before and after harvesting as a percentage of the total inputs. Uncertainties are shown as subscripts (from Wallbrink *et al.* 1997). Snig tracks are skidder or extraction tracks. Reproduced with the permission of CSIRO.

harvest area and filter strips. Coarse sediment was trapped up-slope of the filter strips and fine sediment was preferentially retained in the filter strips.

The Cooperative Research Centre for Catchment Hydrology has also conducted unreplicated rainfall simulator and other field experiments in integrated logged tall open eucalypt forests in southern New South Wales and eastern Victoria (Croke *et al.* 1997, 1999a, 1999b, 1999c, 2001; Croke & Mockler 2001). Unsealed forest roads generated an order of magnitude more sediment than recently disturbed snig or extraction tracks, which in turn, generated an order of magnitude more sediment than the adjacent general harvest area. Rapid sediment exhaustion limited the amount of material available for mobilization from all sources. Well-used roads produced between five and eight times higher sediment concentrations than abandoned roads because the greater intensity traffic dislodged larger amounts of loose material. Soil erosion on snig tracks declined greatly with time since disturbance with about 5 years required for recovery. Therefore, effective sediment control measures are needed during and immediately after disturbance to reduce offsite impacts (Croke *et al.* 1999a, 2001) and such measures are contained in the relevant licence (Anonymous 1999). Roads and snig tracks are significant source areas of runoff and sediment but the general harvest area is an important dispersal area minimizing the transfer of water and sediment from harvested areas to streams (Croke *et al.* 1999a, 1999b, 1999c; Lacey 2000; Wallbrink *et al.* 2002). A 5-m flow path was sufficient to deposit over 50% of the fine-grained sediment (Croke *et al.* 1999b). Lacey (2000) found sediment trap efficiencies of over 90% for 10-m long, undisturbed flow paths.

Current forest management practices aim to transfer water and sediment from disturbed (tracks, roads and log dumps) to less disturbed parts of the forest (general harvest area), increasing sediment storage on hill slopes (Croke *et al.* 1999a; Lacey 2000). Gully initiation at road drainage outlets should be minimized to reduce road-to-stream connection and to prevent increasing drainage density (Croke & Mockler 2001). Receiving area slope and contributing road area are the most significant factors determining whether gully initiation will be induced by road drainage structures (Croke & Mockler 2001).

Channel erosion

Cohen & Brierley (2000) quantified the amount of sand supplied to the Genoa River by incision, widening and straightening of Jones Creek (Fig. 30.3), following base level lowering of the main stream during the 1971 flood. Of the 171,100 m^3 of sediment removed by channel erosion between 1972 and 1997, 99,200 m^3 was sand; about 33% was stored in the channel of Jones Creek and the other 67% was delivered to the Genoa River. Webb *et al.* (2001) found that about 2300 years BP a single flood or series of cataclysmic floods completely stripped the *T. laurina* forest from the sandy floodplain of Bruces Creek (Fig. 30.3) in Nadgee State Forest (Wallagaraugh River basin). Clearly channel erosion by exceptionally large floods was also important for generating sand in the Genoa River basin before the commencement of forestry activities. These results further highlight that channel erosion, even in forested areas, is an important sand source in highly flood-variable watersheds (Erskine 1993, 1996, 1999; Webb *et al.* 2001).

Sand sources, storages, sinks and fluxes are a complex issue that generate much debate, especially in relation to forestry activities (Erskine 1992). Nevertheless, a combination of sediment tracing studies, rainfall simulator experiments, field measurements, historical studies and analyses of air photographs all indicate that channel erosion is the main contemporary sand source in the Genoa River. In relation to fish population effects, it seems most likely that over-exploitation of *A. butcheri* by recreational fishermen during spawning, combined with commercial fishing in the bottom lake of Mallacoota Inlet, have reduced stocks. Channel erosion caused by the 1971 flood reduced the available habitat of *M. novemaculeata*. Habitat recovery is now well advanced and the *M. novemaculeata* population should be responding.

Recommendations for improvement in forestry practices and standards

The following recommendations should be adopted to make forestry practices more 'fish-friendly' in Australia:

- Wherever possible, forest roads should not be located in areas which become saturated and generate storm runoff, and should not be built on the immediate landward side of riparian buffer strips.
- Active thinning of harvesting-induced, regrowth eucalypt forests should be practised to increase water yields and hence available fish habitat.
- Road drainage structures should discharge onto undisturbed forest areas that are at least 5 m and preferably 10 m long to avoid gully initiation and consequent road-to-stream connection.
- Riparian buffers protecting fish habitat (basin areas >2 km² in southeastern Australia) should be at least 30 m wide.
- Riparian tree species should be protected for a distance equivalent to at least one tree height (30 m) away from the river bank.
- Natural large woody debris loadings and recruitment processes should be maintained in Australian streams.
- All types of road crossings should be designed according to principles for fish passage, such as:
  - free air/water surface maintained for all but high flows
  - no vertical discontinuity in bed profile
  - no vertical headlosses at upstream or downstream openings
  - low or no gradient through culverts, pipes, etc.
  - maintain stream depth through whole crossing
  - maximize amount of natural light, minimize pipe length, maximize diameter, and use open systems wherever possible
  - continue natural stream substrate through bed of pipes and culverts
  - maintain road crossings so that they remain clear of debris.
- The principle of minimum specific head should be abandoned as an engineering design approach for **all** road crossings in fish habitat.

Acknowledgements

For their constructive comments on the manuscript, we thank the editors, Peter Bayliss, Peter Unmack and Mike Bullen. Ashley Webb and Tom Rayner are thanked for their continuing collaboration on fish habitat research. State Forests of NSW support for Wayne Erskine's research is gratefully acknowledged.

References

Allen, G.R. (1989) *Freshwater Fishes of Australia*. TFH Publications, Neptune City.

Allen, G.R., Midgley, S.H. & Allen, M. (2002) *Field Guide to the Freshwater Fishes of Australia*. Western Australian Museum, Perth.

Anonymous (1999) *Forestry and National Park Estate Act 1998 Integrated Forestry Operations Approval for Eden Region*. Department of Urban Affairs and Planning, Sydney.

Bayly, I.A.E. & Williams, W.D. (1973) *Inland Waters and their Ecology*. Longman, Hawthorn.

Boubee, J., Jowett, I., Nichols, S. & Williams, E. (1999) *Fish Passage at Culverts. A Review, with Possible Solutions for New Zealand Indigenous Species*. Department of Conservation, Wellington.

Brooks, A., Erskine, W.D. & Finlayson, B.L. (2001) *Report of the Genoa Expert Panel*. East Gippsland Catchment Management Authority, Bairnsdale.

Cohen, T.J. & Brierley, G.J. (2000) Channel instability in a forested catchment: a case study from Jones Creek, East Gippsland, Australia. *Geomorphology*, **32**, 109–28.

Cornish, P.M. & Vertessy, R.A. (2001) Forest age-induced changes in evapotranspiration and water yield in a eucalypt forest. *Journal of Hydrology*, **242**, 43–63.

Croke, J. & Mockler, S. (2001) Gully initiation and road-to-stream linkage in a forested catchment, southeastern Australia. *Earth Surface Processes & Landforms*, **26**, 205–17.

Croke, J., Hairsine, P., Fogarty, P., Mockler, S. & Brophy, J. (1997) Surface runoff and sediment movement on logged hillslopes in the Eden Management Area of south eastern NSW. *Cooperative Research Centre for Catchment Hydrology Report 97/2*.

Croke, J., Wallbrink, P., Fogarty, P., *et al.* (1999a) Managing sediment sources and movement in forests: the forest industry and water quality. *Cooperative Research Centre for Catchment Hydrology Industry Report 99/11*.

Croke, J., Hairsine, P. & Fogarty, P. (1999b) Sediment transport, redistribution and storage on logged forest hillslopes in south-eastern Australia. *Hydrological Processes*, **13**, 2705–20.

Croke, J., Hairsine, P. & Fogarty, P. (1999c) Runoff generation and re-distribution in logged eucalyptus forests, south-eastern Australia. *Journal of Hydrology*, **216**, 56–77.

Croke, J., Hairsine, P. & Fogarty, P. (2001) Soil recovery from track construction and harvesting changes in surface infiltration, erosion and delivery rates with time. *Forest Ecology & Management*, **143**, 3–12.

Davies, P.E. & Nelson, M. (1993) The effect of steep slope logging on fine sediment infiltration into the beds of ephemeral and perennial streams of the Dazzler Range, Tasmania, Australia. *Journal of Hydrology*, **150**, 481–504.

Davies, P.E. & Nelson, M. (1994) Relationships between riparian buffer widths and the effects of logging on stream habitat, invertebrate community composition and fish abundance. *Australian Journal of Marine & Freshwater Research*, **45**, 1289–305.

Dignan, P. (2001) *A functional approach to stream buffer design in mountain ash forest*. PhD thesis, University of Melbourne.

Erskine, W.D. (1992) *An Investigation of Sediment Sources in the Genoa River Catchment*. East Gippsland River Management Board, Mallacoota.

Erskine, W.D. (1993) Erosion and deposition produced by a catastrophic flood on the Genoa River, Victoria. *Australian Journal of Soil & Water Conservation*, **6**, 35–43.

Erskine, W.D. (1994) River response to accelerated soil erosion in the Glenelg River catchment, Victoria. *Australian Journal of Soil & Water Conservation*, **7**, 39–47.

Erskine, W.D. (1996) Response and recovery of a sand-bed stream to a catastrophic flood. *Zeitschrift für Geomorphologie*, **40**, 359–83.

Erskine, W.D. (1999) Oscillatory response versus progressive degradation of incised channels in southeastern Australia. In: *Incised River Channels*, (eds S.E. Darby & A. Simon), pp. 67–95. John Wiley & Sons, Chichester.

Erskine, W.D. (2001) Water yield response to integrated native forest management in southeastern Australia. In: *Integrated Water Resources Management*, (eds M.A. Marino & S.P. Simonovic), pp. 249–55. International Association of Hydrological Sciences, Wallingford, UK.

Erskine, W.D. & Livingstone, E.A. (1999) In-channel benches: the role of floods in their formation and destruction on bedrock-confined rivers. In: *Varieties of Fluvial Form*, (eds A.J. Miller & A Gupta), pp. 445–75. John Wiley & Sons, Chichester.

Erskine, W.D. & Saynor, M.J. (1996) Effects of catastrophic floods on sediment yields in southeastern Australia. In: *Erosion & Sediment Yield: Global & Regional Perspectives*, (eds D.E. Walling & B.W. Webb), pp. 381–8. International Association of Hydrological Sciences, Wallingford, UK.

Gentilli, J. (1986) Climate. In: *Australia – A Geography, Vol. One. The Natural Environment*, (ed. D.N. Jeans), pp. 14–48. Sydney University Press, Sydney.

Gehrke, P.C. & Harris, J.H. (2001) Regional-scale effects of flow regulation on lowland riverine fish communities in New South Wales, Australia. *Regulated Rivers: Research & Management*, **17**, 369–91.

Grant, G.E., Swanson, F.J. & Wolman, M.G. (1990) Pattern and origin of stepped-bed morphology in high gradient streams, Western Cascades, Oregon. *Geological Society of America Bulletin*, **102**, 340–52.

Grant, T.C. (1989) *History of Forestry in New South Wales*. Forestry Commission of New South Wales, Sydney.

Grayson, R.B., Haydon, S.R., Jayasuriya, M.D.A. & Finlayson, B.L. (1993) Water quality in Mountain Ash forests – separating the impacts of roads from those of logging operations. *Journal of Hydrology*, **150**, 459–80.

Harris, J.H. (1984) Impoundment of coastal drainages of south-eastern Australia, and a review of its relevance to fish migrations. *Australian Zoologist*, **21**, 235–50.

Harris, J.H. (1985) Age of Australian Bass, *Macquaria novemaculeata* (Perciformes: Percichthyidae), in the Sydney Basin. *Australian Journal of Marine & Freshwater Research*, **36**, 235–46.

Harris, J.H. (1986) Reproduction of the Australian Bass, *Macquaria novemaculeata* (Perciformes: Percichthyidae) in the Sydney Basin. *Australian Journal of Marine & Freshwater Research*, **37**, 209–35.

Harris, J.H. (1987) Growth of Australian Bass, *Macquaria novemaculeata* (Perciformes: Percichthyidae) in the Sydney Basin. *Australian Journal of Marine & Freshwater Research*, **38**, 351–61.

Harris, J.H. (1988) Demography of Australian Bass, *Macquaria novemaculeata* (Perciformes: Percichthyidae) in the Sydney Basin. *Australian Journal of Marine & Freshwater Research*, **39**, 355–69.

Harris, J. (2001) Fish passage in Australia: experience, challenges and projections. In: Plenary address to the *Third Australian Technical Workshop on Fishways*, (eds R.J. Keller & C. Peterken), pp. 1–17. Monash University, Clayton.

Harris, J.H. & Gehrke, P.C. (1997) *Fish and Rivers in Stress: The NSW Rivers Survey*. NSW Fisheries Office of Conservation and the Cooperative Research Centre for Freshwater Ecology, Cronulla.

Harris, J.H. & Mallen-Cooper, M. (1994) Fish-passage development in rehabilitation of fisheries in mainland south-eastern Australia. In: *Rehabilitation of Freshwater Fisheries*, (ed. I.G. Cowx), pp. 185–93. Fishing News Books, Oxford.

Harris, J.H. & Silveira, R. (1999) Large-scale assessments of river health using an Index of Biotic Integrity with low-diversity fish communities. *Freshwater Biology*, **41**, 235–52.

Jackson, P.D. (1997) Australian Threatened Fishes – 1997 Supplement. *Australian Society for Fish Biology Newsletter*, Sydney.

Jennings, J.N. & Mabbutt, J.A. (1986) Physiographic outlines and regions. In: *Australia – A Geography, Vol. One. The Natural Environment*, (ed. D.N. Jeans), pp. 80–96. Sydney University Press, Sydney.

Kuczera, G. (1987) Prediction of water yield reductions following a bushfire in ash-mixed species eucalypt forest. *Journal of Hydrology,* **94**, 215–36.

Lacey, S.T. (2000) Runoff and sediment attenuation by undisturbed and lightly disturbed forest buffers. *Water Air & Soil Pollution,* **122**, 121–38.

Lane, P.N.J. & Mackay, S.M. (2001) Streamflow response of mixed-species eucalypt forests to patch cutting and thinning treatments. *Forest Ecology & Management,* **143**, 131–42.

Lunney, D. & Moon, C. (1987) The Eden woodchip debate (1969–86). *Search,* **18**, 15–20.

McDowall, R.M. (1996) *Freshwater Fishes of South-Eastern Australia,* 2nd edn. Reed Books, Sydney.

Mackay, S.M., Long, A.C. & Chalmers, R.W. (1985) Erosion pin estimates of soil movement after intensive logging and wildfire. In: *Drainage Basin Erosion and Sedimentation, Conference and Review Papers,* Vol. 2, (ed. R.J. Loughran), pp. 15–22. University of Newcastle, Newcastle.

McMahon, T.A., Finlayson, B.L., Haines, A.T. & Srikanthan, R. (1992) *Global Runoff: Continental Comparisons of Annual Flows & Peak Discharges.* Catena Paperback, Cremlingen-Destedt.

Mallen-Cooper, M. (1994) How high can a fish jump? *New Scientist,* **142**, 32–7.

Mallen-Cooper, M., Stuart, I., Hides-Pearson, F. & Harris, J. (1997) Fish migration in the River Murray and assessment of the Torrumbarry Fishway. In: *1995 Riverine Environment Research Forum,* (eds R.J. Banens & R. Lehane), pp. 33–8. Murray-Darling Basin Commission, Canberra.

MPIG – Montreal Process Implementation Group for Australia (1997) *Australia's First Approximation Report for the Montreal Process.* Commonwealth of Australia, Canberra.

Mussared, D. (1997) *Living on Floodplains.* Cooperative Research Centre for Freshwater Ecology, Canberra.

Nicholson, E. (1999) Winds of change for silvicultural practice in NSW native forests. *Australian Forestry,* **62**, 223–35.

Northcote, T.G. (1978) Migration strategies and production in freshwater fishes. In: *Ecology of Freshwater Fish Production,* (ed. S.D. Gerking), pp. 326–59. Blackwell Scientific, Oxford.

Northcote, T.G. (1998) Migratory behaviour of fish and its significance to movement through fish passage facilities. In: *Fish Migration and Fish Bypasses,* (eds M. Jungwirth, S. Schmutz & S. Weiss), pp. 3–18. Blackwell Scientific, Oxford.

O'Connor, N.A. & Lake, P.S. (1994) Long-term and seasonal large-scale disturbances of a small lowland stream. *Australian Journal of Marine & Freshwater Research,* **45**, 243–55.

O'Connor, W.D. & Koehn, J.D. (1997) Spawning of the Broad-finned Galaxias, *Galaxias brevipinnis* Gunther (Pisces Galaxiiidae) in coastal streams of southeastern Australia. *Ecology of Freshwater Fish,* **7**, 95–100.

Ollier, C.D. (1986) Early landform evolution. In: *Australia – A Geography, Vol. One. The Natural Environment,* (ed. D.N. Jeans), pp. 97–116. Sydney University Press, Sydney.

Ovington, J.D. & Thistlethwaite, R.J. (1976) The woodchip industry: environmental effects of cutting and regeneration practices. *Search,* **7**, 383–92.

Pethebridge, R., Lugg, A. & Harris, J. (1998) *Obstructions to Fish Passage in New South Wales South Coast Streams.* NSW Fisheries Final report Series No. 4.

Raadik, T.A. (1992) Distribution of freshwater fishes in East Gippsland, Victoria, 1967–1991. *Proceedings of the Royal Society of Victoria,* **104**, 1–22.

Raadik, T.A. (1994) *The Distribution of Australian Bass (Macquaria novemaculeata) in Victoria. A preliminary report to the Scientific Advisory Committee.* Freshwater Ecology Section, Flora and Fauna Branch, Department of Conservation and Natural Resources, Victoria.

Reinson, G.E. (1977) Hydrology and sediments of a temperate estuary – Mallacoota Inlet, Victoria. *Bureau of Mineral Resources, Geology and Geophysics Bulletin,* **178**.

Reynolds, L.F. (1983) Migration patterns of five fish species in the Murray-Darling River system. *Australian Journal of Marine & Freshwater Research,* **34**, 857–71.

Riley, S.J. (1988) Soil loss from road batters in the Karuah State Forest, eastern Australia. *Soil Technology*, **1**, 313–32.

Roberts, S., Vertessy, R. & Grayson, R. (2001) Transpiration from *Eucalyptus sieberi* (L. Johnson) forests of different age. *Forest Ecology & Management*, **143**, 153–61.

Roughley, T.C. (1966) *Fish and Fisheries of Australia*. Angus & Robertson, Sydney.

Saynor, M.J., Loughran, R.J., Erskine, W.D. & Scott, P.F. (1994) Sediment movement on hillslopes measured by caesium-137 and erosion pins. In: *Variability in Stream Erosion and Sediment Transport*, (eds L.J. Olive, R.J. Loughran & J.A. Kesby), pp. 87–93. International Association of Hydrological Sciences, Wallingford, UK.

Schiller, C.B. & Harris, J.H. (2001) Native and alien fish. In: *Rivers as Ecological Systems: The Murray-Darling Basin*, (ed. W.J. Young), pp. 229–58. Murray-Darling Basin Commission, Canberra.

Swain, E.H.F. (1912) The forests of the Bellinger River. *Department of Forestry, NSW Bulletin No. 5*.

Thorncraft, G. & Harris, J.H. (2000) *Fish Passage and Fishways in New South Wales: A Status Report*. Cooperative Research Centre for Freshwater Ecology Technical Report 1/2000.

Vertessy, R., Watson, F., O'Sullivan, S., *et al.* (1998) Predicting water yield from Mountain ash forest catchments. *Cooperative Research Centre for Catchment Hydrology Industry Report 98/4*.

Wallbrink, P.J. & Murray, A.S. (1996) Determining soil loss using the inventory ratio of excess lead-210 to cesium-137. *Soil Science Society of America Journal*, **60**, 1201–8.

Wallbrink, P.J., Roddy, B.P. & Olley, J.M. (1997) Quantifying the redistribution of soils and sediments within a post-harvested forest coupe near Bombala, New South Wales, Australia. *CSIRO Land & Water Technical Report No. 7/97*.

Wallbrink, P.J., Roddy, B.P. & Olley, J.M. (2002) A tracer budget quantifying soil redistribution on hillslopes after forest harvesting. *Catena*, **47**, 179–201.

Warner, R.F. (1986) Hydrology. In: *Australia – A Geography, Vol. One. The Natural Environment*, (ed. D.N. Jeans), pp. 49–79. Sydney University Press, Sydney.

Warren, M.L., Jr & Pardew, M.G. (1998) Road crossings as barriers to small fish-stream fish movement. *Transactions of the American Fisheries Society*, **127**, 637–44.

Watson, F.G.R., Vertessy, R.A., McMahon, T.A., Rhodes, B.G. & Watson, I.S. (1999) The hydrologic impacts of forestry on the Maroondah catchments. *Cooperative Research Centre for Catchment Hydrology Report 99/1*.

Webb, A.A. & Erskine, W.D. (2001) Large woody debris, riparian vegetation and pool formation on sand-bed, forest streams in southeastern Australia. In: *Third Australian Stream Management Conference, The Value of Healthy Streams, 27–29 August 2001*, Vol. 2, (eds I. Rutherfurd, F. Sheldon, G. Brierley & C. Kenyon), pp. 659–64. Cooperative Research Centre for Catchment Hydrology, Clayton.

Webb, A.A., Erskine, W.D. & Dragovich, D. (2001) Cataclysmic erosion, *Tristaniopsis laurina* colonization and the late Holocene stability of Bruces Creek, Nadgee State Forest, NSW. In: *Third Australian Stream Management Conference, The Value of Healthy Streams, 27–29 August 2001*, Vol. 2, (eds I. Rutherfurd, F. Sheldon, G. Brierley & C. Kenyon), pp. 653–8. Cooperative Research Centre for Catchment Hydrology, Clayton.

Wilson, C.J. (1999) Effects of logging and fire on runoff and erosion on highly erodible granitic soils in Tasmania. *Water Resources Research*, **35**, 3531–46.

Part VIII
Effecting Better Fish–Forestry Interactions

Chapter 31
Guidelines, codes and legislation

K. MOORE AND G. BULL

Introduction

Earlier chapters in this book describe the impacts that many types of forestry activities have on fish and fish habitat in freshwater and near-shore marine environments around the world. The causal factors include changes to sediment and nutrient regimes, stream flow, water temperature and chemistry, large organic debris and riparian vegetation caused by logging, road construction and forest management practices. In an attempt to reduce or eliminate the impacts on fish habitat, international organizations, national and regional governments, and various other organizations have adopted or proposed codes of forest practices that promote good forest practices and discourage bad ones. Several studies have summarized codes that guide forest practices in selected jurisdictions around the world (Belt *et al.* 1992; Westland 1995; Dykstra & Heinrich 1996; Bull 1999a, 1999b; Young 2000; Applegate & Andrewartha 2002).

Types of codes

Dykstra and Heinrich (1996) describe two basic approaches to establishing guidelines or codes of practice and summarize the benefits and limitations of each. First, codes can be in the form of guidelines that suggest or promote good practices to achieve an objective but do not actually require that those specific practices be carried out. Compliance with these codes is voluntary and the guidelines are often prepared to assist forest enterprises achieve compliance with more general legislative provisions. Many of these types of codes are prepared by teams representing different interest groups and are developed with the understanding that those who endorse them are making a commitment to follow them. This approach has been used by international organizations, such as the Food and Agriculture Organization (FAO), International Tropical Timber Organization (ITTO), and Asia Pacific Forestry Commission (APFC) in numerous countries and by many states in the USA.

The second type of code is established in legislation that sets out very specific enforceable rules that must be followed. Failure to comply with the rules can result in fines or other administrative penalties. This approach has been used by governments in Tasmania, Chile and Russia and especially in the five jurisdictions along the Pacific coast of North America where there are well-documented interactions between forestry activities and fish.

In this chapter we provide examples of these two different approaches in different jurisdictions and summarize some of the provisions that provide protection for fish and fish habitat in those jurisdictions. We have also identified three additional types of codes that are refinements on the two basic types and describe those briefly. In addition to providing a global overview we also provide a detailed analysis of one jurisdiction – British Columbia – that has had experience over a long period of time with both of the two main approaches to regulating forest practices to reduce the impacts on fish habitat.

The third type of code blends detailed recommendations established in guidelines with broad requirements established in legally enforceable regulations. This approach is found in the Best Management Practices Guidelines in the US states of Montana and Maine (Montana Department of State Lands 1991; State of Montana 1991; Maine Forest Service 2002) and in Cambodia's Code of Practice for forest harvesting (Cambodia Department of Forestry and Wildlife 1999).

A fourth type of code takes the shape of legislation that establishes general provisions and then mandates the preparation of forest management plans that contain a code or set of required practices for the specific management plan area. Once the code or the specific standards are adopted in a plan, they acquire the force of law and must be followed. This approach has been used in New Zealand and Malaysia (Awang 2002; New Zealand Ministry of Agriculture and Forestry 2002).

Finally, a fifth type of code is emerging in the standards for certifying forestry enterprises or forest management areas as 'well managed'. These codes promote voluntary compliance with an agreed standard that contains a set of principles, criteria, indicators and, in some cases, verifiers of performance. Certificates are awarded when a third party confirms compliance with the standard. Certification schemes such as the Forest Stewardship Council, Sustainable Forestry Initiative and Canadian Standards Association all contain such standards (Canadian Standards Association 1998; American Forest and Paper Association 2000; Forest Stewardship Council 2000).

Although fish and fish habitat are directly affected by logging, road construction and various post-logging forest management activities, few of the codes around the world specifically link restrictions on forestry activities to the protection of fish habitat. The dominant objectives in most codes relate more generally to protection of the environment, including the prevention of erosion and downstream sedimentation, or to the maintenance of a secure supply of clean water for domestic, irrigational and industrial purposes. In most jurisdictions, codes of logging practices are considered part of environmental protection efforts. They are implemented as a part of sustainable forest management, not specifically as a means to protect fish and fish habitat within the forest. There are some notable exceptions in western North America and one in Tasmania, which are examined in some detail below.

Prior to 1994, most of the developed codes were in industrialized countries and regions, particularly in North America (Dykstra & Heinrich 1996). The National Code of Logging Practice in Fiji was a notable exception (Applegate & Andrewartha 2002). However, since 1994, codes have been developed in a number of tropical and developing countries, particularly in the Asia-Pacific region (APFC 2000). Codes, of one form or another, now exist in Europe, Asia, Oceania, South and Central America, North

America and Africa. These areas include most of the world's largest forested jurisdictions where there are important forestry–fisheries interactions in boreal, temperate and tropical forests – Canada, Russia, USA, China, Indonesia and Brazil.

Guidelines that promote good practices

The FAO model code

Outside the industrialized world, the development of codes to guide forest practices was promoted by the Food and Agriculture Organization of the United Nations (FAO). In 1996, it produced a 'Model Code of Forest Harvesting Practice' (Dykstra & Heinrich 1996) that was intended to stimulate the development of forest codes in FAO member countries around the world. This code is a good example of guidelines that promote good practices to achieve an objective but do not actually require that those specific practices be carried out. These types of codes are intended to achieve results through cooperation, rather than through an enforced regime, and encourage flexibility to achieve objectives (Dykstra & Heinrich 1996). Often they are developed cooperatively by groups of people representing different interests.

The FAO Code followed pioneering work in the development of voluntary guidelines to improve performance done by the International Tropical Timber Organization (ITTO 1992). The ITTO guidelines provided some basic principles to stimulate the development of national policies regarding the protection of the environment in tropical forestry operations, including the establishment of buffer zones, avoiding soil displacement, limiting the season of operations and developing logging plans that minimize disturbance to streams.

The FAO model code provides a more thorough overview of guiding principles and objectives for regulating road construction, timber harvesting and log transport. It makes recommendations about the type of forestry activities that should be avoided to reduce or prevent impacts associated with these practices. These include establishing buffer zones and special management zones along streams where cutting is restricted or prohibited. It recommends that trees should not be felled across streams and, when felled, should be directed away from the buffer zone. It describes the impacts of roads and landings and suggests that roads and landings should be located away from streams and outside streamside buffers. It recommends that culverts should be installed to minimize disturbance to stream channels and stream flow and that skidding should use designated skid trails and be suspended during wet weather.

The FAO model code makes no mention of preventing impacts on fish or fish habitat. It is directed at a more general level of implementing sound forestry practices and protecting the environment. It does not suggest any specific width for buffer zones or any criteria for determining their width but it serves as a basic outline for the types of standards that should be in national or regional codes to prevent impacts on fish and fish habitat.

The Asia-Pacific code

In 1999, the Asia Pacific Forestry Commission (APFC) followed the lead of the FAO model code and developed a code of forest practices for the Asia-Pacific Region (APFC 1999). Like the FAO code, the APFC code was developed to provide a model for the development of national forest practices codes within the Asia-Pacific region. It incorporated provisions from several national codes that existed or were being developed in the region and included a review of codes from around the world. Like the FAO code, it makes no specific reference to fish or fish habitat and provides no special measures specifically tailored to protect any of the attributes of fish habitat. It is also directed at a more general level of environmental protection as part of sustainable forest management.

After an elaborate 2-year process, the APFC 'Code of Practice for Forest Harvesting in Asia-Pacific' was endorsed by the 29 APFC member countries in 1998. It served as an interim set of guidelines for improved harvesting practices in those countries as they developed their national codes. It provided much more detailed guidelines than the FAO code but is a voluntary, not an enforceable, system.

The APFC code includes a very comprehensive and elaborate set of guidelines for planning forestry operations, constructing roads and watercourse crossings, cutting trees, transporting and storing logs, maintaining equipment and storing fuel. It also includes guidelines for camp hygiene and waste disposal, worker safety and fire protection.

The code identifies the need to designate parts of the forest as 'harvest exclusion areas'. These are intended to protect areas of cultural significance, important biodiversity sites, rare and endangered species, and sites of ecological importance including swamps, wetlands and mangroves. Within the remaining 'production forest', it sets out elaborate guidelines for establishment of buffer zones around lakes, lagoons, shorelines and water storage areas and along designated watercourses. These are designed to protect soil, water and riparian vegetation from the impacts of harvesting but there is no reference to protection of fish or fish habitat.

The APFC code defines watercourses to include all areas that receive and conduct concentrated overland flow for some period in most years. It classifies watercourses into streams, gullies and waterways and prescribes a width and composition of the buffer zone along them according to the classification. Five classes are defined based on the permanency of water flow, bed material, width and bank slope. Three of the five classes are streams in which water flows for more than 2 months in most years and in which the stream bed is composed of clean, water-washed stone, gravel or exposed bedrock. The three stream classes are based on stream width. Buffer zones are established on each side of the stream depending on its classification. Streams are:

- class 1: >20 m wide and have a buffer of at least 30 m
- class 2: 10–20 m wide and have a buffer of at least 20 m
- class 3: <10 m wide and have a buffer of at least 10 m.

The Code recommends that no trees should be removed from the buffer zones of class 1, 2 or 3 streams.

The other two classes are:

- class 4: gully, a channel with at least one steep bank (having a slope >25%), where water flows <2 months of the year, and the bed is soil or covered with bark, branches or leaf litter, with a 10-m buffer;
- class 5: waterway, a channel having water <2 months of the year, side slopes <25%, and usually <2 hectares, with a 5-m buffer zone.

The code suggests that merchantable trees may be felled within the buffer areas of the class 4 and class 5 channels.

This classification system does suggest some protection for small, as well as large, streams and for areas that carry water for only short periods of the year. These streams are sensitive to damage and provide habitat for fish in critical periods and may influence downstream areas.

Buffer zones are also required around lakes, lagoons, shorelines and water storage areas. The code prescribes a minimum width for these buffers based on the slope of the adjacent land. Where the slope is <17%, the minimum buffer width is 50 m; where the slope exceeds 17%, the buffer is 100 m.

The APFC code also establishes a number of general restrictions for managing buffer zones along all the classes of watercourses and around lakes, lagoons, shorelines and water storage areas. Machine access is prohibited, except where watercourse crossings are permitted. No spoil from earthworks or debris from logging is to fall within a buffer zone and if any trees inadvertently fall into a watercourse, all debris is to be removed without disturbance to the bank.

Buffer zones are to be established whether or not the watercourses appear on maps and are to be marked in the field before harvesting commences. The buffer width is measured from the high water mark, or the edge of mangrove vegetation if this occurs above the high water mark.

At least 14 countries in the Asia-Pacific region now have national or state codes or reduced impact logging guidelines that have very similar provisions because they have been developed under the APFC framework. These include Australia, Fiji, New Zealand, Papua New Guinea, Samoa, Solomon Islands, Vanuatu, Indonesia, Laos, Malaysia, Myanmar, Vietnam and Japan. In Cambodia, the APFC code has been enacted as the Cambodian Code of Practice for Forest Harvesting (Cambodia Department of Forestry and Wildlife 1999).

There is an ambitious strategy for implementing the APFC Code throughout the region (APFC 2000). Indonesia, for example, developed a set of Principles and Practices for Forest Harvesting in 2000 (Indonesia Ministry of Forestry and Estate Crops 2000). This was quickly followed by the development of a very comprehensive set of Reduced Impact Logging Guidelines (Elias *et al.* 2001) that encourage operators to use environmentally sound practices (Bull *et al.* 2001; Pulkki *et al.* 2001). These principles and guidelines use the same classification system developed in the APFC code and suggest the same widths of buffer zones along watercourses.

United States (selected eastern and central states)

In the United States, several states with forest land have developed 'Best Management Practices' which have served as guidelines for forest practices. Alabama, North Carolina and Florida provide good examples of these types of codes.

In Alabama, the Best Management Practices for Forestry set out non-regulatory guidelines (Alabama Forestry Commission 1993) for the maintenance and protection of water quality. They recommend a minimum streamside management zone of 35 feet (10 m) on both sides of all streams, with the width extended to account for erodible soils, steep slopes or other values such as wildlife. Partial cutting is permitted within the streamside zone, but on perennial streams, the guidelines recommend that 50% of the original crown cover should be retained. On intermittent streams, they recommend that permanent tree cover is not required as long as vegetation and organic debris is left to protect the forest floor along the stream bank. These Best Management Practices do not mention fish or fish habitat.

In North Carolina, the Best Management Practices are also voluntary guidelines and recommend measures to protect water quality. The width of the recommended stream management zone is based on a stream classification system that separates perennial and intermittent streams by stream width and side slope. The recommended management zone ranges from 50 to 100 feet in width in which 80% of the vegetative cover is protected along perennial streams and 60% is protected along intermittent streams. In addition, 75% of the pre-harvest shade is recommended for protection on all stream channels (North Carolina Division of Forest Resources 1989; North Carolina Department of Environment, Health and Natural Resources 1990).

In Florida, the Best Management Practices prescribe a Special Management Zone (SMZ) adjacent to streams, lakes, wetlands and other features such as sinkholes to protect water quality and wildlife habitat values. As with other Best Management Practices there is no reference to protecting fish and fish habitat.

The width of the SMZ and amount of cutting allowed within it is based on the size and type of water body involved, and on a Site Sensitivity Class that indicates the general potential for erosion and sedimentation. The more erodible the soil and the steeper the slope, the higher the site sensitivity class and the wider the SMZ that is prescribed.

On perennial streams, lakes and wetlands, the SMZ varies in width from 35 to 300 feet (10–65 m) per side, depending on the type and size of the water body and the site sensitivity class. Clear-cutting is not permitted within 35 feet (10 m) of the stream but selective harvesting may be conducted anywhere in the SMZ as long as 50% of a fully stocked stand is maintained. The widest SMZs are applied on the most sensitive site classes.

For intermittent streams and lakes and sinkholes with intermittent water an SMZ at least 35 feet wide is also designated but unrestricted selective harvesting and clear-cut harvesting are both permitted in the SMZ adjacent to these streams.

Great Britain

In Great Britain, the Forestry Commission developed a set of guidelines to assist foresters and landowners meet the requirements of the Water Act and other legislation. These Forestry and Water Guidelines (Forestry Commission 1991) do recognize the habitat requirements of fish and establish a number of specific provisions to protect fish habitat.

The UK guidelines require riparian strips of vegetation with a minimum width of 5 m on each bank of small headwater streams. Larger streams require a strip two or three times as wide as the stream. These strips are expected to act as seepage zones, to protect stream banks, to give intermittent shade and protective cover, and to provide nutrients to the stream system.

The UK guidelines also specify road construction and maintenance practices to ensure that culverts are not barriers to fish movement. They recommend that any in-stream work avoids periods when fish are spawning and when salmonid eggs and fry are in the gravel. Guidelines for road drainage suggest that drainage should be discharged through a streamside buffer strip sufficiently wide to prevent coarse sediment from reaching the stream. In areas of high risk of erosion, sumps, settling pools or silt traps are recommended.

The UK guidelines, in contrast to other guidelines, which stress the importance of retaining trees along riparian areas, stress the importance of open ground beside the stream. They state that at least 50% of the stream should be open to sunlight with the remainder under intermittent shade from light foliaged trees and shrubs. Shade-casting trees should be kept sufficiently far back to allow sunlight to reach the stream when the trees are fully grown. Heavy foliaged trees, whether conifer or broad-leaf, must be pruned or cut periodically to maintain open areas and ground vegetation.

This is done because studies have shown that forest canopies scavenge pollutants (particularly gaseous sulphur and nitrogen) from the atmosphere, mist or cloud water. This may increase the acidity of the stream water to levels that are harmful to fish. In the wetter west of Britain, and particularly above 300 m, the scavenging of clouds and mist by forest stands is thought to be particularly important. The guidelines suggest that powdered limestone should be applied to streams if any significant planting of trees is planned near them.

Legislated codes that require compliance

In several jurisdictions, very specific forest practices codes have been implemented as legislation.

Tasmania

The Forest Practices Code of Tasmania (Forest Practices Board of Tasmania 2000) regulates forest practices in native forests and plantations and monitors compliance through periodic audits. The Tasmanian code establishes four classes of streams and determines the width of required streamside reserves and machine exclusion zones based on these classes. The four classes are based primarily on watershed size and are not related to the presence or absence of fish anywhere in the watershed. The classes are:

- class 1: rivers, lakes, artificial storages and tidal waters that are named on 1: 100,000 topographical maps, with a 40-m reserve;
- class 2: creeks and streams that have a catchment area >100 ha, with a 30-m reserve;
- class 3: watercourses that carry water most of the year in catchment areas between 50 and 100 ha, with a 20-m reserve;
- class 4: all other watercourses with running water for all or part of most years, with no reserve but a 10-m machine exclusion zone.

Wider streamside reserves, including reserves on class 4 streams, can be required in plans where necessary to protect fish spawning or nursery areas. In most situations, all native vegetation including trees must be retained in the streamside reserves and trees are not permitted to be felled into reserves. However, there are provisions for the removal of trees on a selective basis as long as no more than 30% of the canopy is removed and trees are not felled in the 10 m adjacent to the stream. Machines are not permitted to operate in the streamside reserves except on approved skid trails and crossings. The boundaries of streamside reserves must be marked in the field before harvesting begins.

China

The People's Republic of China has developed a draft National Code of Practice for Forest Harvesting (People's Republic of China Department of Forest Resources Management 2001). This code will apply to all forest harvesting and forest road construction on forest lands. It requires buffer zones on all streams (including intermittent streams), lakes, wetlands and reservoirs and no harvesting is allowed in these buffer zones. The buffer zone width is dependent on the width of the stream and ranges from 8 m for a stream <10 m wide to 30 m for a stream >50 m wide. The Chinese code is simpler than other codes since there is no distinction between steep streams and low gradient streams. Log landings need to be 40 m from buffer zones.

Russia

In Russia, Decree #1404 requires the protection of riparian zones along all streams, rivers, lakes and water reservoirs (Russian Federation 1996). The law applies to all manner of land use – including industrial plants, farms and feedlots, roads and community grounds – not just to forestry operations. The Russian law determines the width of reserve zones along water bodies based on the distance of the portion of the watercourse affected by the land use from its source and does have special provisions for important fish habitats. For example, a stream within 10 km of its source, presumably a relatively small stream, requires a reserve zone of 50 m, but this is expanded to 100 m if the stream has a high fishing value. A stream >10 km but <50 km from its source requires a reserve of 100 m. A river >500 km from its source, presumably a very large river, requires a protective zone of 500 m.

The width of reserve zones around lakes and wetlands is based on the size of the lake or wetland. For lakes of <200 ha, the protection zone is 300 m; for all lakes >200 ha, the zone is 500 m.

North America

The most extensive use of legislated forestry codes to regulate forestry activities is in five jurisdictions on the west coast of North America – Alaska, British Columbia, Washington, Oregon and California. In all five jurisdictions, there are legally binding codes which provide specific provisions to protect fish habitats. These codes include requirements to classify streams based on fish presence and physical features, and to refrain from many types of practices that have been shown to negatively impact habitat. The use of these legally binding codes probably reflects several factors, including the economic and social importance of salmonids and their dependence on freshwater habitat in forested watersheds. There has also been extensive research in these jurisdictions that has documented the impacts of logging and forest management practices on fish and their habitats. In at least one of these jurisdictions, British Columbia, earlier attempts to use voluntary guidelines to protect habitat failed.

These jurisdictions have developed codes of practice that are similar in some ways but quite different in others. Young (2000) provides an analysis of the different approaches to riparian zone protection in British Columbia, Washington, Oregon and California.

In Alaska, the Alaska Forest Resources and Practices Act of February 2000 requires riparian protection for all streams and water bodies on private, state and other public lands that have anadromous or high value resident fish species that are used for commercial, recreational or subsistence purposes (Alaska Department of Natural Resources 2000). On state and other public lands, no harvest is permitted within 100 feet (30 m) of these water bodies unless it is determined that adequate protection remains. All streams with a gradient of 8% are assumed to be anadromous waters if there is no documentation of a blockage. Additional protection may be imposed through the adoption of land use plans.

On private land in Alaska, a more complex classification system is used to determine the width of the required riparian area on four classes of streams and the required buffer zones are not as wide.

The Washington Forest Practices Rules (Washington Department of Natural Resources 1995) divide streams into five classes, based on fish use, width and substrate. The first three classes are based on high, moderate or low fish use, or domestic water use and are subdivided into subclasses based on width. Classes 4 and 5 are streams with no fish use. None of these classes of streams requires a no-harvest area along the stream but the three fish-bearing classes are required to have riparian management zones (RMZ) ranging in width from 7.5 m to 30 m where practices are restricted. Within all the RMZs operators must leave a number of representative trees on each side of the stream to provide shade. The required level of retention is expressed as a required number of trees per 300 m and is determined by the class and width of the stream and its elevation. It ranges from 25 to 100 trees per 300 m of RMZ.

Oregon classifies streams into three classes – fish use, no fish use but domestic water use, and no fish or domestic water use – which are further sudivided into small, medium and large streams. The Oregon Forest Practice Administration rules require that a Riparian Management Area be established with a width that varies depending on the class and subclass and also depending on their location within the state (Oregon Department of Forestry 1995). Within the RMA a minimum 6-m no-harvest zone must be left along all streams except the smallest non-fish-bearing streams. Outside the no-harvest zone within a wider RMA, a number of large trees must also be retained along the fish-bearing streams to meet a specified level of basal area retention. The required level of retention is determined by the size of the stream and the type of logging planned.

British Columbia

British Columbia provides an excellent example of a jurisdiction that has used both of the main approaches in an attempt to protect fish and fish habitat from the impacts of forestry activities. The province has extensive logging operations in thousands of watersheds that support important recreational and commercial fisheries. There has been a long history of conflict between the two resources, and British Columbia has used a variety of approaches in attempting to manage the interactions and reduce the impacts of forestry operations on fish habitat. British Columbia's experience demonstrates a progression from rigid guidelines developed by government in 1972 and implemented through mandatory clauses in permits, to voluntary guidelines cooperatively developed with the forest industry in 1986, to a complex legislated Forest Practices Code with mandatory compliance in 1995. Each of these approaches included measures that were specifically developed to protect fish habitat and each brought different results. In 2000, the government adopted a fundamentally different approach for the small area of privately owned forest land in the province and, in 2001, made a commitment to fundamentally change the application of the 1995 Forest Practices Code on public lands. Work is now under way to develop yet another approach to managing fisheries–forestry interactions.

The first set of guidelines to address the impact of forestry activities on other forest resources were distributed by the Chief Forester of the province in 1972. The 'Planning Guidelines for Coast Logging Operations' (British Columbia Forest Service 1972) included provisions for the protection of water quality and the protection of fish and fish habitat by leaving strips or blocks of trees along stream banks and lakeshores. Using the words 'shall' and 'must', these guidelines established a limit on cut-block size and a pattern of alternate cut and leave blocks in coastal watersheds. Blocks had to be logged in a checkerboard pattern with temporary leave blocks in between and could not face each other across a stream. Logging practices adjacent to the stream had to be conducted in a way that protected the stream bed and banks. These guidelines led, in 1974, to the development of a 'P1 clause' (protection clause) that was inserted into all permits that allowed companies to log (British Columbia Forest Service 1974). The P1 clause required the retention of all immature, non-merchantable trees and trees leaning over the stream in a 1-chain (20-m) wide strip along the stream edge. The specified width was quickly removed but the concept of a fixed-width streamside buffer remained in place. Another P clause required that no trees be felled into or yarded over streams. Compliance with the clause was mandatory and enforceable.

The intent of these guidelines was noble and the mandatory 'P clause' was a first attempt to restrict logging practices to afford some protection to streams and fish habitat. But this approach was soon widely criticized as arbitrary, impractical and expensive. The guidelines led to watersheds with large numbers of roads and, when the intervening leave blocks were cut, large clear-cut areas were left on both sides of streams. The cut and leave pattern led to extensive blow-down and, in practice, very little vegetation was left along stream edges. A different approach was needed.

In response, the Ministry of Forests moved away from the approach of using mandatory clauses in permits and began to develop other approaches to managing the interaction between fisheries and forestry. They abandoned the cut and leave pattern and maximum cut block size of the 1972 guidelines and reduced enforcement of the P1 clause requirements.

In the southern coastal area, they supported an alternative approach, based on the premise that the measures incorporated for stream protection should be determined on a site-specific basis and should reflect the characteristics and values in the streams (Moore 1980). The decision-making procedure recognized that different widths of buffer strips and different falling and yarding practices should be used on streams with different physical characteristics and fish populations. The decision-making procedure introduced the first stream classification system for the province, with three classes of low gradient streams, based on fish presence, and one class of steep gradient streams. The width and composition of streamside leave strips and the falling and yarding practices allowed near the streams were based on an assessment of site-specific features and values. In contrast to the Planning Guidelines, this was a voluntary system and addressed the retention of trees in specific streamside areas, and the logging practices including falling and yarding of trees and removal of debris in those areas. It did not address rates or pattern of harvest within a watershed. The decision-making procedure was implemented by some companies and in some forest districts on the coast for a number of years.

On the Queen Charlotte Islands, a similar voluntary approach to protect fish habitat and water quality was developed in the 'Streamside Management Methods for the Queen Charlotte Islands' (British Columbia Forest Service 1978). These recommended methods were developed to encourage flexibility and site-specific decision-making and were explicitly not guidelines that applied equally to all sites. They included a stream classification system that distinguished low gradient and steep gradient streams but also distinguished between single channel and multiple channel streams and streams in steep gullies. The streamside management methods recommended that logging should be confined to one side of a watershed as a way of reducing blow-down. Because of concern for blow-down, these methods stressed removal of large trees and suggested leaving deciduous trees and shrubs and coniferous understory trees along the edge of fish-bearing streams. No trees were to be felled into or yarded across fish-bearing streams. The suggested pattern of harvest was combined with a 'rate-of-cut' guideline that emerged from a study of the impact of the cumulative effects of extensive harvesting within watersheds on the islands (Toews & Wilford 1978). That study recommended that cutting in a watershed should be limited to one-third of a watershed over a 25-year period in order to minimize changes to stream hydrology and fish habitat.

During the same time period, major research projects to study the impacts of forest practices on fish habitat were under way at Carnation Creek on Vancouver Island (Narver & Chamberlin 1976) and on the Queen Charlotte Islands (Poulin 1984). In 1977, Canada amended the Fisheries Act to provide much greater legal protection for fish habitat and to prevent the 'harmful alteration, disruption or destruction of fish habitat'. The Department of Fisheries and Oceans produced a handbook that described the many effects of forest activities on fish habitat and suggested measures to avoid them (Toews & Brownlee 1981).

By 1983, important information about the interactions between forest harvesting and fish habitat was emerging from Carnation Creek and the Queen Charlotte Islands studies (Hartman 1982; Hartman & Scrivener 1983; Rood 1984; Tripp & Poulin 1986; Chamberlin 1987). Several high profile charges had been laid under the federal legislation, and there was increasing concern about the lack of adequate measures to protect fish habitat from the impacts of forest practices. The federal and provincial governments jointly assembled a team to develop a comprehensive set of guidelines for the protection of fish habitat from impacts of forestry activities in coastal British Columbia. After much negotiation, the Coastal Fisheries and Forestry Guidelines were completed in 1987 and endorsed for use by ministries of both governments and representatives of the forest industry (British Columbia Ministry of Forests and Lands *et al.* 1987; British Columbia Ministry of Forests *et al.* 1988). These formal guidelines (referred to as the CFFG) replaced the other approaches in coastal British Columbia and on the Queen Charlotte Islands.

Because the CFFG dealt with specific practices in specific cutting areas, the cumulative and hydrological effects of harvesting large areas of a watershed were still a concern. A watershed assessment procedure was developed to identify past hydrological impacts and assess the sensitivity of a watershed to more harvesting (Wilford 1987). This watershed assessment procedure was also incorporated into the CFFG and replaced the arbitrary rate of cut guideline developed on the Queen Charlotte Islands.

The CFFG were a classic example of the type of guidelines described in the FAO model code (Dykstra & Heinrich 1996). They recommended good practices to use and poor practices to avoid. They were developed by a team reflecting different interests. They promoted voluntary compliance and provided a considerable amount of flexibility for interpretation and application.

The CFFG depended on a classification system that distinguished four classes of streams based on fish presence and gradient. Streams were classified as follows:

- class 1: a stream with anadromous fish or high numbers of resident fish and a gradient generally less than 8%;
- class 2: a stream with low numbers of resident species that were large enough to be legally caught by sports fishermen and a gradient between 8% and 12%;
- class 3: a stream with resident non-sport fish;
- class 4: a stream with no fish present and a gradient generally greater than 20%.

Guidelines for constructing roads and landings, falling and yarding trees and undertaking silvicultural treatment were established for each class. A streamside management zone was recommended for all classes of stream but the width and composition of the zone was not specified and was left to site-specific determination. Restrictions on falling and yarding across class 1 and 2 streams and large class 3 and 4 streams were recommended. These guidelines were the product of intense negotiation between government and industry. Many were prefaced by the words 'consider' or 'generally' and the recommended practices were qualified by ' where necessary', 'where practical', 'avoid … if possible', and 'reasonable'.

The guidelines were jointly developed by government and industry, and endorsed by the coastal forest industry. They were expected to be used in all forestry operations on the coast. However they had no legal basis and, because of the qualified language, were essentially unenforceable. Concern grew that the guidelines provided too much latitude and were not being followed in many locations (Moore 1991). Even where they were followed, they were thought to be ineffective in protecting habitat. A series of audits was commissioned to look at the use and effectiveness of the CFFG in protecting fish habitat in cut-blocks logged since the guidelines came into effect (Tripp *et al.* 1992; Tripp 1994, 1995).

These audits provided compelling evidence that the guideline approach had failed. The guidelines were actually implemented in only a few operations and had apparently not changed forestry practices along streams (Tripp 1995). Half the streams inspected in the audits had been affected by logging activities. Roads had been assumed to be the main source of problems, but the audits clearly identified logging, and particularly logging practices along streams, as the main cause of most of the stream damage. The auditors concluded that 'in the absence of site specific recommendations, or strictly defined limits, compliance with the guidelines is very poor. The more room left for interpretation, the more likely minimum standards will be selected' (Tripp 1998).

In response to the audits, the parties moved quickly to revise the guidelines and urged compliance with them (British Columbia Ministry of Forests *et al.* 1992, 1993a). At the same time, many other guidelines to address wildlife and biodiversity concerns, road construction practices and many other aspects of forestry operations were being developed in the province. In the interior of the province, the Interior Fish, Forestry and Wildlife Guidelines (British Columbia Ministry of Forests *et al.* 1993b) were being developed to protect fish habitat as well as wildlife and terrestrial biodiversity. These guidelines were the first to provide a specified width for a streamside management zone. They recommended a minimum 30-m wide streamside management zone (SMZ) on both sides of all continuous and intermittent watercourses. Streams were classified in a three-class system based on fish presence. The extent of cutting within the SMZ was based on the classification, with no cutting being permitted in the SMZ for the class A streams that have anadromous fish present. Wider SMZs were recommended on steep slopes, areas with high wind-throw potential or complex channels. These guidelines also included restrictions on the size of cut-blocks, the extent of watersheds that could be cut, the timing of logging and road construction and on silvicultural activities, including use of herbicides and fertilizers.

It was clear by this time, however, that the guideline approach with voluntary compliance had not been widely implemented, was difficult to enforce and had not sufficiently improved practices. The Forest Resources Commission (1991) had already recommended a single all-encompassing code of forest practices that would set a clear and enforceable minimum standard of practices. They recommended that a code be created through the introduction of a Forest Practices Act.

Thus, in 1995, British Columbia abandoned the guideline approach and adopted a regulatory approach with the passage of the Forest Practices Code of British Columbia Act by the provincial legislature (British Columbia 1995). The legislation included an Act and 18 regulations that set very detailed mandatory minimum standards that had to be met in every forest operation on public lands in British Columbia. The code covered all aspects of forest management, from strategic and operational level planning, to road construction and logging operations, range management, post-harvesting road deactivation and silviculture treatments, fire, insect and disease management. Its scope included a broader range of forest values – recreation, drinking water and cultural values – than the fisheries resource addressed in previous guidelines but there were numerous measures specifically directed to the protection of fish habitat.

The code provided a clear set of legally enforceable minimum standards for all operations across the province. It also recognized the wide variety of sites and conditions, and incorporated the site-specific decision-making approach of earlier guidelines by giving government managers considerable discretion to approve deviations from the legal requirements if they were presented and approved in a plan. The legislation set penalties for failing to comply with the requirements and allowed officials to stop operations that appeared not to comply. To address cumulative impacts, the code replaced the *Watershed Workbook* approach (Wilford 1987) with two guidebooks outlining watershed assessment procedures for coast and interior situations (BC Ministry of Forests 1999).

To protect fish habitat values, the code established a 6-class riparian zone classification system along streams (S1 to S6), as well as a classification system for lakes and wetlands (L1 to L4 and W1 to W5). Riparian classes S1–S4 were low gradient streams (<20%) that provided habitat for fish. As shown in Table 31.1, the classes were based on width. Classes 5 and 6 were steep gradient streams that did not provide fish habitat. In many cases, these streams flowed down into fish habitat and could transport sediment and debris from upstream operations into fish-bearing water.

The code also included many provisions that regulated machine use near streams, yarding of logs across streams, placement of culverts and bridges, locations of roads

Table 31.1   Riparian classes and minimum legal widths of riparian reserve zones (RRZ) and riparian management zones (RMZ)

Riparian class*	Average channel width (m)	Reserve zone width (RRZ) (m)	Management zone width (RMZ) (m)
S1 large rivers	>100	0	100
S1 not large rivers	>20	50	20
S2	5–20	30	20
S3	1.5–5	20	20
S4	<1.5	0	30
S5	>3	0	30
S6	<3	0	20

*Riparian classes S1–S4 are known fish-bearing streams, streams with a gradient of <20% or streams in a community watershed. Riparian class 5 and 6 streams do not contain fish and are not in a community watershed.

and silvicultural practices. The most important provisions, however, involved the protection of riparian areas.

The code required that riparian management areas (RMAs) be established on all classes of streams (and on all classes of lakes and wetlands as well). By law, the RMA was composed of a riparian reserve zone (RRZ) beside the stream and a riparian management zone (RMZ) further from the stream. As shown in Table 31.1, the required minimum width of the RRZ and the RMZ for each stream class was set out in the legislation.

Collectively, the riparian reserve zone (RRZ) and the riparian management zone (RMZ) form the legally required riparian management area (RMA). No trees may be cut in the riparian reserve zones (RRZ) but trees may be cut within the riparian management zone (RMZ). Thus, as shown in Table 31.1, the total width of a riparian management area (RMA) on an S3 stream is 40 m. The 20 m beside the stream must be an undisturbed reserve, and the outer 20 m is a riparian management zone where no machines may operate but where trees may be cut and removed. On the very small, S4, fish-bearing streams and on the non-fish-bearing streams (S5 and S6) no riparian reserves are required so the RMA consists only of a 20- or 30-m management zone (RMZ) where trees may be removed but where machine use is restricted.

The code did not set a minimum standard on how many trees must be retained within the management zone (RMZ). Non-binding recommendations about the average basal area retention for each class of stream, described as best management practices, were provided in a guidebook (BC Ministry of Forests & BC Environment 1995). For small fish-bearing S4 streams, the recommended average retention was 25% of the basal area and for S6 streams, it was only 5%. Thus, although the legislation required that an RMZ be established, in practice, the management zone could be entirely clear-cut. For small streams, including small fish-bearing S4 streams, this meant that all trees along the stream edge could be legally cut down. On larger streams higher retention levels are recommended within the management zone and the reserve zone is 20–50 m wide.

The limited amount of protection for the very small streams led to a report that was highly critical of the code's riparian provisions for fish habitat protection and their implementation (Sierra Legal Defence Fund, SLDF 1997). Based on a review of a sample of cut-blocks logged since the introduction of the code in 1995, SLDF reported that most of the streams within cutting areas were small S4, S5 and S6 and had no reserve zones. The code provision that allowed logging within the RMZ of these small fish-bearing streams and steep streams that flow into fish-bearing streams meant that the riparian areas of these streams were legally clear-cut to the banks. The report also stated that many streams were either not classified or were misclassified and therefore reserve zones were either smaller than required or were logged. A subsequent and much larger study by the Forest Practices Board reached similar conclusions about the extensive cutting of management zones of small steep gradient streams, misclassification and inappropriate reserve zones on small streams (Forest Practices Board 1998).

The latter study, however, found high levels of compliance with other provisions of the code relating to reserves on S1 and S2 streams, falling and yarding practices on all stream classes, removal of any introduced debris and removal of culverts. On many S5 streams, intact reserves were retained even though not required by the code. The study found that the impact of logging on streams was significantly less than found in the audits undertaken before the code (Tripp *et al.* 1992; Tripp 1994). The Forest Practices Board study concluded that, while there was room for improvement, the code had been effective in significantly reducing the impacts of forest harvesting on coastal streams.

Audits conducted by the Forest Practices Board since the 1998 study confirm that some of the classification problems and the retention of trees in management zones along streams have been addressed. Compliance with code requirements has improved each year, and damage to streams from logging has been rarely observed (Forest Practices Board 2001). In 2001, a separate study looked at practices along very small (<1.5-m wide) fish-bearing streams (S4 streams) in the British Columbia interior. This study concluded that the practices implemented to comply with the code along small fish-bearing S4 streams have been effective in protecting fish habitat (Chatwin *et al.* 2001).

The implementation of the regulatory regime in the Forest Practices Code greatly improved practices along streams in the province and led to a significant reduction in the impacts of forestry activities on fish habitat. Public concerns remain about the effectiveness of the measures to protect fish habitat in the code, and the absence of measures to protect small headwater streams. There is also concern about the cumulative effects of harvesting on fish habitat in watersheds. However, the clear, legally required and enforceable standards in the code were a much more successful way of protecting fish habitat from the impacts of forestry operations than the Coastal Fish Forestry Guidelines and other earlier versions of voluntary guidelines and decision-making procedures.

Despite the apparent effectiveness of the code, British Columbia is moving to implement yet another type of regime for managing fish–forestry interactions. In 2002, a new provincial government announced a major initiative to replace the Forest Practices Code with a more 'results-based' approach to regulating forest practices (British Columbia Ministry of Forests & BC Ministry of Water, Land and Air Protection 2002). This is to address concerns that the code is too expensive, too restrictive and too focused on process and regulation. The new approach will streamline the existing legislation and replace many of its provisions. The 'results-based code' will establish objectives for specific forest and environmental values, including riparian areas, and will provide professional foresters and forest companies with greater latitude to implement forest practices and greater responsibility to achieve the required results. The government has stated that it will maintain high environmental standards and that the new code will incorporate the existing provisions for riparian reserve and management zones.

Initially, the new proposals have not been well received by environmental organizations, forest companies or the public (Hoberg 2002) and it is unclear how the provisions for fish habitat protection will be changed in the new legislation. However, it is clear that British Columbia is entering yet another phase in its search for an effective and efficient type of code to manage the interactions between forest harvesting and other forest resource values, including fish and fish habitat.

There is already one example of a 'results-based' approach to fish habitat protection in British Columbia. It is used on the small area of privately owned forest land in the province. Private land is exempt from the riparian protection provisions of the Forest Practices Code and regulated instead by a separate private forest land regulation that was developed by government and the private land owners (British Columbia 1999). This regulation is intended to provide 'results-oriented' environmental standards that allow private landowners latitude to use innovation and local knowledge to protect four key environmental values that include water quality and fish habitat. The regulation provides that a landowner must ensure that stream banks are protected, soil erosion is minimized, machine tracks along the stream edge do not lead to sedimentation and accumulations of debris in the stream do not cause harm to fish habitat. Understory vegetation and non-commercial trees within 5 m of the stream edge must be retained. For fish-bearing streams, there is a requirement that a landowner must retain at least 40 trees evenly distributed along each 200 m of any fish stream that is more than 3 m wide. The trees must be retained in the same range of diameter classes and same proportion of coniferous to deciduous trees that were in the pre-harvest stand so that at least 40 trees are retained. For fish streams that are between 1.5 and 3 m wide, the retention requirement is at least 20 trees along each 200 m.

These requirements are supplemented by a handbook that describes best management practices guidance to achieve the fish habitat standards (Private Forest Landowners Association 1997).

Conclusions

Codes of practice that regulate the interactions between forest harvesting activities and fish habitat are in place in many parts of the world, including Europe, Asia, Oceania,

South and Central America, North America and Africa. This includes most of the world's largest forested jurisdictions where there are important forestry–fisheries interactions in boreal, temperate and tropical forests – Canada, Russia, United States, China, Indonesia and Brazil.

The codes of practice represent a broad spectrum of approaches ranging from very loose guidelines with no legal basis to very detailed prescriptive regulations established within law. We have described provisions in codes that conform to two basic types but have also identified another three types of codes that are used to regulate forest practices.

In most cases, the objective of the codes of forest practices is to protect water quality. Protection of water quality serves as a surrogate for protecting the fish and other organisms that live in the aquatic habitats but in most codes there are no explicit requirements or measures that protect fish or fish habitat directly. The most notable exceptions are in the five jurisdictions on the west coast of North America – Alaska, British Columbia, Washington, Oregon and California – where there is a long and well-documented history of interactions between forestry and fisheries. In these jurisdictions, there are very complex, legally enforceable codes that include provisions to protect fisheries values.

Similarly, most codes classify streams and rivers and identify appropriate practices based on physical features such as width, seasonality, soil susceptibility or slope rather than on the presence of fish. Again, the exceptions are on the west coast of North America where streams are classified and practices prescribed according to the presence or absence of particular species of fish.

The lack of any consistency in the approaches to developing codes and in the types of measures that they prescribe makes it impossible to translate experiences from one jurisdiction to another or to evaluate which approaches and measures are most effective. Independent monitoring of compliance with the codes of practice is generally lacking and many countries that do have codes of practice have poor governance structures, leading to problems with compliance.

References

Alabama Forestry Commission (1993) *Alabama's Best Management Practices for Forestry*. State of Alabama.

Alaska Department of Natural Resources (2000) *Alaska Forest Resources and Practices Act*. Division of Forestry, Department of Natural Resources, Juneau, Alaska.

American Forest and Paper Association (2000) *Sustainable Forestry Initiative Standard*, 2001 edn. Washington, DC.

Applegate, G. & Andrewartha, R. (2002) *Development of Codes of Practice for Tropical Forests in Asia-Pacific*. Unpublished paper. Jaakko Pöyry Consulting (Asia-Pacific) Canberra, Australia.

Asia Pacific Forestry Commission (1999) *Code of Practice for Forest Harvesting in Asia-Pacific*. Food and Agriculture Organization of the United Nations Regional Office for Asia and the Pacific, RAP Publication:1999/12. Bangkok, Thailand.

Asia Pacific Forestry Commission (2000) *Regional Strategy for Implementing the Code of Practice for Forest Harvesting in Asia-Pacific*. Center for International Forestry Research, Jakarta, Indonesia.

Awang, D.I. (2002) *Sustainable Forest Management in Malaysia: The Way Forward*. Malaysia Timber Council, Kuala Lumpur, Malaysia. (http://www.mtc.com.my/publication/speech/sustainable.htm) [accessed 7 September 2002].

Belt, G.H., O'Laughlin, J. & Merrill, T. (1992) *Design of Forest Riparian Buffer Strips for the Protection of Water Quality: Analysis of the Scientific Literature*. Idaho Forest, Wildlife and Range Policy Analysis Group. Report No. 8. University of Idaho, Moscow, ID.

British Columbia (1995) *Forest Practices Code of British Columbia Act*. Victoria, BC.

British Columbia (1999) *Private Land Forest Practices Regulation*. Victoria, BC.

British Columbia Forest Service (1972) *Planning Guidelines for Coast Logging Guidelines*. 29 September 1972. Chief Forester, Forest Service, Victoria, BC.

British Columbia Forest Service (1974) *Vancouver District Circular Letter VR74–245 Re: Administration of P Clauses*. 9 July 1974. BC Forest Service, Vancouver, BC.

British Columbia Forest Service (1978) *Streamside Management Methods for the Queen Charlotte Islands*. Queen Charlotte Islands Forest District, Queen Charlotte City, BC.

British Columbia Ministry of Forests and Lands, BC Ministry of Environment, Federal Department of Fisheries and Oceans & Council of Forest Industries (1987) *British Columbia Coastal Fisheries Forestry Guidelines*. Ministry of Forests and Lands, Victoria, BC.

British Columbia Ministry of Forests, BC Ministry of Environment, Federal Department of Fisheries and Oceans & Council of Forest Industries (1988) *British Columbia Coastal Fisheries Forestry Guidelines*, 2nd edn. Ministry of Forests, Victoria, BC.

British Columbia Ministry of Forests, BC Ministry of Environment, Lands and Parks, Federal Department of Fisheries and Oceans & Council of Forest Industries (1992) *British Columbia Coastal Fisheries/Forestry Guidelines*, 3rd edn. October 1992. Ministry of Forests, Victoria, BC.

British Columbia Ministry of Forests, BC Ministry of Environment, Lands and Parks, Federal Department of Fisheries and Oceans & Council of Forest Industries (1993a) *British Columbia Coastal Fisheries/Forestry Guidelines*, revised 3rd edn. July 1993. Research Branch, Ministry of Forests, Victoria, BC.

British Columbia Ministry of Forests & BC Ministry of Environment, Lands and Parks (1993b) *Interior Fish Forestry and Wildlife Guidelines for the Sub-Boreal Mountains and Central Interior Plateau Ecoprovinces*. Draft August 1993. Ministry of Forests, Victoria, BC.

British Columbia Ministry of Forests & BC Environment (1995) *Riparian Management Area Guidebook*. December 1995. Ministry of Forests, Victoria, BC.

British Columbia Ministry of Forests & BC Ministry of Water, Land and Air Protection (2002) *A Results-Based Forest and Range Practices Regime for British Columbia*. Discussion paper for public review and comment. Government of British Columbia, Victoria, BC.

Bull, G. & Associates (1999a) *A Review of Forest Practices Legislation, Regulations and Guidelines of Selected Canadian Jurisdictions*. Report 1. Prepared for British Columbia Ministry of Forests, Victoria, BC.

Bull, G. & Associates (1999b) *A Review of Forest Practices Legislation, Regulations and Guidelines of Selected International Jurisdictions*. Report 2. Prepared for British Columbia Ministry of Forests, Victoria, BC.

Bull, G.Q., Pulkki, R., Killmann, W. & Schwab, O. (2001) An investigation of the costs and benefits of reduced impact logging. *Tropical Forest Update*, **11**, 1–5. International Tropical Timber Organization, Yokohama, Japan.

Cambodia Department of Forestry and Wildlife (1999) *The Cambodian Code of Practice for Forest Harvesting*. Ministry of Agriculture, Forestry and Fisheries, Cambodia.

Canadian Standards Association (1998) *CSA Sustainable Forestry Management Standards* CAN/CSA-Z808–96 and CAN/CSA-Z809–96. Special Publication PLUS 9015. Mississauga, Ontario.

Chamberlin, T.W. (1987) *Proceedings of a Workshop: Applying 15 Years of Carnation Creek Results*. Workshop held in Nanaimo, BC, 13–15 January 1987.

Chatwin, S., Tschaplinski, P., McKinnon, G., Winfield, N., Goldberg, H. & Scherer, R. (2001) *Assessment of the Condition of Small Fish-bearing Streams in the Central Interior Plateau of British Columbia in Response to Riparian Practices Implemented under the Forest Practices Code*. Working Paper 61. Research Branch, BC Ministry of Forests, Victoria, BC.

Dykstra, D.P. & Heinrich, R. (1996) *FAO Model Code of Forest Harvesting Practice*. Food and Agriculture Organization of the United Nations, Rome, Italy.

Elias, Applegate, G., Kartawinata, K., Machfudh, & Klassen, A. (2001) *Reduced Impact Logging Guidelines for Indonesia*. Center for International Forestry Research, Jakarta, Indonesia.

Forestry Commission (UK) (1991) *Forests and Water Guidelines*. Forestry Commission, London.

Forest Practices Board of British Columbia (1998) *Forest Planning and Practices in Coastal Areas with Streams*. Technical Report. Forest Practices Board, Victoria, BC.

Forest Practices Board of British Columbia (2001) *Backgrounder on Streamside Protection*. Forest Practices Board, Victoria, BC.

Forest Practices Board of Tasmania (2000) *Forest Practices Code*. Hobart, Australia.

Forest Resources Commission (1991) *The Future of Our Forests*. Forest Resources Commission, Victoria, BC.

Forest Stewardship Council (2000) *FSC Principles and Criteria*, revised February 2000. Forest Stewardship Council, Oaxaca, Mexico.

Hartman, G.F. (1982) *Proceedings of the Carnation Creek Workshop, A 10 Year Review*. Malaspina College, Nanaimo, BC.

Hartman, G.F. & Scrivener, J.C. (1983) *Some Implications of Carnation Creek Research Results to the Process of Forest Planning Guideline and Protection Clause Review*. Unpublished Manuscript. Pacific Biological Station, Nanaimo, BC.

Hoberg, G. (2002) *Finding the Right Balance*. Report of Stakeholder Consultations on a Results-Based Forest and Range Practices Regime for British Columbia. Department of Forest Resources Management, University of British Columbia, Vancouver, BC.

International Tropical Timber Organization (1992) *ITTO Guidelines for the Sustainable Management of Natural Tropical Forests*. ITTO Policy Development Series 1. Yokohama, Japan.

Indonesia Ministry of Forestry and Estate Crops (2000) *Principles and Practices for Forest Harvesting in Indonesia*. Jakarta, Indonesia.

Maine Forest Service (2002) *Forestry Best Management Practices Use and Effectiveness in Maine*. Department of Conservation, Augusta, ME.

Montana Department of State Lands (1991) *Montana Forestry Best Management Practices: Forest Stewardship Guidelines for Water Quality*. Publication no. EB0096. Missoula, MT.

Moore, M.K. (1980) *Streamside Management – A Decision-Making Procedure for South Coastal British Columbia*. Land Management Handbook 1. Research Branch, Ministry of Forests, Victoria, BC.

Moore, M.K. (1991) *A Review of the Administrative Use and Implementation of the Coastal Fisheries/Forestry Guidelines*. Moore Resource Management Report to Department of Fisheries and Oceans, Canada and BC Ministry of Environment, Victoria, BC.

Narver, D.W. & Chamberlin, T.W. (1976) *Carnation Creek – An Experiment Towards Integrated Resource Management*. Circular 104. Pacific Biological Station, Nanaimo, BC.

New Zealand Ministry of Agriculture and Forestry (2002) *Standards and Guidelines for Sustainable Management of Indigenous Forests*. Auckland, New Zealand.

North Carolina Department of Environment, Health and Natural Resources (1990) *Best Management Practices for Forestry in the Wetlands of North Carolina*. Durham, NC (http://www.dfr.state.nc.us/source_files/best_man.pdf) [accessed 7 September 2002].

North Carolina Division of Forest Resources (1989) *Forestry Best Management Practices Manual*. Durham, NC (http://www.dfr.state.nc.us/source_files/bmpmanual.pdf) [accessed 7 September 2002].

Oregon Department of Forestry (1995) *Forest Practice Administration Rules*. Oregon Department of Forestry, Salem, OR.

People's Republic of China Department of Forest Resources Management (Draft) (2001) *National Code of Practice for Forest Harvesting in China*. State Forestry Administration, Beijing, China.

Private Forest Landowners Association (1997) *A Handbook of Best Management Practices*. Private Forest Landowners Association, Victoria, BC.

Poulin, V.A. (1984) *A Research Approach to Solving Fish/Forestry Interactions in Relation to Mass Wasting on the Queen Charlotte Islands*. Land Management Report 27. Ministry of Forests, Victoria, BC.

Pulkki, R., Schwab, O. & Bull, G.Q. (2001) *Reduced Impact Logging in Tropical Forest: Literature Synthesis, Analysis and Prototype Statistical Framework*. Forest Products Division Working Paper Series: Working Paper No. 8. FAO, Rome, Italy (www.fao.org/forestry/FOP/FOPH/harvest/x0001e/x0001E00.htm).

Rood, K.M. (1984) *An Aerial Photograph Inventory of the Frequency and Yield of Mass Wasting on the Queen Charlotte Islands, British Columbia*. Land Management Report 34. BC Ministry of Forests, Victoria, BC.

Russian Federation (1996) *The rule of water protection zones of water bodies and their riverbank protective strips*. Decree #1404, 23 November 1996 [unofficial translation].

Sierra Legal Defence Fund (1997) *Stream Protection Under the Code: The Destruction Continues*. Sierra Legal Defence Fund, Vancouver, BC.

State of Montana (1991) *Streamside Management Act*. Cited in Montana Annotated Code (2001) Chapter 5. Timber Resources. 77–05–302 (www.data.opi.state.mt.us/bills/mca/77/5/77-5-302.htm) [accessed 7 September 2002].

Toews, D.A.A. & Brownlee, M.J. (1981) *A Handbook for Fish Habitat Protection on Forest Lands in British Columbia*. Fisheries and Oceans Canada, Vancouver, BC.

Toews, D.A.A. & Wilford, D.J. (1978) *Watershed Management Considerations for Operational Planning on TFL 39 (Block 6A), Graham Island*. Fisheries and Marine Service Manuscript Report No 1473. Department of Fisheries and Oceans, Vancouver, BC.

Tripp, D. (1994) *The Use and Effectiveness of the Coastal Fisheries-Forestry Guidelines in Selected Forest Districts of Coastal British Columbia*. Tripp Biological Consultants report to BC Ministry of Forests, Victoria, BC.

Tripp, D. (1995) *The Use and Effectiveness of the Coastal Fisheries-Forestry Guidelines in the Chilliwack and Mid-Coast Forest Districts of Coastal British Columbia*. Tripp Biological Consultants report to BC Ministry of Forests, Victoria, BC.

Tripp, D. (1998) Problems, prescriptions and compliance with the coastal fisheries-forestry guidelines in a random sample of cutblocks in coastal British Columbia. In: *Carnation Creek and Queen Charlotte Islands Fish/Forestry Workshop: Applying 20 Years of Coastal Research to Management Solutions*, (eds D.L. Hogan, P.J. Tschaplinski & S. Chatwin), pp. 245–55. Land Management Handbook 41. BC Ministry of Forests, Victoria, BC.

Tripp, D.B. & Poulin, V.A. (1986) *The Effects of Mass Wasting on Juvenile Fish Habitats in Streams on the Queen Charlotte Islands*. Land Management Report 45. BC Ministry of Forests, Victoria, BC.

Tripp, D., Nixon, A. & Dunlop, R. (1992) *The Application and Effectiveness of the Coastal Fisheries-Forestry Guidelines in Selected Cutblocks on Vancouver Island*. Tripp Biological Consultants report to BC Ministry of Forests, Victoria, BC.

Washington Department of Natural Resources (1995) *Forest Practice Rules*. Washington Department of Natural Resources, Olympia, Washington.

Westland Resource Group (1995) *A Review of the Forest Practices Code of British Columbia and Fourteen Other Jurisdictions*. Background Report. Prepared for BC Ministry of Forests, Victoria, BC.

Wilford, D.J. (1987) *Watershed Workbook – Forest Hydrology Sensitivity Analysis for Coastal British Columbia Watersheds*. BC Ministry of Forests and Lands, Prince Rupert Forest Region, Smithers, BC.

Young, K.A. (2000) Riparian zone management in the Pacific Northwest: who's cutting what? *Environmental Management*, 26, 131–44.

Chapter 32

Forest management and watershed restoration: repairing past damage is part of the future

G.F. HARTMAN

Introduction

Since late in the nineteenth century, logging has disrupted fish habitat in North America. Beginning in the 1960s, British Columbia (BC) and the western states of the USA actively attempted to improve management practices. However, responsible management of the forested watersheds requires that we continue to improve forest practices, and also effectively restore watersheds that have been damaged by past forestry activities. The cost and scale of required restoration work show us the consequences of inadequate management in the past and foreshadow the expense of cutting management corners in the future.

Chapters 13, 14 and 15 review the effects of forestry activities on streams, lakes and estuaries, respectively. Chapter 13 (Table 13.1) outlines the array of potential physical habitat changes that may occur in streams with forestry activities, and lists the potential changes to fish habitat, particularly that of salmonids. The array of habitat impacts includes change in solar irradiation and temperature increase, increase in sediment transport and deposition, changes in litter input, loss of large woody material, and related changes in channel structure. In response to these changes, restoration work tends to focus on three categories of activity:

(1) work that deals with slope processes and attempts to control landslides and sediment input to streams;
(2) riparian work that may restore vegetation, shade and litter input;
(3) work within the stream channel to provide structural complexity in the channel, and cover and spawning habitat.

The various structures and procedures used in these activities will be listed later in the text.

While restoration of streams and watersheds is vital, these efforts have had a history of varied success. This chapter considers stream and watershed restoration and discusses planning and administrative elements that must be integrated with science to make the process more effective. It supports the perspective on 'ecological restoration' offered by Kauffman *et al.* (1997) and the planning approach by Kondolf (1995). It emphasizes the necessity of understanding processes within rivers before we work in them (Kellerhals & Miles 1996).

There are a number of overlapping definitions for work in watersheds or streams to repair damage or improve conditions. The definition may depend upon the reason why the work is being done. In one situation, construction of new off-channel habitat may be considered to be 'enhancement', in another it may be classed as 'restoration'. Restoration in this chapter refers to the re-establishment of the structure and function of ecosystems. Although the focus is on restoration, there is no substitute for preventing habitat damage from occurring and stopping it once it has begun.

Forest land management and restoration

Early land use legacy

In North America, vast amounts of stream habitat have been affected by land use. Slaney & Martin (1997a) estimate that as much as 20,000 km of BC streams may require channel, riparian or slope restoration. From 70 to 90% of all riparian areas in the USA have been 'extensively altered' by land use, particularly forestry and agriculture (Kauffman *et al.* 1997). The scale of fish habitat damage is so large because it has been ongoing for a long period, and because resource management early in the 'European' history of North America was not sensitive to environmental concerns.

The historical approach to resource use in North America is, in large measure, responsible for the present-day needs for restoration. The pattern has been for early users to claim and exploit natural resources with little concern for the impacts of such use. This approach still prevails and is being expanded in regard to logging, hunting and agriculture in some developing areas of the world (Sizer & Plouvier 2000). In North America, following European colonization, various single uses (e.g. logging, mining, agriculture) began by simply exploiting resources as needed. They expanded with the primary management concern being allocation and regulation for 'free use'. Demarchi & Demarchi (1987) identified this approach as it applied to BC. Its social legacy continues in many places, and in the BC forest sector, expectations for continuing or expanding employment make it difficult to stop impacts.

Resource management: its increasing complexity

During the last 40 years of BC history, interactions among users have led to conflicts and then to a sequence of improving land use approaches. These have evolved from single use allocation, to planning for multiple use, and later, for integrated use. Currently, many agencies and groups discuss or attempt to apply 'ecosystem management'. The phrase implies different things to different people (Lackey 1998, 1999, 2000; Schramm & Hubert 1996). Importantly, some definitions include restoration as part of ecosystem management (Lackey 1998).

To be most meaningful, ecosystem management must involve consideration of all values and resources. However, the application of a pre-determined annual allowable cut, as done in BC, constrains managers in the protection of other resource sectors. For effecting better fish–forestry interactions, level of use, e.g. annual allowable cut, must become an output of the planning process rather than an input to it (Anonymous 1995a). Indeed, in any situation where there is a complex of many land uses, levels of use in various sectors should become outputs of the planning process rather than dominating inputs. Furthermore, as requirements for management and protection of other resources are added to an existing complex of uses, the level of use in all, or most other sectors, must be reduced if the stated management paradigm is to include sustainability. Such balanced use is not possible if human populations continue to increase. As demands for jobs and use of most urgently required resources continue to rise, it will become harder to sustain the complex interconnected environments that are required for many species of fish. This will occur because there is a differential in what society views as necessary to sustain. Energy, food and employment will take precedence over such things as wild fish, which require complex and 'healthy' ecosystems. However, we can ill afford to lose the vast protein production and recreational potential that functional freshwater systems offer. Therefore, in the face of the needs of expanding human populations, and cumulative impacts from satisfying them, it is urgent that degraded watersheds, and the fish habitats within them, are restored to more natural functional conditions. Restoring watersheds and fish habitat, if done correctly, is a route for regaining ground, and as such is an important element of future fish–forestry interaction. The cost of repairing past damage is, however, high.

Funding ranges from 'millions of dollars' spent annually in individual river basins (Roni *et al.* 2002) to small projects of a few thousand dollars each (Hartman & Miles 1995). $428 million was spent in the Watershed Restoration Program (WRP) in BC (Anonymous 2001), but less than half of this funding was applied to work to improve water quality and fish habitat (P. Slaney, pers. comm.). The variation in projects in terms of size, time of execution, groups involved, cause of damage and political control has resulted in inconsistency in application of scientific information, choice of measures applied, and quality of planning, monitoring, maintenance and evaluation.

Changing approaches to restoration work

Efforts at restoration and enhancement of streams over the past 75 years have ranged from small projects involving placement of a few structures in the channel of a small stream, to multiple project programmes to restore whole systems. Stream habitat improvement work began about 1930 in North America with biologists and laymen modifying stream channels. Between 1933 and 1935, 31,084 stream structures were installed on 406 streams in the west and mid-west of the USA. Many of them failed because of the lack of knowledge of stream processes (Koski 1992). As early as the 1930s there was debate about the benefits of constructing cover for greater numbers of fish than the food resources might support. Today there are fisheries biologists and resource specialists (Anonymous 1988; Beschta *et al.* 1995) and many public groups that believe that physical alteration of a stream provides a simple mechanism for improving or restoring fish habitat. Early stream improvement handbooks reflect the emphasis on 'in-channel' work.

The first US Forest Service handbook on stream improvement was published in 1952 (USDA Forest Service 1952), the second in 1985 (Seehorn 1985) and the third (improved edition) in 1992 (Seehorn 1992). Other handbooks are: White & Brynildson (1967), National Research Council (1992) – cited in Beschta *et al.* (1995) – Newbury & Gaboury (1993), House *et al.* (1988), Adams & Whyte (1990), Lowe (1996), Slaney & Zaldokas (1997), Epps *et al.* (1998) and Anonymous (1998). Donat (1995) reviewed bioengineering techniques for restoration of European running waters. Rutherfurd *et al.* (2000) prepared a two-volume manual to guide stream rehabilitation work in Australia. It emphasized that procedures which helped to understand processes in streams were critical components of rehabilitation.

Koski (1992) discussed the requirements for effective restoration. These ranged from understanding limiting factors of fish populations at the outset, to doing proper evaluation during and at the end of the work. Beschta *et al.* (1995) indicated concern that the sequence of manuals did not expand on things learned in the past about technique, or benefit/cost ratios. Kauffman *et al.* (1997), considering recent history, indicated the need for handbooks to have a strong geomorphic perspective (see Kellerhals & Miles 1996) in which changes and their historical causes should be analysed and understood as a prerequisite to restoration work (M. Miles, pers. comm.).

Bibliographic publications have been produced on stream habitat improvement (Wydoski & Duff 1978; Duff *et al.* 1995), and other publications have reviewed monitoring effectiveness – Anonymous (1995b), Hartman & Miles (1995), Sterling (2000), Smokorowski *et al.* (1998). Major workshops and planning documents include Hassler (1985), Miller (1986) and Slaney & Zaldokas (1997).

Diversity of projects and programmes

Stream projects vary widely in size, scope and nature. Activities on valley slopes, intended to reduce slide risk and slope sediment input, include deactivation of roads, returning of water to pre-disturbance routes and channels, and stabilizing slides by planting vegetation and/or building terraces (Underhill 2000). Activities and initiatives in riparian zones may involve planting vegetation and excluding further disturbance. Restoring the connections among habitat elements is critical. Such re-connection of habitat elements can include replacing culverts that block fish access (Wilson *et al.* 2002), or joining channels to riverine ponds (Peterson 1985). Replacement of spawning or rearing habitat, outside of the channel, may be accomplished by digging new channels (Bonnell 1991) or ponds (Cederholm *et al.* 1988) either connected to the main stream or fed by ground water.

In-stream works (guidebooks listed earlier in the text) may include spawning gravel placement, and cover construction with boulders, woody material or gabions. Channel configuration may be changed by boulder placement, rock spur construction, weir construction with logs, and by use of structures that catch and hold large woody material (Wilson *et al.* 2002). Nutrient budgets in streams may be improved by adding fertilizer or restoring riparian vegetation and litter input. These measures are particularly applicable to coastal streams of western North America. Structures that are less durable, e.g. brush bundles, brush mats, half-logs and bank platforms, were used in Wisconsin streams that are characterized by stable flows and low gradient (Hunt 1988).

The organizational nature of programmes varies greatly. They may range from small, one-structure projects, up to multi-year programmes with projects spread over a whole region and executed by an array of different groups and agencies. Spawning gravel placement at a lake outlet is an example of the former. The work in Fish Creek, Oregon, in which over 410 structures were built over several years is an example of a large, multi-year programme (Reeves *et al.* 1990). A complex, whole-system restoration programme in the Keogh River, in coastal BC, has involved placement of wood and boulder structures, addition of nutrients and treatment of valley slopes (McCubbing & Ward 2000).

Diversity of objectives

The number and diversity of programme objectives vary widely. Small projects, such as the placement of spawning gravel in the outlet stream of a lake, may have the single objective of establishing a self-sustaining population of fish in the lake. Large complex programmes, at the 'upper' end of the scale, may have a primary objective of repairing forestry impacts. However, under such an umbrella, they may have several desired outcomes that range from increasing returns to the fishery to simply providing short-term employment. In BC, the WRP (Anonymous 2000) included multiple benefits:

- restoration of disturbed areas in watersheds;
- improvement of fish stocks (i.e. salmonid smolt output and adult returns);
- development of technical forestry planning skills;
- provision of technical community employment;
- provision of support for courses, workshops and conferences;
- provision basis for growth of stream restoration consulting firms;
- provision of short-term employment for displaced forest workers.

Past performance

Ranges in success and failure

There is a wide range in success among the sample of projects reviewed. Chapman (1996) found 'little reliable evidence of benefits' from in-stream works in tributaries of the Columbia River. He suggested the need to shift emphasis toward watershed husbandry. In projects in eight regions or project areas in BC and Pacific Northwest states of the USA, overall success rates ranged from 19% to 94% (Table 32.1). There was no single reason for success or failure. Structures failed because the conditions in the watersheds were not understood. Structures in southwest Alberta streams did poorly in cases where the channels were laterally and vertically unstable, and where bank heights were less than 2 m (Pattenden *et al.* 1998). In Washington and Oregon, failure rates were high in watersheds that were damaged by logging and landslides and

Table 32.1 Ratings of success of restoration programmes* in western Canada and the Pacific Northwest of USA

Region or project area	Structure or procedure used	Time from construction to evaluation, or recurrence interval (years)	Success rating (different terms used by different authors)	References and comments
Alaska	Spawning channels, riparian repair, access repair, bank stabilization	2–12 years (time)	67% success (interpreted from comments in reference)	Parry & Seaman (1994) Explicit ratings not used
Alberta (southwestern)	Boulder structures, log structures	>100 year (recurrence interval)	63% of structures in place before flood event, 19% in place after it	Pattenden *et al.* (1998)
Alberta (Oldman River drainage)	Logs, boulders, pools, and complexes of all	>100 year (recurrence interval)	'Almost all' or 'majority' lost	Goltz & Allan (2001)
Alberta (Oldman River drainage)	Logs, boulders, pools, and complexes of all	>100 year (recurrence interval)	44% damaged	Alberta Environmental Protection (1998)
British Columbia	Boulders, log sills and weirs, spawning gravel placement, spawning channels	22% > 20 years, 20% for 11–20 years, 58% for >10 years (time)	55% in all three groups combined were 'successful'	Hartman & Miles (1995)

Region	Structure	Time/recurrence	Result	Reference
British Columbia	Sills, debris jams, boulder groups	1–5 years (time)	81% functioning as designed for	Wilson et al. (2002)
	Access restoration (culverts and reconnection)	1–5 years (time)	94% functioning as designed for	Wilson et al. (2002)
	Off-channel pool or spawning channel	1–5 years (time)	92% functioning as designed for	Wilson et al. (2002)
Southwest Oregon	Log weirs, deflectors and jams	2 years (recurrence interval)	7–55% 'failure', 27–83% damaged	Frissell & Nawa (1992)
	Log weirs, deflectors and jams	5–10 years (recurrence interval)	40–100% 'failure', 40–100% damaged	Frissell & Nawa (1992)
Southwest Washington	Log weirs, deflectors and jams	2–5 years (recurrence interval)	0–20% 'failure', 0–89% damaged	Frissell & Nawa (1992)
Washington and Oregon	Log, and log and boulder	<15 year (recurrence interval)	94% in place or some movement	Roper et al. (1998)
	Log, and log and boulder	>64 year (recurrence interval)	74% in place or some movement	Roper et al. (1998)
Oregon, Mt Hood	Weirs, log or boulder, log deflectors	15–20 year (recurrence interval)	93% fully functional or damaged but functional	Higgins & Forsgren (1987)

*Types of structures and procedures used are indicated, but space does not permit detailed listing and description of them. Different methods were used in the original papers to give the duration of the evaluation period. The criteria for success are not identical in the various studies. Long recurrence intervals indicate that structures were subject to a major flood event, but not necessarily a long period of evaluation.

there was a trend of more widespread damage and project failure in the low gradient zones in alluvial valleys (Frissell & Nawa 1992). Structures failed where cables and attachment arrangements were too weak, or where they produced changes in the immediate vicinity of the works so that bed load was accumulated or banks were eroded away (Frissell & Nawa 1992; Hartman & Miles 1995). In a high energy BC stream, boulder groups and spurs tumbled into scour holes and were buried or they were carried away (Miles 1998). Structures that are least likely to fail are those that do not form hard resistance to channel-forming processes and that enhance the stability of existing debris collections (Frissell & Nawa 1992). Some enhancement activities such as culvert improvement and provision of access (Parker 1999; Wilson *et al.* 2002), and placement of spawning gravel near to the outlets of lakes (reviewed in Hartman & Miles 1995), may also have low failure rates.

Projects may fail because of a lack of understanding of what was needed. In severely disturbed streams the limiting factor(s) on fish populations may be obvious, but in those in better condition they may be less clear. Understanding limiting factors is crucial to project design. Everest & Sedell (1984a) and Bisson (1989) provided an excellent discussion of the concept of limiting factors and the pitfalls in identifying them.

The factors that prevent full expression of the population production potential may change seasonally. There may be little value in increasing fish numbers during summer by enhancing feeding conditions if the winter habitat is limiting. During any one season fish may use different habitats during the day and night. There may be limited value in improving 'day habitat' if the limitation occurs at night. Such issues are not superfluous. Everest & Sedell (1984b) identified 38 potential limiting factors which, alone or in combination, may limit four salmonid species and age groups within them. Bisson (1989) listed five pitfalls in identifying limiting factors. Four of the key pitfalls are:

(1)   drawing inferences and placing excessive reliance on professional judgement without full examination of the assumptions that formulated it;
(2)   extrapolation of information from a limited number of sites to the whole watershed;
(3)   oversimplification of complex ecological situations;
(4)   focusing exclusively on one aspect of life history.

In some cases it may be very difficult or cost-ineffective to try to determine limiting factors. However, restoration workers must consider and determine them if possible.

Planning

Both socio-economic and ecological conditions affect what restoration may be done and they must be considered fully during planning.

Societal background

The protection of fisheries resources, and the costs, benefits and public's acceptance of restoration work, must be viewed within an overall societal background of competing priorities for resource use. Expanding human numbers will exacerbate this competition. Furthermore, while a majority of people may indicate a desire to maintain salmon as a resource and support restoration programmes, they are unwilling to make the massive changes in lifestyles and human population growth that would permit maintenance or full restoration of the Pacific salmon resource (Lackey 2000). In developing nations, the most simple elemental needs for food, shelter and employment may be so pressing that concerns about restoring damage from forestry activities may be remote or absent. In the broadest ecological sense, regional and global population processes and potential climate change impacts are such that maintenance may not be possible (Hartman *et al.* 2000). Such societal elements indicate the setting within which technical parts of watershed restoration must be carried out.

Ecological background

Watershed assessment is a critical prerequisite to restoration planning. A watershed, whether large or small, is a complex system that includes the plants, animals, stream networks and the landscape in which they occur. The natural system changes slowly except when it undergoes periodic natural disturbance. On a shorter timescale, land use activities such as logging may alter an array of conditions (Chapters 13, 14 and 15). In order to carry out restoration work, it is necessary to understand the natural processes within a watershed. It is also necessary to understand the changes brought about by all land use activities from the top of the valley slopes to the stream mouth.

The WRP for BC proposed a hierarchy of restoration steps in which disturbed hillslopes and unstable roads should be treated first from upstream downward. Once the slopes were stable, riparian areas were to be considered next for restoration. If after these two stages of restoration, in-channel work was required, it should be done (Moore 1994). This approach was not regarded as necessary in all situations. Slaney & Martin (1997b) suggested that in some cases stream rehabilitation might proceed before hill-slope restoration was complete. They considered that in some cases completion of hill-slope work did not assure successful in-stream work.

Programme planning: major needs

Restoration activities must be set within a framework so that they are done in an effective sequence (Moore 1994). Roni *et al.* (2002) proposed a hierarchical strategy for restoration work that stressed reconnecting habitat units (fixing culverts and reconnecting off-channel waters), restoring processes (long-term, e.g. road repair), and restoring habitat (short-term, e.g. restoring riparian processes).

Elements within the plan

There are several important requirements for design of a programme or project. The emphasis on these requirements may change depending on whether a simple small project or a large complex programme is being planned. To guide decision-making it is important to:

- Obtain sound baseline data and understand the system and the problems within it.
- Know what the system was like, before disturbance, and have a clear understanding of what the programmes should accomplish.
- Be aware of natural disturbance patterns, and incorporate elements from them into planning.
- Understand the limiting factors and consider the potential effects, on other species, that may be caused by improving conditions for the target species. It is also critical to understand what the effects of habitat alteration, intended to benefit one life stage of a species, will be for other life stages of the same species.

If a decision is made to carry out active restoration, the following steps should be:

- Incorporate the complementary strategies proposed in Moore (1994) and Roni *et al.* (2002).
- If geomorphic and vegetation processes are being dealt with, involve appropriate specialists in a team approach.
- Design projects with appropriate spatial and temporal controls, and duration of evaluation (Hall & Baker 1982; Hall 1984).
- Design the project or programme to include firm commitments for funding for monitoring, maintenance and evaluation.
- Design programmes, in which fish production is the objective, so that benefit/cost analysis can be carried out.
- Treat causes and integrate short-term restoration, e.g. providing cover or spawning habitat as needed.

Implementation

All of the listed elements of the programme must come together during implementation. The timing of work must be chosen so that impacts on the stream are minimized. The deactivation of roads or placement of large boulders and pieces of wood during actual construction requires skilled and knowledgeable machine operators. These considerations are dealt with in Adams & Whyte (1990), 'Planning the Project'.

Integrating activities of different groups

The environmental effectiveness of large restoration programmes that are driven by socio-political as well as environmental concerns may be diluted by mixed objectives. If the programme is so inclusive that funding must be allocated to different regions, applied to different watersheds within those regions, spent by different groups and used for different objectives, there is a risk that it may lack enough focus to provide effective improvement to fish habitat anywhere.

If public groups are being funded, it is necessary that they have the scientific and planning capacity to design sound programmes. In past reviews that I have done on about 200 completed works or proposals, most lacked sound project design and few made use of relevant published information. This does not imply a total lack of value of such programmes because there are benefits of public involvement, education, and job training. It is critical that public groups involved in restoration work understand and accept the value of sound planning and the use of relevant published research.

Evaluation

Support for evaluation

Evaluating present projects provides information that is essential for future works. This and the matter of accountability are very important where public funds are spent. However, programmes often omit evaluation because it is short-sightedly regarded as an expensive luxury in comparison with project construction. Hall (1984) noted progress away from this, but evaluation is still a weak element in restoration work. In all, 34% of respondents in a survey for Washington State listed lack of funding as a barrier to monitoring, and 14% and 13% listed lack of time and personnel, respectively, as barriers (Bash & Ryan 2002). The large complex BC Watershed Restoration Program started in 1994 is only 8 years old, with a shorter time for actual project implementation. There is no publication that evaluates, in an integrated fashion, the effectiveness of all of the road deactivation and the in-stream and near-stream works. This multi-million dollar programme has been terminated by government and many of the valuable lessons that might have been learned from it will be lost.

Matching objectives and evaluation

In large complex programmes involving projects of many types, the objectives must be stated clearly and the evaluation must match the objectives. If the objective of a project is to provide temporary employment for displaced workers, the evaluation should document employment results. If the project is to deactivate roads, then a measure of kilometres of road done may be adequate. These types of evaluations may be uncomplicated. However, if the object of road deactivation is to reduce sedimentation in a stream, a properly designed study, with 'before' and 'after project' data from the stream and a control is necessary. If the objectives are compound, then both aspects must be evaluated by appropriate means.

 Projects that treat a combination of a number of types of disturbance, and are designed to increase production of adult fish, returning from the sea, require long-term complex monitoring and analysis over a period of at least 10 years. Effects of treatments may vary inter-annually within the streams, and differential effects of ocean conditions or fishing may confound results. There is uncertainty in restoration work and, in some cases, a reluctance to report failures (Hall & Baker 1982; Hamilton 1989; Kondolf

1995). Failure is not necessarily a poor reflection on people or projects, and those that are reported and analysed provide useful information to guide future work.

If complex, long-term projects that require extended monitoring and analysis are planned, funding commitments for such work should be assured. If the costly and long-term evaluation commitments cannot be met, plans should be set up in advance to evaluate strategically selected parts of the programme.

Physical success vs biological effectiveness

Where structural enhancement has been done, physical success may be used as a surrogate for biological utility. In many projects the primary evaluation information involves physical performance (Table 32.1). Biologists surveyed regarding projects in British Columbia (Hartman & Miles 1995) rated physical and biological performance at almost the same level, 55% and 50% respectively.

Structures that are physically sound may not provide functional fish cover. Following a 100-year return period flood, 19% of the structures in streams in southwestern Alberta remained intact (Table 32.1). Underwater examination of 43 of these revealed that 29 provided little or no deep-water refuge for large trout (Pattenden *et al.* 1998). Some of the structures that were damaged did provide cover.

Although it is necessary to observe the function of structures, evaluation of project performance must be based on a measure of change in fish production. Redistribution of fish may occur following in-stream habitat enhancement work. Such redistribution may create an impression of a greater production increase than is actually attributable to increased growth or survival resulting from the enhancement work (Gowan *et al.* 1994).

Duration of evaluation

All physical works, like natural structures, have a particular life expectancy. Also the length of the life cycle of different species of fish varies. The duration of evaluation work must be long enough to extend through at least three life cycles of fish species involved. Most recently reported programmes (Table 32.1) do not have the benefit of a long period of evaluation even though some, during their short duration, experienced floods with long recurrence periods. A short evaluation period may produce misleading results. Hunt (1976) evaluated a project for an initial 3-year period, following construction, and then for a second 3-year period. The number of brook trout (3-year means) in the study section rose from 1746 to 2881 in the first evaluation interval and from 2881 to 4226 in the second.

Quality and evaluation work

In large, high-profile restoration programmes, evaluation work must not become blurred by public relations activities. When politics, restoration management, public group activity and science come together, there are risks of public relations overtaking science with good stories being put foremost. A part of the public education process, in restoration programmes, should be directed at ensuring that people understand the role and benefits of sound evaluation.

Concluding comments

There is need for project designs that incorporate a whole system perspective. If individual elements of a system are manipulated, the full role of such efforts and their effects on the system should be understood. 'Ecological restoration is a holistic approach not achieved through isolated manipulations of individual elements but through approaches ensuring that natural ecological processes occur' (Kauffman *et al.* 1997). This perspective does not exclude any particular kind of restoration work, given that it is scientifically sound and developed within a framework such as that proposed by Moore (1994).

Large programmes with multiple components should be recorded in central databases so that progress may be easily tracked and so that cost/benefit analyses may be carried out. Databases should include a listing of objectives and evaluation steps, description of work performed, annual accounting of costs, annual reporting of results, and names and addresses of all key individuals involved.

This chapter has focused on salmonids. However, such information has some applicability in other regions of the world because trout and salmon have been transplanted widely around the world. Furthermore, there should be universal desirability in the requirements that programmes incorporate a holistic approach; restore processes; apply sound science; evaluate; maintain accountability; and separate science, public relations and politics.

In a world where forests are rapidly being lost or modified (Chapter 1), it is critical to protect and carefully manage what is left and to restore, if possible, that which is damaged. These are the fundamental elements of effecting better interactions between fish management and protection, and forestry.

Acknowledgements

I am very grateful to G. Miller and G. Pattern of the Pacific Biological Station Library, Nanaimo, for much help received. Special thanks to P. Slaney, R. White, R. Pattenden and T. Northcote for their most helpful reviews. A. Thompson assisted with manuscript preparation, and C. Scrivener, W. Hartman and H. Hartman checked the manuscript; I appreciate their efforts.

References

Adams, M.A. & Whyte, I.W. (1990) *Fish Habitat Enhancement: A Manual for Freshwater, Estuarine and Marine Habitats.* Department of Fisheries and Oceans, Canada.

Alberta Environmental Protection (1998) *Environmental Monitoring of the Oldman River Dam: Eight Years of Progress.* Natural Resources Service, Alberta Environmental Protection, Information Centre, Edmonton, Alberta.

Anonymous (1988) *A Strategy for Fisheries Mitigation in the Oldman River Basin,* Vol. 1, *Upstream Component.* Prepared by Dominion Ecological Consulting Ltd, for Alberta Environment, Planning Division and Alberta Forestry, Lands and Wildlife, Fish and Wildlife Division.

Anonymous (1995a) *Sustainable Ecosystem Management in Clayoquot Sound: Planning and Practices.* Prepared for the Government of British Columbia by the Clayoquot Sound Scientific Panel.

Anonymous (1995b) *Ecological Restoration: A Tool to Manage Stream Quality: Annotated Bibliography.* EPA, United States Environmental Protection Agency, Office of Water, EPA841-F-95-007.

Anonymous (1998) *Stream Corridor Restoration: Principles, Processes, and Practices.* Federal Interagency Stream Working Group from eight US agencies or departments.

Anonymous (2000) *Forest Renewal Program Reports.* Ministry of Environment, Lands and Parks Annual Report to Forest Renewal, BC.

Anonymous (2001) *Investing in B.C.'s Forests. Seven Years at a Glance 1994–2001.* Forest Renewal BC, British Columbia Government, Ministry of Forests, Victoria, BC.

Bash, J.S. & Ryan, C.M. (2002) Stream restoration and enhancement projects: is anyone monitoring? *Environmental Management,* **29,** 877–85.

Beschta, R.L., Platts, W.S., Kauffman, J.B. & Hill, M.T. (1995) Artificial stream restoration – money well spent or an expensive failure? In: *Proceedings: Environmental Restoration,* Sponsored by University Council on Water Resources, pp. 76–104. Montana State University, and Universities of Montana, Idaho and Wyoming, August 2–5, 1994.

Bisson, P.A. (1989) *Importance of identification of limiting factors in an evaluation program.* Manuscript, Olympia Forestry Science Laboratory, SW Olympia, WA.

Bonnell, R.G. (1991) Construction, operation, and evaluation of ground-water fed side channels for chum salmon in British Columbia. *Fisheries Bioengineering Symposium, American Fisheries Society Symposium,* **10,** 109–24.

Cederholm, C.J., Scarlett, W.J. & Peterson, N.P. (1988) Low-cost enhancement technique for winter habitat of juvenile coho salmon. *North American Journal of Fisheries Management,* **8,** 438–41.

Chapman, D.W. (1996) Efficacy of structural manipulations of instream habitat in the Columbia River basin. *Rivers,* **5,** 279–93.

Demarchi, D.A. & Demarchi, R.A. (1987) Wildlife habitat – the impacts of settlement. In: *Our Wildlife Heritage: 100 Years of Wildlife Management,* (ed. A. Murray), pp. 159–77. Centennial Wildlife Society of British Columbia.

Donat, M. (1995) *Bioengineering Techniques for Streambank Restoration: A Review of Central European Practices.* British Columbia Watershed Restoration Program, Watershed Restoration Project Report No. 3.

Duff, D., Wullstein, L.H., Nackowski, M., Wilkins, M., Hreha, A. & McGurrin, J. (1995) *Indexed Bibliography on Stream Habitat Improvement.* US Department of Agriculture, Forest Service, Ogden, UT.

Epps, D., Feduk, M. & Gaboury, M. (1998) *Habitat Restoration Prescription Guidebook.* Watershed Restoration Program, British Columbia Ministry of Environment, Lands and Parks, Nanaimo, BC.

Everest, F.H. & Sedell, J.R. (1984a) *Evaluation of Fisheries Enhancement Projects on Fish Creek and Wash Creek, 1982 and 1983.* Project 83–385. Pacific Northwest Forest and Range Experiment Station, Forestry Sciences Laboratory, Corvallis, OR.

Everest, F.H. & Sedell, J.R. (1984b) Evaluating effectiveness of stream enhancement projects. In: *Pacific Northwest Stream Habitat Management Workshop,* (ed. T.J. Hassler), pp. 246–56. California Cooperative Fishery Research Unit, Humboldt State University, Arcata, CA.

Frissell, C.A. & Nawa, R.K. (1992) Incidence and causes of physical failure of artificial habitat structures in streams of western Oregon and Washington. *North American Journal of Fisheries Management*, **12**, 182–97.

Goltz, C. & Allan, J.H. (2001) *Oldman River Dam Fisheries Mitigation Program Habitat Enhancement Program Summary Report*. Prepared by Pisces Environmental Consulting Services Ltd for Alberta Infrastructure, Edmonton, Alberta.

Gowan, C., Young, M.K., Fausch, K.D. & Riley, S.C. (1994) Restricted movement in resident stream salmonids: a paradigm lost? *Canadian Journal of Fisheries and Aquatic Sciences*, **51**, 2626–37.

Hall, J.D. (1984) Evaluating fish response to artificial stream structures: problems and progress. In: *Pacific Northwest Stream Habitat Management Workshop*, (ed. T.J. Hassler), pp. 214–21. California Cooperative Fishery Research Unit, Humboldt State University, Arcata, CA.

Hall, J.D. & Baker, C.O. (1982) *Rehabilitating and Enhancing Stream Habitat*: 1. *Review and Evaluation*. US Forest Service General Technical Report PNW-138.

Hamilton, J.B. (1989) Response of juvenile steelhead to instream deflectors in a high gradient stream. In: *Practical Approaches to Riparian Resources Management*, (eds R.E. Gresswell, B.A. Barton & J.L. Kershner), pp. 149–57. American Fisheries Society, Montana Chapter, Bethesda, MD.

Hartman, G.F. & Miles, M. (1995) *Evaluation of Fish Habitat Improvement Projects in BC and Recommendations for the Development of Guidelines for Future Work*. Fisheries Branch, BC Ministry of Environment, Lands and Parks.

Hartman, G.F., Groot, C. & Northcote, T.G. (2000). Science and management in sustainable fisheries: the ball is not in our court. In: *Sustainable Fisheries Management: Pacific Salmon*, (eds E.E. Knudsen, C.R. Steward, D.D. MacDonald, J.E. Williams & D.W. Reiser), pp. 31–50. CRC, Lewis Publishers, Boca Raton, FL.

Hassler, T.J. (1985) *Pacific Northwest Stream Habitat Management Workshop*, October 10–12, 1984, Humboldt State University, Arcata, CA. American Fisheries Society, Humboldt Chapter, Western Division.

Higgins, B. & Forsgren, H. (1987) *Monitoring and Evaluation of Mt. Hood National Forest Stream Habitat Improvement and Rehabilitation Projects*. 1986 Executive Summary. Mt Hood National Forest, Pacific Northwest Region, Forest Service, US Department of Agriculture.

House, R., Anderson, J., Boehne, P. & Suther, J. (1988) *Stream Rehabilitation Manual Emphasizing Project Design – Construction – Evaluation*. Training Session, 7–8 February 1988, Inn of the 7th Mountain Bend, OR. Oregon Chapter American Fisheries Society.

Hunt, R.L. (1976) A long-term evaluation of trout habitat development and its relation to improving management-related research. *Transactions of the American Fisheries Society*, **105**, 361–4.

Hunt, R.L. (1988) *A Compendium of 45 Trout Stream Habitat Development Evaluations in Wisconsin during 1953–1985*. Wisconsin Department of Natural Resources Technical Bulletin, 162.

Kauffman, J.B., Beschta, R.L., Otting, N. & Lytjen, D. (1997) An ecological perspective of riparian and stream restoration in the western United States. *Fisheries*, **22**, 12–24.

Kellerhals, R. & Miles, M. (1996) Fluvial geomorphology and fish habitat: implications for river restoration. In: *Proceedings of the Second IAHR Symposium on Habitats Hydraulics, Ecohydraulics 2000*, (eds M. Leclerc *et al.*), A261–A279.

Kondolf, G.M. (1995) Five elements for effective evaluation for stream restoration. *Restoration Ecology*, **3**, 133–6.

Koski, KV. (1992) Restoring stream habitats affected by logging activities. In: *Restoring the Nation's Marine Environment*, (ed. G.W. Thayer), pp. 343–403. A Maryland Sea Grant Book, College Park, MD.

Lackey, R.T. (1998) Seven pillars of ecosystem management. *Landscape and Urban Planning*, **40**, 21–30.

Lackey, R.T. (1999) Radically contested assertions in ecosystem management. *Journal of Sustainable Forestry*, **9**, 21–34.

Lackey, R.T. (2000) Restoring wild salmon to the Pacific Northwest: chasing an illusion? In: *What We Don't Know about Pacific Northwest Fish Runs? An Inquiry into Decision-Making*, (eds P. Koss & M. Katz), pp. 91–143. Portland State University, Portland, OR.

Lowe, S. (1996) *Fish Habitat Enhancement Designs, Typical Structure Designs*. Alberta Environmental Protection, River Engineering Branch, Edmonton, Alberta.

McCubbing, D.J. & Ward, B.R. (2000) *Stream Rehabilitation in British Columbia's Watershed Restoration Program: Juvenile Salmonid Response in the Keogh and Waukwaas Rivers 1998*. Project Report No. 12. British Columbia Forest Renewal, Watershed Restoration Program, Ministry of Environment, Lands and Parks, and Ministry of Forests.

Miles, M.J. (1998) Restoration difficulties for fishery mitigation in high-energy gravel-bed rivers along highway corridors. In: *Gravel-Bed Rivers in the Environment*, (eds P.C. Klingman, R.L. Beschta, P.D. Komar & J.B. Bradley), pp. 393–414. Water Resources Publications, LLC, Highlands Ranch, CO.

Miller, J.G. (1986) *Fifth Trout Stream Habitat Improvement Workshop*, 12–14 August 1986, Lock Haven University, Lock Haven, PA. Pennsylvania Fish Commission, Harrisburg, PA.

Moore G.D. (1994) *Resource Road Rehabilitation Handbook: Planning and Implementing Guidelines (Interim Methods)*. British Columbia Ministry of Environment, Lands and Parks, and Ministry of Forests, Watershed Restoration Technical Circular No. 3.

National Research Council (1992) *Restoration of Aquatic Ecosystems*. National Academy Press, Washington, DC.

Newbury, R.W. & Gaboury, M.N. (1993) *Stream Analysis and Fish Habitat Design: A Field Manual*. Manitoba Heritage Corporation, Manitoba Fisheries Branch.

Parker, M.A. (1999) *Fish Passage – Culvert Inspection Procedures*. Watershed Restoration Program, British Columbia Ministry of Environment, Lands and Parks, and Ministry of Forests, Technical Circular 11.

Parry, B.L. & Seaman, G.A. (1994) *Restoration and Enhancement of Aquatic Habitats in Alaska: Case Study Reports, Policy Guidance, and Recommendations*. Alaska Department of Fish and Game, Habitat and Restoration Division, Technical Report No. 94–3.

Pattenden, R., Miles, M., Fitch, L., Hartman, G. & Kellerhals, R. (1998) Can instream structures effectively restore fisheries habitat? In: *Fish-Forest Conference: Land Management Practices Affecting Aquatic Ecosystems*, (eds M.K. Brewin & D.M. Monita), pp. 1–11. Canadian Forest Service, Northern Forestry Centre 1998, Information Report NOR-X-356.

Peterson, N.P. (1985) *Riverine Pond Enhancement Project: October 1982–December 1983*. Progress Report 233. Washington Department of Fisheries and Washington Department of Natural Resources, Olympia, WA.

Reeves, G.H., Everest, F.H. & Sedell, J.R. (1990) *Influence of Habitat Modifications on Habitat Composition and Anadromous Salmonid Populations in Fish Creek, Oregon, 1983–88*. Project No. 84–11. Pacific Northwest Forest and Range Experiment Station, Forestry Sciences Laboratory, Corvallis, OR.

Roni, P., Beechie, T.J., Bilby, R.E., Leonetti, F.E., Pollock, M.M. & Pess, G.R. (2002) A review of stream restoration techniques and a hierarchical strategy for prioritizing restoration in Pacific Northwest watersheds. *North American Journal of Fisheries Management*, **22**, 1–20.

Roper, B.B., Konnoff, D., Heller, D. & Wieman, K. (1998) Durability of Pacific Northwest instream structures following floods. *North American Journal of Fisheries Management*, **18**, 686–93.

Rutherfurd, I.D., Jerie, K. & Marsh, N. (2000) *A Rehabilitation Manual for Australian Streams*, Vols 1 & 2. Land and Water Resources Research and Development Corporation, Canberra ACT 2601, and Cooperative Research Centre for Catchment Hydrology, Department of Civil Engineering, Monash University, Clayton, Victoria.

Schramm, H.L. & Hubert, W.A. (1996) Ecosystem management: implications for fisheries management. *Fisheries*, **21**, 6–11.

Seehorn, M.E. (1985) *Stream Habitat Improvement Handbook*. Southern Region Technical Publication R8-T-7. USDA Forest Service, Atlanta, GA.

Seehorn, M.E. (1992) *Stream Habitat Improvement Handbook*. USDA Forest Service, Southern Region, Atlanta, GA.

Sizer, N. & Plouvier, D. (2000) *Increased Investment and Trade by Transnational Logging Companies in ACP-Countries: Implications for Sustainable Forest Management and Conservation*. A joint report of WWF-Belgium WRI & WWF International. EC Project B7–6201/96–161/VIII/FOR.

Slaney, P.A. & Martin, A.D. (1997a) The watershed restoration program of British Columbia: accelerating natural recovery processes. *Water Quality Research Journal of Canada*, **32**, 325–46.

Slaney, P.A. & Martin, A.D. (1997b) Planning fish habitat rehabilitation: linking to habitat protection. In: *Fish Habitat Rehabilitation Procedures*, (eds P.A. Slaney & D. Zaldokas), pp. 1.1–1.23. British Columbia Ministry of Environment, Lands and Parks, Watershed Restoration Technical Circular No. 9.

Slaney, P.A. & Zaldokas, D. (1997) *Fish Habitat Rehabilitation Procedures*. British Columbia Ministry of Environment, Lands and Parks, Watershed Restoration Program, Technical Circular 9.

Smokorowski, K.E., Withers, K.J. & Kelso, J.R. (1998) Does habitat creation contribute to management goals? An evaluation of literature documenting freshwater habitat rehabilitation or enhancement projects. *Canadian Technical Report of Fisheries and Aquatic Sciences*, **2249**.

Sterling, M. (2000) *Annotated Bibliography of Watershed Restoration Effectiveness Monitoring*. Prepared for Forest Renewal, British Columbia, Victoria, BC.

Underhill, D. (2000) *Annual Compendium of Aquatic Rehabilitation Projects for the Watershed Restoration Program 1999–2000*. British Columbia Ministry of Environment, Lands and Parks, Watershed Restoration Project Report No. 18.

USDA Forest Service (1952) *Stream Habitat Improvement Handbook*. USDA Forest Service, Washington, DC.

White, R.J. & Brynildson, O.M. (1967) *Guidelines for Management of Trout Stream Habitat in Wisconsin*. Wisconsin Department of Natural Resources Technical Bulletin Number 39. Madison, WI.

Wilson, A., Slaney, P. & Deal, H. (2002) Evaluating the performance of channel and fish habitat restoration projects in British Columbia's Watershed Restoration Program. *Streamline*, **6**, 3–7.

Wydoski, R.S. & Duff, D.A. (1978) *Indexed Bibliography on Stream Habitat Improvement*. Technical Note 322. US Department of the Interior, Bureau of Land Management.

Chapter 33

Better and broader professional, worker and public education in fish–forestry interaction

T.G. NORTHCOTE AND J.D. HALL

Introduction

From the material presented in the previous chapters it should be clear that the subject of fish–forestry interaction is highly complex. It involves an in-depth appreciation of ecological processes in the functioning of forests, watersheds (streams, rivers, lakes, reservoirs, estuaries), and the communities of fishes within them (life histories, stock diversities, migrations, feeding and reproduction). It also includes the wide range of forest harvesting and processing activities that bring about a host of both short-term and long-term effects on these aquatic ecosystems and their fishes. We have also seen the range and severity of forestry activities that have occurred and still are occurring in many regions of the world, some focused on very poorly known aquatic habitats and fish communities. In addition we have examined the attempts to effect better fish–forestry interactions through development of forest harvesting and processing guidelines, codes and legislation, along with watershed management and restoration. But all of this complexity and breadth demands that professional foresters of all specialties receive in-depth coverage of the subject areas outlined above in lecture, laboratory and field exposure. Furthermore, freshwater fish and fisheries professionals should have some coverage in their training, both of how forestry activities are being and should be conducted to prevent or greatly minimize harmful effects on fish and fish habitat, and of how to facilitate meaningful dialogue with forestry professionals.

Appropriate educational opportunities and on-the-job training in fish–forestry interactions are also required by both forestry and fishery technical workers. They are the ones at the front lines, where unexpected but quick decisions sometimes have to be made to prevent serious loss of fish and their habitat. Errors in such decisions occasionally cause loss of habitat. These technical workers must also understand the difficulties that their co-workers may face in the field. Some attempts have been made to provide useful information to both technical groups, but often these efforts have been insufficient.

Finally, the general public and their political representatives must be presented with meaningful information on fish–forestry interactions. They must also have opportunities to discuss, question, see directly, and understand the complexities of fish–forestry interactions, and to realize clearly how the long-term solutions to some of the problems may impinge on their way of life and work in the watersheds involved.

This is the range of educational subjects that we will cover briefly in this penultimate chapter, along with comments and suggestions on how best to effect fish–forestry educational transfer across such a broad front at so many different levels.

Professional forestry educational centres

Reviews of forestry education have been made in Canada and the USA; see for example Dana & Johnson (1963), Garratt (1971), Godbout (1997), Miller & Lewis (1999).

The most comprehensive worldwide listing of information on professional forestry education is that developed by the Food and Agriculture Organization (FAO) of the United Nations. The latest available in late 2001 is their 1996 Directory of Forestry Education and Training Institutions (Anonymous 1996). This directory gives information on about 800 centres from over 100 countries. No analysis of the listings, regionally or by subject area of training, is provided. Information for this directory was obtained from a six-page questionnaire sent by FAO to all forestry education and training centres listed in their 1994 directory, as well as to many others not included in that directory. In spite of this extensive coverage, there are a number of centres not listed in the 1996 directory, either because of failure to receive the questionnaire or failure to complete it. Plans to continue updating the directory every 2 years have not materialized.

The questionnaire identified some 28 main forestry or related teaching areas, up to 10 of which were to be selected by respondents in order of priority. This ranking was intended to identify the main focus of their educational programme. The listed teaching areas potentially relevant to fish–forestry interaction, as indicated by FAO descriptors in parentheses, were (1) environmental conservation (conservation of wildlife resources, nature conservation/conservation of natural resources), and (2) watershed management (hydrology). A third area – wildlife (described as parks and recreation management, range management, wildlife management, game management, wildlife biology) – seemed to imply a teaching area dealing mainly with terrestrial game animals and human recreation. Although no explicit mention in the questionnaire was made of fish, fish habitat or fisheries as a teaching area, the possibility that 'wildlife' may be considered by some to include fish leads to possible problems in our interpretation of the questionnaire results.

Furthermore it was difficult to compare numbers of institutions in the FAO Directory listed for different geographical areas, owing to the inconsistent interpretation of the term 'institution' by the responding groups. For example, the listings for Germany included as separate institutions the many Departments, Institutes, Chairs, etc. within a larger institution of higher learning. In the USA and Canada, and in many other countries, only one listing was provided for each institution (except for a few duplications). In the following analysis, we have attempted standardization by combining multiple educational units into one parent centre, at least where it was clearly possible to do that.

The over 700 forestry educational centres that responded to the survey are distributed among 114 countries in the world (Table 33.1), with strongest representation in

Table 33.1   A summary by major world regions of forestry training centres[a] assembled from information in the FAO Directory (Anonymous 1996)

World region[b] (number of countries)	Forestry training centres	Centres giving main teaching areas	Number (%) giving teaching areas relevant to fish–forestry interaction
North America (3)	114	97	50 (52)
Central America & Caribbean (9)	23	20	9 (45)
South America (11)	93	81	28 (35)
Europe (32)	198	150	62 (41)
Africa (30)	83	60	24 (40)
Asia (17)	136	93	32 (34)
Oceania (12)	64	48	24 (50)
Totals (114)	711	549	229 (42)

[a]Includes those located within universities, colleges and technical schools, but combines centre subdivisions (e.g. departments, institutes, etc.) within these, so numbers are less than given in FAO Directory; see text for details.
[b]As in *The Times Atlas of the World 1999*.

Europe, followed by Asia (mostly in China, Japan and India), with fewer than 10 in the 15 other Asian countries covered. The 114 listed in North America are mostly in the USA (67), followed by Mexico (27) and Canada (20). The 93 in South America are concentrated in Brazil and Chile, next in Argentina and Peru, with fewer than 10 in the remaining South American countries with such centres. The 83 in Africa, spread among 30 countries, are most numerous in Nigeria and Cameroon, with five or fewer in all others. For Oceania (insular Southeast Asia and Pacific oceanic islands, the Antipodes), there are 18 in the Philippines and 15 in Australia, 12 in Indonesia, and five or fewer in each of the remaining countries with centres in that area (Malaysia, Papua New Guinea and New Zealand).

Over half of the North American forestry training centres listed in the 1996 FAO Directory provided teaching in areas relevant to fish–forestry interaction (Table 33.1). Slightly lower percentages (40–50%) did so for centres in Oceania, Central America and the Caribbean, Europe and Africa. Just over a third did so in South America and Asia.

Modern forestry education is about two centuries old. In the first half of the nineteenth century at least six forestry training centres were established in Europe (Table 33.2), one of the first being in Hungary (1808), followed by Poland (1816), Belgium (1817), France (1824), Germany (1831) and Spain (1848). During the later half of that century another 21 were started in Europe and three in Asia. Most other major regions of the world started such training centres in the first two decades of the twentieth century, and numbers in all grew over the following six decades up to 1980 (Table 33.2). Since then the rate of addition of forestry training centres has decreased sharply in all world areas except South America.

Table 33.2 Sequential establishment of forestry training centres[a] in major world regions, mainly assembled from corrected information[b] in Anonymous (1996)

World region (number of countries)	1800–1850	1851–1899	1900–1920	1921–1940	1941–1960	1961–1980	1981–2000	Total
North America (3)			24	11	20	48	9	112
Central America & Caribbean (5)					3	7	1	11
South America (10)			4	2	14	28	20	68
Europe (28)	6	21	18	18	36	41	17	157
Africa (30)			1	5	11	36	15	68
Asia (14)		3	7	6	32	26	11	85
Oceania (8)			2	2	3	37	3	47
Total	6	24	56	44	119	223	76	548

[a]See footnote in Table 33.1.
[b]Many starting dates given in the 1996 FAO Directory were those for the umbrella university/college/institution, not for the forestry training centre; those given have been corrected as far as possible from information given in appropriate websites and publications, and others have been added.

Levels of training in fish–forestry interaction

We made a concerted effort to contact about 350 forestry centres that seemed likely (FAO Directory) to include training relevant to fish–forestry interaction in their teaching areas. They were sent an explanatory letter and a one-page questionnaire (see Appendix). The rationale was to obtain up-to-date information on the extent to which fish–forestry subject areas are now included in the curricula of these forestry training centres, and to determine the date of their introduction.

Analysis of questionnaire response has to be done cautiously. Of 353 questionnaires sent out, 97 (27%) were returned (Table 33.3). Only four came back from the 45 sent to centres in Caribbean and Central and South American countries. Low returns (18) came from the 114 sent to institutions in Asia, Africa and Oceania.

There are marked differences among major world regions in the levels of fish–forestry interaction training now offered by their forestry educational centres (Table 33.3). Most of those that responded from North America provide at least some such training, whereas about 36% and 50% of those responding from Europe and from the rest of the world, respectively, gave no fish–forestry interaction training. Nearly all North American centres responding provided at least Level 1 training, and about 60% also gave training in Levels 2–5. About 60% of European centres responding offered Level 1 training, but much lower percentages had Levels 2–5. Of those centres that responded in other world regions, few offered third to fifth level training in fish–forestry interaction (Table 33.3).

Table 33.3 Comparison among major world regions in the level of fish–forestry training[a] provided by forestry centres offering some courses with an aquatic environmental focus: data from responses to Appendix questionnaire; percentages rounded to whole numbers

	Number of centres		Level of training courses relevant to fish–forestry interaction					
World region	contacted	responded	0	1	2	3	4	5
North America	87	42	1	38	29	25	24	23
Central & South America, Caribbean	45	4	2	2	1	1	0	0
Europe	107	33	12	20	11	6	6	6
Africa, Asia	74	10	6	4	3	0	0	0
Oceania	40	8	3	5	3	3	2	2
Total	353	97	24	69	47	35	32	31
Percentage		27	25	71	48	36	33	32

[a]Training levels: 0 = virtually none; 1 = some aspects of hydrology, stream, river, lake, reservoir, or estuary conditions; 2 = fish biology, ecology, habitat needs; 3 = effects of forestry practices on 1 & 2; 4 = means of reducing harmful effects under 3; 5 = restoration of fish habitat degradation or loss under 3.

A few responding centres in North America and Europe started training in first level fish–forestry interaction before 1941. The number rose quickly thereafter for North America in the 1960s to 1980s, for Europe after 1980, and started then for those in other parts of the world (Fig. 33.1). Training at Levels 2 and 3 did not seem to get well under way in North America until the period 1961–1980, and still remains low in Europe and other areas. Levels 4 and 5 training, low in most European and other responding institutions, reached higher levels in North American institutions between 1961 and 1980, and more have started such training there in the most recent 20 years (Fig. 33.1).

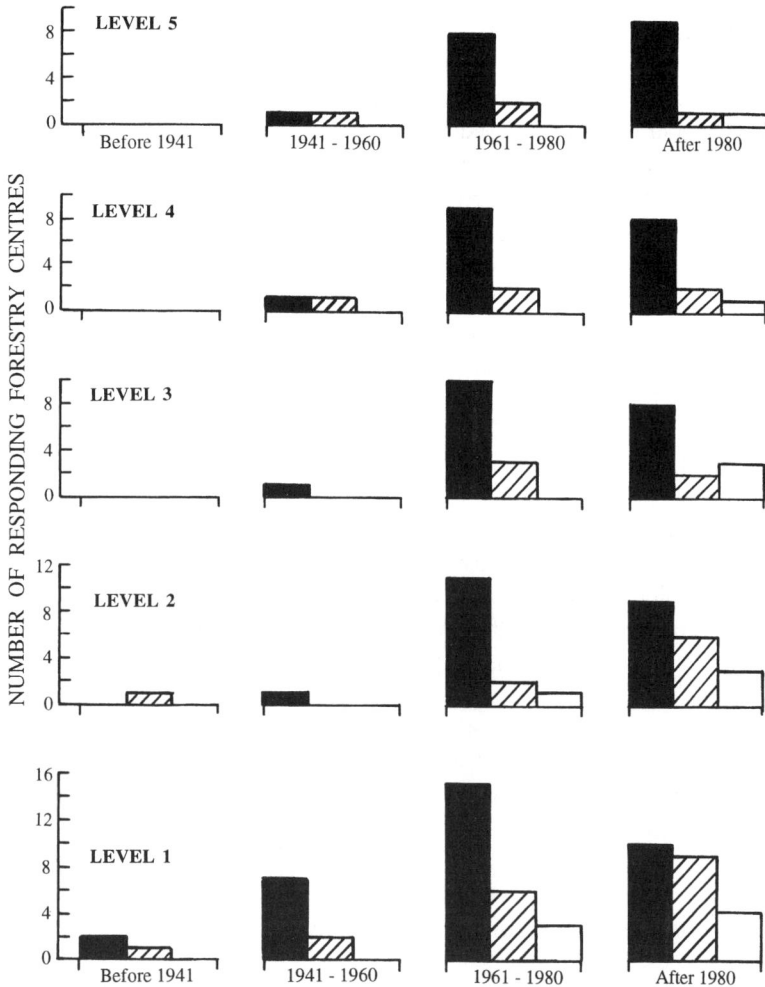

Fig. 33.1 Initiation period[a] for North American (black shading), European (hatched shading) and other (no shading) forestry educational centres to start different training levels[b] of fish–forestry interaction subjects in their programme. Data obtained from responses to Appendix questionnaire. [a]Slightly fewer than given in Table 33.3 because some respondents were uncertain of the initiation year for training levels. [b]As given in Table 33.3.

Respondents from several European institutions claimed that forestry there had relatively little impact on freshwater fishes and their habitat, at least in comparison to dams, weirs, agricultural and industrial pollution, and other human activities. That may be true in certain areas, but not in all (Mills 1980, 1986; Smith 1980). A number of respondents, especially from Latin American and African countries and some from Oceania countries, noted that they did not, at present, give much or any training in fish–forestry interaction. However, after receiving the questionnaire, they realized the need for such education, as there were many problems between fish and forestry in their area.

Focus of education

At least in the United States, there has always been tension in forestry education regarding the proper role of the profession. The debate reaches back to the very beginnings of professional forestry education, just prior to the turn of the twentieth century (Miller & Lewis 1999). In a general sense, the dichotomy might be characterized as between utilitarian timber management and a broader perspective of natural resource management. Although the University of Michigan established a School of Forestry and Conservation in 1927, the predominant emphasis of most forestry schools until the middle of the twentieth century remained on technical subjects that focused on timber and its management (Miller & Lewis 1999). By 1960, a number of educational units responsible for forestry training had broadened their emphasis to include other resources, particularly wildlife (Dana & Johnson 1963). There was a similar evolution of instructional philosophy among forestry programmes in Canada (Garratt 1971; Godbout 1997).

In our experience of teaching in fish–forestry interaction since about 1970, there has been a significant shift in focus in the past 30 years. This shift was prompted, at least in part, by changing interests of students, both in fisheries and in forestry. As the environmental movement gained momentum after the initial Earth Day in 1970, many students sought a broader emphasis in their education, including ecosystem perspectives and environmental protection. This trend has continued to the present. With increasing regulation affecting both fishery and forest management, curricula have broadened to improve this aspect of training. However, much remains to be done, particularly in the specific area of fish–forestry interaction. Instruction in this topic is still not widely available.

Professional fish–fishery training institutions

Professional training relevant to fish–forestry interactions should include not only that provided at forestry educational institutions in subject areas such as hydrology and freshwater ecology, fish biology and habitat ecology, effects of forestry practices on these, and means of reducing or mitigating these effects along with habitat restoration. Also needed is some training at fish and fishery educational institutions in forest ecology, forest harvesting and silvicultural practices, and also timber and wood processing, with their relevant environmental effects. Meaningful communication in the broad fields of fisheries and forestry requires that professionals in both have a general understanding of key processes in these two areas of natural resource use. Mutual understanding is essential for development of sound protective and management practices.

So far we have been considering mainly the forestry training. It would also be useful to consider briefly the extent to which forestry-related subjects have been covered in fisheries institutional programmes in North America, a region where interaction between these two natural resource fields has long been intense.

In 1998 the American Fisheries Society issued a listing of North American institutions offering fisheries degrees, recognizing that others may offer programme options under different names – wildlife, natural resources or forestry, for example (Anonymous 1998). This listing included 152 USA institutions (in 49 states), 16 in Canada (8 provinces), 1 in Puerto Rico and 2 in Mexico. Of these 171 centres for fish and fisheries training, 10 were within forestry educational faculties/departments/centres, where there should be strong opportunities for liaison, and 16 were in environmental or natural resource science schools. The remaining 145, however, were in separate fisheries, fisheries and wildlife, biology, or zoology departments, where opportunities for interaction between fisheries and forestry students probably would be much less. In some institutions where forestry and fisheries are taught in separate colleges (Oregon State University, for example), linkages have been established that provide at least an opportunity for mutual understanding of each other's field.

Recently there have been several conferences in North America to draw together forestry and fisheries workers. Nevertheless, of 76 responses to evaluate a 1996 forest–fish conference on land management practices affecting aquatic ecosystems in North America, only 11 (14.5%) mentioned requirements for additional training and education in this area (Monita 1998), pointing to the need for better information exchange between the two fields.

One indication of an increasing visibility for forestry–fisheries interaction was given by the 2-day symposium held at the 2000 Annual Meeting of the American Fisheries Society: 'Reflections on forest management: can fish and fiber coexist?' Symposium speakers, 32 in number, reviewed historical and current influences of forestry on fish and fish habitat, from coast to coast in North America. There was considerable emphasis on ways in which conflicts could be minimized.

More recently, an American Fisheries Society questionnaire was sent to some 111 senior fishery managers to seek their view of training needs for fishery professionals (Rassam & Eisler 2001). Surprisingly, forestry or fish–forestry interaction was not included in a list of 12 skills and techniques that managers thought needed periodic upgrading, nor in a list of 67 technical courses set out for selection in the questionnaire sent to them, although several topics of relevance to fishery–forestry interaction were highly ranked (watershed processes, riparian and stream ecology, hydrology, for example).

In its section on fisheries education issues and needs, the advisory booklet 'A Crisis in Fisheries Education', released by the Pacific Fisheries Resource Council (Anonymous 2001), noted that 'An obvious requirement will be for students in fisheries study programs to gain exposure to other fields, and to pursue more interdisciplinary research and studies. The initiative of the Fisheries Centre at the University of British Columbia to form links with the Faculty of Forestry and the Liu Centre for the Study of Global Issues is an example in the right direction'. Just such a linkage was made there in 1972 with the Faculty of Forestry when, as a professor in Zoology and Fisheries, one of us (TGN) sought and obtained cross-appointment in Forestry to start training courses in fish–forestry interaction. These I carried on for 20 years (see course outline in Northcote 1992), and they have been expanded further since. Some of my students in zoology and fisheries took my fish–forestry interaction courses, and some in forestry took courses in fisheries.

Over several decades there have been some concerted efforts made through special workshops, symposia, and other types of sessions to make available to forestry and fishery professionals the results coming from relevant research work in both these fields. See for example Hall & Lantz (1969), Mills (1969), Krygier & Hall (1971), Chamberlin *et al.* (1974), Hartman (1982), Salo & Cundy (1987), Chamberlin (1988) and especially Hartman (1988) therein, Meehan (1991), Hartman & Hicks (1992), Hayes & Davis (1992), Laursen (1996), Brewin & Monita (1998) and Anonymous (2000).

Forestry worker training

In addition to training of professional forestry and fishery personnel in fish–forestry interaction, there is a need to provide meaningful training for workers involved in day-to-day field operations, be it in forest road construction, tree felling and yarding, hauling, storing, processing, or related activities. Since the early 1970s, special field and workshop sessions have been arranged for forestry workers by the University of British Columbia Faculty of Forestry in 2–4-day sessions involving short seminars, question and answer sessions, and on-site fish and forestry demonstrations (Northcote 1992). These have been presented by faculty specialists in forest ecology, watershed hydrology, and fish ecology, at field locations convenient to the workers. Training for forest workers is now provided by a number of government and non-government agencies in various countries. Forest Renewal BC, Canada, did offer such training periodically through its Forestry Continuing Studies Network. Unfortunately this agency has now been phased out in government cutbacks.

General public information transfer

Information transfer on fish–forestry interaction must also extend to the general public and to their political representatives, if understanding of the processes involved, of the problems and shortcomings, and of means of improvement is to be achieved. Too often the messages that are widely disseminated through the many media outlets are weak in process content or are greatly oversimplified, glossing over problems as being unimportant. They present statements or pictures indicating that all is well in watersheds subject to various forestry-related activities. To be fair, sometimes all is well, but sometimes it surely is not. Particularly in times of economic downturn, coupled with globalized industrial competition, the view being made more forcefully is that attention to minimizing forestry effects on fish and their habitats can no longer be justified or afforded. Habitat restoration efforts are being similarly impacted.

Of course there have been, and in some cases still are, careful attempts to present to the general public unbiased information and meaningful review of the complexities involved in fish–forestry interactions. One of the more outstanding examples in this area is the series of Westland television programmes in British Columbia.

Approaches for improvement

The following brief list (see also Chamberlin 1988) provides some possibilities for improvement of information transfer in fish and forestry interaction – some resurrected, others innovative.

- Encourage forest/forestry and fish/fisheries training centres to strengthen or develop integrative courses in fish–forestry interaction.
- Encourage usually separate government agencies to strongly support the above with funded chairs, positions, scholarships and awards.
- Encourage the forest industry and the fisheries industry (recreational as well as commercial) to do the same, perhaps via stumpage and/or tax rebate incentives that need not be large to be effective.
- Provide the general public with unbiased, interesting, yet realistic information (not down-playing the complexities involved) via TV, video, brochure and magazine material developed with input and support from both forestry and fishery sectors.
- Provide the above, with appropriate modifications and approvals, for schoolchildren (primary and secondary levels), following the successful 'Scientists & Innovators in Schools' programme in British Columbia, Canada (www.scienceworld.bc.ca).
- Develop a series of rotating (time, place, subject focus) short courses for forestry and fisheries personnel (professional and worker level) using combined workshop and field demonstrations to illustrate key fish–forestry concepts, processes, interactions, problems and corrective methodologies.

Summary and conclusion

There is minimal interaction between fish, fisheries and forestry education where given at universities, colleges, schools and technical institutes. There is also a general lack of an integrated approach in these subject areas where presented not only to forestry but also to fishery students. Furthermore, teaching areas basic to even a rudimentary understanding of the complexities involved in fish–forestry interactions, such as environmental conservation and watershed management, apparently are given at less than a third and about a sixth, respectively, of forestry educational institutions throughout the world (data from 1996 FAO directory).

Although some forestry educational centres in Europe started in the early 1800s, and in North America about a century later, those in both areas have shown a rapid increase in their numbers up to the 1980s.

Training in fish–forestry interaction for forestry professionals in North America extends over a broad front of subject areas and has done so since the late 1960s or early 1970s at many institutions, whereas that in Europe only recently seems to be providing such coverage. In other parts of the world such training is still at a low level.

There is still a great need to provide appropriate and meaningful information exchange in fish–forestry interaction for forestry workers in the field and in timber processing operations. This need also extends to the general public and to their political representatives.

Acknowledgements

The many colleagues who provided contacts with forestry training centres in various parts of the world include Dr S. Andreasson, Dr D.T. Crisp, Dr T. Eriksson, Dr W. Erskine, Dr G. Glova, Dr N. Jonsson, Dr J. Lobon-Cervia, Dr J.A. McLean, Dr A. Peter, Dr M. Rask, Dr J. Reynolds, Dr A. Sánchez Vélez and Dr F.J. Ward. Dr G.F. Hartman reviewed an early draft of the manuscript as well as the questionnaire sent to selected forestry training centres in North America, the Caribbean and Central America, South America, Europe, Africa and Oceania. We sincerely thank all those who completed the questionnaire and returned it to us.

References

Anonymous (1996) *Directory of Forestry Education and Training Institutions.* (FAO) Food and Agriculture Organization of the United Nations, Rome (available at www.fao.org).

Anonymous (1998) *Fisheries Programs and Related Courses at North American Colleges and Universities.* American Fisheries Society, Bethesda, MD.

Anonymous (2000) *International Conference on Wood in World Rivers.* Abstracts. Oregon State University, Corvallis, OR (http://riverwood.orst.edu).

Anonymous (2001) *A Crisis in Fisheries Education.* Advisory September 2001. Pacific Fisheries Resource Conservation Council.

Brewin, M.K. & Monita, D.M.A. (1998) *Forest–Fish Conference: Land Management Practices Affecting Aquatic Ecosystems.* Information Report NOR-X-356. Canadian Forest Service, Edmonton, Alberta, Canada.

Chamberlin, T.W. (1988) *Proceedings of the Workshop: Applying 15 Years of Carnation Creek Results.* Pacific Biological Station, Nanaimo, British Columbia, Canada.

Chamberlin, T., Chestnut, C.W., Drew, G.A., *et al.* (1974) *Stream Ecology Symposium,* Parksville, British Columbia, Canada. Association of BC Professional Foresters, Faculty of Forestry, University of BC, and Centre for Continuing Education, University of BC.

Dana, S.T. & Johnson, E.W. (1963) *Forestry Education in America: Today and Tomorrow.* Society of American Foresters, Washington, DC.

Garratt, G.A. (1971) *Forestry Education in Canada.* Canadian Institute of Forestry, Macdonald College, Quebec.

Godbout, C. (1997) Forestry education in Canada. *Forestry Chronicle,* **73**, 341–7.

Hall, J.D. & Lantz, R.L. (1969) Effects of logging on the habitat of coho salmon and cutthroat trout in coastal streams. In: *Symposium on Salmon and Trout in Streams,* (ed. T.G. Northcote), pp. 355–75. H.R. MacMillan Lectures in Fisheries, Institute of Fisheries, University of British Columbia, Canada.

Hartman, G.F. (1982) *Proceedings of the Carnation Creek Workshop: a 10 Year Review.* Pacific Biological Station, Nanaimo, BC, Canada.

Hartman, G.F. (1988) Research and forestry – fisheries management: institutional voids in technology transfer. In: *Proceedings of the Workshop: Applying 15 Years of Carnation Creek Results,* (ed. T.W. Chamberlin), pp. 225–7. Pacific Biological Station, Nanaimo, BC, Canada.

Hartman, G.F. & Hicks, B.J. (1992) Application of fisheries/forestry research. In: *Proceedings of the Fisheries/Forestry Conference, Christchurch, New Zealand,* (eds J.W. Hayes & S.F. Davis), pp. 78–86. Ministry of Agriculture and Fisheries, New Zealand Freshwater Fisheries Report No. 136.

Hayes, J.W. & Davis, S.F. (1992) *Proceedings of the Fisheries/Forestry Conference, Christchurch, New Zealand.* New Zealand Freshwater Fisheries Report No. 136.

Krygier, J.T. & Hall, J.D. (1971) *Forest Land Uses and Stream Environment.* Oregon State University, Corvallis, OR.

Laursen, S.B. (1996) *At the Water's Edge: The Science of Riparian Forestry.* Conference Proceedings, Minnesota Extension Service, University of Minnesota, St Paul, USA.

Meehan, W.R. (1991) *Influences of Forest and Rangeland Management on Salmonid Fishes and their Habitats.* American Fisheries Society, Special Publication 19, Bethesda, MD.

Miller, C. & Lewis, J.G. (1999) A contested past: forestry education in the United States, 1898–1998. *Journal of Forestry,* **97**, 38–43.

Mills, D.H. (1969) The survival of juvenile Atlantic salmon and brown trout in some Scottish streams. In: *Symposium on Salmon and Trout in Streams,* (ed. T.G. Northcote), pp. 217–28. H.R. MacMillan Lectures in Fisheries, Institute of Fisheries, University of BC, Vancouver, Canada.

Mills, D.H. (1980) *The Management of Forest Streams.* Forestry Commission Leaflet 78. Her Majesty's Stationery Office, London.

Mills, D.H. (1986) The effects of afforestation on salmon and trout rivers and suggestions for their control. In: *Effects of Land Use on Fresh Waters,* (ed. J. Solbe), pp. 422–31. Ellis Horwood, Chichester, UK.

Monita, D.M.A. (1998) Summary of the evaluations following the forest–fish conference, 1996. In: *Forest–Fish Conference: Land Management Practices Affecting Aquatic Ecosystems,* (eds M.K. Brewin & D.M.A. Monita), pp. 523–30. Information Report NOR-X-356. Canadian Forest Service, Edmonton. Alberta.

Northcote, T.G. (1992) Getting the message across to foresters. In: *Proceedings of the Fisheries/ Forestry Conference, Christchurch, New Zealand,* (ed. J.W. Hayes & S.F. Davis), pp. 70–4. Ministry of Agriculture and Fisheries, New Zealand Freshwater Fisheries Report No. 136.

Rassam, G.N. & Eisler, R. (2001) Continuing education needs for fisheries professionals: a survey of North American fisheries administrators. *Fisheries* **26**, 24–8.

Salo, E.O. & Cundy, T.W. (1987) *Streamside Management: Forestry and Fisheries Interactions.* Institute of Forest Resources, University of Washington, Seattle.

Smith, B.D. (1980) The effects of afforestation on the trout of a small stream in southern Scotland. *Fisheries Management,* **11**, 39–58.

Appendix: Questionnaire on fish–forestry interaction educational opportunities

1. Institution name and affiliation:
2. Year started:
3. Training levels: (1) technical ... (2) university undergraduate ... (3) university graduate [masters ...; doctorate ...] (4) other ...
4. Breadth covered:
   (1) Forest resources management [a] timber ... [b] grazing ... [c] wildlife ... [d] recreation & aesthetics ... [e] water & fish ... [f] others ...
   (2) Forestry operations [a] forest road selection, design, construction ... [b] logging system planning, supervision ... [c] site protection & rehabilitation ... [d] computer applications for harvesting ... [e] others ...
   (3) Forest science [a] genetics ... [b] soils ... [c] weather & climate ... [d] tree form & function ... [e] forest ecology ... [f] entomology ... [g] pathology ... [h] silvics ... [i] silviculture ... [j] fire ... [k] water & hydrology ... [l] others ...
   (4) Wood products & processing [list major subjects] ...
   (5) Others [list major subjects] ...
5. Courses offered (**& year started**) relevant to fish-forestry interaction with some coverage of:
   (1) [a] ecology of streams & rivers ...
       [b] ecology of lakes and reservoirs ...
       [c] ecology of estuaries & coastal foreshores ...
   (2) fish biology, habitat needs, ecology ...
   (3) effects of forestry practices on (1), (2) ...
   (4) means of reducing any harmful effects to fish & fish habitat under (3) ...
   (5) means of mitigation or restoration of fish habitat degradation or loss under (3) ...
6. Courses under (5) available in other departments, faculties, institutions associated with or nearby yours **and recommended** to your forestry students to take as part of their training programme (**give year use started by your institution of these courses**) ...
7. Additional comments, observations, suggestions (e.g. is there a need for fish–forestry training in your area; are there any special fish–forestry problems in your area?) ...

Chapter 34
Towards a new fish–forestry interaction in the world's watersheds

T.G. NORTHCOTE AND G.F. HARTMAN

Opening remarks

Fish–forestry interactions around the world, alone or in conjunction with other issues, pose challenges and risks that must be addressed. Forestry effects that jeopardize fish occur primarily in the volumetrically small but important freshwater environments of the world. Freshwater in lakes, rivers and streams makes up only 0.7% of the world's water, but provided, on average between 1996 and 2000, some 9% of the world's fish production (FAO 2000a). Furthermore freshwater of relatively low salinity is essential for much of the world's agricultural production as well as for direct human consumption. The stakes for overall freshwater protection are high.

The issues of forestry impacts on fishes, as we see them, are even more critical when considered as part of the complex of major environmental problems that humanity faces. Water availability, per capita, is falling. Water quality in many areas of the world is being degraded by chemical pollution and soil erosion, the latter linked to agricultural practices and deforestation. Human numbers continue to rise exponentially, even if more slowly. As such, they drive the complex of developing environmental problems, one of which is forestry effects on fishes.

We visualize human population increase as a primary force that demands higher fuel and timber production. The consequent forest loss and associated degradation of aquatic environments, products of population increase, are at the foundation of fish–forestry interactions (Fig. 34.1).

Given these perceptions, our discussion of the foregoing chapters and of future fish–forestry interactions must be set against considerations of human population growth and projected status of freshwater supplies. The manner in which fish–forestry matters are managed in the future will depend on how societies guide themselves in regard to population increase and their overall treatment of freshwater resources.

The reality and relevance of human population limits

The notion of limits to human population growth dates back to the 1660s with works by Graunt, and then later by Malthus, Verhulst, Lotka and others. Hutchinson (1978) reviewed this background and outlined the consequences of exponential population growth that so often have been derided up to the present day by many economists and politicians. Several of the above population ecologists considered supply of potable

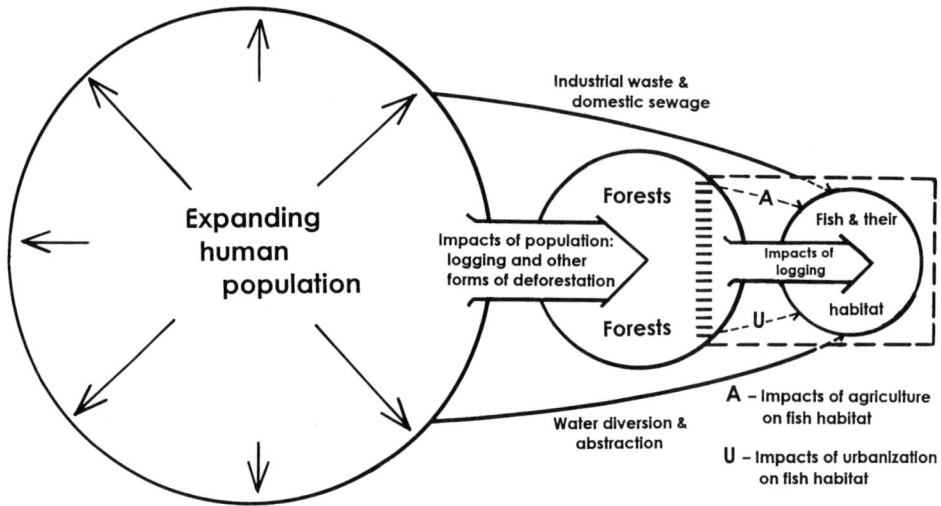

Fig. 34.1 Human population increase as a driving force in deforestation through logging and other activities that have impacts on fishes and their habitat. These effects are compounded by other population-driven impacts such as agriculture, urbanization, water diversions and pollution.

fresh water to set the upper limit for human population size. More recently, Postel *et al.* (1996), Ayensu *et al.* (1999), Gleick (2000, 2001), Vorosmarty *et al.* (2000), Johnson *et al.* (2001) and Leshner (2002) have updated the impending reality of limitations by fresh water (see also Pimentel & Morse 2003).

By the year 2025 Johnson *et al.* (2001) project that there will be some 53 river basins of the world where annual per capita water use for domestic, agricultural and industrial purposes will be <2500 m³, classified as water scarcity. If present use levels continue, they estimate that nearly half (48%) of the world's population will live in water-stressed river basins in 2025 (Table 34.1). Five such basins are in North America (three of these entirely in Mexico), and one in the Buenos Aires area of South America. Nine occur in Europe and eight in Africa. Western Asia has 17 such river basins, five of them where the annual renewable water supply is projected to be <500 m³ per person per year. In eastern Asia 9 of the basins will be severely stressed for water in just over two decades (see also Holden 2003).

Humans are distributed very unevenly across the lands of the world (Fig. 34.2). Over a quarter of its solid surface is either uninhabited or at low densities of less than 10 inhabitants per km² (Anonymous 1999a). The three largest concentrations of people are in eastern Asia, the Indian subcontinent, and Europe, together making up over half the present over six billion world total at densities well over 100 per km², and in places such as Bangladesh exceeding 800 per km². Not surprisingly in such areas there is little room for trees let alone forests! Smaller areas of human high densities (over 200 per km²) occur in Jawa (Java), in parts of northern, central and southeastern Africa, and in South and North America. Annual growth rates of 1.5% or more (1995–2000) are common in the southern half of Asia, Africa and Latin America. India and China

Table 34.1 Renewable water supply projected to 2025 in world river basins where human populations will reach at least 10 million residents and where there will be <2500 m³ of water per person per year

Region	River basin	Annual renewable water supply (m³ per person per year)
North America	Rio Grande	<500
	Grande de Santiago	500–1000
	Balsas, Colorado, Panuco	1000–1700
South America	Buenos Aires[1]	<500
Europe	Meuse, Seine, Thames	500–1000
	Don, Elbe, Odra, Rhine, Wistla	1000–1700
	Dnepr	1700–4000
Africa	Northern Algeria[1]	<500
	Galana, Jubba, Limpopo, Orange	500–1000
	Nile, Volta	1000–1700
	Niger	1700–4000
Asia (western)	Cauweri, Indus, Penner, Shamal Sina[1], Southern Saudi Arabia[1]	<500
	Krishna, Syr Darya, Tapti Farah, Gordavari, Kura-Araks, Narmada, Tehran[1], Tigris &	500–1000
	Euphrates	1000–1700
	Amu Darya, Ganges, Mahanadi	1700–4000
Asia (eastern)	Hai Ho, Han, Huang He, Hong, Huai, Ibaraki (Japan)[1], Jawa[2]-Jakarta[1], Jawa[2]-Timur[1], Liao	<500
	Several in SE China	500–1000
	Chao Phraya, Fuchun Jiang	1000–1700
	Yangtze	1700–4000

Data from figure in Johnson *et al.* (2001); antipodes coverage not included.
[1]River basins unnamed by Johnson *et al.* (2001), so a name to designate the nearby location is used.
[2]Formerly Java.

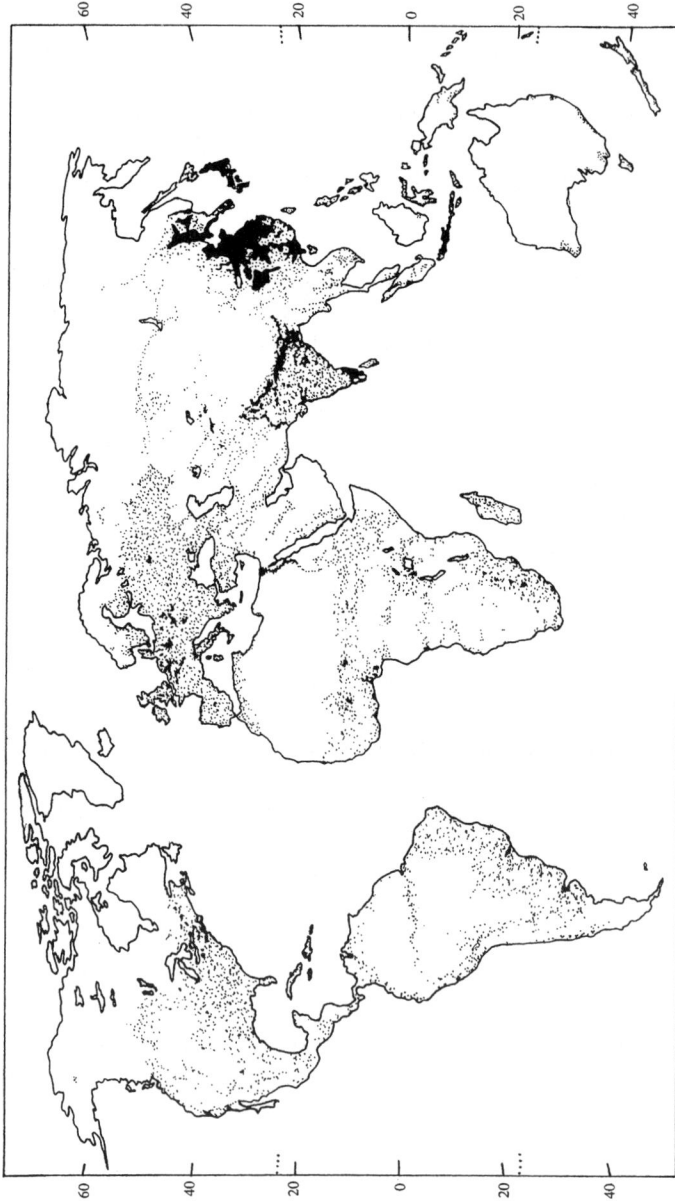

Fig. 34.2 World population density in 1998. Black, >100 inhabitants per km²; stippled, 10–100 per km²; blank, <10–0 per km². Latitudes shown on ordinates; upper and lower dotted partial lines on ordinates show Tropic of Cancer and Tropic of Capricorn, respectively. Adapted from Anonymous (1999a), Winkel Tripel Projection.

account for over half the world's current population growth, and India will probably take top rank before 2050. Nevertheless low to moderate human population densities (10–100 per km^2) occur in many countries of the world, often in forested areas, although the latter also can be found in broad regions with very low human densities (cf. Figs. 1.1, 34.2).

Two books should be required reading, and re-reading, not only for the general populace but also for economists, politicians, and their many entrepreneuristic advocates. First, Garrett Hardin's *Living Within Limits* (*Ecology, Economics and Population Taboos*) (Hardin 1993), and then most recently E.O. Wilson's *The Future of Life* (Wilson 2002), especially its Chapter 2 'The Bottleneck'. A few brief quotations from the latter are appropriate here.

> They [leading economists and public philosophers] have mostly ignored the numbers that count. Consider that with the global population past six billion and on its way to eight billion or more by mid-century, per-capita fresh water and arable land are descending to levels resource experts agree are risky. … If soil erosion and withdrawal of groundwater continue at their present rates until the world population reaches (and hopefully peaks) at 9 to 10 billion, shortages of food seem inevitable. There are two ways to stop short of the wall. Either the industrialized populations move down the food chain to a more vegetarian diet, or the agricultural yield of productive land worldwide is increased by more than 50 percent.

Smil (1993) drew together the causes, facts and figures for the current environmental crisis in China, arising from its approach then to reaching a fifth of the world population and its inability to produce an adequate food supply for its present population level, still increasing. Wilson (2002) updates China's present situation, whereby food production has been supplemented by massive irrigational, domestic and industrial use of both surface and groundwater. But the US National Intelligence Council predicts that China will have to import some 175 million tons of grain annually by 2025, and by 2030 this will rise to 200 million tons, the total amount now exported annually throughout the world (Wilson 2002).

Furthermore, the dominant effect on world biodiversity is habitat degradation or destruction, and the most consequential cause is forest clearing (Wilson 2002), either directly for timber and pulp production, or for agricultural, housing and other purposes. About 70% of tropical dry forest, over 60% of temperate hardwood and mixed forest, 45% of tropical rainforest and 30% of coniferous forest coverages have been lost (Wilson 2002). Chemical pollution is another major cause of aquatic biodiversity loss. For example in China with 50,000 km of major rivers, some 80% no longer support fish (United Nations Food and Agriculture Organization), as noted by Wilson (2002). See also Northcote (1996) and the publications therein dealing with inland water quality and quantity problems in China, as well as in other countries around the Pacific rim in the context of human population growth.

In the summer of 2002 a group of 11 social ecologists and world conservationists from Austria, Mexico, UK, Switzerland, USA and Norway put forward an insightful analysis of human demands on the regenerative capacity of the biosphere (Wackernagel *et al.* 2002) which caught world attention, even to headlines in daily newspapers of the Okanagan Valley in British Columbia, Canada. The group conservatively assembled and summated demands of six human activities that use the biologically productive space of the world: (1) growing crops for food and other human uses; (2) grazing animals for these uses; (3) harvesting timber for wood, fibre and fuel; (4) harvesting fish in marine and inland waters; (5) accommodating infrastructure for housing, transportation, industrial production and hydroelectric power; and (6) burning fossil fuel. They demonstrated that these combined demands on the earth's total ecological capacity, expressed as the number of earths required to support it, had risen from about 0.70 in 1961, reaching 1.00 in 1977, and overshot it by 1.20 in 1999. Factoring in as 12% the bioproductive area required to protect other species, they showed that the demand line crossed the supply line in the early 1970s, not the 1980s.

One of the more recent coverages of human population growth (Birdsall *et al.* 2001) concludes that rapid population growth is generally adverse to economic development. In his review of the book, Sachs (2002) points out that it has too little discussion of linkages to environmental degradation, noting the many studies showing that deforestation, soil depletion and other environmental problems are increased by high population growth coupled with low income.

Additional geographic coverage of fish–forestry interaction

As noted in the Preface, for various reasons we were unable to have complete geographical coverage of several land areas of the world where important fish–forestry interactions occur. Here we provide synopses of the subject for seven widely different regions not covered in Parts VI and VII of the book.

Southeastern USA

This area covers nearly 1.5 million km², including the states of Alabama, Arkansas, Florida, Georgia, Kentucky, Louisiana, Mississippi, North Carolina, South Carolina, Tennessee, Virginia, and parts of Oklahoma and Texas. It contains the Appalachian Mountain Range, eastern coast river drainages, and the lower Mississippi River basin.

The area has over 866,000 km² of forest land, representing 91% of such cover there in 1907 (Wear & Greis 2002). Some 89% of the land is in private ownership among about five million people. There are about 2370 km² of 'old growth' forest left, much of which is in the 84,986 km² of public lands (Wear & Greis 2002). The five major categories of forest are upland hardwoods, lowland hardwoods, mixed oak forest, natural pine and plantation pine. From 1953 to 1999, natural pine area decreased from 291,380 to 133,549 km² while plantation pine area increased from 8094 to 121,408 km². Forest harvesting is done by ground-based machines, and cable logging systems are only used in the Appalachian Mountains. At least 3 of the 13 states have developed

'Best Management Practices' that include use of streamside management zones and protection of water quality (Chapter 31).

There are about 1,504,400 km of rivers and streams in southeastern USA and about 5790 km/year are considered to become impaired by non-point pollution from silvicultural activities. However, waterways are affected by agriculture, heavy industry and urban developments as well as by forestry. The impacts of forestry on stream ecosystems include effects on water yield, peak flows, water table level, sediment input, water temperature and dissolved oxygen (Wear & Greis 2002).

Southeastern USA contains a fish fauna of high diversity. Numerically, nearly half of the North American fish fauna is located there. Four species are now extinct and 28% of the 662 freshwater and diadromous species are classed as 'jeopardized'; 165 of the species are given 'rare' status. However, 90% of the rare species, the 'narrow endemics', have very restricted ranges. The species most at risk are those that nest in bottom crevices, feed by sight or feed on loose material on stream bottoms, depend on surface algal films on rocks for food, require loose gravel to stir their eggs down into the bottom, or reside in small streams or warm-water ponds.

Expanding forest plantation operation will create further impacts on fish faunas because of specific habitat requirements of some species and the restricted ranges of others. Forestry effects will occur among other impacts, and may not be the ones most risky for fish. The low percentage of public lands, the diversity of ownerships among private landholders, and continuing development with population growth will make it difficult to protect species that use small streams and ponds and exist in narrow distributions.

Central America

Seven small countries – Guatemala, Belize, El Salvador, Honduras, Nicaragua, Costa Rica, Panama – comprise Central America, with a total area of 544,700 km². Topographically all are mountainous with sharp climatic differences vertically and between western and eastern lowland seaboards. Highlands are largely forested with pines, oak and other trees, with broad-leafs lower down and into the tropical rainforests. The total population in 1970 was over 16 million, increasing 3.2% annually, and now over 37 million. Even today much of the economy relies on agricultural crops (major contributors being bananas, cotton, sugar, coffee, cocoa) starting in the mid-1800s with major forest clearing. Forestry itself began three centuries ago in some areas such as Belize. Ecotourism now is important throughout Central America, but especially so in Costa Rica. Freshwater fishes are common and locally abundant in the many streams, rivers and rather few lakes of Central America. Sharks and other species of fish abound in Lake Nicaragua and its San Juan river outlet, as well as elsewhere close to the seas. Some six species of freshwater fishes including introduced rainbow trout provide sport fishing in Costa Rica (Baker 2001; see also Bussing 1987).

Among the earlier studies covering estuarine and strictly freshwater fishes in Central America, two of the more recent are those by Villa (1982) and Bussing (1987). Together these list some 328 species. As a further update we made a search (http://ichthyonb.fr/Country/CountrySearchList.cfm) which gave a total of 389 species included in 19 orders. Central American freshwater fishes are dominated by four groups – characins, catfishes, cyprinodonts and percids – as is the case for three of them in the world overall (see Table 1.2). Over 7% of them have been introduced, mainly percids (15 species) and cyprinids (7 species). Over 14% (mostly catfishes and percids) occur mainly in estuarine waters. The remainder are difficult to separate between those using both estuarine and inland waters, and those that are heavily dependent on the latter either entirely or for reproductive and juvenile rearing habitat (anadromous forms). A number of the strictly freshwater species, for example, some catfishes, percids (*Gobiesox* spp.) and others can live in high gradient headwater streams where they may be most vulnerable to effects of deforestation. Throughout Central America there is a high diversity of fishes living from upper reaches of small streams down into middle and lower reaches, and on to estuarine habitats. All of this fish diversity is sensitive to watershed impacts including those of forestry. Members of at least 12 of the 19 orders of Central American freshwater fish have importance for human food, fisheries and other purposes.

Deforestation is said to be a serious problem in Central America, notably in Guatemala (Brooks 1978), in Honduras (McGaffey 1999), and even in Costa Rica where about 25% of its area is under some form of protection for forest cover and which is said to lead the way in moving Central America away from 'the soil-leaching deforestation that plagues the isthmus' (Baker 2001). Nevertheless, between 1981 and 1985 Costa Rican deforestation loss, as a percentage of the remaining stocks, was at 4.0% and rose to 7.6% between 1987 and 1990 (Anonymous 1999b). Despite Costa Rican government attempts with reserves, refuges and parks to protect large areas of forested habitats with their rich biotic diversity, and aid by outside scientific and governmental support, the loss of such lands to logging and ranching continues, pushed by population increase pressures still at 2.5% annually, and by ineffective forestry management (Baker 2001). Lack of forestry training would not seem to be a strong cause for this problem. There are at least 19 forestry training centres in Central America (Anonymous 1996), distributed among six of the seven countries, almost all of which have forestry management as a main teaching area, and six of which offer courses in environmental conservation and/or watershed management. None apparently give courses in fish–forestry interaction. There is little doubt that inland fish populations, especially those in the many watersheds draining into both Pacific and Atlantic coasts of Central America, have been seriously impacted by the several centuries of large-scale deforestation that have occurred in the region. The breadth and extent of such effects must await more quantitative study, but that should not delay a move toward better forest management and habitat restoration in the region. One of the more promising developments is that of a project to link protected areas from southernmost Mexico to Panama (Kaiser 2001), an approach that also is gaining interest in North America, Brazil, Europe, western Australia and elsewhere.

Temperate South America

This region is mainly made up of mid- to southern Chile west of the Andes Mountains, and east of them by southern Argentina. The Chilean land cover of the region is largely mixed forest with smaller areas of closed shrublands and natural vegetation mosaic (Anonymous 1999a). That for the larger Argentinian side is a combination of open and closed shrublands apparently with little extensive forest cover. A myriad of east to west flowing watersheds occur on the Chilean side with far fewer but larger ones flowing west to east into Argentina. A varied freshwater fish fauna occurs in the region and includes native galaxiids and introduced salmonids of both commercial and recreational importance. Forestry, after mining, is Chile's second resource-based export sector with some 60 species of hardwoods, along with plantation-based pine softwoods that supply its pulp and paper industry. In contrast, forestry is not of great importance in temperate Argentina with agriculture and livestock production providing most of the country's foreign exchange earnings.

Russia and Siberia

The Russian Federation contains one of the largest forest areas in the world. Notwithstanding the size and remoteness of parts of the area, the future of its forest resources is threatened by ineffective management policies, uncertain property rights, deteriorating forest conditions, pollution, and the prospect of massive cutting in Siberia (WRI 1996).

Russian forests include about 70% of the world's boreal forest within a forested area estimated at 7.5–7.7 million km² in some of the harshest climates on earth. About 65% of Siberia's forests are located in the permafrost zone, lying north of a line extending from near the base of the Kamchatka Peninsula to the south end of Lake Baikal, and thence northwest to the Arctic Ocean southwest of Novaya Zemlya (Kotlyakov & Khromova 2002). It includes the headwaters of the rivers that run into the Arctic Ocean from the eastern end of Russia to Novaya Zemlya. Much of Russia's boreal forest lies north of this line. The forests are unproductive and composed of larch, pine, spruce and birch species. In the region surrounding the lower Amur River, the Korean pine-broad-leaved forests are complex, support forestry and contain valuable salmon rivers (Owston *et al.* 2000).

At present, 39% of Siberia's eastern forest zone and 30% of its western forest zone is within 'intact forest landscapes'. Only 9% of the European forest zone is described as 'intact' by Dobrynin *et al.* (2002). In European Russia only 14% or 320,000 km² remains in relatively undisturbed blocks of 500 km² or more (Yaroshenko *et al.* 2002).

There are about 39 families and 352 species of fishes within the inland waters of the Russian Federation (Froese & Pauly 2002), with the species distributed unevenly among the families. Twenty of the 39 families are represented by one or two species, while six of the major families (Cyprinidae, Abyssocottidae, Cottidae, Gobiidae, Salmonidae, Acipenseridae) are represented by 14 or more species each.

Russian inland fish catches ranged from 212,874 to 307,823 metric tonnes, in 1991–2000, and made up 60.2% of the total European inland waters catch, with Azov Sea sprat (*Alosa tanaica*) contributing a major part (FAO 2000b). Salmonids and smelt made up from 59,206 to 83,964 metric tonnes of the European freshwater catch annually from 1994 to 2000. From 42 to 52%, by weight, of this group were Pacific salmon and sturgeons, mostly caught in Russia.

There are two major elements that provide a more positive outlook for fisheries protection in the face of forestry expansion in Russia:

(1)   A strong body of capable scientists and basic, relevant information exists there. Effects of logging on water yield, peak flows, water temperature and snowmelt are understood (Efremov & Morin 2001). Prohibition of log driving, requirements for skid trail dimensions, protective measures for pollution prevention, permafrost protection, leave strip and bridge requirements indicate that Russian foresters understand the needs for protecting fish habitat (V. Morin, pers. comm.).

(2)   Russia has forestry legislation. The 'Russian Federation Forest Code', adopted by the State Duma, January 1997, is a detailed and strong code. It established criteria and categories for protection of shores of lakes and rivers. Harvesting operation rules for the far east of Russia require leave strips from 0.25 to 4.0 km along streams and rivers, depending on their widths. To protect water quality, there are requirements for preserving or creating optimum forest cover within watersheds (V. Morin, pers. comm.). The glimpses that we get into forestry–fish matters in Russia indicate that scientists and foresters know what should be done.

The risks that are faced regarding sound forest management and fish habitat protection revolve around three elements:

(1)   Support for and maintenance of the management structures have been weak following the break-up of the USSR. At least 20% of the timber logged in Russia is taken illegally or in severe violation of existing legislation (see website: Forest.ru 2002).

(2)   Russia has large volumes of wood and international companies are seeking to establish operations there. Given the difficult economic situation in the country and the desire of the companies to extract logs for processing elsewhere, there will be pressure on the country to open up large areas. Management agencies in Russia find it difficult to balance conflicting environmental and development concerns (WRI 1996).

(3)   About 65% of Siberia's forests are located in the permafrost zone and are sensitive to disturbance. Once permafrost soils are exposed to sunlight, the top layer may melt, making areas wet or swampy, which precludes reforestation (WRI 1996). The implications of such disturbance for water quality and temperature, and fish spawning habitat, may be very serious if rigorous planning and controls are not implemented.

Risks surrounding the fate of boreal forests in Russia should be recognized as equal to those facing the tropical forests of the world. To the extent that poor forest management is a surrogate for negative impacts on fish, we are concerned about the future of Russia's boreal forests.

China

Total forest cover in China in 1950 was only 5.2% of its land mass (Zhang *et al.* 2000). By the mid-1970s its forest cover had risen to 12.7% (Smil 1993), but still large areas were unforested, in part because they were deserts, semi-deserts, or at high elevations where only grasses and a few other small plants could survive. Furthermore, timber harvesting more than trebled from 20 million m³/year in the mid-1950s to over 60 million m³/year in the mid-1990s. By then natural forest cover had dropped to less than a third of the total forest area (Zhang *et al.* 2000). Official claims of the area afforested were often more than three-fold above actual tree survival from 1960 to 1990 (Smil 1993). In 1998 a new forest policy (Natural Forest Conservation Program – NFCP) was started in China (Zhang *et al.* 2000) to try to address 50 years of forestry overexploitation which had resulted in forest cover degradation, massive soil erosion, catastrophic flooding and loss of biodiversity (Smil 1992, 1993; Zhang *et al.* 2000; Liu *et al.* 2001; Dudgeon 2002). The NFCP includes 18 inland provinces and autonomous regions covering middle to upper reaches of the major rivers, especially the Yellow and Yangtze where environmental degradation had been highest. These and other problems related to human population growth around inland waters were the subject of a large session at the 1995 Pacific Science Congress in Beijing, China (Northcote 1996), and among the 16 published presentations dealing with China, those of Tang & Bi (1996), Bi (1996) and Dudgeon (1996) are especially relevant here. Major thrusts of the NFCP extend to increased technical training, land management planning, mandatory conversion of marginal farmlands to forest lands, resettlement and retraining of forest dwellers (human population has increased five-fold in forested areas since 1950, twice that in China overall), shares in private ownership of forest lands and expanded forest research. At least China has the potential to extend the first priority of the NFCP – that of increased technical training – as there are some 43 forestry education and training institutions spread widely throughout the country, a number of them having teaching areas in environmental conservation, forest management and watershed management (Anonymous 1996). Not surprisingly, shortcomings of NFCP have been pointed out (Xu *et al.* 2000) and response given (Zhao *et al.* 2000). Forest biomass carbon storage has increased significantly in China since the mid-1970s (Fang *et al.* 2001), suggesting that alteration of forest management practices could help offset industrial carbon dioxide emissions. Nevertheless these emissions also contain large quantities of soot (black carbon) which may alter climate and thereby increase the tendency for greater summer floods in south China and more drought in north China (Mennon *et al.* 2002).

China has long been recognized as a region of high species diversity for freshwater fishes, especially of cyprinids, as documented by Nichols (1943). Of the 567 species listed then, 403 (71%) were cyprinids. Our assemblage of 957 species (http.// ichthyonb.fr/Country/CountrySearchList.cfm) shows that 674 (70%) are cyprinids, virtually all of which occupy inland waters rather than brackish or estuarine habitats. Many are found in middle to uppermost reaches of rivers and small mountain streams. Furthermore over 100 species of catfishes occur mainly in inland waters, as well as 38 percid species, 22 osmerid and salmonid species, and a few members of eight other major groups. Another 77 species can be found in both inland and estuarine habitats, and 19 species occupy only estuarine waters. Some 22 species, mainly cyprinids, cat-fishes and percids, have been introduced. Even more so than in Central America, the estuarine and inland water fishes of China, along with their key habitats, are highly vulnerable to environmental alterations coming from forestry as well as other human activities. Members of 17 orders of fishes have significant benefits to the high human population density in China for food, fisheries and other inputs. Four species of native carp in China rank first, second, fourth and seventh in world aquaculture fish production (Lu *et al.* 1997) and more than 75% of China's total freshwater fish production. The Yangtze River contains the most important natural populations of these species, essential even for fish cultural purposes. Results of the study by Lu *et al.* (1997) imply that carp of the Yangtze River cannot be managed as a unit because human over-exploitation and habitat change have differential impacts on fish abundance for different parts of the river.

Dudgeon (2002) lists five main threats to Asian rivers, including those of China: (1) flow regulation or modification (in part effected by forestry activities), (2) pollution, (3) habitat degradation by deforestation, (4) over-harvesting, and (5) synergistic effects of introduced fish species and climate change. All of this is happening in a country apparently fourth highest in the world for freshwater fish biodiversity, exceeded only marginally by Venezuela and Indonesia, but greatly by Brazil (Dudgeon 2002). On the island of Taiwan, where now there is no climax forest cover remaining, 11 out of the 32 total freshwater species of fish are threatened with extinction or already extinct (Dudgeon 1996).

Africa

Africa has about 17% (c. 5.2 million km²) of the world's forest area, mostly in its central regions. This is receiving progressively more use. Twenty percent of the region's forest area is in the Democratic Republic of Congo (Zaire) (UNEP 2000).

The rate of forest loss across the continent is high and has almost doubled since 1970. About 90% of Africans depend on the forest for firewood and charcoal for fuel, and the volume used for these purposes has risen from 250 to 502 million m³/year (UNEP 2000). From 1980 to 1995, Africa has lost about 0.7 million km² of forest to the combined effects of mining, forestry, agriculture and elephant damage. Natural forest cover has declined from 2.16 to 1.98 million km² from 1980 to 1995. The forests of the Congo basin are in better condition than those in other parts of Africa. Afforestation is improving, but cannot keep up with the losses (Sizer & Plouvier 1997).

Fish resources of the central African countries are enormously diverse. Minimal numbers of families and species (see Table 1.2 for Africa as a whole and Nelson (1994) for further information) are as follows:

- Cameroon – 38 families and 407 species. First four families, in order of species numbers, are given here and below in parentheses (Aplocheilidae, Cyprinidae, Cichlidae, Mormyridae).

- Central African Republic – 17 families and 59 species (Aplocheilidae, Cyprinidae, Cichlidae, Clupeidae).
- Congo Republic – 33 families and 350 species (Cyprinidae, Aplocheilidae, Alestiidae, Cichlidae).
- Democratic Republic of Congo – 37 families and 1041 species (Cichlidae, Cyprinidae, Alestiidae, Mormyridae).
- Equatorial Guinea – 21 families and 74 species (Aplocheilidae, Cichlidae, Gobiidae, Cyprinidae).
- Gabon – 30 families and 203 species (Aplocheilidae, Mormyridae, Alestiidae, Cyprinidae).

It has been very difficult to make contact with anyone who is doing, or has done, work on the effects of forestry activities on fish in tropical Africa. The task of designing and carrying out fish–forestry research such as that presented in Chapter 18 would be extremely difficult in areas such as the central African countries. The faunas are complex and vary from region to region, the infrastructure is limited and skilled technical help may be absent. In addition, the forest industry is expanding rapidly with an influx of companies from Asia. There are, for example, 295 companies operating in Cameroon, 270 being foreign. Many of the trans-national companies are either corrupt or are prepared to corrupt nationals within the developing countries (Sizer & Plouvier 1997). Forestry operations of one company are described by them as follows: 'There is no management plan or significant effort to ameliorate environmental impacts through, for example, skid trail planning or low impact logging.' Although we have little direct information on fish–forestry interaction in central Africa, the state of governance, the complexity of the fish faunas, the large geographical areas involved, would all indicate strongly that fish and fish habitat will not fare well as forest industry expands in the central African countries.

Caribbean, Oceania islands

In the Caribbean there are some 11 islands or island groupings totalling 220,000 km² in area, with elevations rarely over 1000 m, and with a combined population of well over 31 million people (nearly 85% on just three islands – Cuba, Dominican Republic and Haiti). Most of the larger islands support some forested areas. Four forestry training centres operate in the Caribbean area, one in Cuba, two in the Dominican Republic and one in Trinidad & Tobago, all of which give forest management, watershed management and environmental conservation as main teaching areas (Anonymous 1996).

Small islands or island groupings in Oceania total some 82,000 km² in area, many of them under 1000 m in elevation, and with a combined population of 2.3 million people have a much lower overall human density than on the Caribbean islands (28.4 versus 143.6 inhabitants per km², respectively). Many of the Oceania islands shown on land cover mapping by satellite imagery (Anonymous 1999a) indicate considerable areas with forest cover. Monterey pine (*Pinus radiata*) plantations are common on Fiji, for example, and may reach commercial size in under two decades.

There are six forestry training centres on the small Oceania islands, one on each of New Caledonia, the Solomon Islands, Tonga, Western Samoa, and two on Fiji. Main teaching areas include forest management at the Solomon Islands centre and at one of the Fiji centres, where watershed management and environmental conservation are also given. Streams are evident on almost all of the islands, and small lakes occur on a few such as Guam, Kiritimati, Nauru, New Caledonia and Tahiti. Eels (*Anguilla* spp.) occur in streams of some Oceania islands and mudskippers (Gobiidae) in their estuarine mouths. Deforestation and reforestation with exotics probably have had effects on the fish fauna of Oceania's small island streams but too little is known of the species present, their population dynamics and habitat requirements to make judgement.

Contrasts and perceptions of fish–forestry interaction

When a book on fish–forestry interactions attempts to cover an extremely complex topic on a wide geographical scale, and when it includes the knowledge and experience of many people, two important elements emerge:

(1)   there will be wide variability in the kinds of interactions and in the degree of their understanding;
(2)   the interactions will be viewed in widely differing contexts.

Such a book may lack the cohesiveness of one written by one or two authors, but it may be more valuable for its perceptual diversity. At this stage we review some of the perceptions that have emerged from the Parts and chapters, and indicate their contrasts.

In the 'ecology of the systems' Part, the thoughtful and provocative chapter on forest ecology examines the elements of forest ecosystems, and especially the role and significance of disturbance. It does not refer to a key paper by Reeves *et al.* (1995) who pointed out that timber harvest disturbance does not necessarily have the same spatial or temporal characteristics as that caused by stand-setting wildfire. The chapter has an interesting philosophical approach, which is captured in part of the concluding statement: 'Management and conservation must respect nature as it is and not as we might wish it to be for one or more of convenience, simplicity or emotion-based reasons.' The chapters on stream, lake and estuary ecology take a strong ecosystem approach and stress the basic elements of ecosystem function in those watershed components.

Streams are dynamic environments in which logging impacts often are shown quickly and clearly. They are environments that are interconnected through the movement of water, and impacts are transmitted down them, but dampened as the size and tributary summing effects increase. Lakes, especially large ones, may respond more slowly to watershed impacts of forestry, and records of long-term watershed changes may be available in the paleolimnology of their bottom sediments. Nevertheless lake shorelines are their most complex, productive and vulnerable areas – sensitive to many forms of human activities including forestry. Most land use impacts within a watershed are summed within the estuary, again a highly dynamic, productive and sensitive region where the forestry effects of logging, log storage, pulp mills and other infrastructure may be combined.

The Part on fish biology indicates the large range in patterns of life history, migration, reproduction and feeding behaviours among some of the 10,000 species of freshwater fishes. Migration may link the use of spawning, feeding and refuge habitats. It is a crucial life history behaviour during which fish may be vulnerable to forestry-related impacts that alter habitat or block their access, both upstream and downstream. The range of environments and the movement patterns to and from them during reproduction are very wide. The physiology and behaviour of developing and hatched larvae are highly attuned to the features of the environments in which spawning occurs. The chapter on foraging provides insight into the different components of feeding behaviour, and points out the importance of forests and arboreal food resources for tropical fish such as those in the Amazon River basin.

Three chapters on 'forestry activities' provide an important view into pivotal forestry activities of locating and designing roads, choosing and employing harvesting systems, developing harvesting and reforestation plans, and designing and operating pulp mills. These should be of particular interest to non-engineers and non-foresters, and show how far recent developments have come in reducing impacts of these activities.

Several perceptions emerge from the Part on forestry effects on aquatic systems and fishes. Streams exhibit quick and severe responses to poorly planned logging. Small tributaries are particularly vulnerable. Impacts are transmitted downstream but may be dampened in lakes and larger-sized lower reaches of rivers. The chapter reviewing forestry impacts within streams showed that there may be different effects on different species and life stages of fish, that effects may be delayed in expression up to decades after initial disturbance, and that they may be confounded by non-forestry impacts. Some of the impacts within a river system may be tracked by multi-decade air photograph series, while those in lakes may be studied by analysis of long-term sediment records. The chapter on pulp mill effects shows how sensitive and complex fish response to components of pulp mill effluent may be, and how highly sophisticated the study of such effluent effects has become.

Chapters 17 to 30 provide overviews of conditions and effects of forestry in North America and other parts of the world. The concentration on North American coverage reflects the large amount of fish–forestry research that has been done there. Within this group of chapters there is a wide range of very different situations wherein fish–forestry interactions occur. While there is risk in generalizing, in North America we see:

- relatively simpler fish faunas, particularly in the Pacific Northwest;
- a long history of generally benign use of forests and fish by low populations of aboriginal people;
- a relatively short period of forest removal along with simple combinations of other land use activities within watersheds where logging occurs;
- a large body of fish–forestry research after the mid-1900s;
- high institutional capacity for good governance and regulation.

In Australia and New Zealand four points emerge:

- ancient and in places heavy use of forests, for up to 40,000 years in Australia, and removal by forest burning in parts of New Zealand starting in the ninth century;
- relatively recent (post-1800) European influence on forests and fishes;
- important impacts of forestry including plantation use of non-native tree species;
- well-developed research programmes, good governance and regulations relevant to fish–forestry interactions.

In Europe and Japan the situations are different:

- there are long histories of forest removal starting as early as 1300 AD in the UK, and even earlier in other parts of Europe and in Japan (700 AD);
- the fish faunas influenced are more complex and contain some highly valued species;
- important forestry effects involve afforestation as well as logging removal of forests, and include studies that focus on soil chemistry;
- while there is excellent research (e.g. Norway, Sweden, and recently in Japan), it has been less fish-focused than in North America;
- multiple land uses occur in watersheds and confound fish–forestry interactions;
- there is sound governance and capacity for regulation.

In South America, southeastern Asia, and equatorial Africa fish–forestry interactions occur in a much more complex and problematic arena:

- the areas are large and contain complex forest systems and fish faunas, with as many as 4000 fish species in the Amazon basin, over 3000 in southeast Asia, and over 1000 in the Zaire (Congo) basin;
- freshwater fish form a critical protein source for people;
- high levels of precipitation, leached soils and erosion potential occur;
- logging activities are relatively recent, but there is, and has been, a heavy dependence of indigenous people on forests for fuel wood;
- the countries have limited financial resources, trained technical staff, and established fish-forestry research;
- governance may be weak or corrupt;
- many recently emerging forestry companies from wood-starved Asian countries are inclined to be corrupt and log with destructive forestry practices;
- there is wide divergence between forestry codes, planning (on paper) and the practices that occur in the forests.

Other important perceptions emerge from the chapters dealing with worldwide coverage of fish–forestry interactions. Based on North American experience, the tendency is to consider the effects of forestry as a major causal factor affecting fishes and their habitat. However, in Indonesia, fish farming developments destroy mangrove forests, even though many species of native fish depend on intact mangrove habitat. In Mexico, fish–forestry matters were considered in a historical and broad societal perspective, and presented as part of overall land use involving population, agriculture, power generation and domestic water supply. In brief consideration of fish–forestry matters in Russia, we raised the concern that the huge tracts of boreal forest and fisheries resources associated with them might be as much at risk as were those of tropical rain forests.

In the final Part dealing with means to effect better fish–forestry interactions, four main points arise from the chapter on guidelines, codes and legislation:

- there are two categories of codes – those from international organizations that establish guidelines to achieve objectives but don't carry the force of law, and those that set out specific enforceable rules that must be followed;
- guidelines, codes, and regulations of many countries require some form of buffer strips along waterways;
- very few of the codes specifically link restrictions on forestry activities to the protection of fish habitat, so environmental and water quality protection are surrogates for fish protection;
- many countries that do have codes have poor governance structures, so there are problems with compliance.

An extremely important element of future fish–forestry work involves education, and the worldwide review of the subject showed that there is great need for inclusion and improvement in this subject for forestry managers and the general public. Fish–forestry topics have received little attention in major international conferences. There are few citations of relevant fish–forestry literature in publications on forest management from such organizations as FAO and the International Tropical Timber Organization.

Closing comments

Realities and world views

Parts VI and VII of this book cover very diverse regions of the world. They include countries that have a strong history of forest use, affluence, reasonably good governance, and well-established forestry and fisheries science. At the opposite end of the spectrum they include countries that have been ravaged by recent war, which have weak or corrupt government agencies, and lack strong science and management capacity. In some countries, people have been able to develop a high level of sensitivity about environmental degradation, water quality loss and impacts on fish habitat. In other parts of the world people may be pressed by hunger, lack of employment, and social and economic disruption to the point where environmental concerns, important though they should be, are not understood, or are pushed aside by more pressing problems of day-to-day keeping alive. While there are major national differences among societies, there are also wide differences in the attitudes of different groups or interests within nations.

For developing countries there are different levels of appreciation and concern for forest, wildlife and fish values that influence people's behaviour. In central Africa, logging roads facilitate hunting, and as road networks expand, hunters move in to obtain wildlife for meat supply to city dwellers. Many such areas now suffer from the 'empty forest' syndrome (Minnemeyer 2002). National in-migrants that exploit forest resources may have little concern or appreciation for the sustainability of wildlife, fish or environmental quality.

In regions such as Southeast Asia there are different sectors of society that have very different levels of interest in forestry effects on environment or fish habitat. Owners of many major forestry corporations are prepared to engage in corrupt practices (Sizer & Plouvier 1997) without concern for well-planned logging or measures for fish protection. However, within these same regions, where aid organizations have proposed plans for selective logging by residents within the immediate forest area, these people have requested lower levels of selective logging. They advise very conservative cutting because they use a wide array of the forest resources. Ideal models for forestry and fishery management should reflect not only the requirements for sustainability, but also the cultural needs of people living within the forested area.

Community forestry or corporate forestry?

Over much of North America forest harvesting is carried out on private land by landowners, or on public lands by forest companies operating under some form of cutting permission or tenure. Operation of forestry programmes within an area by the local community does occur, but it is not the predominant approach. The chapters on Mexico, Indonesian mangrove forests, and forestry in Borneo and Cambodia suggested the need to consider implications of having forestry management placed into the hands of local communities. We do not offer any conclusion about the suitability of forestry approaches for different parts of the world, but must suggest different management approaches because for many parts of the forested world there is little or no information on local effects of forestry activities on fishes. In such cases we can only indicate the probable impacts on fish, along with some consideration of the overall needs for appropriate forest management.

Corporate forestry may be carried out with the benefit of knowledge of trained foresters, engineers and biologists if the companies involved have the interest, needs and funding to employ such people. They may provide management expertise that a financially limited country may not be able to afford within government structures. The limitations of corporate forestry rest strongly upon the appetites and behaviour of the company, along with the government legislation and the people involved.

The scenarios for corporate and community forestry in Southeast Asia have different features. Corporate operations bring in crews of non-local workers and set up large camps. The first creation of road access may occur with logging (Minnemeyer 2002). The camps may be fed in part by use of wildlife and fish from within the area. Conflicts may arise with local populations that depend on the wildlife and fish. The companies cut for profit and try to maximize harvest from road networks they established. Many company logging operations are tarnished with corruption, bribery and unsustainable forestry (Sizer & Plouvier 1997), displacement of indigenous people who have only 'customary tenure' (Greenpeace 2002), and complex influential connections with government officials and politicians (Kartodihardjo 1999).

People involved in community forestry live within the area and use the forest for many purposes including fishing, hunting, food-gathering, obtaining medicinal plants and housing materials, and use of specific sites for spiritual purposes. For some forest people who draw their livelihood from the forest in Cameroon, 'those forests symbolize their civilization' (Ekoko 1999). As a result, the level of timber removal from community forestry will be much lower than that from corporate forestry.

It may be difficult to significantly shift emphasis from corporate forestry to community or cooperative programmes. The situations in countries such as Cameroon (Ekoko 1999), Indonesia (Kartodihardjo 1999) and Papua New Guinea (Filer & Kalit 1999) involve complex entanglements of laws, bureaucratic and political manipulation, and corporate influences, all set against a background of control and influence by the World Bank. Community forest management may offer more hope for consideration of fish and their habitat. However, in the milieu of international financial organizations, government policy, bureaucracy, corporate interests and changing community interests, it is unlikely that fish–forestry interactions would be matters of high priority.

Ranges of complexity in management issues

Throughout this book, readers will see wide ranges in the complexity of management situations. These differences in complexity are caused by a number of elements, five of which are notable and are listed below:

- wide diversity in fish faunas, from situations where there are less than ten species of management concern, to those where there may be hundreds of important fish species in faunas containing thousands;
- great diversity of land ownership ranging from situations where most land is held by the government (e.g. much of Canada), to those in which there may be hundreds of landowners within a single watershed (e.g. southeastern USA);
- broad range in the number of human activities and impacts within a watershed or part of it;
- complexity in the number and roles of organizations, agencies and legislation involved;
- wide range in time during which forests first began to be cut, and when forests and fish started to be managed.

The result of these diverse elements is that there are very different situations in which management of fish–forestry interaction occurs. In British Columbia, Canada, there is a relatively simple situation in which to manage forestry impacts on fish. Forest land is largely owned by the government, there may be only one forestry company involved, the primary impacts of concern (aside from over-fishing) are forestry-related, the streams may contain only a few species of interest, and there is a reasonably good foundation of science and management experience.

In southeastern USA the fish faunas are complex, some fish species may be of concern because of endangered status, and others because of fishing values. There may be hundreds of landowners, along with multiple impacts and agencies involved within a single watershed. On the positive side, there is good governance, a strong body of science, and very knowledgeable managers.

In tropical ecosystems, management of fish–forestry interactions may be very problematic. Fish faunas can be extremely complex with thousands of species, many of them may be valuable, and all may require protection in the interests of maintaining biodiversity. Governance may be weak, societies impoverished and suffering from recent wars, scientific information limited or absent, and only an extremely limited management capability may exist.

A series of case studies over Latin America, Asia and Africa that examine single-factor up to five-factor causation as driving forces in tropical deforestation suggests that there is no universal link between causes and effects (Geist & Lambin 2002). Instead, tropical forest decline results from many complex pressures, local and regional, which act in various combinations in different geographical locations. These must be understood in detail before any policy intervention is invoked.

In these situations involving great complexity in physical, biotic and socio-economic conditions, we offer four observations:

- Among the chapters on Mexico, Indonesia, and South America, the authors indicate concern about erosion, increased sediment loads and water temperatures. There is universality about the negative impacts of these elements.
- There is universality about the manner in which hill-slopes and streams will respond to poorly planned road construction and logging in areas of high precipitation.
- It is widely understood that watersheds must remain structurally intact to protect fisheries values.
- There is limited knowledge about the specific measures that should be taken to protect particular species or life stages of fish.

Into the unknown future

Although beyond the scope of this book to discuss in detail, there are two important circumstances that will alter future forest use and its impacts on fish:

- Climate warming will affect the distribution and density of forests, particularly in the northern hemisphere above 40° north latitude where vegetated areas already show sizable increases in density (Holden 2001). There also may be effects on fire frequency and insect outbreak patterns. As these and other changes occur, and government and industrial responses to them develop, the nature of forestry and other impacts on fish and their habitat will also change (Northcote 1992).
- Human population numbers will continue to rise in most parts of the world, creating an array of cumulative pressures that will affect fishes. Some of these pressures will be related to forestry and others will be caused by agriculture, fishing and urbanization (Hartman *et al.* 2000).

We believe that there are many positive measures that should be taken, and many of these are implicit in the national and regional chapters, and in other parts of this book. Major forestry and fisheries organizations of the world must begin to put more emphasis on the spheres of interaction between their two primary resource sectors. This effort must extend to conference structure, internationally funded fisheries and forestry programmes, and to broad educational opportunities. There is a great need to begin to evaluate and understand the differences between actual forestry practices and those indicated in government policy and planning documents. There is need for education programmes that can prepare the nationals of all countries to contribute in this context. And there is need for international conference structure that avoids domination by politicians and senior political appointees. Finally there is urgent need for people throughout the world to consider the consequences of poorly managed forestry on inland fishes and fisheries that provide an important part of the world's protein supply.

But there are glimmerings of light to be seen when peering into the tunnel of the future. Advances in scientific methodology are providing better tools and networks to document forest cover and its loss worldwide (for examples see Kaiser 2000, 2002, Nelson *et al.* 2000; www.globalforestwatch.org). Although some 40–50% of land cover on earth has been irreversibly transformed or degraded by human actions, with high rates of natural forest disappearance (Ayensu *et al.* 1999, Williams 2003), ecosystem assessment and correctives are now being put forward at an international level. In Brazilian national forests of the Amazon there are means developing to protect large areas from deforestation (Veríssimo *et al.* 2002). Establishment of protected areas for maintaining biodiversity in 22 tropical countries showed that most were successful in stopping land clearing and to a lower degree at mitigating logging and other damaging human activities (Bruner *et al.* 2001). Congo basin forests are becoming a high conservation priority and since 1999 coverage of protected areas has shown a 36% increase (40,607 km²) over the region's forests (Kamdem-Toham *et al.* 2003). Over a decade ago means were suggested to provide local people with economic benefits from forests without the inhabitants having to undertake severe deforestation by logging (Alper 1993). Positive results of such practices via ecotourism are evident in some high elevation forests of Mexico (T.G. Northcote & J.D. Hall, personal observations). Rainforest conservation can generate significant benefits over logging and agriculture at both local and global scales (Kremen *et al.* 2000). Canada's Model Forest Program (Anonymous 1999c) offers effective local, provincial and national means to achieve sustainable forest management within the context of other resource uses. And in the USA, science may become an important player in formulation of new forest management plans (Paul 2000). As a result of scientific intervention a key region of old-growth forest in Baja California of Mexico recently has been spared for the present (Shouse 2003).

Nevertheless, as convincingly pointed out at the world level by Suzuki (2003), "forestry must switch from its current focus on resource extraction – what to take – to ecosystem-based management – what to leave". ... "having a strategy is just the first of three critical components that are needed for conservation to succeed. Success also requires authority and accountability ... most well-intended conservation measures start and **end** with an action plan." All too true, too often, we are sorry to say!

Acknowledgements

Moira S. Greaven provided invaluable assistance in obtaining freshwater fish species lists for several countries of the world. J.S. Nelson was most helpful in providing information on freshwater fish biology at a worldwide level. A.C. Dolloff gave much help in obtaining coverage of fish–forestry interactions in southeastern USA. D. Efremov, V. Morin and V. Harberger (Far East Scientific Forestry Research Institute, Russia) provided help in obtaining information on forestry in Russia.

References

Alper, J. (1993) How to make the forests of the world pay their way. *Science*, **260**, 1895–6.

Anonymous (1996) *Directory of Forestry Education and Training Institutions*. Food andAgriculture Organization of the United Nations, Rome.

Anonymous (1999a) *The Times Atlas of the World*. Random House, New York.

Anonymous (1999b) *Oxford Atlas of the World*, 7th edn. Oxford University Press, New York.

Anonymous (1999c) *Achieving Sustainable Forest Management through Partnership*. Natural Resources Canada, Canadian Forest Service, Model Forest Secretariat, Ottawa.

Ayensu, E., Claasen, D., Collins, M. *et al.* (1999) International ecosystem assessment. *Science*, **286**, 685–6.

Baker, C.P. (2001) *Costa Rica*, 4th edn. Avalon Travel, Emeryville, CA.

Bi, J. (1996) New water quality indices for sustainable development in China. *GeoJournal*, **40**, 9–15.

Birdsall, N., Kelly, A.C. & Sinding, S.W. (2001) *Population Matters. Demographic Change, Economic Growth, and Poverty in the Developing World*. Oxford University Press, New York.

Brooks, J. (1978) *The 1979 South American Handbook*. Trade & Travel Publications, Bath, UK.

Bruner, A.G., Gullison, R.E., Rice, R.E. & da Fonseca, G.A.B. (2001) Effectiveness of parks in protecting tropical biodiversity. *Science*, **291**, 125–8.

Bussing, W.A. (1987) *Peces de las Aguas Continentales de Costa Rica*. Editorial de la Universidad de Costa Rica. San Jose, Costa Rica.

Dobrynin, D., Isaev, A. & Laestadius, L. (2002) *Atlas of Russia's Intact Forest Landscapes*. World Resources Institute, Washington, DC.

Dudgeon, D. (1996) Anthropogenic influences on Hong Kong streams. *GeoJournal*, **40**, 53–61.

Dudgeon, D. (2002) The most endangered ecosystems in the world? Conservation of riverine biodiversity in Asia. *Verhandlungen Internationale Vereinigung für theoretische und angewandte Limnologie*, **28**, 59–68.

Efremov, D.F. & Morin, V.A. (2001) Percentage of forest land – an instrument of hydrological regimes of mountain basins regulation. *River Basin Management*, **11–13**, Abstract, September 2001, Cardiff, UK.

Ekoko, F. (1999) Environmental adjustment in Cameroon: challenges and opportunities for policy reform in the forestry sector. *Workshop on 'Environmental Adjustment: Opportunities for Progressive Policy Reform in the Forestry Sector?'*, World Resources Institute, April 1999.

Fang, J., Chen, A., Peng, C., Zhao, S. & Ci, L. (2001) Changes in forest biomass carbon storage in China between 1949 and 1998. *Science*, **292**, 2320–2.

FAO (2000a) *FAO Yearbook, Fishery Statistics: Capture Production*, Vol. 90/1. Food and Agriculture Organization of the United Nations, Rome, Italy.

FAO (2000b) *Yearbook of Fishery Statistics – Summary Tables*. FAO, Rome, Italy.

Filer, C. & Kalit, K. (1999) Papua New Guinea case study. *Workshop on 'Environmental Adjustment: Opportunities for Progressive Policy Reform in the Forest Sector?'*, World Resources Institute, April 1999.

Forest.ru (2002) *Survey of Illegal Forest Felling Activities in Russia* (http://www.forest.ru/).

Froese, R. & Pauly, D. (eds) (2002) *FishBase*. Worldwide electronic publication (www.fishbase.org).

Geist, H.J. & Lambin, E.F. (2002) Proximate causes and underlying driving forces of tropical deforestation. *BioScience*, **53**, 143–50.

Gleick, P.H. (2000) *The World's Water 2000–2001. The Biennial Report on Freshwater Resources*. Island Press, Washington, DC.

Gleick, P.H. (2001) Our water – making every drop count. *Scientific American*, **February**, 39–55.

Greenpeace (2002) *Greenpeace Forests: The Paradise Forest* (http://www.paradiseforest.org/).

Hardin, G. (1993) *Living within Limits*. Oxford University Press, New York.

Hartman, G.F., Groot, C. & Northcote, T.G. (2000) Science and management in sustainable salmonid fisheries: the ball is not in our court. In: *Sustainable Fisheries Management: Pacific Salmon*, (eds E.E. Knudsen, C.R. Steward, D.D. MacDonald, J.E. Williams & D.W. Reiser), pp. 31–50. CRC, Lewis Publishers, Boca Raton, New York.

Holden, C. (2001) Greenhouse is here. *Science*, **293**, 1987.

Holden, C. (2003) Global water dilemma. *Science*, **299**, 1657.

Hutchinson, G.E. (1978) *An Introduction to Population Ecology*. Yale University Press, New Haven, CT.

Johnson, N., Revenga, C. & Echeverria, J. (2001) Managing water for people and nature. *Science*, **292**, 1071–2.

Kaiser, J. (2000) Network to log world forest loss. Network, *Science*, **293**, 2196–9.

Kaiser, J. (2002) Satellites spy more forest than expected. *Science*, **297**, 919.

Kamdem-Toham, A., Adeleke, A.W., Burgess, N.D., *et al.* (2003) Forest conservation in the Congo Basin. *Science*, **299**, 346.

Kartodihardjo, H. (1999) Toward an environmental adjustment: structural barriers to forest policy reform in Indonesia. *Workshop on 'Environmental Adjustment: Opportunities for Progressive Policy Reform in the Forest Sector?'* World Resources Institute, April 1999.

Kotlyakov, V. & Khromova, T. (2002) Maps of permafrost and ground ice. In: *Land Resources of Russia*, CD-ROM, (V. Stolbovoi & I. McCallum), Laxemburg, Austria; International Institute for Applied Systems Analysis and the Russian Academy of Sciences.

Kremen, C., Niles, J.O., Dalton, M.G., *et al.* (2000) Economic incentives for rain forest conservation across scales. *Science*, **288**, 1828–32.

Leshner, A. (2002) Science and sustainability. *Science*, **297**, 897.

Liu, J., Linderman, M., Ouyang, Z., An, L., Yang, J. & Zhang, H. (2001) Ecological degradation in protected areas: the case of Wolong Nature Reserve for giant pandas. *Science*, **292**, 98–101.

Lu, G., Li, S. & Bernatchez, L. (1997) Mitochondrial DNA diversity, population structure, and conservation genetics of four native carps within the Yangse River, China. *Canadian Journal of Fisheries and Aquatic Sciences*, **54**, 47–58.

McGaffey, L. (1999) *Honduras*. Marshall Cavendish, New York.

Mennon, S., Hansen, J., Nazarenko, L. & Luo, Y. (2002) Climate effects of black carbon aerosols in China and India. *Science*, **297**, 2250–3.

Minnemeyer, S. (2002) *An Analysis of Access to Central Africa's Rainforests*, (ed. K. Holmes), World Resources Institute.

Nelson, J.S. (1994) *Fishes of the World*. John Wiley & Sons, New York.

Nelson, R.F., Kimes, D.S., Salas, W.A. & Routhier, M. (2000) Secondary forest age and tropical forest biomass estimation using thematic mapper imagery. *BioScience*, **50**, 419–31.

Nichols, J.T. (1943) *The Fresh-Water Fishes of China. Natural History of Central Asia*, Vol. 9. American Museum of Natural History, New York.

Northcote, T.G. (1992) Prediction and assessment of potential effects of global environmental change on freshwater sport fish habitat in British Columbia. *GeoJournal*, **28**, 39–49.

Northcote, T.G. (1996) Effects of human population growth on freshwater quality, quantity and biotic systems. *GeoJournal*, **40**, 1–2.

Owston, P.W., Schlosser, W.E., Efremov, D.F. & Miner, C.L. (2000) Korean pine-broadleaved forests of the far east. *Proceedings from the International Conference*. General Technical Report PNW-GTR-487. USDA Forest Service, Pacific Northwest Research Station.

Paul, E. (2000) Science could play starring role in new forest management plans. *BioScience*, **50**, 108.

Pimentel, D. & Morse, J. (2003) Malnutrition, disease, and the developing world. *Science*, **300**, 251.

Postel, S.L., Daily, G.C. & Ehrlich, P.R. (1996) Human appropriation of renewable fresh water. *Science*, **271**, 785–8.

Reeves, G.H., Benda, L.E., Burnett, K.M., Bisson, P.A. & Sedell, J.R. (1995) A disturbance-based ecosystem approach to maintaining and restoring freshwater habitats of evolutionarily significant units of anadromous salmonids in the Pacific Northwest. In: *Evolution and the Aquatic Ecosystem: Defining Unique Units in Population Conservation*, (ed. J.L. Nielsen), pp. 334–49. *American Fisheries Society Symposium*, **17**.

Sachs, J.D. (2002) Rapid population growth saps development. *Science*, **297**, 341.

Shouse, B. (2003) Old-growth forest spared for now. *Science*, **299**, 802.

Sizer, N. & Plouvier, D. (1997) *Increased Investment and Trade by Transnational Logging Companies in APC-Countries: Implications for Sustainable Forest Management and Conservation*. A joint report to WWF-Belgium WRI & WWF International. EC-Project B7–6201/96–161.VIII/For.

Smil, V. (1992) China's environment in the 1980s: some critical changes. *Ambio*, **21**, 431–6.

Smil, V. (1993) *China's Environmental Crisis*. M.E. Sharpe, Amonk, New York.

Suzuki, D. (2003) A look at world parks. Editorial. *Science*, **301**, 1289.

Tang, Y. & Bi, J. (1996) Spatial and temporal analysis of water pollution in China. *GeoJournal*, **40**, 3–7.

UNEP (2000) *United Nations Environment Programme, Global Environment Outlook 2000*. Produced by UNEP GEO Team.

Veríssimo, A., Cochrane, M.A. & Souza, C. (2002) National forests in the Amazon. *Science*, **297**, 1478.

Villa, J. (1982) *Peces Nicaraguensis de Agua Dulce*. Serie Geografía y Naturaleza. No.3. Coleccion Cultural. Fondos de Promocion Cultural. Banco de America. Managua, Nicaragua.

Vorosmarty, C.J., Green, P., Salisbury, J. & Lammers, R.B. (2000) Global water resources: vulnerability from climate change and population growth. *Science*, **289**, 284–8.

Wackernagel, M., Schultz, N.B., Deumling, D. *et al.* (2002) Tracking the ecological overshoot of the human economy. *Proceedings of the National Academy of Sciences, USA*, **99**, 9266–71.

Wear, D.N. & Greis, J.G. (2002) *Southern Forest Resource Assessment*. General Technical Report SRS-53. Asheville, NC, US Department of Agriculture, Forest Service, Southern Research Station (http://www.srs.fs.fed.us/).

Williams, M. (2003) *Deforesting the Earth*. University of Chicago Press, Chicago.

Wilson, E.O. (2002) *The Future of Life*. Knopf, New York.

WRI (1996) *World Resources 1996–97: A Guide to the Global Environment*. 9. Forests and Land Cover. World Resources Institute (http://www.wri.org/wri/biodiv/temperat.html).

Xu, M., Qi, Y. & Gong, P. (2000) China's new forest policy. *Science*, **289**, 2049.

Yaroshenko, A.Y., Potapov, P.V. & Turubanova, S.A. (2002) *The Last Intact Forest Landscapes of Northern European Russia*. World Resources Institute, Washington, DC.

Zhang, P., Shao, G., Zhao, G., *et al.* (2000) China's forest policy for the 21st century. *Science*, **288**, 2135–6.

Zhao, G., Shao, G., Zhang, P. & Bai, G. (2000) Response. *Science*, **288**, 2049–50.

Index